Lennette's Laboratory Diagnosis of Viral Infections

INFECTIOUS DISEASE AND THERAPY

Series Editor

Burke A. Cunha
Winthrop-University Hospital
Mineola, New York
and
State University of New York School of Medicine
Stony Brook, New York

1. Parasitic Infections in the Compromised Host, *edited by Peter D. Walter and Robert M. Genta*
2. Nucleic Acid and Monoclonal Antibody Probes: Applications in Diagnostic Methodology, *edited by Bala Swaminathan and Gyan Prakash*
3. Opportunistic Infections in Patients with the Acquired Immunodeficiency Syndrome, *edited by Gifford Leoung and John Mills*
4. Acyclovir Therapy for Herpesvirus Infections, *edited by David A. Baker*
5. The New Generation of Quinolones, *edited by Clifford Siporin, Carl L. Heifetz, and John M. Domagala*
6. Methicillin-Resistant *Staphylococcus aureus*: Clinical Management and Laboratory Aspects, *edited by Mary T. Cafferkey*
7. Hepatitis B Vaccines in Clinical Practice, *edited by Ronald W. Ellis*
8. The New Macrolides, Azalides, and Streptogramins: Pharmacology and Clinical Applications, *edited by Harold C. Neu, Lowell S. Young, and Stephen H. Zinner*
9. Antimicrobial Therapy in the Elderly Patient, *edited by Thomas T. Yoshikawa and Dean C. Norman*
10. Viral Infections of the Gastrointestinal Tract: Second Edition, Revised and Expanded, *edited by Albert Z. Kapikian*
11. Development and Clinical Uses of Haemophilus *b* Conjugate Vaccines, *edited by Ronald W. Ellis and Dan M. Granoff*
12. *Pseudomonas aeruginosa* Infections and Treatment, *edited by Aldona L. Baltch and Raymond P. Smith*
13. Herpesvirus Infections, *edited by Ronald Glaser and James F. Jones*
14. Chronic Fatigue Syndrome, *edited by Stephen E. Straus*
15. Immunotherapy of Infections, *edited by K. Noel Masihi*
16. Diagnosis and Management of Bone Infections, *edited by Luis E. Jauregui*
17. Drug Transport in Antimicrobial and Anticancer Chemotherapy, *edited by Nafsika H. Georgopapadakou*
18. New Macrolides, Azalides, and Streptogramins in Clinical Practice, *edited by Harold C. Neu, Lowell S. Young, Stephen H. Zinner, and Jacques F. Acar*
19. Novel Therapeutic Strategies in the Treatment of Sepsis, *edited by David C. Morrison and John L. Ryan*
20. Catheter-Related Infections, *edited by Harald Seifert, Bernd Jansen, and Barry M. Farr*
21. Expanding Indications for the New Macrolides, Azalides, and Streptogramins, *edited try Stephen H. Zinner, Lowell S. Young, Jacques F. Acar, and Harold C. Neu*
22. Infectious Diseases in Critical Care Medicine, *edited by Burke A. Cunha*

23. New Considerations for Macrolides, Azalides, Streptogramins, and Ketolides, *edited by Stephen H. Zinner, Lowell S. Young, Jacques F. Acar, and Carmen Ortiz-Neu*
24. Tickborne Infectious Diseases: Diagnosis and Management, *edited by Burke A. Cunha*
25. Protease Inhibitors in AIDS Therapy, *edited by Richard C. Ogden and Charles W. Flexner*
26. Laboratory Diagnosis of Bacterial Infections, *edited by Nevio Cimolai*
27. Chemokine Receptors and AIDS, *edited by Thomas R. O'Brien*
28. Antimicrobial Pharmacodynamics in Theory and Clinical Practice, *edited by Charles H. Nightingale, Takeo Murakawa, and Paul G. Ambrose*
29. Pediatric Anaerobic Infections: Diagnosis and Management, Third Edition, Revised and Expanded, *Itzhak Brook*
30. Viral Infections and Treatment, *edited by Helga Ruebsamen-Waigmann, Karl Deres, Guy Hewlett, and Reinhotd Welker*
31. Community-Aquired Respiratory Infections, *edited by Charles H. Nightingale, Paul G. Ambrose, and Thomas M. File*
32. Catheter-Related Infections: Second Edition, *edited by Harald Seifert, Bernd Jansen, and Barry Farr*
33. Antibiotic Optimization: Concepts and Strategies in Clinical Practice (PBK), *edited by Robert C. Owens, Jr., Charles H. Nightingale, and Paul G. Ambrose*
34. Fungal Infections in the Immunocompromised Patient, *edited by John R. Wingard and Elias J. Anaissie*
35. Sinusitis: From Microbiology To Management, *edited by Itzhak Brook*
36. Herpes Simplex Viruses, *edited by Marie Studahl, Paola Cinque and Toms Bergström*
37. Antiviral Agents, Vaccines, and Immunotherapies, *Stephen K. Tyring*
38. Epstein-Barr Virus, *edited by Alex Tselis and Hal B. Jenson*
39. Infection Management for Geriatrics in Long-Term Care Facilities, Second Edition, *edited by Thomas T. Yoshikawa and Joseph G. Ouslander*
40. Infectious Diseases in Critical Care Medicine, Second Edition, *edited by Burke A. Cunha*
41. Infective Endocarditis: Management in the Era of Intravascular Devices, *edited by John L. Brusch*
42. Fever of Unknown Origin, *edited by Burke A. Cunha*
43. Rickettsial Diseases, *edited by Didier Raoult and Philippe Parola*
44. Antimicrobial Pharmacodynamics in Theory and Clinical Practice, Second Edition, *edited by Charles H. Nightingale, Paul G. Ambrose, George L. Drusano, and Takeo Murakawa*
45. Clinical Handbook of Pediatric Infectious Disease, Third Edition, *Russell W. Steele*
46. Anaerobic Infections: Diagnosis and Management, *Itzhak Brook*
47. Diagnosis of Fungal Infections, *edited by Johan A. Maertens and Kieren A. Marr*
48. Antimicrobial Resistance: Problem Pathogens and Clinical Countermeasures, *edited by Robert C. Owens, Jr. and Ebbing Lautenbach*
49. Lyme Borreliosis in Europe and North America, *edited by, Sunil Sood*
50. Lennette's Laboratory Diagnosis of Viral Infections, Fourth Edition, *edited by Keith R. Jerome*

Lennette's Laboratory Diagnosis of Viral Infections

Fourth Edition

Edited by
Keith R. Jerome
University of Washington
Fred Hutchinson Cancer Research Center
Seattle, Washington, U.S.A.

CRC Press
Taylor & Francis Group
Boca Raton London New York

CRC Press is an imprint of the
Taylor & Francis Group, an **informa** business

First published in 2010 by Informa Healthcare,

Published 2023 by CRC Press
Taylor & Francis Group
6000 Broken Sound Parkway NW, Suite 300
Boca Raton, FL 33487-2742

CRC Press is an imprint of Taylor & Francis Group, an Informa business

ISBN-13: 978-1-4200-8495-5 (hbk)

Visit the Taylor & Francis Web site at
http://www.taylorandfrancis.com

and the CRC Press Web site at
http://www.crcpress.com

A CIP record for this book is available from the British Library.

Library of Congress Cataloging-in-Publication Data available on application

Edwin H. Lennette: A Tribute

Ed Lennette was born in Pittsburgh, Pennsylvania, on September 11, 1908, and died of respiratory failure on October 1, 2000, following surgery. These cold, hard facts in no way describe or tell us anything useful about this man and his accomplished life.

He earned a B.S. degree at the University of Chicago in 1931 and a Ph.D. degree in 1935. His Ph.D. degree is believed to be the first awarded specifically in the field of virology. Ed then completed an M.D. degree at Rush Medical College (also at the University of Chicago) in 1936, and following his internship, he spent brief periods at the Pathology Department of Washington University School of Medicine in St. Louis, Missouri, and at the Rockefeller Foundation laboratories in New York City.

At that time the Rockefeller Foundation was interested in yellow fever and its International Health Division (IHD) assigned him to Brazil, where he spent most of World War II, working on yellow fever and encephalitis viruses.

In 1944, the IHD transferred him to their laboratory in Berkeley, California, to work on hepatitis and encephalitis. When that laboratory was transferred to the California Department of Public Health in 1947, Ed became its Director, following a year as Chief of the Medical-Veterinary division of the U.S. Army facility at Fort Detrick, Frederick, Maryland.

For the next 31 years he molded this laboratory into a world-renowned training laboratory, as well as a, perhaps the, leading laboratory for the diagnosis of viral and rickettsial diseases. This Viral and Rickettsial Diseases Laboratory, or VRDL as it was known, conducted substantial programs of research on Q-fever and arthropod-borne encephalitis, on polioviruses and other infections, and on the role of viruses in causing human cancer. Many of the people trained at the Berkeley VRDL went on to become leading scientists and administrators of laboratories and health agencies worldwide. He also served as a consultant to many government agencies and participated in numerous advisory committees.

Ed published many scientific papers and edited several books that became classics in their field. Among them was "Diagnostic Procedures for Viral, Rickettsial, and Chlamydial Infections," coedited with Drs. Nathalie J. Schmidt and Richard W. Emmons.

After he retired from public service in 1978, Ed became the President of the California Public Health Foundation, and in 1981 and 1982, served as Acting Director of the W. Alton Jones Cell Science Center in Lake Placid, New York.

Of course, I had heard of Dr. Lennette and his legendary accomplishments, but it was only in the 1980s that we met for the first time, although I had participated in writing chapters for earlier editions of his book on diagnostic virology. Awed by meeting such a senior person and eminent scientist, Ed put me at ease immediately. To this day, I remain in awe of him.

His encyclopedic recall and brilliant, practical insights were remarkable. To the end he strongly supported gaining experience that leads us to knowledge, as opposed to exclusively technical procedures, relied on by some in the rapidly moving field of diagnosis. Whenever, as Moderator for Virus Diseases of ProMED-mail, I would comment harshly (but fairly!) about an organization misinterpreting or overinterpreting its data and suggest to them a different, albeit "old-fashioned," method, or if I simply said someone was wrong, I could expect a telephone call or e-mail from Ed saying, essentially, "Right on, baby."

At least as much as I enjoyed hearing about his experiences and rereading his early papers with Hilary Koprowski, Bill Hammon, and many more of the founding fathers of virology, I enjoyed his company. Ed was tough; he didn't take any guff from anyone, although he usually was diplomatic and always polite. He had a marvelous sense of humor, and he was socially adventurous, generous of his time, and patient with young people. Through sorrows and disappointments, Ed Lennette maintained his love of life. He was one of the great people in my life and in the lives of many, many others, and his influence lives on, as witness this book.

Charles H. Calisher

Preface

One of the great joys of editing this latest version of *Laboratory Diagnosis of Viral Infections* has been learning more about the original editor, Dr. Edwin H. Lennette. Although I never had the privilege of meeting Ed, the stories and anecdotes cheerfully offered by his many friends and colleagues have made me feel as if I knew him. At the same time, this has presented a problem—to try to meet the impossibly high bar Ed set as an editor who also had a distinguished career as a scientist and diagnostician. Although it would be futile to try to fill Ed's shoes, I hope that this text fulfills the promise to inform the field he loved so well. To honor Ed, then, it is altogether fitting that this series has been renamed *Lennette's Laboratory Diagnosis of Viral Infections.*

A major challenge in assembling this text was finding a niche not already occupied by any of the other excellent books touching upon diagnostic virology. This inspired a significant reorganization. As in previous editions, the work is divided into two parts. Part one is similar in scope to that of the previous editions and provides a detailed description of the various techniques forming the foundation of modern diagnostic virology. Part two, on the other hand, presented a greater challenge. A simple listing of virus families and their various clinical manifestations was clearly the easiest option, but this approach has already been well traveled. Instead, we have taken a syndromic approach, an idea originally suggested by my colleague, Dr. Yi-Wei Tang. Thus, if presented with a patient having symptoms of viral encephalitis, for example, readers can now refer to the chapter on CNS infections, where they will find a differential diagnosis of potential causative agents, along with suggestions for the appropriate diagnostic approach. While this reorganization has brought its own challenges in avoiding redundancy and omissions, I believe this unique approach will make the book particularly valuable to students of infectious disease as well as laboratorians.

Clinical virology has changed at an astounding pace in the 10 years since publication of the previous edition, and this edition has been completely rewritten to reflect this new reality. Molecular techniques continue to grow in importance and are covered in depth by new chapters on a variety of topics, including the design of molecular tests, the importance of genotyping and viral sequence analysis, and the use of microarrays in diagnostic virology. Another emerging theme is the increased awareness of global health issues, reflected here by a new chapter regarding viral testing in resource-limited settings. Finally, new associations continue to be made between clinical disease and viruses, and these are discussed in the chapters on respiratory infections, polyomavirus infections, hemorrhagic fevers, and elsewhere throughout the book.

The process of bringing this edition to reality owes much to Maria Lorusso at Informa, who initially brought the project to my attention, and Aimee Laussen, also of Informa, who has taken care of innumerable logistical issues since the early days of the project. I would also like to thank my colleague, Dr. Rhoda Morrow, for advice and support at many stages along the way.

In his preface to the first edition, Ed stated that the book was directed toward the laboratorian who needs a ready reference source to assist in reaching a laboratory diagnosis of a viral infection. This remains the goal of the new edition; no easy task given the rapid changes in technology, the continuing emergence of new viruses, and newly described viral etiologies for clinical syndromes. I hope that readers of this new edition will find the book useful and will gain a little of Ed's enthusiasm for this ever changing and endlessly fascinating field.

Keith R. Jerome

Contents

Contributors

Robert L. Atmar Department of Medicine and Department of Molecular Virology and Microbiology, Baylor College of Medicine, Houston, Texas, U.S.A.

Jeffrey M. Bender Department of Pediatrics, University of Utah School of Medicine, Salt Lake City, Utah, U.S.A.

Annemarie Berger Institute for Medical Virology, Johann Wolfgang Goethe University Frankfurt/Main, Germany

Karen C. Bloch Departments of Medicine and Preventive Medicine, Vanderbilt University School of Medicine, Nashville, Tennessee, U.S.A.

D. S. Bowden Victorian Infectious Diseases Reference Laboratory, Victoria, Australia

James L. Caruso United States Navy and Duke University Medical Center, Durham, North Carolina, U.S.A.

Michael Chang Department of Medicine, Portland VA Medical Center, Portland, Oregon, U.S.A.

John M. Childs United States Navy and Duke University Medical Center, Durham, North Carolina, U.S.A.

Paola Cinque Clinic of Infectious Diseases, San Raffaele Hospital, Milan, Italy

Robert W. Coombs Department of Laboratory Medicine, University of Washington, Seattle, Washington, U.S.A.

Rosella Creed Center for Clinical Studies, Houston, Texas, U.S.A.

Oliver Donoso-Mantke Robert Koch-Institut, Berlin, Germany

Alexis Dumoulin Transplantation Virology and Division of Diagnostics, Institute for Medical Microbiology, Department of Biomedicine, University of Basel, Basel, Switzerland

Mary K. Estes Department of Medicine and Department of Molecular Virology and Microbiology, Baylor College of Medicine, Houston, Texas, U.S.A.

Bagher Forghani Viral Immunoserology Section, Viral and Rickettsial Disease Laboratory, California Department of Public Health, Richmond, California, U.S.A.

Julie Fox Department of Microbiology and Infectious Diseases, University of Calgary and Provincial Laboratory for Public Health (ProvLab), Calgary, Alberta, Canada

Patti E. Gravitt Department of Epidemiology, Johns Hopkins Bloomberg School of Public Health, Baltimore, Maryland, U.S.A.

Stephen B. Greenberg Department of Medicine and Department of Molecular Virology and Microbiology, Baylor College of Medicine, Houston, Texas, U.S.A.

Hans H. Hirsch Transplantation Virology and Division of Diagnostics, Institute for Medical Microbiology, Department of Biomedicine, University of Basel, and Division of Infectious Diseases and Hospital Epidemiology, University Hospital of Basel, Basel, Switzerland

Richard L. Hodinka Departments of Pathology and Pediatrics, Clinical Virology Laboratory, Children's Hospital of Philadelphia and University of Pennsylvania School of Medicine, Philadelphia, Pennsylvania, U.S.A.

Noah Hoffman Department of Laboratory Medicine, University of Washington, Seattle, Washington, U.S.A.

David N. Howell Duke University Medical Center and Veterans Affairs Medical Center, Durham, North Carolina, U.S.A.

Peter B. Jahrling NIAID Integrated Research Facility at Fort Detrick (IRF-Frederick), Fort Detrick, Frederick, Maryland, U.S.A.

Holly Janes Division of Public Health Sciences and Program in Biostatistics and Vaccine and Infectious Disease Institute, Fred Hutchinson Cancer Research Center, Seattle, Washington, U.S.A.

Keith R. Jerome Department of Laboratory Medicine, University of Washington, and Vaccine and Infectious Disease Institute, Fred Hutchinson Cancer Research Center, Seattle, Washington, U.S.A.

Uwe Kühl Medizinische Klinik II, Cardiology und Pneumonology, Charité—Universitätsmedizin Berlin, Berlin, Germany

Harald H. Kessler Molecular Diagnostics Laboratory, Institute of Hygiene, Microbiology, and Environmental Medicine, Medical University of Graz, Austria

David Kimberlin Department of Pediatrics, University of Alabama at Birmingham, Birmingham, Alabama, U.S.A.

Jens H. Kuhn NIAID Integrated Research Facility at Fort Detrick (IRF-Frederick), Fort Detrick, Frederick, Maryland, and Tunnell Consulting, Inc., King of Prussia, Pennsylvania, U.S.A.

Marie Louise Landry Department of Laboratory Medicine, Yale University School of Medicine, New Haven, Connecticut, U.S.A.

Diane Leland Department of Pathology and Laboratory Medicine, Indiana University School of Medicine, Indianapolis, Indiana, U.S.A.

Thomas J. Liesegang Mayo Clinic, Jacksonville, Florida, U.S.A.

Amalia Magaret Department of Laboratory Medicine, University of Washington, and Program in Biostatistics and Vaccine and Infectious Disease Institute, Fred Hutchinson Cancer Research Center, Seattle, Washington, U.S.A.

Natalia Mendoza Center for Clinical Studies, Houston, Texas, U.S.A.

Sara E. Miller Department of Pathology, Duke University Medical Center, Durham, North Carolina, U.S.A.

Rhoda A. Morrow Department of Laboratory Medicine, University of Washington, and Seattle Children's, Seattle, Washington, U.S.A.

Matthias Niedrig Robert Koch-Institut, Berlin, Germany

Andreas Nitsche Robert Koch-Institut, Berlin, Germany

Steven A. Pergam Department of Medicine, University of Washington, and Vaccine and Infectious Disease Institute, Fred Hutchinson Cancer Research Center, Seattle, Washington, U.S.A.

Cathy A. Petti Department of Pathology, University of Utah School of Medicine, Associated Regional University Pathologists (ARUP) Laboratories, and Department of Medicine, University of Utah School of Medicine, Salt Lake City, Utah, U.S.A.

Holger F. Rabenau Institute for Medical Virology, Johann Wolfgang Goethe University Frankfurt/Main, Germany

Sheli R. Radoshitzky United States Army Medical Research Institute of Infectious Diseases (USAMRIID), Fort Detrick, Frederick, Maryland, U.S.A.

Reinhard B. Raggam Molecular Diagnostics Laboratory, Institute of Hygiene, Microbiology, and Environmental Medicine, Medical University of Graz, Austria

Parisa Ravanfar Department of Dermatology, Dartmouth-Hitchcock Medical Center, Lebanon, New Hampshire, U.S.A.

P. A. Revill Victorian Infectious Diseases Reference Laboratory, Victoria, Australia

Julie E. Reznicek Department of Medicine, Division of Infectious Diseases, Vanderbilt University School of Medicine, Nashville, Tennessee, U.S.A.

Brian Rha Department of Pediatrics, University of Alabama at Birmingham, Birmingham, Alabama, U.S.A.

Misha Roshal Department of Laboratory Medicine, University of Washington, Seattle, Washington, U.S.A.

Anita Satyaprakash Center for Clinical Studies, Houston, Texas, U.S.A.

Heinz-Peter Schultheiss Medizinische Klinik II, Cardiology und Pneumonology, Charité—Universitätsmedizin Berlin, Berlin, Germany

John Scott Department of Medicine, University of Washington, Seattle, Washington, U.S.A.

Rosemary C. She Department of Pathology, University of Utah School of Medicine, Salt Lake City, Utah, U.S.A.

Theo P. Sloots Queensland Paediatric Infectious Diseases Laboratory, Sir Albert Sakzewski Virus Research Centre, Queensland Children's Medical Research Institute, Children's Health Service District, Brisbane, Queensland, Australia

Yi-Wei Tang Departments of Pathology and Medicine, Vanderbilt University School of Medicine, Nashville, Tennessee, U.S.A.

Stephen Tyring Center for Clinical Studies, Houston, Texas, U.S.A.

Anna Wald Departments of Medicine, Laboratory Medicine, Epidemiology, University of Washington, and Vaccine and Infectious Disease Institute, Fred Hutchinson Cancer Research Center, Seattle, Washington, U.S.A.

David M. Whiley Queensland Paediatric Infectious Diseases Laboratory, Sir Albert Sakzewski Virus Research Centre, Queensland Children's Medical Research Institute, Children's Health Service District, Brisbane, Queensland, Australia

Richard Whitley Department of Pediatrics, University of Alabama at Birmingham, Birmingham, Alabama, U.S.A.

P. A. White School of Biotechnology and Biomolecular Sciences (BABS), University of New South Wales, Sydney, Australia

Atif Zaman Department of Medicine, Oregon Health and Sciences University, Portland, Oregon, U.S.A.

Danielle M. Zerr Department of Pediatrics, University of Washington and Seattle Children's Hospital, Seattle, Washington, U.S.A.

1 Verification and Validation of Virological Laboratory Tests in the Routine Diagnostic Laboratory

Holger F. Rabenau
Institute for Medical Virology, Johann Wolfgang Goethe University Frankfurt/Main, Germany

Harald H. Kessler and Reinhard B. Raggam
Molecular Diagnostics Laboratory, Institute of Hygiene, Microbiology, and Environmental Medicine, Medical University of Graz, Austria

Annemarie Berger
Institute for Medical Virology, Johann Wolfgang Goethe University Frankfurt/Main, Germany

INTRODUCTION

Routine viral diagnostics includes techniques for both indirect and direct detection of viruses. Indirect detection of viruses is performed by serological studies. Techniques for direct detection of viruses include detection of viral antigens, viruses, or viral components by isolation of viruses on cell cultures (or through animal experiments), and detection of viral nucleic acids is also referred to as nucleic acid testing (NAT). Furthermore, viral morphologic structures can be investigated by means of transmission electron microscopy.

Today, NAT is having a major impact on viral diagnostics. Molecular assays are used in many if not most virological laboratories. Technological improvements, from automated sample preparation to real-time amplification technology, provide the possibility to develop and introduce assays for most viruses of clinical interest. The risk of contamination has been reduced significantly and the turnaround time to generate results shortened. In contrast, standardization and quality assurance/quality control issues have often remained underemphasized, requiring urgent improvement.

Moreover, it must be taken into consideration that reliable viral diagnostics depend on additional preanalytical issues, such as choice of the correct sample material, optimal sampling time with regard to the course of disease, and the duration and conditions of sample transport to the laboratory.

QUALITY ASSURANCE/QUALITY CONTROL

In the international standard ISO 15189, special requirements for medical laboratories have been established. Among several issues, this standard demands certain verification and validation procedures. For laboratories in the United States, the FDA has established regulations based on existing ISO standards (1).

The European Union's Directive on In Vitro Diagnostic (IVD) Medical Devices (98/79/EC) requires data demonstrating that an IVD achieves the stated performance and will continue to perform properly after it has been shipped, stored, and put to use at its final destination (2).

Quality control systems have been implemented in the majority of routine diagnostic laboratories. In contrast to certification that is mainly based on the supervision, description, and conformity of processes, accreditation additionally focuses on the competence of the laboratory providing reliable test results and their correct interpretation.

Quality assurance requires careful documentation in the routine diagnostic laboratory. For each newly implemented test or test system, a standard operating procedure must be available. Additionally, verification or validation data must be available for each test.

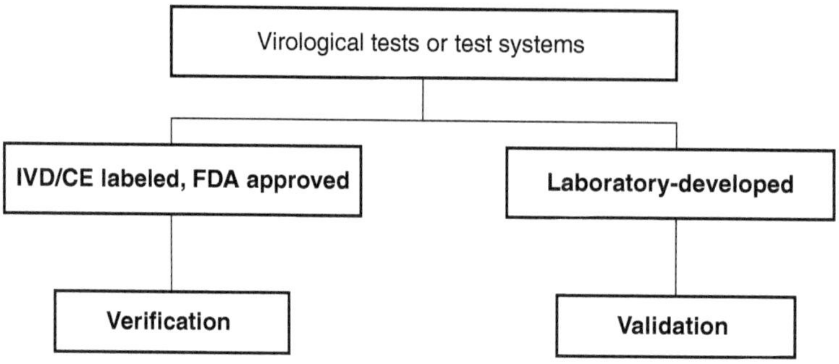

Figure 1 Verification/validation of virological tests or test systems.

Verification and Validation of Tests or Test Systems Employed in the Routine Laboratory

Suitability of a technique does not necessarily mean that it is performed correctly and provides valid results. The ISO 15189, the IVD Directive 98/79/EC, and the FDA regulations (clearly described in the *Code of Federal Regulations*) require verification or validation of each investigational procedure in order to prove both the correct application and the correct performance of a diagnostic test. The complexity and the extent of the verification or validation procedure depend on whether an IVD/CE-labeled and/or FDA-approved test or a "home-brewed" laboratory-developed test or test system is involved. For a laboratory-developed test or test system, "analyte-specific reagents" (ASRs), medical devices that are regulated by the FDA, should be used preferentially. Implementation of any reagent labeled "research use only" (RUO) is not permitted for any test or test system in the United States routine laboratory. In Europe, one or more RUO reagents may be implemented following validation of the test or test system. Both terms "in vitro diagnostic medical device" as used in the IVD Directive 98/79/EC and "device" as used in the FDA regulations do not only mean "test" but also "test system" if more than a single component is required to generate a diagnostic result. For instance, molecular test systems based on PCR usually consist of a combination of different reagents and instruments for nucleic acid extraction, amplification, and detection of amplification products.

Verification or validation work has to be done if a new test or test system is introduced in the routine diagnostic laboratory (Fig. 1). Additionally, any change of an existing test procedure requires further validation work (3).

For a commercially available IVD/CE-labeled and/or FDA-approved test or test system, the manufacturer is responsible that the IVD achieves the performance as stated. Nevertheless, the user must verify that performance characteristics, such as accuracy and precision, are achieved in the laboratory (Table 1). The accuracy (or "trueness" in the recent nomenclature) is defined as the degree of conformity of a measured or calculated quantity to its actual (true) value and can be estimated by analyses of reference materials or comparisons of results with those obtained by a reference method (Fig. 2). These are the only accepted approaches to trueness. When neither is available, other evidence is required to record the ability of the method to measure the analyte. The imprecision is defined as the level of deviation of the individual test results within a single run (intraassay imprecision) and from one run to another (interassay imprecision) (Fig. 2). Imprecision is usually characterized in terms of the standard deviation of the measurements and relative standard variation (variation coefficient). In case of a

Table 1 Minimum Requirements for Verification or Validation of a Test or a Test System in Clinical Virology

Verification	Validation
Accuracy	Accuracy
Imprecision (intra- and interassay)	Recovery
Linearity (if quantitative)	Selectivity
	Imprecision (intra- and interassay)
	Linearity (if quantitative)

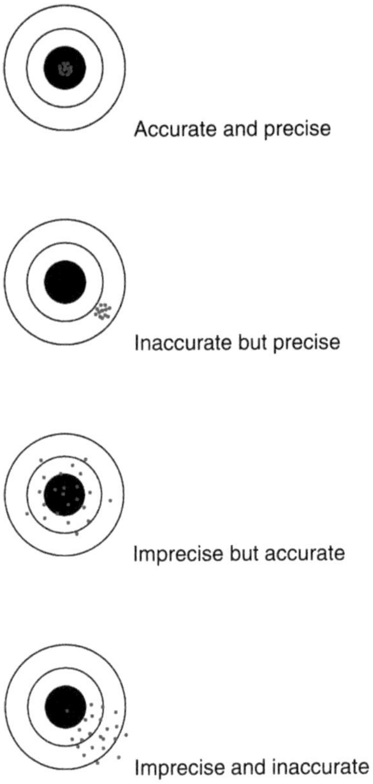

Figure 2 Accuracy ("trueness") and imprecision.

quantitative test or test system, the linearity must be evaluated additionally (Table 1). The linearity is defined as the determination of the linear range of quantification. Data for linearity studies should be subjected to linear regression analysis with an ideal regression coefficient of 1. In case of a nonlinear curve, any objective, statistically valid method may be used (4).

In contrast, the clinical laboratory that uses laboratory-developed test or test systems, or combines different IVD/CE-labeled and/or FDA-approved tests or test systems without recommendation of the manufacturer, is acting as manufacturer of a medical device and thus responsible for both the suitability and the correct performance of the test. Those tests or test systems must be validated including accuracy, recovery, selectivity, imprecision, and, if quantitative, linearity (Table 1). Recovery (also known as "analytical sensitivity") studies involve analyses after known amounts of analyte are added to the biological matrix on which the determination will be performed. Selectivity (also known as "analytical specificity") testing reflects the ability of an analytical method to detect an analyte (and quantify it in case of a quantitative test or test system) in complex mixtures of biological sample material also referred to as matrix. For selectivity testing, cross-reactivity with any other analyte has to be excluded. Furthermore, interference studies must be performed to assess the effects of possible interferents including, for instance, hemoglobin, rheumatoid factor, and autoantibodies, and those of exogenous materials, such as ingredients of blood collection containers and commonly used or coadministered drugs. It is important to mention that the introduction of an internal control (IC; see below) checks for a possible matrix-induced effect and ensures the reliability of a NAT test or test system.

Minimum requirements for verification and validation procedures for virological tests or test systems are described in the following sections. A more simplified validation procedure may be applied if calibrators are not commonly accessible or if a test or test system for validation is based on a scientific publication. In general, reference material, patient samples, or pooled sera may serve as calibrators for a verification or validation experiment. If patient samples or pooled sera are used, they must have been tested earlier with the existing "gold standard," as far as available and/or defined. Calibrators are classified into positive, low-positive, and negative controls. For detection of virus-specific antibodies and viral antigens, positive controls are defined as having concentrations more than threefold above the limit of detection (LOD; see below) or the limit of quantitation (LOQ; see below) of the test or test system, and within the upper limit of linearity, while low-positive controls are defined as having concentrations up to threefold over the LOD or the LOQ of the test or test system. For NAT, positive controls are defined as having concentrations more than 1 $\log_{10}$ over the LOD or the LOQ of the test or test system and within the upper limit of linearity for detection of virus-specific antibodies and viral antigens, while low-positive controls are defined as having concentrations up to 1 $\log_{10}$ over the LOD or LOQ of the test or test system. If more than one positive control is necessary to complete testing for certain performance characteristics, they should always contain different concentrations (within the linearity range as defined above) of the parameter to be tested.

Minimum requirements outlined in this chapter are valid for all verification and validation procedures in clinical virology. However, tests or test systems for pathogens included in List A of Annex II to Directive 98/79/EC (human immunodeficiency virus type 1 and 2, human T-cell lymphotropic virus type I and II, hepatitis B, C, and D viruses) are not covered here because of special regulations (Directive 98/79/EC, Article 9). Common technical specifications enforced for tests or test systems on those parameters are outlined in the Commission Decision of May 7, 2002, on common technical specifications for IVD medical devices (5).

Minimum Requirements for Verification of IVD/CE-Labeled and/or FDA-Approved Tests or Test Systems for Detection of Virus-Specific Antibodies, Viral Antigens, or NAT

If a new IVD/CE-labeled test or test system for detection of virus-specific antibodies, viral antigens, or NAT is introduced in the routine diagnostic laboratory, verification experiments are performed to verify accuracy, imprecision, and, in case of a quantitative test or test system, linearity (Table 2). For determination of the accuracy, three positive, three low-positive, and three negative samples are used. In case of a qualitative test or test system, one positive and one low-positive sample are used for determination of intraassay imprecision. Each sample is tested three times within a run. For interassay imprecision, one positive and one low-positive sample are used. Each sample is tested one time on three different days. In case of a quantitative test or test system for detection of virus-specific antibodies or viral antigens, four positive and three low-positive samples are used for determination of intraassay imprecision, and two positive and one low-positive sample for determination of interassay imprecision. The corresponding recommendations for a quantitative NAT test or test system are three positive and three low-positive samples each for determination of intraassay imprecision and one positive and

Table 2 Verification of IVD/CE-Labeled and/or FDA-Approved Tests or Test Systems for Detection of Virus-Specific Antibodies, Viral Antigens, or Viral Nucleic Acid Testing

		No. of samples required			
		Detection of antibodies or antigens		Nucleic acid testing	
	Calibrator (sample)	Qualitative	Quantitative	Qualitative	Quantitative
Accuracy	Positive[a]	3	3	3	3
	Low positive[b]	3	3	3	3
	Negative	3	3	3	3
Intraassay imprecision	Positive[a]	1	4	1	3
	Low positive[b]	1	3	1	3
Interassay imprecision	Positive[a]	1	2	1	1
	Low positive[b]	1	1	1	1
Linearity	Positive[a]	0	1	0	1

[a]More than 1 $\log_{10}$ over the limit of detection (LOD) or the limit of quantification (LOQ) and within the upper limit of linearity of the test or test system.

[b]Up to 1 $\log_{10}$ over the LOD or the LOQ of the test or test system.

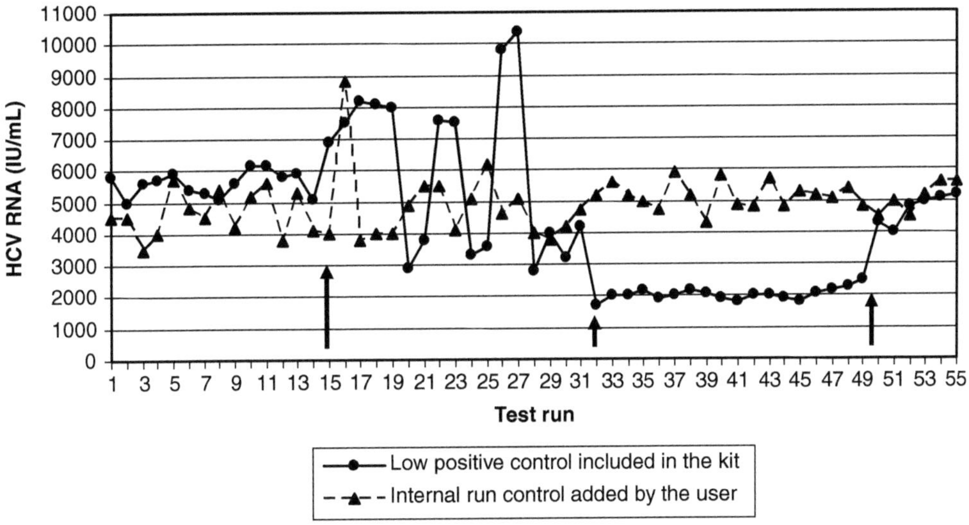

Figure 3 Performance of an internal run control (IRC) implemented in NAT (*arrows* indicating introduction of a new test lot).

low-positive sample each for determination of interassay imprecision. In order to optimize the verification workflow, it may be useful to take the first result of intraassay imprecision testing as first result of interassay imprecision testing thus allowing a reduction of the number of further runs for interassay imprecision testing to two. In case of a quantitative test or test system, linearity must be verified additionally by analyzing a serial dilution (tenfold dilution series with at least three dilution steps) of one positive sample in duplicate.

Additionally, it is recommended to survey the correctness of a test result obtained by an IVD/CE-labeled and/or an FDA-approved test or test system continuously after implementation in the routine diagnostic laboratory. This is achieved by introduction of an internal run control (IRC), which is independent from the positive control(s) included by the manufacturer of the test or test system and may be implemented either in each test run or within defined intervals. When introducing a new test lot, comparison of the results obtained by the IRC with those obtained by the positive control(s) included by the manufacturer of the test or test system enables identification of relevant aberrations at an early stage (Fig. 3). Statistical analysis of results obtained by both the IRC and the positive control(s) may also be helpful.

Minimum Requirements for Validation of a Laboratory-Developed Test or Test System for Detection of Virus-Specific Antibodies, Viral Antigens, or NAT

If a laboratory-developed test or test system for detection of virus-specific antibodies, viral antigens, or NAT is introduced in the routine diagnostic laboratory, validation experiments are performed to validate accuracy, recovery, selectivity, imprecision, and, in case of a quantitative test or test system, linearity (Table 3). For determination of the accuracy, three positive, three low-positive, and three negative samples are used. For recovery, 10 positive and 10 low-positive samples are tested. The selectivity of a test or test system for detection of virus-specific antibodies is determined by analyzing 10 negative samples including samples containing antibodies that may lead to cross-reactivity. For tests or test systems detecting viral antigens or NAT, 10 samples testing positive for antigens or viruses of the same family and samples spiked with reference material that may lead to cross-reactivity are analyzed. Each potentially cross-reactive analyte must be present in a high concentration (at least 10^5 $TCID_{50}$/mL or 10^5 genome equivalents/mL). Additionally, selectivity testing requires 10 low-positive samples including, for instance, samples with elevated hemoglobin levels, testing positive for rheumatoid factor, and/or containing auto-antibodies. Determination of intra- and interassay imprecision are similar to those for verification procedures except for an extension in the validation of quantitative tests or test systems regarding positive samples (use of six positives instead of three for

Table 3 Validation of a Laboratory-Developed Test or Test System for Detection of Virus-Specific Antibodies, Viral Antigens, or Viral Nucleic Acid Testing

		No. of samples required			
		Detection of antibodies or antigens		Nucleic acid testing	
	Calibrator (sample)	Qualitative	Quantitative	Qualitative	Quantitative
Accuracy	Positive[a]	3	3	3	3
	Low positive[b]	3	3	3	3
	Negative	3	3	3	3
Recovery	Positive[a]	10	10	10	10
	Low positive[b]	10	10	10	10
Selectivity	Negative[c]	10	10	10	10
	Low positive[b,d]	10	10	10	10
Intraassay imprecision	Positive[a]	1	6	1	6
	Low positive[b]	1	3	1	3
Interassay imprecision	Positive[a]	1	2	1	2
	Low positive[b]	1	1	1	1
Linearity	Positive[a,e]	0	2	0	2

[a]More than 1 $\log_{10}$ over the limit of detection (LOD) or the limit of quantification (LOQ) and within the upper limit of linearity of the test or test system.
[b]Up to 1 $\log_{10}$ over the LOD or the LOQ of the test or test system.
[c]Samples that may lead to cross-reactivity.
[d]Samples including possible interferents.
[e]Serial dilutions (at least four dilution steps) in duplicate on two different days.

determination of intraassay imprecision and two instead of one for determination of interassay imprecision). In the case of a quantitative laboratory-developed test or test system, linearity must be validated additionally by analyzing serial dilutions (at least four dilution steps) of two positive samples in duplicate on two different days.

Issues Regarding Introduction of a Laboratory-Developed NAT Assay

When establishing a laboratory-developed NAT assay, primer and probe sequences must be checked carefully by use of a genome sequence databank. It is advisable to verify the amplification product by means of sequencing and to use a primer pair that has already been published in a highly recognized journal. The latter helps to avoid testing of a more or less extended specificity panel. However, the published sequences should always be subjected to an alignment analysis by means of a genome sequence databank to ensure that the correct sequence has been published.

Moreover, several issues including the molecular technique employed, the detection format, introduction of an IC, and quantitation must be addressed. With regard to the molecular technique employed, it must be taken into consideration that automation reduces hands-on work and thus helps avoid human error. To ensure analyte-specific results, introduction of a probe detection format is required while melting curve analysis without probe detection format does not provide sufficient specificity. Because amplification may fail in a reaction due to interference from inhibitors, an IC must be incorporated in every NAT assay to exclude false-negative results. To ensure an accurate control of the entire NAT assay, the IC should be added to the sample before the start of the nucleic acid extraction procedure. Either a homologous or a heterologous IC can be employed. The homologous IC is a DNA sequence (for DNA amplification targets) or an in vitro transcript (for RNA targets) consisting of primer-binding regions identical to those of the target sequence, a randomized internal sequence with a length and base composition similar to those of the target sequence and a unique probe-binding region that differentiates the IC amplification product from the target amplification product. Either a single IC or multiple ICs for a set of NAT assays can be generated (6,7). In contrast to the homologous internal control, the heterologous internal control represents a second amplification system within the same reaction vessel. The control must have the same or similar extraction and amplification efficiencies as the target. Plasmids or housekeeping genes can be used as heterologous internal controls (8). Any IC (homologous or heterologous) must be added at a suitable concentration to prevent extreme competition with the target template for reagents. When PCR-based NAT assays are introduced, quantitation by end-point analysis should be avoided; instead, log-phase analysis is preferable.

Further Considerations Regarding Validation of a Laboratory-Developed Test or Test System

When employing a laboratory-developed test or test system, it is mandatory to determine either the LOD or the LOQ. The LOD is defined as the lowest concentration or quantity of an analyte that can be reliably detected as being qualitatively present in the sample, while the LOQ is defined as the lowest concentration or quantity of an analyte that can be reproducibly quantified in a sample. The operational definition of those limits must be stated clearly in the validation protocol. A partly nonparametric approach for determining the LOD and the LOQ has been published recently (9). If there is no reference material available, both the determination of the LOD and LOQ are impossible. In this case, for the introduction of a laboratory-developed NAT assay, the application of the real-time PCR technique may be the best approach for yielding at least relatively quantitative results (10,11).

Furthermore, diagnostic accuracy must be included in the evaluation process, especially if an existing test or test system is modified or replaced. In studies of diagnostic accuracy, the outcome from a test or test system under evaluation is compared with the outcome from the reference test or test system. Proposed items to include in determination of diagnostic accuracy have been published recently (12). Diagnostic accuracy includes diagnostic sensitivity (the ability of an assay to detect individuals with the condition of interest in a group) and diagnostic specificity (the ability of an assay to correctly identify an individual who does not have the condition of interest). In clinical virology, a minimum requirement is the comparison of results obtained by the new test or test system with those obtained by the existing test or test system. To fulfill this, 20 samples (seven positives, six low positives, and seven negatives) must be tested in parallel.

Validation of Isolation of Viruses on Cell Cultures

Virus isolation on cell cultures is a technique that is difficult to standardize, thus validation is particularly demanding. First of all, the suitability of the cells for the detection of a certain virus must be proved. During the implementation of a new cell line as an indicator system, the cell line should be tested for its susceptibility with two concentrations of both a reference virus strain and a wild-type isolate. After titration of the virus stock, the inoculums should contain a multiplicity of infection of 0.1 (positive) and 0.01 (low positive). Tests must be done in triplicate on three days. Determination of imprecision is performed by using 20 wild-type samples that must be tested in parallel on the existing and the newly introduced cell line (Table 4). The viability of the cells and the influence of the sample matrix must be monitored and recorded carefully.

Table 4 Validation of Isolation of Viruses on Cell Cultures

	Sample requirements	No. of samples required
Susceptibility	Positive[a]	1
	Low positive[b]	1
Imprecision	Wild type	20

[a]Multiplicity of infection = 0.1.
[b]Multiplicity of infection = 0.01.

CONCLUSION

Implementation of a new test or test system in the routine diagnostic virological laboratory demands verification or validation procedures in compliance with a quality management system and according to ISO 15189 and/or the FDA regulatory framework. While CE/IVD-labeled and FDA-approved tests or test systems require verification, laboratory-developed tests or test systems demand validation. However, both verification and validation procedures are no guarantee of constant correctness of test results requiring continuous quality control measures in the routine diagnostic laboratory.

REFERENCES

1. U.S. Food and Drug Administration Website. http://www.fda.gov. Accessed August 25, 2009.
2. Directive 98/79/EC of the European Parliament and of the Council of 27 October 1998 on in vitro diagnostic medical devices. Official J Eur Communities 1998; L331:1–37.
3. Rabenau HF, Kessler HH, Kortenbusch M, et al. Verification and validation of diagnostic laboratory tests in clinical virology. J Clin Virol 2007; 40:93–98.
4. Linnet K, Boyd JC. Selection and analytical evaluation of methods—With statistical techniques. In: Burtis CA, Ashwood ER, Bruns DE, eds. Tietz Textbook of Clinical Chemistry and Molecular Diagnostics, 4th ed. St. Louis, MO: Saunders, 2006:353–407.
5. Commission Decision of 7 May 2002 on common technical specifications for in vitro-diagnostic medical devices. Official J Eur Communities 2002; L131:17–30.
6. Raggam RB, Leitner E, Mühlbauer G, et al. Rapid detection of *Legionella* species in bronchoalveolar lavages and induced sputa by automated DNA extraction and real-time polymerase chain reaction. Med Microbiol Immunol 2002; 191:119–125.
7. Raggam RB, Leitner E, Berg J, et al. Single-run, parallel detection of DNA from three pneumonia-producing bacteria by real-time polymerase chain reaction. J Mol Diag 2005; 7:133–138.
8. Koidl C, Bozic M, Mossböck G, et al. Rapid diagnosis of adenoviral keratoconjunctivitis by a fully automated molecular assay. Ophthalmology 2005; 112:1521–1527.
9. Linnet K, Kondratovich M. Partly nonparametric approach for determining the limit of detection. Clin Chem 2004; 50:732–740.
10. Rasmussen R. Quantification on the LightCycler. In: Meuer S, Wittwer C, Nakagawara K, eds. Rapid Cycle Real-time PCR, Methods and Applications. Berlin: Springer-Verlag, 2001:21–34.
11. Koidl C, Bozic M, Berg J, et al. Detection of transfusion transmitted virus DNA by real-time PCR. J Clin Virol 2004; 29:277–281.
12. Bossuyt PM, Reitsma JB, Bruns DE, et al. The STARD statement for reporting studies of diagnostic accuracy: explanation and elaboration. Clin Chem 2003; 49:7–18.

2 Statistical Evaluation of Diagnostic Tests

Holly Janes

Division of Public Health Sciences and Program in Biostatistics and Vaccine and Infectious Disease Institute, Fred Hutchinson Cancer Research Center, Seattle, Washington, U.S.A.

Amalia Magaret

Department of Laboratory Medicine, University of Washington, and Program in Biostatistics and Vaccine and Infectious Disease Institute, Fred Hutchinson Cancer Research Center, Seattle, Washington, U.S.A.

INTRODUCTION

A useful medical test must satisfy a variety of criteria, including identifying disease that is serious, prevalent in the target population, and treatable; and that the test is not harmful to the individual and is accurate (1–4). An exception is the need for accurate diagnosis of some infectious diseases without serious morbidity in the host or effective treatments, as this diagnosis may aid in the prevention of transmission to others in whom morbidity may vary. This chapter describes the last criterion alone, the accuracy of the test.

We consider evaluation of three types of laboratory tests: diagnostic, screening, and prognostic tests. Diagnostic tests, such as serology, are aimed at diagnosing disease in symptomatic individuals. Screening tests, such as HIV tests applied in routine medical exams, are intended to detect preclinical disease. Prognostic tests, for example those used to determine infection subtype, are used to identify patients with good and poor prognosis. The statistical evaluation of these three types of tests has a common theme: the key question is how well the test discriminates between two groups of individuals. For simplicity, we refer to all tests as "diagnostic tests" and the two groups as "diseased" and "nondiseased" subjects, but the approaches are understood to apply equally well to prognostic and screening tests.

This chapter is one of two focusing on evaluation of laboratory tests. Chapter 1 focuses on laboratory validation of the assay, including assessment of test reproducibility. This chapter concerns the clinical evaluation of the test, namely its ability to distinguish between diseased and nondiseased individuals. We present methods for evaluating diagnostic accuracy that are appropriate for both commonly used types of study designs: a case-control design, where fixed numbers of diseased and nondiseased individuals are enrolled and then tests are performed on each group; or a cohort design where the test is applied to a population of interest and then true disease status is determined (4). Basic statistical methods for evaluating binary and continuous tests are described, approaches to handling indeterminate test results are discussed, and fundamental concepts in study design are introduced. The last section draws attention to more complex issues beyond the scope of the chapter and provides references for further reading.

EVALUATING BINARY TESTS

Consider a binary test (Y) used to diagnose disease (D). For example, enzyme-linked immunosorbent assays (ELISA) are used to test for HIV infection. Among individuals with disease ($D = 1$), the test result is either positive ($Y = 1$), called a true positive result, or negative ($Y = 0$), a false negative result (Table 1). Similarly, among nondiseased individuals ($D = 0$), true negative ($Y = 0$), and false positive ($Y = 1$) results may occur.

Classification Probabilities

The accuracy of a diagnostic test is typically characterized using a pair of classification probabilities. The true positive rate (TPR), or *sensitivity*, is the proportion of diseased subjects who are classified as positive by the test. The false positive rate (FPR) is the proportion of disease-free subjects who are classified as positive by the test. The FPR is equivalent to 1 minus the *specificity*

Table 1 Tabulation of the Outcomes of a Binary Diagnostic Test (Y) by True Disease Status (D)

	Disease status (D)		
Test (Y)	**Diseased (D = 1)**	**Nondiseased (D = 0)**	**Total**
Positive (Y = 1)	a	b	a + b
Negative (Y = 0)	c	d	c + d
Total	a + c	b + d	a + b + c + d

of the test, the proportion of disease-free subjects who are classified as negative. A perfect test detects all diseased individuals (TPR = 1) and no nondiseased individuals (FPR = 0); and a useless test, no better than a coin toss, is one for which diseased and nondiseased subjects have equal chance of testing positive (TPR = FPR). These parameters are relevant to public health practitioners, as they describe the value of the test when applied in the population. But they are also relevant to individual patients and clinicians, in helping to decide whether to use the test at all (5). If I am diseased, what is the chance that the test will detect this? If I am not diseased, what is the chance I will have a false-positive result? These are questions an individual will have before the test is performed.

The TPR and FPR can be estimated using data derived from cohort or case-control studies. The estimates, using notation from Table 1, are as follows:

$$\mathrm{TPR} = p(Y = 1|D = 1) = \frac{a}{a + c} \tag{1}$$

$$\mathrm{FPR} = p(Y = 1|D = 0) = \frac{b}{b + d}$$

Confidence intervals should be provided in order to characterize the precision of the estimates. While standard binomial-based confidence intervals can be used and are available in most statistical software packages, confidence intervals based on logit-transformations of the TPR and FPR often have better performance and are easy to calculate (4). The formula for a 95% confidence interval (CI) for the TPR is

$$\frac{\exp\left(\log[\mathrm{TPR}/(1-\mathrm{TPR})] \pm 1.96\sqrt{1/[\mathrm{TPR}(1-\mathrm{TPR})(a+c)]}\right)}{1+\exp\left(\log[\mathrm{TPR}/(1-\mathrm{TPR})] \pm 1.96\sqrt{1/[\mathrm{TPR}(1-\mathrm{TPR})(a+c)]}\right)},$$

where the natural log is used. The corresponding formula for 95% CI for the FPR is:

$$\frac{\exp\left(\log[\mathrm{FPR}/(1-\mathrm{FPR})] \pm 1.96\sqrt{1/[\mathrm{FPR}(1-\mathrm{FPR})(b+d)]}\right)}{1+\exp\left(\log[\mathrm{FPR}/(1-\mathrm{FPR})] \pm 1.96\sqrt{1/[\mathrm{FPR}(1-\mathrm{FPR})(b+d)]}\right)}.$$

To illustrate, we consider the following example. We quantify the accuracy of the Focus ELISA test (Y) for diagnosing HSV-2 infection (D) (Table 2) using data from a representative

Table 2 Example: The Accuracy of Focus ELISA for Diagnosing HSV-2 Infection Among Subjects Undergoing HIV Testing at STD Clinics in Kampala, Uganda[a]

	HSV-2 Status		
Focus ELISA result	**Infected (D = 1)**	**Uninfected (D = 0)**	**Total**
Positive (Y = 1)	142	27	169
Negative (Y = 0)	1	75	76
Total	143	102	245

[a]True HSV-2 status is determined using Western Blot.

sample of subjects undergoing HIV testing at STD clinics in Kampala, Uganda (6). True HSV-2 status is determined using the Western Blot test, which is considered the gold standard test for HSV-2 seropositivity. We estimate that the Focus ELISA test detects 142/143 = 99.3% of HSV-2 infected subjects (95% CI: 95.2–99.9%). However, 27/102 = 26.5% of HSV-2 uninfected subjects also test positive (95% CI: 18.8–35.9%). While this high FPR may be acceptable in HIV-infected populations where a false-positive HSV-2 result is relatively inconsequential, in other populations the high FPR would likely make use of the test impractical. It turns out that the Focus test has much better performance in the US population (7).

Predictive Values

Predictive values are different measures of test performance that describe how well the test results reflect true disease status. Whereas the classification probabilities describe test results for diseased and nondiseased subjects, the predictive values describe disease outcomes for those who test positive and those who test negative. Therefore, they are most relevant to individual patients and clinicians who have test results in hand, and can provide guidance in making subsequent treatment decisions. The positive predictive value (PPV) is the proportion of subjects found positive by the test who are in fact diseased. The negative predictive value (NPV) is the proportion of subjects classified as negative by the test who are in fact disease-free. A perfect test is one for which all subjects who test positive are diseased (PPV = 1) and all subjects who test negative are disease-free (NPV = 1). A test is useless when the chance of being diseased is the same regardless of test result (PPV = 1 − NPV).

Predictive values can also be interpreted as refinements of the pretest probability of disease or the disease prevalence, P(D = 1), based on the test result. The PPV is the updated probability of disease in populations among those who test positive, and 1 minus the NPV is the updated probability of disease for those who test negative.

The predictive values are directly estimable from data collected under a cohort study design. Referring to Table 1, the estimates are given by:

$$\mathrm{PPV} = \mathrm{p}(\mathrm{D}=1|\mathrm{Y}=1) = \frac{\mathrm{a}}{\mathrm{a}+\mathrm{b}}$$

$$\mathrm{NPV} = \mathrm{p}(\mathrm{D}=0|\mathrm{Y}=0) = \frac{\mathrm{d}}{\mathrm{c}+\mathrm{d}}.$$

Confidence intervals are of the same form as those for TPR/FPR. For PPV, the formula is:

$$\frac{\exp\left(\log[\mathrm{PPV}/(1-\mathrm{PPV})] \pm 1.96\sqrt{1/[\mathrm{PPV}(1-\mathrm{PPV})(\mathrm{a}+\mathrm{b})]}\right)}{1+\exp\left(\log[\mathrm{PPV}/(1-\mathrm{PPV})] \pm 1.96\sqrt{1/[\mathrm{PPV}(1-\mathrm{PPV})(\mathrm{a}+\mathrm{b})]}\right)},$$

and for NPV the formula is:

$$\frac{\exp\left(\log[\mathrm{NPV}/(1-\mathrm{NPV})] \pm 1.96\sqrt{1/[\mathrm{NPV}(1-\mathrm{NPV})(\mathrm{c}+\mathrm{d})]}\right)}{1+\exp\left(\log[\mathrm{NPV}/(1-\mathrm{NPV})] \pm 1.96\sqrt{1/[\mathrm{NPV}(1-\mathrm{NPV})(\mathrm{c}+\mathrm{d})]}\right)}.$$

We illustrate estimation of predictive values using the HSV-2 example. Recall that the data were collected under a cohort study design. We estimate that 142/169 = 84.0% of subjects who test positive are in fact HSV-2 infected (95% CI: 77.7–88.8%); 16.0% of these individuals are actually not infected. A negative test result is more reliable; 75/76 = 98.7% of subjects who test negative are in fact HSV-2 seronegative (95% CI: 91.2–99.8%).

Predictive values cannot be directly estimated from case-control data because under a case-control design the proportion of subjects with disease (the prevalence) is fixed in the study by design. However, there is a direct relationship between predictive values, disease prevalence, and classification probabilities that can be exploited:

$$\text{PPV} = \frac{\text{TPRp(D = 1)}}{\text{TPRp(D = 1)} + \text{FPR}[1 - \text{p(D = 1)}]}$$

$$\text{NPV} = \frac{(1 - \text{FPR})(1 - \text{p(D = 1)})}{(1 - \text{FPR})[1 - \text{p(D = 1)}] + (1 - \text{TPR})\text{p(D = 1)}}, \qquad (2)$$

where p(D = 1) is the population prevalence. Expression (2) can be used to estimate the PPV and NPV where the TPR and FPR are calculated from case-control data and a prevalence estimate is obtained from an external source.

Expression (2) also shows explicitly how the values of PPV and NPV depend on prevalence, and thus vary between populations with different prevalences. In the HSV-2 example shown in Table 2, the prevalence was 143/245 = 58.4%. However, if the Focus ELISA were applied in a population with a lower prevalence of 10%, the same TPR and FPR (99.3% and 26.5%, respectively) would result in a PPV of 29.4% and a NPV of 99.9%. In this setting, the PPV is much lower, since few people are truly infected. This example illustrates that predictive values are population-specific; and tests that discriminate well based on the TPR and FPR can have low PPVs in low prevalence settings (8).

Other Measures of Test Performance

In epidemiologic studies, the odds ratio is commonly used to describe the association between two binary variables. While a useful measure of association, the odds ratio does not characterize classification accuracy of a diagnostic test (9). The odds ratio relating a binary test result to disease status is a function of the TPR and FPR of the test (9). Using notation from Table 1:

$$\text{Odds ratio} = \frac{\text{TPR}(1 - \text{TPR})}{\text{FPR}(1 - \text{FPR})} = \frac{\text{ad}}{\text{bc}}.$$

The odds ratio combines the TPR and FPR into a single number, and therefore many (TPR, FPR) pairs are consistent with the same odds ratio. More specifically, the odds ratio does not distinguish between a high TPR and a low FPR. For example, the Focus ELISA test shown in Table 2 has an estimated odds ratio of 394. That is, the odds of a positive Focus test are 394 times higher in those with HSV-2 infection than in those without. If the numbers of false-positive and false-negative test results (27 and 1) were switched, the TPR rate would decrease from 99.3% to 84.0% and the FPR would decrease from 28.7% to 1.3%, but the odds ratio would still be 394. Yet these new operating characteristics would have very different implications. The two components of test accuracy, the TPR and the FPR, must be reported separately (9).

It is also worth noting that tests with good classification accuracy have odds ratios much higher than those usually reported in studies measuring association (9). We saw above that a TPR of 99.3% and FPR of 28.7% yields an odds ratio of 394. Therefore, demonstrating that an odds ratio is statistically significantly greater than 1 is inadequate for concluding that the test has good classification accuracy.

Another common single summary measure of test performance is the misclassification rate, or the proportion of test results that are inconsistent with true disease status. Using notation from Table 1:

$$\text{Misclassification rate} = \frac{\text{b} + \text{c}}{\text{a} + \text{b} + \text{c} + \text{d}} = (1 - \text{TPR})\text{p(D = 1)} + \text{FPRP(D = 0)}$$

Note that this parameter is also a function of the TPR and FPR of the test, as well as the prevalence. In the HSV-2 example shown in Table 2, the estimated misclassification rate is 11.4% regardless of whether the numbers of false-positive and false-negative test results are swapped with one another (or whether the numbers of true-negative and true-positive test

Table 3 Example Data: Focus and Kalon ELISA Tests for Diagnosing HSV-2 Infection

	Kalon ELISA		
Focus ELISA	**Positive**	**Negative**	**Total**
HSV-2 Infected (D = 1)			
Positive	138	4	142
Negative	0	1	1
Total	138	5	143
HSV-2 Uninfected (D = 0)			
Positive	5	22	27
Negative	0	75	75
Total	5	97	102

results are swapped with one another). As with the odds ratio, the total misclassification rate does not distinguish TPR and FPR and therefore does not provide a complete summary of test performance.

Comparing Binary Tests

Comparisons between medical tests may be based on cost/resources, ease of use, speed of results, risk to the patient, or accuracy (1, 2). Here we focus on comparisons of test accuracy.

It is generally helpful when comparing tests to use a paired study design, that is to evaluate both tests on the same set of subjects. Such a design reduces variability and avoids confounding associated with comparisons between populations (4). A paired design, however, requires statistical methods that can account for correlation between multiple test results on the same subject. We refer the reader to Pepe (4) and Zhou et al. (10) for reviews of these methods.

To illustrate, we use the same HSV-2 example. The Focus ELISA test is compared with the Kalon ELISA (Table 3). Both tests are applied to the same sample of individuals in a paired design. We estimate that the relative TPR for Focus versus Kalon, that is, the ratio of the TPRs, is 1.0 (95% CI: 1.0 to 1.1), which suggests that the Focus and Kalon tests detect equal proportions of HSV-2 seropositives. The estimated relative FPR is 5.4 (95% CI: 2.5 to 11.9), which implies that the Focus test has a substantially larger FPR, 5.4 times higher than the Kalon test.

EVALUATING CONTINUOUS TESTS

With the development of more precise quantitation methods, such as polymerase chain reaction (PCR) and certain immunological assays, many medical tests provide continuous test results. For example, the Focus ELISA test described in our data example uses the optical density of serum binding to HSV-2–specific antigen gG-2 as read by a spectrophotometer to assess HSV-2 seropositivity. Larger values of the response are more indicative of disease or infection, and quantitative levels often function as thresholds for positivity determination. In this section, we describe methods for assessing the accuracy of these continuous tests (Y) in determining disease status (D).

The ROC Curve

The classification accuracy of a continuous test is typically summarized using the ROC curve (4,10). This is based on dichotomizing the continuous test result at a threshold, c, and plotting the TPR versus the FPR for the binary rule "$Y > c$." The threshold is then varied over all possible values to generate a curve (Fig. 1). A perfect test has an ROC curve that contains the point at the top left corner of the plot, where for some threshold TPR = 1 and FPR = 0. The ROC curve for a useless test is the 45 degree line where TPR = FPR.

The ROC curve has several important attributes. First, it puts all tests on a common scale and thus facilitates the comparison of tests measured in different units or of results obtained

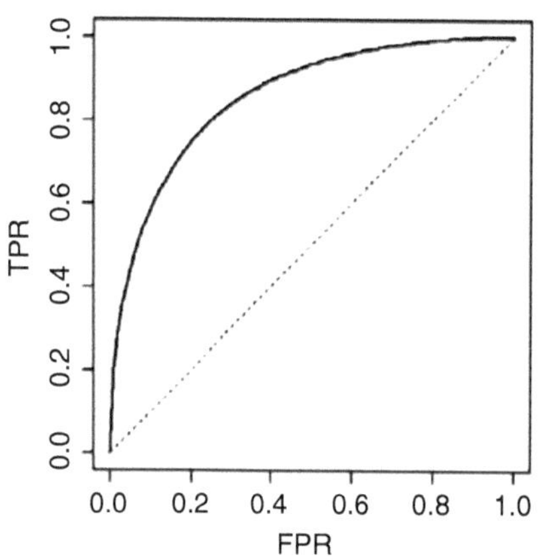

Figure 1 A sample ROC curve.

across studies. The curve also shows the range of possible operating characteristics associated with dichotomizing the test result at all possible thresholds.

We revisit the HSV-2 example to illustrate the ROC curve. The Focus ELISA test yields a continuous result called an index. We previously applied the standard threshold of 1.1 to generate binary test results. Here we summarize the accuracy of the test across all possible thresholds. Figure 2A shows the distribution of Focus test results in HSV-2 seropositive and HSV-2 seronegative individuals. By applying a series of thresholds to these distributions, we generate the ROC curve, shown in Figure 2B. We estimate that a TPR of 90.2% and an FPR of 2.0% can be achieved using a threshold of 3.3. The standard threshold of 1.1 results in more HSV-2 seropositives being classified as positive (TPR = 99.3%) but more seronegatives test positive as well (FPR = 26.5%).

Choosing the test threshold that is to be used in practice involves weighing the costs and benefits of false and true positive designations. These valuations are specific to the clinical context. For example, correctly diagnosing a life-threatening viral infection will receive substantially greater weight if the treatment is relatively benign. In contrast, if the infection is not life threatening and testing positive involves invasive work-up, onerous treatment, or psychological stress, false positive test results will typically receive greater weight than false negatives. Proposals have been made for choosing thresholds based on statistical criteria, for example identifying the threshold that corresponds to the point on the ROC curve that is closest to the upper left corner of the plot (11,12). However, these methods ignore the fact that the

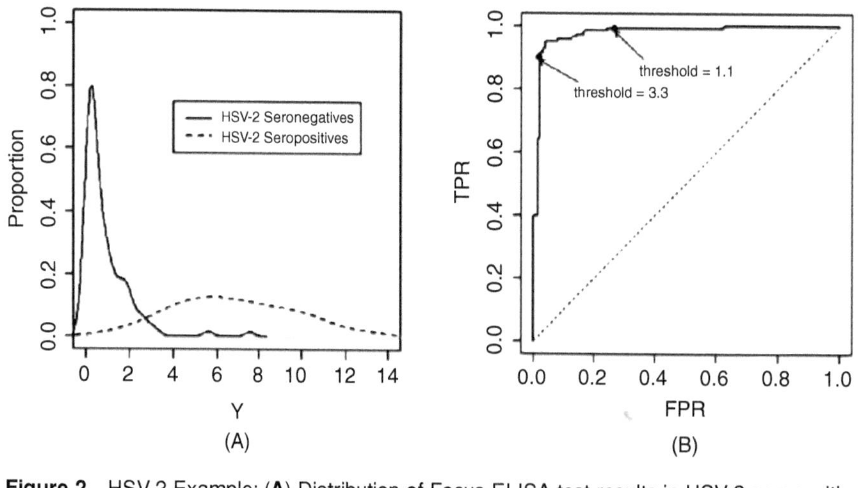

Figure 2 HSV-2 Example: (**A**) Distribution of Focus ELISA test results in HSV-2 seropositive and seronegative individuals. The proportion of subjects with a given test result, Y, is plotted. (**B**) The ROC curve for the Focus ELISA test.

relative importance of the TPR and FPR depends entirely on the clinical context. These statistical approaches cannot replace careful thought about the context in which the test is to be applied.

There are a number of different approaches to estimating ROC curves, ranging from nonparametric to fully parametric (see Refs. 4 and 10 for summaries). Most statistical software packages contain programs for ROC estimation. Two recent articles summarize the advantages and disadvantages of the available software (13,14). Another pair of articles describes a new suite of programs in Stata for estimating ROC curves (15,16).

Summary Measures of the ROC Curve

ROC summary indices are used to compare continuous tests. The most clinically relevant summaries are the points on the ROC curve. The FPR at a fixed TPR of interest is a useful summary measure in diagnostic studies, where maintaining a high TPR is often the priority. In our HSV-2 example, if we specify that 99.0% of HSV-2 seropositives are to be detected, we find that the corresponding FPR is 23.5% for the Focus ELISA. Alternatively, the TPR can be calculated at a fixed FPR of interest. This is appropriate in contexts, such as screening studies, where maintaining a low FPR is paramount.

The area under the ROC curve (AUC) is a commonly used summary index. It can be interpreted as the probability that the test result for a randomly chosen diseased individual is higher than that for a randomly chosen nondiseased individual (17,18). However, the clinical relevance of this measure is questionable (5,19). Subjects do not present at the clinic in pairs. In addition, it is often true that only a portion of the ROC curve is of practical interest, commonly the left-most part where FPRs are low. The partial area under the curve (pAUC) is the area under a limited region of the ROC curve, e.g., between FPR = 0 and FPR = .1. Again, however, this measure lacks clinical relevance. We recommend that measures with direct clinical interpretations be used to summarize test performance.

Other Measures of Test Performance

The ROC curve is the generalization of classification probabilities (TPR, FPR) to the continuous test setting. Various proposals have been made for generalizing predictive values to continuous tests (20,21), but at this time there is no standard approach. When there is a specific threshold of interest, the PPV and NPV can be calculated for the binary rule "Y > c."

Comparing Continuous Tests

Continuous tests can be compared using any of the ROC summary indices described above. As in the binary test setting, statistical methods that take into account correlation between test results on the same subject are necessary (4,22,23).

We illustrate comparisons of continuous tests by revisiting the HSV-2 example. The Focus and ELISA tests are examined on their original continuous scales in order to compare performance at all possible thresholds. Figure 3 shows the ROC curves for the two tests. We estimate that in order to correctly diagnose 99.0% of HSV-2 seropositives, Focus has an FPR of 23.5%

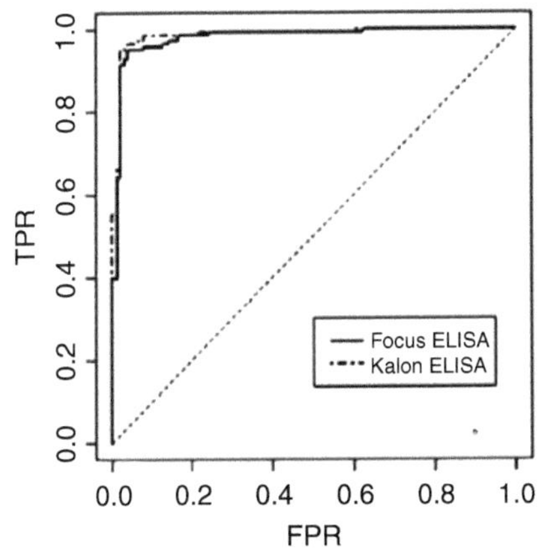

Figure 3 HSV-2 Example: ROC curves for Focus and Kalon ELISA tests for diagnosing HSV-2 infection.

Table 4 Example: Frequency of Focus Positive, Negative, and Equivocal Test Results by HSV-2 Serostatus

	HSV-2 Status		
Focus ELISA result	Infected (D = 1)	Uninfected (D = 0)	Total
Positive (>1.1)	142	27	169
Equivocal (0.9–1.1)	0	8	8
Negative (<0.9)	1	67	68
Total	143	105	245

(95% CI: 10.9–69.6%) whereas Kalon has an FPR of 19.6% (95% CI: 2.9–67.6%). Therefore, there is some suggestion that the Kalon test is superior, but there is substantial uncertainty in the FPR estimates.

INDETERMINATE TEST RESULTS

Many tests produce indeterminate test results. Some indeterminate results are still somewhat informative, such as responses whose values fall outside the limits of quantitation and are therefore known to be particularly low or particularly high. Others may contain few clues as to proper categorization, such as those resulting from a specimen that was insufficient in volume or responses corresponding to standards or controls that were out of range. Indeterminate results should not be ignored when calculating test accuracy. A thorough description of a test's performance will include the number of indeterminate results and reasons for their inadequacy, and may also include multiple measures of the test's accuracy that vary depending on the treatment of these indeterminate findings.

To illustrate, we consider the HSV-2 example. The Focus test package insert recommends that 1.1 be used as the threshold for a "positive" test result, 0.9 be used as the threshold for a "negative" test result, and results between 0.9 and 1.1 are neither "positive" nor "negative" and should be characterized as "equivocal". In previous sections we have described the performance of the Focus test using a single threshold of 1.1, and therefore the "equivocal" results were grouped with the negative test results. However, if the two thresholds are used to define an equivocal range a complete description of test performance would provide the frequency of positive, negative, and equivocal test results by HSV-2 serostatus (Table 4).

STUDY DESIGN ISSUES

Phases of Study

As the development of new therapeutics follows a phased framework, so too has a phased approach been proposed for the development of new diagnostic tests. See Pepe et al. (24), Zhou et al. (10), and Baker et al. (25) for variations on this theme. Elements common to these frameworks include a progression from exploratory studies, to case-control studies used to evaluate classification accuracy, to prospective cohort studies which evaluate predictive values in a prospective context. We overview the basic elements of study design.

Components of Study Design

Proper study design involves first identifying the clinical context in which the test is to be used. The study population should then be randomly sampled from the target population of interest (5,26). A classic source of bias in case-control studies arises when cases and controls are sampled from different populations, for example cases being treated in the clinic and healthy clinic employees as controls. This leads to confounding, where cases and controls differ in many ways other than disease status.

Random sampling of study subjects is important to avoid selection bias (27,28) and spectrum bias (4), where subjects included in the study do not represent the population of interest. Enrolling participants at multiple sites also helps to ensure generalizability.

Other potential sources of bias may be introduced in the laboratory. Prospective collection of samples for diagnostic testing helps to ensure uniform collection, processing, and storage of specimens (5). In addition, testing should be performed by technicians who are blinded to true disease status in order to avoid bias in test integrity (5).

Finally, a study should be large enough that conclusions can be drawn from it. The goal is to determine whether the test meets minimally acceptable performance standards. Therefore, sample size calculations should be based on: (*i*) identifying measures of test performance, and (*ii*) specifying values of these measures that constitute a minimally useful test (5). See Ref. 4 for sample size calculations for both continuous and binary tests.

ISSUES BEYOND THE SCOPE OF THIS CHAPTER

This chapter has introduced basic methods for evaluating diagnostic test accuracy. Here we briefly mention more complex situations and the corresponding methods that have been developed to deal with them.

A common problem occurs when disease is diagnosed using a "gold standard" test which itself is inaccurate. Various statistical techniques have been proposed to deal with this issue (29–31). In reality this is not a problem that statistics can overcome (4,32,33). A fundamental task in any study is defining disease; however, when there is no definitive test a composite reference standard is often a useful solution (33).

In some settings, it is infeasible or unethical to obtain true disease status on all subjects. When disease verification is selective, for example determined on the basis of the test result, verification bias is incorporated. As an extreme example, failure to verify negative test results severely biases estimation of both TPR and FPR as all subjects will be positive for the test regardless of true disease status, and therefore both measures will be estimated at 100% regardless of test performance. Statistical methods for adjusting for verification bias have been developed (34,35). Wherever possible, verification bias should be avoided by determining disease status on all subjects.

In many contexts, test results may be impacted by factors other than disease status, for example patient characteristics or aspects of the specimen collection, processing, or storage procedure. Statistical methods have been developed to incorporate this type of covariate information into the evaluation of diagnostic test accuracy (36–40).

Finally, in some contexts, interest lies in combining results from several diagnostic tests, to improve discriminatory accuracy. The general principles behind combining test results are summarized in McIntosh and Pepe (41) and Pepe (4). It is important that development and evaluation of the combination of tests not be performed using the same data; doing so is well known to induce overfitting bias. The dataset can be split into two portions: one for developing the combination and one for evaluating its performance, or statistical procedures such as cross-validation and bootstrapping (42) can be used if the steps for developing the combination can be defined a priori and automated.

REFERENCES

1. Wilson JMG, Jungner YG. Principles and practice of screening for disease. Public Health Papers 34. World Health Organization, Geneva, Switzerland, 1968.
2. Cole P, Morrison AS. Basic issues in population screening for cancer. J Natl Cancer Inst 1980; 64:1263–1272.
3. Obuchowski NA, Granam RJ, Baker ME, et al. Ten criteria for effective screening: Their application to multislice CT screening for pulmonary and colorectal cancers. Am J Roentgenol 2001; 176:1357–1362.
4. Pepe MS. The Statistical Evaluation of Medical Tests for Classification and Prediction. Oxford: Oxford University Press, 2003.
5. Pepe MS, Feng Z, Janes H, et al. Pivotal evaluation of the accuracy of a classification biomarker. J Natl Cancer Inst 2008; 100:1432–1438.
6. Lingappa J, Nakku-Joloba E, Magaret A, et al. HSV-2 serology testing in Uganda: Sensitivity and specificity of various testing methods among HIV-infected and uninfected persons. Manuscript in preparation.
7. Ashley RL. Performance and use of HSV type specific serology test kits. Herpes 2002; 9:38–45.
8. Steurer J, Fischer JE, Bachmann LM, et al. Communicating accuracy of tests to general practitioners: A controlled study. Br Med J 2002; 324:824–826.
9. Pepe MS, Janes H, Longton G, et al. Limitations of the odds ratio in gauging the performance of a diagnostic, prognostic, or screening marker. Am J Epidemiol 2004; 159:882–890.
10. Zhou XH, McClish DK, Obuchowski NA. Statistical Methods in Diagnostic Medicine. New York: Wiley, 2002.

11. Youden WJ. Index for rating diagnostic tests. Cancer 1950; 3:32–35.
12. Moses LE, Shapiro D, Littenberg B. Combining independent studies of a diagnostic test into a summary ROC curve: Data-analytic approaches and some additional considerations. Stat Med 1993; 12:1293–1316.
13. Stephan C, Wesseling S, Schink T, et al. Comparison of eight computer programs for receiver-operating characteristic analysis. Clin Chem 2003; 49(3):433–439.
14. Lasko TA, Bhagwat JG, Zou KH, et al. The use of receiver operating characteristic curves in biomedical informatics. J Biomed Inform 2005; 38:404–415.
15. Pepe MS, Longton G, Janes H. Estimation and comparison of receiver operating characteristic curves. Stata J 2009; 9:1–16.
16. Janes H, Longton G, Pepe MS. Accommodating covariates in ROC analysis. Stata J 2009; 9:17–39.
17. Bamber D. The area above the ordinal dominance graph and the area below the receiver operating characteristic graph. J Math Psychol 1975; 12:387–415.
18. Hanley JA, McNeil BJ. The meaning and use of the area under an ROC curve. Radiology 1982; 143:29–36.
19. Cook NR. Statistical evaluation of prognostic versus diagnostic models: Beyond the ROC curve. Circulation 2008; 54:17–23.
20. Moskowitz CS, Pepe MS. Quantifying and comparing the predictive accuracy of continuous prognostic factors for binary outcomes. Biostatistics 2004; 5:113–127.
21. Huang Y, Pepe MS, Feng Z. Evaluating the predictiveness of a continuous marker. Biometrics 2007; 63:1181–1188.
22. DeLong ER, DeLong DM, Clarke-Pearson DL. Comparing the areas under two or more correlated receiver operating characteristic curves: A nonparametric approach. Biometrics 1988; 44:837–845.
23. Wieand S, Gail MH, James BR, et al. A family of nonparametric statistics for comparing diagnostic markers with paired or unpaired data. Biometrika 1989; 76:585–592.
24. Pepe MS, Etzioni R, Feng Z, et al. Phases of biomarker development for early detection of cancer. J Natl Cancer Inst 2001; 93:1054–1061.
25. Baker SG, Kramer BS, McIntosh M, et al. Evaluating markers for the early detection of cancer: Overview of study designs and methods. Clin Trials 2006; 3:43–56.
26. Ransohoff DF. How to improve reliability and efficiency of research about molecular markers: Roles of phases, guidelines, and study design. J Clin Epidemiol 2007; 60:1205–1219.
27. Van Der Shouw YT, Verbeek ALM, Ruijs SHJ. Guidelines for the assessment of new diagnostic tests. Invest Radiol 1995; 30:334–340.
28. Sica GT. Bias in research studies. Radiology 2006; 238:780–789.
29. Walter SD, Irwig LM. Estimation of test error rates, disease prevalences, and relative risk from misclassified data: A review. J Clin Epidemiol 1988; 41:923–937.
30. Kosinski AS, Flanders WD. Evaluating the exposure and disease relationship with adjustment for different types of exposure misclassification: A regression approach. Stat Med 1999; 18;2795–2808.
31. Qu Y, Tan M, Kutner MH. Random effects models in latent class analysis of revaluating accuracy of diagnostic tests. Biometrics 1996; 52:797–810.
32. Pepe MS, Janes H. Insights into latent class analysis of diagnostic test performance. Biostatistics 2007; 8:474–484.
33. Alonzo TA, Pepe MS. Using a combination of reference tests to assess the accuracy of a new diagnostic test. Stat Med 1999; 18:2987–3003.
34. Begg CB, Greenes RA. Assessment of diagnostic tests when disease verification is subject to selection bias. Biometrics 1983; 39:207–215.
35. Alonzo TA, Pepe MS. Assessing accuracy of a continuous screening test in the presence of verification bias. J R Stat Soc Ser C Appl Stat 2005; 54:173–190.
36. Tosteson AAN, Begg CB. A general regression methodology for ROC curve estimation. Med Decis Making 1988; 8:204–215.
37. Pepe MS. An interpretation for the ROC curve and inference using GLM procedures. Biometrics 2000; 56:352–359.
38. Alonzo TA, Pepe MS. Distribution-free ROC analysis using binary regression techniques. Biostatistics 2002; 3:421–432.
39. Janes H, Pepe MS. Adjusting for covariate effects on classification accuracy using the covariate-adjusted receiver operating characteristic curve. Biometrika 2009; 96(2):371–382.
40. Janes H, Pepe MS. Adjusting for covariates in studies of diagnostic, screening, or prognostic markers: An old concept in a new setting. Am J Epidemiol 2008; 168:89–97.
41. McIntosh M, Pepe MS. Combining several screening tests: Optimality of the risk score. Biometrics 2002; 58:657–664.
42. Harrell FE. Regression Modelling Strategies: With Applications to Linear Models, Logistic Regression, and Survival Analysis. New York: Springer-Verlag, 2001.

3 Molecular Amplification Methods in Diagnostic Virology

David M. Whiley and Theo P. Sloots

Queensland Paediatric Infectious Diseases Laboratory, Sir Albert Sakzewski Virus Research Centre, Queensland Children's Medical Research Institute, Children's Health Service District, Brisbane, Queensland, Australia

INTRODUCTION

Effective clinical management and possible treatment of patients with viral infections relies on the rapid and specific identification of the causative organism. The early recognition of an infectious agent allows clinicians to make sound therapeutic decisions and avoid the indiscriminate use of antibiotics. Traditionally, these methods have included virus isolation by cell culture, detection of viral products, or the detection of antibodies produced as a result of infection. However, in many cases these methods can be laborious, time-consuming, and may lack sensitivity, thereby prolonging or denying definitive diagnosis and subsequent treatment of the patient. Rapid molecular diagnostic tools and detection methods, such as nucleic acid amplification, are used increasingly in the clinical microbiology laboratory to enhance the identification of viral pathogens and to assist physicians in the diagnosis and management of a variety of viral diseases.

Nucleic acid amplification strategies and advances in the detection of amplification products have been key aspects in the progress of molecular microbiology. Sophisticated new amplification–detection combinations are resulting in many new applications in laboratory testing for infectious diseases. These applications include qualitative detection, subspecies-level DNA fingerprinting, molecular resistance testing, genotyping, and quantitative (viral load) testing. When applied selectively in the laboratory, these applications can enhance diagnostic approaches and clinical management and will most likely evolve into standard laboratory and point-of-care testing protocols in the near future.

A variety of nucleic acid amplification techniques were developed in the mid- to late 1980s, including the polymerase chain reaction (PCR) (1), ligation-mediated amplification (2), and transcription-based amplification (3). Since then, these techniques have been improved and alternative approaches for the amplification of target sequences have been developed [e.g., transcription-mediated amplification (TMA), nucleic acid sequence-based amplification (NASBA), ligase chain reaction (LCR), strand displacement amplification (SDA), and linear linked amplification (4)]. However, none of these techniques has achieved the same widespread application as PCR, most likely due to the simplicity and cost-effectiveness of the PCR methodology.

CYCLING AMPLIFICATION TECHNOLOGIES

The Polymerase Chain Reaction

PCR originated in 1983 as a means of in vitro DNA amplification using DNA polymerase (1,5,6). Briefly, PCR is performed in a reaction mixture containing the target DNA, a heat-stable DNA polymerase, an excess of deoxynucleoside triphosphates, and forward and reverse oligonucleotide primers that flank the particular target DNA sequence of interest. The PCR process is then facilitated by repeated cycles of heating and cooling of the reaction mixture. Double-stranded target DNA (dsDNA) is heat denatured (94–97°C) and then cooled (50–65°C) enabling forward and reverse primers to anneal to complementary sequences on each target DNA strand. The primers are extended by the DNA polymerase enzyme (60–72°C) creating new double-stranded copies of the target DNA, which can then act as further template for DNA amplification (Fig. 1).

Since its inception, PCR technology has been at the forefront of revolutionizing viral diagnostics, and has facilitated the rapid and sensitive detection of a broad range of clinically

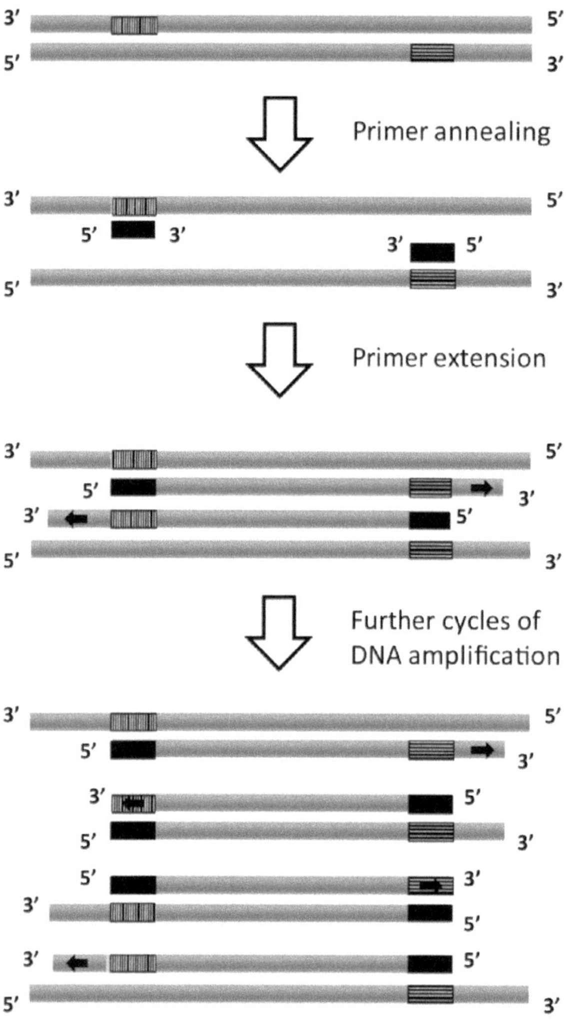

Figure 1 Polymerase chain reaction. In step 1 of the reaction, the two strands of DNA are separated by heat denaturation and primers anneal to the target sequences in a complementary manner. Next *Taq* DNA polymerase initiates extension at the 3′-end of each primer to synthesize a complementary strand of DNA containing the primer-binding sites. In each subsequent cycle of heat denaturation and extension, the new DNA strands can then act as further template for primer annealing and extension initiating further DNA synthesis.

relevant viruses, including RNA viruses via reverse transcription PCR (RT-PCR). This success has largely been driven by significant advancements in PCR detection technologies. Many PCR protocols have now been published and commercial assays are available for a number of important viruses including human immunodeficiency virus (HIV) (7), hepatitis B and C viruses (8), and cytomegalovirus (CMV) (9). However, the number of commercial assays available is still quite limited and this has led to the extensive use and development of "in-house" or "home-brew" PCR protocols.

Conventional PCR Detection Methods

PCR product detection was traditionally performed by direct visualization of the product using agarose gel electrophoresis with DNA-binding fluorescent dyes such as ethidium bromide. Gel-based methods, although laborious and dependent on subjective result interpretation, are still widely used in diagnostic virology, particularly for PCR-based sequencing and typing (10). In addition, gel-based visualization of PCR product remains an effective way for troubleshooting problems encountered using alternative detection methods. Thus, most diagnostic laboratories maintain gel-based methods to some extent.

Enzyme immunoassays, including the enzyme-linked immunosorbent assay (ELISA), provided the first key advancement in PCR detection methodology. Briefly, the PCR-ELISA system used a colorimetric microtiter plate probe-based capture system whereby a 5′ biotinylated oligonucleotide probe targeting a DNA sequence internal to the primers was used to capture

PCR product to streptavidin-coated wells (11). A major advantage of the PCR-ELISA over gel-based techniques is that it is objective, minimizing the potential for interpretation errors. In addition, the technology is particularly suitable for multiplex PCR reactions (discussed below) (12). PCR-ELISA was extensively used in "home-brew" assays as well as in several commercial assays targeting a variety of important viral pathogens, including human immunodeficiency virus (7) and hepatitis C virus (13). However, the PCR-ELISA technology has now been largely superseded by real-time PCR.

Real-Time PCR Detection

The advent of real-time PCR probably represents the greatest leap in DNA amplification technology since the development of PCR itself, and real-time PCR has been the key advancement in revolutionizing diagnostic virology. Briefly, real-time PCR is achieved through the use of fluorescent detection technology. Fluorescent molecules are added to a PCR reaction mix and interact with the PCR product to produce an increase in fluorescent signal when PCR amplification occurs. Monitoring of the fluorescent output is achieved through real-time PCR instrumentation, which measures the fluorescent signal during or following each thermal cycle. Thus, PCR amplification is monitored in "real-time," providing numerous advantages over conventional detection methods. From a practical perspective, real-time PCR removes the need for a separate detection step, which significantly reduces PCR result turnaround times and decreases staff hands on time. Also, the system is closed (i.e., reactions do not need to be opened for detection), reducing the potential for carryover contamination. The technology also provides an additional key performance characteristic, in that it has an extremely broad dynamic range for virus detection making it highly suitable for viral quantification (discussed below). For these reasons, numerous real-time PCR protocols have been described for almost every human viral pathogen and the technology has been widely embraced in diagnostic laboratories (14–16).

There are two main types of real-time PCR fluorescent technology, including DNA intercalating dyes, such as SYBR green, which bind nonspecifically to dsDNA, and sequence-specific oligonucleotide chemistries such as dual-labeled probes (17,18). Upon intercalation into dsDNA, SYBR green emits fluorescent signal at 522 nm, which can then be readily observed using real-time PCR instrumentation. Thus, the use of SYBR green offers a very simple and quick means of producing real-time PCR methods. The main disadvantage of SYBR green is that it will bind to any dsDNA, including nonspecific PCR products such as primer dimer, and so may rely on additional analyses, such as melting curve analysis, for assay specificity. For these reasons, sequence-specific probe chemistries are favored over intercalating dyes in diagnostic virology as they are specific to the target DNA sequence of interest, and so offer superior result resolution (15,16). The most commonly used sequence-specific oligonucleotide probe format used in diagnostic real-time PCR has been the dual-labeled TaqMan® probe. TaqMan® probes consist of a reporter fluorescent dye (e.g., FAM) covalently coupled to the 5′-end and a quenching dye (e.g., TAMRA) at the 3′-end. When the probe is intact, the close proximity of the quenching dye to the reporter dye prevents emission of fluorescent signal from the reporter dye by Förster resonance energy transfer (FRET), more commonly known as fluorescence resonance energy transfer. However, during PCR primer extension the DNA polymerase enzyme digests any bound TaqMan® probe, separating the two dyes. The reporter dye is no longer suppressed by the quencher dye and so may now emit fluorescent signal. There are many different dye-primer–based signaling systems used in real-time PCR, including the simple light upon extension (LUX; Invitrogen) system, scorpion primers, and the more recent Plexor system (Promega). A number of labeled probe-based systems are commonly used in addition to hybridization and hydrolysis probes, such as molecular beacons, minor groove binding (MGB), and locked nucleic acid (LNA) probes.

Nested PCR Formats

Although technically capable of detecting very low copy numbers of target nucleic acid per reaction, the detection limit of "single round" PCR can in fact be improved by a nested PCR format. In this approach, a small aliquot of reaction mix from a primary PCR reaction is transmitted to a second PCR reaction containing a second set of primers, which target sequences that are internal to the primers used in the primary reaction (Fig. 2). The overall result of this

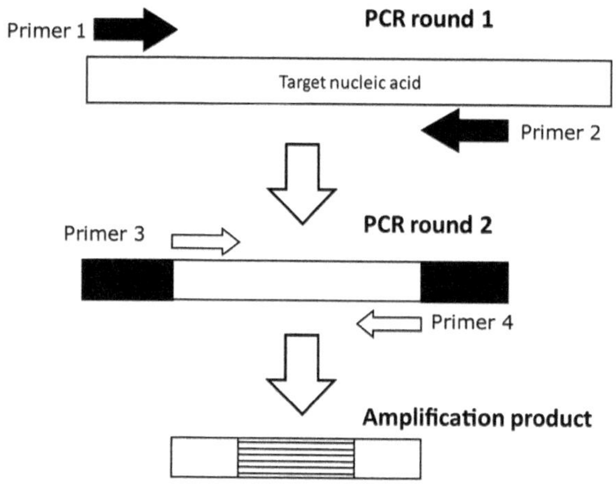

Figure 2 Nested PCR. In this modification of PCR, two separate reactions, or "rounds," of PCR amplification are used. Following first round amplification, a small aliquot of reaction mix from the primary PCR reaction is transferred to a second PCR reaction. The primers used in the second reaction mix target sequences that are internal to the target sequences used by the primers in the initial reaction.

approach is twofold. First, the sensitivity of an assay can be significantly improved and on average may improve the detection limit of a PCR by 10- to 100-fold (15). This is particularly useful when trying to detect viruses that are at low load (19) or when trying to detect viruses using suboptimal specimens, such as dried blood spots (20). The added benefit of nested PCR is that it provides a "clean" DNA template for the second PCR reaction. This not only improves the specificity of the PCR for the target organism but also improves result resolution, particularly when using nonspecific detection methods including gel electrophoresis or SYBR green-based real-time PCR. Despite these advantages, nested PCR protocols are generally not regarded suitable for routine diagnostics, mainly due to the substantial increase in risk of PCR carryover contamination. Questions have also been raised over the clinical significance of extremely low viral loads that are detectable only by nested PCR methods (19).

Multiplex PCR Formats

Although offering increased sensitivity and rapid result turnaround times, an inherent limitation of PCR is that it is specifically directed for detection of an organism containing the appropriate primer targets. In contrast, traditional techniques, including cell culture, allow for a more *pan-viral* approach. This puts PCR at a disadvantage when a particular clinical question may implicate a variety of viral pathogens. Respiratory viruses provide a key example of this type of problem as similar clinical symptoms may be observed for a range of respiratory pathogens. Thus, using conventional PCR, a respiratory sample may need to be tested by multiple individual PCR reactions to cover all potential viral agents. This can make the technology prohibitively labor-intensive, expensive, and low throughput by sequestering valuable space on PCR thermocycling instrumentation.

Fortunately, these limitations can be overcome by using multiplex PCR. The multiplex PCR format is a significant improvement over conventional PCR protocols when multiple viruses are in question, and is achieved by incorporating multiple primer sets for simultaneous detection of several viruses within a single PCR reaction. The different amplification products may then be differentiated in a number of ways, including band size using gel electrophoresis, by sequence-specific oligonucleotide probes using technologies such as PCR-ELISA or real-time PCR, or by melting temperature using SYBR green-based real-time methods. Numerous multiplex PCR assays, including commercial tests, have been described for respiratory viruses, with up to 19 viral agents successfully detected and differentiated within a single PCR (12,21–23). Multiplex PCR has also proved useful for detecting the common viral causes of central nervous system disease, including herpes and enteroviruses (24). The only disadvantage of multiplex PCR methods is that the sensitivity of these reactions can be compromised, when compared to their monoplex counterparts, as a result of nonspecific reactions between the large number of oligonucleotides within the reaction mix or by competitive inhibition caused by competition between the specific reactions. Nonetheless, it is likely that the huge potential offered by multiplex PCR technology will see it continue to grow in diagnostic virology, particularly with the advent of newer commercial detection systems, including liquid-array (23), which provide greater flexibility for multiplex PCR.

Liquid arrays use tiny color-coded beads, called microspheres, that are grouped into distinct sets. Each bead set can be coated with a reagent specific to a particular bioassay, allowing the capture and detection of specific analytes from a sample. Within the analyzer, lasers excite the internal dyes that identify each microsphere particle, and also any reporter dye captured during the assay. Many readings may be made on each bead set, further validating the results. In this way, the technology allows detection of multiple targets within a single sample, both rapidly and precisely.

Quantitative PCR

A significant benefit of PCR, particularly real-time PCR, is that it can readily be adapted to quantify the viral load in clinical specimens. The principle behind quantitative PCR (qPCR) is that during thermocycling PCR amplification will begin sooner in specimens with higher viral load compared with specimens with lower viral load. In real-time PCR, this will be observed as earlier generation of fluorescent signal (or earlier cycle threshold values). Using qPCR, results can be expressed in absolute terms (e.g., copies per mL) with reference to quantified standards, or in relative terms compared to another target sequence present in the sample. In diagnostic virology, qPCR offers considerable advantages over *qualitative* PCR, as it enables the possibility to determine the dynamics of viral proliferation, monitor the response to treatment, and distinguish between latent and active infections. Prominent examples include HIV, hepatitis B and C viruses, and CMV (8,9,25).

OTHER TARGET AMPLIFICATION METHODS

Nucleic Acid Sequence-Based Amplification

The first non-PCR–based target amplification system was described in 1989 by Kwoh et al. (3). This technique, originally known as transcription-based amplification (TAS), was based on the amplification of a target sequence by in vitro transcription. This method was subsequently refined to an isothermal transcription-based amplification technique that exploited the simultaneous enzymatic activities of three enzymes in a process initially called self-sustaining sequence replication (3SR) or nucleic acid sequence-based amplification (NASBA). NASBA amplifies RNA from an RNA target and utilizes a dual function reverse transcriptase/DNA polymerase [avian myeloblastosis virus (AMV) reverse transcriptase], a T7 RNA polymerase, the enzyme RNaseH, and a T7 promoter-labeled target-specific primer (26). The reaction comprises continuous cycles of reverse transcriptase and RNA transcription to replicate the target nucleic acid sequence via a double-stranded cDNA intermediate (Fig. 3).

Briefly, in the reaction a RNA:DNA hybrid is produced containing a T7 promoter. The RNA in this hybrid is degraded by RNaseH enzyme, and the DNA is extended by DNA polymerase to form a double-stranded DNA molecule. This acts as a template for the production of multiple RNA transcripts by T7 RNA polymerase utilizing the T7 promoter sequence. These transcripts can subsequently be used for the production of additional DNA fragments containing T7 promoters, and act as transcription templates. This process continues in a self-sustained cyclic reaction at 42°C until reagents are exhausted or inactivated. NASBA produces 100 to 1000 copies per target per cycle as compared to PCR and LCR that produce only two copies per cycle. This results in a 10 billion-fold increase of target RNA copies within about 15 to 30 minutes.

One major advantage of this technique is that it is not affected by DNA contamination of the test samples, which means that the quantification of template RNA can be achieved even on crude cell extracts. NASBA has demonstrated equivalent or improved sensitivity to PCR-based methods (27,28), and has the potential advantage of being easier to optimize than conventional PCR (29). To date, NASBA has been most commonly used for the detection of RNA viruses, typically using commercial kits and probe-based chemiluminescent detection of the amplified RNA (30–33). bioMérieux has combined NASBA and molecular beacons into a test system called EasyQ (bioMérieux, Durham, NC, USA) for monitoring the generation of amplification product in real time (34), which has been applied to quantify viruses (35–38). In addition to real-time

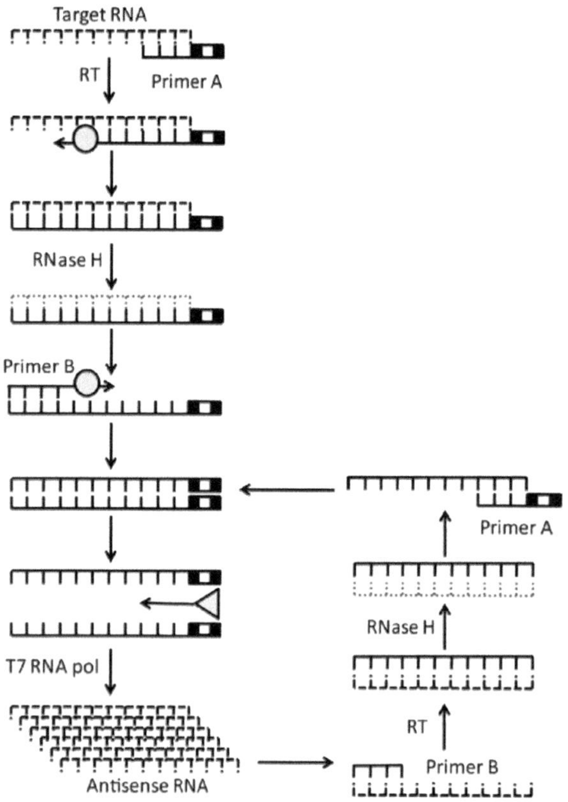

Figure 3 Nucleic acid sequence-based amplification (NASBA). NASBA comprises continuous cycles of RNA transcription to duplicate target nucleic acid through a double-stranded cDNA intermediate. Primer A contains a T7 promoter site, and binds to the target RNA strand. Reverse transcriptase (RT) extends the primer and introduces a functional T7 promoter in the cDNA strand. RNaseH degrades the RNA in the RNA:DNA hybrid. Primer B binds to the cDNA strand and reverse transcriptase extends the primer to produce a double-stranded DNA intermediate containing T7 promoters. The T7 RNA polymerase then produces multiple copies of antisense RNA transcripts. These are immediately converted to T7 promoter-containing double-stranded cDNA, which acts as a further transcription template.

detection, NASBA has also been combined with liposome signal amplification technology to develop biosensors for the detection of dengue virus (39).

Transcription-Mediated Amplification

Transcription-mediated amplification (TMA) is a variation of NASBA that uses the RNaseH activity of the reverse transcriptase in the reaction, rather than a separate enzyme (RNAseH). Like NASBA, it also utilizes isothermal amplification conditions and can target either DNA or RNA. TMA uses RNA transcription by RNA polymerase and DNA synthesis by reverse transcriptase to produce an RNA amplification product from the target nucleic acid. The possibility of carryover contamination in the laboratory is diminished because of the more labile nature of the RNA molecule compared to DNA. TMA has gained popularity in the clinical laboratory with the development of commercial assays including the APTIMA tests (Gen-Probe Incorporated, San Diego, CA, USA) for the detection of human immunodeficiency virus 1 (HIV-1) and hepatitis C virus (HCV) (40,41).

Strand Displacement Amplification

Strand displacement amplification (SDA) was first described in 1992 (42). It is another isothermic amplification method based on the ability of DNA polymerases to initiate DNA synthesis at a break within a single-stranded target DNA molecule and to displace the nicked strand during DNA synthesis (Fig. 4). The key technology behind SDA is the generation of site-specific nicks by a restriction endonuclease. Normally, endonuclease enzymes cleave double-stranded DNA, which then cannot act as a template for SDA. However, in the SDA reaction, alpha-thio-substituted nucleotides are incorporated into newly synthesized DNA creating a DNA:hemi-phosphorothioated DNA hybrid. The amplification reaction mix incorporates exonuclease-deficient DNA polymerase, a restriction endonuclease, an alpha-thio-substituted deoxynucleotide to allow the synthesis of hemi-phosphorothioated DNA, and two sets of primers.

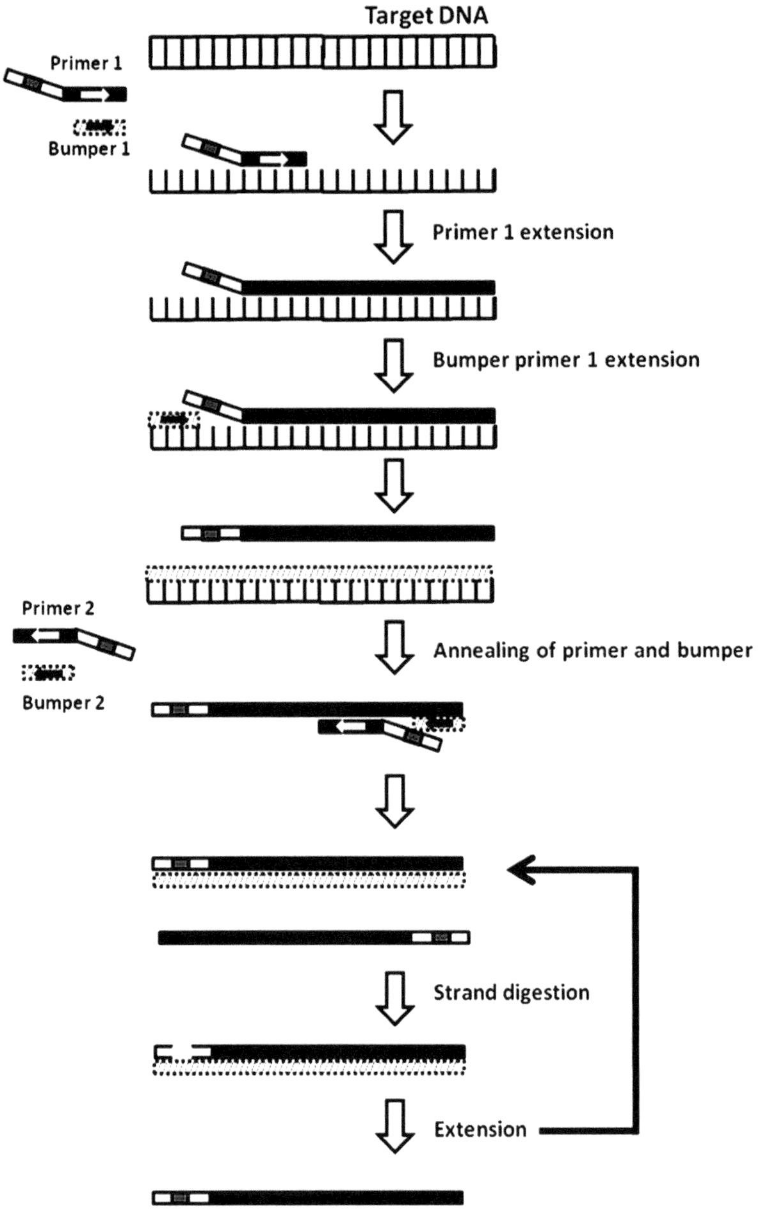

Figure 4 Strand displacement amplification (SDA). After denaturation of the nucleic acid, the primer binds to a single-stranded target sequence. The primer contains a recognition site at the 5′-end for the *Bso*B 1 restriction enzyme. Both primer and target are extended by DNA polymerase lacking 5′ to 3′ exonuclease activity in the presence of three deoxynucleoside triphosphates (dGTP, dUTP, dATP) and dCTP that contains an alpha-thiol group (dCTPαS). The resultant DNA synthesis generates a double-stranded *Bso*B 1 recognition site, one strand of which contains the hemiphosphorothioate linkages. The restriction enzyme nicks the nonthiolated strand only, and DNA polymerase extends the nucleic acid from the nick, displacing the original DNA strand.

The first set of primers act in the same way as forward and reverse primers used in PCR, but they have a restriction enzyme recognition site inserted at their 5′-ends. The second set of primers is known as "bumper" primers and these are designed to bind immediately 5′ of the forward and reverse primers. After denaturation of the target DNA, the forward and reverse primers promote the synthesis of hemi-phosphorothioated DNA, creating a DNA:hemi-phosphorothioated DNA hybrid. These strands are separated by extension of the bumper primers, which displace the newly synthesized hemi-phosphorothioated DNA strand. The resulting ssDNA is converted to dsDNA by primer extension using the respective forward or reverse primer. The hemi-phosphorothioated dsDNA contains a restriction site and forms the template for SDA. This template is cut by a restriction enzyme to introduce a single stranded nick at the restriction site, which promotes the synthesis of a new strand of DNA by the DNA polymerase. The synthesis of the new strand of DNA results in the displacement of the old one. Once the hemi-phosphorothioated template has been produced, the process is self-sustaining.

A limitation with SDA is the potential for mis-priming, which has the potential to increase the background signal. This can be largely overcome by using highly stringent operating temperatures or by incorporating the single-stranded binding protein from gene 32 of bacteriophage T4 (43). This protein also enhances the ability to amplify longer target sequences.

Becton Dickinson Technologies (North Carolina, USA) has combined SDA with fluorogenic reporter probes that permit real-time, sequence-specific detection of amplification product (44). The new probes possess the single-strand half of a *Bso*BI recognition sequence flanked on opposite sides by a fluorophore and a quencher. The probes also contain target-binding sequences located 3′ to the *Bso*BI site. Fluorophore and quencher are maintained in sufficiently close proximity such that fluorescence is quenched in the intact single-stranded probe. If target is present during SDA, the probe is converted into a fully double-stranded form and is cleaved by the restriction enzyme *Bso*BI, which also serves as the nicking agent for SDA. Fluorophore and quencher separate upon probe cleavage, causing increasing fluorescence. Target replication may thus be followed in real time during the SDA reaction. Probe performance may be enhanced by embedding the fluorogenic *Bso*BI site within the loop of a folded hairpin structure. This new probe designs permit detection of as few as 10 target copies within 30 minutes in a closed-tube, real-time format, minimizing the possibility of carryover contamination.

This technology is the basis for the commercial BD ProbeTec tests (Becton Dickinson Technologies, North Carolina, USA) that are used for the clinical diagnosis of bacterial infections. The application of SDA to virology has been slow although some viral SDA assays have been described in a research environment (45,46).

Rolling Circle Amplification

Rolling circle amplification (RCA) involves the isothermal amplification of a circular target molecule by the extension of a single forward primer by DNA polymerase for many rounds. During the reaction, the polymerase displaces upstream sequences, generating a long single-stranded DNA containing multiple repeat copies of the target sequence. Linear amplification kinetics occur during the reaction which may run at constant temperature for several hours or days, producing millions of copies of the small circle sequence. RCA was first described in the mid-1990s (47,48) and has been applied for diagnostic purposes in the direct or indirect detection of DNA or RNA using various detection mechanisms.

A modification of the technique involves the use of two primers and is called exponential (49), hyperbranched (50), ramification (51), or cascade RCA (52). One primer is complementary, and hybridizes to the circular target sequence, whereas the second primer targets the DNA product of the first primer and initiates hyper-branching during DNA replication, creating as many as 10^{12} copies/hour (50–52). The kinetics of this reaction is exponential.

Special instrumentation for the performance of RCA is not needed as the reaction proceeds at a constant temperature, and temperature cycling is not required. Also, RCA is more resistant to contamination and, unlike some other isothermal technologies, requires little or no assay optimization. Applications of RCA for the diagnosis of infectious disease have been discussed (53), and even though the application of RCA to the detection of viruses has been explored (54), practical application of this technique to diagnostics is still not widely evident.

Loop-Mediated Isothermal Amplification

In 2000, Notomi and coworkers developed a novel DNA amplification method called loop-mediated isothermal amplification (LAMP) that rapidly amplifies DNA with high specificity and efficiency under isothermal conditions, thereby obviating the need for expensive thermal cyclers (55). The method is a single tube technique that makes use of four primers that are homologous to six distinct sequences on the target DNA, with an inner primer, containing sequences of the sense and antisense strands of the target, which initiates the LAMP reaction. Then an outer primer initiates strand-displacement DNA synthesis releasing a single-stranded DNA molecule. This DNA serves as a template for further DNA synthesis primed by the second inner and outer primers that hybridize to the other end of the target, producing a stem-loop structure. In subsequent cycles, one inner primer binds to the loop of the progeny DNA and initiates further displacement DNA synthesis in the form of the original stem-loop and a new stem-loop DNA of twice the original length. The reaction produces more than 10^9 copies of target DNA in less than one hour of cycling time.

This method is highly specific, because LAMP recognizes the target by six distinct sequences in the initial step, and by four distinct sequences in subsequent steps. It may be combined with a reverse transcription step to allow the detection of RNA. Detection of amplification products can be by the addition of SYBR green and the detection of fluorescence. LAMP has the potential to be used as a simple screening assay in the field or at the point of care by clinicians.

PROBE AMPLIFICATION METHODS

Ligase Chain Reaction

Ligase Chain Reaction (LCR) is another nucleic acid amplification method developed shortly after PCR and uses two pairs of probes that are complementary to each other (Fig. 5) (56). Unlike PCR, a pair of probes does not flank the target sequence; instead they cover the target DNA immediately adjacent to one another, typically leaving a gap of 1–3 bases. The gap between the probes acts as a template for ligation by thermostable DNA ligase that joins the two probes only if they match exactly to the template sequence. DNA ligase is highly specific and intolerant of base mismatches, a property that has been exploited for use in the real-time detection of single nucleotide polymorphism (SNP) (57). Following ligation, the reaction mixture is heated to 95°C to separate the ligation product and target DNA. On cooling, further copies of the probes can anneal to the target and the complementary probes can anneal to the ligation product produced by the first set of probes. Successive rounds of denaturing/annealing/ligation will result in the exponential generation of ligation products. The advantages of this technology is its sensitivity

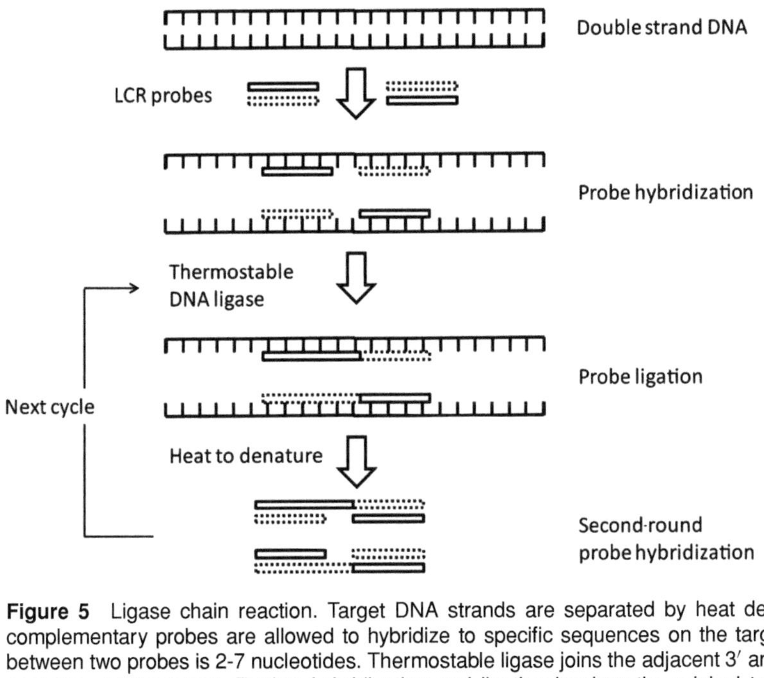

Figure 5 Ligase chain reaction. Target DNA strands are separated by heat denaturation, and two pairs of complementary probes are allowed to hybridize to specific sequences on the target molecule so that the gap between two probes is 2-7 nucleotides. Thermostable ligase joins the adjacent 3′ and 5′ ends to form a duplicate sequence to the target. Further hybridization and ligation involves the original target as well as ligated DNA fragments produced during the reaction.

to single base pair changes in the target and its potential for automated detection through labeling of the probes.

Uptake of LCR to the diagnostic environment has been limited and only preliminary reports on the use of LCR for the identification and/or detection of viruses have been published. A nonisotopic ligase-based DNA amplification assay using oligonucleotides targeting part of the gag region of HIV-1 has been described (58) and LCR technology has also been applied for the nonradioactive detection of herpes simplex virus and human papilloma virus (HPV) and allowed rapid detection of these viruses as compared to traditional detection methods using cell culture techniques (59,60). So far, the commercial application of LCR appears to be restricted to the detection of bacterial genomes (Abbott Laboratories, Abbott Park, IL, USA) with varying success (4).

Cycling Probe Technology

Cycling probe technology (CPT) was developed by ID Biomedical (Vancouver, Canada) in 1999 (61) and subsequently licensed to the Takara Biomedical Group (Takara Shuzo Company, Tokyo, Japan). It is a signal amplification system that allows detection of nucleic acid target sequences without target amplification (Fig. 6). The technique uses an RNA–DNA chimeric probe that consists of an RNA sequence that hybridizes to a complementary target DNA sequence and is flanked by two DNA sequences. The RNA in the probe becomes a substrate for RNaseH. Once hybridized, the internal RNA part of the probe is cleaved by RNaseH at the RNA internucleotide linkages, and results in dissociation of the probe from the target, thereby making it available for the next probe molecule. Once the probe is cleaved, the reporter dye and quencher dye on each side of the probe are separated and fluorescence is emitted. The fluorescence signal increases proportionally as the probe is cleaved, allowing for measurement of the amplified product. Probe amplification is linear and not exponential, thus eliminating carryover contamination, and also gives a quantitative assessment of viral or bacterial load. Because a single cleavage step is involved, the test is easy and cheap to produce and can be automated.

A refinement of the method was described by Bhatt et al. (61), who attached chimeric probes to magnetic particles, thereby creating an effective method of separating the cleaved

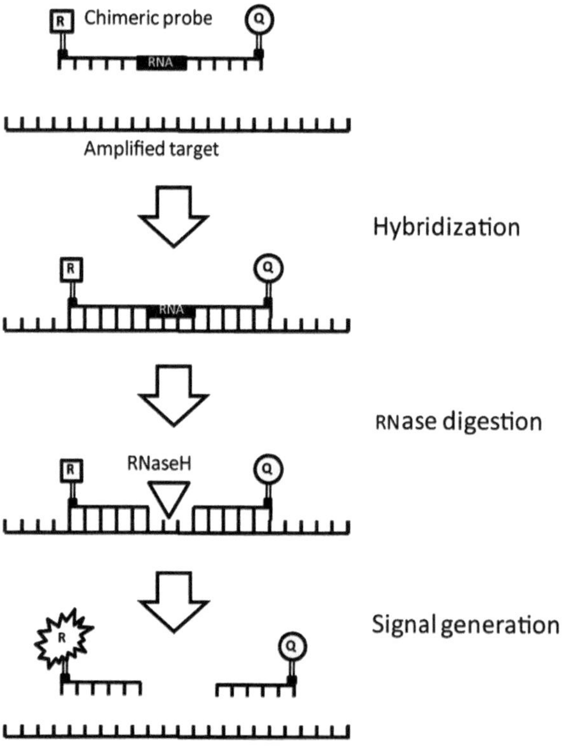

Figure 6 Cycling probe technology. Cycling probe amplification utilizes a DNA:RNA:DNA probe usually containing a reporter (R) at the 5′-end and a quencher molecule (Q) at the 3′-end. The RNA sequence binds in a complementary manner to the target DNA, and is subsequently hydrolyzed by RNaseH, releasing the noncomplementary DNA portions of the probe. Separation of the reporter and quencher generates an appropriate signal.

probe from noncycled probe. By capturing the target DNA on particles and separating it from the extraneous nonspecific DNA, they were able to reduce background signal, and better discriminate between samples of positive and negative targets. So far this technology has been used for the rapid identification of methicillin-resistant staphylococci (62) but has not been widely applied in virology.

SIGNAL AMPLIFICATION METHODS

Branched DNA

Another technique that utilizes signal rather than target amplification is called branched DNA (bDNA). Chiron Corporation (Emeryville, CA, USA) first described this technique in 1987 (63), and it has proven to be one of the most versatile signal-amplification systems to date. Signal amplification by bDNA incorporates several simultaneous hybridization steps involving several different types of oligonucleotide probes (Fig. 7). These are capture probes, a series of target probes, a novel branched secondary probe, and short enzyme-linked tertiary probes. First, a set of target-specific target probes bind to the target nucleic acid and hybridize with capture probes that are immobilized on a solid support. A second set of target-specific target probes hybridize to the immobilized target nucleic acid molecule and serve as binding sites (via 5′ extensions) for the branched secondary probe. The branched probe typically contains 15 or more branches that are complementary to sequences on the enzyme-labeled tertiary probes. As many as 300 to 3000 enzyme labels can be incorporated onto each target molecule in this manner. Following hybridization and stringency washing, a chemiluminescent substrate is used to generate a signal. It is estimated that the sensitivity of this system is in the range of 10^3 to 10^5 target molecules.

The bDNA method has been enhanced through oligonucleotide probe redesign by the inclusion of the novel nucleotides, isoC and isoG and reagent modifications, resulting in increased sensitivity and a reduction in background signal (64). Other modifications include the use of shorter overhang sequences of target probes for capture, the design of target probes for amplification, and the addition of preamplifier molecules. The bDNA technology has been

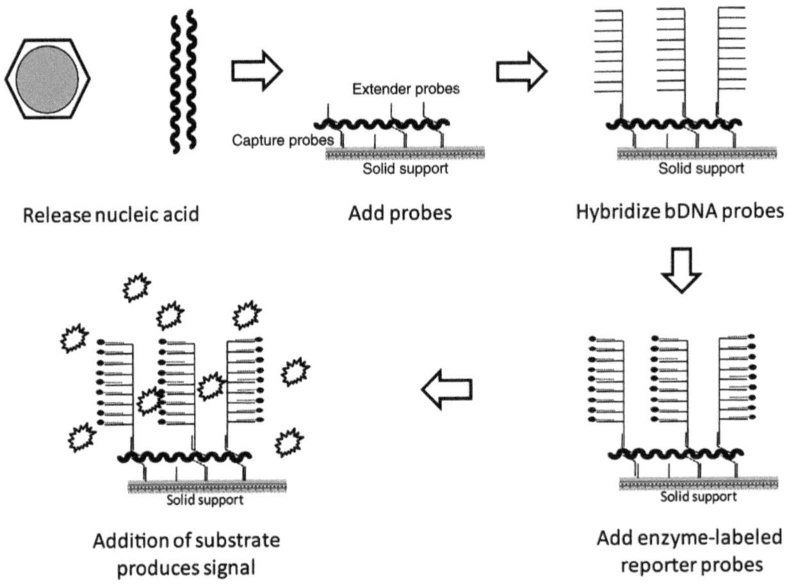

Figure 7 Branched chain DNA signal amplification. The target nucleic acid is captured onto a solid support via multiple capture probes. Contiguous extender probes bind to several target sequences. The distal ends of these probes are complementary to one section of the branched secondary probes. Enzyme-labeled reporter probes hybridize to the multiple arms of the branched probes. All hybridization reactions occur simultaneously. Upon addition of an appropriate substrate, a signal is generated as chemiluminescence.

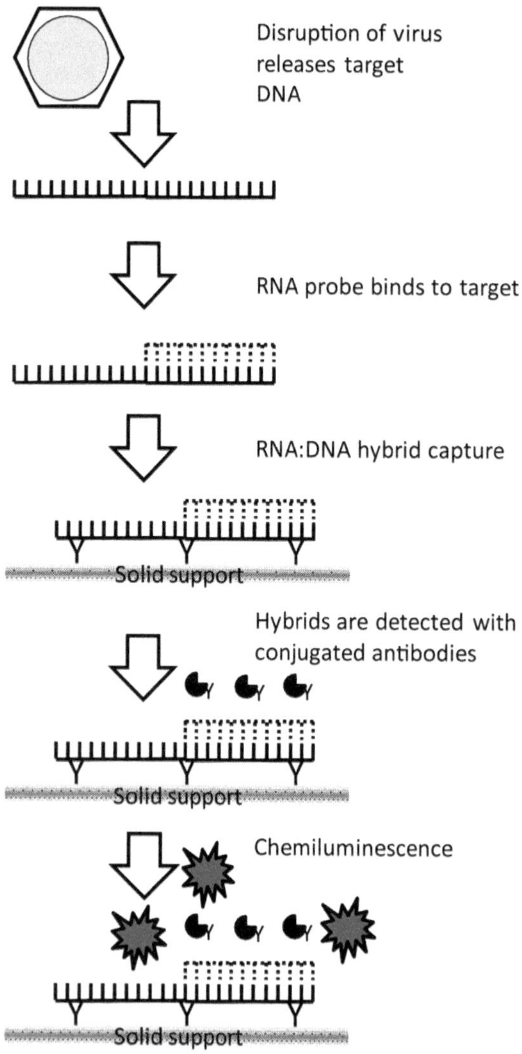

Figure 8 Hybrid capture assay. Clinical specimens are treated to release viral target DNA. Target DNA combines with RNA probes to form RNA:DNA hybrids. Multiple RNA:DNA hybrids are captured onto a solid phase coated with universal capture antibodies specific for RNA:DNA hybrids. The captured RNA:DNA hybrids are detected with multiple antibodies conjugated to an enzyme. Presence of target DNA is detected by the addition of a chemiluminescent dioxetane substrate, which is hydrolyzed to produce light which is measured on a luminometer. The resultant signal is amplified at least 3000-fold.

commercially applied by Siemens Healthcare Diagnostics (Deerfield, IL, USA) in the Versant™ 440 Molecular System to determine viral load, for the management of patients undergoing antiviral therapy against Hepatitis C virus and HIV-1 (65).

Hybrid Capture

Signal amplification is also the basis for Hybrid Capture (HC) assays. HC technology detects nucleic acid targets directly and uses signal amplification to provide a sensitivity that is comparable to target amplification methods (Fig. 8) (66). In the reaction, target DNA and specific RNA probes combine to form a RNA:DNA hybrid that is captured to a solid support by antibodies specific to RNA:DNA hybrids. The captured hybrids are detected by a secondary antibody conjugated to an enzyme that cleaves a chemiluminescent substrate to release light. Each hybrid combines with many conjugate antibody molecules thereby amplifying the resultant signal. A major benefit of the technology is that amplification products are not produced in the laboratory, thus reducing the possibility of cross-contamination. HC has great sensitivity, speed, and ease-of-use, and the ability to measure viral loads. Commercial Hybrid Capture Tests (Digene Corporation, Gaithersburg, MD, USA) are available to detect HPV and the blood-borne viruses, hepatitis B virus, and CMV.

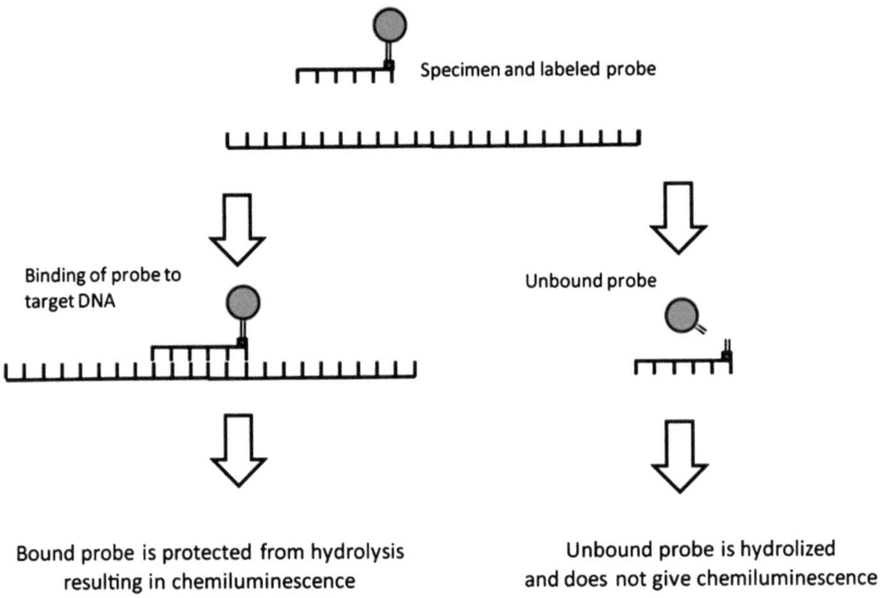

Figure 9 Hybridization protection assay. An acridinium-ester-label is covalently attached to a target-specific DNA probe via an acid-sensitive ether linkage. If present, the probe hybridizes to the target nucleic acid, protecting the bond from acid hydrolysis and generating a signal upon the addition of an appropriate substrate. Nonhybridized probes are hydrolyzed and become inactive.

Hybridization Protection Assay

The Hybridization Protection Assay (HPA) utilizes a chemiluminescent acridinium ester as a detector molecule bound to a DNA probe after binding to specific target RNA sequences (Fig. 9). The label is covalently bound to the oligonucleotide probes via an acid-sensitive ether bond (67). Probes bound to the target are luminescent and protected from acid hydrolysis, while unbound probes are readily hydrolyzed to be rendered permanently nonluminescent. The technology has been commercially developed by Gen-probe (AccuProbe; San Diego, CA, USA) primarily for the detection of bacterial pathogens.

DISCOVERY OF UNKNOWN VIRUSES

Until recently almost all new viruses were discovered by traditional methods such as isolation in cell culture or by detection in clinical specimens using electron microscopy. However, in 1989, hepatitis C virus (HCV) was the first virus to be identified by strictly molecular methods, and since then these techniques have become the primary tools for viral discovery (68). The discovery of HCV was followed by the molecular identification of human herpesvirus-8 (69) and hepatitis G virus (70) using target amplification methods. Molecular techniques are now more widely applied to detect new viruses in samples collected from various body compartments, particularly respiratory, stool, and blood samples. Perhaps the greatest activity in this area has been in the discovery of new viruses associated with the human respiratory tract. In particular, since the discovery of human metapneumovirus (HMPV) in 2001 (71), six previously undescribed viruses have been identified by molecular analysis of clinical specimens from the human respiratory tract. These include three new human coronaviruses (HCoV): the severe acute respiratory syndrome (SARS) associated coronavirus in 2003 (72), coronavirus NL63 (NL63) in 2004 (73), coronavirus HKU1 (HKU1) in 2005 (74), as well as human bocavirus (HBoV) in 2005 (75) and the recently described human polyomaviruses KI (KIV) and WU (WUV) in 2007 (76,77).

These new viral agents were detected by novel molecular methods such as VIDISCA (73), pan-viral DNA microarrays (78), and high-throughput sequencing (76,77). These and other methods were comprehensively reviewed by Ambrose and Clewley (79).

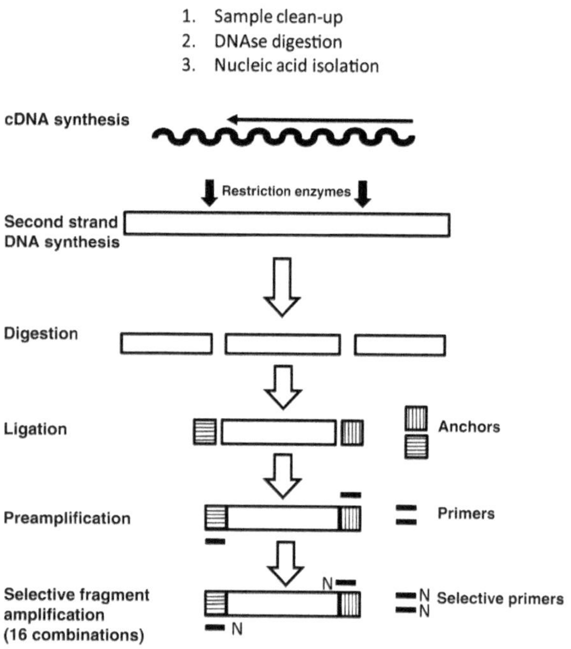

Figure 10 VIDISCA. Following cDNA synthesis, DNA is digested with two restriction enzymes, for example *MseI* and *HinP1I*, producing DNA molecules with *MseI* and *HinP1I* overhangs at either end. Adapters (or anchors) bind to one specific end of the DNA fragment, according to its complementary overhang. Two primers specific to each adapter are then used in an exponential amplification reaction by PCR. A second selective nested PCR amplification is then used to simplify the resultant PCR products from a DNA smear to specific bands.

Virus-Discovery-cDNA-Amplified Fragment Length Polymorphism

In 2004, van der Hoek and colleagues used a modification of a sequence-independent primer amplification technique, called Virus-Discovery-cDNA-AFLP (VIDISCA), to detect a new human coronavirus, HCoV-NL63, in the human respiratory tract (Fig. 10). This technique employs two primers in the PCR amplification step, and includes an amplified fragment length polymorphism (AFLP) method previously described (73).

DNA is digested with two restriction enzymes, for example *MseI* and *HinP1I*, both of which have four base pair recognition sites. This produces DNA molecules with *MseI* and *HinP1I* overhangs at either end, as well as some with *MseI–MseI* and *HinP1I–HinP1I* overhangs. Only the MseI and HinP1I fragments are amplified in the subsequent PCR as each adapter binds to one specific end of the DNA fragment, according to its complementary overhang. Two primers specific to each adapter are then used in an exponential amplification reaction by PCR. A second selective nested PCR amplification can be used to simplify the resultant PCR products from a DNA smear to specific bands. By extending the 3′-end of the primers by one to three nucleotides, a subset of the PCR products is generated, which are subject to further characterization by nucleotide sequencing (80).

Pan-Viral DNA Microarrays

Wang et al. (81) has designed comprehensive DNA microarrays for viral discovery and applied these in the identification of the novel coronavirus associated with severe acute respiratory syndrome (SARS) (78) and the discovery of human polyomavirus WU (77). These arrays consist of 70-mer oligonucleotides representing highly conserved viral sequences, derived from reference sequences of existing viral families obtainable from public sequence databases (78). Ten 70-mers were used for each virus, totaling approximately 10,000 oligonucleotides from about 1000 viruses. Wang et al. (81) used these pan-viral arrays to identify and characterize SARS coronavirus after it had been isolated and cultured in Vero cells from a patient suffering from SARS. Viral sequences hybridized to the individual array elements were recovered and sequenced, to identify this novel coronavirus.

Other viral-specific microarrays have been developed to detect PCR amplicons from sequence-independent amplification reactions. Boriskin et al. (82) developed a diagnostic DNA microarray specific for central nervous system viral infections and applied it to the examination of Cerebrospinal fluid (CSF) and non-CSF specimens. The array contains 38 gene targets for

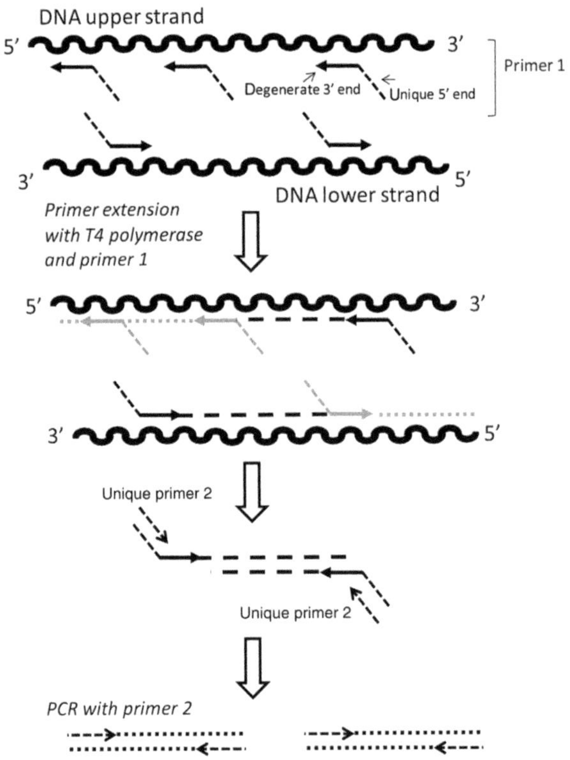

Figure 11 Random PCR. This PCR is initially performed using a single primer with a unique nucleotide universal sequence at the 5′-end, and a degenerate hexa- or heptamer sequence on the 3′-end for random amplification. A second primer is used in a subsequent PCR amplification which is complementary to the 5′ universal region of the first random primer.

13 viral causes of meningitis and encephalitis. Other arrays have been described for the rapid detection and serotyping of acute respiratory disease–associated adenoviruses (83), and for the simultaneous detection of herpesviruses, enteroviruses, and flaviviruses (84). Comprehensive microarrays representing the most up-to-date sequence information for all viral families have much promise for the detection of previously unidentified viruses, provided these have sufficient homology to the known viral sequences (85).

Random PCR Amplification and High-Throughput Sequencing

In some instances, it is advantageous to amplify viral nucleic acids by random PCR amplification before these can be identified using microarrays (81). Generally, random PCR uses one primer with a unique nucleotide universal sequence at the 5′-end (Fig. 11). This sequence contains restriction enzyme sites for subsequent cloning. On the 3′-end this primer contains a degenerate hexa- or heptamer sequence (80,86). A second primer is used in subsequent PCR amplification which is complementary to the 5′ universal region of the first random primer. PCR products are then cloned and DNA sequenced (Fig. 12). Random PCR can be used to detect both DNA and RNA viral genomes (87).

LIMITATIONS OF MOLECULAR AMPLIFICATION METHODS IN DIAGNOSTIC VIROLOGY

While nucleic acid–based assays offer many advantages for the clinical laboratory, care must be exercised when using these tests, particularly those that involve amplification of target nucleic acid sequences, and contamination prevention and quality control must be rigorously implemented. Theoretically, in a nucleic acid amplification test, one copy of a target gene can be amplified. Therefore, if the one copy is from a laboratory contaminant or previous experiment, a false-positive result will be observed. Conversely, inhibitors in clinical specimens or nucleic acid degradation can lead to false-negative results. False-negative results may also occur where the nucleic acid extraction step has failed (14,16).

Nevertheless, false-negative results may still occur even where the very best of quality control measures are implemented. This is because for many viruses it may be difficult to

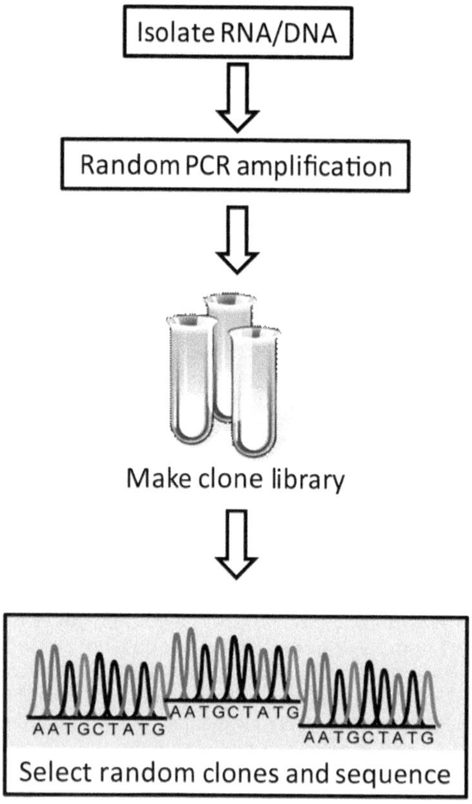

Figure 12 Random PCR and high-throughput sequencing. Following random PCR, PCR products are cloned and then DNA sequenced.

identify sufficiently conserved sequences for diagnostic assays, and so false-negative results may still arise through sequence variation in primer or probe targets. There are several reasons for this, including sequence polymorphism of the viral genome as well as a lack of sequence information. Rapidly evolving RNA viruses such as respiratory syncytial virus (RSV) and the parainfluenza viruses can be particularly problematic as each new season can bring a new variant. Newly characterized or emerging viruses present the greatest challenge in terms of limited sequence data. For instance, we do not know what the actual nucleotide sequence of a pandemic H5N1 strain would be, if such a pandemic were to occur. Therefore, assays designed on the basis of currently circulating H5N1 sequences offer the most effective approach for pandemic preparedness. Multitarget NAT methods have recently been proposed as means of avoiding sequence-related false-negative results (88).

The impact of sequence variation on molecular amplification methods is not just limited to false-negative results, but in certain circumstances may have more subtle effects. Rather than completely preventing nucleic acid amplification, mismatches in *primer* targets may sometimes simply delay amplification. In real-time PCR assays, this may be observed as an increase in cycle threshold (Ct) value. For *quantitative* real-time PCR assays, this delay in Ct value can introduce error and may lead to an underestimation of viral load by several logs. For purely *qualitative* assays, this delay in Ct value can reduce the sensitivity of an assay up to 1000-fold (89). Likewise, sequence variation in *probe* targets can decrease fluorescent signal of positive specimens to a point where it may be difficult to distinguish the signal from that of negative specimens (90). Probe-based genotyping can also be impeded by sequence variation within probe targets (91).

Another disadvantage of this technology is that NATs detect nucleic acids but do not indicate viability of the pathogen. Yet, nucleic acid amplification does provide a sensitive alternative for the diagnosis of noncultivatable or slowly growing pathogens.

CONCLUSION

The advances in molecular techniques witnessed over the last 20 years have revolutionized the diagnosis of viral disease and have provided the tools for the detection and characterization of previously unknown viruses. So far, PCR has been most widely applied in the diagnostic laboratory, but more recently the commercial application of alternative technologies has gained significant momentum. New instrumentation and the development of kit-based systems has introduced a much needed level of standardization and simplicity that will see the implementation of molecular methods in most laboratories over the next few years.

With the development of new molecular technology, our ability to detect and characterize new viral agents has greatly improved. As a result, genome sequences have been described for new viruses that are associated with the human respiratory tract and gastrointestinal tract, as well as new blood-borne viruses. Some of these are recognized as significant human pathogens causing disease in certain population groups. Others can be found in clinical specimens without definitive evidence for their role as the causative agent of disease, and yet others, like TT-(torqueteno) virus (92,93) and mimivirus (94,95) have only been loosely associated with respiratory disorders in humans.

Still, for a significant proportion of clinical infectious disease of suspected viral origin, a pathogen cannot be identified. Although new molecular methods are increasingly used to investigate these unknown causes of disease, they remain technically challenging and prone to the amplification of nonviral related sequence artifacts. However, with continuing advances in molecular technology and the development of more reliable, robust, and reproducible molecular techniques, it seems certain that new potential viral pathogens of humans will continue to be discovered.

With the wider acceptance of molecular technologies, physicians involved in the care of patients can expect another "quantum" leap in the understanding of the epidemiology and genetic aspects of viral disease and its diagnosis. Although conventional clinical microbiology techniques will still occur in other areas of microbiology, it is expected that viral diagnosis will become predominantly molecular. Significant progress can be expected in the next decade in the rapid molecular diagnoses of significant childhood viral disease, with genetic antiviral drug resistance and virulence determinants provided in four to six hours following admission. Also, these techniques will increase our knowledge of the molecular epidemiology of common viral diseases of childhood, particularly those concerning infections of the respiratory and gastrointestinal tracts. The accurate detection and identification of new and known viruses in children and the immunocompromised will continue to improve with these latest molecular techniques, and in combination with advances in cellular biology will lead to the development of novel antiviral and immunologic therapies.

REFERENCES

1. Mullis KB, Faloona FA. Specific synthesis of DNA in vitro via a polymerase-catalyzed chain reaction. Methods Enzymol 1987; 155:335–350.
2. Wu DY, Wallace RB. The ligation amplification reaction (LAR)—Amplification of specific DNA sequences using sequential rounds of template-dependent ligation. Genomics 1989; 4:560–569.
3. Kwoh DY, Davis GR, Whitfield KM, et al. Transcription-based amplification system and detection of amplified human immunodeficiency virus type 1 with a bead-based sandwich hybridization format. Proc Natl Acad Sci USA 1989; 86:1173–1177.
4. Monis PT, Giglio S. Nucleic acid amplification-based techniques for pathogen detection and identification. Infect Genet Evol 2006; 6:2–12.
5. Saiki RK, Gelfand DH, Stoffel S, et al. Primer-directed enzymatic amplification of DNA with a thermostable DNA polymerase. Science 1988; 239:487–491.
6. Saiki RK, Scharf S, Faloona F, et al. Enzymatic amplification of beta-globin genomic sequences and restriction site analysis for diagnosis of sickle cell anemia. Science 1985; 230:1350–1354.
7. Mulder J, McKinney N, Christopherson C, et al. Rapid and simple PCR assay for quantitation of human immunodeficiency virus type 1 RNA in plasma: application to acute retroviral infection. J Clin Microbiol 1994; 32:292–300.
8. Berger A, Braner J, Doerr HW, et al. Quantification of viral load: Clinical relevance for human immunodeficiency virus, hepatitis B virus and hepatitis C virus infection. Intervirology 1998; 41: 24–34.

9. Boeckh M, Boivin G. Quantitation of cytomegalovirus: Methodologic aspects and clinical applications. Clin Microbiol Rev 1998; 11:533–554.
10. Lee WM, Kiesner C, Pappas T, et al. A diverse group of previously unrecognized human rhinoviruses are common causes of respiratory illnesses in infants. PLoS ONE 2007; 2:e966.
11. Landgraf A, Reckmann B, Pingoud A. Direct analysis of polymerase chain reaction products using enzyme-linked immunosorbent assay techniques. Anal Biochem 1991; 198:86–91.
12. Syrmis MW, Whiley DM, Thomas M, et al. A sensitive, specific, and cost-effective multiplex reverse transcriptase-PCR assay for the detection of seven common respiratory viruses in respiratory samples. J Mol Diagn 2004; 6:125–131.
13. Roth WK, Lee JH, Rüster B, et al. Comparison of two quantitative hepatitis C virus reverse transcriptase PCR assays. J Clin Microbiol 1996; 34:261–264.
14. Niesters HG. Molecular and diagnostic clinical virology in real time. Clin Microbiol Infect 2004; 10: 5–11.
15. Ratcliff RM, Chang G, Kok T, et al. Molecular diagnosis of medical viruses. Curr Issues Mol Biol 2007; 9:87–102.
16. Espy MJ, Uhl JR, Sloan LM, et al. Real-time PCR in clinical microbiology: Applications for routine laboratory testing. Clin Microbiol Rev 2006; 19:165–256.
17. Livak KJ, Flood SJ, Marmaro J, et al. Oligonucleotides with fluorescent dyes at opposite ends provide a quenched probe system useful for detecting PCR product and nucleic acid hybridization. PCR Methods Appl 1995; 4:357–362.
18. Wittwer CT, Herrmann MG, Moss AA, et al. Continuous fluorescence monitoring of rapid cycle DNA amplification. Biotechniques 1997; 22:130–131, 134–1348.
19. Brainard JA, Greenson JK, Vesy CJ, et al. Detection of cytomegalovirus in liver transplant biopsies. A comparison of light microscopy, immunohistochemistry, duplex PCR and nested PCR. Transplantation 1994; 27:1753–1757.
20. Yamamoto AY, Mussi-Pinhata MM, Pinto PC, et al. Usefulness of blood and urine samples collected on filter paper in detecting cytomegalovirus by the polymerase chain reaction technique. J Virol Methods 2001; 97:159–164.
21. Schützle H, Weigl J, Puppe W, et al. Diagnostic performance of a rapid antigen test for RSV in comparison with a 19-valent multiplex RT-PCR ELISA in children with acute respiratory tract infections. Eur J Pediatr 2008; 167:745–749.
22. Kehl SC, Henrickson KJ, Hua W, et al. Evaluation of the Hexaplex assay for detection of respiratory viruses in children. J Clin Microbiol 2001; 39:1696–1701.
23. Lee WM, Grindle K, Pappas T, et al. High-throughput, sensitive, and accurate multiplex PCR-microsphere flow cytometry system for large-scale comprehensive detection of respiratory viruses. J Clin Microbiol 2007; 45:2626–2634.
24. Read SJ, Kurtz JB. Laboratory diagnosis of common viral infections of the central nervous system by using a single multiplex PCR screening assay. J Clin Microbiol 1999; 37:1352–1355.
25. Jung R, Soondrum K, Neumaier M. Quantitative PCR. Clin Chem Lab Med 2000; 38:833–836.
26. Guatelli JC, Whitfield KM, Kwoh DY, et al. Isothermal, in vitro amplification of nucleic acids by a multienzyme reaction modelled after retroviral replication. Proc Natl Acad Sci USA 1990; 87:1874–1878.
27. Loeffler J, Hebart H, Cox, P, et al. Nucleic acid sequence-based amplification of *Aspergillus* RNA in blood samples. J Clin Microbiol 2001; 39:1626–1629.
28. Wacharapluesadee S, Hemachudha T. Nucleic-acid sequence based amplification in the rapid diagnosis of rabies. Lancet 2001; 358:892–893.
29. Mahony JB, Song X, Chong S, et al. Evaluation of the NucliSens Basic Kit for detection of *Chlamydia trachomatis* and *Neisseria gonorrhoeae* in genital tract specimens using nucleic acid sequence-based amplification of 16S rRNA. J Clin Microbiol 2001; 39:1429–1435.
30. Lanciotti RS. Molecular amplification assays for the detection of flaviviruses. Adv Virus Res 2003; 61:67–99.
31. van Gemen B, van Beuningen R, Nabbe A, et al. A one-tube quantitative HIV-1 RNANASBA nucleic acid amplification assay using electrochemiluminescent (ECL) labelled probes. J Virol Methods 1994; 49:157–167.
32. Witt DJ, Kemper M, Stead A, et al. Analytical performance and clinical utility of a nucleic acid sequence-based amplification assay for detection of cytomegalovirus infection. J Clin Microbiol 2000; 38:3994–3999.
33. Wu SJ, Lee EM, Putvatana R, et al. Detection of dengue viral RNA using a nucleic acid sequence-based amplification assay. J Clin Microbiol 2001; 39:2794–2798.
34. Leone G, van Schijndel H, van Gemen B, et al. Molecular beacon probes combined with amplification by NASBA enable homogeneous, real-time detection of RNA. Nucleic Acids Res 1998; 26:2150–2155.

35. Ayele W, Pollakis G, Abebe A, et al. Development of a nucleic acid sequence-based amplification assay that uses gag-based molecular beacons to distinguish between human immunodeficiency virus type 1 subtype C and C′ infections in Ethiopia. J Clin Microbiol 2004; 42:1534–1541.
36. Capaul SE, Gorgievski-Hrisoho M. Detection of Enterovirus RNA in cerebrospinal fluid (CSF) using NucliSens EasyQ Enterovirus assay. J Clin Virol 2005; 32:236–240.
37. Gulliksen A, Solli L, Karlsen F, et al. Real-time nucleic acid sequence-based amplification in nanoliter volumes. Anal Chem 2004; 76:9–14.
38. Moore C, Hibbitts S, Owen N, et al. Development and evaluation of a real-time nucleic acid sequence based amplification assay for rapid detection of influenza A. J Med Virol 2004; 74:619–628.
39. Zaytseva NV, Montagna RA, Lee EM, et al. Multianalyte single-membrane biosensor for the serotype-specific detection of Dengue virus. Anal Bioanal Chem 2004; 380:46–53.
40. Walensky RP. APTIMA assay approved for HIV testing. AIDS Clin Care 2006; 18:109–110.
41. Morishima C, Morgan TR, Everhart JE, et al. HCV RNA detection by TMA during the hepatitis C antiviral long-term treatment against cirrhosis (Halt-C) trial. Hepatology 2006; 44:360–367.
42. Walker GT, Fraiser MS, Schram JL, et al. Strand displacement amplification—An isothermal, in vitro DNA amplification technique. Nucleic Acids Res 1992; 20:1691–1696.
43. Walker GT, Linn CP, Nadeau JG. DNA detection by strand displacement amplification and fluorescence polarization with signal enhancement using a DNA binding protein. Nucleic Acids Res 1996; 24:348–353.
44. Nadeau JG, Pitner JB, Linn CP, et al. Real-time, sequence-specific detection of nucleic acids during strand displacement amplification. Anal Biochem 1999; 6:177–187.
45. Nycz CM, Dean CH, Haaland PD, et al. Quantitative reverse transcription strand displacement amplification: Quantitation of nucleic acids using an isothermal amplification technique. Anal Biochem 1998; 259:226–234.
46. Nuovo GJ. In situ strand displacement amplification: An improved technique for the detection of low copy nucleic acids. Diagn Mol Pathol 2000; 9:195–202.
47. Fire A, Xu SQ. Rolling replication of short DNA circles. Proc Natl Acad Sci U S A 1995; 92(10): 4641–4645.
48. Liu D, Daubendiek SL, Zillman MA, et al. Rolling circle DNA synthesis: Small circular oligonucleotides as efficient templates for DNA polymerases. J Am Chem Soc 1996; 118(7):1587–1594.
49. Mothershed EA, Whitney AM. Nucleic acid-based methods for the detection of bacterial pathogens: Present and future considerations for the clinical laboratory. Clin Chim Acta 2006; 363:206–220.
50. Lizardi PM, Huang X, Zhu Z, et al. Mutation detection and single-molecule counting using isothermal rolling-circle amplification. Nat Genet 1998; 19:225–232.
51. Zhang DY, Brandwein M, Hsuih TCH, et al. Amplification of target-specific, ligation-dependent circular probe. Gene 1998; 211:277–285.
52. Thomas DC, Nardone GA, Randall SK. Amplification of padlock probes for DNA diagnostics by cascade rolling circle amplification or the polymerase chain reaction. Arch Pathol Lab Med 1999; 123:1170–1176.
53. Demidov VV. 10 years of rolling the minicircles: RCA assays in DNA diagnostics. Expert Rev Mol Diagn 2005; 5:477–478.
54. Cheglakov Z, Weizmann Y, Basnar B, et al. Diagnosing viruses by the rolling circle amplified synthesis of DNAzymes. Org Biomol Chem 2007; 5:223–225.
55. Notomi T, Okayama H, Masubuchi H, et al. Loop-mediated isothermal amplification of DNA. Nucleic Acids Res 2000; 28:E63.
56. Barany F. Genetic disease detection and DNA amplification using cloned thermostable ligase. Proc Natl Acad Sci USA 1991; 88:189–193.
57. Chen X, Livak KJ, Kwok PY. A homogeneous, ligase-mediated DNA diagnostic test. Genome Res 1998; 8:549–556.
58. Carrino JJ, Laffler TG. Detection of HIV DNA sequences using the ligase chain reaction (LCR). Clin Chem 1991; 37:1059.
59. Rinehardt L, Hampl H, Laffler TG. Ultrasensitive non-radioactive detection of herpes simplex virus by LCR, the ligase chain reaction. In: 20th Annual Meeting of the Keystone Symposia on Molecular and Cellular Biology, March 8–15, 1991:101. Lake Tahoe, CA.
60. Bond S, Carrino J, Hampl H, et al. New methods of detection of HPV. In: Monsonego J, ed. Serono Symposia. Paris, France: Raven Press, 1990:425–434.
61. Bhatt R, Scott B, Whitney S, et al. Detection of nucleic acids by cycling probe technology on magnetic particles: High sensitivity and ease of separation. Nucleosides Nucleotides 1999; 18:1297–1299.
62. Bekkaoui F, McNevin JP, Leung CH, et al. Rapid detection of the mecA gene in methicillin resistant staphylococci using a colorimetric cycling probe technology. Diagn Microbiol Infect Dis 1999; 34: 83–90.

63. Urdea MS, Running JA, Horn T, et al. A novel method for the rapid detection of specific nucleotide sequences in crude biological samples without blotting or radioactivity; application to the analysis of hepatitis B virus in human serum. Gene 1987; 61:253–264.
64. Kern D, Collins M, Fultz T, et al. An enhanced-sensitivity branched-DNA assay for quantification of human immunodeficiency virus type 1 RNA in plasma. J Clin Microbiol 1996; 34:3196–3202.
65. Elbeik T, Dalessandro R, Loftus RA, et al. HIV-1 and HCV viral load cost models for bDNA: 440 molecular system versus real-time PCR AmpliPrep/TaqMan test. Expert Rev Mol Diagn 2007; 7:723–753.
66. Black CM, Morse SA. Diagnosis and epidemiologic study of sexually transmitted infections. Current Infectious Disease Reports 2000; 2:31–43.
67. Arnold LJ Jr, Hammond PW, Wiese WA, et al. Assay formats involving acridinium-ester-labeled DNA probes. Clin Chem 1989; 35:1588–1594.
68. Choo QL, Kuo G, Weiner AJ, et al. Isolation of a cDNA clone derived from a blood-borne non-A, non-B viral hepatitis genome. Science 1989; 244:359–362.
69. Chang Y, Cesarman E, Pessin MS, et al. Identification of herpesvirus-like DNA sequences in AIDS-associated Kaposi's sarcoma. Science 1994; 266:1865–1869.
70. Birkenmeyer LG, Desai SM, Muerhoff AS, et al. Isolation of a GB virus-related genome from a chimpanzee. J Med Virol 1998; 56:44–51.
71. van den Hoogen BG, de Jong JC, Groen J, et al. A newly discovered human pneumovirus isolated from young children with respiratory tract disease. Nat Med 2001; 7:719–724.
72. Ksiazek TG, Erdman D, Goldsmith CS, et al. A novel coronavirus associated with severe acute respiratory syndrome. N Engl J Med 2003; 348:1953–1966.
73. van der Hoek L, Pyrc K, Jebbink MF, et al. Identification of a new human coronavirus. Nat Med 2004; 10:368–373.
74. Woo PC, Lau SK, Chu CM, et al. Characterization and complete genome sequence of a novel coronavirus, coronavirus HKU1, from patients with pneumonia. J Virol 2005; 79:884–895.
75. Allander T, Tammi MT, Eriksson M, et al. Cloning of a human parvovirus by molecular screening of respiratory tract samples. Proc Natl Acad Sci USA 2005; 102:12891–12896.
76. Allander T, Andreasson K, Gupta S, et al. Identification of a third human polyomavirus. J Virol 2007; 81:4130–4136.
77. Gaynor AM, Nissen MD, Whiley DM, et al. Identification of a novel polyomavirus from patients with acute respiratory tract infections. PLoS Pathog 2007; 3:595–604.
78. Wang D, Urisman A, Liu YT, et al. Viral discovery and sequence recovery using DNA microarrays. PLoS Biol 2003; 1:257–260.
79. Ambrose HE, Clewley JP. Virus discovery by sequence-independent genome amplification. Rev Med Virol 2006; 16:365–383.
80. Vos P, Hogers R, Bleeker M, et al. AFLP: A new technique for DNA fingerprinting. Nuc Acids Res 1995; 23:4407–4414.
81. Wang D, Coscoy L, Zylberberg M, et al. Microarray-based detection and genotyping of viral pathogens. Proc Natl Acad Sci USA 2002; 99:15687–15692.
82. Boriskin YS, Rice PS, Stabler RA, et al. DNA microarrays for virus detection in cases of central nervous system infection. J Clin Microbiol 2004; 42:5811–5818.
83. Lin B, Vora GJ, Thach D, et al. Use of oligonucleotide microarrays for rapid detection and serotyping of acute respiratory disease-associated adenoviruses. J Clin Microbiol 2004; 42:3232–3239.
84. Korimbocus J, Scaramozzino N, Lacroix B, et al. DNA probe array for the simultaneous identification of herpesviruses, enteroviruses, and flaviviruses. J Clin Microbiol 2005; 43:3779–3787.
85. Clewley JP. A role for arrays in clinical virology: Fact or fiction? J Clin Virol 2004; 29:2–12.
86. Bachem CWB, van der Hoeven RS, de Bruijn SM, et al. Visualization of differential gene expression using a novel method of RNA fingerprinting based on AFLP: Analysis of gene expression during potato tuber development. Plant J 1996; 9:745–753.
87. Stang A, Korn K, Wildner O, et al. Characterization of virus isolates by particle-associated nucleic acid PCR. J Clin Microbiol 2005; 43:716–720.
88. Whiley DM, Lambert SB, Bialasiewicz S, et al. False-negative results in nucleic acid amplification tests-do we need to routinely use two genetic targets in all assays to overcome problems caused by sequence variation? Crit Rev Microbiol 2008; 34:71–76.
89. Lengerova M, Racil Z, Volfova P, et al. Real-time PCR diagnostics failure caused by nucleotide variability within exon 4 of the human cytomegalovirus major immediate-early gene. J Clin Microbiol 2007; 45:1042–1044.
90. Whiley DM, Sloots TP. Sequence variation can affect the performance of minor groove binder TaqMan probes in viral diagnostic assays. J Clin Virol 2006; 35:81–83.

91. Whiley DM, Sloots TP. Melting curve analysis using hybridisation probes: Limitations in microbial molecular diagnostics. Pathology 2005; 37:254–256.
92. Maggi F, Pifferi M, Fornai C, et al. TT virus in the nasal secretions of children with acute respiratory diseases: Relations to viremia and disease severity. J Virol 2003; 77:2418–2425.
93. Pifferi M, Maggi F, Caramella D, et al. High torqueteno virus loads are correlated with bronchiectasis and peripheral airflow limitation in children. Pediatr Infect Dis J 2006; 25:804–808.
94. Khan M, La SB, Lepidi H, et al. Pneumonia in mice inoculated experimentally with *Acanthamoeba polyphaga* mimivirus. Microb Pathog 2007; 42:56–61.
95. Raoult D, La SB, Birtles R. The discovery and characterization of mimivirus, the largest known virus and putative pneumonia agent. Clin Infect Dis 2007; 45:95–102.

4 Viral Genotyping and the Sequencing Revolution

P. A. Revill and D. S. Bowden
Victorian Infectious Diseases Reference Laboratory, Victoria, Australia

P. A. White
School of Biotechnology and Biomolecular Sciences (BABS), University of New South Wales, Sydney, Australia

INTRODUCTION

Virus characterization has undergone continual refinement as a consequence of the development of new technologies. Most of the human viruses we know today were first identified by observing the consequences of viral replication in laboratory animals, embryonated eggs, or cell cultures. Although not all viruses grow in culture and not all viruses produce a cytopathic effect, infection of cultures can result in reproducible and characteristic changes in cell morphology. This is in a crude sense a form of phenotyping. With the development of serology technology in the 1970s, some classes of viruses could then be identified using specific antibodies capable of neutralizing their infectivity. Serotyping, using neutralizing antibodies raised to type-specific antigens, has been an important tool for classifying viruses, including poliovirus (types 1, 2, and 3), hepatitis B virus (adw, adr, ayw, and ayr), dengue virus (types 1, 2, 3, and 4), and many other virus groups. Serotyping, however, is not suitable for all viruses and the dawn of the molecular biology era allowed viruses to be classified genotypically, at first by using nucleic acid hybridization technology and more recently with the aid of amplification technology, such as the polymerase chain reaction (PCR) followed by DNA sequencing and analysis.

VIRAL MUTATION

Genotypes are a product of virus evolution and brought about by mutation of their genetic material. Viruses mutate rapidly within a narrow sequence space (1). In comparison, the rate of mammalian evolution can be considered slow, owing to a faithful genomic replication process and a low genetic turnover. Viruses exploit all known mechanisms of genetic variation to explore their functional sequence space. Mechanisms attributed to viral evolution include mutation, recombination, inversion, and reassortment. In general, viral mutation rates are related to the fidelity of their respective polymerases. RNA viruses are thought to have high mutation rates ranging between $u = 1 \times 10^{-3}$ and 1×10^{-4} [u is the per-nucleotide mutation rate and is given by $u = M/N$, where M is the number of mutations and N is total copying events (2)]. Drake estimated the RNA genomic mutation rate (U) to be between 1 and 0.1 for most RNA viruses, where $U = G \times u$ [G is the genome size in nucleotides (2)]. In other words, this represents 0.1 and 1 mistakes for every genome copied. Viruses such as hepatitis C virus (HCV) and human immunodeficiency virus (HIV) have amongst the highest mutation rate, with $U = {\sim}1$. Mutation rates substantially higher than 1 per genome per replication cycle (U=1) cannot be tolerated, thus RNA viruses exploit their mutational limit (1). When mutation rates are this high, multiple viral sequences, termed quasispecies, exist within a single host at any given time. Quasispecies can be considered as a group of self-replicating RNA or DNA molecules, which are different, yet closely related to each other, and evolve as a single unit when adapting to changes in the environment (3). The generation of quasispecies in HCV and HIV infection and the extremely high genetic variation of these viruses are dependent upon the combination of an error-prone RNA polymerase and the lack of proofreading ability during polymerization (4). In contrast, DNA viruses, which replicate using a DNA polymerase, have error rates around 100-fold lower ($u = 1 \times 10^{-5}$ to 1×10^{-6}) than RNA viruses. One exception to this rule is hepatitis B virus (HBV). Although HBV is a DNA virus, it encodes a pregenomic RNA (pgRNA) intermediate that is integral to viral replication. The pgRNA is reverse-transcribed into viral DNA using a viral encoded reverse transcriptase that lacks proofreading ability, resulting in error rates similar to HIV and HCV.

The lower mutation rate of most DNA viruses can be largely explained by the presence of the proofreading and mismatch repair functions of their polymerase, which is lacking for the polymerases of RNA viruses and retroviruses. One result of a lower mutation rate in DNA viruses is that longer viral genomes can be replicated (as viruses can only tolerate $U<1$), providing enough sequence space to encode other functions such as mechanisms of immune evasion and enzymes for RNA transcription. Again HBV is an exception, with the high mutation rate associated with reverse transcription of the pgRNA providing one explanation for HBV having one of the smallest genomes of all DNA viruses, being only 3.2 kilobases (kb) in length.

Random mutations from copying errors, whether in RNA or DNA viruses, can lead to phenotypic changes, which may in turn confer a selective advantage—termed positive selection. The advantages conferred by positive selection may range from the ability to replicate in the presence of antiviral drugs, to the ability to evade the cellular immune response to infection. In more extreme cases, these changes may enable the virus to invade and replicate in a new host. In HIV and HBV, mutations in the reverse transcriptase gene confer resistance to a range of antiviral drugs (see later), whereas mutations in the hemagglutinin (H) and neuraminidase (N) genes of influenza virus have enabled it to expand its host range from birds to humans (see excellent reviews in Refs. 5–7). However, not all mutations confer a selective advantage. Those mutations that are deleterious to the virus will be selected against and removed from the population. However, one paradigm argues that the majority of sequence changes may have no significant effect on phenotype and become fixed in the population purely by chance (8). This is known as the "neutral" theory. Genetic variability is maintained in a finite population due to mutational production of neutral and nearly neutral isoalleles. For a virus like HCV, this neutral sequence drift likely accounts for the diversity seen with HCV genotypes, where geographically isolated virus strains accumulate mutations over a long period of time while they phenotypically remain largely unchanged (9). Estimates of genotype divergence in HCV by more conventional methods estimate a range from 500 to 1000 years, although it is likely this is substantially longer (9).

Although neutral mutations do not directly induce phenotypic change, they may still play a very important role in viral evolution and function. Silent mutations may alter noncoding transcription factors and promoter sequences, or RNA secondary structure, thereby affecting RNA synthesis, genome stability, and protein synthesis.

DEFINITION OF GENOTYPES

In virology, genetic sequences within species are subdivided into one or more groups. These classifications are commonly termed genogroups, genotypes, or clades. The degree of variation seen between these classifications is species specific and no standardization in nomenclature exists between viral species. In humans, HIV and HCV probably demonstrate mutation rates among the highest so far recorded, while other viruses, including RNA viruses such as measles and mumps, demonstrate little genetic variation and exist as one or a few genetic clusters or genotypes. In HIV-1, viruses from group M are the most common globally and they share around 50% to 80% nucleotide identity in the envelope region with viruses from other HIV-1 groups (Table 1). HIV groups are subdivided into clades, which share around 30% nucleotide identity within the envelope region and viruses within a clade vary by around 15% (Table 1).

In HCV genetic diversity is classified on three levels. The first level defines genotypes which differ by >35% in nucleotide sequence over the whole 9.5 kb genome (Table 1; Ref. 10–13). The second level divides genotypes into subtypes whose nucleotide sequence differs by approximately 27%. The third level defines isolates, within a subtype, which differ by approximately 8%. In HBV, eight different genotypes have been identified worldwide (A to H), based on intergroup divergence of ≥8% nucleotide sequence variability over the complete genome (14–17). Sequence divergence of up to 4% has been identified within genotypes (18–21), and numerous subgenotypes are now defined for each HBV genotype, although the clinical relevance of many subgenotype classifications is unknown. In HBV, genotypic separation is based on the sequences within the "a determinant" of the surface antigen gene (13), and more recently the promoter and coding sequence of the precore gene that encodes the secreted hepatitis B e-antigen (HBeAg) (22). A computer program has recently been developed that enables identification of the

Table 1 Genetic Distance Within Each Subclassification Level

		Level 1		Level 2		Level 3		
Virus	**Gene**	**Nomenclature**	**% Diversity**	**Nomenclature**	**% Diversity**	**Nomenclature**	**% Diversity**	**Reference**
HIV-1		Inter-group (M vs N)		Inter-clade (A vs B)		Intra-clade (A vs A)		10,11
	gag		~30%		~15%		4–10.5%	
	env		20–52%		~30%		10–15%	
NoV		Inter-genogroup (GI vs GII)		Inter-genotype (GII.4 vs GII.6)		Intra-genotype (G11.4 vs GII.4)		12
	Genome		~46%		~18%		~6%	
HCV	Genome	Inter-genotypes (1 vs 2)		Inter-subtypes (1a vs 1b)		Intra-subtypes (1a vs 1a)		This chapter
			~33%		~23%		~9%	
HBV	s gene	Genotypes		Subgenotypes				13
			>8%		~4%			

genotype of any submitted HBV sequence that encompasses the "a determinant," as well as all mutations present throughout the HBV genome (23).

High degrees of genotypic variability are also observed in other viral groups, with full length sequencing analysis of norovirus (NoV) genogroup II leading to the classification of 17 genotypes that differ by approximately 18% (Table 1; Ref. 12).

Recombination

Genotypic classification is further complicated by the presence of viral recombinants. RNA recombination is one of the major driving forces of viral evolution (reviewed in Ref. 24). Viral recombination can affect phylogenetic groupings, increase the virulence of the virus, confuse molecular epidemiological studies, and have major implications for vaccine design. Recombination occurs when the genomes of two viruses infecting the same host recombine during viral replication, resulting in a new virus capable of autonomous replication (24). Recombination is common in viruses such as HIV, HBV, and NoV (25), but not in HCV. The most common subtype of HBV genotype B circulating in Asia is actually a recombinant between genotypes B and C (21,26), and numerous other recombinants harboring mixtures of other genotypes have also been identified. Recombination can lead to taxonomic confusion, as demonstrated by the recent identification of a recombinant virus in Vietnam harboring genomes of HBV genotypes A, B, and D that led to the proposal that the virus represented a new genotype, designated genotype I (27). This classification has subsequently been challenged (28) and suggests great care need be taken when classifying viruses based on a small number of sequences. Classification difficulties caused by viral recombination are not restricted to HBV taxonomy. A recombinant form of HIV subtype A was initially identified as subtype E (29,30) until reclassified as an A subtype (reviewed in Ref. 31), and HCV recombinants containing the structural genes of a genotype 2k with the nonstructural genes of genotype 1b have also been identified (32). The actual frequency of virus recombination is likely to be underestimated because it is uncommon to sequence full viral genomes.

Phylogenetic Analysis

Once a sequence has been generated, phylogenetic analysis may be used to determine relationships between viruses, establish clusters of related sequences, determine rates of evolution, trace infections, and even assist with vaccine design. Early phylogenetic analysis simply involved comparing the sequence with one or two close relatives using a simple alignment program. But as sequence data increased in volume and complexity, this soon became impractical. Sophisticated analysis programs such as neighbor-joining, maximum likelihood, parsimony, and more recently Bayesian analysis (33) enables production of phylogenetic trees that graphically demonstrated sequence relationships and in some cases makes use of time-stamped viral sequences. With very little training, it is possible for researchers with basic computer skills to generate a phylogenetic tree. However, this itself is not without problems. While it is relatively simple to perform phylogenetic analysis, if performed incorrectly the conclusions may be completely erroneous. The prerequisite for meaningful phylogenetic analysis is accurate alignment of nucleotide or amino acid sequences. The ease of using alignment programs, such as CLUSTAL_X (34), and tree drawing programs, such as TreeView (35), means that phylogenetic trees can be generated, without necessarily any fundamental understanding of the analysis performed. It is imperative that all alignments are checked manually before they are used in phylogenetic analyses, as the quality and relevance of the final tree is totally dependent on the input alignment data. It is also important to provide an estimate of the reliability of the tree, using methods such as bootstrapping (36). Consideration should also be given to generating trees using different methods (i.e., neighbor-joining and maximum likelihood) and comparing the resultant phylogenies. The reader is directed to two excellent texts for detailed explanations of phylogenetic analysis (37,38).

Although the most common usage for phylogenetics in virological terms has probably been to simply determine genotypic or phenotypic relationships among viruses, phylogenetics is much more than just generating trees. It is also a powerful tool that can be used to monitor molecular evolution and trace the origin of viral infections (33). Programs such as HyPhy (39) available at the DataMonkey website (www.datamonkey.org) and BEAST (33) enable calculation of the rate of positive, neutral, and negative selection within a gene by calculating the number of synomonous (silent) and nonsynomonous (nonsilent) mutations. This is useful for determining the regions of a genome under selection pressure and has ramifications for fields such as vaccine design. A study of poliovirus epitopes concluded that epitopes under negative selection may be better choices for vaccine targets, as they are less likely to mutate to a vaccine-avoiding phenotype than sites under positive selection (40).

GENOTYPING METHODOLOGY: TRADITIONAL TECHNIQUES

Genotyping is typically carried out by some form of sequence interrogation. This may entail full genome sequencing, the analysis of a discriminatory region of the genome, or the identification of signature nucleotides. For viruses with small genomes or genomes of modest size, full genome sequencing followed by phylogenetic analysis remains the gold standard. In some instances, the sequencing and phylogenetic analysis of single genes may prove sufficient for genotyping. This will depend on the degree of sequence homology and size of the gene. Comparative sequence programs such as BLASTn are useful for searching sequences in GenBank that have the highest degree of similarity, although few submissions list the virus genotype. Large-scale sequence analysis can be a cumbersome method of determining genotype and numerous methods have been developed to simplify the process and improve throughput (Table 2). Nevertheless, many of these techniques rely on sequence knowledge for their design and implementation.

Traditional Sequencing Methods

While a number of methodologies have been reported in the past, the most common technique initially adopted was based on dideoxynucleotide sequencing (9). The method utilizes the ability of dideoxynucleotides, corresponding to the four naturally occurring nucleotides, to terminate the growing DNA strand initiated from a primer designed to the target DNA strand (hence, these are also known as dideoxyterminators). By radiolabeling either the primer or the dideoxyterminators, four separate reactions can be carried out and the terminated oligonucleotides of varying lengths separated by high-resolution gel electrophoresis. After autoradiography, this provides a nucleotide ladder and nucleotide-by-nucleotide sequence. Unfortunately, this process was cumbersome and individual sequence reads were limited to 200 to 300 bases. More recently, sequencing technology has been refined by using PCR (cycle sequencing) to incorporate deoxyterminators each labeled respectively with different fluorescent dyes. By either gel electrophoresis or capillary electrophoresis, the terminated oligonucleotides pass a laser scanner where the individual fluorescent dye is detected and recorded, enabling the sequence to be determined. Read lengths approaching 1000 bases can be achieved which in many instances may be sufficient to allow viral genotyping.

Direct sequencing of PCR products also meant that viral genomes could be sequenced without the need for cloning, with the obvious caveat that the sequences obtained represented the dominant sequence in the viral population. PCR and automated sequencing technologies resulted in a rapid increase in the number of genomic-length and partial sequences submitted to public databases. GenBank release note 162 (41) states that since its inception in 1982, the GenBank database has doubled in size every 18 months and currently contains over 61 billion nucleotides representing 61 million sequences. PCR and automated sequencing technologies also opened the door to large-scale phylogenetic and evolutionary analysis. It is now possible to routinely amplify and sequence viral genomes from different geographic regions, hosts, or different time periods, and compare the sequences using one of the many analysis programs available over the internet.

Restriction Fragment Length Polymorphism

Restriction fragment length polymorphism (RFLP) is a common method used for genetic fingerprinting and can also be applied to viral genotyping. The technique usually involves prior amplification of a region or regions of the viral genome by PCR and subsequent digestion by restriction endonucleases. After gel electrophoresis, the fragment sizes form a characteristic pattern for a certain genotype. The selection of restriction endonuclease(s) relies on the analysis of different genotypic sequences to find suitable sites for digestion and discrimination. A deficiency of the method is that single nucleotide polymorphisms can result in a change within the

Table 2 Advantages and Disadvantages of Different Genotyping Assays

	Comparison of genotyping assays	
	Advantages	**Disadvantages**
Sequencing and phylogenetics	Current gold standard. Produces the greatest amount of information, allowing classification of new genotypes and identification of recombinants	Time consuming, labor intensive, requires expertise in use of phylogeny computer programs. Reference sequence data needed.
Restriction fragment length polymorphism	Restriction enzyme digestion after PCR amplification is inexpensive, simple and quick. Identification of known genotype patterns is straightforward.	Sequence information is needed to identify restriction enzyme sites. Single nucleotide polymorphisms in restriction enzyme sites or incomplete digestion can make interpretation difficult. A single mutation could result in discordant results.
Type-specific PCR	Simple procedure. Can give an indication of mixed infection with different genotypes.	Sequence information required to deduce PCR primers. Mutations or single nucleotide polymorphisms in primer-binding sites may reduce efficiency of PCR. Cross-reactivity of primers may affect results. Mixed genotype infections may not be identified.
Reverse-phase hybridization	Some commercial assays available that improves quality control and reproducibility. Assays designed to suit most laboratories that do not have extensive experience in genotyping.	Expensive. Results do not always conform to manufacturer's guidelines.
Heteroduplex mobility analysis	Consensus PCR primers can be used. No further sequence knowledge required. Simple and inexpensive technique.	Interpretation can be difficult. Panel of reference sequences needed.
Restriction fragment mass polymorphism	Has the ability to genotype large numbers of samples quickly. Very sensitive at detection of minor species.	Equipment expensive. Expertise and sequence knowledge required to set up initial assay conditions.
Microarray	Multiple specific primers available, flexible design, reproducible.	Requires sequence knowledge, array instrumentation can be expensive, usually a single sample per chip.
Mass sequencing	Can provide full genomic sequencing. Very sensitive.	Expensive instrumentation and start-up costs. Inexpensive bioinformatic programs capable of evaluating the large amount of data generated are not readily available.

restriction site that affects the digestion and thus the reliability of the genotyping for the viruses with less conserved genomes. This method has been successfully applied for the genotyping of numerous viruses including HCV (42).

PCR and Genotype-Specific Primers

Often, genotype-specific PCR primers can be designed based on nucleotide differences or a lack of sequence homology between genotypes. The primers also need to correspond to a sufficiently conserved region within a genotype. Most of these PCR-based assays use a multiplex approach (primers are added to the one reaction mix) with the primers designed to amplify products of different sizes. After electrophoresis, the genotypes can then be easily identified by the size of the amplicons. For example, in clinical diagnostics settings the highly conserved 5′-untranslated region (5′-UTR) of HCV is almost exclusively used for routine reverse transcription–polymerase chain reaction (RT-PCR) detection of HCV. The 5′-UTR also exhibits specific polymorphisms between types and subtypes, which allows classification into six genotypes, but not all subtypes (43). HCV genotyping assays that have utilized type-specific PCRs include those targeting NS5b (44) or core regions (45).

Reverse-Phase Hybridization

For HBV and HCV, one of the commonly used genotyping assays is the commercial line probe assay (LiPA), originally developed by Innogenetics (Belgium). The LiPA is a reverse-phase hybridization assay in which denatured PCR product is hybridized to genotype-specific oligonucleotides bound to a nitrocellulose strip (46,47). The biotinylated primers on the hybridized PCR product allow for a conventional EIA format of conjugate/substrate reaction and after color development, the pattern of reactivity is compared to a supplied template to determine genotype.

Heteroduplex Mobility Analysis

Heteroduplex mobility analysis (HMA) has been applied successfully for the genotyping of HIV (48), HCV (49) and others. HMA relies on the formation of mismatches when two divergent DNA molecules (usually PCR products) are mixed, denatured, and allowed to reanneal. This results in the formation of homoduplexes and heteroduplexes that migrate at different speeds by polyacrylamide gel electrophoresis (PAGE). The mismatches reduce the mobility of the heteroduplexes, which are retarded roughly in proportion to the divergence between the two sequences. Unpaired nucleotides produce larger shifts compared to mismatched nucleotides (40,42). Genotyping by HMA involves mixing a PCR product of unknown genotype separately with a panel of reference products of each genotype and the resultant heteroduplexes are then separated by PAGE (Fig. 1). Ideally, the sequence of the subtypes in the panel should adhere as closely as possible to the consensus sequence for each subtype. Genotype determination relies on the identification of heterologous genotypes in lanes that contain heteroduplexes with reduced mobility (40,43).

GENOTYPING METHODOLOGY: NEW TECHNIQUES

Mass Spectrometry

Once the almost exclusive domain of protein chemists, mass spectrometry has recently been developed as a sensitive tool for virologists. This technology has been adapted to enable analysis of viral genotypes and also enables the identification of mixed viral sequences within a quasispecies pool, as well as the detection of mutations associated with drug resistance (see later).

Kim et al. have recently adapted mass spectrometry to differentiate HCV genotypes (50). They have developed a novel sensitive technique termed restriction fragment mass polymorphism (RFMP), which is based on PCR amplification of the HCV 5′UTR, using primers that introduce sequences recognized by type IIS restriction endonucleases. Importantly, these enzymes cleave outside the restriction sites, resulting in a large array of short amplicons of defined length, unencumbered by the risk that polymorphic amplicons may encode restriction endonuclease recognition sites (Fig. 2). Three variable regions within the HCV 5′UTR were used in

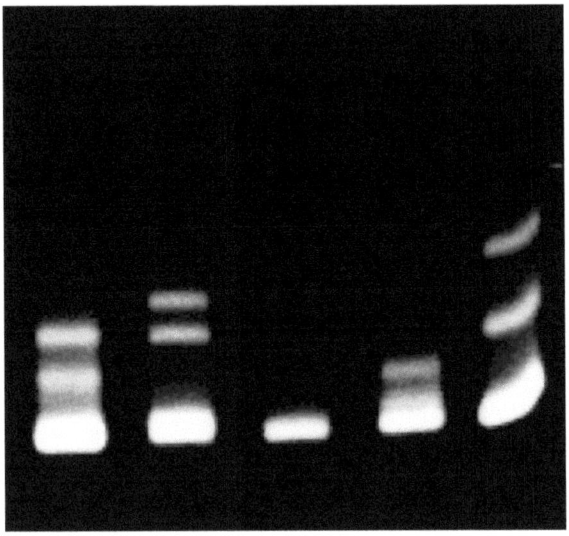

Figure 1 Ethidium bromide stained 8% polyacrylamide gels showing genotyping by HMA. The reference panel of subtypes used is shown along the bottom and the test sample is genotype 3a. Heteroduplexes are observed in all lanes except the reference 3a lane, which therefore indicates the genotype of the test sample.

the analysis, and following type IIS restriction endonuclease digestion, generated fragments of known length ranging from 7 to 19 mer depending on the region analyzed. Mass spectrometry, or more specifically matrix-assisted laser desorption/ionization time-of-flight mass spectrometry (MALDI-TOFF) was then used to identify the variant sequences within the amplicon pool. Using this approach, Kim et al. (50) identified six major genotypes and 27 subtypes, although it is noted that initial results did not match reference sequences for approximately 12% (38/318) of samples. However, subsequent reanalysis of the data enabled classification of an additional 21 samples and it is also possible that further improvements could be achieved by analyzing different regions of the HCV genome. Importantly, Kim et al. (50) reported that the method detected minor genotypes that represented only 0.5% of the quasispecies pool, which is much greater sensitivity than is possible with standard genotyping methods. The sensitivity of the RFMP method suggests that mass spectrometry has an important place in the modern genotyping toolkit.

Microarrays

Oligonuleotide microarrays are another useful method for virus genotyping. In the microarray method, amplicons derived from viral sequence are generally used and tagged in some way, usually with a fluorescent dye. They are then annealed to a microarray chip which is then read. There are numerous ways of labeling viral sequences, for example, Jaaskelainen et al. transcribed RNA from a PCR template and hybridized it to short detection primers on a microarray (51) and used reverse transcription to add fluorescent nucleotides to the hybridized RNA template, while Sengupta et al. used products labeled with cyanine dyes (52). The advantages of microarray are that, a large number of specific primers can be used without additional effort and high costs. For this reason, microarray is particularly suitable for detection and subtyping of a panel of viruses based on their diversity. Microarrays have already been used for the detection and genotyping of viruses such as rotavirus (53), astrovirus and NoV (51), HBV (53), influenza (52), and others.

Mass Sequencing by Synthesis

Powerful new methods in DNA sequencing also offer exciting opportunities for viral genotyping, with degrees of sensitivity that were unimaginable only a few years ago. A large-scale whole genome ultra-deep sequencing approach can provide a means to detect genetic changes associated with selective pressures in more detail than any other current methodology. There are currently two platforms for such an approach: the GS20^{TM} (Roche 454 Life Science) and the Illumina Genome Analyzer System. Both systems utilize a "sequencing-by-synthesis" technology,

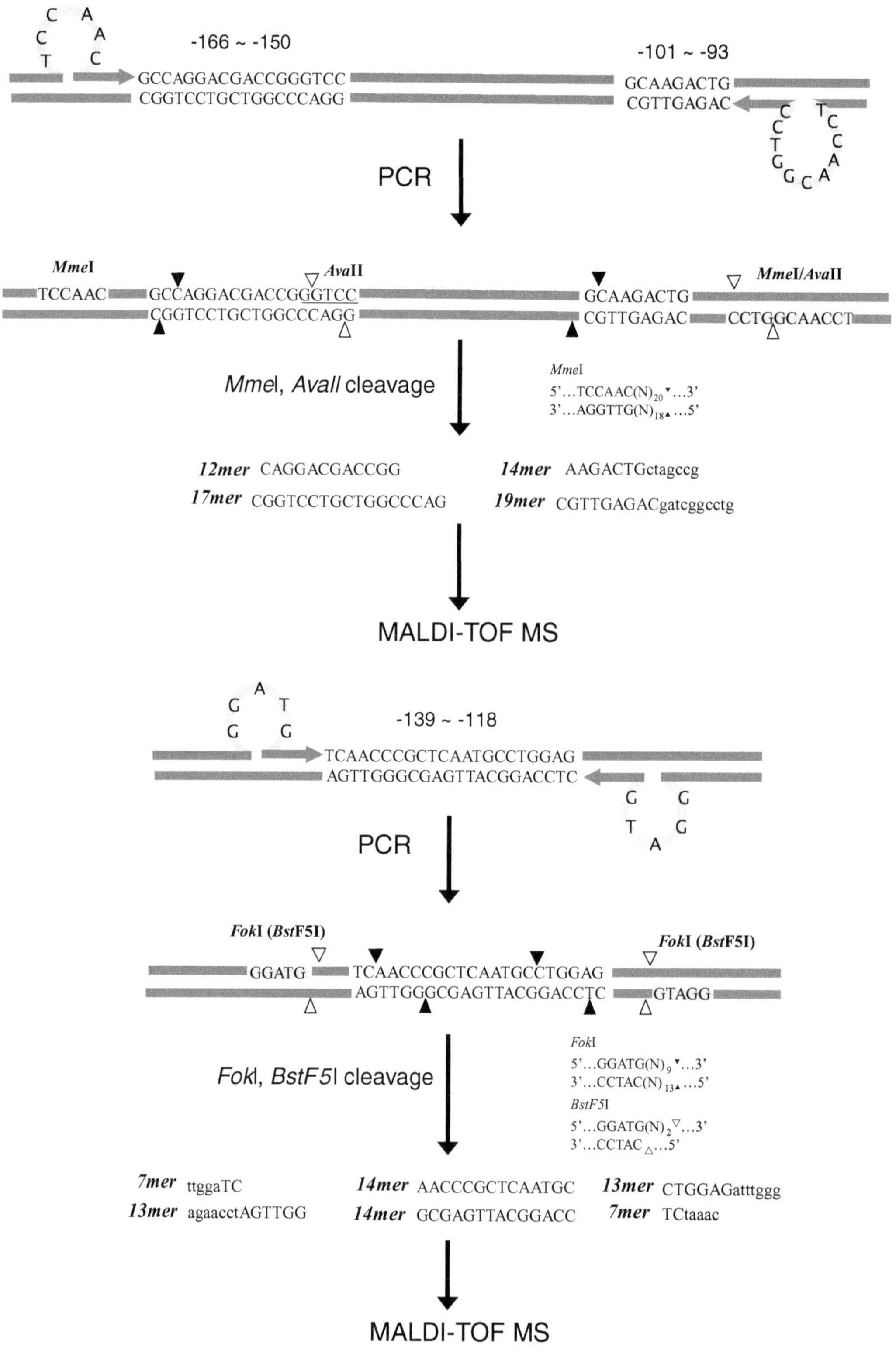

Figure 2 The method used by Kim et al. (50) to generate PCR products for analysis by mass spectrometry.

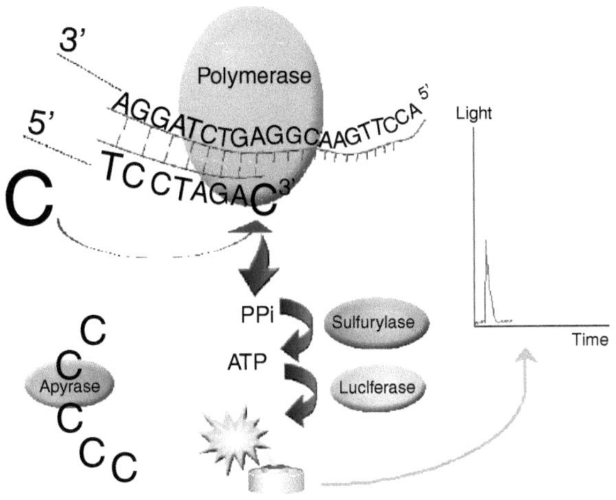

Figure 3 (*See color insert*). The pyrosequencing reaction, demonstrating release of pyrophosphate (PPi) accompanying the addition of each nucleotide. This in turn releases ATP which drives luciferase expression.

allowing for direct single-molecule sequencing without cloning the target sequences into bacteria. They also provide much greater coverage to detect rare sequence changes, although the methods by which the sequence is obtained differs substantially for each platform. The GS20 uses pyrosequencing technology, while the Illumina system is based on Sanger dideoxy sequencing. These technologies will enable in-depth analysis of viral quasispecies with hitherto impossible degrees of sensitivity and will have important ramifications for the treatment of chronic viral infections, as well as evolutionary and epidemiological studies of viral infection.

Pyrosequencing

Pyrosequencing technology is a method for sequencing-by-synthesis in real time (54). It is based on an indirect bioluminometric assay of the pyrophosphate (PPi) that is released from each deoxynucleotide (dNTP) upon DNA chain elongation (54,55). A DNA template/primer complex is presented with a dNTP in the presence of exonuclease-deficient Klenow DNA polymerase (Fig. 3). The four nucleotides, including the dATP analogue dATPalpha-S to avoid background signal, are sequentially added to the reaction mix in a predetermined order. If the nucleotide is complementary to the template base and thus incorporated, PPi is released and used as a substrate, together with adenosine 5′-phosphosulfate (APS), for ATP sulfurylase, which results in the formation of adenosine triphosphate (ATP). Luciferase then converts the ATP, together with luciferin, to oxy-luciferin, AMP, PPi, and visible light that is detected by a luminometer or charge-coupled device. The light produced is proportional to the number of nucleotides added to the extended primer chain. Excess nucleotide is digested by apyrase present in the reaction mixture, before the addition of the next nucleotide. Further improvements on the initial method have enabled extended and more robust read-lengths, for example through the use of single-stranded DNA binding protein (SSB) to reduce secondary structure in DNA templates (56) and the use of purified Sp-Isomer form of dATPalphaS to increase read-length (57).

Until relatively recently, pyrosequencing had been used mainly to identify single nucleotide polymorphisms (SNPs) in the human genome (58,59). However, SNP pyrosequencing has also been used for viral genotyping. Although the basic principles of pyrosequencing are the same in the SNP and ultradeep methods, the SNP method differs in that specific PCR and sequencing primers are used to first amplify and then sequence the PCR product of interest, whereas ultradeep pyrosequencing requires no prior knowledge of the target sequence. Pyrosequencing technology for SNP analysis has been commercialized by Biotage AB, Sweden, for processing up to 96 post-PCR samples in parallel use solid-phase (60). Normally, SNP analysis using pyrosequencing technology involves sequencing less than 10 bases, meaning that 96 samples can be genotyped in approximately 10 minutes. The system also supports multiplexing of SNP, or mutation detection in different templates or positions, detection of multiple SNPs in one template, analysis of insertions and deletions, allele frequency quantification, as well as sequencing of short stretches of typically 20 to 40 bases.

Ultradeep Pyrosequencing

Ultradeep pyrosequencing performed with the Roche GS20™ uses an emulsion-based pyrosequencing platform. Initially, genomic DNA is fragmented into 2.5 kb pieces, methylated and blunt ended, following which an adaptor DNA oligo is ligated onto both ends. The fragments are then digested with the restriction endonuclease *Eco*RI and ligated to form circular molecules. As the adaptor DNA contains two *MmeI* restriction endonuclease sites, digestion with *MmeI* cleaves the circularized DNA, generating small DNA fragments that have the adaptor DNA in the middle and 20 nucleotides of genomic DNA on each end. These small DNA fragments are termed "paired end" fragments and since they are biotinylated, they may be purified from the genomic DNA using streptavidin beads.

Clonal amplification is achieved by incorporating one fragment onto a bead that is then encapsulated into a microreactor by forming a water droplet in an oil background. This microreactor contains all the reagents for DNA amplification. The one fragment is amplified and millions of copies are ultimately attached to the one bead. This process is called emulsion PCR. The clonally amplified bead is then removed from the oil background, beads are enriched for only those that have DNA and then are placed into a PicoTiterPlate for sequencing. Only one bead can fit into one well of the plate and clonal amplification is carried out directly from each DNA-containing bead (one bead is equivalent to one clone). Up to 16 samples can be sequenced in one run. This system is capable of sequencing both PCR amplified products (amplicons) and viral cDNA. Target molecules are attached to the bead with the ratio of two copies (paired end) per bead. The exact number of samples will depend on the desired depth of coverage. Advances in the technology will soon enable reads of up to 500 bp, with over 1 million reads per instrument run (T. Harkin, personal communication, August 2008).

Illumina Genome Analyser

An alternative method of sequencing-by-synthesis is offered by the Illumina (originally Solexa) Genome Analyzer System, which uses a flow-cell (chip) platform. Target molecules are also ligated to an adaptor (one or two types depending on whether bidirectional sequencing is desired) and these molecules are later attached to the flow-cell surface. Each molecule occupies a position on the flow-cell. The sequencing-by-synthesis reactions can then be carried out directly using fluorescently labeled nucleotides, with the fluorescent signal calculated relative to background from each occupied position. Up to eight samples can be sequenced in one run and the system is capable of generating over a billion bases of DNA sequence per run. In theory, the Illumina system can also sequence target cDNA directly. The system currently generates sequence reads of 35 to 50 bases, with up to 100 million base reads routinely obtained for each sequence.

It is likely that massively parallel sequencing-by-synthesis will supersede microarray studies in their current form. The emergence of the aforementionned technologies, as well as other methods such as the ABI SOLID system, means that researchers are no longer restricted to identifying mRNA or DNA sequences using known probe sequences. We are now able to undertake discovery projects unencumbered by the need for prior sequence information.

APPLICATIONS OF GENOTYPING

There may be no observable phenotypic differences between the genotypes of certain viruses. However, for others, virus genotype has been shown to correlate with disease pathogenesis, infectivity, transmission properties, and response to antiviral agents.

Disease Pathogenesis

Genotypic differences may be important in disease pathogenesis. The severity of HBV-related liver disease appears at least in part to be genotype dependent, with genotype C generally resulting in more severe liver disease than genotype B in Asian countries (61–68) whereas genotype D may result in more severe disease than genotype A in Western societies (69,70). The influence of genotype on HBV pathogenesis is confounded by the identification of subtypes within HBV genotypes, some of which are caused by intergenotypic recombination. A subgenotype of genotype B (Ba) is a genotype C recombinant that responds less well to lamivudine therapy and has been linked to the development of more severe liver disease than the nonrecombinant genotype B (Bj) virus (71,72). Three subtypes of genotype A have also been identified, with the A1 (or Aa) subtype prevalent in southern Asia and sub-Saharan Africa; the A2 (Ae) common in the USA and Europe; and the A3 genotype present in West Africa (73–75). The African (A1) subtype is associated with rapid disease progression and a higher incidence of HCC (76). The reasons for the differences in pathogenic response of different genotypes remain unclear, although there is increasing evidence that mutations in the promoter region and the gene encoding the HBV precore protein, which is subsequently processed into the secreted hepatitis B antigen (HBeAg), are associated with disease severity. For example, the African A1 subtype encodes a mutation in the N-terminal signal sequence of the precore protein that results in intracellular retention of HBeAg (77). It remains to be determined if this retention of viral protein is directly related to the more rapid progression to HCC associated with this genotype.

It appears that HCV genotype does not generally influence progression to chronicity or severity of disease. However, HCV infection has also been associated with a variety of clinical disorders, including metabolic disorders. Steatosis, the accumulation of fat within hepatocytes, has been found to be a common feature of chronic hepatitis C infection. Interestingly, in patients infected with HCV genotype 3 there is some evidence to suggest that hepatic steatosis may be a genotype-induced lesion (78). This suggests that a viral protein produced during a genotype 3 infection is involved in the steatogenic process, while the same protein produced during infection by other genotypes is not.

Response to Therapy

One clear association of HCV genotype is its role as a major predictor of outcome of interferon-based therapy (79,80). With recent improvements in the efficacy of antiviral treatment, up to 50% of patients infected with HCV genotype 1 and 80% of those infected with HCV genotypes 2 and 3 achieve a sustained viral response (SVR) six months posttreatment (81). Treatment is generally recommended for 48 weeks in patients with HCV genotype 1 and genotype 4 infections, while patients infected with HCV genotypes 2 and 3 are recommended to have 24 weeks of therapy.

HBV genotypes also respond differently to interferon. HBV genotype A is more sensitive and responds earlier to interferon treatment than all other genotypes, although the reasons for this are unknown (82,83). Differences in response to interferon are also observed in other HBV genotypes, with genotype B responding better to interferon treatment than genotype C, in HBeAg positive individuals (84).

Viral Quasispecies Analysis

Sensitive sequencing analysis methods are extremely useful for analysis of viral quasispecies. These techniques enable the detection of minor populations of infecting virus that may otherwise go undetected due to the lack of sensitivity of some methodologies. Therefore, infections involving two different isolates of the same viral species will be more readily detected, and the clinical implications of dual infection can now be explored. Techniques such as real-time PCR (85), ultradeep pyrosequencing (86), and SNP pyrosequencing (87) have been used to determine the abundance of mutant viral genomes within populations, without the need to clone large numbers of viral genomes. Although time consuming, data generated by cloning is still valid, as demonstrated by a recent study that used PCR and cloning to elegantly demonstrate high levels of viral quasispecies diversity prior to seroconversion in persons with chronic HBV infection (88). However, it is likely that large-scale cloning studies will fall out of favor as more rapid and automated techniques evolve.

Antiviral Resistance Detection

One of the earliest applications of sequence analysis in the clinical situation was for the detection of HIV drug resistance. A broad spectrum of antiviral drugs is available for the treatment of HIV, and resistance testing is an important component of patient management (89). The complexity of resistance changes for HIV is such that some of the methods applicable for detection of changes associated with resistance for other viruses are unsuitable. The most common method used for detection of HIV resistance is direct sequencing of PCR product and studies have shown that in experienced laboratories a high concordance can be obtained. Resistance testing for HIV provides many challenges. Other than the sequencing itself, interpretation of what can be complicated patterns of changes can be subjective. Several databases are available to help infer antiviral resistance from HIV sequence, including the Stanford Database (90) and CREST algorithm (91).

As there is some potential for both inter- and intralaboratory variation, several countries have introduced external quality assurance schemes to assess accuracy and reliability of testing (92). Commercial PCR and sequencing-based assays, such as the Viroseq HIV-1 genotyping system (Celera Diagnostics) and TruGene HIV-1 genotyping kit (Bayer HealthCare), supply reagents and software to help identify resistance changes in the HIV protease and the reverse transcriptase domain. The ViroSeq HIV-1 genotyping system has been shown to have high sensitivity and specificity, even when samples with low viral load are used (93).

HCV, HBV, and HIV all exist as a quasispecies pool of viral genomes. The lack of proof-reading in the respective HCV RNA polymerase, and HBV and HIV reverse transcriptases means that many errors are generated with each round of replication. Thus, mutant viruses that are more resistant to antiviral agents used to treat infections emerge. Although most mutant viruses are not as fit as wild-type virus, many mutants are able to replicate at low levels and under certain selection pressures emerge as the dominant virus. For example, preexisting mutations in the HBV polymerase gene confer resistance to nucleoside analogues such as lamivudine (94–101), adefovir (102,103), and entecavir (97–99) Fig. 4). Determining the abundance of viral genomes harboring these mutations prior to the commencement of therapy may enable better targeting of therapy and improved treatment outcomes for patients. Numerous methods have been used to identify HBV mutations associated with drug resistance. These include real-time PCR (85), LiPA (100,101), amplicon sequencing (102,103), mass spectrometry (100,104–106), SNP pyrosequencing (87), and most recently ultradeep pyrosequencing (86).

It is likely that methods that are suitable for automation and scale-up into 96 well or 384 well platforms will find most favor in the foreseeable future. In this regard, mass spectrometry, real-time PCR, and pyrosequencing are most suited to large-scale analysis. Kim et al. (104) used a 384-well RFMP mass spectrometry assay to genotype HCV variants in patient serum. RFMP analysis has also been used to determine the relative abundance of lamivudine-resistant (100,104,105) and adefovir-resistant (105) HBV mutants. RFMP detected lamivudine-resistant mutant viral genomes at a sensitivity of 1% in mutant/wild-type mixtures, compared to LiPA technology that detected mutants to 4% in the same population (104). In the setting of HBV mono-infection, Hong et al. (106) showed that mass spectrometry could detect as few as 100 copies of HBV DNA per mL of serum.

Ultradeep pyrosequencing has recently been used to identify mutations associated with drug resistance in HIV (86) and HBV. Shafer et al. have used Roche ultradeep pyrosequencing

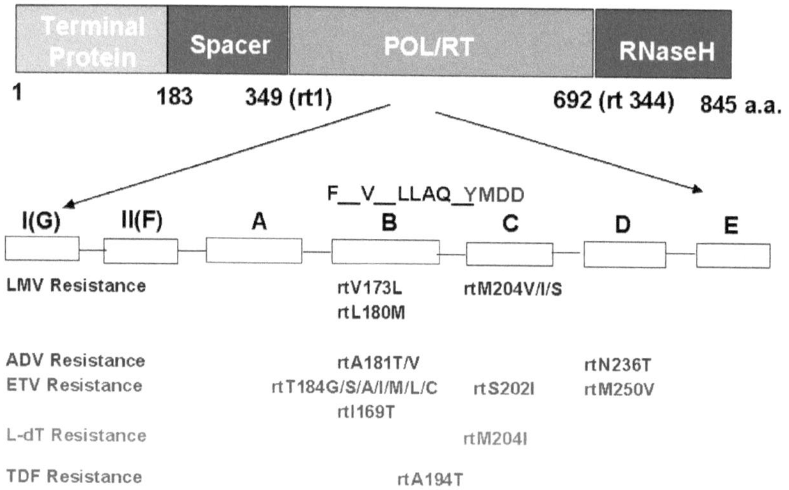

Figure 4 (*See color insert*). Primary HBV polymerase mutations (domains A to G) associated with antiviral drug resistance in chronic HBV infection. *Abbreviations*: LMV, lamivudine; ADV, adefovir; ETV, entecavir; L-dT, telbivudine; TDF, tenofovir. The YMDD motif associated with lamivudine and telbivudine resistance is located at residues 203 to 206.

to identify mutations associated with antiviral drug resistance with higher degrees of sensitivity than previously possible with amplicon sequencing (R. Shafer personal communication, September 2008)[a]. In samples from nucleoside treated patients, mutations were identified at a level of 2% or greater, with this degree of sensitivity confirmed by cloning. Although there are challenges ahead to sort out the true mutations from background noise such as PCR error and G to A hypermutation (107), the potential for ultradeep pyrosequencing and other sensitive methods to analyze the abundance of drug resistance mutations prior to nucleoside analogue therapy should be realized.

SNP pyrosequencing has also been used to analyze HBV mutations associated with antiviral drug resistance (87). Lindstrom et al. (87) showed that pyrosequencing was faster and more accurate than direct sequencing of PCR amplicons and more amenable to scale up. It also enabled detection of the relative abundance of mutant and wild-type genomes with greater accuracy than traditional PCR and direct sequencing. However, Yang et al. (85) compared amplicon-sequencing, pyrosequencing, and real-time PCR to detect YMDD mutants associated with lamivudine resistance, in patients with chronic HBV infection and showed that real-time PCR was the most sensitive and cost-effective assay. Ultradeep pyrosequencing was not included in the comparison and it is questionable whether real-time PCR would be more sensitive than ultradeep pyrosequencing in a direct comparison.

Interferon-based therapy for chronic HCV infection is intended to enhance the immune-mediated eradication of virus, and with this treatment there is no evidence of the development of viral resistance. The adverse side effects of interferon and the low sustained viral response rate for individuals infected with the common HCV genotype 1 have provided momentum for the development of specifically targeted antiviral therapy for HCV (STAT-C), principally to inhibit the HCV serine protease and RNA polymerase. Preliminary clinical data using STAT-C was encouraging, showing significant decreases in viral load; however, when used as monotherapy resistance was rapidly selected limiting their use in this capacity (108). In HCV, detection of the resistance mutations is generally carried out using type-specific PCR primers and subsequent direct sequencing and/or cloning. The emergence of more sensitive technologies amenable to mass analysis suggests it is only a matter of time before some of these methods will also be used to monitor HCV drug resistance.

Viral Discovery

Traditional viral discovery methods were often painstakingly slow, requiring initial identification by electron microscopy, culturing, viral and nucleic acid purification, cloning and, finally, sequencing. However, despite the success of these techniques, many viruses have eluded electron microscopists and virologists alike and proved recalcitrant to purification and subsequent characterization. New strategies using instruments such as the Roche FLX ultradeep pyrosequencer and the Illumina Genome Analyser, enable sequencing of complete viral genomes without the need for prior viral purification or cloning. These methods may result in a rapid increase in the discovery of previously unidentified viruses.

The power of sequencing-by-synthesis for viral discovery was recently demonstrated with the identification of a lethal arenavirus in three transplant recipients using ultradeep pyrosequencing (109). These patients received organs from a person that had died of a febrile illness, and they all died within six weeks of transplantation. Culture, microarray, PCR, and serological analyses for a range of bacterial and viral pathogens were uninformative; however, analysis of over 103,000 sequences generated by ultradeep pyrosequencing identified 14 sequences with similarity to old world arenaviruses at the amino acid level. Specific PCR primers based on these sequences enabled amplification of the remainder of the genome by traditional PCR. This clearly demonstrated the power of ultradeep pyrosequencing technology to identify a previously unknown viral pathogen. The massively parallel "sequencing-by-synthesis" dideoxy platform used by the Illumina Genome Analyser also requires no prior sequence knowledge. However, even these exciting new technologies still have their limitations. The aforementioned

[a] Margeridon-Thermet S, Shulman N, Ahmed A, Shahriar R, Liu T, Wang C, Holmes S, Babrzadeh F, Gharizadeh B, Hanczaruk B, Simen B, Egholm M, and Shafer R. 2009. Ultra-Deep Pyrosequencing of Hepatitis B Virus Quasispecies from Nucleoside and Nucleotide Reverse-Transcriptase Inhibitor (NRTI)–Treated Patients and NRTI-Naive Patients. Journal of Infectious Diseases, 199:1275–85.

novel arenavirus was identified because of its similarity to the deduced sequences encoded by previously identified viruses. If arenavirus sequences had not already been placed on GenBank, the 14 novel sequences identified by Palacios et al. (109) may have been overlooked. It is important to note that the nucleotide sequences themselves bore no similarity to sequences on GenBank. It was only when the sequences were translated, using the BlastX algorithm available on the NCBI website (110) that the similarity to old world arenaviruses became apparent. However, we could still be faced with the dilemma of how to identify a completely novel viral sequence, which bears no relationship to currently identified viruses, if sequence data is the only information available.

Molecular Epidemiology

Molecular epidemiological techniques have provided an important new approach to the study of virus transmission and have often been used to complement traditional epidemiological investigations. In particular for the blood-borne viruses HIV and HCV, molecular techniques have been useful in the investigation of virus evolution (9,111,112), characterizing past epidemics (113–115), following viral transmission on a global scale (116,117), outbreak investigation (118,119), and tracing individual sources of transmission (120–123). More recently, in the era of antiviral therapy, molecular epidemiological techniques have been used to trace the transmission of drug-resistant HIV (124,125) and drug-resistant HBV (126).

CONCLUSION

We are at the dawn of new age in viral genotyping and DNA sequencing analysis that will profoundly alter virological research over the next decade. Techniques such as mass spectrometry, massively parallel Sanger dideoxy sequencing, and ultradeep pyrosequencing will revolutionize DNA and RNA sequencing and enable discovery of viruses that have proved undetectable by traditional methods. The challenge for each of us is how to manage and analyze the large amount of data that is generated, to maximize the potential of these exciting tools.

ACKNOWLEDGMENTS

The authors thank Soo Ok Kim (Genematrix) for providing Figure 2, Phillip Hudson (Millennium Science Pty Ltd) and Nigel Tooke (Biotage) for providing Figure 3 and some of the text on pyrosequencing, and Tim Harkins, Roche Applied Sciences, Indianapolis for information on Ultradeep pyrosequencing. We thank Stephen Locarnini, Lilly Yuen, and Anna Ayres for providing Figure 4 and Robert Shafer and Nancy Shulman for providing access to their data prior to publication. We also thank Doris Chibo and Yi-Mo Deng for information on HIV and influenza viruses, respectively.

REFERENCES

1. Holland J, Spindler K, Horodyski F, et al. Rapid evolution of RNA genomes. Science 1982; 215:1577.
2. Drake JW. Rates of spontaneous mutation among RNA viruses. Proc Natl Acad Sci U S A 1993; 90(9):4171–4175.
3. Domingo E. Quasispecies and the implications for virus persistence and escape. Clin Diagn Virol 1998; 10(2–3):97–101.
4. Manrubia SC, Escarmis C, Domingo E, et al. High mutation rates, bottlenecks, and robustness of RNA viral quasispecies. Gene 2005; 347(2):273–282.
5. Cheung TK, Poon LL. Biology of influenza a virus. Ann N Y Acad Sci 2007; 1102:1–25.
6. Ghedin E, Sengamalay NA, Shumway M, et al. Large-scale sequencing of human influenza reveals the dynamic nature of viral genome evolution. Nature 2005; 437(7062):1162–1166.
7. Lipatov AS, Govorkova EA, Webby RJ, et al. Influenza: Emergence and control. J Virol 2004; 78(17):8951–8959.
8. Kimura M. Genetic variability maintained in a finite population due to mutational production of neutral and nearly neutral isoalleles. Genet Res 1968; 11(3):247–269.
9. Simmonds P. Genetic diversity and evolution of hepatitis C virus—15 years on. J Gen Virol 2004; 85(Pt 11):3173–3188.
10. Zagury JF, Franchini G, Reitz M, et al. Genetic variability between isolates of human immunodeficiency virus (HIV) type 2 is comparable to the variability among HIV type 1. Proc Natl Acad Sci U S A 1988; 85:5941–5945.

11. Spira S, Wainberg MA, Loemba H, et al. Impact of clade diversity on HIV-1 virulence, antiretroviral drug sensitivity and drug resistance. J Antimicrob Chemother 2003; 51:229–240.
12. Katayama K, Shirato-Horikoshi H, Kojima S, et al. Phylogenetic analysis of the complete genome of 18 Norwalk-like viruses. Virology 2002; 299:225–239.
13. Norder H, Courouce AM, Coursaget P, et al. Genetic diversity of hepatitis B virus strains derived worldwide: Genotypes, subgenotypes, and HBsAg subtypes. Intervirology 2004; 47(6):289–309.
14. Arauz-Ruiz P, Norder H, Robertson BH, et al. Genotype H: A new Amerindian genotype of hepatitis B virus revealed in Central America. J Gen Virol 2002; 83(Pt 8):2059–2073.
15. Chu CJ, Keeffe EB, Han SH, et al. Hepatitis B virus genotypes in the United States: Results of a nationwide study. Gastroenterology 2003; 125(2):444–451.
16. Norder H, Courouce AM, Magnius LO. Molecular basis of hepatitis B virus serotype variations within the four major subtypes. J Gen Virol 1992; 73(Pt 12):3141–3145.
17. Okamoto H, Tsuda F, Sakugawa H, et al. Typing hepatitis B virus by homology in nucleotide sequence: Comparison of surface antigen subtypes. J Gen Virol 1988; 69(Pt 10):2575–2583.
18. Kramvis A, Weitzmann L, Owiredu WK, et al. Analysis of the complete genome of subgroup A' hepatitis B virus isolates from South Africa. J Gen Virol 2002; 83(Pt 4):835–839.
19. Kimbi GC, Kramvis A, Kew MC. Distinctive sequence characteristics of subgenotype A1 isolates of hepatitis B virus from South Africa. J Gen Virol 2004; 85(Pt 5):1211–1220.
20. Mbayed VA, Barbini L, Lopez JL, et al. Phylogenetic analysis of the hepatitis B virus (HBV) genotype F including Argentine isolates. Arch Virol 2001; 146(9):1803–1810.
21. Sugauchi F, Kumada H, Sakugawa H, et al. Two subtypes of genotype B (Ba and Bj) of hepatitis B virus in Japan. Clin Infect Dis 2004; 38(9):1222–1228.
22. Kramvis A, Arakawa K, Yu MC, et al. Relationship of serological subtype, basic core promoter and precore mutations to genotypes/subgenotypes of hepatitis B virus. J Med Virol 2008; 80(1):27–46.
23. Yuen LK, Ayres A, Littlejohn M, et al. SeqHepB: A sequence analysis program and relational database system for chronic hepatitis B. Antiviral Res 2007; 75(1):64–74.
24. Worobey M, Holmes EC. Evolutionary aspects of recombination in RNA viruses. J Gen Virol 1999; 80:2535–2543.
25. Bull RA, Tanaka MM, White PA. Norovirus recombination. J Gen Virol 2007; 88(Pt 12):3347–3359.
26. Sugauchi F, Orito E, Ichida T, et al. Hepatitis B virus of genotype B with or without recombination with genotype C over the precore region plus the core gene. J Virol 2002; 76(12):5985–5992.
27. Tran TT, Trinh TN, Abe K. New complex recombinant genotype of hepatitis B virus identified in Vietnam. J Virol 2008; 82(11):5657–5663.
28. Kurbanov F, Tanaka Y, Kramvis A, et al. When should "I" consider a new hepatitis B virus genotype? J Virol 2008; 82(16):8241–8242.
29. Carr JK, Salminen MO, Koch C, et al. Full-length sequence and mosaic structure of a human immunodeficiency virus type 1 isolate from Thailand. J Virol 1996; 70(9):5935–5943.
30. Gao F, Robertson DL, Morrison SG, et al. The heterosexual human immunodeficiency virus type 1 epidemic in Thailand is caused by an intersubtype (A/E) recombinant of African origin. J Virol 1996; 70(10):7013–7029.
31. Taylor BS, Sobieszczyk ME, McCutchan FE, et al. The challenge of HIV-1 subtype diversity. N Engl J Med 2008; 358(15):1590–1602.
32. Kalinina O, Norder H, Mukomolov S, et al. A natural intergenotypic recombinant of hepatitis C virus identified in St. Petersburg. J Virol 2002; 76(8):4034–4043.
33. Drummond AJ, Rambaut A. BEAST: Bayesian evolutionary analysis by sampling trees. BMC Evol Biol 2007; 7:214–221.
34. Thompson JD, Gibson TJ, Plewniak F, et al. The CLUSTAL_X windows interface: flexible strategies for multiple sequence alignment aided by quality analysis tools. Nucleic Acids Res 1997; 25(24):4876–4882. http://www.uk.plbio.kvl.dk/bioinfo.htm. Accessed June 2008.
35. Page RD. TreeView: An application to display phylogenetic trees on personal computers. Comput Appl Biosci 1996; 12(4):357–358. http://www.uk.plbio.kvl.dk/bioinfo.htm. Accessed February 2008.
36. Felsenstein J. Confidence limits on phylogenies: An approach using the bootstrap. Evolution 1989; 39:783–791.
37. Hall BG. Phylogenetic Trees Made Easy—A How-To Manual for Molecular Biologists. Sunderland, MA: Sinauer Associates, Inc., 2001.
38. Salemi M, A-M Vandamme. The Phylogenetic Handbook. A Practical Approach to DNA and Protein Phylogeny. Cambridge, U.K.: Cambridge University Press, 2003.
39. Pond SL, Frost SD, Muse SV. HyPhy: Hypothesis testing using phylogenies. Bioinformatics 2005; 21(5):676–679.
40. Suzuki Y. Negative selection on neutralization epitopes of poliovirus surface proteins: Implications for prediction of candidate epitopes for immunization. Gene 2004; 328:127–133.

41. http://www.ncbi.nlm.nih.gov/Genbank/index.html. Accessed August 2008.
42. Pohjanpelto P, Lappalainen M, Widell A, et al. Hepatitis C genotypes in Finland determined by RFLP. Clin Diagn Virol 1996; 7(1):7–16.
43. Smith DB, Davidson F, Simmonds P. Hepatitis C virus variants and the role of genotyping. J Hepatol 1995; 23(Suppl 2):26–31.
44. Chayama K, Tsubota A, Arase Y, et al. Genotypic subtyping of hepatitis C virus. J Gastroenterol Hepatol 1993; 8(2):150–156.
45. Holland PV, Barrera JM, Ercilla MG, et al. Genotyping hepatitis C virus isolates from Spain, Brazil, China, and Macau by a simplified PCR method. J Clin Microbiol 1996; 34(10):2372–2378.
46. Stuyver L, Rossau R, Wyseur A, et al. Typing of hepatitis C virus isolates and characterization of new subtypes using a line probe assay. J Gen Virol 1993; 74(Pt 6):1093–1102.
47. Teles SA, Martins RM, Vanderborght B, et al. Hepatitis B virus: Genotypes and subtypes in Brazilian hemodialysis patients. Artif Organs 1999; 23(12):1074–1078.
48. Delwart EL, Herring B, Rodrigo AG, et al. Genetic subtyping of human immunodeficiency virus using a heteroduplex mobility assay. Genome Res 1995; 4(5):S202–S216.
49. White PA, Li Z, Zhai X, et al. Mixed viral infection identified using heteroduplex mobility analysis (HMA). Virology 2000; 271(2):382–389.
50. Kim YJ, Kim SO, Chung HJ, et al. Population genotyping of hepatitis C virus by matrix-assisted laser desorption/ionization time-of-flight mass spectrometry analysis of short DNA fragments. Clin Chem 2005; 51(7):1123–1131.
51. Jaaskelainen AJ, Maunula L. Applicability of microarray technique for the detection of noro- and astroviruses. J Virol Methods 2006; 136(1–2):210–216.
52. Sengupta S, Onodera K, Lai A, et al. Molecular detection and identification of influenza viruses by oligonucleotide microarray hybridization. J Clin Microbiol 2003; 41(10):4542–4550.
53. Pas SD, Tran N, de Man RA, et al. Comparison of reverse hybridization, microarray, and sequence analysis for genotyping hepatitis B virus. J Clin Microbiol 2008; 46(4):1268–1273.
54. Ronaghi M, Uhlen M, Nyren P. A sequencing method based on real-time pyrophosphate. Science 1998; 281(5375):363, 365.
55. Alderborn A, Kristofferson A, Hammerling U. Determination of single-nucleotide polymorphisms by real-time pyrophosphate DNA sequencing. Genome Res 2000; 10(8):1249–1258.
56. Ronaghi M. Improved performance of pyrosequencing using single-stranded DNA-binding protein. Anal Biochem 2000; 286(2):282–288.
57. Gharizadeh B, Nordstrom T, Ahmadian A, et al. Long-read pyrosequencing using pure 2′-deoxyadenosine-5′-O′-(1-thiotriphosphate) Sp-isomer. Anal Biochem 2002; 301(1):82–90.
58. Neve B, Froguel P, Corset L, et al. Rapid SNP allele frequency determination in genomic DNA pools by pyrosequencing. Biotechniques 2002; 32(5):1138–1142.
59. Gruber JD, Colligan PB, Wolford JK. Estimation of single nucleotide polymorphism allele frequency in DNA pools by using Pyrosequencing. Hum Genet 2002; 110(5):395–401.
60. http://www.biotagebio.com. Accessed July 2008.
61. Kao JH, Chen PJ, Lai MY, et al. Hepatitis B genotypes correlate with clinical outcomes in patients with chronic hepatitis B. Gastroenterology 2000; 118(3):554–559.
62. Lee CM, Chen CH, Lu SN, et al. Prevalence and clinical implications of hepatitis B virus genotypes in southern Taiwan. Scand J Gastroenterol 2003; 38(1):95–101.
63. Ding X, Mizokami M, Ge X, et al. Different hepatitis B virus genotype distributions among asymptomatic carriers and patients with liver diseases in Nanning, southern China. Hepatol Res 2002; 22(1):37–44.
64. Kobayashi M, Arase Y, Ikeda K, et al. Clinical characteristics of patients infected with hepatitis B virus genotypes A, B, and C. J Gastroenterol 2002; 37(1):35–39.
65. Lindh M, Horal P, Dhillon AP, et al. Hepatitis B virus DNA levels, precore mutations, genotypes and histological activity in chronic hepatitis B. J Viral Hepat 2000; 7(4):258–267.
66. Orito E, Ichida T, Sakugawa H, et al. Geographic distribution of hepatitis B virus (HBV) genotype in patients with chronic HBV infection in Japan. Hepatology 2001; 34(3):590–594.
67. Sakugawa H, Nakasone H, Nakayoshi T, et al. Preponderance of hepatitis B virus genotype B contributes to a better prognosis of chronic HBV infection in Okinawa, Japan. J Med Virol 2002; 67(4):484–489.
68. Sugauchi F, Chutaputti A, Orito E, et al. Hepatitis B virus genotypes and clinical manifestation among hepatitis B carriers in Thailand. J Gastroenterol Hepatol 2002; 17(6):671–676.
69. Janssen HL, van Zonneveld M, Senturk H, et al. Pegylated interferon alfa-2b alone or in combination with lamivudine for HBeAg-positive chronic hepatitis B: A randomised trial. Lancet 2005; 365(9454):123–129.
70. Thakur V, Guptan RC, Kazim SN, et al. Profile, spectrum and significance of HBV genotypes in chronic liver disease patients in the Indian subcontinent. J Gastroenterol Hepatol 2002; 17(2):165–170.

71. Akuta N, Kumada H. Influence of hepatitis B virus genotypes on the response to antiviral therapies. J Antimicrob Chemother 2005; 55(2):139–142.
72. Sugauchi F, Orito E, Ichida T, et al. Epidemiologic and virologic characteristics of hepatitis B virus genotype B having the recombination with genotype C. Gastroenterology 2003; 124(4):925–932.
73. Bowyer SM, van Staden L, Kew MC, et al. A unique segment of the hepatitis B virus group A genotype identified in isolates from South Africa. J Gen Virol 1997; 78(Pt 7):1719–1729.
74. Sugauchi F, Kumada H, Acharya SA, et al. Epidemiological and sequence differences between two subtypes (Ae and Aa) of hepatitis B virus genotype A. J Gen Virol 2004; 85(Pt 4):811–820.
75. Hannoun C, Soderstrom A, Norkrans G, et al. Phylogeny of African complete genomes reveals a West African genotype A subtype of hepatitis B virus and relatedness between Somali and Asian A1 sequences. J Gen Virol 2005; 86(Pt 8):2163–2167.
76. Kew MC, Kramvis A, Yu MC, et al. Increased hepatocarcinogenic potential of hepatitis B virus genotype A in Bantu-speaking sub-saharan Africans. J Med Virol 2005; 75(4):513–521.
77. Chen CY, Crowther C, Kew MC, et al. A valine to phenylalanine mutation in the precore region of hepatitis B virus causes intracellular retention and impaired secretion of HBe-antigen. Hepatol Res 2008; 38(6):580–592.
78. Koike K, Moriya K. Metabolic aspects of hepatitis C viral infection: Steatohepatitis resembling but distinct from NASH. J Gastroenterol 2005; 40(4):329–336.
79. McHutchison JG, Gordon SC, Schiff ER, et al. Interferon alfa-2b alone or in combination with ribavirin as initial treatment for chronic hepatitis C. Hepatitis Interventional Therapy Group. N Engl J Med 1998; 339(21):1485–1492.
80. Poynard T, Marcellin P, Lee SS, et al. Randomised trial of interferon alpha2b plus ribavirin for 48 weeks or for 24 weeks versus interferon alpha2b plus placebo for 48 weeks for treatment of chronic infection with hepatitis C virus. International Hepatitis Interventional Therapy Group (IHIT). Lancet 1998; 352(9138):1426–1432.
81. Di Bisceglie AM, Hoofnagle JH. Optimal therapy of hepatitis C. Hepatology 2002; 36(5 Suppl 1):S121–S127.
82. Erhardt A, Blondin D, Hauck K, et al. Response to interferon alfa is hepatitis B virus genotype dependent: genotype A is more sensitive to interferon than genotype D. Gut 2005; 54(7):1009–1013.
83. Hou J, Schilling R, Janssen HL, et al. Genetic characteristics of hepatitis B virus genotypes as a factor for interferon-induced HBeAg clearance. J Med Virol 2007; 79(8):1055–1063.
84. Wai CT, Chu CJ, Hussain M, et al. HBV genotype B is associated with better response to interferon therapy in HBeAg(+) chronic hepatitis than genotype C. Hepatology 2002; 36(6):1425–1430.
85. Yang ZJ, Tu MZ, Liu J, et al. Comparison of amplicon-sequencing, pyrosequencing and real-time PCR for detection of YMDD mutants in patients with chronic hepatitis B. World J Gastroenterol 2006; 12(44):7192–7196.
86. Wang C, Mitsuya Y, Gharizadeh B, et al. Characterization of mutation spectra with ultra-deep pyrosequencing: Application to HIV-1 drug resistance. Genome Res 2007; 17(8):1195–1201.
87. Lindstrom A, Odeberg J, Albert J. Pyrosequencing for detection of lamivudine-resistant hepatitis B virus. J Clin Microbiol 2004; 42(10):4788–4795.
88. Lim SG, Cheng Y, Guindon S, et al. Viral quasi-species evolution during hepatitis Be antigen seroconversion. Gastroenterology 2007; 133(3):951–958.
89. Shafer RW. Genotypic testing for human immunodeficiency virus type 1 drug resistance. Clin Microbiol Rev 2002; 15(2):247–277.
90. http://hivdb.stanford.edu/pages/algs/HIVdb.html. Accessed August 2008.
91. Birch C, Middleton T, Hales G, et al. Limited evolution of HIV antiretroviral drug resistance-associated mutations during the performance of drug resistance testing. J Acquir Immune Defic Syndr 2003; 32(1):57–61.
92. Sayer DC, Land S, Gizzarelli L, et al. Quality assessment program for genotypic antiretroviral testing improves detection of drug resistance mutations. J Clin Microbiol 2003; 41(1):227–236.
93. Eshleman SH, Crutcher G, Petrauskene O, et al. Sensitivity and specificity of the ViroSeq human immunodeficiency virus type 1 (HIV-1) genotyping system for detection of HIV-1 drug resistance mutations by use of an ABI PRISM 3100 genetic analyzer. J Clin Microbiol 2005; 43(2):813–817.
94. Allen MI, Deslauriers M, Andrews CW, et al. Identification and characterization of mutations in hepatitis B virus resistant to lamivudine. Lamivudine Clinical Investigation Group. Hepatology 1998; 27(6):1670–1677.
95. Bartholomew MM, Jansen RW, Jeffers LJ, et al. Hepatitis-B-virus resistance to lamivudine given for recurrent infection after orthotopic liver transplantation. Lancet 1997; 349(9044):20–22.
96. Chayama K, Suzuki Y, Kobayashi M, et al. Emergence and takeover of YMDD motif mutant hepatitis B virus during long-term lamivudine therapy and re-takeover by wild type after cessation of therapy. Hepatology 1998; 27(6):1711–1716.
97. Honkoop P, Niesters HG, de Man RA, et al. Lamivudine resistance in immunocompetent chronic hepatitis B. Incidence and patterns. J Hepatol 1997; 26(6):1393–1395.

98. Ling R, Mutimer D, Ahmed M, et al. Selection of mutations in the hepatitis B virus polymerase during therapy of transplant recipients with lamivudine. Hepatology 1996; 24(3):711–713.
99. Ogata N, Fujii K, Takigawa S, et al. Novel patterns of amino acid mutations in the hepatitis B virus polymerase in association with resistance to lamivudine therapy in Japanese patients with chronic hepatitis B. J Med Virol 1999; 59(3):270–276.
100. Tipples GA, Ma MM, Fischer KP, et al. Mutation in HBV RNA-dependent DNA polymerase confers resistance to lamivudine in vivo. Hepatology 1996; 24(3):714–717.
101. Yeh CT, Chien RN, Chu CM, et al. Clearance of the original hepatitis B virus YMDD-motif mutants with emergence of distinct lamivudine-resistant mutants during prolonged lamivudine therapy. Hepatology 2000; 31(6):1318–1326.
102. Angus P, Vaughan R, Xiong S, et al. Resistance to adefovir dipivoxil therapy associated with the selection of a novel mutation in the HBV polymerase. Gastroenterology 2003; 125(2):292–297.
103. Villet S, Pichoud C, Billioud G, et al. Impact of hepatitis B virus rtA181V/T mutants on hepatitis B treatment failure. J Hepatol 2008; 48(5):747–755.
104. Kim HS, Han KH, Ahn SH, et al. Evaluation of methods for monitoring drug resistance in chronic hepatitis B patients during lamivudine therapy based on mass spectrometry and reverse hybridization. Antivir Ther 2005; 10(3):441–449.
105. Yeon JE, Yoo W, Hong SP, et al. Resistance to adefovir dipivoxil in lamivudine resistant chronic hepatitis B patients treated with adefovir dipivoxil. Gut 2006; 55(10):1488–1495.
106. Hong SP, Kim NK, Hwang SG, et al. Detection of hepatitis B virus YMDD variants using mass spectrometric analysis of oligonucleotide fragments. J Hepatol 2004; 40(5):837–844.
107. Noguchi C, Ishino H, Tsuge M, et al. G to A hypermutation of hepatitis B virus. Hepatology 2005; 41(3):626–633.
108. Harrison SA. Small molecule and novel treatments for chronic hepatitis C virus infection. Am J Gastroenterol 2007; 102(10):2332–2338.
109. Palacios G, Druce J, Du L, et al. A new arenavirus in a cluster of fatal transplant-associated diseases. N Engl J Med 2008; 358(10):991–998.
110. http://blast.ncbi.nlm.nih.gov/Blast.cgi. Accessed August 2008.
111. Ndjomou J, Pybus OG, Matz B. Phylogenetic analysis of hepatitis C virus isolates indicates a unique pattern of endemic infection in Cameroon. J Gen Virol 2003; 84(Pt 9):2333–2341.
112. Paraskevis D, Lemey P, Salemi M, et al. Analysis of the evolutionary relationships of HIV-1 and SIVcpz sequences using bayesian inference: implications for the origin of HIV-1. Mol Biol Evol 2003; 20(12):1986–1996.
113. Kim YB, Cho YK. Monophyletic clade of HIV-1 subtype B in Korea: evolutionary pressure or single introduction? AIDS Res Human Retroviruses 2003; 19:619–623.
114. Pybus OG, Cochrane A, Holmes EC, et al. The hepatitis C virus epidemic among injecting drug users. Infect Genet Evol 2005; 5(2):131–139.
115. Yoshii E, Shinzawa H, Saito T, et al. Molecular epidemiology of hepatitis C virus infection in an area endemic for community-acquired acute hepatitis C. Tohoku J Exp Med 1999; 188(4):311–316.
116. Baskar P, Narayan O, McClure HM, et al. Simian immunodeficiency virus SIVsmmPBj 1.9 induces multinucleated giant cell formation in human peripheral blood monocytes. AIDS Res Hum Retroviruses 1994; 10(1):73–80.
117. Cochrane A, Searle B, Hardie A, et al. A genetic analysis of hepatitis C virus transmission between injection drug users. J Infect Dis 2002; 186(9):1212–1221.
118. Smith DB, Simmonds P. Review: Molecular epidemiology of hepatitis C virus. J Gastroenterol Hepatol 1997; 12(7):522–527.
119. Visco-Comandini U, Cappiello G, Liuzzi G, et al. Monophyletic HIV type 1 CRF02-AG in a nosocomial outbreak in Benghazi, Libya. AIDS Res Hum Retroviruses 2002; 18(10):727–732.
120. Allander T, Gruber A, Naghavi M, et al. Frequent patient-to-patient transmission of hepatitis C virus in a haematology ward. Lancet 1995; 345(8950):603–607.
121. Birch CJ, McCaw RF, Bulach DM, et al. Molecular analysis of human immunodeficiency virus strains associated with a case of criminal transmission of the virus. J Infect Dis 2000; 182(3):941–944.
122. Ou CY, Ciesielski CA, Myers G, et al. Molecular epidemiology of HIV transmission in a dental practice. Science 1992; 256(5060):1165–1171.
123. Tallis GF, Ryan GM, Lambert SB, et al. Evidence of patient-to-patient transmission of hepatitis C virus through contaminated intravenous anaesthetic ampoules. J Viral Hepat 2003; 10(3):234–239.
124. Briones C, Perez-Olmeda M, Rodriguez C, et al. Primary genotypic and phenotypic HIV-1 drug resistance in recent seroconverters in Madrid. J Acquir Immune Defic Syndr 2001; 26(2):145–150.
125. Kaye M, Chibo D, Birch C. Phylogenetic investigation of transmission pathways of drug-resistant HIV-1 utilizing Pol sequences derived from resistance genotyping. J Acquir Immune Defic Syndr 2008; 49(1):9–16.
126. Besisik F, Karaca C, Akyuz F, et al. Occult HBV infection and YMDD variants in hemodialysis patients with chronic HCV infection. J Hepatol 2003; 38(4):506–510.

5 Design of Molecular Virologic Tests

Noah Hoffman and Misha Roshal
Department of Laboratory Medicine, University of Washington, Seattle, Washington, U.S.A.

INTRODUCTION

Molecular methods play an ever-expanding role in many areas of the clinical laboratory. Despite their growing familiarity, the application of diagnostic assays involving nucleic acid sequencing, amplification, or hybridization still presents unique challenges in the realm of clinical virology. The primary challenge in assay design arises from the potentially extreme genetic heterogeneity of viruses, not only among isolates infecting different human populations, but even within an individual. This chapter briefly reviews the biological basis for viral genetic heterogeneity, demonstrates the importance of matching assay technology to test objective, describes general methods for using viral genetic sequence data to direct assay design, and explores strategies for accommodating unavoidable sequence heterogeneity.

BIOLOGICAL SOURCES OF SEQUENCE VARIABILITY

Many viruses are remarkably well adapted to tolerate a large burden of mutations in at least some genomic regions while maintaining fitness. Relatively high error rates of viral polymerases, high replication rate, and frequent recombination events all contribute to accumulation of genomic diversity that is remarkably rapid even on human time scales. Genetic heterogeneity within circulating pools of viruses is shaped by factors including drift, natural selection, and, in modern times, artificial selection due to introduction of antiviral therapy. In the past, the relative geographic isolation of human populations resulted in genotype-defining diversification of the circulating viral pools through both drift and separate zoonotic transmission events (1–3). Natural selection, both positive for increased virus propagation within the human hosts and negative (purifying selection) against deleterious virus mutants, has continuously molded the existing viral pools. Human mobility and demographic changes have reshaped the distribution of viruses and led to the emergence of recombinant virus strains and new zoonotic transmissions (1,4–6). Recent introduction of antiviral drugs has caused the evolution of drug resistance mutations.

In the era of widespread availability of sequencing technology, classification based on nucleic acid sequences has largely replaced serotyping for epidemiological grouping of virus strains. Historically, subgroups within a viral species (common terms include "strains," "genotypes," "subtypes," and "clades") were mostly important for epidemiologic surveillance. Recently, however, the genotype of certain viruses, notably HCV (7,8) and HBV (9–11), has been shown to correlate with natural history of infection and/or response to therapy. Thus, genotyping of these viruses has become a part of the clinical laboratory mission.

By convention, classification schemes differ among viruses. HIV-1, for instance, is classified first into groups (M, N, O) then further into subtypes and sub-subtypes (12). On the other hand, HCV (2) and HBV (13,14) are subdivided into genotypes and subgenotypes. The classification of the virus strains is fluid and revisions of the group definitions occur fairly frequently as more sequence information becomes available. Furthermore, the genotypes themselves are not rigid biological categories. Geographic mixing of the viral pools has led to increased inter-genotype recombinant strains in many viruses, including HCV (2,15), HBV (16), and HIV-1 (17). Continuous reshuffling of influenza virus Hemagglutinin and Neuraminidase segments is another example of the fluidity of viral nucleic acid content.

For many viruses, genotypic diversity (that is, differences between viruses belonging to different lineages) accounts for the bulk of systematic sequence divergence. For example, hepatitis B virus is subdivided into eight genotypic groups, each showing greater than 8% nucleotide divergence between groups, but less than 4% genetic divergence within most individual genotypes (18,19). Still others, like HIV-1, show significant variation within clades with variation of 25% to 35% between subtypes and 15% to 20% within an individual subtype (20). Even within a single individual, viral populations continue to evolve during infection. Phenotypic evidence of this continuing evolution is seen, for example, in the emergence of CXCR4 tropic viruses during HIV infection and drug-resistant variants during chronic HBV and HIV therapy. Both ongoing evolution and co-infection with different strains can lead to existence of multiple clinically important viral subpopulations within a single individual.

Introduction of antiviral therapy represents novel evolutionary pressures resulting in the emergence of drug-resistant mutants. Resistance often arises as a multistep process with primary mutations allowing ongoing replication in the presence of the drug, at a cost of reduced enzyme activity or stability, and secondary mutations that increase the fitness by masking the deleterious effect of the drug-resistant mutant (21–23). The accumulation of secondary mutations leads to progressively higher levels of resistance (lower apparent replication inhibition) and is therefore clinically important to detect and report. As new drugs are introduced and more experience is gained with patterns of resistance, the number of possible resistance mutations for a given viral target or to newer drugs will undoubtedly grow.

GOALS OF TESTING AND TESTING METHODOLOGY

Over the last two decades, molecular testing has become an important modality in diagnosis, prognostication, and epidemiologic surveillance of most human viral disease (24,25). The diversity of applications for molecular testing is ever-expanding. Table 1 lists some of the common clinical virology questions that can be answered with molecular testing, as well as their utility and intrinsic requirements. It is evident that certain clinical questions require test designs that impose specific constraints on the test performance. Some of the tests, particularly quantitative PCR and RT-PCR, have very strict requirements for conservation of primer and probe-binding sequences. Others, such as genotype determination, involve the identification of regions that are sufficiently divergent between viral strains to support confident differentiation. Addressing the design of individual virologic tests is well beyond the scope of this chapter. Rather than focusing on any individual test or technology, this chapter emphasizes i) bioinformatic tools and approaches that may be used to describe the genetic heterogeneity of medically important viruses for the purpose of informing assay design; and ii) the identification of appropriate methodologies to address specific clinical questions that pertain to viral infections given the underlying variability of molecular targets.

MOLECULAR TESTING CHALLENGES

Most clinical molecular genetic applications—no matter the platform, the ultimate diagnostic question, or downstream data processing—share the need to specifically amplify the viral DNA or RNA of interest using polymerase chain reaction (PCR), or a variant thereof. PCR requires at least two short oligonucleotides (primers) that are complementary to viral nucleic acid sequences. Many applications also use one or two additional fluorescent probes. Because of the requirements for perfect or near-perfect complementarity of the primers and probes, nucleotide variability in the primer sites poses a significant challenge in test design for many viruses. These problems are related to the uniquely high rates of genetic variation in many of the most commonly encountered human viral infectious agents. Some DNA viruses, notably VZV (26), HSV-1 (27), and parvovirus B19 (28) are relatively well conserved with overall nucleotide sequence conservation of 98% or more. Others, typically RNA viruses and retroviruses including such common pathogens as HIV-1 (20), HCV, and norovirus (29) show tremendous genomic variation with inter-strain differences of as much as 40% overall, and up to 70% in surface proteins. The overall higher variability of RNA and reverse-transcribing viruses can mainly be attributed to the generally lower fidelity of RNA-dependent polymerases.

The likelihood of such extensive diversity calls for careful selection of the primers and probes to ensure that the assay is both sensitive for the virus tested, and specific enough to prevent cross-reactivity with undesired nucleic acid sequences. Often, the task of identifying a combination of two primers and a probe within a reasonable distance of each other that amplify representatives of all strains of a highly diverse virus is not trivial. One simplifying observation is that because different areas of the genome have different tolerance for sequence variation, genetic diversity of viruses is not evenly distributed across their genomes. For example, viral

Table 1 Examples of Molecular Testing Modalities and Clinical Applications

Clinical questions	Examples of testing modalities	Test requirements	Clinical utility examples
Is the patient infected?	Qualitative PCR/RT-PCR, multiplexed PCR, other amplification-based technologies, oligonucleotide array microchip	Very high sensitivity and specificity for all the virus strains tested	Blood component testing, diagnosis of viral meningitis/encephalitis, diagnosis of respiratory viral disease, HCV infection status in patients with anti-HCV core Ab
What is the viral load?	Quantitative PCR/RT-PCR based tests	Sensitivity requirement dictated by clinical decision-making cutoffs. Requires good precision and accuracy; equal amplification efficiency for all virus strains tested; absence of unwanted cross-reactivity.	Decision support for initiation and monitoring of antiviral therapy. Monitoring of disease activity and treatment efficacy
What is the genotype?	Line probe, genotype-specific multiplex PCR, restriction fragment length polymorphism, PCR followed by informative region DNA sequencing	Adequate genome sampling for confident genotypes assignment. Sensitivity to mixed genotypes/recombinants	Prognosis of disease progression and response to therapy. Epidemiologic surveillance
Is this virus resistant to a drug or are there clinically significant mutations?	Amplification followed by DNA sequencing of the region of interest. Line probe assays. Mutation-specific PCR probes[KJI] Oligonucleotide ligation-based assays	Coverage of all known primary and compensatory resistance mutations for a given target or drugs of interest. Sensitivity to resistant/mutated subpopulations	Determination of drug resistance profiles, determination of core/precore mutation status in HBV

genes encoding products under strong positive selection, such as envelope proteins, are likely to display increased variability across isolates. Conversely, loci under negative selection due to stringent functional constraints are more likely to be well conserved: examples include sequences encoding enzymes, particularly the active sites or substrate-binding regions; nucleic acid binding sites of regulatory proteins or viral structural components; ribozymes; and highly structured regions of RNA.

In addition to careful selection of primers with respect to genomic diversity, genotyping and drug resistance tests pose another set of challenges. An ideal genotyping test design requires adequate genome sampling to provide both confident genotype discrimination and the ability to detect both recombinant strains and mixtures of subpopulations. A drug resistance test should detect both primary and compensatory resistance mutations and be able to discern subpopulations with differing drug resistance profiles. Sophisticated interpretive algorithms may be required to detect and describe the patterns of resistance. In selecting the platform for the genotyping or mutation testing, the need for flexibility to rapidly adjust to changes in phylogenetic nomenclature and to the discovery of novel mutations must also be taken into consideration. This includes the ability to collect relevant information arising from new research and to interpret it in accordance with the rapid changes in the field. In some cases, the need for flexibility may need to be balanced with considerations of assay sensitivity. Thus, while bulk sequencing of an entire coding region of a viral polymerase would likely detect all the possible mutations that lead to resistance to polymerase inhibitors, this approach will not be sensitive enough to detect resistant subpopulations comprising less than 20% of the total viral burden. Conversely, a ligation-based assay testing for a defined set of point mutations may be highly sensitive, but will require extensive revalidation as new resistance targets are discovered. Table 2 provides a brief summary of considerations for appropriate target selection and the anticipated challenges associated with each of the test modalities.

One other consideration for the selection of the target of an assay should be kept in mind, particularly in clinical laboratories that charge fees to perform assays: intellectual property protections may place restrictions on the choice of specific genomic regions or sequences that can be used as assay targets. Some patents contain claims that are very broad, and severely limit the available sequences that can be used without obtaining licensing agreements with the patent holder. Others are much narrower and are unlikely to prove an obstacle to assay design (e.g., hepatitis C virus genotype, and its use as prophylactic, therapeutic, and diagnostic agent. US Patent No. #7196183). In general, broad claims relevant to assay development are becoming more difficult to obtain, and are more likely to apply to specific viral variants or recently identified genotypes. Unfortunately, the determination of the patent landscape may require expertise that can only be provided by intellectual property professionals (for reasons of both practicality and liability); assistance in deciding whether to perform an extensive search for applicable patents may, for example, be provided by the Technology Transfer office of research institutions.

STRATEGIES FOR SELECTION OF POTENTIAL TARGET SEQUENCES

Many considerations related to assay design are either method-specific or do not differ significantly between virologic assays and other contexts, such as molecular genetics or molecular microbiology. As discussed above, the challenges unique to molecular virology arise largely from heterogeneity of potential target sequences, and the suitability of a locus as a target for a primer-binding site depends on the purpose of the assay: viral detection and quantitation rely on the availability of well-conserved sequences, whereas genotyping assays might require analysis of regions of above-average diversity (often with the additional requirement that flanking sequences be sufficiently conserved to serve as targets for primers used for amplification or sequencing). We will therefore focus much of our discussion on general approaches for assessing sequence heterogeneity.

Conceptually, two strategies for identifying appropriate viral genomic targets may be described. As noted above, the first involves consideration of the underlying biological function of the potential probe target. The second, of course, is to directly examine sequence data. Relative heterogeneity across viral genomes can certainly be measured without prior knowledge of biological function (indeed, patterns of variability among isolates may provide evidence

Table 2 Examples and Applicability of Test Methods

Kinds of tests	Genomic targets	Technical considerations
Qualitative virus infection testing	Conserved regions of the viral genome	Requires a highly conserved site for optimal primer/oligonucleotide binding that is also sufficiently different from related virus or human genomic DNA sequences. Need high sensitivity across viral genotypes and absence of cross-reactivity with similar viruses. Shorter amplicon lengths provide for more efficient and faster amplification reactions. Multiplexing increases complexity of the test design.
Quantitative viral load testing	Conserved regions of the viral genome	Same as above with stricter requirements on primer- and probe-binding site conservation to ensure equal binding efficiency across genotypes. Shorter amplicon lengths are required for real-time assays. Need for real-time detection hardware and fluorescence probes.
Genotype determination	Regions that are sufficiently divergent between all the genotypes to allow confident genotype assignment. For downstream line probe assays and sequencing need primer-binding sites to be conserved between the genotypes to allow amplification. For genotype-specific PCR need primer-binding sites that discriminate between different genotypes.	Longer amplicons may be required for sufficient genome sampling to allow confident genotype assignment. Potential for coinfection with multiple strains and recombinant viruses should be kept in mind. Sequencing-based assays may require potentially complex data analysis procedures. If sampling point mutations, sufficient representation of polymorphic sites must be ensured to allow confident genotype determination.
Tests for drug resistance or the presence of clinically significant mutations	Genomic regions that contain the sequence of interest. Consider including flanking regions to anticipate discovery of additional resistance mutations.	Similar to genotype determination. Longer amplicons may be needed to cover both primary and compensatory mutation regions for drug resistance. Primer-binding sites need to be sufficiently conserved to allow amplification. Consider suitability of technology for sensitivity requirements. For example, bulk sequencing will not pick up small resistant subpopulations.

of the underlying role). In practice, these strategies are often used in concert. For example, if a well-conserved region is required, candidate viral genomic loci for sequencing might be restricted to loci known to be under strong negative selective pressure.

SOURCES OF VIRAL SEQUENCE DATA

Preexisting sequences representing a given virus are frequently available; often these sequences are sufficient to permit confident identification of appropriate primer or probe targets, since a vast and growing body of sequence data resides in public databases. Nevertheless, in some cases additional data must be generated experimentally. In either case, as discussed below, assembly of an appropriate data set is essential for a meaningful analysis.

Curated Viral Sequence Databases

Specialized public repositories of viral sequences, typically accessible via websites, have become critical resources for investigators studying viral biology, evolution, and epidemiology (Table 3). These sites should usually be the starting place for a search for sequence data if an appropriate resource exists for the virus of interest. Sequences included in specialized collections are often represented in GenBank (see below), but may have undergone additional screening for the quality of the submission and completeness of accompanying information compared to those from other sources (though this should be verified, not assumed).

Curated databases often provide valuable accompanying information, such as genotype, clinical information like viral load or the host's history of exposure to antiviral drugs, place of origin, or viral phenotype. Sophisticated tools for sequence retrieval, visualization, and analysis may also be available from these websites. The Los Alamos HIV sequence database (http://www.hiv.lanl.gov) is one of the best-established examples of such a resource. This site provides an interface for downloading aligned HIV sequences corresponding to the user's choice of genomic region, and selected according to criteria such as those listed above. The HIV sequence database also provides a variety of tools and tutorials that illustrate nearly all of the approaches to sequence analysis discussed below. Table 3 is a partial list of publicly available viral sequence databases available online.

Other Public Sequence Databases

If there can be said to be a "master repository" of the world's biological sequence data, it is the International Nucleotide Sequence Database Collaboration, whose members include the DNA DataBank of Japan (DDBJ), the European Molecular Biology Laboratory (EMBL), and GenBank at NCBI. For researchers in the United States, the most familiar of these resources is the NCBI web portal to GenBank (http://www.ncbi.nlm.nih.gov/). The importance and ubiquity of these databases cannot be overstated, but their enormity may be daunting. In the absence of a dedicated resource providing sequence data representing a given virus or other organism, how should one approach the task of assembling an appropriate set of sequences for comparison?

A number of search strategies should be considered. An exhaustive search might employ all of these methods:

1. Searches may be performed using NCBI Taxonomy ID. The Taxonomy Browser (http://www.ncbi.nlm.nih.gov/taxonomy) supports identification and retrieval of sets of sequences according to their organization within the NCBI taxonomic hierarchy. Alternatively, a search may be performed directly from the main search interface to the NCBI nucleotide database (http://www.ncbi.nlm.nih.gov/nucleotide) using the identifier for a taxonomic group. For example, a search using the term "txid11102[Organism:exp]" will identify all available HCV sequences. The classification depends on annotation provided by either the submitter or the database maintainers. This can often be the single best approach for an exhaustive search, but misclassification of records may still result in an imperfect data set.
2. A text- or keyword-based database query of the complete nucleotide database can also be a useful starting point. This approach is most fruitful when search terms are chosen carefully and applied only to a subset of database fields (http://www.ncbi.nlm.nih.gov/Sitemap/samplerecord.html). A naive approach can easily result in the inclusion of

Table 3 A Partial List of Public Viral Sequence Databases

Database name	Viruses	URL	Host organization	Notes
HIV sequence database	HIV-1, HIV-2, SIV	http://www.hiv.lanl.gov	LANL	Site also contains extensive literature, tutorials, and software tools
HPV Sequence Database	HPV	http://hpv-web.lanl.gov/	Los Alamos National Laboratory (LANL)	
VBRC	Arenaviridae, Bunyaviridae, Flaviviridae, Filoviridae, Paramyxoviridae, Poxviridae, Togaviridae (see note)	http://www.vbrc.org	Viral Bioinformatics Resource Center (VBRC)[a]	Curated sequence data representing listed viral families as well as a suite of online bioinformatic tools and other resources; the VBRC also hosts more specialized virus databases listed separately.
Hepatitis C Virus Database (HCVdb)	HCV	http://www.hcvdb.org	VBRC	Shares funding and resources with the LANL HCV database (http://hcv.lanl.gov)
Dengue Virus Database (DengueDB)	Dengue virus	http://www.denguedb.org	VBRC	
Poxvirus Bioinformatics Resource Center	Poxviruses	http://www.poxvirus.org	VBRC	
BioHelathBase Influenza Database	Influenza	http://www.biohealthbase.org	The National Institute of Allergy and Infectious Diseases (NIAID)	Provides online tools for multiple sequence alignment creation and visualization
NCBI Virus Variation	Influenza, Dengue	http://www.ncbi.nlm.nih.gov/genomes/VirusVariation/	NCBI	Extension of Influenza Virus Resource (http://www.ncbi.nlm.nih.gov/genomes/FLU/FLU.html) to other organisms; addition of West Nile Virus is planned.

[a]A collaboration of the University of Alabama at Birmingham and the University of Victoria.

inappropriate records that incidentally match the search term, or the omission of records in which alternative names are used to describe the organism of interest. For example, results of a taxonomy search for "BK Polyomavirus" (Taxonomy ID 10629) contained records with at least seven variations in the "Definition" field, including "BK virus BKV," "BK polyomavirus," "Human polyomavirus BK," "Human BK virus," "Papovavirus BKV," "Human Polyomavirus (BKV)," and "Human papovavirus BK." Clearly, a relatively sophisticated search would be required to capture all of these records using the description field.
3. Searches using sequence similarity, typically performed using the BLAST search tool, can be useful to quickly identify only those records that overlap the region of interest. Similarity and match-length parameters should be set to exclude distantly related or very short matches. This approach should be used with caution, as it may fail to identify less closely related, but still relevant, genotypes or variants.
4. The GenBank PopSet database (http://www.ncbi.nlm.nih.gov/popset) contains predefined sets of records "that have been collected to analyze the evolutionary relatedness of a population." These collections of sequences are frequently useful, but a single set may contain representatives of more than one species or other taxonomic classification.
5. The GenBank Viral genomes Database (http://www.ncbi.nlm.nih.gov/genomes/GenomesHome.cgi) identifies a full-length viral genome as a reference sequence for recognized viral species.

Successful search strategies should be carefully documented so that they may be reproduced later as additional sequences become available. Regardless of the search strategy used, the results may be downloaded in a variety of formats for subsequent analysis. Sequences downloaded in FASTA format (http://www.ncbi.nlm.nih.gov/blast/fasta.shtml) can be used as input to nearly any sequence-analysis utility, but contains only limited accompanying information.

SEQUENCE ANALYSIS

It is rarely critical that a search be absolutely exhaustive, since in all but the smallest data sets each additional sequence may add only incrementally to the available information. A higher priority is to exclude sequences of low quality, those that do not actually represent members of a viral group of interest, highly passaged lab strains that have accumulated substitutions not represented in natural populations, and closely related groups of sequences. In addition, ensuring that clinically relevant subpopulations are sufficiently (and proportionately) represented is of equal importance. Strategies for performing these tasks are described below.

Assessing Sequence Quality

In the absence of primary sequence data (e.g., capillary gel electrophoresis chromatograms), the researcher must rely on surrogate markers of sequence quality. The most obvious indication of low-quality sequence data is the presence of a high proportion of ambiguous positions, typically represented by "N" or other IUPAC ambiguity codes. Though the presence of ambiguous characters should not be taken as evidence of lack of care on the part of the authors of the submission, they may reflect limitations in the methods used to collect the sequence, such as bulk-sequencing of heterogeneous viral populations. In other cases, generation of high-quality sequences may not have been an objective of the study for which they were generated. Setting a threshold for the maximum proportion of ambiguous characters can be an effective screen for low-quality sequences. Another approach, particularly relevant for GenBank sequences but very conservative, is to exclude records not accompanied by a peer-reviewed publication (i.e., "direct submissions"). Finally, alignments with sequences of known quality may reveal insertions or deletions that are likely to be artifactual, such as those causing nonsense mutations in genes known to be required for viral replication.

Multiple Sequence Alignment

A multiple sequence alignment (MSA) consists of a position-by-position comparison of a set of related biological sequences. Alignment of a set of sequences all of the same length in which each starts and ends at the same relative position is the most trivial case, and can be performed "by hand" without the assistance of alignment software. More often, however, sequences are not the same length, or contain insertions and deletions relative to one another, and some positions must be occupied by "gaps" to maintain the register of the alignment. The optimal placement of gaps becomes less certain as the divergence among the sequences increases.

An MSA reflects the underlying assumption that nucleotides or amino acids in a column (a "position" in the alignment) share a common evolutionary origin; likewise, the addition of gaps implies a hypothesis that insertion and deletion events have occurred in the evolutionary history of the sequences. Thus, multiple alignment tools commonly use evolutionary models to achieve an alignment that is biologically plausible.

The implementation, selection, and evaluation of multiple sequence alignment algorithms have been extensively discussed in the bioinformatics literature (30). Multiple alignment algorithms have undergone great refinement since the earliest implementations. For the purpose of most tasks in viral assay design, however, it may be that the most important criteria for software selection may be ease of use, speed of execution, ability to handle the length and quantity of sequences to be compared, and compatibility with available computing platforms. Because the most frequent use of the MSA in assay design is to compare sequences in regions of relatively high sequence similarity where gaps are rare, alignments in these regions should be relatively insensitive to differences between alignment algorithms. Thus, for most purposes, the ubiquitous ClustalW, used from a terminal command line, or ClustalX, which provides a graphical user interface, will suffice (software available at http://www.clustal.org/) (31). A systematic comparison between ClustalW and other algorithms using a database of prealigned reference sequences demonstrated that though other programs performed better than ClustalW at certain tasks, the benefit may have been marginal, and the improved performance was often at the cost of significantly increased computational time (32). That said, users are encouraged to explore other options. For example, the creation of multiple alignments containing very long or very many sequences using progressive alignment algorithms not specifically optimized for large alignments (such as ClustalW) is computationally intensive and likely to be prohibitively slow. Tools that are more specialized should be used for this purpose. The LANL HIV Sequence Database uses the HMMER software package to create alignments of tens of thousands of HIV sequences using profile hidden Markov models (or alignment "profiles") created from smaller reference alignments (33). Useful guidelines for the selection of multiple alignment software are provided in a review by Edgar et al. (34).

An MSA created by any computer program is not guaranteed to be optimal, and may require additional manipulation by hand. On a practical note, use of word-processing software to edit multiple sequence alignments (or any file containing sequence data) is likely to be an exercise in frustration and a source of difficult-to-identify errors in subsequent steps of the analysis. Instead, we highly recommend using any of a large number of free or commercial sequence alignment editors. Because the universe of sequence alignment editors is extremely dynamic, any set of suggestions provided here would be immediately out of date; lists of alignment editors are maintained in a variety of locations, such as Wikipedia (http://en.wikipedia.org/wiki/List_of_sequence_alignment_software).

Creating a Representative Sequence Set

Most molecular virologic assays are intended either to sensitively and uniformly detect isolates of a given virus or group of viruses, to discriminate between viruses or viral subtypes, or to detect polymorphisms associated with phenotypic characteristics. In either case, assay design may be compromised by the failure to describe background genetic variability unless relevant subpopulations are appropriately represented. Once a set of candidate sequences are assembled and aligned, additional steps are necessary to ensure that the data set accurately reflects sequence heterogeneity in target populations.

Outliers

Outliers of a trivial sort can be introduced through annotation errors or insufficiently specific search strategies; in some cases, these might be detected simply by reviewing the annotations of the sequence records. A category of outlier that is more difficult to identify might be termed "biological outliers"; that is, isolates that are biologically dissimilar to most isolates of the intended viral target. Highly passaged laboratory isolates deserve a particular note of caution, especially as they are often enshrined as "reference" strains in type collections: often the earliest isolates of a viral pathogen have accumulated substitutions, insertions, deletions, or rearrangements through prolonged passage in cell culture, and are no longer representative of primary clinical isolates. A well-known example is the HIV-1 reference strain HXB2 (GenBank K03455), which is the numbering standard for HIV-1 despite having an unusual pattern of amino acid insertions and deletions in an important functional region of Env. Detection of outliers using phylogenetic analysis is discussed below.

Accurately Representing Population Structure

Perhaps the most challenging and time-consuming task in assembling a set of viral reference sequences is to ensure adequate representation of distinct subpopulations without over-representing groups of closely related isolates. Viral subgroups may be highly correlated with some combination of geography, epidemiology, or the ethnic background of the human subject of origin, and one must ensure that the viral sequences included for analysis reflect the genetic variability circulating within the target human population for the assay. It is impossible to overstate the importance of the observation that viral sequence databases cannot be assumed to proportionately represent the full range of genetic heterogeneity. For example, strains or subtypes that are prevalent in Europe and North America are often better represented than those prevalent in other areas. Another extremely common sampling artifact is the presence of large sets of viral sequences derived from viruses collected from a single individual or group of individuals. Including a set of closely related sequences that are not identified as such can be greatly misleading.

Estimating the representation of various subpopulations can be nontrivial when primary sequence data is used, or when sequences are retrieved from a repository that does not indicate group assignments (35). An even more significant difficulty exists for more recently identified pathogens, for which the population structure may not be well characterized at the time the search is performed, and named subgroups may not yet have been defined. In the latter case, it is essential to experimentally estimate the extent of heterogeneity by sequencing isolates from representative populations before attempting to design an assay that will be deployed in a clinical setting.

Visualizing population structure can in some cases be performed by simply inspecting an MSA. Inspection is greatly facilitated by software that highlights nonconsensus nucleotide substitutions [software for calculating or editing MSAs such as ClustalX can often accomplish this, as can a variety of online tools, such as the Sequence Alignment Publishing Tool (http://www.hiv.lanl.gov/content/sequence/SeqPublish/seqpublish.html), Highlighter Tool (http://www.hiv.lanl.gov/content/sequence/HIGHLIGHT/highlighter.html), or Jalview (36) (http://www.jalview.org)]. On a practical note, it is useful to name reference sequences so that assignment to subtype or clade can be readily determined during the inspection of the alignment.

In some cases, a phylogenetic tree can aid in the visualization of population structure. Outliers, in particular, are best identified using this approach. Even more than multiple sequence alignment, phylogenetic reconstruction is a complex and specialized topic, and an even more extensive array of software tools are available. But again, our needs in the setting of viral assay design are relatively modest, and even a rudimentary phylogenetic analysis can rapidly identify outliers and highlight high-level population structure.

A reasonable place to start is the neighbor-joining tree option in ClustalW or ClustalX. More sophisticated calculations, such as meaningful estimates of evolutionary distance or site-specific rates of nucleotide substitution, require more finesse (at the level of both MSA creation and phylogenetic analysis), and are well outside the scope of this chapter. Visualization of phylogenetic trees can be performed using TreeView (http://taxonomy.zoology.gla.ac.uk/rod/treeview.html) among other tools. An extensive list of software resources is maintained by Joe Felsenstein (the author of the widely used PHYLIP phylogenetics software package) at the University of Washington (http://evolution.genetics.washington.edu/phylip/software.html).

If all subgroups are not well represented, the first consideration is whether the underrepresented groups are sufficiently divergent to be of concern in the target regions. For example, if all known subgroups are extremely well conserved in a region of interest, one might decide that assay design can proceed without additional sequence information. Otherwise, it may be necessary to design the assay with the understanding that its performance for members of the underrepresented groups cannot be predicted. If underrepresented groups are prevalent in the human population served by the laboratory, it may be necessary to sequence additional isolates before the assay can reasonably be deployed.

In summary, therefore, assuming that subgroups have been defined and representative sequences are available, a typical process for placing additional unclassified sequences in subgroups and selecting a representative data set might be as follows:

1. Identify a set of reference sequences of known classification spanning the region of interest, for example from a database of reference strains or as classified in the literature. For practical reasons, the classification of each sequence should be reflected in the sequence name.
2. If additional sequences are required, search other public databases.
3. Assess the sequences for adequate length, quality (e.g., proportion of ambiguous bases), and coverage of the genomic region of interest. Examine sequence annotations for evidence that groups of records are closely related (e.g., from the same individual) and exclude all but one or a few representatives.
4. Create a multiple sequence alignment.
5. Assign unclassified sequences to subgroups using some combination of inspection and/or phylogenetic analysis by noting similarity between unclassified sequences and reference sequences representing each subgroup.
6. Determine if subpopulations are appropriately represented. Consider removing sequences from highly overrepresented subgroups or clusters of very closely related isolates.

Interpreting Sequence Heterogeneity

The ultimate purpose for the painstaking assembly of the sequence data described in the steps above is to provide a guide for the placement of primers or probes within the viral sequence region of interest. Because the approach required for assay design is determined by methodology, integration between primer or probe design and assessment of heterogeneity in the target sequences is rarely seamless.

A few options do exist for using the multiple sequence alignment as the primary input for PCR primer design. For example, Primaclade (37) and PriFi (38) are two free web-based tools that can identify conserved regions while also assessing primer characteristics such as length, melting temperature, GC content, etc. Most primer design software, however, expects a single sequence as input, such as Primer3 (http://primer3.sourceforge.net/), a popular primer design program developed at the Whitehead Institute and Howard Hughes Medical Institute. In this case, selection of candidate regions based on sequence conservation (given an MSA) and sequence characteristics (given a single nucleotide sequence) is an iterative process. For example, one might use primer design software to perform an initial search for candidate primers by providing sequence coordinates defining regions of relatively high conservation. Output primers can then be compared one-by-one to the MSA (or another representation of sequence heterogeneity as described below) to eliminate those that are likely to be affected by individual positions with many polymorphisms.

It may be convenient to use a "consensus" sequence as the template for the primer design software. A consensus sequence is calculated from a multiple sequence alignment, and is composed of the most frequently observed nucleotide at each alignment position. Positions without a clear consensus may be represented by ambiguity codes as mixtures of bases. One obvious advantage of using the consensus rather than an arbitrary reference sequence as the basis for primer design is that candidate primers will be more likely to match the majority of sequences. A less obvious benefit is that nucleotide positions in the consensus sequence correspond exactly to the MSA from which it was calculated, facilitating the mapping of primers back to the aligned sequences.

Two general classes of nucleotide polymorphisms can be described. The first comprises those substitutions that are highly correlated with subgroup, probably having first occurred in a common ancestor after the subgroup's divergence from other viral populations. The second class includes polymorphisms introduced into the population after the time of divergence of the major subgroups, and are not highly correlated with group membership. From the perspective of assay design, the polymorphisms in the former class are the most important to take into account, as one can be assured that nearly every member of a given subgroup will be affected by a mismatch with a primer or probe at that position.

A technique for highlighting group-specific polymorphisms is to examine an alignment of consensus sequences, each constructed using only members of a given subgroup (thus there is one consensus for each subgroup). It is useful as well to include an overall consensus from this alignment. Candidate primers can be rapidly compared to this condensed representation of group-specific variability, and only those compatible with all groups need be compared to the sequences of individual isolates.

Other representations of site-specific nucleotide variability have been described. Many alignment editors (such as ClustalX) can display a plot of overall conservation at each position. Qiu et al. (39) described the construction of a position weight matrix spanning the entire HCV genome to use as a guide for primer design. Zhang et al. (40) described a simple script called BxB in the setting of primer design for the amplification of HBV sequences. BxB filters regions of low sequence conservation and generates output that may be visualized in ClustalX.

Other Bioinformatic Tools

The proliferation of task-specific software tools is evidence of the complexity of the individual steps involved in completing the analyses described above. In some cases, the complexity of the analysis is best addressed using flexible packages of sequence analysis tools in a scripted or programming environment. Those experienced in complex sequence analysis tasks often turn to command-line oriented tools such as the free EMBOSS (http://emboss.sourceforge.net/) package, which allows one to construct a scriptable workflow by combining a series of single purpose software tools. It should also be noted that a web-interface to the EMBOSS software is available (http://bips.u-strasbg.fr/EMBOSS/). Though the initial learning curve may be steep, libraries for popular computer languages such as Java (http://biojava.org), Perl (http://www.bioperl.org), Python (http://biopython.org), and R (http://cran.r-project.org/and http://www.bioconductor.org/) permit rapid development of tools for performing complex sequence analyses that can make otherwise nearly impossible tasks (such as the analysis of tens of thousands of sequences) routine.

ADDRESSING UNAVOIDABLE GENETIC HETEROGENEITY IN THE ASSAY DESIGN

In some cases, it may prove impossible or undesirable to use primers and probes fully complimentary to all sequences of interest. Reasons may include an absence of suitable conserved targets in the region of interest of a highly diverse virus or the need to amplify regions with high genetic diversity. In such a case, four possibilities have to be considered. These include allowing primer mismatches, use of degenerate primers, use of universal bases, or a combination of these approaches.

Mismatches near the 5′ end of a primer are usually not fatal to amplification, but decrease the melting temperature of the primer and thereby its specificity. In contrast, even a single mismatch at or near the 3′ end can severely destabilize the primer/target complex and prevent amplifications (41,42). Therefore, if mismatches are inevitable, restricting them to 5′ end of the primer offers an advantageous, though by no means a perfect solution (43,44). The use of degenerate primers with multiple conventional bases at a given position, or "universal" bases that allow decreased specificity of base-pairing may in some cases improve the sensitivity of the test. Both strategies have been successfully used alone or in combination by multiple investigators, though both may lead to significant complications in test development. The use of degenerate bases can allow one to include only the nucleotides that are required to result in complementarity to the sequence in question. Thus, two bases can be represented at a specific position in an oligonucleotide, as opposed to a universal base that will pair with all four bases. This in theory improves the primer specificity. On the other hand, degenerate bases in multiple sites result in a complex mixture of primers into the reaction (three different bases at four different positions results in 81 different primers) and decrease the concentration of each primer in proportion, which may lead to primer depletion and inefficient amplification. Moreover, each of the primers will have different patterns of cross-reactivity and melting temperatures, which may contribute to nonspecific amplification and make quantitative assays very difficult to standardize. In addition, optimization of a PCR reaction containing multiple primers may be complex. It is therefore recommended to use degenerate bases sparingly.

The most commonly used alternative base is deoxyinosine. Deoxyinsosine (I) is a universal base that can pair with any of the naturally occurring bases, although with unequal efficiencies; for example, pairing between I and G or T is less efficient than with A or C (45). Inclusion of I can also reduce the total number of primers in the reaction and therefore prevent primer exhaustion. The most common use of deoxyinosine has historically been for primer design for sites with conserved amino acid, but degenerate or not fully known nucleic acid sequence (46). Successful use of I at the 3′ terminus has also been reported for population sampling applications where fixing a base at the 3′ end position is impossible (47). Conceptually, this provides an extra "insurance" against total amplification failure due to an unexpected 3′ end mismatch. However, this strategy seems risky due to increased risk of mispriming with the 3′ degenerate base, as well as a reported decrease in efficiency of amplification (48). The disadvantages of using I include decreases in specificity, annealing temperature (49), and the inability to use some proofreading polymerases with I-containing primers (50,51). The lowering of the annealing temperature due to duplex destabilization is perhaps the most confounding effect and low annealing temperature has to be used for primers with high inosine content.

A combination of both degenerate bases and I is a common strategy that seems to avoid the pitfalls of excessive use of either approach. In one recent example of this strategy, a single pair of primers was used for successful amplification of three divergent flavivirus subgroups comprising 65 different virus species (52). The use of any of these strategies in the setting of a quantitative clinical test however may have a profound confounding effect due to unequal efficiency of primer binding to different sequence combinations (44,53,54). As with any clinical test, careful verification of the performance parameters with a variety of expected templates is required.

CONCLUSION

As with other applications of molecular testing modalities in the clinical laboratory, the landscape of tools and technologies available for molecular virologic assay design, test performance, and data analysis is dynamic. Some of the software or applications of specific technologies described in this chapter may soon be out of date. However, we believe that the general approaches we have outlined for assessing and accommodating viral heterogeneity will remain valid, barring the introduction of new technology that is less dependent on the conservation of biological sequences than are PCR and its relatives. Beyond the familiar peer-reviewed literature, the best resources for keeping up to date on approaches and software for analyzing viral sequences may be the viral sequence databases described in Table 3. As an exceptionally dynamic field in the age of genomics, clinical molecular virology will continue to draw its most valuable insights and tools from the basic research (and researchers) that spawned the field; and perhaps more than other areas of the clinical laboratory, molecular virology must be prepared to rapidly integrate new findings and techniques in pursuit of its rapidly evolving biological targets.

REFERENCES

1. Hahn BH, Shaw GM, De Cock KM, et al. AIDS as a zoonosis: Scientific and public health implications. Science 2000; 287:607–614.
2. Simmonds P. Genetic diversity and evolution of hepatitis C virus—15 years on. J Gen Virol 2004; 85:3173–3188.
3. Bowden R, Sakaoka H, Ward R, et al. Patterns of Eurasian HSV-1 molecular diversity and inferences of human migrations. Infect Genet Evol 2006; 6:63–74.
4. Bryant JE, Holmes EC, Barrett AD. Out of Africa: A molecular perspective on the introduction of yellow fever virus into the Americas. PLoS Pathog 2007; 3:e75.

5. Alter MJ. Epidemiology of hepatitis C virus infection. World J Gastroenterol 2007; 13:2436–2441.
6. Kyle JL, Harris E. Global spread and persistence of dengue. Annu Rev Microbiol 2008; 62:71–92.
7. Manns MP, McHutchison JG, Gordon SC, et al. Peginterferon alfa-2b plus ribavirin compared with interferon alfa-2b plus ribavirin for initial treatment of chronic hepatitis C: A randomised trial. Lancet 2001; 358:958–965.
8. Lamarre D, Anderson PC, Bailey M, et al. An NS3 protease inhibitor with antiviral effects in humans infected with hepatitis C virus. Nature 2003; 426:186–189.
9. Chu CJ, Lok AS. Clinical significance of hepatitis B virus genotypes. Hepatology 2002; 35:1274–1276.
10. Sumi H, Yokosuka O, Seki N, et al. Influence of hepatitis B virus genotypes on the progression of chronic type B liver disease. Hepatology 2003; 37:19–26.
11. Tsubota A, Arase Y, Ren F, et al. Genotype may correlate with liver carcinogenesis and tumor characteristics in cirrhotic patients infected with hepatitis B virus subtype adw. J Med Virol 2001; 65: 257–265.
12. Taylor BS, Sobieszczyk ME, McCutchan FE, et al. The challenge of HIV-1 subtype diversity. N Engl J Med 2008; 358:1590–1602.
13. Bowyer SM, Sim JG. Relationships within and between genotypes of hepatitis B virus at points across the genome: Footprints of recombination in certain isolates. J Gen Virol 2000; 81:379–392.
14. Purdy MA, Gonzales AC, Dimitrova Z, et al. Supragenotypic groups of the hepatitis B virus genome. J Gen Virol 2008; 89:1179–1183.
15. Kalinina O, Norder H, Mukomolov S, et al. A natural intergenotypic recombinant of hepatitis C virus identified in St. Petersburg. J Virol 2002; 76:4034–4043.
16. Kurbanov F, Tanaka Y, Fujiwara K, et al. A new subtype (subgenotype) Ac (A3) of hepatitis B virus and recombination between genotypes A and E in Cameroon. J Gen Virol 2005; 86:2047–2056.
17. Perrin L, Kaiser L, Yerly S. Travel and the spread of HIV-1 genetic variants. Lancet Infect Dis 2003; 3:22–27.
18. Okamoto H, Tsuda F, Sakugawa H, et al. Typing hepatitis B virus by homology in nucleotide sequence: Comparison of surface antigen subtypes. J Gen Virol 1988; 69(Pt 10):2575–2583.
19. Schaefer S. Hepatitis B virus taxonomy and hepatitis B virus genotypes. World J Gastroenterol 2007; 13:14–21.
20. Hemelaar J, Gouws E, Ghys PD, et al. Global and regional distribution of HIV-1 genetic subtypes and recombinants in 2004. AIDS 2006; 20:W13–W23.
21. Nijhuis M, Schuurman R, de Jong D, et al. Increased fitness of drug resistant HIV-1 protease as a result of acquisition of compensatory mutations during suboptimal therapy. AIDS 1999; 13:2349–2359.
22. Nijhuis M, van Maarseveen NM, Boucher CA. Antiviral resistance and impact on viral replication capacity: Evolution of viruses under antiviral pressure occurs in three phases. Handb Exp Pharmacol 2009; 189:299–320.
23. Canducci F, Sampaolo M, Marinozzi MC, et al. Dynamic patterns of human immunodeficiency virus type 1 integrase gene evolution in patients failing raltegravir-based salvage therapies. AIDS 2009; 23:455–460.
24. Scott JD, Gretch DR. Molecular diagnostics of hepatitis C virus infection: A systematic review. JAMA 2007; 297:724–732.
25. Mangia A, Antonucci F, Brunetto M, et al. The use of molecular assays in the management of viral hepatitis. Dig Liver Dis 2008; 40:395–404.
26. Norberg P, Liljeqvist JA, Bergstrom T, et al. Complete-genome phylogenetic approach to varicella-zoster virus evolution: Genetic divergence and evidence for recombination. J Virol 2006; 80:9569–9576.
27. Norberg P, Bergstrom T, Rekabdar E, et al. Phylogenetic analysis of clinical herpes simplex virus type 1 isolates identified three genetic groups and recombinant viruses. J Virol 2004; 78:10755–10764.
28. Gallinella G, Venturoli S, Manaresi E, et al. B19 virus genome diversity: Epidemiological and clinical correlations. J Clin Virol 2003; 28:1–13.
29. Phan TG, Kaneshi K, Ueda Y, et al. Genetic heterogeneity, evolution, and recombination in noroviruses. J Med Virol 2007; 79:1388–1400.
30. Abecasis A, Vandamme A, Lemey P. Sequence Alignment in HIV Computational Analysis. In: Leitner T, Foley B, Hahn B, et al., eds. HIV Sequence Compendium 2006/2007: Theoretical Biology and Biophysics Group. Los Alamos, NM: Los Alamos National Laboratory, LA-UR 07–4826, 2007:2–16.
31. Larkin MA, Blackshields G, Brown NP, et al. Clustal W and Clustal X version 2.0. Bioinformatics 2007; 23:2947–2948.
32. Lassmann T, Sonnhammer EL. Quality assessment of multiple alignment programs. FEBS Lett 2002; 529:126–130.
33. Eddy SR. Profile hidden Markov models. Bioinformatics 1998; 14:755–763.
34. Edgar RC, Batzoglou S. Multiple sequence alignment. Curr Opin Struct Biol 2006; 16:368–373.

35. Hoffman NG, Cook L, Atienza EE, et al. Marked variability of BK virus load measurement using quantitative real-time PCR among commonly used assays. J Clin Microbiol 2008; 46:2671–2680.
36. Waterhouse AM, Procter JB, Martin DM, et al. Jalview Version 2—A multiple sequence alignment editor and analysis workbench. Bioinformatics 2009; 25:1189–1191.
37. Gadberry MD, Malcomber ST, Doust AN, et al. Primaclade—A flexible tool to find conserved PCR primers across multiple species. Bioinformatics 2005; 21:1263–1264.
38. Fredslund J, Schauser L, Madsen LH, et al. PriFi: Using a multiple alignment of related sequences to find primers for amplification of homologs. Nucleic Acids Res 2005; 33:W516–W520.
39. Qiu P, Cai XY, Wang L, et al. Hepatitis C virus whole genome position weight matrix and robust primer design. BMC Microbiol 2002; 2:29.
40. Zhang Q, Wu G, Richards E, et al. Universal primers for HBV genome DNA amplification across subtypes: A case study for designing more effective viral primers. Virol J 2007; 4:92.
41. Huang MM, Arnheim N, Goodman MF. Extension of base mispairs by Taq DNA polymerase: Implications for single nucleotide discrimination in PCR. Nucleic Acids Res 1992; 20:4567–4573.
42. Ayyadevara S, Thaden JJ, Shmookler Reis RJ. Discrimination of primer 3′-nucleotide mismatch by taq DNA polymerase during polymerase chain reaction. Anal Biochem 2000; 284:11–18.
43. Christopherson C, Sninsky J, Kwok S. The effects of internal primer-template mismatches on RT-PCR: HIV-1 model studies. Nucleic Acids Res 1997; 25:654–658.
44. Bru D, Martin-Laurent F, Philippot L. Quantification of the detrimental effect of a single primer-template mismatch by real-time PCR using the 16S rRNA gene as an example. Appl Environ Microbiol 2008; 74:1660–1663.
45. Martin FH, Castro MM, Aboul-ela F, et al. Base pairing involving deoxyinosine: Implications for probe design. Nucleic Acids Res 1985; 13:8927–8938.
46. Kilpatrick DR, Nottay B, Yang CF, et al. Group-specific identification of polioviruses by PCR using primers containing mixed-base or deoxyinosine residue at positions of codon degeneracy. J Clin Microbiol 1996; 34:2990–2996.
47. Ben-Dov E, Shapiro OH, Siboni N, et al. Advantage of using inosine at the 3′ termini of 16S rRNA gene universal primers for the study of microbial diversity. Appl Environ Microbiol 2006; 72:6902–6906.
48. Zheng L, Gibbs MJ, Rodoni BC. Quantitative PCR measurements of the effects of introducing inosines into primers provides guidelines for improved degenerate primer design. J Virol Methods 2008; 153:97–103.
49. Seela F, Chen Y. Oligonucleotides containing fluorescent 2′-deoxyisoinosine: Solid-phase synthesis and duplex stability. Nucleic Acids Res 1995; 23:2499–2505.
50. Knittel T, Picard D. PCR with degenerate primers containing deoxyinosine fails with Pfu DNA polymerase. PCR Methods Appl 1993; 2:346–347.
51. Fujiwara H, Fujiwara K, Hashimoto K. PCR with deoxyinosine-containing primers using DNA polymerases with proofreading activity. PCR Methods Appl 1995; 4:239–240.
52. Maher-Sturgess SL, Forrester NL, Wayper PJ, et al. Universal primers that amplify RNA from all three flavivirus subgroups. Virol J 2008; 5:16.
53. Bracho MA, Garcia-Robles I, Jimenez N, et al. Effect of oligonucleotide primers in determining viral variability within hosts. Virol J 2004; 1:13.
54. Whiley DM, Sloots TP. Sequence variation in primer targets affects the accuracy of viral quantitative PCR. J Clin Virol 2005; 34:104–107.

6 Microarray Detection of Viruses

Julie Fox

Department of Microbiology and Infectious Diseases, University of Calgary and Provincial Laboratory for Public Health (ProvLab), Calgary, Alberta, Canada

This chapter describes the use of microarray technologies for the diagnosis of viral infections. In the majority of cases, such diagnosis relates to detection of taxonomically related viruses or those associated with similar symptomatology. When analysis of a particular virus strain or genotype is useful in management, this may also be considered diagnostic in some circumstances. Thus, examples have been included in this chapter where arrays have been used to provide added value beyond initial detection of a virus. Additional uses for microarrays beyond management of an individual include vaccine efficacy studies, investigation of epidemiologically linked cases, and surveillance.

The majority of this chapter focuses on microarrays for nucleic acid detection and analysis of viruses, this is because there are many examples of this type of application already in routine use. However, array-based diagnosis of viruses may also be achieved through antigen-based formats. This chapter does not cover the wealth of data gathered from expression profiling arrays, both to investigate the transcription profile of the host in the case of a viral infection and to undertake detailed investigation of viral transcription and interaction with the host. Both of these uses for expression arrays are important in enhancing our understanding of viral pathogenesis and to identify potential targets for anti-viral therapy and vaccine development. Use of microarrays for vaccine quality control is an important area but beyond the scope of this review as the intention of this approach is only to analyze cultured viruses and high-titre preparations.

Inevitably, this review cannot provide an exhaustive list of cited references. Those quoted are either examples where arrays have already proved useful in diagnosis or are articles providing more detailed methodology and information on a particular virus or virus group that is beyond the scope of this review.

INTRODUCTION TO NUCLEIC ACID MICROARRAYS

Nucleic acid amplification tests (NATs) are emerging as the preferred (gold standard) approach for diagnosis of many viral infections, either as an adjunct to other testing or as a replacement. Despite the enhancements provided by NATs, the broad range of pathogens that can cause similar symptomatology makes it difficult to apply individual (monoplex) or small multiplex assays targeting DNA or RNA to a situation where comprehensive diagnosis is needed. Undertaking multiple NATs can be cost prohibitive and may not even be possible if specimen quantity is limiting. Attempts to build diagnostic capacity by introduction of multiple primer and probe sets in a real-time assay have not been entirely successful as there is a tendency for a loss in amplification efficiency when targets are analyzed in such a complex mix. One way to increase the capacity to test for multiple possible pathogens in a single specimen is to separate the nucleic acid amplification away from the hybridization/detection reaction. Using this approach, it is easier to broaden the amplification without compromising the diagnostic sensitivity for each target. The advantage of such an approach is multiple pathogen detection in a single assay. Thus, the convenience of individual real-time NATs is replaced by the enhanced capacity of separate amplification and detection. Microarrays potentially have the benefit of being able to resolve very complex amplified product mixtures. Thus, redundancy can be built into the assay with multiple gene targets represented for each potential pathogen, and inclusion of genotyping information for critical targets such as influenza virus (IFV), human papillomavirus (HPV), human immunodeficiency virus (HIV), and other variable DNA and RNA viruses.

The initial setups of nucleic acid arrays were adaptations of original Southern or dot blots and were the precursors of line-probe and other low density hybridization assays in use in many laboratories. In Southern and dot blots, the complex nucleic acid mixture (from the sample, with or without prior nucleic acid amplification) is generally applied to the solid surface (nitrocellulose or nylon) and a mix of labeled probes applied in solution to query this sample material. In the case of modern arrays, the "probe" (generally a short oligonucleotide or cDNA sequence) is attached to the support matrix (which can be a solid surface or suspension beads as described below) and the "target" (starting material, generally amplified by PCR) is then applied to the matrix containing an "array" of probes.

The original "nucleic acid array" formats were the line-probe/blot assays that are well established for HIV, hepatitis B (HBV), hepatitis C (HCV), and HPV genotyping. The term "nucleic acid microarray" is generally applied when the number of probes is higher than these simple formats but many of the principles are similar.

In short, a microarray is an array with enhanced capacity for detection and/or typing of a wide range of viruses. In general, such microarrays have either been applied to broad-range detection of viruses (such as in the examples of respiratory virus detection) or for detailed analysis of a viral group (such as detection and analysis of IFV types and strains).

MICROARRAY FORMATS USED FOR DIAGNOSTICS

Until recently, application of microarrays to viral detection and diagnosis from clinical specimens was limited by complexity of design, cost, and a lack of sensitivity when compared with conventional or real-time NATs. Many studies have demonstrated the utility of microarrays for detection of amplified products when cultured (high-titre purified) viruses are used as template. While culture of a virus prior to microarray analysis may be appropriate where detailed epidemiological study of a virus group or strain is to be undertaken, the delay and lack of sensitivity of culture would limit the use of such an approach severely for front-line individual patient diagnosis.

In the last few years we have seen enhancements in microarray technology with adaptations and customization for "in-house" use as well as commercialization and regulation of some diagnostic and typing assays. The intent is that these assays can be applied directly to amplified products produced from an original sample without compromise in sensitivity and specificity compared with alternative methods.

Details of different assay formats applied to specific viral detection and analysis are described below. Table 1 gives examples of microarray assay formats that have already been applied to virus diagnosis together with some example protocols and references.

For nucleic acid–based microarrays to be useful in a diagnostic setting, enhancements to nucleic acid amplification procedures, labeling methods, probe synthesis, and hybridization formats have been necessary.

Amplification and Labeling Methods

Template (target) nucleic acid needs to be amplified prior to hybridization on a microarray. The majority of diagnostic virology protocols utilize multiplex PCR. Where a single gene is to be analyzed with low-density array detection, generic primers may be used to amplify across the variable region to be queried (32). Random amplification methods give the broadest approach and have been combined with high-density array detection (48,51). A combination amplification approach has been suggested with random priming to allow unbiased amplification of all templates with the addition of virus-specific primers to enrich for important targets present at low level or which may amplify inefficiently (56). In order to ensure maximum sensitivity for analysis of primary specimens, some protocols utilize a nested amplification procedure prior to hybridization (31,32). The amplification reaction may be "skewed" in an asymmetric manner to enhance the relative amount of the cDNA strand available to bind to the probe (8,41). RNA polymerase promoter sequences can be incorporated into PCR products as a 5′ "tail" on one of the primers. Incorporation of these sequences allows run-off RNA transcripts to be prepared for application to the array (25,26,30). Amplified cDNA or run-off transcripts may be fragmented, or sheared, prior to hybridization on the array to facilitate binding to the available probes.

Label may be incorporated directly in to an amplified product as a component of the primer or by using a labeled nucleoside triphosphate. Such label can also be added during a secondary strand-specific labeling reaction (68–71,73), during a second-round (nested) PCR (32),

Table 1 Microarray Formats and Example Protocols for Virus Detection and Analysis

Microarray format	Advantages	Disadvantages	Example protocols		
			Target or group	**Application**	**Reference (s)**
Solid-phase low–medium density arrays (tens to hundreds of probes)	Easy to design and set up "in house." May not require complex equipment for set up and reading. Can be adapted and changed easily. Commercialized applications are already available for diagnostics. Tends to be a cost-effective means of broad diagnosis or analysis compared with real-time NATs.	Difficult to standardize "in-house" formats and compare data between laboratories. The format limits the number of targets that can be queried in a single reaction.	HPV	Detection and genotyping	(1–11)
			Blood-borne viruses	Detection of a panel of viruses	(12)
				HBV analysis	(13–15)
				HCV analysis	(16,17)
				HIV analysis	(17–20)
			Respiratory viruses	Detection of a panel of viruses	(21–23)
				IFVA analysis	(24–29)
				IFVB analysis	(27,28,30)
				hCoV analysis	(31)
				ADV analysis	(32)
				Hantavirus analysis	(33)
			Picornaviruses	Detection and analysis of EV71 and CA16	(34)
				Detection and analysis of polioviruses	(35)
			CNS pathogens	Detection of a panel of viruses.	(36)
			Herpesviruses	Detection and differentiation	(37–39)
				VZV genotyping	(40)
			Orthopoxvirus and Herpesvirus	Detection and differentiation	(41)
			Measles virus	Genotype analysis	(42)
			Enteric viruses	Detection of a panel of enteric viruses	(43)
				Norovirus analysis	(43)
				Astrovirus analysis	(43,44)
				Rotavirus analysis	(45–47)

High-density arrays (generally many thousands of probes, including resequencing arrays)	Capacity is much higher than other formats with ability to identify new sequence variants. Provides the only close to "catch all" array approach to viral diagnostics.	Arrays of this nature tend to be expensive and inflexible. Sophisticated computational analysis is needed to deal with the high complexity data. High-density arrays tend to have lower sensitivity for analysis of primary specimens. QC of such high-complexity assays for primary diagnosis is very difficult.	Broad virus detection	Panel testing for all viruses	(48–51)
			Bioterrorism agents	Detection of a panel of viruses	(52)
				Smallpox analysis	(53)
			Blood-borne viruses	HIV analysis	(19,20,54)
			Respiratory viruses	Detection of a panel of viruses.	(55–58)
				IFVA and IFVB analysis	(50,55,56)
				hCoV analysis	(51,59,60)
				Rhinovirus analysis	(51)
Flow-through and 3D/4D microarrays (applications currently limited to less than 100 probes but more may be added)	Fast hybridization kinetics. Potentially very sensitive as more probes can be addressed during the reaction.	Requires specialized equipment that is not compatible with other array formats. Not yet in routine diagnostic use.	Respiratory viruses	IFVA detection and analysis	(61)
				IFVB detection and analysis	(61)
Suspension microarrays (current applications have up to 100 probes but more may be added)	Flexible in design with new targets added easily. Method uses detection equipment that is common in many laboratories. Equipment is "open access" for flexibility in use.	Requires equipment and software for reading.	HPV	Detection and genotyping.	(62–66)
			Respiratory viruses	Detection of a panel of viruses.	(67–74)
				IFVA analysis.	(68,69)

Abbreviations: HPV, human papillomavirus; CNS, central nervous system; HBV, hepatitis B; HCV, hepatitis C; HIV, human immunodeficiency virus; IFV, influenza virus; hCoV, human coronavirus; ADV, adenovirus; CA16, coxsackievirus A16; E71, enterovirus 71; NATs, nucleic acid amplification tests.

or by chemical labeling of run-off ssRNA transcripts (40,41). For added specificity and to control the amount of label applied, two probes may be utilized: one as part of the microarray for capture of target sequence and a second target-specific labeled probe for detection of bound target (25–27,30). Primer extension to identify single nucleotide polymorphisms (SNPs), sometimes termed mini-sequencing, may be undertaken "on chip" with labeled nucleotides incorporated during this extension (47). Enhancement of labeling may be achieved by using an indirect labeling method such as that described for use with the GreeneChip systems (50,56). In this case, more than 300 fluorescent reporter molecules are incorporated into the probe-target hybridization. The majority of labeling and detection methods already utilized for viral detection and analysis use fluorescence (e.g., Cy5, Cy3) or chemiluminescence for detection of hybridized products (see Table 1 for example methods). However, simple colorimetric procedures (24) or electrochemical detection (ECD) methods (21) may also be used.

Microarray Substrates and Probe Synthesis

The array (or chip) substrate may be nylon, membrane, glass, silicon, or polystyrene microbeads of variable density (numbers of specific probes and thus targets to be queried). The hybridization probes on an array may be oligonucleotides or longer nucleic acid sequences (such as cDNAs produced as PCR products and cloned). Oligonucleotide probes can be synthesized "on chip" or linked to the array surface after synthesis. For presynthesized probes, attachment to the surface may be by simple "spotting" or may make use of microelectrodes or covalent attachment methods.

Probe design and hybridization conditions can be adjusted in an array to allow some mismatch of sequences, enabling possible identification of novel viruses, sequence variants, or pathogens not well represented in current sequence databases. Where specific sequences are available, dedicated software may be used to help probe design. In one procedure, an amplicon retrieval software was designed to detect all possible amplicons from 2 primers directed against a given set of FASTA format sequences (32). This approach was utilized in low-density array analysis of adenoviruses (ADV) amplified using generic hexon gene primers, but could be applied more widely for "in-house" development of microarrays.

Solid-Phase Microarray Detection

Solid-phase microarrays were the first to be made available to diagnostic laboratories. In their simplest form, they are the well-recognized line-probe/hybridization assays already in routine use for HIV, HCV, HBV, and HPV genotyping. These utilize nylon or nitrocellulose membrane as the solid phase to which probes are applied. The number of probes that can be applied is limited by the porous nature of the membrane. However, for some viruses where genotyping is necessary for assessment of risk and management of a patient these have well-demonstrated diagnostic utility. The equipment required is not complex and blots may be read by eye as shown in the example line-probe assay for HPV given in Figure 1 (6).

Probes arrayed on to a slide or other non-porous solid-phase format can be spotted in a well-defined nonoverlapping manner and can be relatively long, ensuring flexibility in melt temperatures used for hybridization. Glass is often used in such a system as it is easily activated for covalent attachment chemistry and can be used with low-hybridization volumes. Solid-phase arrays may be low–medium density customized "chips" prepared "in house" or may be high-density manufactured arrays using automated systems and sold commercially.

Figure 2 shows an example of a simple solid-phase diagnostic array used for detection and differentiation between two important zoonotic viruses, Newcastle disease virus (NDV), and avian influenza virus (AIV) (24). This array utilized conserved primers and probes based on the matrix (M) gene of AIV and the hemagglutinin (HA) gene of different AIV subtypes. The amplified PCR products were biotinylated by incorporation of the label at the 5′ end of the primers. Each oligonucleotide probe was prepared with a 19 T base tail through which it was UV linked to a polymer substrate. The hybridization reaction was undertaken using kit-based reagents with streptavidin-alkaline phosphate and a colorimetric substrate reaction. The results are clearly readable without further imaging equipment making this an easy to implement low-complexity system for detection and typing of multiple viruses using oligonucleotide arrays.

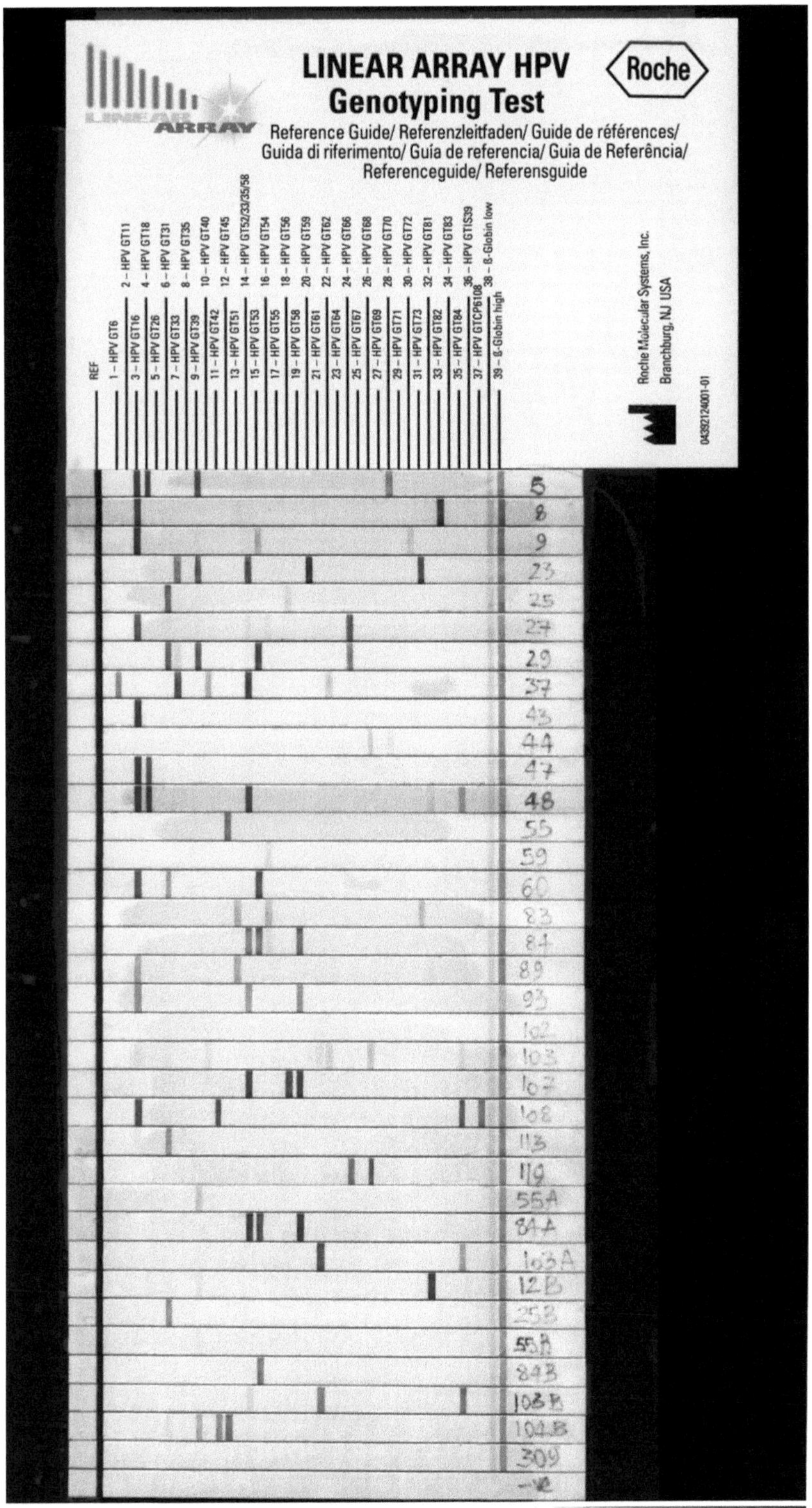

Figure 1 Linear array strips for detection and typing of HPV. A range of HPV types are shown using the Linear array system (Roche). The method allows detection and differentiation of high- and low-risk HPV genotypes and will prove useful for monitoring vaccine efficacy and genotype replacement. *Source*: From Ref. 6.

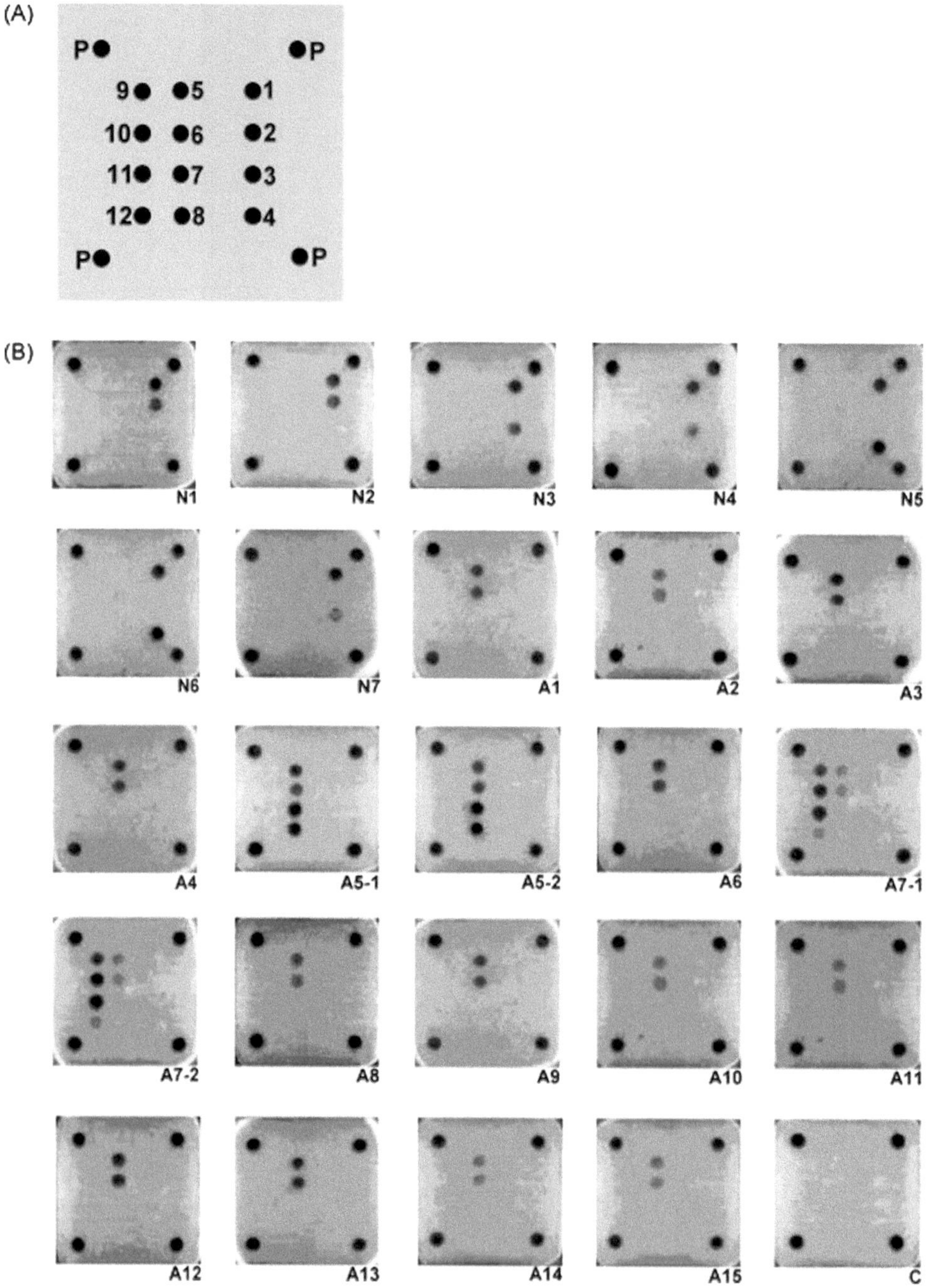

Figure 2 Detection and typing of Newcastle disease and avian influenza viruses using oligonucleotide array. (**A**) Array map. Each dot indicates the spotted position of each probe. *Abbreviation*: P, positive control. (**B**) Detection and typing results for Newcastle disease virus (NDV) strains 1–7. Avian influenza virus (AIV) hemagglutinin subtypes are given as A1–A15. The strain in A5-1 and A5-2 is Influenza A/Chicken/Taiwan/1209/03 (H5N2) and Influenza A/Black duck/New York/184/1988 (H5N2), respectively. The strain shown in A7-1 and A7-2 is Influenza A/Mallare/Ohio/322/1998 (H7N3) and DK/TPM/A45/03 (H7N7), respectively. *Abbreviation*: C, negative control. *Source*: From Ref 24.

Figure 3(A) shows an example methodology for detection of biotinylated target products in a semiconductor-based oligonucleotide microarray format that can be adapted to utilize fluorescence or ECD (21). An example of results relating to the ECD of respiratory viral and bacterial products using this type of array is given in Figure 3(B) (21). The complexity of this solid-phase array is higher than the examples shown in Figures 1 and 2, having approximately 850 unique probes (with many replicates of each) per array. The ECD system uses an enzymatic reaction to create electrical signals, which can be read directly without the need for image analysis and optical scanning equipment. The evaluation utilized the ElectraSense® microarrays and scanner (CombiMatrix Corp.). With more than 12,000 potential addressable electrodes, this format has the potential for high throughput for broad viral detection and analysis.

A high-density microarray was developed containing the most highly conserved 70mer sequences from every fully sequenced reference viral genome in GenBank (48,49,51,59). Combined with a random amplification procedure, the prototype of this array proved useful in identifying a previously uncharacterized human coronavirus (hCoV) in a viral isolate cultivated from a patient with severe acute respiratory syndrome (SARS) (59) as well as for later analysis of SARS hCoV variation (60). Different versions of this Virochip array have been used as a final "catch all" approach to identification of viruses (especially respiratory viruses), even when divergent from the prototype sequences. The sensitivity of this random amplification and array approach may not yet be quite as good as individual NATs but the feasibility of the approach has been shown for direct detection of rhinoviruses in clinical samples (51).

A panmicrobial oligonucleotide array was reported for broad detection of infectious disease (GreeneChipPm) (50). This array comprised 29,455 60-mer oligonucleotides that were directed against vertebrate viruses, bacteria, fungi, and parasites. Combined with a random amplification procedure, this method is extremely powerful for the broadest identification of pathogens. The authors suggest a staged algorithm with highly multiplexed PCR being used initially to attempt to identify the causative agent, and the GreeneChipPm being used in a second line procedure for further analysis. A minimum of 3 gene targets were included for every family or genus, with one highly conserved target and at least two more variable genes queried with an aim to identify the known and the novel. The reported sensitivity of this array was 10,000 copies for a representative DNA virus (ADV) and 1,900 copies for respiratory RNA viruses. In further adaptations of this system, an array with 14,795 probes with special focus on respiratory viruses and influenza virus array (IFVA) subtyping was developed with a reported sensitivity of 1000 copies (56).

Resequencing Arrays

Resequencing arrays were originally designed for SNP analysis and microbial typing in a high-density format, and provide high-quality sequence for all or part of a pathogen. Such arrays, when applied to viral diagnostics or as a viral epidemiological tool, have been developed largely as part of the custom array program using the GeneChip® scanner, fluidics station, workstation, and analysis software (Affymetrix). Resequencing arrays tend to be developed using short oligonucleotide probes (typically less than 25 bases). They potentially have very high capacity for identification and speciation of pathogens, but the short-hybridization region means that any sequence variation will disrupt the signal. This technology is already beginning to impact viral diagnostics with for example assays developed for broad viral detection as well as analysis of viral strains within a family (detailed below).

Application of resequencing arrays for epidemiological investigation of outbreaks or emergence of novel viruses is well established. This approach can also be used for primary diagnosis, particularly where novel viruses are sought or there may be sequence divergence. For diagnostic purposes, however, there is a need for re-design of components of the array regularly to reflect virus sequence variation. There is some evidence that a random amplification approach with resequencing array detection of products may not be sensitive enough for direct detection and analysis of samples directly from clinical samples, despite the reported analytical sensitivity of 10 to 1000 copies of target (58). Alternative methodologies have utilized multiplexed target-specific PCRs for amplification prior to analysis of products from clinical samples using resequencing formats (52,58).

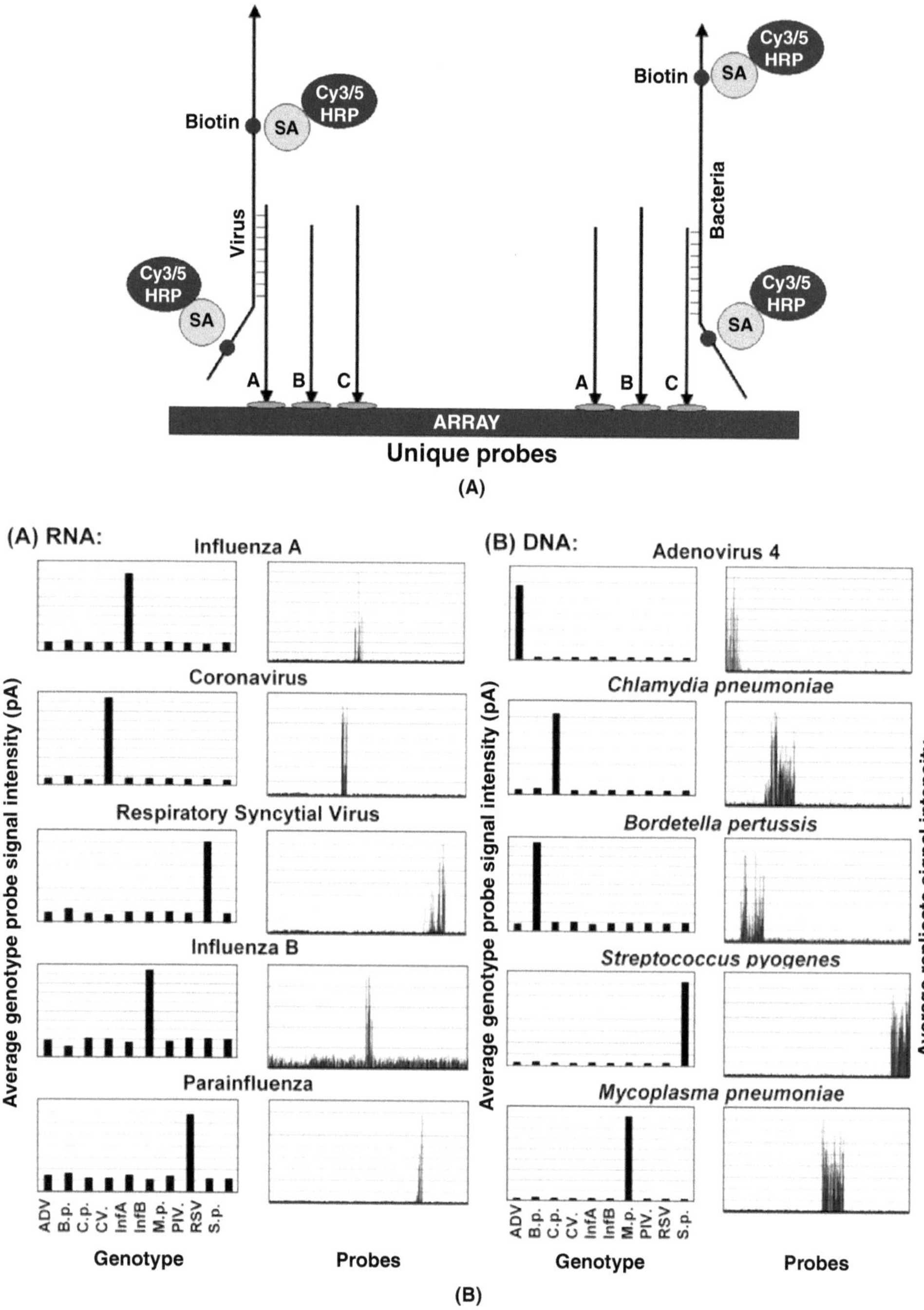

Figure 3 (**A**) Detection methodology for an oligonucleotide array designed to detect viral and bacterial targets. Biotinylated, single-strand target is hybridized to complementary probes on the microarray (**A**, **B**, and **C**) and labeled by the addition of streptavidin (SA)–horseradish peroxidise (HRP) (for electrochemical detection) or SA-Cy3/Cy5 (for fluorescent detection). (**B**) Respiratory pathogen validation studies using an oligonucleotide array with electrochemical detection. Results show average RNA genome (**A**) and DNA genome (**B**) target signal intensity for 10 upper respiratory pathogens studied. The left panels (GENOTYPE) show the average genotype-specific probe signal in picoamps that is used to determine a genome identity. The right panels (PROBES) illustrate the average probe signal intensities of 12 replicates for each probe and also illustrate signal specificity for both positive and negative probes. *Abbreviations*: ADV, adenovirus; B.p., *Bordetella pertussis*; C.p., *Chlamydia pneumoniae*; CV, coronavirus; infA, influenza A virus; infuB, influenza B virus; PIV, parainfluenzvirus; RSV, respiratory syncytial virus; S.p., *Streptococcus pyogenes*. *Source*: From Ref. 21.

Example results using a resequencing array for detection and differentiation between IFVs are given in Figure 4 (55). Part of an Affymetrix respiratory pathogen microarray (RPM) developed to identify a range of bacteria and viruses (57,58) is shown to which random RT-PCR products were applied. In addition to identification of IFVA, this procedure provided primary sequence information illustrating the distinct viral lineages (IFVA and IFVB) circulating in each season.

Flow-Through and 3D/4D Microarrays

These have the advantage above solid-phase microarrays of allowing kinetic binding studies in an array-based format with the possibility for enhanced sensitivity, rapid hybridization, and high capacity because of the relatively large probe binding surface area.

The Flow-thru Chip™ (from Xceed Molecular, previously MetriGenix) is a high-performance 3D microarray platform that requires smaller amounts of sample and reagents than some other systems. Application of this commercial biochip platform to IFV diagnosis was successful and demonstrated the potential of this approach for detection and analysis of IFVA types, subtypes, and strains in a single reaction (61).

The BioFilmChip™ Microarray (Autogenomics) is a novel, film-based microarray, which consists of multiple layers of hydrogel matrices giving a 3D matrix. This type of array has been combined with a high level of automation (INFINITI™ Analyzer) giving a continuous flow, random access microarray platform that integrates sample handling, hybridization, and detection in a self-contained system. The analyzer features a built-in confocal microscope with two lasers, a thermal stringency station, and a temperature cycler for denaturing nucleic acids for allele-specific primer extension. This technology has significant potential for diagnostic application because of its high level of automation. Assays suitable for viral diagnosis using this procedure (but currently for research use only) include those for broad spectrum detection of respiratory viruses and for detection and genotyping of HPV.

Suspension Microarray Detection

These employ a liquid-phase bead-conjugated array technology known as Luminex® xMap™ for detection of amplified products. Such suspension microarrays exhibit rapid hybridization kinetics, flexibility in assay design, and low cost. Spectrally distinct fluorescent-labeled polystyrene microspheres (beads) are utilized in the suspension microarray. Up to 100 different beads are available in the original format but many more labels are becoming available. The Luminex® detection system can be utilized for detection of many different molecules. In the case of diagnostic NATs, a different nucleic acid probe is conjugated to each bead type and a mixture of beads is used in the array for detection and differentiation between amplified products. Microspheres are interrogated by two different lasers in the Luminex® analyzer to identify the specific bead (and thus conjugated probe) and to identify any binding of the product (which is labeled during amplification or in a separate reaction) to this probe. The equipment is open and can be used for analysis of commercially available suspension microarray assays as well as for development of "in-house" assays. The Luminex® technology is very flexible as new beads (and probes) can be added or others replaced without having to reformat and print new arrays (a disadvantage for high-density solid-phase arrays). The methodology is also suitable, and has been widely applied, to detection of polymorphisms in human genetics or for strain differentiation of pathogens.

The xTAG™ Respiratory Viral Panel (RVP from Luminex Molecular Diagnostics) is a comprehensive assay for the detection of multiple viral strains and subtypes and is the first FDA approved assay of its type. Two other commercial assays utilizing the Luminex® analyzer have also been developed for respiratory pathogen detection (discussed in more detail in the respiratory virus section below). Comparison of assay strategies utilizing the Luminex suspension array system is given in Figure 5. The labeling and detection formats for these assays are different but the basic principles are the same. The amplification is highly multiplexed and either has a biotin incorporated at this point (ResPlex from Qiagen) (67,72,74) or during a second-stage strand-specific target extension [Luminex RVP (68,69) and EraGen RMA/MultiCode PLx (70,71,73)]. The specificity of these approaches is either assured during a second-primer–directed extension [Luminex RVP (68,69) and EraGen RMA/MultiCode PLx (70,71,73)] or by having a

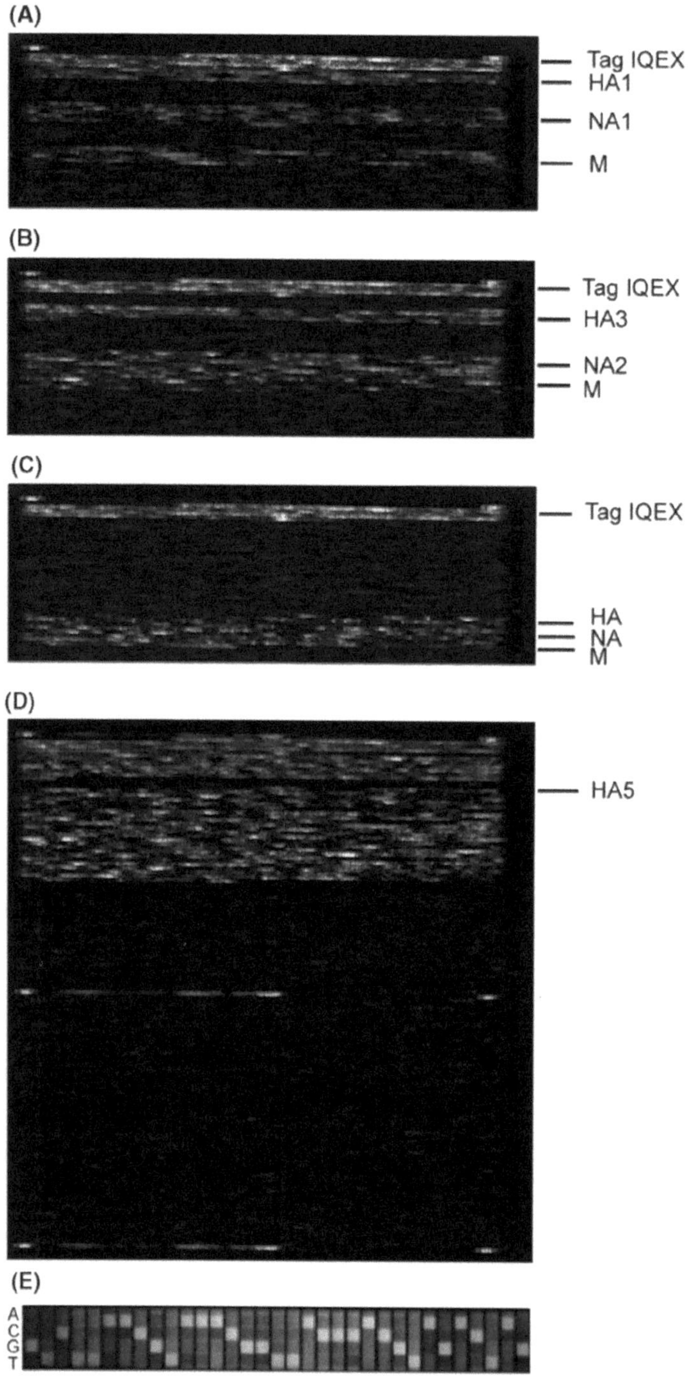

Figure 4 Hybridization images for the respiratory pathogen microarray (RPM) version 1. Analysis of three influenza virus (IFV) isolates and trivalent FluMist vaccine using the RPM version 1 is shown. (**A**) IFVA H1N1, (**B**) IFVA H3N2, (**C**) IFVB, and (**D**) FluMist vaccine. Only the IFV-specific regions of the RPM are shown for **A**, **B**, and **C**. Hybridization positive identifications are shown on the right. The image for **D** is the whole RPM and, in this case, only the negative portion of the hybridization (for H5) is quoted on the right. (**E**) Magnified portion of panel **B** showing an example of the primary sequence data generated by the hybridization of randomly amplified targets to the RPM version 1 HA3 probe set. The primary sequence generated can be read from left to right. *Abbreviations*: HA, hemagglutinin; NA, neuraminidase; IQEX, internal positive hybridization control (Affymetrix); M, matrix. *Source*: From Ref. 55.

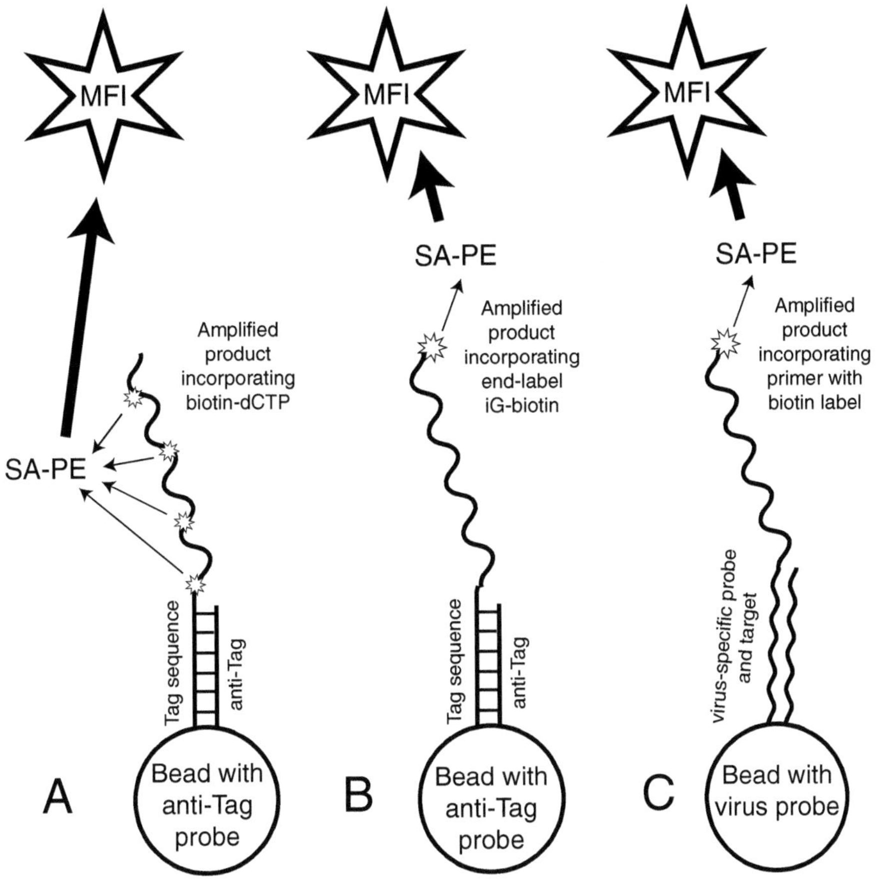

Figure 5 Example detection formats for respiratory virus suspension arrays. (**A**) xTag™ RVP from Luminex Molecular Diagnostics, (**B**) MultiCode® PLx respiratory panel (RMA) from EraGen, (**C**) ResPlex I and II from Qiagen Molecular Diagnostics. *Abbreviations*: MFI, median fluorescence intensity; SA-PE, streptavidin-phycoerythrin.

target-specific probe on the bead (Qiagen ResPlex) (67,72,74). The format of these assays will no doubt be expanded to other targets and broad detection of pathogen groups. In particular, HPV genotyping using suspension microarray formats are being developed and evaluated for routine use (62–66).

Universal Microarrays

Re-designing the probes on microarrays to accommodate changing needs can be expensive and laborious. Universal microarrays make use of standard sequences for detection, the complementary sequence for which is usually included in the amplification or labeling reaction as a tag or tail sequence on one of the target-specific primers. The advantage of using a universal array is that the hybridization conditions in the array can be optimized and design of the probes is not constrained by sequence variation in the viral target. Also, it means that arrays can be designed and kept constant despite the wide range of assay targets or required changes to the diagnostic testing repertoire. Universal arrays have been used in two respiratory virus suspension microarray formats (68–71,73) and may be more cost-effective than resequencing arrays for analysis of mutations and sequence variation for divergent RNA viruses.

REGULATORY APPROVAL OF DIAGNOSTIC ASSAYS UTILIZING MICROARRAY

Some NAT microarrays are already being marketed and have regulatory approval (e.g., FDA, Health Canada, and CE Mark). Further evaluation of microarray methods in prospective diagnostic studies is necessary as transition from reference or research laboratories to front-line diagnostics takes place. Sensitivity and specificity need to be compared with alternative diagnostic formats (particularly individual and multiplex NATs for all targets) in order for clinical interpretation to be simplified. The range of formats being developed means that each laboratory will be able to select the method that suits their particular testing needs, with automation, turn-around time, throughput, and range of pathogens analyzed being key deciding factors. Approval and regulation of diagnostic assays by appropriate agencies, with concurrent availability of suitable quality control and proficiency panel materials, will likely establish amplification methods, combined with array-based detection, as the next "gold standard" for many areas of virus diagnosis. The main comparator for such broad spectrum approaches will be high-throughput sequencing and amplification and detection of complex mixtures of products by methods such as capillary electrophoresis.

QUALITY CONTROL, ASSURANCE, AND PROFICIENCY

The highly multiplexed analyses that are inevitably part of diagnostic microarrays present particular challenges in the area of quality control, assurance, and proficiency. Where arrays are printed or prepared by a manufacturer, they must assure identity of each probe and placement. The verification of such an array would then be undertaken by the test laboratory, although this would be a relatively complex process for all but the lowest-density arrays. Procedures for specimen handling, amplification, and labeling must be adequately controlled and the interpretation algorithms must also be included in assessment of microarray (and complete assay) reproducibility. Many viral diagnostic laboratories are used to having controls for every target in each test run. With microarray procedures it is difficult to control for every possible viral target in each test and a compromise must be made. For many commercially available tests, internal controls are included which can be useful to ensure efficient sample extraction and to control, generically, for the reverse-transcriptase (RT) and DNA amplification steps. In the case of such kits, lots can be tested with a panel of targets to ensure suitability of the lot. After this, each laboratory has to develop a strategy for inclusion of enough controls to ensure quality and reproducibility of diagnostic results. Guidelines are being developed (e.g., Clinical and Laboratory Standards Institute. Diagnostic Nucleic Acid Microarrays: Approved Guideline MM12-A) to address the problems of laboratory validation, proficiency, and control of highly complex viral microarrays.

EXAMPLES WHERE MICROARRAYS HAVE PROVED USEFUL IN VIRAL DETECTION AND DIAGNOSIS

In most cases, when microarrays are used for detection of viruses, a close to "catch all" approach is necessary. As NATs are becoming part of the routine for blood and organ donor screening, broad spectrum approaches incorporating microarrays could impact this area. Much of the initial work in the area of nucleic acid–based microarrays for detection of viruses has revolved around respiratory pathogen detection because such a broad range of viruses and bacteria can present with similar symptomatology. In a similar way, viruses causing central nervous system (CNS) disease or gastroenteritis can present similarly, and diagnostic strategies benefit from a broad approach to detection. Examples are provided below where customized (in-house) and commercial assays have been developed specifically for viral diagnostics.

Blood-Borne Virus Screening

Microarrays may have impact on the screening of blood and other products for blood-borne viruses. In a proof-of concept study, a low-density solid-phase oligonucleotide microarray for detection of HIV, HBV, and HCV multiplex-PCR products was developed (12). The assay was very sensitive, detecting 1-20 international units of each target with added capacity to identify mixed infections.

Broad Respiratory Virus Detection

Microarrays have been applied to broad respiratory virus detection (sometimes alongside bacterial detection). There is a need for RPMs to be adaptable to include additional targets as they are identified. Recently identified viruses that have yet to be incorporated into available microarrays include human bocavirus and the polyomaviruses WU and KI.

Assays Using Solid-Phase Arrays

Initial setup of broad respiratory virus detection came from direct application of multiplex PCR to array-based detection. Low-density formats gave excellent results for 14 different respiratory viruses with sensitivity close to that for nested PCR procedures (23). As described above, Figure 3(A) shows an example medium-density methodology for detection of a range of respiratory pathogens using a semiconductor-based oligonucleotide microarray format that can be adapted to utilize fluorescence or ECD. Utility has been demonstrated for detection of upper respiratory tract pathogens (nine viruses and four bacteria in this initial study) (21).

Solid-phase microarray hybridization of randomly amplified PCR products from respiratory cultures and clinical samples has also been successful using high-density arrays (48,50,56). These reports suggest that a syndromic approach to respiratory pathogen diagnosis may be possible using such technology and that we can get close to a "catch all" approach. The reported sensitivity of these arrays (GreenChip and ViroChip), combined with random-amplification PCR, was good enough for direct detection of some respiratory viruses in clinical specimens but may not be as sensitive as individual real-time NATs (56). The very powerful broad spectrum detection of viruses using these high-density arrays is ideally suited to search for the novel (especially from viral cultures), but this approach may not be a front-line procedure for diagnostics because of continuing concerns about sensitivity as well as cost.

Original resequencing array formats for respiratory virus detection utilized random amplification procedures. However, on application to clinical studies, the authors concluded that an optimized multiplex-target–specific PCR combined with a respiratory pathogen resequencing array provided the best sensitivity for diagnosis of respiratory pathogens in clinical samples (57,58). In their study, the authors were able to show correct sequence and strain identification using an array targeting 57 genes for 26 respiratory pathogens. Resequencing arrays have been applied to further detailed analysis of respiratory viruses, as detailed below.

Assays Using Suspension Microarray

Commercial assays for respiratory virus detection utilizing multiplex amplification with detection using the suspension microarray Luminex® system have been reported (67–74). In one format, multiplex PCR products are detected and discriminated using template-specific probes conjugated to different microspheres (67,72,74). In an alternative methodology, a multiplex PCR is used in a first step followed by primer-directed (and target-specific) strand extension and labeling. Each target-specific primer used in this labeling reaction incorporates a unique capture sequence. It is these capture sequences that are used for detection of amplified products in the universal suspension microarray (68–71,73). Thus, one of the main differences between these two approaches is use of target-specific hybridization (67,72,74) versus primer extension (68–71,73) for specificity. The MultiCode-PLx assay (Eragen) utilizes a novel base in their design that simplifies the steps in the process of amplification and primer-directed strand extension (70,71,73). Evaluation of all three of these commercially available assays, which utilize suspension-bead microarray, has confirmed they have good sensitivity and specificity compared with antigen- and culture-based procedures for detection of respiratory viruses. The MultiCode PLx Respiratory assay (RMA) from EraGen is currently the fastest to a reportable result, but each would give a result within an 8-hour shift (including nucleic acid extraction). Most clinical data has been published for the Luminex xTAG™ RVP assay, which is now FDA and Health Canada approved (and also carries the CE mark). This assay is very easy to set up in a routine viral diagnostic laboratory. Example results for selected viral targets are given in Figure 6 with raw data and interpretation in Table 2. The median fluorescent intensity (MFI) is very easy to interpret and, in general, there is a clear distinction between positive and negative results. The assay includes interpretation software for ease of analysis.

Interpretation of Microarray Results for Respiratory Viruses

The ability to analyze samples for multiple respiratory viruses concurrently in a microarray format has allowed us to appreciate the frequency of dual and triple viral infections (72). Most laboratories have decided to report all viral infections identified, but as more viruses are added to the diagnostic investigation, it is becoming clear that such mixed infections are very common. Whether these infections have a more severe outcome, or just reflect testing in an individual who is sicker, remains to be confirmed.

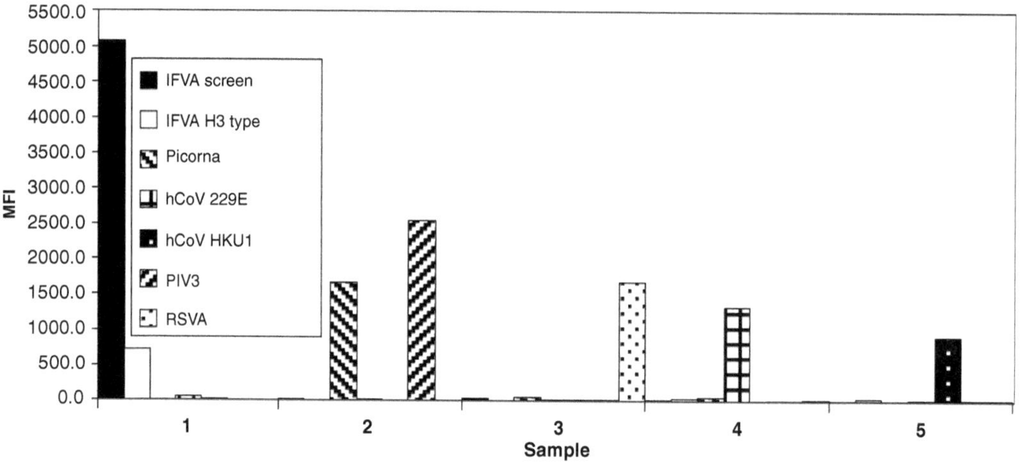

Figure 6 Example results for respiratory specimens analyzed by Luminex xTag™ RVP. Example results for five samples with a limited number of targets included in the assay are shown. Median fluorescence intensity (MFI) clearly falls above (>150) or below the cut-off for each target making interpretation straightforward. See Table 2 for raw data and interpretation for these targets. *Abbreviations*: IFV, influenza virus; Picorna, picornavirus; hCoV, human coronavirus; PIV, parainfluenzavirus; RSV, respiratory syncytial virus.

Viral Central Nervous System Infections

Despite the use of sensitive NATs for viral detection in cases of meningitis, encephalitis, and other neurological symptoms, most cases of presumed viral CNS infections go undiagnosed. This is thought to be because of the large number of potential viral causes with a similar presentation as well as the usually very limited amount of cerebrospinal fluid (CSF) for analysis using individual assays. Many laboratories have moved to a panel-testing approach for viral causes of CNS infections, which inevitably involves multiplexed amplification. One example low-density solid-phase microarray for 13 viral causes of meningitis and encephalitis gave very good results compared with single-virus PCR (36).

Detection of and Differentiation Between Herpesviruses

Herpesviruses are an important cause of CNS disease as well as complications in the immunocompromised. Microarray-based technology has been applied specifically to the detection of and differentiation between human herpesviruses. Low-density oligonucleotide arrays were described to differentiate 7 human herpesviruses from multiplex PCR product mixtures. The methods have been applied successfully to CSF and blood (37–39).

Table 2 Interpretation of Results for Respiratory Specimens Analyzed by Luminex xTag™ RVP

	Median fluorescence intensity for selected targets							
Sample	hCoV 229E	hCoV HKU1	Picorna	IFVA H3 type	IFVA screen	PIV3	RSVA	Result interpretation
1	41.0	20.0	0.0	**715.0**	**5083.0**	8.0	7.0	IFVA H3
2	18.5	1.0	**1675.0**	5.5	14.0	**2544.0**	1.0	Picorna/PIV3
3	18.0	18.0	50.0	0.5	25.0	22.0	**1690.0**	RSVA
4	**1324.5**	0.0	44.0	27.0	13.0	6.0	12.0	hCoV 229E
5	13.5	**906.0**	0.0	29.0	3.0	23.0	21.0	hCoV HKU1

Raw data and interpretation of results for the 5 samples and targets illustrated in Figure 6 are given. Median fluorescence intensity (MFI) clearly falls above (>150) or below the cut-off for each target making interpretation straightforward. Abbreviations as for Figure 6.

Detection and Differentiation Between Poxviruses and Herpesviruses

As part of emergency preparedness, methods have been developed to identify potential agents of bioterror, which would include variola virus (smallpox). A low-density solid-phase microarray was developed to detect and differentiate between orthopoxviruses as well as to address a potential need to differentiate poxvirus infections from those caused by herpes simplex virus (HSV) types 1 and 2 or varicella-zoster virus (VZV) (41). Amplification was undertaken by multiplex asymmetric PCR to enhance the amount of the strand needed to bind to the array. The single-stranded DNA was then chemically labeled with wither Cy3 or Cy5 before binding to oligonucleotide probes against orthopoxviruses and the human alphaherpesviruses.

Enteric Virus Infections

The divergence of viruses such as noroviruses and astroviruses means that multiplexed approaches are necessary to pick up a broad range of naturally circulating variants. A microarray-based detection method was developed for detection of noroviruses and astroviruses (combined with genotyping) (43). Future development of broad-spectrum arrays for diagnosis of viral gastroenteritis would include rotaviruses, sapoviruses, and those ADV serotypes found in individuals with gastrointestinal symptoms.

ANALYSIS OF VIRAL SEROTYPES, VARIANTS, AND GENOTYPES BY ARRAY

Nucleic acid–based arrays have enormous potential for identification of novel viruses and analysis of divergent viral strains and clusters. For many viral infections, the extra information provided by typing is critical to evaluate clinical status and inform treatment.

Human Papillomavirus-Type Differentiation

HPV infections are associated with an increased risk of developing a high-grade cervical intraepithelial lesion. Although array methods are not necessary for HPV screening, they are useful to assign an individual woman's risk of developing cervical carcinoma for genotyping and epidemiological studies. Also, with HPV vaccines being utilized more routinely, it will be important to monitor the circulating types to determine whether type replacement occurs.

A number of different format arrays have been developed for typing of HPV. These include commercially available line-probe/blot assays where oligonucleotides against an increasing number of HPV types are immobilized onto a membrane for analysis of PCR products (Fig. 1 and example Refs. 1–6). The utility of the PathogenMip Assay was demonstrated with an initial set of 24 probes in a microarray targeting the most clinically relevant HPV genotypes associated with cervical cancer progression, with one of the comparative formats being a low-density microarray procedure (7). Low-density "in-house" developed solid-phase microarrays have also been reported by other groups with excellent results for detection of single and mixed HPV-type infections (8–10). One solid-phase procedure for detection of HPV types made use of a novel integrated photodiode array chip with detection of biotinylated, amplified DNA using gold nanoparticle-promoted silver development (11). The microarray was sensitive and has the potential to be cost-effective and miniaturized as the hybridization is measured using electrical signal in each photodiode, obviating the need for expensive scanners.

As HPV diagnostics is proving to be such an important area, other formats for HPV microarray genotyping are being developed. These include automated versions of solid-phase microarrays (e.g., those developed by Autogenomics) and suspension-based microarrays using the Luminex® system, some of which have already been subjected to peer review (62–66). As nucleic acid–based tests are the only approaches available for detection and analysis of HPV; further, detailed comparative studies are needed to ensure that vaccine and epidemiological study data produced from difference microarray systems can be compared between laboratories. As commercially based assays are subjected to regulation (e.g., FDA, Health Canada, CE Mark), this will make standardization of HPV genotyping easier for epidemiological studies.

Analysis of Blood-Borne Viruses by Microarrays

Genotype-specific information, as well as analysis of mutations associated with anti-viral drug resistance, is critical in ensuring appropriate management of HBV, HCV, and HIV-infected patients. The familiar line-probe systems are still the most common ways of assessing genotype and sequence divergence for these important blood-borne pathogens. As the gene targets and mutations associated with resistance expand, the formats in which such analysis is undertaken need to be adapted.

Analysis of HBV

Line-probe assays are still the most common hybridization procedures for analysis of HBV mutations (13,14). A recent study reported the use of an updated line-probe assay for detection of an expanded number of mutations (at 11 positions within the RT region of the polymerase) associated with HBV resistance (Innogenetics). Although results were highly concordant with direct sequencing, the sensitivity was not optimal for all positions (15).

Analysis of HCV

Genotyping HCV is important for patient management and predicting outcome. Such genotyping is usually undertaken in diagnostic laboratories by line-probe assays or sequence-based procedures. The most convenient target for both approaches is the 5′ non-coding region as this is the amplicon used in the majority of diagnostic testing (16). Although this region is useful for assigning genotype by linear array, methods utilizing other regions of the genome are required for accurate sub-typing (16,17). However, microarray procedures are not yet in routine use for this type of HCV multi-gene analysis.

Analysis of HIV

Analysis of mutations that confer resistance to antiretroviral therapy is important for management of HIV-infected individuals. Currently, most laboratories undertake such testing by sequencing procedures, although there are line-probe assays targeting mutations in RT and protease (18–20) as well as some resistance-based microarrays in development (17,19).

The Affymetrix GeneChip HIV assay was one of the first high-density commercial microarrays available to clinical and research laboratories (54). This is a resequencing array with 16,000 unique oligonucleotide probes complementary to the viral RT and protease genes. In one study, the HIV-1 RT line-probe assay (Innogenetics) and the HIV-1 GeneChip were compared with a sequencing system for HIV genotyping using plasma from treatment naïve patients or those failing combination therapy (20). The assays gave good concordance for the codons analyzed by each. Although there was some difference in results for each patient, in general each assay was suitable for identifying clinically relevant mutations in the genes analyzed. However, further enhancement and validation of the GeneChip resequencing array will be necessary in order to ensure accurate analysis of non-clade B viruses (54).

Respiratory Virus Analysis and Tracking Using Microarrays

IFV Subtyping and Strain Identification

Microarrays have been used to provide detailed IFV typing, subtyping, and strain-specific information that will be important for predicting potential zoonotic events and emergence of novel viruses in humans. A low-density solid-phase microarray was developed based on 4 HA, 3 neuraminidase (NA), and 2 M gene targets from IFVA and IFVB. cDNA clones of 500 bp were linked to the solid glass support through covalent binding with a 5′ amino tag, and PCR products were labeled with Cy3 and Cy5 dyes to hybridize to the array. This was one of the earlier developed arrays showing feasibility of this approach for sensitive and specific detection and analysis of IFVs (28). Figure 2 shows a good example of a low-density solid-phase array used for detection and differentiation between subtypes of AIV (24). This array has wide utility for screening studies but has been applied mainly to viruses propagated in embryonated eggs.

Generally, subtyping of IFVA involves direct analysis of HA and NA genes and such analysis is usually separated from the broad detection of all IFVA subtypes using conserved gene detection. In an interesting adaptation of microarray-based methods it was found that one low-density format assay targeting the M gene of IFVA could be used for surveillance of viruses and to infer subtype (25). An illustration of the method used for detection of run-off, fragmented RNA in a low-density microarray is given in Figure 7. An artificial neural network was utilized to automate microarray image interpretation in this procedure (25). Although use of the MChip

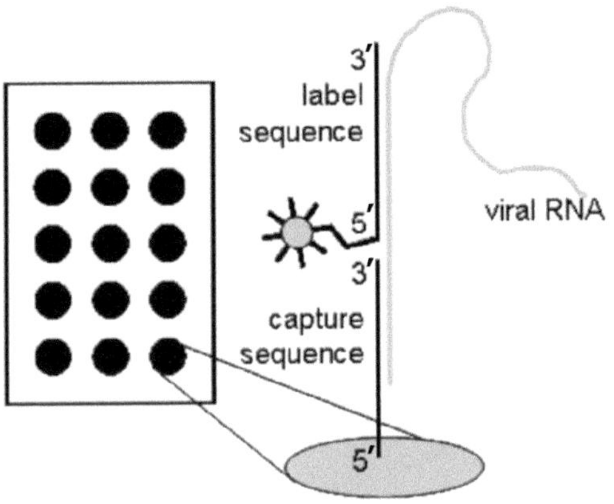

Figure 7 Hybridization method used for detection and differentiation between IFVs. The non-labeled viral RNA hybridizes to a DNA capture sequence on a microarray and the binding event is visualized by a second hybridization step using a specific 5′-Quasar Q570 dye containing a DNA label sequence. Example results using this method are given in Figure 8. *Source*: From Ref. 30.

would not be the best front-line method to identify new recombinant viruses, it could have great utility for simplified surveillance alongside arrays addressing the strain drift within HA and NA included in the original broader IFVA described by the same group (27). Incorporation of amplification and detection methods for anti-viral resistance would be a natural adjunct to IFVA low-density array testing. This has been achieved for two of the most common mutations in the M2 protein associated with adamantane resistance in IFVA: V27A and S31N (29).

An initial study using resequencing arrays demonstrated excellent specificity for detection and analysis of IFVA using cultured viruses (including subtype specificity for IFVA). However, it was clear that the sensitivity of the amplification/array approach may not be good enough to apply to clinical samples directly, and further modifications of the amplification procedure were undertaken for clinical samples. The application of resequencing arrays to tracking of influenza genetic variation confirmed the utility of this approach to inform vaccine development (55).

The high-density GreeneChipResp oligonucleotide array was combined with an amplification protocol with additional amplification primers facilitating detection of IFVA and IFVB, with subtyping of IFVA by recognition of HA types 1–16 and NA types 1–9 (56). The use of modified primers enhanced the sensitivity of this broad-spectrum array for IFV detection and typing on clinical specimens above the random amplification methods, with a reported sensitivity close to that of individual real-time NATs. A third of the 14,795 oligonucleotides in this array are directed towards IFVA subtyping with the advantage that this microarray procedure would be the most useful described to date for identifying a novel (potentially pandemic) virus. Sensitivity of the amplification and array procedure was reported as 1000 copies.

A Flow-Thru Chip (MetriGenix, now Xceed Molecular) was developed for typing and subtyping of IFVs (61). Oligonucleotide probes were immobilized in the microchannels of a silicon wafer. Probes were directed against the IFVA M and H1, H3, H5, N1, and N2 subtypes of the HA and NA genes. Biotinylated amplicons resulting from either multiplex or random reverse transcription-PCR were hybridized to arrayed oligonucleotides with a chemiluminescence detection system.

Influenza B Strain Identification

Many of the IFVAs described above have the added capacity for detection of IFVB. Determination of IFVB lineage and strain-specific information, however, is included in only a small number of low-density arrays and as part of some of the high-density panel virus and resequencing array approaches. As only one of the IFVB lineages is included in the current trivalent seasonal inactivated influenza vaccine, such discrimination is important for vaccine efficacy studies. One part of the low-density IFV microarray assay developed for detection and analysis of IFVs gave clear-cut results to separate the important IFVB lineages. The array detection methodology used is the same as shown in Figure 7. Figure 8 gives example results for discrimination between the Yam88 and Vic87 lineages using probes in the HA gene (30). The array can be clearly read by eye but an artificial neural network was also developed and evaluated for more automated, subjective, and quantitative reading of results.

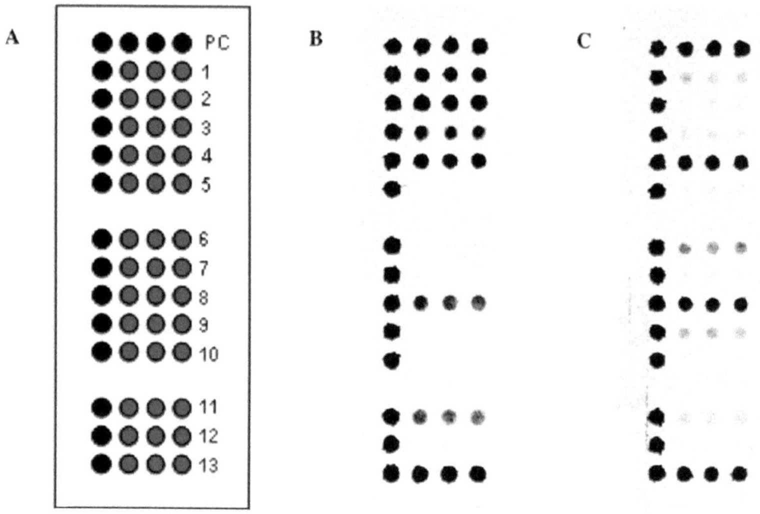

Figure 8 Discrimination between the two major IFVB virus lineages using the HA section of the BChip. (**A**) Microarray layout of the HA section, (**B**) sample B/Johannesburg/5/99 (Yam-88-like), (**C**) sample B/South Carolina/4/2003 (Vic-87-like). For the fluorescence images, darker shades represent higher fluorescence. *Source*: From Ref. 30.

Analysis of Respiratory Virus "Serotype" by Nucleic Acid Array

Culture-based procedures with serotyping of virus isolates using specific antibodies are a long and laborious process. The advent of NATs has led to many diagnostic laboratories changing their culture procedures for viruses such as ADVs, enteroviruses (EVs), and rhinoviruses (RVs) to give a more sensitive, rapid turn around of results. The lack of serotype data, however, makes investigation of potential outbreaks and epidemiologically linked cases problematic. The high capacity of microarrays allows for assignment of "serotype" for viruses based on probe hybridization of amplified products. Such an approach is convenient and may be easier for diagnostic laboratories to perform than formal sequencing reactions. Some examples of microarrays providing "serotype" data are given below.

Picronavirus Serotyping

Picornaviruses are a very diverse virus family with variation from SNPs as well as from recombination. Low-density arrays have been developed and used to detect and differentiate between the EVs coxsackievirus A16 and enterovirus 71, both of which are associated with hand-foot-and-mouth disease as well as outbreaks of neurological symptoms (34). In another study using microarrays, analysis of polioviruses was undertaken to show feasibility of such an approach for tracking of divergent vaccine-derived viruses. The authors reported that this low-density oligonucleotide array had great utility for identifying minority sequences in a mixture, which would be difficult to confirm using other procedures (35).

The high-density Virochip was able to pick up many more RVs from respiratory specimens than viral culture and gave a sensitivity of 97% and specificity of 98% compared with PCR (51). The Virochip hybridization signatures for RVs represented 16 RVA serotypes, 8 RVB serotypes, and a novel third set of 5 divergent RVs. High-density arrays, such as the Virochip, have the potential to enhance our understanding of viral diversity by providing more detailed analysis information than the original serotyping methods.

ADV Serotyping

Distinction of ADV serotypes is based on hemagglutination inhibition assays targeting the surface (hexon) protein, and to date 52 serotypes are recognized. There is some evidence that the original serotype designation of ADVs provided information relevant to severity of infection and was useful in epidemiological assessment of possibly linked cases. A simple low-density microarray procedure was described to address the need for more information beyond detection of ADV in clinical samples (32). The microarray, and associated software to help design the probes, was developed to detect and differentiate the common ADV serotypes associated with respiratory infection and disease in children and adults (serotypes 1, 2, 3, 4, 5, 7, 14, and 21).

Analysis of SARS and Other hCoVs

A low-density microarray was developed and evaluated for generic detection of all hCoVs (31). Probes were included in the microarray for each hCoV group and for individual virus detection using a proprietary technology (Chipron). When applied directly to clinical samples, the method lacked sensitivity for detection of hCoVs 229E, NL63, OC43, and HKU1 in respiratory specimens, but this was resolved by a change to a nested PCR, with results then equivalent to individual real-time RT-PCRs (31).

Random amplification procedures combined with hybridization to the Virochip microarray (48) led to identification of hCoVs in respiratory specimens that were missed by viral culture (51). However, the authors did not undertake parallel individual RT-PCR assays to assess the relative sensitive of the array approach in this context.

The SARS resequencing GeneChip from Affymetrix was developed to interrogate 29,724 bases of SARS hCoV in a single hybridization. The chip comprises eight unique 25-mer probes per base position allowing detection of both known as well as novel SNPs. This proved useful for tracking different genomic changes and serves as a model for other analysis of viruses over time and geography (60). Although such an approach is an extremely powerful method for viral analysis, validation utilized cultured virus and thus the method is not likely, as yet, to be used for front-line viral diagnosis.

Detection and Analysis of Hantaviruses

Detection and analysis of hantaviruses was undertaken using 500-nucleotide overlapping and 250-nucleotide nonoverlapping fragments. It was possible to detect and distinguish between isolates of virus correctly despite 90% sequence similarity (33).

Analysis of Enteric Virus Infections by Microarray

Astroviruses

Human astroviruses are an important cause of gastroenteritis in young children. Microarrays can be useful for analysis of PCR products to assign "serotype" for epidemiological studies as shown in one study where RT-PCR products produced using degenerate primers were hybridized against a microarray consisting of short oligonucleotide probes in a solid-phase format (44).

Rotaviruses

Group A rotaviruses are important causes of diarrhea in young children and infants. Two rotaviruses vaccines have now been licensed and it is important to be able to monitor vaccine efficacy and changes in rotavirus epidemiology. Although originally undertaken by serotyping, the lack of availability of suitable antisera means that epidemiological study of rotaviruses now generally utilizes nucleic acid–based procedures. Classification of rotaviruses by NATs is based on analysis of the surface VP7 and VP4 genes, and the sequence variation and complexity means that such analysis is ideal for array procedures. In one early study, analysis of VP7 G1–G4 and G9 genotypes of rotaviruses was undertaken using a low-density oligonucleotide array (46). In an expanded version of this array, additional targets were added to the original design to define four different VP4 (P) genotypes as well as five additional VP7 genotypes (G5, G6, G8, G10, and G12) (45). This new array allowed sensitive and specific detection of and discrimination between rotavirus genotypes in human stool samples. The method assigned mixed rotavirus infections appropriately and could be used for more detailed analysis of polymorphisms in the VP4 and VP7 genes. The authors compared their results with a previously reported array-based method for rotavirus genotyping that utilized a multiplex low-density oligonucleotide capture array combined with type-specific primer extension (47). In this latter methodology, Cy5 label is only incorporated where there is a match between the PCR product and the 3′ end of the capture oligonucleotide. Such a match results in primer extension and labeling of the array spot that can be read by eye.

Measles Genotyping

An oligonucleotide microarray was developed for detection of and differentiation between measles virus genotypes with analysis of virus directly from clinical samples (42). The low-density oligonucleotide microarray included 71 pairs of probes directed against the nucleoprotein (N) gene sequence. This study reported good sensitivity (90.7%), specificity (100%), and genotype agreement (91.8%) for the new method compared with sequencing. In addition, the microarray demonstrated the ability to identify potential new genotypes based on the similarity of the hybridization patterns with known genotypes.

VZV Genotyping

A rapid and sensitive microarray-based method was used to distinguish the three major circulating genotypes of VZV. Pairs of short oligonucleotide probes with sequences corresponding to all of the observed SNPs in open reading frame 22 were designed and utilized to detect labeled RNA transcripts produced from a PCR process with T7 RNA polymerase sequence on the primers (40).

Smallpox Analysis

Resequencing arrays were designed for detailed analysis of the complete genome sequences of 24 strains of smallpox virus (53). Twenty-two overlapping segments were amplified by long-range PCR to cover the whole genome and the seven GeneChips were used to provide sequence-specific information for 14 different smallpox virus strains. This type of detailed analysis of complete virus genomes by high-density resequencing arrays demonstrates the power of this technology and its possible application to complete, rapid characterization of a smallpox genome in the case of bioterrorism event.

CONCLUSION

Individual real-time NATs are ideal for viral diagnosis if one has a clear idea what pathogen is suspected. However, in some cases it is useful to know more than just whether a virus is present in order to interpret results and for appropriate patient management. Discovery of new, clinically relevant viruses means we can enhance diagnosis, but adding more targets gets logistically more difficult and expensive if using individual or small multiplex assays. Microarray technology is beginning to have an impact on viral diagnostics, particularly where a "syndromic" approach to testing is needed. Broad-spectrum respiratory virus microarrays are now available in multiple formats, some of which already have FDA and other regulatory approvals for diagnostic use. Those that have been incorporated into routine testing have proven utility for individual patient diagnosis and for assessment of viral etiology in respiratory outbreaks.

In the next few years we will see further application of high-density broad virus microarrays for a "catch all" approach, although at the moment the cost of such high-density arrays means they will likely be a final analysis after individual NATs or low-density arrays have not identified a viral cause. Eventually, microarray-based methods for virus discovery, detection, and analysis will overlap, and such technology will become more accessible. This will provide the diagnostic laboratory with extremely powerful virus detection and analysis tools to inform vaccine and anti-viral developments as well as clinical and patient management protocols.

REFERENCES

1. Castle PE, Gravitt PE, Solomon D, et al. Comparison of linear array and line blot assay for detection of human papillomavirus and diagnosis of cervical precancer and cancer in the atypical squamous cell of undetermined significance and low-grade squamous intraepithelial lesion triage study. J Clin Microbiol 2008; 46(1):109–117.
2. Castle PE, Porras C, Quint WG, et al. Comparison of two PCR-based human papillomavirus genotyping methods. J Clin Microbiol 2008; 46(10):3437–3445.
3. Koidl C, Bozic M, Hadzisejdic I, et al. Comparison of molecular assays for detection and typing of human papillomavirus. Am J Obstet Gynecol 2008; 199(2):144–146.

4. Sandri MT, Riggio D, Salvatici M, et al. Typing of human papillomavirus in women with cervical lesions: Prevalence and distribution of different genotypes. J Med Virol 2009; 81(2):271–277.
5. Stevens MP, Garland SM, Tabrizi SN. Validation of an automated detection platform for use with the roche linear array human papillomavirus genotyping test. J Clin Microbiol 2008; 46(11):3813–3816.
6. Woo YL, Damay I, Stanley M, et al. The use of HPV Linear Array Assay for multiple HPV typing on archival frozen tissue and DNA specimens. J Virol Methods 2007; 142(1–2):226–230.
7. Akhras MS, Thiyagarajan S, Villablanca AC, et al. PathogenMip assay: A multiplex pathogen detection assay. PLoS One 2007; 2(2):e223.
8. Albrecht V, Chevallier A, Magnone V, et al. Easy and fast detection and genotyping of high-risk human papillomavirus by dedicated DNA microarrays. J Virol Methods 2006; 137(2):236–244.
9. Lee HS, Kim KM, Kim SM, et al. Human papillomavirus genotyping using HPV DNA chip analysis in Korean women. Int J Gynecol Cancer 2007; 17(2):497–501.
10. Choi BS, Kim O, Park MS, et al. Genital human papillomavirus genotyping by HPV oligonucleotide microarray in Korean commercial sex workers. J Med Virol 2003; 71(3):440–445.
11. Baek TJ, Park PY, Han KN, et al. Development of a photodiode array biochip using a bipolar semiconductor and its application to detection of human papilloma virus. Anal Bioanal Chem 2008; 390(5):1373–1378.
12. Hsia CC, Chizhikov VE, Yang AX, et al. Microarray multiplex assay for the simultaneous detection and discrimination of hepatitis B, hepatitis C, and human immunodeficiency type-1 viruses in human blood samples. Biochem Biophys Res Commun 2007; 356(4):1017–1023.
13. Libbrecht E, Doutreloigne J, Van De Velde H, et al. Evolution of primary and compensatory lamivudine resistance mutations in chronic hepatitis B virus-infected patients during long-term lamivudine treatment, assessed by a line probe assay. J Clin Microbiol 2007; 45(12):3935–3941.
14. Olivero A, Ciancio A, Abate ML, et al. Performance of sequence analysis, INNO-LiPA line probe assays and AFFIGENE assays in the detection of hepatitis B virus polymerase and precore/core promoter mutations. J Viral Hepat 2006; 13(6):355–362.
15. Degertekin B, Hussain M, Tan J, et al. Sensitivity and accuracy of an updated line probe assay (HBV DR v.3) in detecting mutations associated with hepatitis B antiviral resistance. J Hepatol 2009; 50(1):42–48.
16. Nolte FS, Green AM, Fiebelkorn KR, et al. Clinical evaluation of two methods for genotyping hepatitis C virus based on analysis of the 5′ noncoding region. J Clin Microbiol 2003; 41(4):1558–1564.
17. Domiati-Saad R, Scheuermann RH. Nucleic acid testing for viral burden and viral genotyping. Clin Chim Acta 2006; 363(1–2):197–205.
18. Jlizi A, Ben Ammar El GA, Slim A, et al. Profile of drug resistance mutations among HIV-1-infected Tunisian subjects failing antiretroviral therapy. Arch Virol 2008; 153(6):1103–1108.
19. Sen S, Tripathy SP, Paranjape RS. Antiretroviral drug resistance testing. J Postgrad Med 2006; 52(3): 187–193.
20. Wilson JW, Bean P, Robins T, et al. Comparative evaluation of three human immunodeficiency virus genotyping systems: The HIV-GenotypR method, the HIV PRT GeneChip assay, and the HIV-1 RT line probe assay. J Clin Microbiol 2000; 38(8):3022–3028.
21. Lodes MJ, Suciu D, Wilmoth JL, et al. Identification of upper respiratory tract pathogens using electrochemical detection on an oligonucleotide microarray. PLoS One 2007; 2(9):e924.
22. Causse M, Garcia-Mayorgas AD, Gutierrez JB, et al. Preliminary evaluation of a multiplex reverse transcription-PCR assay combined with a new DNA chip hybridization assay for detecting respiratory syncytial virus. J Clin Microbiol 2007; 45(11):3811–3813.
23. Coiras MT, Lopez-Huertas MR, Lopez-Campos G, et al. Oligonucleotide array for simultaneous detection of respiratory viruses using a reverse-line blot hybridization assay. J Med Virol 2005; 76(2):256–264.
24. Wang LC, Pan CH, Severinghaus LL, et al. Simultaneous detection and differentiation of Newcastle disease and avian influenza viruses using oligonucleotide microarrays. Vet Microbiol 2008; 127(3–4): 217–226.
25. Dawson ED, Moore CL, Smagala JA, et al. MChip: A tool for influenza surveillance. Anal Chem 2006; 78(22):7610–7615.
26. Dawson ED, Moore CL, Dankbar DM, et al. Identification of A/H5N1 influenza viruses using a single gene diagnostic microarray. Anal Chem 2007; 79(1):378–384.
27. Townsend MB, Dawson ED, Mehlmann M, et al. Experimental evaluation of the FluChip diagnostic microarray for influenza virus surveillance. J Clin Microbiol 2006; 44(8):2863–2871.
28. Li J, Chen S, Evans DH. Typing and subtyping influenza virus using DNA microarrays and multiplex reverse transcriptase PCR. J Clin Microbiol 2001; 39(2):696–704.
29. Townsend MB, Smagala JA, Dawson ED, et al. Detection of adamantane-resistant influenza on a microarray. J Clin Virol 2008; 42(2):117–123.
30. Dankbar DM, Dawson ED, Mehlmann M, et al. Diagnostic microarray for influenza B viruses. Anal Chem 2007; 79(5):2084–2090.

31. de Souza Luna LK, Heiser V, Regamey N, et al. Generic detection of coronaviruses and differentiation at the prototype strain level by reverse transcription-PCR and nonfluorescent low-density microarray. J Clin Microbiol 2007; 45(3):1049–1052.
32. Lopez-Campos G, Coiras M, Sanchez-Merino JP, et al. Oligonucleotide microarray design for detection and serotyping of human respiratory adenoviruses by using a virtual amplicon retrieval software. J Virol Methods 2007; 145(2):127–136.
33. Nordstrom H, Johansson P, Li QG, et al. Microarray technology for identification and distinction of hantaviruses. J Med Virol 2004; 72(4):646–655.
34. Chen TC, Chen GW, Hsiung CA, et al. Combining multiplex reverse transcription-PCR and a diagnostic microarray to detect and differentiate enterovirus 71 and coxsackievirus A16. J Clin Microbiol 2006; 44(6):2212–2219.
35. Cherkasova E, Laassri M, Chizhikov V, et al. Microarray analysis of evolution of RNA viruses: Evidence of circulation of virulent highly divergent vaccine-derived polioviruses. Proc Natl Acad Sci U S A 2003; 100(16):9398–9403.
36. Boriskin YS, Rice PS, Stabler RA, et al. DNA microarrays for virus detection in cases of central nervous system infection. J Clin Microbiol 2004; 42(12):5811–5818.
37. Jaaskelainen AJ, Piiparinen H, Lappalainen M, et al. Improved multiplex-PCR and microarray for herpesvirus detection from CSF. J Clin Virol 2008; 42(2):172–175.
38. Jaaskelainen AJ, Piiparinen H, Lappalainen M, et al. Multiplex-PCR and oligonucleotide microarray for detection of eight different herpesviruses from clinical specimens. J Clin Virol 2006; 37(2):83–90.
39. Zheng ZB, Wu YD, Yu XL, et al. DNA microarray technology for simultaneous detection and species identification of seven human herpes viruses. J Med Virol 2008; 80(6):1042–1050.
40. Sergeev N, Rubtcova E, Chizikov V, et al. New mosaic subgenotype of varicella-zoster virus in the USA: VZV detection and genotyping by oligonucleotide-microarray. J Virol Methods 2006; 136(1–2):8–16.
41. Ryabinin VA, Shundrin LA, Kostina EB, et al. Microarray assay for detection and discrimination of Orthopoxvirus species. J Med Virol 2006; 78(10):1325–1340.
42. Neverov AA, Riddell MA, Moss WJ, et al. Genotyping of measles virus in clinical specimens on the basis of oligonucleotide microarray hybridization patterns. J Clin Microbiol 2006; 44(10): 3752–3759.
43. Jaaskelainen AJ, Maunula L. Applicability of microarray technique for the detection of noro- and astroviruses. J Virol Methods 2006; 136(1–2):210–216.
44. Brown DW, Gunning KB, Henry DM, et al. A DNA oligonucleotide microarray for detecting human astrovirus serotypes. J Virol Methods 2008; 147(1):86–92.
45. Honma S, Chizhikov V, Santos N, et al. Development and validation of DNA microarray for genotyping group A rotavirus VP4 (P[4], P[6], P[8], P[9], and P[14]) and VP7 (G1 to G6, G8 to G10, and G12) genes. J Clin Microbiol 2007; 45(8):2641–2648.
46. Chizhikov V, Wagner M, Ivshina A, et al. Detection and genotyping of human group A rotaviruses by oligonucleotide microarray hybridization. J Clin Microbiol 2002; 40(7):2398–2407.
47. Lovmar L, Fock C, Espinoza F, et al. Microarrays for genotyping human group a rotavirus by multiplex capture and type-specific primer extension. J Clin Microbiol 2003; 41(11):5153–5158.
48. Wang D, Coscoy L, Zylberberg M, et al. Microarray-based detection and genotyping of viral pathogens. Proc Natl Acad Sci U S A 2002; 99(24):15687–15692.
49. Wang D, Urisman A, Liu YT, et al. Viral discovery and sequence recovery using DNA microarrays. PLoS Biol 2003; 1(2):E2.
50. Palacios G, Quan PL, Jabado OJ, et al. Panmicrobial oligonucleotide array for diagnosis of infectious diseases. Emerg Infect Dis 2007; 13(1):73–81.
51. Kistler A, Avila PC, Rouskin S, et al. Pan-viral screening of respiratory tract infections in adults with and without asthma reveals unexpected human coronavirus and human rhinovirus diversity. J Infect Dis 2007; 196(6):817–825.
52. Wilson WJ, Strout CL, DeSantis TZ, et al. Sequence-specific identification of 18 pathogenic microorganisms using microarray technology. Mol Cell Probes 2002; 16(2):119–127.
53. Sulaiman IM, Tang K, Osborne J, et al. GeneChip resequencing of the smallpox virus genome can identify novel strains: A biodefense application. J Clin Microbiol 2007; 45(2):358–363.
54. Vahey M, Nau ME, Barrick S, et al. Performance of the Affymetrix GeneChip HIV PRT 440 platform for antiretroviral drug resistance genotyping of human immunodeficiency virus type 1 clades and viral isolates with length polymorphisms. J Clin Microbiol 1999; 37(8):2533–2537.
55. Wang Z, Daum LT, Vora GJ, et al. Identifying influenza viruses with resequencing microarrays. Emerg Infect Dis 2006; 12(4):638–646.
56. Quan PL, Palacios G, Jabado OJ, et al. Detection of respiratory viruses and subtype identification of influenza A viruses by GreeneChipResp oligonucleotide microarray. J Clin Microbiol 2007; 45(8): 2359–2364.

57. Lin B, Wang Z, Vora GJ, et al. Broad-spectrum respiratory tract pathogen identification using resequencing DNA microarrays. Genome Res 2006; 16(4):527–535.
58. Lin B, Blaney KM, Malanoski AP, et al. Using a resequencing microarray as a multiple respiratory pathogen detection assay. J Clin Microbiol 2007; 45(2):443–452.
59. Rota PA, Oberste MS, Monroe SS, et al. Characterization of a novel coronavirus associated with severe acute respiratory syndrome. Science 2003; 300(5624):1394–1399.
60. Sulaiman IM, Liu X, Frace M, et al. Evaluation of affymetrix severe acute respiratory syndrome resequencing GeneChips in characterization of the genomes of two strains of coronavirus infecting humans. Appl Environ Microbiol 2006; 72(1):207–211.
61. Kessler N, Ferraris O, Palmer K, et al. Use of the DNA flow-thru chip, a three-dimensional biochip, for typing and subtyping of influenza viruses. J Clin Microbiol 2004; 42(5):2173–2185.
62. Antonishyn NA, Horsman GB, Kelln RA, et al. The impact of the distribution of human papillomavirus types and associated high-risk lesions in a colposcopy population for monitoring vaccine efficacy. Arch Pathol Lab Med 2008; 132(1):54–60.
63. Jiang HL, Zhu HH, Zhou LF, et al. Genotyping of human papillomavirus in cervical lesions by L1 consensus PCR and the Luminex xMAP system. J Med Microbiol 2006; 55(Pt 6):715–720.
64. Nazarenko I, Kobayashi L, Giles J, et al. A novel method of HPV genotyping using Hybrid Capture sample preparation method combined with GP5+/6+ PCR and multiplex detection on Luminex XMAP. J Virol Methods 2008; 154(1–2):76–81.
65. Oh Y, Bae SM, Kim YW, et al. Polymerase chain reaction-based fluorescent Luminex assay to detect the presence of human papillomavirus types. Cancer Sci 2007; 98(4):549–554.
66. Schmitt M, Bravo IG, Snijders PJ, et al. Bead-based multiplex genotyping of human papillomaviruses. J Clin Microbiol 2006; 44(2):504–512.
67. Brunstein J, Thomas E. Direct screening of clinical specimens for multiple respiratory pathogens using the genaco respiratory panels 1 and 2. Diagn Mol Pathol 2006; 15(3):169–173.
68. Mahony J, Chong S, Merante F, et al. Development of a respiratory virus panel test for detection of twenty human respiratory viruses by use of multiplex PCR and a fluid microbead-based assay. J Clin Microbiol 2007; 45(9):2965–2970.
69. Pabbaraju K, Tokaryk KL, Wong S, et al. Comparison of the Luminex xTAG respiratory viral panel with in-house nucleic acid amplification tests for diagnosis of respiratory virus infections. J Clin Microbiol 2008; 46(9):3056–3062.
70. Nolte FS, Marshall DJ, Rasberry C, et al. MultiCode-PLx system for multiplexed detection of seventeen respiratory viruses. J Clin Microbiol 2007; 45(9):2779–2786.
71. Lee WM, Grindle K, Pappas T, et al. High-throughput, sensitive, and accurate multiplex PCR-microsphere flow cytometry system for large-scale comprehensive detection of respiratory viruses. J Clin Microbiol 2007; 45(8):2626–2634.
72. Brunstein JD, Cline CL, McKinney S, et al. Evidence from multiplex molecular assays for complex multipathogen interactions in acute respiratory infections. J Clin Microbiol 2008; 46(1):97–102.
73. Marshall DJ, Reisdorf E, Harms G, et al. Evaluation of a multiplexed PCR assay for detection of respiratory viral pathogens in a public health laboratory setting. J Clin Microbiol 2007; 45(12):3875–3882.
74. Li H, McCormac MA, Estes RW, et al. Simultaneous detection and high-throughput identification of a panel of RNA viruses causing respiratory tract infections. J Clin Microbiol 2007; 45(7):2105–2109.

7 Virus Isolation

Diane Leland

Department of Pathology and Laboratory Medicine, Indiana University School of Medicine, Indianapolis, Indiana, U.S.A.

Marie Louise Landry

Department of Laboratory Medicine, Yale University School of Medicine, New Haven, Connecticut, U.S.A.

INTRODUCTION

Viruses are obligate intracellular parasites that proliferate only within living cells. Early virus isolation and identification work required human volunteers, laboratory animals, or embryonated eggs (Table 1). Although animals remain useful for detection of certain groups of viruses, they are cumbersome and expensive to maintain in clinical practice. Suckling mice should be inoculated via the intracranial or intraperitoneal route at 24 to 48 hours of age for the most sensitive results. Thus, ready access to colonies of pregnant mice is required. Embryonated hens' eggs are less expensive than laboratory animals but also require inoculation at specified ages, and unused eggs must be discarded before they hatch. Depending on the virus, allantoic or amniotic cavities or chorioallantoic membrane can be inoculated. Details for use of these systems for virus isolation have been published previously (1). Currently, these methods are reserved for specialized public health, reference, or research laboratories, or in the case of embryonated eggs, for vaccine production. However, they are too expensive and inconvenient for use in the routine diagnostic virology laboratory.

With the discovery in the early 1900s that human cells could be propagated in vitro, a new source of large numbers of cells suitable for virus isolation was identified. As early as 1913, vaccinia virus (2) was grown in cell cultures, and in the 1930s both smallpox (3) and yellow fever virus (4) were propagated in cell cultures for the purpose of vaccine production. Interest in using cell cultures for virus isolation expanded in the 1950s when it was discovered that polioviruses would proliferate in cell cultures that were not of neural origin (5,6). By the early 1960s, virus isolation in cell cultures was employed by research laboratories and in some major medical centers.

Addition of antibiotics to cell culture media, the development of chemically defined culture media, and the use of cell-dispensing equipment for preparing replicate cultures also contributed to the increased interest in the use of cell culture (7). With the commercial availability of cell lines in the early 1970s, diagnostic virology expanded dramatically (8). Virus isolation in cell cultures in monolayers soon became the "gold standard" method for virus detection.

CONVENTIONAL CELL CULTURE

Although cell cultures can be purchased or prepared in a variety of containers, the 16 × 125 mm glass round-bottom screw-capped tube is standard, with the cell monolayer adhering from the midpoint to the bottom of one side of the tube. At this writing, there are two manufacturers of ready-to-use cell cultures in the United States: Diagnostic Hybrids, Inc., Athens, OH and ViroMed Laboratories, Minnetonka, MN. Many diagnostic virology laboratories purchase all of their cell cultures, although in other settings, virologists still prepare some or most of their cell cultures by subpassaging some cell lines in house.

Types of Cell Cultures

Viruses infect cells by attachment to receptors on the cell surface. The type of receptor needed for this attachment varies from virus to virus and among different cell types. Clinical virology laboratories carry a limited number of cell lines, usually three to six, depending on the viruses targeted. Typically, several different cell lines are inoculated with each clinical sample. Basic categories of cell lines include primary, diploid, and heteroploid or continuous cells (Table 2).

Table 1 Utility of Embryonated Eggs and Newborn Mice for Virus Isolation from Clinical Specimens

Culture method	Viruses isolated from clinical specimens
Embryonated hens' eggs amniotic[a] and/or allantoic cavity	Influenza viruses and mumps virus
Embryonated hens' eggs chorioallantoic membrane (CAM)	Poxviruses and herpes simplex viruses
Newborn (suckling) mice	Arboviruses, coxsackievirus groups A and B, herpes simplex viruses

[a]Inoculation of amniotic cavity is preferable to allantoic cavity for primary isolation of influenza virus.

Primary cells are the first generation of cells that grow from the tissues of origin and can usually be subcultured for only one or two passages. An example of primary cells used routinely in virology laboratories in the United States is rhesus monkey kidney cells (RhMK). Diploid cell lines are usually derived from human fetal or newborn tissues, maintain their diploid chromosome number during passaging, and can be subcultured 20 to 50 times before senescence. MRC-5 and human foreskin fibroblasts are examples of diploid cell lines. Heteroploid or continuous cell lines may originate from human or animal tissues, and usually arise from tumors or infection with a transforming virus, but may result from spontaneous transformation of normal tissues. These have a heteroploid chromosome number and can be subcultured indefinitely. A549 and HEp-2 cells are examples of heteroploid cell lines. The number and types of cell culture tubes inoculated for each clinical specimen depend on the specimen source and the viral suspects indicated by the ordering physician.

Specimen Collection, Transport, and Processing

The collection and transport of clinical samples to the laboratory are of critical importance. Improper site, collection method, or transport conditions can inactivate viruses, resulting in falsely negative viral culture results. Specimens may be collected from the site of viral pathology or from a site where the virus may be replicating and shed asymptomatically. For example, in enteroviral meningitis, a sample of cerebrospinal fluid—the site of pathology—may be useful in detecting the infecting enterovirus; however, samples such as stool from the gastrointestinal tract—where the virus replicates and is shed in higher titers—may also be excellent sources of virus. Swabs and biopsies are placed in viral transport medium to prevent drying, with antibiotics to prevent overgrowth of microbial flora. The latter is particularly important for samples collected from body sites such as skin, respiratory tract, and genital or gastrointestinal tract, which are contaminated with microbial flora. Upon arrival in the laboratory, swabs, biopsies, stools, and aspirates may be further processed by centrifugation, addition of antibiotics, or filtering. In contrast, body fluids collected from routinely sterile sites such as cerebrospinal fluid, urine, and blood are not placed in transport medium with antibiotics. While CSF requires little if any processing prior to inoculation into cell cultures, urine and blood often require pH

Table 2 Common Cell Lines Used in Clinical Virology Laboratories

Type	Examples
Primary (1 to 3 passages)	Rhesus monkey kidney, cynomolgus monkey kidney[a], African green monkey kidney, rabbit kidney
Heteroploid (continuous passage)	
tumor derived	Hep-2, A549, RD
spontaneous transformation	MDCK, LLCMK2, Vero[b], BGMK
Diploid, embryonic or newborn (20–50 passages)	MRC-5, WI-38, HDF, HELF, human foreskin (e.g., HFF, HNF)

Note: Cell lines from different sources and at different passage levels can vary in their sensitivity to virus infection.
[a]Two recently isolated viruses, human metapneumovirus and NL-63 coronavirus, were first isolated in tertiary cynomolgus monkey kidney cells.
[b]Vero E6 cells are used to isolate filoviruses, arenaviruses, tick-borne flaviviruses, bunyaviruses, henipaviruses, and SARS coronavirus.

adjustment (urine) or dilution to reduce toxicity. Rapid transport of samples to the virology laboratory is important in keeping viruses viable. Ideally, transport time should not exceed four hours. Samples other than peripheral blood samples should be kept cold (on wet ice or cold packs or refrigerated at 2–8°C) and moist between collection and arrival in the laboratory if the transport process takes three or fewer days. If the sample cannot be processed in the virology laboratory within three days of collection, the sample should be frozen at −70°C or colder and transported on dry ice in a sealed container (9). Guidelines for specimen collection, transport, and processing have been published previously (10).

Inoculation and Incubation

After appropriate preparation of the clinical sample, the processed inoculum is added to the cell culture tube, either by simply dispensing the sample into the tube containing media or by first decanting the cell culture medium and then applying the inoculum directly to the cell culture monolayer and allowing it to adsorb. In the adsorption method, after one hour of incubation of the inoculated tube in a horizontal position at 35°C to 37°C, excess inoculum can either be discarded or left in place, and fresh cell culture medium added (11). The adsorption method is more time consuming but allows enhanced recovery of low levels of virus.

Inoculated cell culture tubes are incubated at 35°C to 37°C, although many rhinoviruses may proliferate more efficiently at 33°C. A rotating rack (roller drum) provides motion that has been shown to enhance the viral replication process. Tubes are held at a slight angle of 5° to 7° to keep the top of the tubes higher than the bottom. This keeps the cell culture medium in the bottom half of the tube where it is needed for nourishing the cell monolayer and prevents the medium from collecting in or near the cap of the tube where it can serve as a route for entry of contaminants into the tubes. Rotation speeds of 0.2 to 2 revolutions per minute have been shown to be acceptable (12). The rotation aerates the cell monolayers and disperses newly released virus particles into the cell culture medium.

Stationary slanted racks may be used for cell culture incubation if roller drums are not available. Like the rotating rack of the roller drum, the stationary rack is slanted to keep the cell culture medium in the bottom half of the tubes. When stationary racks are used, cell culture tubes must be positioned in the rack with the cell monolayer on the lower surface of the tube to ensure that the cell culture medium covers the cells. Accidental mispositioning of the monolayer on the upper surface will result in drying and eventual death of the monolayer cells. Commercially purchased cell culture tubes bear an insignia or label on the side of the tube opposite the monolayer. Tubes should be positioned so that the insignia faces upward, thus ensuring that the cell monolayer is in the correct position and will be covered with medium.

An uninoculated cell culture tube from each lot of cell cultures should be incubated along with the inoculated cultures to serve as "control" tubes. Any changes in the appearance of the cells in the uninoculated tubes may signal the presence of endogenous viruses or contaminants or indicate that the lot of cell cultures is of poor quality.

The length of the incubation period and the frequency of microscopic examination of tubes may vary depending on the target virus(es). For routine cultures with no specified viral suspect, culture tubes are routinely incubated for 10 to 14 days and examined daily for the first 5 to 7 days and on alternate days thereafter. If herpes simplex virus (HSV) alone is targeted, tubes can be incubated for a shorter period of seven days because HSV proliferates rapidly, and the tubes should be examined daily. Cytomegalovirus (CMV) may take 7 to 21 days or longer to produce visible signs of proliferation, and thus requires an extended incubation period. CMV cultures may be examined on alternate days during the first two weeks and only once per week for weeks 3 and 4 of incubation.

Microscopic Examination of Cell Culture Monolayers

The microscopic exam of unstained cell culture monolayers has long been the standard approach for detecting viral proliferation. Degenerative changes ranging from swelling, shrinking, and rounding of cells to clustering and syncytia formation, and, in some cases, to complete destruction of the monolayer may be produced by viruses. These changes are collectively called the cytopathogenic or cytopathic effect (CPE) of the virus. The rate and patterns of CPE induced by various viruses are dependent upon (*i*) the type of cell cultures used, (*ii*) the concentration of virus in the specimen, and (*iii*) the properties of the individual virus. The types of CPE described below are observed most commonly in unstained preparations.

Vacuoles—large, frothy, bubble-like areas usually in the cytoplasm of infected cells.

Syncytia—large cell masses which may contain up to 100 nuclei. They are sometimes called "giant cells" and result from fusion of virus-infected cells, which facilitate cell-to-cell spread of the virus.

General morphologic changes—rounding, swelling, shrinking, or forming grape-like clusters.

Loss of adherence—loss of adherence of cells to the surface of the culture vessel. Cells may float free in the culture medium, leaving clear areas or fine prolongations.

Cellular granulation—cells have a dark, rough, finely speckled appearance. This granulation may be confused with nonspecific degeneration or aging of the cell culture.

The extent of monolayer involvement by CPE is routinely estimated and scored as follows: 4+ = 100% of cells in monolayer affected; 3+ = 75% of cells in monolayer affected; 2+ = 50% of cells in monolayer affected; 1+ = 25% of cells in monolayer affected (localized areas); or the term "focal" should be used to describe CPE that is restricted to infrequent small individual areas of involvement that represent less than 25% of the monolayer.

Hemadsorption

An alternative approach for detecting viral presence is hemadsorption (HAD) testing. HAD is useful only for viruses such as influenza, parainfluenza, and mumps that express their hemagglutinating proteins on the plasma membranes of virus-infected cells. These proteins can be detected by their affinity for erythrocytes. HAD testing is routinely performed if CPE appears and a hemadsorbing virus is suspected or at days 3 and 7 of incubation and at the end of the incubation period for cell cultures that fail to produce CPE (13). Hemadsorbing foci have been found in human fetal lung diploid cell cultures within 12 hours after inoculation with influenza viruses A and B (14). In HAD testing, the cell culture medium is removed and replaced with a dilute suspension of erythrocytes, usually guinea pig erythrocytes, and the cell culture tubes are incubated at 4°C for 30 minutes (11). For parainfluenza type 4, hemadsorption at room temperature is recommended. Tubes are then rinsed and examined microscopically (1). If a hemadsorbing virus is present, erythrocytes will adhere in clumps to the infected areas of the cell monolayer (Fig. 1). Erythrocytes will not adhere to uninfected cells or to cells infected by nonhemadsorbing viruses. Nonadherent erythrocytes float free when the cell culture tube is tapped or rotated. Nonspecific scattered adsorption can be seen when aged erythrocytes are used. Although only a few human viral pathogens produce a positive HAD result, confirmatory testing of all HAD-positive cell cultures is required to differentiate among the hemadsorbing viruses.

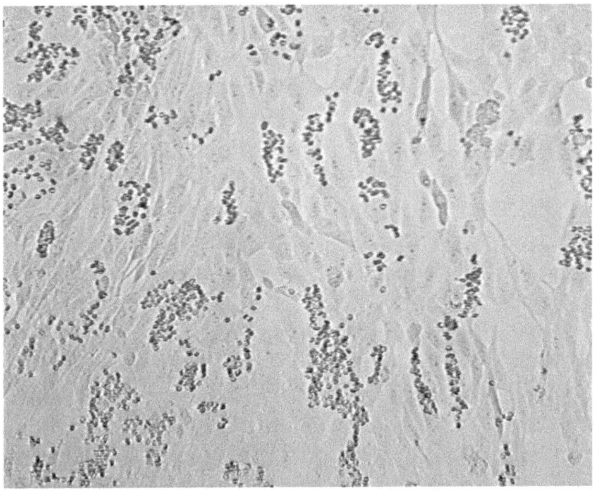

Figure 1 Positive hemadsorption result in parainfluenza-infected RhMK cells (100×).

Definitive Identification of Viral Isolates

An experienced virologist can usually predict which virus is present based on the characteristics of the CPE, the HAD result, the cell line involved, the length of incubation, and the type of clinical specimen, but confirmatory testing is needed to make a definitive viral identification. Confirmatory testing of viruses detected by CPE or HAD has traditionally been based on the reaction of antibodies of known specificity with viral antigens expressed in the infected cells (e.g., neutralization or hemadsorption inhibition tests). At present, virus identification is largely confined to immunofluorescence (IF) techniques that use fluorescein isothiocyanate (FITC)-labeled monoclonal antibodies (Mabs). The cells are scraped from the infected monolayer and placed on a microscope slide. The preparation is fixed in acetone and then flooded with Mabs of known specificity. Binding of antibodies to viral proteins is signaled by the presence of fluorescence when the preparation is viewed using the fluorescence microscope. The type of fluorescence (e.g., speckled vs. confluent) and the location of the fluorescence in the cell (e.g., nuclear and/or cytoplasmic) is also useful in differentiating certain viruses. This process takes only one to two hours and gives a sensitive and specific viral identification. At present, the following viruses can be identified by IF methods when detected in cell cultures: adenovirus (group only, not type), enteroviruses (groups only, not type), HSV types 1 and 2, influenza types A and B, measles virus, mumps virus, parainfluenza virus types 1, 2, 3, and 4, respiratory syncytial virus, human metapneumovirus, cytomegalovirus (CMV), and varicella zoster virus (VZV).

Unfortunately IF reagents are not available for definitively identifying certain viruses. These, notably the coxsackie, polio, and echoviruses—which are closely related and have numerous serotypes—may be identified as "enterovirus" by IF, but serotype identification requires neutralization. In neutralization testing, a standardized amount of the unknown virus is incubated with antibodies of known specificity; then an aliquot of the mixture is inoculated into susceptible cell cultures, and the cell cultures are observed for evidence of viral proliferation. If the virus produces CPE, the antibodies did not bind and inactivate or neutralize the virus. Conversely, if the virus is unable to produce CPE, the virus infectivity is neutralized. The identity of the virus is then established by the specificity of the antibody used. This is a cumbersome procedure, requiring virus titration prior to the start of the procedure to select a challenge dose and a lengthy incubation after inoculation. Although neutralization testing may be used in identifying a wide variety of viruses, it is used only when less cumbersome, more rapid methods are not available, or when serotype identification is required.

Viruses Isolated in Cell Cultures

A limited number of common human viral pathogens will produce CPE in standard cell cultures. These are adenovirus, CMV, enteroviruses except many coxsackievirus group A serotypes, HSV 1 and 2, influenza A and B, parainfluenza 1, 2, 3, and 4, respiratory syncytial virus, rhinovirus, and VZV. Characteristic CPE of some of these is shown in Figure 2. The cell lines that routinely show CPE, characteristics of the typical CPE, and the time to appearance of CPE for each of these viruses is shown in Table 3 (15).

Measles and mumps viruses will also produce CPE in cell cultures. These viruses are seldom seen in laboratories in the United States due to the success of vaccine programs in dramatically reducing the incidence of infections with these viruses. However, recent outbreaks have been reported in the United States for both measles (16) and mumps virus (17). Measles virus will produce syncytia and generalized deterioration in Vero or primary monkey kidney cells in 7 to 10 days but proliferates most effectively in monolayers of B95a cells. These are Epstein–Barr virus-transformed B lymphoblastoid cells derived from marmoset lymphocytes and are estimated to be 10,000 times more sensitive than other cell lines for measles virus isolation (18). Mumps virus proliferates in traditional cell cultures of primary monkey kidney, human neonatal kidney, HeLa, and Vero, characteristically showing rounding of cells and multinucleated giant cells in six to eight days (19). However, the B95a cell line has been shown to be as sensitive as primary monkey kidney cells for mumps isolation (20).

Several viruses that are seldom seen in the United States will proliferate in standard cell cultures, and U.S. laboratories have been involved in identifying these agents. Chikungunya virus, a mosquito-borne virus common in Africa, India, and the Indiana Ocean islands, has been

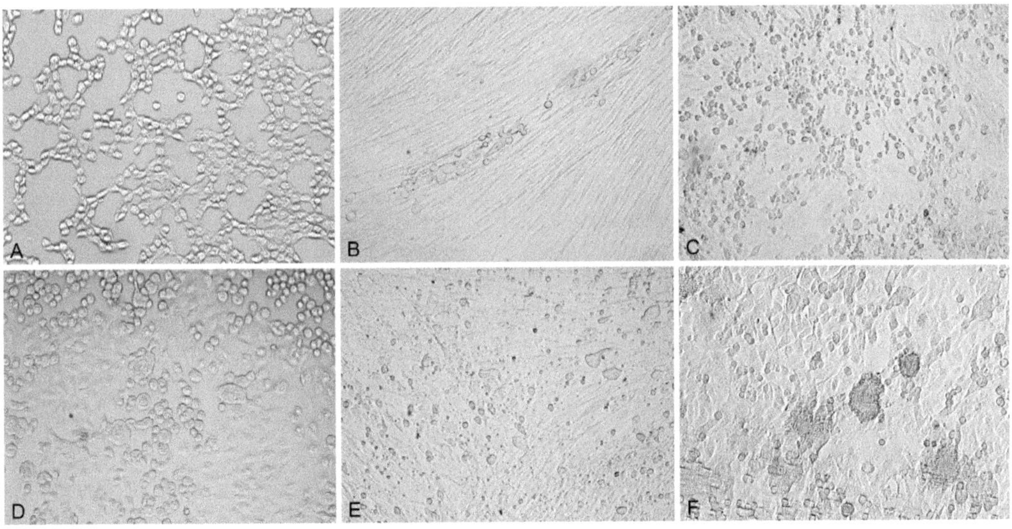

Figure 2 Typical CPE (100×) of (**A**) adenovirus in A549 cells, (**B**) CMV in MRC-5 cells, (**C**) enterovirus in RhMk cells, (**D**) HSV type 2 in A549 cells, (**E**) rhinovirus in MRC-5 cells, and (**F**) RSV in HEp-2 cells.

seen in the United States, with more than 26 cases reported in 2006 (21). This virus proliferates in cultures of primary monkey kidney cells and is classified as a BSL-3 agent.

Monkeypox, a poxvirus seen in animals and transmitted from animals to humans, was seen in the United States in 2003. Monkeypox was transmitted from imported Gambian rats to prairie dogs housed together by an exotic pet dealer (22). Members of many families who purchased the infected prairie dogs were infected. Monkeypox virus proliferates in several established cell culture lines including Vero, BSC-1, CV-1, and others. It can be handled at BSL-2 only if laboratory personnel have received smallpox vaccine within the previous 10 years.

Other viruses that should not be cultivated in a BSL-2 laboratory also replicate in common cell lines (e.g., avian influenza, tick-borne encephalitis viruses). The laboratory should be alert to this possibility and avoid culture inoculation or quarantine or destroy inoculated cultures if indicated.

Adventitious Agents Contaminating Cell Cultures

Primary cell cultures and passaged cell lines can become contaminated with adventitious agents or mycoplasma. Such contaminants can affect the interpretation of results (23–25). CPE induced by adventitious agents may mimic the changes induced by common human pathogens, leading to a false-positive report, or infection with these agents may be inapparent yet affect the ability of the cells to grow viruses in the clinical specimens. Furthermore, some endogenous animal viruses, such as herpes B virus, can pose a safety risk to laboratory personnel. Primary cell cultures prepared from monkey kidneys can contain endogenous simian polyomavirus (SV40), adenoviruses, CMV, enteroviruses, reoviruses, parainfluenza virus (SV5), and foamy virus (a retrovirus). Virus infection can affect the tissues and blood products of nonprimate species as well (26). For example, calf serum contaminated with infectious bovine rhinotracheitis virus led to CPE similar to HSV type 2 in commercially distributed A549 cells (27). Mycoplasmas have been a significant contaminant of passaged cell lines, necessitating periodic testing (28). Conversely, inhibitory substances and/or antibodies in calf serum used in the cell culture media can reduce the isolation of certain viruses, especially of the orthomyxo- and paramyxovirus groups (29). Although animal colonies and products are screened when commercially prepared and problems are infrequent, they can still occur. Thus, some laboratories use only diploid or continuous cell lines and avoid primary cells, especially of primate origin.

Table 3 Cytopathogenic Effect in Standard Cell Cultures of Human Viral Pathogens Common in the United States[a]

Virus	Cytopathogenic effect: Fibroblasts	A549	RhMK	Other	Final identification of isolates
Adenovirus	Some produce clusters	Grape-like clusters or "lacy" pattern, 2–8 days	Some produce clusters (if endogenous SV40 infection)	HEK: grape-like clusters 5–7 days	IF for group, Neut for type
CMV	Foci of contiguous rounded cells, 10–30 days	None	None	Use shell vials for rapid detection	IF or pass into shell vial, CPE alone if shell vial positive
Enterovirus	Some produce CPE, same as RhMK, 2–5 days	Infrequent, degenerative	Small, round cells with cytoplasmic tails 2–5 days		IF for groups, Neut for type
Herpes simplex	Rounded large cells, 2–6 days	Rounded large cells, 1–4 days	Some produce CPE, same as A549, 4–8 days	RK or HEK: rounded large cells, 1–4 days	IF
Influenza	None	None	Undifferentiated CPE, cellular granulation 4–8 days	HAD positive with GP RBC	IF
Parainfluenza	None	None	Rounded cells, some syncytia 4–8 days	HAD positive with GP RBC	IF
Rhinovirus	Degeneration, rounding, 7–10 days	None	None	Incubate fibroblasts at 33°C	Acid sensitivity test CPE can be difficult to differentiate from enteroviruses
RSV	Infrequent, granular, degeneration	Infrequent	Syncytia 4–10 days	HEp-2: syncytia 4–10 days	IF
Varicella-zoster	Some CPE, small, round cells 4–8 days	Small, round cells 4–8 days	Small, round cells 6–8 days if VZV high titer	HEK: small, round cells 6–8 days	IF

[a]Measles, mumps, and rubella virus are seldom encountered in the United States at present. Measles (large syncytia in RhMK cells, 7–10 days, HAD positive with Rhesus erythrocytes, identification confirmed by IF) and mumps (rounded cells with large syncytia in RhMK, 6–8 days, HAD positive with GP RBC, identification confirmed by IF). Rubella requires special cultures such as African Green Monkey Kidney, RK, or BSC-1 cells and does not produce CPE; special detection by interference challenge or other method is needed.

Abbreviations: A549, human lung carcinoma; CMV, cytomegalovirus; CPE, cytopathogenic effect; GP RBC, guinea pig erythrocytes; HAD, hemadsorption; HEp-2, human laryngeal carcinoma; HEK, human embryonic kidney cells; IF, immunofluorescence; Neut, neutralization; RhMK, primary monkey kidney cells; RK, rabbit kidney cells; RSV, respiratory syncytial virus.

Source: From Ref. 15.

Discovery of Novel Agents

In principle, conventional cell culture is more "open minded" than methods that target specific viral proteins or genetic sequences, and thus has the ability to reveal an unanticipated pathogen or detect a novel agent. Occasionally, CPE is observed that cannot be identified by common reagents. In such cases, electron microscopy can be extremely useful in identifying the virus family by morphology; then molecular methods can be used to sequence and characterize the unknown agent. In recent years, human metapneumovirus (30) and coronavirus NL-63 (31,32) were discovered after isolation in tertiary monkey kidney cells. Likewise, the recognition and characterization of Severe Acute Respiratory Syndrome coronavirus (SARS-CoV) was greatly facilitated by the ability to grow the virus in Vero E6 cells (33).

Advantages and Disadvantages of Virus Isolation in Traditional Cell Culture Tubes

There are both advantages and disadvantages associated with the use of traditional cell culture tubes in the diagnostic virology laboratory. Several of these are listed below:

Advantages:

1. A variety of viruses can be isolated. This is especially important in the following situations: when there is no specified viral suspect, when the sample may contain more than one virus, and when a virus appears that is unsuspected (in an unusual geographic location, outside the usual season, or as an emerging or reemerging pathogen).
2. Isolation is more sensitive and specific than viral antigen detection methods for many viruses.
3. Isolation can differentiate viable virus from nonviable viral antigen or nucleic acid.
4. An isolate is available if additional studies are needed.

Disadvantages:

1. Technical expertise is needed in evaluating cell culture monolayers microscopically.
2. An incubation period of 1 to 21 days is required for viruses to produce CPE.
3. Cell culture tubes are expensive to purchase and maintain.
4. Unanticipated isolation of dangerous viruses, such as SARS-CoV, influenza H5N1, tick-borne encephalitis viruses, and Ebola virus, is possible.
5. Many viruses of clinical importance cannot be cultivated in routine cell cultures (e.g., noroviruses, hepatitis viruses, parovovirus B19).

RAPID CULTURE METHODS

Centrifugation Culture (Shell Vial Technique)

Conventional virus isolation typically involves examination of tube cultures for CPE, which can take days to weeks to appear. Thus, the application of centrifugation cultures to rapid diagnosis in the clinical laboratory constituted a significant advance. Although the mechanism remains unclear, low-speed centrifugation of monolayers enhances the infectivity of viruses as well as Chlamydia (34). In 1984 the use of centrifugation cultures followed by staining with a monoclonal antibody at 24 hours postinoculation was first reported for CMV (35) (Fig. 3). In subsequent reports, its usefulness was documented for rapid diagnosis of other viruses, including HSV (36), VZV (37), adenovirus (38–40), respiratory viruses (41), polyomavirus BK (42), and enteroviruses (43). When the inoculum is standardized, semiquantitative results can be obtained by counting the number of virus-positive cells (44).

The shell vial method combines (*i*) centrifugation to enhance viral infectivity, (*ii*) cell culture to amplify virus in the specimen, and (*iii*) early detection of virus-induced antigen (before CPE) by the use of specific antibodies. Viruses with a long replication cycle, such as CMV, can be detected many days before CPE is apparent, especially if viral antigens produced early in the replication cycle are targeted. Viruses that replicate more rapidly, such as HSV, show less time gained for positive results with the shell vial technique. However, labor savings accrue since negative cultures are usually terminated and reported at two days for shell vial cultures, compared to 7 to 14 days for conventional cultures (Table 4) (45).

By this method, flat-bottomed shell vials containing cell culture monolayers on round coverslips are inoculated with sample, then centrifuged for 30 to 60 minutes at $700 \times g$. The

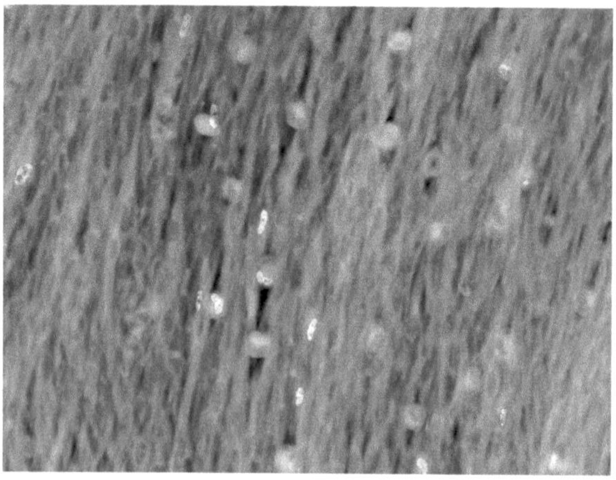

Figure 3 Centrifugation culture: Detection of CMV immediate early antigens in infected nuclei at 16 to 24 hours postinoculation (immunofluorescence stain) (200×).

inoculum is removed if desired, maintenance medium added, and the culture incubated at 35°C to 37°C. At designated days postinoculation, cultures are fixed in acetone or acetone/methanol and stained with virus-specific antibody. In some instances, centrifugation cultures are monitored for CPE and tested by HAD, similar to conventional cultures in roller tubes. To facilitate high volume testing, centrifugation cultures can be performed using 24- or 48-well tissue culture plates, instead of individual shell vials. Either immunofluorescence or immunoperoxidase methods can be used, but the former is more common.

A number of factors influence the sensitivity of the shell vial technique, including the type of specimen (46), the length and temperature of centrifugation (47), the virus sought, the type of cell culture, the antibody employed, and the time of fixation and staining. In general, the use of young cell cultures and inoculation of multiple shell vials enhances the recovery rate (48,49). Toxicity, especially from blood and urine specimens, can lead to monolayer loss. Passage of the specimen or specimen reinoculation may be needed.

It should be noted that rapid techniques that target one specific virus will detect only the virus sought. In contrast, conventional isolation using a spectrum of cell cultures can detect a variety of virus types, including the unexpected (50). When optimal recovery is needed, both conventional culture and centrifugation cultures should be performed in parallel (41,51–53).

Mixed Cell Cultures and Monoclonal Antibody Pools

To apply shell vial cultures to the detection of the spectrum of viruses potentially present in a clinical sample requires multiple cell lines and antibodies. In order to simplify the process and detect more viruses with fewer cell cultures, antibodies to multiple viruses have been pooled in one reagent (41,54,55), and two to three different cell cultures have been combined in one vial

Table 4 Time to Virus Detection by Conventional and Centrifugation Culture

	Time (days) to virus detection	
Virus	**Conventional culture by CPE (avg and range)**	**Shell vial centrifugation culture IF staining**
Respiratory viruses		
Adenovirus	6 (1–14)	2–5 days
Influenza A	2 (1–7)	1–2 days
Influenza B	2 (1–7)	1–2 days
Parainfluenza 1–3	6 (1–14)	1–2 days
Respiratory syncytial	6 (2–14)	1–2 days
Herpesviruses		
Cytomegalovirus	8 (1–28)	1–2 days
Herpes simplex	2 (1–7)	1–2 days
Varicella-zoster	6 (3–14)	2–5 days

Abbreviations: CPE, cytopathic effects; IF, immunofluorescence.
Source: From Ref. 45.

Table 5 Mixed Cell Cultures and Genetically Engineered Cell Lines Commercially Available

Culture	Composition	Viruses routinely targeted	Other recoverable viruses	References
R-mix^{TM}	Mink lung (Mv1Lu) and A549	RSV, Influenza A and B, parainfluenza 1,2,3, adenovirus HMPV	HSV, VZV, CMV, enteroviruses	58, 60–66
R-mix Too^{TM}	MDCK and A549	Same as R-mix, except not susceptible to SARS CoV	HSV, VZV, enteroviruses	
H&V Mix^{TM}	African green monkey kidney (strain CV-1) and MCR-5 cells	HSV-1 and 2, VZV	CMV, mumps, measles, rotavirus, encephalitis viruses, rhinoviruses, adenoviruses, and enteroviruses, RSV	68
Super E-mix^{TM}	BGMK-hDAF and A549	Enteroviruses	HSV, VZV, adenoviruses	71
ELVIS®	BHK cell line with UL39 promoter and *E. coli* lacZ gene	HSV-1 and HSV-2		72,74–78

Abbreviations: A549, human lung carcinoma; MDCK, Madin Darby canine kidney; MRC-5, human diploid fibroblasts; BGMK, Buffalo Green monkey kidney; BHK, baby hamster kidney; RSV, respiratory syncytial virus; HMPV, human metapneumovirus; HSV, herpes simplex virus; VZV, varicella-zoster virus; CMV, cytomegalovirus; SARS CoV, severe acute respiratory syndrome coronavirus.

(56–58). Although experience is needed to differentiate specific from nonspecific patterns, the training time for new personnel to learn interpretation of IF staining of shell vial cultures is much shorter than required for conventional CPE. Mixed cell cultures and corresponding fluorescent reagents are now available commercially, and the cultures have been further enhanced through genetic engineering. With this technology, detection of common respiratory viruses is simplified, labor is reduced, and results are more rapidly reported on both positives and negatives.

If desired, mixed cell cultures can be maintained longer than the one to two days typically employed for IF staining and observed for CPE for one to two weeks. Consequently, some laboratories have eliminated conventional cell culture tubes and converted to shell vials with mixed cells (59).

There are a variety of mixed cell cultures to choose from, according to the viruses sought (Table 5). R-mix^{TM} (Mv1Lu and A549) and R-mix Too^{TM} (MDCK and A549) are used with monoclonal antibody pools to rapidly detect selected respiratory viruses (58,60–65). R-mix Too^{TM} was developed to avoid inadvertently growing SARS-CoV by replacing Mv1Lu with MDCK cells. Human metapneumovirus also can be isolated and detected by IF (66). In addition, other viruses can be detected by observing R-mix cultures for CPE.

Traditional enterovirus detection requires inoculation of three to five different cell lines. That number was reduced with the original E-mix A (RD and H292) and E-mix B (BGMK and A549) cells. Subsequently, E-mix A and B cells were replaced by the more sensitive genetically engineered Super E-mix^{TM} described below under genetically modified cell lines (67).

H&V mix^{TM} (CV-1 and MRC-5) was developed for isolation of HSV 1 and 2 and VZV, but also can detect CMV. Although HSV and CMV are detected by IF staining after one or two days of incubation, optimal detection of VZV may require staining at two days and again at four to five days. Many other viruses also can replicate in H&V mix^{TM} and can be detected by CPE (68).

The protocols for inoculation, incubation, and staining for commercially obtained mixed cell cultures are generally those recommended by the supplier and modified as needed by the user. In general, two to three shell vials are inoculated. For respiratory viruses, inoculation of three R-mix^{TM} or R-mix Too^{TM} shell vials is recommended and the following protocol employed:

1. On day one post-inoculation, one shell vial is fixed and stained with the respiratory virus antibody pool.
2. If the first shell vial is positive, a second shell vial is scraped and spotted onto an eight-well slide to identify the unknown virus by staining with individual antibodies.

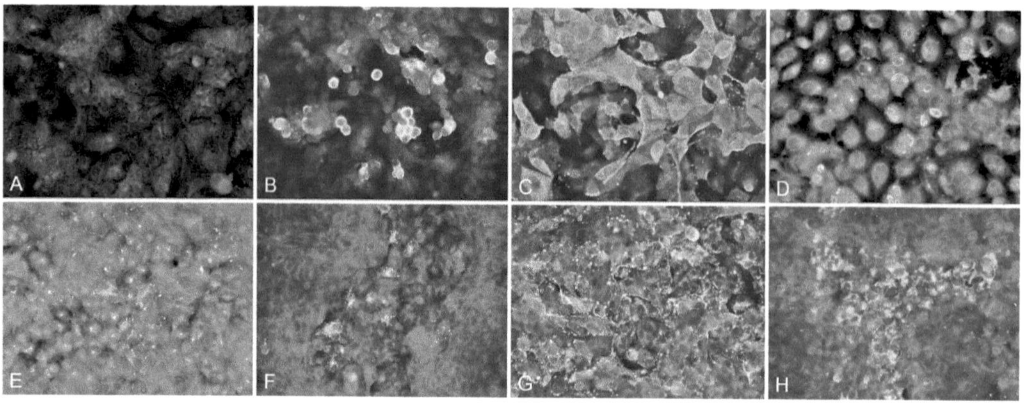

Figure 4 R-mix™ Too cells (200×) in shell vial format stained with pooled respiratory virus immunofluorescence screen reagent at day 1 postinoculation: (**A**) uninoculated, (**B**) adenovirus, (**C**) influenza A, (**D**) influenza B, (**E**) parainfluenza 1, (**F**) parainfluenza 2, (**G**) parainfluenza 3, and (**H**) RSV.

3. If the first vial is negative, a second shell vial is scraped on day two of incubation and spotted onto both a single well and an eight-well slide.
4. The screening reagent is applied to the single-well slide.
5. If the screening reagent is positive, the eight-well slide is then stained with individual antibodies to identify the unknown virus.
6. If the screening reagent is negative, the eight-well slide is discarded.
7. The third shell vial can be observed for CPE for a longer period, since some slower growing and low titered viruses may be detected.
8. Alternatively, the second shell vial can be stained with the screening reagent in situ and if positive, the third shell vial used to prepare an eight-well slide for identification.

Figure 4 shows staining of R-mix Too™ shell vials stained in situ with the pooled reagent. Subsequent staining on eight-welled slides identified infection for the viruses shown.

For enteroviruses, two Super E-mix™ shell vials are inoculated, and staining at two and five days is recommended. Samples that contain high titers of virus, such as stools, are generally positive by day two, but up to five days are required for spinal fluids (67).

Genetically Modified Cell Lines

Genetic modification of cell lines is an emerging technology with great potential for the diagnostic laboratory (69). For example, a cellular receptor for several enteroviruses, human decay–accelerating factor (hDAF) or CD55, was transfected into Buffalo green monkey kidney (BGMK) cells (70,71). BGMK-hDAF cells were then combined with the human colon adenocarcinoma cell line (CaCo-2) in one culture vessel and designated Super E-mix™ cells. These mixed cells were reported to be more sensitive for isolation of enteroviruses than inoculation of three separate conventional tube cultures using primary rhesus monkey kidney, A549, and fetal foreskin (SF) cells (67). The current Super E-mix™ contains A549 instead of the CaCo-2 cells.

Genetic elements derived from viral, bacterial, or cellular sources can be introduced into a cell for a different approach. When the target virus enters the cell, the viral replication cycle triggers the production of a measurable enzyme. The application of a simple histochemical assay results in infected cells staining a characteristic color. Even untrained observers can recognize infected cells stained in the inducible system and the earliest stages of infection can be detected. Using different strategies for enzyme induction, this approach has been applied to both DNA and RNA viruses (72,73).

An inducible system for isolation of HSV, called ELVIS® for enzyme-linked virus inducible system, provides both positive and negative results in 16 hours (74). In this commercially available system, transgenic baby hamster kidney (BHK) cells have been altered to include an HSV-specific promoter and an *Escherichia coli* LacZ reporter gene. The HSV positive cells form a blue precipitate when reacted with a chromogenic substrate (X-Gal). ELVIS® is simple, sensitive, and rapid and can be used for the simultaneous detection, identification, and typing of HSV isolates from clinical specimens (75–78). However, ELVIS remains somewhat less sensitive than the most sensitive of conventional culture systems for detection of very low levels of infectious HSV.

CRYOPRESERVED CELLS

Clinical laboratories commonly receive shipments of ready-to-use cell cultures once or twice a week from suppliers. In the process of shipment, the cultures may be stressed by extremes in temperature, mishandled as they are packed, stacked, and loaded, or compromised by delays in delivery due to bad weather, holiday closures, and many other uncontrollable circumstances (79). In addition, laboratories must determine the number of cell cultures needed in advance. If there is an unexpected surge in demand, such as an outbreak of a viral illness in the hospital or community, the laboratory may not have sufficient cell cultures to deal with the increased specimen volume.

To help address this issue, frozen preparations of a number of cell types are marketed in the United States, which are produced at densities expected to grow to confluency within four days from planting (Frozen FreshCells™, Diagnostic Hybrids, Inc., Athens, OH). These can be stored frozen for up to six months from the date of shipment and prepared for use when there is a need for additional cell cultures or for a type of cell culture that is not routinely kept on hand in the laboratory.

Cryopreserved ready-to-use cell monolayers grown in shell vials, shipped on dry ice, and stored at −70°C with a six-month outdate from the date of manufacture (ReadyCells™, Diagnostic Hybrids Inc., Athens, OH) are also available. These vials are ready to be inoculated with clinical samples after thawing for four minutes in a 35°C to 37°C water bath and refeeding with cell culture medium supplied by the manufacturer. Cryopreserved monolayers for culturing chlamydia (McCoy ReadyCells™), HSV and CMV (Hs27 ReadyCells™), and the various viral respiratory pathogens (R-Mix ReadyCells™) are available. In comparison studies, these frozen monolayers performed with sensitivity comparable to that of standard cell cultures for the detection of HSV and influenza A and B viruses (79).

SUMMARY

Cell culture isolation has long been considered the gold standard for viral diagnosis, but with the advent of more rapid methods, its role has been challenged. Limitations of culture that are cited include the inherent time delay required for virus growth, the expertise needed to maintain cell cultures and recognize CPE, the presence of endogenous viruses or other adventitious agents, variable quality, and decreased sensitivity of cell lines at higher passage levels. Some viruses do not produce identifiable effects in commonly used cell cultures. When samples are shipped a distance or are mishandled, virus may lose viability, leading to falsely negative results. Inadvertent recovery of BSL 3 or 4 agents is a safety concern.

The advantages culture offers include detection of a broad spectrum of viruses at lower cost than similar detection by molecular assays, greater sensitivity and specificity than antigen detection, and ability to differentiate infectious from noninfectious virus. With sensitive molecular assays, clinically irrelevant infections can be detected and lead to unnecessary treatment. Detection of infectious virus in culture may have a better predictive value for clinical disease, and is used by some physicians as a "test of cure." In addition, cell culture can detect the "unexpected" (80) and facilitate the discovery and rapid molecular characterization of new viruses.

Recent cell culture innovations have shortened turnaround times to one to two days in most cases, and significantly reduced the technical expertise, labor, and quality control required. Since most hospitals still have limited in-house molecular diagnostic capability, the time to result with rapid culture can be faster than molecular methods that are not performed daily or are sent to a distant reference laboratory.

For laboratories performing rapid RSV and influenza tests, mixed cell cultures provide a broader diagnosis, enhanced sensitivity and specificity, and results in one to two days. For laboratories that offer direct IF for rapid diagnosis of herpes and respiratory viruses, cell culture is essential to establish and monitor the performance of these tests. Furthermore, culture methods should be used for samples that are inadequate for IF and for lower respiratory tract and tissue biopsy samples to detect additional or unsuspected viruses.

While molecular methods are essential for detection of viruses in spinal fluid and to monitor viral load in blood, virus isolation continues to play an important role in viral diagnosis. The central role of conventional virus isolation with observation for CPE may have diminished. However, a variety of cell culture methods are now available to choose from, and culture remains essential for validating rapid methods and for patient management, especially when performed on-site in hospital laboratories.

REFERENCES

1. Hsiung GD, Fong CKY, Landry ML. Hsiung's Diagnostic Virology, 4th ed. New Haven, CT: Yale University Press, 1994.
2. Stinehardt E, Israeli C, Lambert R. Studies on the cultivation of the virus of vaccinia. J Infect Dis 1913; 13:204–300.
3. Rivers TM, Ward SM. Jennerian prophylaxis by means of intradermal injections of culture vaccine virus. J Exp Med 1935; 62:549–560.
4. Lloyd W, Theiler M, Ricci NI. Modification of the virulence of yellow fever virus by cultivation in tissues in vitro. Trans R Soc Trop Med Hyg 1936; 19:481–529.
5. Enders JF, Weller TH, Robbins FC. Cultivation of the Lansing strain of poliomyelitis virus in cultures of various human embryonic tissues. Science 1949; 109:85–87.
6. Robbins FC, Enders JF, Weller TH. Cytopathogenic effect of poliomyelitis viruses in vitro on human embryonic tissues. Proc Soc Exp Biol Med 1950; 75:370–374.
7. Schmidt NJ. Tissue culture technics for diagnostic virology. In: Lennette EH, Schmidt NJ, eds. Diagnostic procedures for Viral and Rickettsial Infections, 4th ed. New York, NY: American Public Health Association, Inc., 1969:81–178.
8. Hsiung GD. Diagnostic virology: from animals to automation. Yale J Biol Med 1984; 57:727–733.
9. Clinical and Laboratory Standards Institute. Viral Culture: Approved guideline. Document M41-A. Wayne, PA: Clinical and Laboratory Standards Institute, 2006.
10. Forman MS, Valsamakis A. Specimen collection, transport, and processing: Virology. In: Murray PR, Baron EJ, Jorgensen JH, Landry ML, Pfaller MA, eds. Manual of Clinical Microbiology, 9th ed. Washington, DC: American Society for Microbiology, 2007:1284–1296.
11. Leland DS. Clinical Virology. Philadelphia, PA: W. B. Saunders, 1996.
12. Hughes JD. Physical and chemical methods for enhancing rapid detection of viruses and other agents. Clin Microbiol Rev 1993; 6:150–175.
13. Minnich LL, Ray GC. Early testing of cell cultures for detection of hemadsorbing viruses. J Clin Microbiol 1987; 25:421–422.
14. Phipps P, McCulloch HBG, Miller HR, et al. Rapid detection of influenza virus infections in human fetal lung diploid cell cultures. J Infect 1989; 18:269–278.
15. Leland DS, Ginocchio CC. Role of cell culture for virus detection in the age of technology. Rev Clin Microbiol 2007; 20:49–70.
16. Mulholland EK. Measles in the United States, 2006. N Engl J Med 2006; 355:440–443.
17. Bitsko RH, Cortese MM, Dayan GH, et al. Detection of RNA of mumps virus during an outbreak in a population with a high level of measles, mumps, and rubella vaccine coverage. J Clin Microbiol 2008; 46:1101–1103.
18. Kobune F, Sakata H, Sugiura A. Marmoset lymphoblastoid cells as a sensitive host for isolation of measles virus. J Virol 1990; 64:700–705.
19. Leland DS. Parainfluenza and mumps viruses. In: Murray PR, Baron EJ, J Jorgensen JH, Landry ML, Pfaller MA, eds. Manual of Clinical Microbiology, 9th ed. Washington, DC: American Society for Microbiology, 2007:1352–1360.
20. Knowles WA, Cohen BJ. Efficient isolation of mumps virus from a community outbreak using the marmoset lymphoblastoid cell line B95a. J Virol Methods 2001; 96:93–96.
21. Centers for Disease Control and Prevention. Update: Chikungunya fever diagnosed among international travelers—United States, 2006. MMWR Morb Mortal Wkly Rep 2007; 56:276–277.
22. Kyle JC, Fleischauer AT, Beard B, et al. Transmission of monkeypox among persons exposed to infected prairie dogs in Indiana in 2003, 2007.
23. Hsiung GD. Latent virus infections in primate tissues with special reference to simian viruses. Bacteriol Rev 1968; 32:185–205.
24. Smith KO. Adventitious viruses in cell cultures. Prog Med Virol 1970; 12:302–336.
25. Whiteman MD. Scope and practicality of in vivo testing for adventitious agents. Dev Biol (Basel) 2006; 123:147–152; discussion 183–197.
26. Merten OW. Development of serum-free media for cell growth and production of viruses/viral vaccines–safety issues of animal products used in serum-free media. Dev Biol (Basel) 2002; 111:233–257.

27. Fong CKY, Landry ML. An adventitious viral contaminant in commercially supplied A549 cells: Identification of infectious bovine rhinotracheitis virus and its impact on diagnosis of infection in clinical specimens. J Clin Microbiol 1992; 30:1611–1613.
28. Stanbridge E. Mycoplasmas and cell cultures. Bacteriol Rev 1971; 35:206–227.
29. Krizanova O, Rathova V. Serum inhibitors of myxoviruses. Curr Top Microbiol Immunol 1969; 47:125–151.
30. van den Hoogen BG, de Jong JC, Groen J, et al. A newly discovered human pneumovirus isolated from young children with respiratory tract disease. Nat Med 2001; 7:719–724.
31. van der Hoek L, Pyrc K, Jebbink MF,et al. Identification of a new human coronavirus. Nat Med 2004; 10:368–373.
32. Fouchier RA, Hartwig NG, Bestebroer TM, et al. A previously undescribed coronavirus associated with respiratory disease in humans. Proc Natl Acad Sci U S A 2004; 101:6212–6216.
33. Ksiazek TG, Erdman D, Goldsmith CS, et al; SARS Working Group. A novel coronavirus associated with severe acute respiratory syndrome. N Engl J Med 2003; 348:1953–1966.
34. Hudson JB, Misra V, Mosmann TR. Cytomegalovirus infectivity: Analysis of the phenomenon of centrifugal enhancement of infectivity. Virology 1976; 72:235–243.
35. Gleaves CA, Smith TF, Shuster EA, et al. Rapid detection of cytomegalovirus in MRC-5 cells inoculated with urine specimens by using low-speed centrifugation and monoclonal antibody to an early antigen. J Clin Microbiol 1984; 19:917–919.
36. Gleaves CA, Wilson DJ, Wold AD, et al. Detection and serotyping of herpes simplex virus in MRC-5 cells by use of centrifugation and monoclonal antibodies 16 hr postinoculation. J Clin Microbiol 1985; 21:29–32.
37. Gleaves CA, Lee CF, Bustamante CI, et al. Use of murine monoclonal antibodies for laboratory diagnosis of varicella-zoster virus infection. J Clin Microbiol 1988; 26:1623–1625.
38. Espy MJ, Hierholzer JC, Smith TF. The effect of centrifugation on the rapid detection of adenovirus in shell vial. Am J Clin Pathol 1987; 88:358–360.
39. Mahafzah AM, Landry ML. Evaluation of conventional culture, centrifugation culture, and immunofluorescence reagents for the rapid diagnosis of adenovirus infections. Diagn Microbiol Infect Dis 1989; 12:407–411.
40. van Doornum GJJ, DeJong JC. Rapid shell vial culture technique for detection of enteroviruses and adenoviruses in fecal specimens: comparison with conventional virus isolation method. J Clin Microbiol 1998; 36:2865–2868.
41. Olsen MA, Shuck KM, Sambol AR, et al. Isolation of seven respiratory viruses in shell vials: a practical and highly sensitive method. J Clin Microbiol 1993; 31:422–425.
42. Marshall WF, Telenti A, Proper J, et al. Rapid detection of polyomavirus BK by a shell vial cell culture assay. J Clin Microbiol 1990; 28:1613–1615.
43. She RC, Crist G, Billetdeaux E, et al. Comparison of multiple shell vial cell lines for isolation of enteroviruses: a national perspective. J Clin Virol 2006; 37:151–155.
44. Slavin MA, Gleaves CA, Schock HG, et al. Quantification of cytomegalovirus in bronchoalveolar lavage fluid after allogeneic marrow transplantation by centrifugation culture. J Clin Microbiol 1992; 30:2776–2779.
45. Landry ML. Rapid Viral Diagnosis. In: Detrick B, Hamilton RG, Folds JE, eds. Manual of Molecular and Clinical Laboratory Immunology, 7th ed. Washington, DC: American Society for Microbiology, 2006:610–616.
46. Paya C, Wold AD, Smith TF. Detection of cytomegalovirus infections in specimens other than urine by shell vial assay and conventional tube cell cultures. J Clin Microbiol 1987; 25:755–757.
47. Shuster EA, Beneke JS, Tegtmeier GE, et al. Monoclonal antibody for rapid laboratory detection of cytomegalovirus infections: characterization and diagnostic application. Mayo Clin Proc 1985; 60:577–585.
48. Paya C, Wold AD, Ilstrup DM, et al. Evaluation of number of shell vial cultures per clinical specimen for rapid diagnosis of cytomegalovirus infection. J Clin Microbiol 1988; 26:198–200.
49. Fedorko DP, Ilstrup DM, Smith TF. Effect of age of shelf vial monolayers on detection of cytomegalovirus from urine specimens. J Clin Microbiol 1989; 27:2107–2109.
50. Blanding JG, Hoshiko MG, Stutman HR. Routine viral culture for pediatric respiratory specimens submitted for direct immunofluoresence testing. J Clin Microbiol 1989; 27:1438–1440.
51. Rabella N, Drew WL. Comparison of conventional and shell vial cultures for detecting cytomegalovirus infection. J Clin Microbiol 1990; 28:806–807.
52. Espy MJ, Wold AD, Jespersen DJ, et al. Comparison of shell vials and conventional tubes seeded with rhabdomyosarcoma and MRC-5 cells for the rapid detection of herpes simplex virus. J Clin Microbiol 1991; 29:2751–2753.
53. Rabalais GP, Stout GG, Ladd KL, et al. Rapid diagnosis of respiratory viral infections by using a shell vial assay and monoclonal antibody pool. J Clin Microbiol 1992; 30:1505–1508.

54. Brumback BG, Farthing PG, Castellino SN. Simultaneous detection of and differentiation between herpes simplex virus and varicella-zoster viruses with two fluorescent probes in the same test system. J Clin Microbiol 1993; 31:3260–3263.
55. Engler HD, Pruess J. Laboratory diagnosis of respiratory virus infections in 24 hours by utilizing shell vial cultures. J Clin Microbiol 1997; 35:2165–2167.
56. Brumback BG, Wade CD. Simultaneous culture for adenovirus, cytomegalovirus and herpes simplex virus in same shell vial by using three-color fluorescence. J Clin Microbiol 1994; 32:2289–2290.
57. Navarro-Mari JM, Sanbonmatsu-Gamez S, Perez-Ruiz M, et al. Rapid detection of respiratory viruses by shell vial assay using simultaneous culture of HEp-2, LLC-MK2, and MDCK cells in single vial. J Clin Microbiol 1999; 37:2346–2347.
58. Huang YT, Turchek BM. Mink lung cells and mixed mink lung and A549 cells for rapid detection of influenza virus and other respiratory viruses. J Clin Microbiol 2000; 38:422–423.
59. Stevenson D, Hoffman J, Beckman KW, et al. Impact of Converting from Conventional Tube Cultures to Shell Vials: A Community Hospital Laboratory's Experience. Poster M-59, 23rd Annual Clinical Virology Symposium, Clearwater Beach, FL, April 29–May 2, 2007.
60. Fong CKY, Lee MK, Griffith BP. Evaluation of R-mix fresh cells in shell vials for detection of respiratory viruses. J Clin Microbiol 2000; 38:4660–4662.
61. Barenfanger J, Drake C, Mueller T, et al. R-mix cells are faster, at least as sensitive and marginally more costly than conventional cell lines for the detection of respiratory viruses. J Clin Virol 2001; 22:101–110.
62. Dunn JJ, Woolstenhulme RD, Langer J, et al. Sensitivity of respiratory virus culture when screening with R-mix fresh cells. J Clin Microbiol 2004; 42:79–82.
63. St. George K, Patel NM, Hartwig RA,et al. Rapid and sensitive detection of respiratory virus infections for directed antiviral treatment using R-mix cultures. J Clin Virol 2002; 24:107–115.
64. Fader RC. Comparison of the Binax NOW flu A enzyme immunochromatographic assay and R-mix shell vial culture for the 2003–2004 influenza season. J Clin Microbiol 2005; 43:6133–6135.
65. Weinberg A, Brewster L, Clark J, et al. Evaluation of R-mix shell vials for the diagnosis of viral respiratory tract infections. J Clin Virol 2004; 30:100–105.
66. McRae KL, Bocchini JA Jr., Anga A,et al. Human Metapneumovirus Detection Using MGB Eclipse RT PCR and Mixed Cell Culture. Poster TP-45, 23rd Annual Clinical Virology Symposium, Clearwater Beach, FL, April 29–May 2, 2007.
67. Buck GE, Wiesemann M, Stewart L. Comparison of mixed cell culture containing genetically engineered BGMK and CaCo-2 cells (super E-mix) with RT-PCR and conventional cell culture for the diagnosis of enterovirus meningitis. J Clin Virol 2002; 25:S13-S18.
68. Huang YT, Hite S, Duane V, et al. CV-1 and MRC-5 mixed cells for simultaneous detection of herpes simplex viruses and varicella zoster virus in skin lesions. J Clin Virol 2002a; 24:37–43.
69. Olivo PD. Transgenic cell lines for detection of animal viruses. Clin Microbiol Rev 1996; 9:321–334.
70. Lublin DM, Atkinson JP. Decay-accelerating factor: Biochemistry, molecular biology, and function. Ann Rev Immunol 1989; 7:35–58.
71. Huang YT, Yam P, Yan H, et al. Engineered BGMK cells for sensitive and rapid detection of enteroviruses. J Clin Microbiol 2002b; 40:366–371.
72. Stabell EC, O'Rourke SR, Storch GA, et al. Evaluation of a genetically engineered cell line and a histochemical b-galactosidase assay to detect herpes simplex virus in clinical specimens. J Clin Microbiol 1993; 31:2796–2798.
73. Lutz A, Dyall J, Olivo PD, et al. Virus-inducible reporter genes as a tool for detecting and quantifying influenza A virus replication. J Virol Method 2005; 126:13–20.
74. Proffitt MR, Schindler SA. Rapid detection of HSV with an enzyme-linked virus inducible system™ (ELVIS™) employing a genetically modified cell line. Clin Diag Virol 1995; 4:175–182.
75. Patel N, Kauffmann L, Baniewicz G, et al. Confirmation of low-titer, herpes simplex virus-positive specimen results by the enzyme-linked virus-inducible system (ELVIS) using PCR and repeat testing. J Clin Microbiol 1999; 37:3986–3989.
76. Turchek BM, Huang YT. Evaluation of ELVIS™ HSV ID/typing system for the detection and typing of herpes simplex virus from clinical specimens. J Clin Virol 1999; 12:65–69.
77. Kowalski RP, Karenchak LM, Shah C, et al. A new 24-hour culture test for detecting herpes simplex virus from ocular samples. Arch Ophthalmol 2002; 120:960–962.
78. Crist GA, Langer JM, Woods GL, et al. Evaluation of the ELVIS plate method for the detection and typing of herpes simplex virus in clinical specimens. Diagn Microbiol Infect Dis 2004; 49:173–177.
79. Huang YT, Yan H, Sun Y, et al. Cryopreserved cell monolayers for rapid detection of herpes simplex virus and influenza virus. J Clin Microbiol 2002c; 40:4301–4303.
80. Mackenzie JS. Emerging viral diseases: An Australian perspective. Emerging Infect Dis 1999; 5:1–8.

8 Diagnosis by Viral Antigen Detection

Bagher Forghani

Viral Immunoserology Section, Viral and Rickettsial Disease Laboratory, California Department of Public Health, Richmond, California, U.S.A.

INTRODUCTION

Major dynamic advances are occurring in all fields of science and technology, including the emergence of many new and exciting viral diagnostic assay systems. The application of molecular diagnostic assays in the last two decades, the impact of information technology on laboratory information management, and the availability of commercial antigen detection kits have been extremely helpful for the rapid diagnosis of viral infections. The new frontier in diagnostic assays may be the application of nanotechnology in clinical samples. Nanotechnology assay systems combine extremely sensitive detector signals with the high specificity of immunological reactions, with or without the need for virus isolation. Currently, the potential of nanotechnology in diagnostic virology is being vigorously explored. However, in the meantime, conventional viral antigen detection systems play a major role in the clinical management of patients (1–6).

Traditionally, laboratory diagnosis of viral infections was accomplished by the isolation and identification of virus from clinical samples, using cell cultures, laboratory animals, or embryonated eggs. More recently, new genetically modified cell lines have been extremely helpful for rapid isolation of some important human viruses, but this remains a slow process (7–13). The delays inherent in viral isolation in cell culture, the "gold standard" for the detection of viruses, have created an urgent need for rapid, sensitive, specific, and reproducible methods of viral diagnosis. The ability of immunoassays to detect viral antigens directly in clinical specimens, using antibodies labeled with fluorescein, radioisotopes, or enzymes, has helped meet this need.

Compared with isolation in cell culture, immunoassays are "close minded" and have certain limitations. They can detect and identify only the antigens of viruses specified by the labeled antibodies, and not those of new viruses, nor of known but unsuspected viruses. Furthermore, the small amounts of some clinical samples may limit extensive, direct probing for multiple viruses. Virus isolation, on the other hand, is often slow and expensive, and requires the use and maintenance of several cell lines, sterile media preparation, and highly trained personnel. However, viral isolation offers specificity and the ability to detect unsuspected or new viral agents. For viral culture to remain a useful and viable approach for rapid viral diagnosis, it will increasingly utilize emerging cell culture methodologies (e.g., transgenic cell lines) and/or new stem cell technology. Stem cell lines have been established from human tissue (embryonic or adult) and from several animal species, and have been successfully used in several disciplines of medicine and genetics. From such research it may be possible to generate and select cell lines to support rapid virus replication, especially for viruses that have slow growth rates, and to establish cell lines susceptible to viruses that currently cannot be cultivated.

THEORETICAL ASPECTS OF ANTIGEN AND ANTIBODY INTERACTION

Antigen

An antigen is a substance capable of inducing antibody formation in animals, and of binding to the antibodies it has induced. Therefore, the immunogenicity of viral antigens is measured by their capacity to elicit a specific immune response and their ability to combine with antibody. All viral immunoassay systems are based on the exploitation of these principles.

The antigenicity of a substance is dependent on its physiochemical properties, such as molecular weight, chemical composition, secondary and tertiary structures, and the degree of foreignness to the host (14,15). Antibodies bind primarily to chemical structures exposed at the surface of antigens. These attachment regions on the antigen are called *antigenic domains*, and each individual binding site is an antigenic determinant or *epitope*. An antigenic domain may contain one or more epitopes. These epitopes are conformational; they may be continuous, composed of adjacent amino acid sequence, or discontinuous, made up of amino acids brought into juxtaposition by protein folding. The size of an epitope appears to be large enough to accommodate a hexapeptide or pentasaccharide, but may vary. The current view is that an epitope may be flexible or rigid. A rigid configuration may prevent stable antigen–antibody interaction, whereas flexibility allows the epitope to assume a thermodynamically stable configuration in antibody binding. Furthermore, antigen–antibody binding depends both on the chemical structure of the epitope and its ability to form an electrostatic hydrophilic or hydrophobic interaction with the amino acid residues within the binding site of the antibody.

Basic Kinetics of Immunoassays

The binding site of antibody to an antigen is called *paratope*. The binding of an antibody paratope to a particular antigenic epitope constitutes the basis of immunological specificity. The study of the kinetics and stoichiometry of antigen–antibody reactions reveals that antigen–antibody reactions follow the rules of biochemical interactions, and antigen–antibody reactions can be used to quantitate either antigen or antibody. These concepts have led to the development of numerous immunological assays, both for basic investigations of antigen–antibody reactions and for the practical detection of microbial and nonmicrobial antigens in clinical samples. Detection of viral antigen from clinical specimens has been at the frontier in these methodologies (3,6).

Our knowledge of the exact nature of the interaction between antibody and antigen is sketchy, especially where complex antigens such as viruses are concerned. It is presumed that antigen is complexed to antibody by weak, noncovalent, short-range bonds of the electrostatic and *van der Waals* type, and that the formation of these complexes is governed by the kinetics of the *law of mass action*. Therefore, it is generally accepted that the antibody–antigen binding reaction reaches equilibrium but does not go to completion under laboratory testing conditions. The formation of antibody–antigen complexes is proportional to, and dependent upon, the concentration of antibody. In addition, the higher the degree of stereo-complementarity between antigen and antibody, the greater the number of bonds formed. These bonds are enhanced by the hydrophobic interaction between antigen and antibody. Since this reaction is exothermic, binding is enhanced by decreasing temperature, whereas the reaction rate is enhanced by increasing temperature.

An important factor in any immunoassay is the affinity and avidity of an antibody during antigen and antibody interaction. *Affinity* is defined as the attraction force of an antibody in the immunological reaction, whereas *avidity* is defined as the strength of the antibody–antigen bond after the immune complex is formed. High avidity is a prerequisite for specific and sensitive solid-phase immunoassays, as the many washing steps involved will remove immunoreactants if their interactions are weak. High antibody avidity will prevent the elution of an already formed antibody–antigen complex in these assays. Avidity is an important concern in the production and selection of mAbs to be used in solid-phase immunoassay systems for detection of viral antigen in clinical specimens. The antibody concentration affects also antigen and antibody binding. The use of mAbs, all of a desired specificity, is generally the best way to increase antibody concentration (15–17).

DESCRIPTION OF VIRAL ANTIGEN DETECTION ASSAYS

The assays described in this chapter are the most widely used in public health and clinical laboratories for the detection of viral antigen in clinical specimens. Furthermore, special attention is given to commercial rapid viral antigen kits. The sensitivity and specificity of these kits have been evaluated by large number of investigators; only selected studies are presented here. In summary, these studies show that most rapid antigen detection kits have sensitivities of 60% to 90% and specificities of 90% to 100% when compared either with virus isolation or molecular techniques such as PCR (18–22).

Detection of Viral Antigen by Immunofluorescence

Immunofluorescence antibody (IFA) staining has become a standard method in many laboratories for the rapid and direct detection of viral antigens in clinical samples. This is because several fluorochromes can be chemically bound to antibody with high efficiency without compromising the immunologic specificity of the antibody and the fluorescence intensity of the fluorochrome. These labeled antibodies can be used in several immunoassay formats to detect either viral antigen or antibody. In recent years, the purity of the fluorochromes and the design and quality of the optics of fluorescence microscopes have improved tremendously, leading to an increased application of immunofluorescence in virology. Immunofluorescence was introduced in the early 1940s, and was applied in diagnostic virology in the mid-1950s. Its usefulness was later considerably broadened by the nondestructive conjugation of fluorescein isothiocyanate (FITC) to antibody (23). The two most widely used fluorochromes are FITC, which fluoresces yellowish-green (apple green), and tetramethyl rhodamine isothiocyanate (TRITC), which fluoresces reddish-orange. Although both fluorochromes are efficiently and intensely fluorescent, fluorescein offers three advantages: the human eye is more sensitive to the green portion of the spectrum; the background autofluorescence of clinical specimens is more commonly red than green; and nonspecific background staining can be blocked by agents such as Evans blue and Congo red. Recently, antibodies have been labeled with two new fluorochromes, Cy3 (green) and Cy5 (red), with good results. The presence of Evans blue in conjugate solution stains all parts of the cell, and its red fluorescence provides a useful contrast to the green fluorescence of FITC. Evans blue is a carcinogenic and teratogenic agent, which must be handled with care to prevent skin contact.

Direct and Indirect IFA Procedure

The direct IFA (sometime referred to as "DFA") staining procedure is usually the method of choice when examining clinical specimens for viral antigen. It is the simplest and most reliable of the various staining methods, with fewer nonspecific reactions, and is therefore less subject to misinterpretation. Pretitrated conjugate is applied directly to the specimen being examined (infected cultured cells, vesicular fluids, skin scrapings, tissue smears, etc.), which has been previously acetone fixed on a microscope slide, and incubated for about 30 to 45 minutes in a humid atmosphere at 37°C; the unbound conjugate is then removed and the stained preparations are washed, rinsed with distilled water, air-dried, mounted, and examined by a fluorescence microscope (Fig. 1). Although it is necessary to prepare and maintain conjugates for each virus, the greater specificity of the direct technique outweighs this drawback (24–28). Conjugates for direct IFA are available commercially from several sources for Adenovirus, CMV, HSV, influenza A and B, Parainfluenza type 1, 2, and 3, RSV, and VZV (29).

The indirect procedure is slightly more sensitive for antigen detection than the direct method, but may have problems with nonspecificity. For indirect antigen detection, the specimen to be examined is first incubated with antigen-specific primary antiviral antibody (e.g., rabbit anti-HSV) for a period of time (30–45 minutes) in a humid atmosphere at 37°C. The sample is then

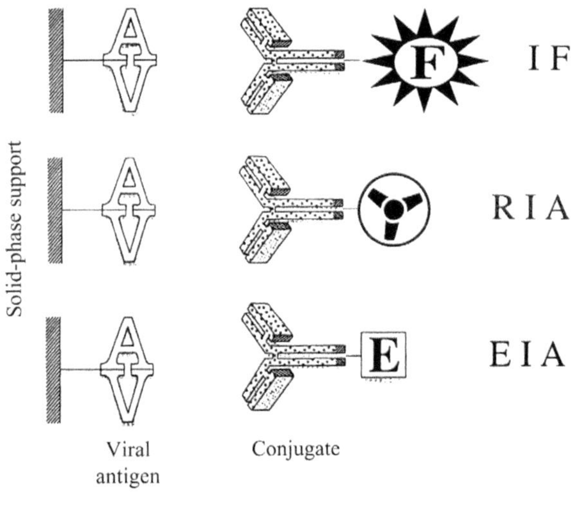

Figure 1 Principle of direct immunoassay systems for detection of viral antigen in tissues.

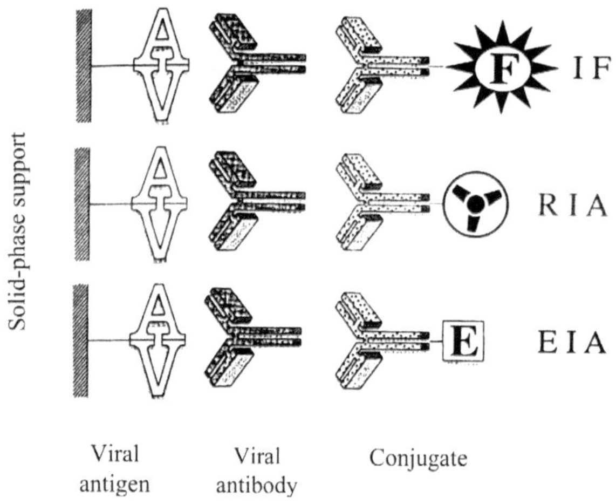

Figure 2 Principle of indirect immunoassay systems for detection of viral antigen in tissues.

washed with PBS, and incubated with diluted conjugate prepared from antiserum against the species of the primary antiviral antibody (e.g., FITC conjugated goat anti-rabbit). The incubation is again done for 30 to 45 minutes in a humid atmosphere at 37°C. The preparations are washed, mounted, and examined as described for direct detection (Fig. 2). Large numbers of polyclonal and monoclonal antibodies are available commercially and can be used for in-house indirect IFA or other assays (29–31).

Detection of Viral Antigen by Radioimmunoassay

RIA in diagnostic virology, first used to detect hepatitis B surface antigen, led to a new generation of assays for detection of other hepatitis virus antigens. However, RIA for detecting viruses has been overtaken by the development of sensitive and specific enzyme immunoassays (EIAs). Other major drawbacks are the short shelf-life of radionuclides, danger working with radioactive reagents, and the need for licenses from regulatory agencies for working with radioisotopes, both for the manufacturers producing these products and for the testing laboratories using these kits.

Historically, in direct RIA, unlabeled antiviral antibodies are adsorbed to a solid-phase support and used to "capture" viral antigens or viruses in the clinical specimen. The nonreactants in the initial incubation are removed by washing. ^{125}I-labeled antibody directed against the suspected virus is then added as *indicator* antibody. The unbound antibody is removed after incubation, and ^{125}I-labeled antibody bound to the solid-phase support is counted. The results are evaluated after comparison with those obtained on the appropriate controls. This assay is also known as *sandwich* RIA. The capture and *indicator* antibodies can be prepared in the same or different species (Fig. 3). The drawback of this system is that a labeled antibody is needed

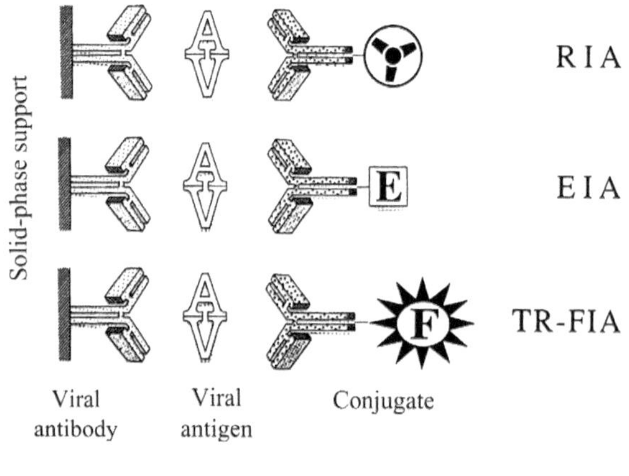

Figure 3 Principle of direct immunoassay systems for detection of viral antigen in body fluids and exudates.

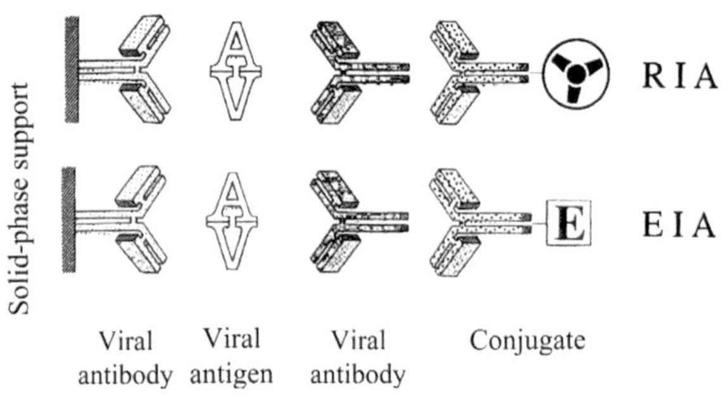

Figure 4 Principle of indirect immunoassay systems for detection of viral antigen in body fluids and exudates.

for each individual virus. In indirect RIA, the first three steps are the same as those for direct RIA, except that the viral antibody is unlabeled (32,33). In the fourth step, ^{125}I-labeled antibody directed against the species of viral antibody donor is used to detect the antigen/antibody complexes (Fig. 4).

Detection of Viral Antigen by Enzyme Immunoassay Systems

Enzyme immunoassay (EIA) methodology offers an almost ideal combination of sensitivity, specificity, and practicality for detection of viral antigens from clinical samples (28). These include the choice of enzymes with diverse physicochemical properties, the sensitivity gained from the amplification effect of the enzyme/substrate reaction, the potential for qualitative and quantitative immunoassays, the potential for automation, the safety of nonradiolabeled reagents, their long shelf-life, and their commercial availability. After decades of research developments, two applications using enzyme-labeled antibody for viral antigen detection have emerged: the cyto-EIA and the solid-phase quantitative EIA (34,35).

Cyto-Enzyme Immunoassay (Cyto-EIA)

Traditionally, fluorescein-conjugated antibody has been used to localize viral antigen in clinical tissues or in virus-infected cells. However, certain enzymes that generate insoluble colored substrate reaction *products* can be conjugated to an antibody and used in an immunoenzymatic staining assay for presence of viral antigen in the fixed tissue or cell. This enzymatic reaction can be visualized by the naked eye or by light microscopy, or electron-dense products can be observed by electron microscopy. The most widely used enzymes are horseradish peroxidase and alkaline phosphatase, which have both been used in qualitative and quantitative EIA. Advantages of horseradish peroxidase are its high enzymatic activity, the availability of several chromogens giving insoluble reaction products, and the ease of visualizing the products of the reaction. Therefore, this assay is sometimes called *immunoperoxidase* staining (35). The cyto-EIA has been widely applied in immunocyto- and histochemistry; however, its application in rapid viral antigen detection has been more limited.

Direct and Indirect Cyto-EIA Procedures

Direct Cyto-EIA staining is commonly the method of choice for rapid diagnosis of viral infections, and results can be obtained within two hours. Clinical specimens, such as vesicle or nasopharyngeal smears, tissue sections, or cell scrapings are fixed in acetone on a microscopic slide. Peroxidase-conjugated antibody is added to the specimen. After proper incubation, the unreacted conjugate is removed by washing in buffer, enzyme–substrate solution is added, and the specimen is incubated at room temperature, depending on the enzyme–substrate system. After development of the colored product, slides are again rinsed, counter-stained (optional), and mounted in a permanent or semipermanent mounting medium before viewing by light microscopy (35) (Fig. 1).

In indirect Cyto-EIA, unconjugated primary antiviral antibody (e.g., rabbit anti-HSV) is incubated with the fixed antigen preparation at 37°C for 30 to 40 minutes. After the unbound antibody is rinsed away, the slides are incubated with enzyme-conjugated antibody directed against the species of origin of the antiviral antibody (e.g., goat anti-rabbit), which binds to the virus–antibody complexes. Unbound conjugate is rinsed away and the sample incubated with enzyme–substrate, rinsed, mounted, and visualized as above (Fig. 2).

Quantitative Enzyme Immunoassay

There are two types of quantitative immunoassay using enzyme-labeled antibodies: the homogeneous and the heterogeneous EIA. In the homogeneous EIA, the inhibition or enhancement of enzymatic activity from the interaction between enzyme-labeled antibody and antigen (or between antibody and enzyme-labeled antigen) is measured. The test does not require removal of nonreactants and the whole assay is performed in minutes. The following enzymes have been utilized: lysozyme, malate dehydrogenase, and more recently, galactosidase. The sensitivity of homogeneous EIAs is relatively low, and they are not used for rapid viral antigen detection in clinical specimens.

In heterogeneous EIAs, the interaction between antigen and enzyme-labeled antibody does not alter the enzymatic activity. A separation step to remove unbound enzyme-labeled antibody and a relatively long incubation time are required. All quantitative EIAs in virology are of this type and this has been the basis of all solid-phase EIAs (34).

Solid-Phase Enzyme Immunoassay Procedures

Most quantitative EIAs are based on solid-phase systems and these assays are also called enzyme-linked immunosorbent assays (*ELISA*). The major advantage of these methods is their versatility: they can be used to detect both viral antigens and viral antibody. The direct and indirect EIAs are based on the same principles as direct and indirect RIAs for detection of viral antigen by capturing the virus by an antiviral antibody-coated solid-phase surface.

In direct (antigen-capture) assays, unlabeled antiviral antibody (e.g., guinea pig antihepatitis surface antigen) is first bound to a solid-phase support (microtiter plates, membranes, tubes, beads, or cuvettes). After adsorption, unbound antibody is removed. Potential nonspecific binding sites of the solid-phase support are blocked by blocking agents (e.g., bovine serum albumin). Next, samples are added, and after proper incubation, the nonreactants are removed, followed by the addition of enzyme-conjugated antiviral antibody (e.g., guinea pig antihepatitis surface antigen). The latter will bind to the antigen captured by the first antibody. The unreacted conjugate is removed and the substrate is added. The enzymatic activity is measured by its hydrolysis or oxidation of the substrate to produce a reaction product (Fig. 3). The amounts of reaction product detected are proportional to the amount of enzyme bound to the antigen retained on the solid phase. The amount of viral antigen present is determined from the degree of enzymatic activity of the test sample, compared with the reactivity of appropriate positive and negative control samples. Reaction products can be measured spectrophotometrically, fluorometrically, or chemoluminescently, depending on the substrate solution used. The assay and modified versions have been extensively applied for detection of many viruses (36–43).

The indirect EIA is useful when higher sensitivity is desired. This increased sensitivity occurs because several antispecies antibody molecules bind to a single molecule of the "*detector*" antiviral antibody. Indirect EIA requires the availability of two antibodies produced in different animal species that do not cross-react with one another. Briefly, the test procedure is as follows: unlabeled antiviral antibody (e.g., guinea pig antihepatitis e antigen) is adsorbed to a solid-phase support and the clinical test specimen is added, followed by unlabeled antiviral antibody prepared in an animal species different from that used for production of the capture antiviral antibody (e.g., rabbit antihepatitis e antigen). Then enzyme-labeled antispecies antibody against the second antibody (e.g., goat anti-rabbit antibody) is added; followed by the substrate solution and measurement of the enzymatic activity as described above (Fig. 4).

Avidin–Biotin Systems

The use of the avidin–biotin complex (ABC) can also increase the sensitivity of the indirect assay. Avidin (from egg white) binds with high affinity to biotin (a low molecular weight vitamin); one molecule of avidin can react with four biotin molecules. Streptavidin (from bacterium *Streptomyces avidinii*) shows less nonspecificity than egg white avidin because it is not glycosylated, and is the preferred substitute. Streptavidin can be labeled with several enzymes, but alkaline phosphatase and horseradish peroxidase are used widely. The ABC reaction occurs

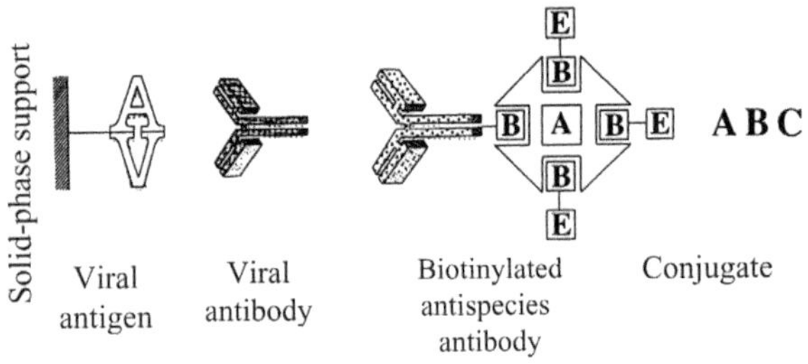

Figure 5 Principle of indirect avidin–biotin immunoassay systems for detection of viral antigen.

independently of any immunological reactions in the assay. Avidin–biotin systems have wide applications and are used in both qualitative and quantitative EIA.

In avidin–biotin systems, biotin-conjugated primary antiviral antibodies (e.g., rabbit anti-HSV) are incubated with test samples, unbound antibodies are washed, and enzyme-conjugated streptavidin is added. The latter binds to the primary antibody through the strong attraction of streptavidin for the biotin conjugated to the antibody. Subsequent incubation with substrate and mounting of slides are as described above (Fig. 5). In the indirect variation, there are three incubations prior to color development to complete the test: (*i*) unlabeled primary antibody with the fixed antigen preparation, (*ii*) biotin-labeled antispecies antibody, (*iii*) biotin–streptavidin enzyme complex. The avidin–biotin system can avoid or reduce the nonspecific reaction due to antispecies antibody. One of the advantages of these assays is that a single streptavidin conjugate can be used for all biotinylated antibodies.

The remaining principles of quantitative biotin–avidin systems are the same as described above for either RIA or quantitative EIA. Avidin labeled with various fluorochromes (e.g., fluorescein), radionuclides, and enzymes are widely available commercially (29).

Latex Agglutination Assay for Detection of Viral Antigen

The viral agglutination assay is based on agglutination of antibody coated microspherical particles in the presence of viral antigens. Specific polyclonal or monoclonal antiviral antibodies are bound either by covalent linkage or adsorbed passively to the particles. Latex particles (microparticles of nanometer size) have most often been used. The most widely used latexes are polystyrene, polyacrylate, polyacrolein, and polyacrylamide. The test is very simple and rapid. However, latex agglutination assays suffer from lower sensitivity and nonspecificity as compared with EIA, because many clinical specimens produce nonspecific agglutination. In the test, antibody coated latex particles are mixed with clarified and/or diluted clinical specimens (e.g., stool specimens or nasopharyngeal secretions) on a microscope slide. The appearance of agglutination is read visually within 10 to 15 minutes. Latex agglutination assays are used for adenovirus, respiratory syncytial virus, and rotavirus. Currently, several slide latex agglutination test kits are available commercially with varying sensitivity and specificity (44–47).

Detection of Viral Antigen by Membrane-Based EIA

The membrane based EIA is also is known as the "cassette" EIA. The cassette EIA has become a popular method for antigen detection, especially for single sample testing (e.g., influenza virus). This is because large amounts of antibody can be bound to nitrocellulose (80 mg/cm^2), nylon membrane (480 mg/cm^2), or other modified membranes, greatly increasing the sensitivity and reducing total assay time to minutes. Generally, a membrane is attached to the bottom of a rigid plastic well which is in turn attached to a cassette containing absorbent material capable of holding all waste fluid generated by the assay. All reagents are added with a dropper. The antiviral antibody and controls are dotted or slotted onto the membrane in one well or separate wells in the same cassette. Alkaline phosphatase is the most frequently used detection enzyme. Nonspecific reactions are blocked by addition of blocking agent as described above.

In practice, the clarified and/or diluted clinical samples and the detection reagents are added as for the solid-phase direct or indirect assays. The reaction product is a colored, insoluble precipitate and generally read visually. Many kits are available commercially for respiratory viruses. Some commercial kits provide color charts to assist determination of a low positive or borderline reaction. Such samples should be retested before reporting results.

Lateral Flow Immunochromatographic Assays

The lateral flow immunochromatographic is a colorimetric assay on a membrane strip primed with antiviral antibody for the direct visual detection of viral antigen in clinical specimens. The basic principle for the detection is the use of colored particles (e.g., red) labeled with antiviral antibody which after addition of extracted clinical samples (e.g., nasal pharyngeal washing) travels through the test strip. The flow carries the mixture of sample and the red particles labeled with antiviral antibody (e.g., Influenza virus) through the membrane containing the viral antibody test line and then through a standard control line. When the viral antigen is present in clinical samples, the fluid phase antiviral antibody binds the antigen and this antigen–antibody complex is then in turn bound at the test line (forming a red line). Blue particles labeled with a standard control line system are also in the label formation. When the clinical sample is applied to the test, the blue particles flow with the sample and will bind directly to the control line to form a blue line. Therefore, when a positive sample is applied to the test, two lines become visible: one red at the Test Line and a second blue at the Control Line. When viral antigens are not present in the clinical samples, only the blue Control Line appears. The assay has been applied for several viruses. Commercial kits are available for RSV and influenza virus A and B and sold under name QuickVue (Quidel Corporation) and BinaxNOW (Inverness Medical). Figure 6 illustrates the Quidel QuickVue tests for influenza virus A and B using either nasal swabs or nasal washes. These assays have been evaluated by many investigators, and show a sensitivity of 54% to 95% and a specificity of 85% to 100% (48–57).

Optical Immunoassay

The basic principle of optical immunoassay (OIA) is the same as membrane-based EIA, except that it uses a gold color silicon thin-film biosensor wafer instead of a membrane. The OIA

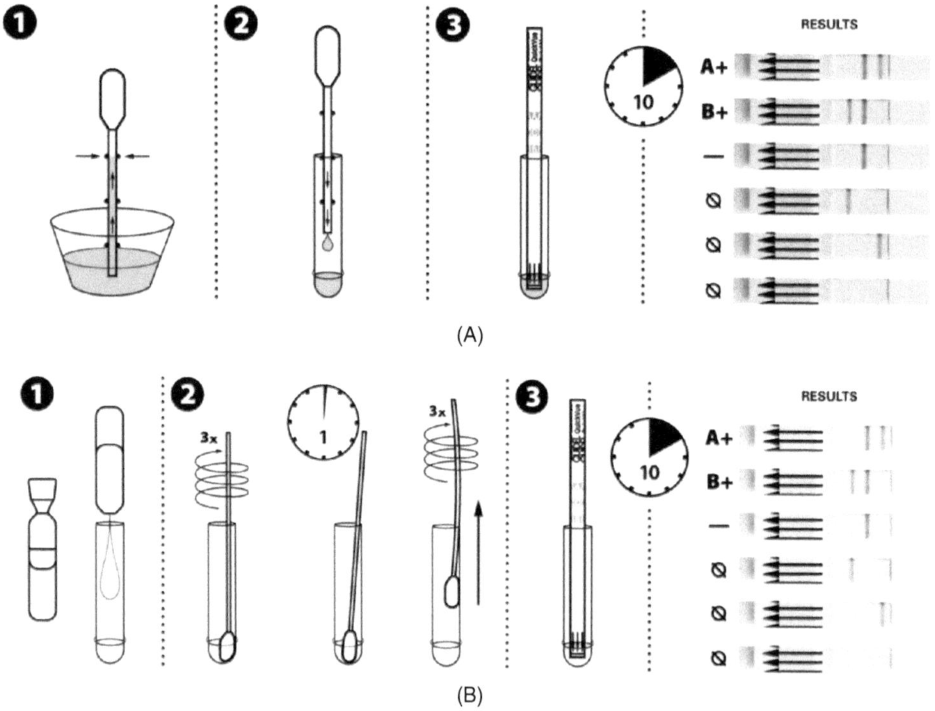

Figure 6 (*See color insert*) Quidel's QuickView test for influenza virus (**A**) and (**B**), using either nasal wash (*top*) or/and nasal swabs (*bottom*).

technology enables the direct visual detection of a physical change in the optical thickness of molecular films. The silicon wafer is coated with antiviral antibody and the optical change is the result of antigen–antibody binding on the surface of a silicon wafer. In the test, clinical samples (nasal wash, nasopharyngeal swab, etc.) are added to the wafer, the unbound compound is removed by washing, followed by conjugate and substrate solution. The enzymatic reaction increases the thickness (mass enhancement) of the molecular thin film. The change of thickness alters the reflected light path and is visually perceived as a color change. A positive result appears as purple dots on the predominant gold color background. For a negative result, the thickness is unchanged and the surface remains the original gold color (49,50,58).

Time-Resolved Fluoroimmunoassay

A very sensitive immunoassay for viral antigen detection in clinical specimens is time-resolved fluoroimmunoassay (TR-FIA). The principle of TR-FIA is based upon measuring the characteristics of fluorescence decay in fluoresceinated substances upon irradiation. Ordinary background autofluorescence of proteins or other compounds found in clinical samples (e.g., human serum albumin) has fluorescence decay times of 3 to 4 ns (nanoseconds). Because the background autofluorescence and FITC (4.5 ns) have approximately similar decay times, they cannot be differentiated by TR-FIA. However, certain earth metals, the lanthanides, have very long decay times of one thousand to one million ns (59). The lanthanide, which has been exclusively used in TR-FIA, is the trivalent europium (Eu3+) that has a decay time on the order of one microsecond and can be clearly separated from the background fluorescence. In addition, the Stoke's shift or the difference between the excitation wavelength (360 nm) and the emission wavelength (613 nm) is very large for europium.

The principle of TR-FIA is the same as direct EIA for detection of viral antigen in clinical samples with the following modification. In TR-FIA, the microtiter plate or strip is coated with purified capture antibody, and the remaining free binding sites of the plate blocked. The clinical specimen and the conjugate are added simultaneously to the appropriate well. After proper incubation time, the unbound components are washed. An enhancement solution is added and the fluorescence measured for one second with a single-photon fluorometer. The assay has been used for detection of several viruses and more recently has been applied in nanotechnology-based assays (59,60).

Protein Arrays

The need for technologies that allow highly parallel quantitation of specific viral proteins in a rapid and extremely small-volume format has become increasingly apparent. Protein array-based assays have great importance as approaches to potential global epidemics of highly lethal viruses such avian influenza and other infectious agents. The ability to measure multiple antigens simultaneously has application in many disciplines including the diagnosis of viral infection. This is because in most instances clinical samples are in limited amounts, precluding multiple probing. Protein arrays have the potential to probe over one thousand analytes in single slide. In these tests, proteins (e.g., antiviral antibodies) are printed on a glass slide as a spot a few millimeters in diameter. Each slide consists of a grid of several hundreds of spots of proteins. The assay can designed either for detection of viral antigen or antiviral antibody. Antibody printed on a microarray format can detect antigens at concentration below 1 ng/mL (61). A multiplex protein microarray for the simultaneous detection of multiple antigens and antibodies to five human hepatitis viruses has been reported (62).

Multiplex Assays

A simplified and/or small-scale version of the protein array is multiplex testing. Multiplex analysis provides the ability to perform multiple discrete assays in a single slide, tube, well, chip, or other format with the same sample at the same time. Although nucleic acid arrays have been applied extensively for many years, protein arrays may have the same application potential because either antigen or antibody can be detected and/or measured.

The current assay format of protein arrays is the same as for quantitative EIA. First, antiviral antibodies to several viruses are applied, followed by the clinical sample, and specific antigen–antibody complexes detected with labeled antibody. Several fluoro-cytometric multiplex platforms are currently on the market with different technologies and detector signals (63–65).

The Luminex Laboratory Multiple Analyte Profiling System is one of the multiplex viral antigen detection assays. The technology is a microsphere-based, multiplexed data acquisition and analysis platform for simultaneous, real-time flow cytometric analysis of up to 100 analytes in a microtiter plate format. Specifically, it consists of 100 distinct same-size (5.5 μm) sets of fluorescent polystyrene microspheres (red, orange, green), a flow cytometer, a 96-well microtiter platform reader, and analysis software. Individual sets of microsphere beads can be modified with reactive compounds such as antigen, antibody, and olignucleotides via amine linkage, and mixed to form a desired multiplex assay set. Currently, Luminex has a commercial kit, the xTag Respiratory Viral Panel (xTag RVP), which detects several respiratory viruses and viral subtypes (63–65).

Nanotechnology-Based Viral Diagnostic Assays

Nanotechnology-based diagnostics could provide a new generation of viral diagnostic assays due to their extremely high degree of sensitivity, high immunological specificity, high potential for multiplexing, and their ability to use different assay configurations, detector signals, and instrumentation. Currently, several nanotechnology-based methods have been described for detection of viral antigen in clinical samples. However, a valid comparison between these assays has not been performed to determine which of these methods or approaches are superior, and therefore a standard procedure does not exist. Such assays have applied for detection of Adenovirus, HBV, HIV, HSV-1, and RSV with various sensitivities and specificities (66–74).

FACTORS AFFECTING IMMUNOASSAYS

Multiple factors affect the sensitivity, specificity, reproducibility, and robustness of all immunoassays for detection of viral antigen in clinical samples. Good Laboratory Practice (GLP) is the foundation of quality control (QC) and quality assurance (QA) practices for obtaining and holding licenses from regulatory agencies [e.g., Clinical Laboratory Improvement Act (CLIA), College of American Pathologists (CAP)]. Each run should have positive and negative controls and exactly follow in-house procedures regarding incubation time, temperature, washing time, etc.

If a commercial kit is in use, the test must strictly follow the procedure as described by the manufacturer. Some of the newly approved FDA rapid virus antigen kits are waived and do not need further regulatory oversight. These assays are simple to perform and can be run in a doctor's office laboratory or at other institutions. However, a recently published survey by the CDC identified major problems, and deficiencies have been reported in terms of staff training, documentation, quality control, and other GLP issues (75).

Specimen Collection

Prerequisites for a successful immunoassay for detection of viral antigen are that a proper clinical sample is collected correctly, at the right time, and that the proper test is done for the probable viral disease diagnosis (76–79). All these factors will affect the final results of an immunoassay. For specific viruses the particular chapter offers more detailed information on collecting and handling of clinical samples and the reader is referred to those chapters.

Antibody

The sensitivity and specificity of all immunoassays depend on the quality of the antibody. The source of purified IgG can be from hyperimmune sera or monoclonal antibody (mAb). The antiserum should have high titer, high affinity and avidity, and should react specifically with the immunizing virus. The purity of the secondary and conjugate antibody is as important as the capture antibody on the solid phase. Optimal working dilutions must be determined for each antibody used in the assay. For example, serial dilutions of each lot of primary antibody must be evaluated against dilutions of the secondary or conjugate antibody used in the assay in a checkerboard fashion. More than one optimum dilution may be found. A lower concentration of primary antiserum, for example, may be acceptable with an increased concentration of conjugate. In addition to cost saving, higher dilutions of some immunoreagents may overcome problems with nonspecificity.

If antisera are prepared in-house, access to an animal colony, seed virus, cell culture, and growth media will be needed. One possibility is that the cell culture and immunizing host can be selected from the same or homologous host. Alternatively, if the immunizing virus is grown in cell culture derived from a heterologous host, a highly purified cell-free immunizing virus must be used. Antihost antibodies may be removed with multiple absorptions with uninfected host cells.

With the advent of highly specific mAbs produced by cell hybridization techniques, many of the problems associated with the production of polyclonal antisera have been eliminated (80). Although hybridoma technology is very labor intensive and requires personnel highly trained in cell culture, once the desired mAb is found and characterized it can be consistently produced in unlimited quantity and in high purity. The drawbacks of some mAbs are low affinity and avidity, with low stability under physicochemical manipulation, but through selection processes mAbs with high avidity can be produced. Low affinity/avidity can allow reagent bound to the solid-phase support to be removed during the washing steps in the antibody capture assay; they can also cause dissociation of antigen–antibody complexes after they are formed. In either case, a weak signal or false negative result can occur.

The sensitivity of an individual mAb preparation may be enhanced by pooling with mAbs directed to different epitopes (poly-mAbs), to obtain sensitivity equal to or greater than that of highly purified hyperimmune IgG. Many high quality mAbs to different viruses are available commercially, although they are often relatively expensive. mAbs have largely replaced polyclonal hyperimmune sera in most viral antigen detection assays (29).

Conjugates

In the past, most conjugates were prepared in laboratories where viral diagnostics were done. Preparing high quality conjugates required highly trained staff and dedicated animal facilities, limiting diagnostic virology to a few public health laboratories and large medical centers. However, currently many manufacturers nationally or internationally sell high-quality fluorochromes, enzymes, and biotin-labeled conjugates. Additionally, a few manufacturers offer polyclonal antibody or mAbs and complete antibody labeling kits and reagents, allowing preparation of the desired conjugates in-house (29).

The quality of the conjugate for detection of viral antigen is determined by the conjugation procedure and the quality of the antibodies. The affinity, avidity, concentration, and specificity of antibody in the conjugate will therefore set the limits of the test sensitivity. The main advantage of enzymes is their amplification of signal, in spite of the less efficient conjugation of the complex enzymes to antibodies. All fluorochromes and radionuclides generate either constant or diminishing signals with time. In contrast, enzymes generate progressively more signal with time, thus increasing the sensitivity attainable.

Fluorescence Microscopy

The first generation fluorescence microscopes used transmitted light. In these microscopes, the light passes through the excitation filter and is reflected by a mirror through a cardioid dark-field condenser up through the specimen. Fluorescence emitted by the specimen passes up through the objective and the barrier filter through the oculars to the observer. Virtually all modern fluorescence microscopes in virus laboratories use incident or epi-illumination systems, and are equipped with Ploem interference filters. In this type of microscope, the light source is positioned above the specimen. The exciting light passes through the exciter filter to a dichroic beam-splitting mirror, which deflects light of selected wavelengths down through the objective to the specimen top surface. Fluorescence light emitted by the specimen is guided through the objective, the dichroic mirror, and a barrier filter through the oculars to the observer. Light sources of sufficient excitation intensity are essential for fluorescence microscopy. The three most common light sources for epi-illumination are mercury arc bulbs, halogen quartz bulbs, and high-pressure xenon arc bulbs, which have a spectrum close to daylight. A common misconception is that ultraviolet light is required for excitation of fluorochrome. However, peak FITC absorption is 495 nm, with emission at 525. With interference filters, up to 85% of transmitted light is between 400 and 500 nm, in the visible spectrum and not in the ultraviolet range. For a CLIA approved IFA test, QC and QA is essential, including continuous maintenance of the microscope, regular replacement of the light source, and careful light alignment.

Solid-Phase Supports

The use of a solid-phase support for immunoassays was first described to eliminate the separation step in RIA for measuring bound and free antigen. Several solid-phase supports are in use, including polystyrene or polyvinyl tubes, beads, cuvettes, various membranes, microparticles, and microtiter plates. Microtiter plates constitute one of the most convenient solid supports, especially when many specimens are to be tested. They eliminate the need for individual racks, tubes, beads, and transfers. They can be used in automated equipment, together with rapid colorimetric or fluorometric readers and computerized data analysis (32,34).

The exact interaction of proteins with solid-phase matrices is not known. Plastics used for preparation of microtiter plates (e.g., polystyrene) are generally hydrophobic in nature and it is presumed that the immobilization of proteins to these plastics occurs predominantly by a hydrophobic reaction. In an effort to reach thermodynamic stability, proteins incubated with the plastic orient their hydrophobic region toward the adsorbing surface. To achieve their energetically favorable conformation on the surface, they may hide or change the epitope conformation normally expressed and exposed on the surface of the protein in solution. The loss of epitopes or the conformational changes of capture antibodies during immobilization are important factors that will affect the sensitivity of immunoassays. Hence, immobilization of proteins depends on the surface matrix, the structure of the protein, and the condition of immobilization.

The influence of pH on adsorption of protein on plastic surfaces remains controversial. Early studies indicated a pH dependence and the most widely used coating buffer is carbonate, but other buffers with lower pH have also been used. It appears that immobilization of proteins to membrane matrices is more pH dependent than to polyvinyl or polystyrene microtiter plates.

The sensitivity of solid-phase immunoassays is dependent on the amount of capture antibody that can be adsorbed on the solid support. Antibody to be immobilized to the microtiter plates should be highly purified with a final concentration of 10 to 12 μg/mL, and generally 50 to 100 μL is added to each well. Usually, over sensitizing the microtiter plates does not increase specific binding and, in some instances, has an adverse effect. Treatment of microtiter plates with poly-L-lysine increases nonspecific protein binding. Another limiting factor is the uneven binding of antigen or antibody to the solid support; this uneven protein binding is more of a problem with microtiter plates, although it is shared with all other solid supports. Chemically treated microtiter plates can reduce, but not eliminate, uneven binding. Antigen and antibody reactions are much faster and more efficient in liquid phase than in solid-phase immunoassays. Therefore, a solid-phase immunoassay requires relatively longer incubation for each step and this increases nonspecific reactions (34).

Enzyme

Enzymes are catalysts that participate in and accelerate chemical and biochemical processes without being consumed. One enzyme molecule can cleave millions of substrate molecules per minute without losing its enzymatic activity. The reaction product generated can be identified visually or microscopically, or can be measured colorimetrically, fluorometrically, or by luminescence. The sensitivity of all EIAs is influenced both by the kinetics of antibody–antigen interaction and the kinetics of enzyme–substrate reaction. Therefore, the ideal enzyme for EIA should have a high turnover rate, be stable under physicochemical manipulation, small in size, easily conjugated to antibody, have stable substrates, and be commercially available. For diagnostic work, well-standardized and stable commercial enzyme conjugates are available. The most commonly used enzymes for immunocytochemical staining and quantitative solid-phase EIA are horseradish peroxidase (HRP), alkaline phosphatase (AP), glucose oxidase (GO), and ß-D-galactosidase (ß-Gal), which fulfill most of these criteria. However, HRP and AP have been most widely used either in cyto-immunostaining or in quantitative EIA (35).

Substrates

The sensitivity of EIA is greatly affected by the purity and shelf-life of the substrates; it also is dependent on the solubility of cleaved products in quantitative EIA and their insolubility in Cyto-EIA, and on the stability of substrate product during measurement. The detectability of the products depends on the molar extinction of the substrate product, which is in the range of 10^{-5} to 10^{-6} M for colored products and 10^{-8} M for fluorescent products. Depending on the methods for detecting the reaction products, enzyme–substrates can be divided into several categories.

Cyto-EIA requires a precipitable chromogenic substrate, which must not diffuse from the site of formation during subsequent steps or under mountant. The substrate for peroxidase is hydrogen peroxide (H_2O_2) which has several chromogenic electron donors, of which the most commonly used are 3,3′-diaminobenzidine tetrahydrochloride (DAB), 3-aminoethyl-carbazole (AEC), and 4-chlorol-naphthol (4-C-1 N). DAB forms a brown, alcohol-soluble precipitate. AEC produces a red to reddish-brown precipitate, and 4-C-1 N a blue-black to blue-gray precipitate; both of these precipitates are soluble in alcohol and organic solvents and require a water-based mounting medium. All immunoperoxidase staining can be enhanced by metallic ions such as osmium tetroxide. Stock solutions of any of the above chromogens can be prepared in advance for daily use. DAB and AEC are potential carcinogens, while 4-C-1 N is toxic; they should all be handled with caution. However, noncarcinogenic and nonmutagenic substrates are available. One safe and versatile alternative is HistoMark^R chromogen (Kiregaard and Perry Laboratory, Gaithersburg, Maryland), which is available in a ready-to-use solution and suitable for various cyto-EIA formats for viral antigen detection. In our mumps virus plaque reduction neutralization test, HistMark^R was used for visualization and enumeration of viral plaques in 48 well plates. HistoMark^R generates dark black-brown spots (plaques) to mumps virus-infected cells, with little or no background staining of uninfected cells, and is more sensitive for plaque visualization than AEC (81). The most frequently used soluble substrate chromogens for HRP in quantitative EIA have been *O*-phenylenediamine dihydrochloride (OPD), 3,3′,5,5′-tetramethyl benzidine, dihydrochloride (TMB), and 2,2′-azinobis(3-ethylbenzothiazoline-6-sulfonic acid) (ABTS).

A number of azo dyes are safe chromogens for AP for light microscopy. The alkaline phosphatase product can be developed with a naphthol salt as a coupling agent in the presence of a diazonium salt as a capture agent (e.g., Fast Blue BBN, blue precipitate; or Fast Red TR, red precipitate). An alternate and more sensitive substrate for dot and transfer blot AP immunostaining is the McGrady reagent, a mixture of 5-bromo-4-chloro-3-indolyl phosphate and nitroblue tetrazolium (BCIP/NBT). The dark blue to purple-brown precipitate provides superior visualization of stained preparations (82). The substrate of choice for AP in quantitative EIA is *p*-nitrophenyl diphosphate (PNPP), which is easily soluble and is available in convenient tablets in many sizes.

Nonspecific Reactions

An effective immunoassay, whether fluorescent, Cyto-IE staining, or quantitative solid phase, should be specific, reproducible, and give a strong signal easily discernible against negligible background. However, the causes of nonspecificity are multifactorial and the amount varies considerably from one assay to another. Some nonspecific reactions are common to all assays, while others are intrinsic to a particular assay. Nonspecificity can severely affect the interpretation of results, so recognizing and controlling the causes are important. Nonspecific reactions can be classified as immunological nonspecificity or method nonspecificity (83).

Immunological Nonspecificity

The use of proper controls, including infected and uninfected cells, and preimmune and postimmune antibody, will detect most problems of nonspecificity at the test level. Most antibody nonspecificity problems can be avoided if purified antigens are used for antibody production. Polyclonal antisera contain mixed populations of antibody that can bind to the clinical specimen nonspecifically, especially at high antibody concentration. Some of these nonspecific reactions can be reduced by using lower concentrations of antibody, or they can be removed by absorption with uninfected tissues or cell pack. In addition, excessive antibody concentration may inhibit immunological reactions, a phenomenon commonly referred to as the "prozone effect." Prozone effects are very critical when a single screening dilution is used.

Tissue-nonspecific reactive components may be inhibited with neutral protein, normal IgG, or normal serum of the same species as the fluorescein–enzyme-labeled antibody. Infection of cell culture, especially with herpesviruses (e.g., CMV), produces nonspecific Fc receptors. Thus, using labeled whole serum antibody conjugate is not advised. Fc receptor binding can be avoided by using labeled F(ab′)2 antibody fragments. Equally effective and less expensive than antibody fragments are goat antispecies-globulin or IgG conjugates, which have been shown to have low Fc receptor binding activity compared to rabbit IgG (84). Virus-induced Fc receptor problems in CMV, HSV, and VZV testing can be avoided by using mAbs to immediate early proteins (IEPs), which stain the nuclear antigens of the virus (24,29,85).

Method Nonspecificity

With a thorough knowledge of the test system, appropriate precautionary steps can be taken to avoid most method nonspecificity problems. For example, binding of immunoreagents to unreacted test sites through ionic charge in Cyto-EIA, as well as to membrane and microtiter plates, can yield nonspecific background staining and lead to difficulties in test interpretation. To eliminate this, incubation with a blocking agent usually precedes each antibody incubation and is also incorporated in the conjugate diluents. A recent refinement includes the use of blocking agent(s) in all antibody diluents, which shortens the total test time significantly. Suitable blocking agents include normal goat, horse, or fetal bovine serum, bovine serum albumin, gelatin, and casein.

In general, fluorescein-labeled antibodies are negatively charged, binding readily to positively charged acidophilic components of the cell cytoplasm and nucleus. Certain fixatives, such as aldehyde derivatives, may increase the positive charge of the specimen, increasing these nonspecific reactions. The presence of residual free fluorescein in the conjugate is another source of nonspecificity. An excessive fluorescein/protein ratio may induce aggregation of antibodies during labeling, also causing nonspecific reaction. Some cells or tissues may autofluorescence under the light spectrum used for fluorescence microscopy. In general, autofluorescence is more yellow-green than the fluorescence of fluorescein. Autofluorescence and some other causes of background staining can be reduced or eliminated by addition of Evans blue as a counterstain (13,28).

Endogenous enzyme activity, though not generally a problem of fixed monolayer cell preparations, is frequently encountered in clinical specimens (tissue scrapings, urine) due to naturally occurring tissue enzymes or heme-containing cells (macrophages, neutrophils, eosinophils, erythrocytes), mucus, bacteria, damaged tissues, neoplastic tissues, or even some normal tissues. Unless inhibited or destroyed, endogenous enzymes will react with the substrate chromogen and lead to false-positive staining. Enzyme inhibitors should be selected to inhibit the unwanted enzymes irreversibly without inhibiting the antibody–antigen reactions. Measures to suppress endogenous peroxidase activity generally include a pretreatment of the specimen with methanol/H_2O_2, sodium azide/H_2O_2, or acid/alcohol prior to incubation with primary antibody. Alkaline phosphatase inhibitors include levamisole or acetic acid. Pretreatment of virus-infected cells or tissues to remove endogenous enzymes may destroy some viral antigens.

False-negative staining may be attributable to the masking of antigenic determinants by overfixation of the tissue specimen, and false-positive results may occur due to endogenous antibody. Proteolytic enzyme pretreatment of tissues has successfully abolished fixation-induced antigenic crosslinking. Protease, pronase, and trypsin appear to free cross-linked antigen molecules, thus allowing antibody to enter and react. Retrieval of viral antigens from formalin-fixed and paraffin-embedded tissues has posed certain difficulties. An antigen retrieval procedure has been described for VZV, based on boiling deparaffinized tissue sections attached to microscopic slides in an acidic buffer, followed by standard IF staining (28).

Other major false negative reactions in assays result from immune complexes, which are present in some clinical samples in certain viral infections. This is especially true during testing for p24 antigen in the sera of HIV infected individuals. In order to prevent false negative reactions, clinical samples are treated with an acidic buffer to dissociate the antigen and antibody complexes before testing. Nearly all commercial kits currently available for testing HIV p24 antigen are based on this principle (86,87).

All solid-phase immunoassays incorporate washing steps, and inadequate washing may cause nonspecific reactions. For example, inadequate washing may produce a very uniform background staining over some or all of the specimen, thereby masking a specific reaction, or it may result in the nonspecific binding of residual immunoreagents to the specimen or solid support. Adequate blocking reagent will inhibit these types of nonspecific binding. All microtiter plate washers must be decontaminated, washed, and calibrated daily prior each run, to prevent nonspecific binding.

Fixation and Fixatives

The ideal fixative should insolublize viral antigens, and should not denature the protein structure or change its immunological reactivity or permeability to antibody. It should also be able to inactivate infectious virions in the tissue or cells. Acetone is the usual fixative, with varying time and temperatures. Acetone will inactivate many viral agents, but some nonlipid viruses (e.g., adenovirus) may require acetone for 30 minutes at 50°C to 60°C or 10% formalin in buffer for a few minutes. EIA-stained slides may also be treated in 2% formalin for two minutes for inactivation before reading with the light microscope (88). The disposal of acetone, formalin, or other fixatives is of great concern environmentally, and there is interest in other means of fixation. For example, rabies virus preparations can be fixed in a microwave oven with good results (89). Viral antigens are generally labile, and thus fixatives must be selected carefully. The minimum time and concentration should be used to prevent changes in antigenic characteristics. Ethanol or methanol in combination with acetone, as well as formaldehyde and glutaraldehyde are used. Cell culture monolayers grown on plastic surfaces have been fixed with absolute ethanol or 80% to 90% cold acetone with good results. Some experimentation may be necessary to find a satisfactory method. Overfixation can result in false negative tests due to denaturation of viral antigen, preventing recognition by antibody. Albumin is sometimes necessary to "glue" a sample to the slide during fixation and testing.

Mounting Media

IFA preparations rapidly fade under prolonged illumination by the intense excitation light of epi-illumination. Initially intense fluorescence and contrast become weak and can no longer be readily observed. This is a drawback for photography with long exposure times. Fading can sometimes be reduced by adding certain substances to the mounting medium. A good general-purpose mounting medium is a polyvinyl alcohol (PVA)/glycerol formulation in Tris buffer, containing 25 μg/mL DABCO(1,4-diazabicyclo[2.2.2]octane) (90). IF preparations mounted in this medium can be examined and photographed with longer exposure times. Because of the solubility of some of the precipitated chromites in organic solvents, mounting medium is carefully chosen to preserve immunoenzymatically stained specimens. The mounting medium causes the coverslip to adhere to the slide, and prevents fading of the chromogen. Slides stained with the peroxidase chromogens DAB or AEC and mounted in PVA retain their color intensity over many years. A number of suitable mounting media are available commercially.

Automation

Automation, especially for the testing of large numbers of samples, has been a goal of virological and serological development since the 1970s. Instruments are continuously being improved to provide rapid, accurate, and reproducible assays, and to relieve the skilled virologist from repetitive operations. Robotic technology is used for liquid handling and transfers. Automatic and semi-automatic pipettors, dispensers, diluters, plate washers, and plate readers with complete computer programming (analysis and printout) are available from numerous manufacturers (91–93). Automation must be cost effective and affordable, and the computer programming should allow upgrading. The space required for the instrument can also be a factor. Initial capital investment can often be reduced by options (e.g., lease) offered by many manufacturers. Upgrades may also be available as manufacturers introduce new instruments. Some companies provide maintenance for their instruments, and some have trained technicians available on call to handle repairs or adjustment problems.

Another major innovation in automation is the application of information technology (IT) for the management of laboratory operations. Since the end product of any laboratory testing is information, having appropriate information management solutions is crucial for effective daily operations. Currently, there are many companies that produce information management system software. For laboratory applications, two approaches have evolved: the Laboratory Information System (LIS) and the Laboratory Information Management System (LIMS). LIS and LIMS perform similar functions. LIS has been primarily used in small or large hospitals and clinical labs, whereas LIMS is targeted toward research analysis and clinical trials, or the pharmaceutical, environmental, and industrial markets. There are two distinct forms of LIMS software: web-based LIMS and web-enabled LIMS. A LIMS should be the central nervous system of the laboratory and be able to perform the following functions: specimen validation and identification, barcode labeling, work scheduling, test initiation, data acquisition, automated or manual entry of results, automatic or manual test result reporting, data analysis, statistical analysis, QC reporting, assay validation, technologist training, specimen/test/technologist tracking, maintenance of standard operating procedures, billing, compliance, and other laboratory management functions (94).

For several years, the California Department of Public Health (CDPH) has used the STARLIMS software in all its laboratories, including the Viral and Rickettsial Disease Laboratory. This is because STARLIMS offers a web-based solution designed to manage the operations of multidisciplinary Public Health Laboratories. Furthermore, the CDC, the Association of Public Health Laboratory (APHL), several State Public Health Laboratories including CDPH and the STARLIMS Company coordinate a nationwide information-based system, allowing rapid communication between labs.

REPORTING OF DIAGNOSTIC IMMUNOASSAY RESULTS

The function of an immunoassay is to provide information and data to confirm a clinical diagnosis of viral infection. Immunoassays are useful only if they produce accurate, sensitive, specific, and reproducible results. An important consideration is how best to report the results of a quantitative assay such as EIA or RIA to the clinician (95,96). The nomenclature and interpretation of data derived from virological assays are quite different from those obtained by the same methodology in other disciplines. For example, in clinical chemistry the concentration of serum insulin should be within a certain "normal" range; values outside of that range are considered abnormal and clinically significant for the patient, and therefore the exact quantitation of insulin is the objective of these assays. However, in certain viral infections (e.g., for detection of hepatitis B surface antigen [HBsAg] in serum) a qualitative result may be sufficient, and in other instances viral quantitation cannot be correlated with clinical disease.

In some instances where a critical diagnosis is required, a combination of other viral antigen markers may provide additional information on the status of the disease and a more precise clinical diagnosis. For example, in addition to HBsAg determination in a serum, another useful marker for hepatitis B infection is hepatitis "e" antigen. Presence of a high level of HBeAg and absence of anti-e in the serum may indicate acute or recent infection, whereas a high level of anti-e in the absence of HBeAg in the assay may indicate either a convalescent condition in the absence of HBsAg, or a persistent carrier state if HBsAg is present (20–22).

In the case of HIV infection, the presence of core p24 antigen in the serum has been a potential prognostic marker for the progression of AIDS. In an HIV infected individual, if p24 antigen is on the rise, the level of anti-p24 is declining, and other immunological data (e.g., $CD4^{+}$ T-lymphocytes) are consistent with these, the progression of full-blown AIDS can be accurately predicted (97,98). These examples illustrate why detailed testing and quantitation of viral antigens are required for accurate viral diagnosis and assessment of clinical status.

In contrast, the majority of routinely used viral diagnostic tests for detection of viral antigen are qualitative in nature, such as IFA or Cyto-EIA staining. These assays provide only positive or negative results and cannot determine the quantity of viral antigen in the clinical samples or the state of infection. In both assays the intensity of staining is commonly read and graded in terms of 1+ to 4+ and reported as antigen detected or not detected, which may be sufficient for rapid diagnosis of most viral infections.

DISCLAIMER

The names of reagents, test kits, laboratory instruments, and other commercial products described in this chapter are intended for information purposes only, and not endorsed by either the author or by the Viral and Rickettsial Disease Laboratory Branch of State of California Department of Public Health.

REFERENCES

1. Schmidt NJ. Advances in the laboratory diagnosis of viral infections. In: de la Maza LM, Peterson EM, eds. Medical Virology. New York, NY: Elsevier, 1981:1–33.
2. Lennette DA. General principles for laboratory diagnosis of viral rickettsial and chlamydial infections. In: Lennette EH, Lennette DA, Lennette ET, eds. Diagnostic Procedures for Viral, Rckettsial and Chlamydial Infections, 7th ed. Washington, D.C.: American Public Health Association, 1995:3–25.
3. Forghani B, Hagens S. Diagnosis of viral infections by antigen detection. In: Lennette EH, Lennette DA, Lennette ET, eds. Diagnostic Procedures for Viral, Rickettsial and Chlamydial infections, 7th ed. Washington DC: American Public Health Association, 1995:79–96.
4. Forghani B. Detection by antigen detection. In: Lennette EH, Smith TF, eds. Laboratory Diagnosis of Viral Infection, 3rd ed. New York, NY: Marcel Dekker, 1999:71–94.
5. Storch GA. Diagnostic virology. Clin Infect Dis 2000; 31:739–751.
6. Leland DS. Clinical virology: Concepts and perspectives. In: Lennette EH, Smith TF, eds. Laboratory Diagnosis of Viral Infection, 3rd ed. New York, NY: Marcel Dekker, 1999:1–20.
7. Leland DS, Ginocchio. Role of cell culture for virus detection in the age of technology. Clin Microbiol Rev 2007; 20:49–78.
8. Olivo P. Application of genetically engineered cell lines to diagnostic virology. In: Lennette EH, Smith TT, eds. Laboratory Diagnosis of Viral Infection, 3rd ed. New York, NY: Marcel Dekker, 1999:159–175.
9. Crist GA, Langer JM, Woods GL, et al. Evaluation of the Elvis plate method for the detection and typing herpes simplex virus in clinical specimens. Diagn Microbiol Infect Dis 2004; 49:173–177.
10. Kowalski RP, Karenchak LM, Shah C, et al. Elvis: A new 24-hour culture test for detecting herpes simplex virus from ocular samples. Arch Ophthal 2002; 120:960–962.
11. Lasola P, Bufton K, Ismail N, et al. Perspective comparison of R-mix shel vial system with direct antigen test and conventional cell culture for respiratory virus detection. J Clin Virol 2007; 38:210–216.
12. Buck GE, Wieseman M, Stewart L. Comparison of mixed cell culture containing genetically engineered BGMK and CaCo-2 cells (super mix E mix) with RT-PCR and conventional cell culture for diagnosis of enterovirus meningitis. J Clin Virol 2002; (24 suppl 1):S13–S18.
13. Forghani B, Hurst JW, Chan CS. Advantages of a human immunodefficiency virus type 1 (HIV-1) persistently infected HeLa T4$^+$ cell line for HIV-1 indirect immunofluorescence serology. J Clin Microbiol 1991; 29:2266–2272.
14. Benjamin DC, Berzofsky JA, East IJ, et al. The antigenic structure of proteins: A reappraisal. Ann Rev Immunol 1984; 2:7–101.
15. Underwood PA. Measurement of the affinity of viral antibodies. Adv Virus Res 1988; 34:283–289.
16. Colman PM, Air GM, Webster RG, et al. How antibodies recognize virus proteins. Immunol Today 1987; 8:323–326.
17. Getzoff ED, Trainer JA, Lerner RA, et al. The chemistry and mechanism of antibody binding to protein antigens. Adv Immunol 1988; 43:1–98.
18. Bonner AB, Monroe KW, Tally LI, et al. Impact of the rapid diagnosis of influenza on physician decision-making and patient management in the pediatric emergency department: Results of a randomized, prospective, control trial. Pediatrics 2003; 112:363–367.
19. Uyeki TM. Influenza diagnosis and treatment in children: A review of studies on clinically useful tests and antiviral treatment for influenza. Pediatr Infect Dis J 2003; 22:164–177.
20. World Health Organization. HIV Assays: Operational characteristics, Report 15, Antigen/Antibody ELISAs. UNAIDS 2004. www.who.int/diagnostic_laboratory/evaluations/hiv/en. Accessed July 2008.
21. World Health Organization. WHO recommendation on the use of rapid testing for influenza diagnosis 2005. www.who.int/csr/disease/influenzanetwork/en/index.html. Accessed July 2008.
22. U.S. FDA, Center for Biologic Evaluation Research. Donner Screening assays for infectious agents and HIV diagnostic assays. 2008; Silver Spring, MD. www.fdw.gov/cbr/products/testkits.htm. Accessed September 2008.
23. Riggs JL. Immunofluorescence staining. In: Schmidt NJ, Emmons RW, eds. Diagnostic Procedures for Viral, Rickettsial, and Chlamydial Infections, 5th ed. Washington, DC: American Public Health Association, 1979:123–133.

24. St. George K, Boyd MJ, Lipson SM,et al. A multisite trial comparing two cytomegalovirus (CMV) antigenemia test kits, Biotest CMV Brite and Bartels/Argene antigenemia. J Clin Microbiol 2000; 38:1430–1433.
25. Landry ML, Ferguson D. SimulFluor respiratory screen for rapid detection of multiple respiratory viruses in clinical specimens by immunofluorescence staining. J Clin Microbiol 2000; 38:708–711.
26. Landry ML, Cohen S, Ferguson D. Real-time PCR compared to Binax Now and cytospinn immunofluorescence for detection of influenza in hospitalized patients. J Clin Virol 2008; 43:148–151.
27. Landry ML, Cohen S, Ferguson D. Perspective study of human metapneumovirus detection in clinical samples by use of light diagnostics direct immunofluorescence reagent and real-time PCR. J Clin Microbiol 2008; 46:1098–1100.
28. Forghani B. Laboratory diagnosis of infection. In: Arvin AM, Gershon AA, eds. Varicella-zoster Virus: Virology and Clinical Management. New York, NY: Cambridge University Press, 2000:351–382.
29. Linscott's directory of immunological and biological reagents. www.linscott'sdirectory.com. Accessed July 2008.
30. NIH AIDS Research and Reference Reagent Program. aidsreagents@thermofisher.com. Accessed July 2008.
31. American Type Culture Collection. www.atcc.org. Accessed July 2008.
32. Forghani B. Radioimmunoassay systems. In: Lennette EH, ed. Laboratory Diagnosis of Viral Infections, 2nd ed. New York, NY: Marcel Dekker, 1992:89–104.
33. Forghani B, Schmidt NJ, Lennette EH. Solid phase radioimmunoassay for identification of herpes hominis types 1 and 2 from clinical materials. Appl Microbiol 1974; 28:661–667.
34. Forghani B. Enzyme immunoassay systems. In: Lennette EH, ed. Laboratory Diagnosis of Viral Infection. New York, NY: Marcel Dekker, 1992:105–125.
35. Forghani B, Dennis J. Immunoenzymatic staining. In: Schmidt NJ, Emmons RW, eds. Diagnostic Procedures for Viral, Rckettsial and Chlamydial Infections, 6th ed. Washington, DC: American Public Health Association, 1989:135–156.
36. Gray JJ, kohli E, Ruggeri FM, et al. European multicenter evaluation of commercial enzyme immunoassys for detection of norovirus antigen in fecal samples. Clin Vaccine Immunol 2007; 14:1349–1355.
37. Jiang X, Wilton N, Zhong WM, et al. Diagnosis of human calciviruses by use of enzyme immunoassays. J Infect Dis 2000; 181(suppl 2):5349–5359.
38. McIver CJ, Palombo EA, Doultree, JC, et al. Detection of astrovirus gastroenteritis in children. J Virol Methods 2000; 84:99–105.
39. Wilhelmi dCI, Revilla A, del Alamo JM. Evaluation of two commercial enzyme immunoassays for detection of norovirus in fecal samples from hospitalized children with sporadic acute gastroenteritis. Clin Microbiol Infect 2007; 13:341–343.
40. Alvinn R, Janz B, Doerr HW. Viral gastroenteritis. An epidemiologic investigation between the period 2001–2006. Med Klin (Munich) 2008; 103:389–395.
41. Chen D, Kaplan LA. Performance of a new-generation chemiluminescent assay for detection of hepatitis B surface antigen. Clin Chem 2006; 52:1592–1598.
42. Leary TP, Guterrez RA, Muehoff AS, et al. A chemiluminescent, magnetic particle-based immunoassay for the detection of hepatitis C virus core antigen in human serum and plasma. J Med Virol 2006; 78:1436–1440.
43. Hamilton MS, Abel DM, Ballam YJ, et al. Clinical evaluation of the ZstatFluII test: A chemiluminescence rapid diagnostic test for influenza virus. J Clin Microbiol 2002; 40:2331–2334.
44. Landry ML, Cohen S, Ferguson D. Comparison of Binax NOW and Directigen for rapid detection of influenza A and B. J Clin Virol 2004; 31:113–115.
45. Raboni SM, Nogueira MB, Hakim, et al. Comparison of latex agglutination with enzyme immunoassay for detection of rotavirus in fecal specimens. Am J Clin Pathol 2002; 117:392–394.
46. Ruest A, Michaud S, Delandes S, et al. Comparison of Directigen Flu A+B test, the QuikVue influenza test, and clinical case definition to viral culture and reverse transcription-PCR for rapid detection of influenza virus infection. J Clin Microbiol 2003; 41:3487–3493.
47. Rodriguez WJ, Schwartz RH, Thorne MM. Evaluation of diagnostic tests for influenza in a pediatric practice. Pediatr Infect Dis J 2002; 21:193–196.
48. Fader RC. Comparison of the Ninax NOW Flu A enzyme immunochromatographic assay and R-Mix shell vial culture for the 2003–2004 influenza season. J Clin Microbiol 2005; 43:6133–6135.
49. Cazacu AC, Demmler GJ, Neuman MA, et al. Comparison of a new lateral-flow chromatographic membrane immunoassay to viral culture for rapid detection and differentiation of influenza A and B in respiratory specimens. J Clin Microbiol 2003; 42:3661–3664.
50. Cruz AT, Cazacu AC, Greer JM, et al. Performance of a rapid assay (Binax NOW) for detection of respiratory syncytial virus at a children's hospital over a 3-year period. J Clin Microbiol 2007; 45:1993–1995.

51. Jonathan N. Diagnostic utility of Binax Now RSV- an evaluation of the diagnostic performance of Binax Now RSV in comparison with cell culture and direct immunofluorescence. Ann Microbiol Antimicrob 2006; 5:13–21.
52. Konstantinos A, Mack K, Bonsu BK, et al. Evaluation of the Quidel QuickVue test for detection of influenza viruses in pediatric emergency medicine stetting by using different specimen collection methods. J Clin Microbiol 2006; 44:2638–2643.
53. Kikuta H, Sakata C, Gamo R, et al. Comparison lateral-flow immunochromatography assay with real-time reverse transcription-PCR for detection of human metapneumovirus. J Clin Microbiol 2008; 46:928–932.
54. Zainah S, Abdul-wahab AH, Mariam M. Performance of commercial rapid dengue N1S1 antigen immunochomatographic test with reference toN1S1 antigen capture ELISA. J Virol Methods 2009; 155:157–160.
55. Khamrin P, Nguyen TA, Phan TG, et al. Evaluation of immunochromatography and commercial enzyme-linked immunosorbent assay for rapid detection of norovirus antigen in stool samples. J Virol Methods 2008; 147:360–363.
56. Hideaki K, Sakata C Gamo, R et al. Comparison of a lateral-flow immunochromatography as time reverse transcription-PCR for detection of hman metapneumovirus. J Clin Microbiol 2008; 46: 928–932.
57. Rahman M, Vandermause MF, Kiel BA, et al. Performance of Binax Now Flu A and B and direct fluorescent assay in comparison with a composite of viral culture or reverse transcription PCR for detection of influenza infection during 2006–2007 season. Diagn Microbiol Infect Dis 2008; 62: 162–164.
58. Hindiyeh M, Goulding C, Morgan H, et al. Evaluation of Biosta Flu OIA assay for rapid detection of influenza A and B viruses in respiratory specimens. J Clin Virol 2000; 17:119–126.
59. Hierholzer JC, Anderson LJ, Halonen P. Monoclonal time-resolved fluoroimmunoassay: Sensitive systems for the rapid diagnosis of respiratory viruses. In: de la Maza LM, Peterson EM, eds. Medical Virology. New York, NY: Plenum Press, 1990:17–46.
60. Valane A, Huopalahti S, Vainiopaa R, et al. Rapid and sensitive HBsAg immunoassay based fluorescent nanoparticle labels and time-resolved detection. J Virol Methods 2005; 129:83–90.
61. Brian Haab B, Dunham MJ, Brown PO. Proten microarrays for highly parallel detection and quantitation of specific protens and antibodies in complex solution. Genome Biol 2001; 2:Research 004.
62. Xu R, Gan X, Fang Y, et al. A simple, rapid, and sensitive integrated protein microarray for simultaneous detection of multiple antigens and antibodies of five human hepatitis viruses (HBV, HCV, HDV, HEV, and HGV). Anal Biochem 2007; 362:69–75.
63. Merate F, Yghoubian S, Janeczko R. Principles of the xTAG respiratory viral panel assay (RVP). J Clin Virol 2007; 40(suppl 1):S31–S35.
64. Krunic N, Yager TD, Himsworth D, et al. xTAG RVP assay: Analytical and clinical performance. J Clin Virol 2007; 40(suppl 1):S39–S46.
65. Nolte FS, Marshall DJ, Rasberry C, et al. MultiCode-Plex system for multiplexed of seventeen respiratory viruses. J. Clin Microbiol 2007; 45:2779–2786.
66. Tripp RA, Alvarez R, Anderson B, et al. Bioconjugated nanoparticle detection of respiratory syncytial virus. Int J Nanomedicine 2007; 2:117–124.
67. Kim EY, Stanton J, Korber BT, et al. Detection of HIV-1 p24 Gag in plasma by a nanoparticle-based bio-barcode-amplification method. Nanomedicine 2008; 3:293–303.
68. Tang S, Zhao J, Storhoff, et al. Nanoparticle-based biobarcode amplification assay (BCA) for sensitive and early detection of human immunodeficiency type1 capsid (p24) antigen. J Acquir Immune Defic Syndr 2007; 46:231–237.
69. Jaiswal JK, Simon SM. Potential and pitfalls of fluorescent dots for biological imaging. Trends Cell Biol 2004; 14:497–504.
70. Takeda S, Ozaki H, Hattori S, et al. Detection of influenza virus hemagglutination with randomly immobilized anti-hemaagglutinin antibody on a carbon nanotube sensor. J Nanosci Nanotechnol 2007; 7:752–756.
71. Yun Z, Zhengatao D, Jiachang Y, et al. using cadmium telluride quantum dots as a proton flux sensor and applying to detect H9 avian influenza virus. Anal Biochem 2007; 364:122–127.
72. Dong C, Huang X, Ren J. Characterization of water-soluble luminescent quantum dots by fluorescence correlation spectroscopy. Ann N Y Acad Sci 2008; 1130:253–261.
73. Ymeti A, Greve J, Lambeck PV, et al. Fast, ultrasensitive detection using Young interferometer sensor. Nano Lett 2007; 7:394–397.
74. Perez JM, Simeone FJ, Saeki Y, et al. Viral-induced self-assemmmbly of magnetic nanoparticles allows the detection of viral particles in biological media. J Am Chem Soc 2003; 125:10192–10193.
75. Center for Disease Control and Prevention. Good laboratory practices for waived testing sites. MMWR Morb Mort Wkly Rep, November 2005.

76. Covalciuc K, Webb K, Carlson CA. Comparison of four clinical specimen types for detection of influenza A and B virus by optical immunoassay (FLU OIA test) and cell culture. J Clin Microbiol 1999; 37:3971–3974.
77. Smith TF. Specimen requirements: Selection, Collection, transport and processing. In: Spector S, Hodinka RL, Young SA, eds. Clinical Virology Manual, 3rd ed. Washington, D.C.: ASM Press, 2000:11–26.
78. Landry ML, Cohen S, Ferguson D. Impact of sample type on rapid detection by cytospin-enhanced immunofluorescence and membrane enzyme-linked immunosorbent assay. J Clin Microbiol 2000; 38:429–430.
79. Loens K, Van Heirstraeten L, Malhotra-Kumar S, et al. Optimal sampling sites and methods for detection of pathogens possibly causing community-acquired lower respiratory tract infection. A mini review. J Clin Microbiol 2009; 47:21–31.
80. Forghani B, Karu AE. Monoclonal antibodies. In: Harper DR, ed. Virology Labfax. Oxford, UK BIOS Scientific Publishers LTD, 1993:187–213.
81. LeBaron CW, Forghani B, Beck C. Persistence of mumps antibodies after 2 doses of measles-mumps-rubella vaccine. J Infect Dis 2009; 199:552–560.
82. McGrady J. A tetrazolium method for non-specific alkaline phosphatses. Histochemie 1970; 23:180–184.
83. Van Noorden S. Problems and solutions. In: Beesley JE, ed. Immunocytochemistry- a practical approach. New York, NY: Oxford University Press Inc., 1993:207–239.
84. Alexander EL, Sanders SK. F(ab′)2 reagents are not required if goat, rather than rabbit, antibodies are used to detect human surface immunoglobulin. J Immunol 1977; 119:1084–1088.
85. Forghani B, Mahalingam R, Vafai A, et al. Monoclonal antibody to immediate early protein encoded by varicalla-zoster virus gene 62. Virus Res 1990; 16:195–210.
86. Schupbach J. Measurement of HIV-1 p24 antigen by signal-amplification-boosted ELISA of heat-denatured plasma is a simple and inexpensive alternative. AIDS Rev 2002; 4:83–92.
87. Knuchel MC, Tomasik Z, Speck RF, et al. Ultrasensitive HIV-1 p24 antigen assay adapted for dried plasma spots to improve treatment monitoring in low-resource settings. J Clin Virol 2006; 36:64–67.
88. Groen G, Kurata T, Mets C. Modification of indirect immunofluorescence tests on Lassa, Marburg, and Ebola material. Lancet 1983; 1:654–655.
89. Davis C, Neill S, Raj P. Microwave fixation of rabies specimens for fluorescent antibody test. J Virol Methods 1997; 68:177–182.
90. Johnson GD, Davidson RS, McNamee KC, et al. Fading of immunofluorescence during microscopy: A study of the phenomenon and its remedy. J Immunol Methods 1982; 55:213–242.
91. Ismail N, Fish GE, Smith MB. Laboratory evaluation of a fully automated chemiluminiscence immunoassay for rapid detection of HBsAg, antibodies to HBsAg, and antibody to hepatitis C virus. J Clin Microbiol 2004; 42:610–617.
92. Kwon JA, Yoon SY, Lee CK, et al. Performance evaluation of tree automated human immunodeficiency virus antigen-antibody combination assays. J Virol Methods 2006; 133:20–26.
93. Taylor P, Pickard G, Gammie A, et al. Comparison of the ADVIA Centaur and Abbott AxSYM immunoassay systems for a routine diagnostic virology laboratory. J Clin Virol 2004; 30(suppl 1):S11–S15.
94. Raszko C, Turner E, Hinton MD. Laboratory Information Management Systems. Boca Raton, FL USA CRC Press, 2002.
95. Lanigan R, Mahony JB. Issues for reporting results. J Clin Virol 2007; 40(suppl 1):S47–S50.
96. Rabenau HF, Kessler HH, Kortenbush M, et al. Verification and Validation of Laboratory tests in clinical virology. J Clin Virol 2007; 40:93–98.
97. Saag MS. Use of virologic markers in clinical practice. J Acquir Immune Defic Syndr Hum Retrovirol 1997; 16(suppl 1):S3–S13.
98. Speers D, Philips P, Dyer J. Combination assay detection both Human Immunodeficiency Virus (HIV) p24 antigen and anti-HIV antibody opens a second diagnostic window. J Clin Microbiol 2005; 43:5397–5399.

9 Serologic Tests in Clinical Virology

Richard L. Hodinka

Departments of Pathology and Pediatrics, Clinical Virology Laboratory, Children's Hospital of Philadelphia and University of Pennsylvania School of Medicine, Philadelphia, Pennsylvania, U.S.A.

INTRODUCTION

Although the current emphasis in clinical virology is on the direct identification of viruses using antigen- or nucleic acid–based assays, antibody detection continues to play a significant role in the diagnosis and management of many viral diseases. This is particularly true for human immunodeficiency virus types 1 and 2 (HIV-1 and -2), the hepatitis viruses A–E, the herpesviruses, measles, mumps, and rubella viruses, parvovirus B19, the arboviruses, and HTLV-I and -II. The performance of viral serologies is useful in the diagnosis of recent or chronic viral infections, for determining the immune status of a person or population to a specific virus, and to verify the immune response to vaccination (Table 1). Rapid determination of immune status in a hospital setting may aid in the prevention of unnecessary spread of certain viruses to nonimmune patients or healthcare workers. Screening for virus-specific antibodies in blood products and organ donors and recipients prior to transplantation is important in preventing transmission of blood-borne viruses to individuals at high risk for severe disease. Knowing the serostatus of the donor and recipient is also helpful in determining the treatment or prophylaxis to be used following transplantation and in considering the type of donor and blood products to be given. Prenatal antibody screening can supply useful information on the risk for contracting certain viral infections during pregnancy. Identification of virus-specific antibodies in a patient's serum may also be the only means of making a viral diagnosis under certain circumstances. A number of viruses are difficult to grow in culture or to detect by other methods. Proper specimens for culture or direct detection assays may be difficult to obtain or may not be obtained. Specimens may also be collected too late in the course of the disease to detect viruses directly. The identified virus may also have an uncertain role in the current disease process, and serology may assist in establishing a causal relationship.

Technological advances over the last decade have led to more rapid, sensitive, and accurate tests for measuring virus-specific antibodies. Synthetic peptides and recombinant antigens are now being incorporated into serological assays to improve their performance over traditional tests based on whole viral lysates. Signal methods for the detection of captured antibody have improved and include highly sensitive fluorescent, phosphorescent, chemiluminescent, and electrochemiluminescent compounds. The continued development and advancement of automated technology and point-of-care devices has also made these tests easier and faster to perform.

ANTIBODY RESPONSE TO VIRAL INFECTIONS

Viral serologies represent an indirect approach to making a diagnosis of viral infection, since antibodies are measured as they develop in response to an invading virus. Exposure to a virus can lead to primary infection in an immuologically susceptible host or to exogenous reinfection in someone who has previously been infected. Primary infection with certain viruses, such as the family of herpesviruses, results in the establishment of a persistent or latent infection. Reactivation of these viruses can occur in response to different stimuli. The classes of immunoglobulins (Ig) produced following a viral infection include IgG, IgM, IgA, and possibly IgD, and IgE (1,2). During primary infection, IgM appears within several days after onset of symptoms, peaks at 7 to 10 days, and normally declines to undetectable levels within 1 to 2 months (Fig. 1). Because of the transient nature of the IgM antibody response, its presence is generally indicative of current or recent viral infection. Following natural viral infection or after successful immunization, IgG antibodies appear several days after the production of IgM, reach higher levels than IgM, and can persist lifelong in lower quantities. A significant rise in IgG antibody concentration over

Table 1 Utility of Serological Determinations in Clinical Virology

Clinical application	Most common virus(es)
Diagnosis of recent or chronic infections	
Hepatitis	HAV-HEV and HGV, CMV, EBV, HSV, VZV, HIV, coxsackievirus B, adenovirus, yellow fever virus
Central nervous system	HSV, CMV, VZV, EBV, HHV-6, enteroviruses, arboviruses, measles virus, mumps virus, rubella virus, rabies virus, HIV, LCMV
Congenital or perinatal	CMV, HSV, VZV, rubella virus, parvovirus B19, HBV, HCV, LCMV
Exanthems	Measles virus, rubella virus, parvovirus B19, HHV-6, HHV-7, arboviruses
Myocarditis or pericarditis	Coxsackievirus B types 1–5, influenza virus types A and B, CMV, parvovirus B19
Infectious mononucleosis	
Heterophile antibody positive	EBV
Heterophile antibody negative[a]	EBV, CMV, HIV, rubella virus
Nonspecific febrile illness	CMV, EBV, HHV-6, HHV-7, parvovirus B19, HIV, dengue virus, Colorado tick fever virus
T-cell leukemia	HTLV-I and HTLV-II
Hemorrhagic fever	Filoviruses, arenaviruses, flaviviruses, bunyaviruses
Hantavirus pulmonary syndrome	Sin nombre virus, other hantaviruses
Sudden acute respiratory syndrome	SARS coronavirus
Screening for immune status	
Preemployment	VZV, measles virus, mumps virus, rubella virus, HBV
Prenatal	Rubella virus, CMV, HSV, VZV, parvovirus B19, HBV, HCV, HIV
Pretransplant	CMV, HSV, EBV, VZV, HBV, HCV, HIV
Blood donation	HIV, HBV, HCV, HTLV-I, HTLV-II
Postexposure	HIV, HAV, HBV, HCV, VZV
Epidemiology/surveillance	All viruses
Verify response to vaccination	HAV, HBV, VZV, measles virus, mumps virus, rubella virus

[a]A comprehensive panel of EBV-specific serologic tests should be performed for patients with heterophile-negative infectious mononucleosis.

Abbreviations: CMV, cytomegalovirus; EBV, Epstein–Barr virus; HSV, herpes simplex virus; VZV, varicella-zoster virus; HIV, human immunodeficiency virus; LCMV, lymphocytic choriomeningitis virus; HAV, hepatitis A virus; HBV, hepatitis B virus; HCV, hepatitis C virus; HEV, hepatitis E virus; HGV, hepatitis G virus; HHV-6, human herpesvirus 6; HHV-7, human herpesvirus 7; HTLV, human T-cell lymphocytic virus.

time is accepted as evidence of a current or recent viral infection. The onset, level, and duration of IgA antibody are less predictable than either IgM or IgG, and serological tests specific for IgA antiviral antibodies are not performed routinely in diagnostic laboratories. The role and level of IgD and IgE produced in viral infections are unclear. During reactivation or exogenous reinfection, an anemnestic response in IgG antibodies will occur and an IgM response may or may not be observed (Fig. 1).

PROCEDURES FOR DETECTING ANTIVIRAL ANTIBODIES

A variety of methods are available for serodiagnosis of viral infections (Table 2). The more common traditional assays include complement fixation (CF), hemagglutination inhibition (HI), neutralization (NT), indirect immunofluorescence (IF), and anticomplement immunofluorescence (ACIF). With the exception of the immunofluorescence tests, these long-established assays involve considerable time, labor, and standardization and are now performed primarily

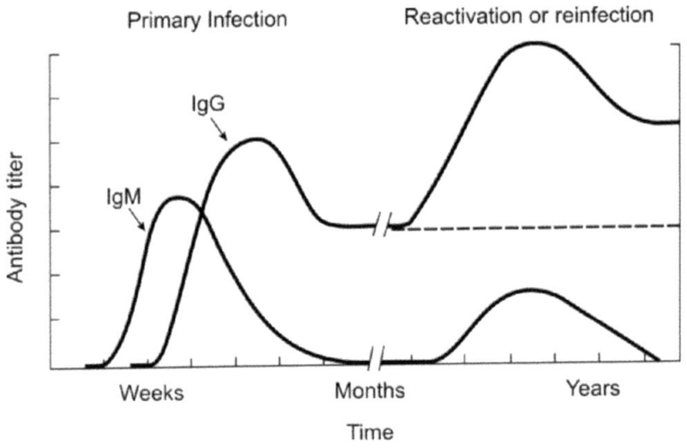

Figure 1 Diagram depicting the typical IgM and IgG antibody response following primary viral infection and reactivation or reinfection. The dotted line represents virus-specific IgG antibodies that can persist lifelong in lower quantities following primary infection and in the absence of reactivation or reinfection.

in public health, research, and specialized reference laboratories and only for a select number of viruses. They have been largely replaced over time by solid-phase immunoassays (SPIA), passive latex agglutination (PLA), passive hemagglutination (PHA) tests, and immunoblotting techniques. The selection of which tests to perform will depend on test availability for a given virus, the patient population and clinical situation, the number of specimens to be tested, turnaround time, equipment needs, ease of testing, and the resources and capabilities of the individual laboratory. Qualitative measurements of virus-induced antibody can be performed when simply knowing that the presence or absence of a specific antibody provides useful information. Quantitation can be performed when knowing the amount of antibody is important; this is most helpful in measuring virus-specific IgG antibodies when attempting to diagnose a primary viral infection or when a value is used to define immunity following vaccination or past exposure. (For additional reviews of the various methodologies, see Refs. 3–5.)

Specimen Collection and Handling

Serum is the specimen of choice for most serological testing, although plasma can be used as an acceptable alternative in some instances (e.g., for serological diagnosis of HIV and hepatitis viruses). A total of 1 to 2 mL of serum or plasma is usually sufficient, but the amount needed will vary depending on the number of tests to be performed. As a general rule, approximately 1 to 2 mL of blood should be collected for every two to three tests ordered. For obtaining serum, collection tubes should not contain anticoagulants or preservatives. Typical anticoagulants used for plasma collection include potassium EDTA, sodium citrate, sodium heparin, and acid-citrate-dextrose. One should remove the serum or plasma from clotted blood or anticoagulated red cells, respectively, as soon as possible to avoid hemolysis. Blood collection tubes with gel barriers are now available and can be used for efficient separation of serum or plasma from other blood components. The best results are generally observed for serum or plasma specimens that are clear and nonhemolyzed. Lipemic, icteric, or hemolyzed specimens should be avoided when possible, and specimens with obvious microbial contamination should not be used. Specimens

Table 2 Common Types of Viral Serological Assays

Solid-phase immunoassays	Agglutination assays
Enzyme immunoassay	Passive latex agglutination
Immunofluorescence immunoassay	Passive hemagglutination
Cemiluminescence immunoassay	Hemagglutination inhibition
Immunofluorescence assays	Complement fixation
Indirect immunofluorescence	Neutralization
Anticomplement immunofluorescence	Immunoblotting

containing unremoved clots, red blood cells, or particulate matter may give inconsistent results and should be clarified by centrifugation before testing. Unprocessed blood specimens should routinely be refrigerated at 4°C shortly after collection and transport to the laboratory, and should not be held at room temperature for longer than eight hours. Processed specimens may be stored at 2°C to 8°C for several days pending the completion of tests. If an extended delay in transport or testing of a specimen is anticipated (e.g., holding acute-phase serum until the convalescent-phase serum is collected), it should be frozen at −20°C or colder. If space permits, frozen specimens may be stored indefinitely to facilitate retrospective testing as patients are evaluated and for future epidemiological studies or evaluation of newly developed serological assays. Repeat freezing and thawing of specimens can lead to antibody degradation and should be avoided, and specimens should not be stored in frost-free freezers. Heat inactivation of specimens is not recommended unless otherwise specified when using a particular serological assay. A single serum specimen is required to determine the immune status of an individual or for the detection of IgM-specific antibody. With few exceptions [e.g., Epstein-Barr virus (EBV), hepatitis B virus (HBV), and parvovirus B19], paired sera specimens, collected 10 to 14 days apart, are required for the diagnosis of current or recent viral infections when specimens are tested for IgG antibody. The acute-phase serum should be obtained as soon as possible during the course of the illness and no later than five to seven days after onset. The most useful results are obtained by submitting acute- and convalescent-phase sera together to be tested simultaneously. Depending on the virus, the timing may vary for when IgM and IgG antibodies are produced and can be detected. Serum specimens from mother, fetus, and newborn can be submitted for the detection of prenatal, natal, or postnatal viral infections with cytomegalovirus (CMV), herpes simplex virus (HSV), rubella virus, parvovirus B19, HIV, HBV, hepatitis C virus (HCV), and others.

Serological testing is not usually available for body fluids other than serum or plasma. However, whole blood collected by finger stick or venipuncture and dried blood on filter paper have been studied as practical and effective substitutes to obtain serum, especially when screening for antibody to HIV (6–10). In patients with viral neurological disease, cerebrospinal fluid (CSF) may be tested for antiviral antibody if paired with a serum specimen from the same date (11). Although many hospital laboratories no longer perform antibody testing on CSF samples, it may be beneficial to do so for certain viruses, including the arboviruses, measles, mumps, and rabies viruses, lymphocytic choriomeningitis virus (LCMV), and herpes B virus. Also, whole saliva (12–16), oral mucosal transudates rich in gingival crevicular fluid (17–19), and urine (20–23) have been advocated as noninvasive alternatives to the collection of blood for the detection of antibodies to a number of different viruses. Particular attention has been given to the value of oral fluids (e.g., unstimulated and stimulated saliva and oral mucosal transudates) for the diagnosis of infections with HIV (for review, see Ref. 19). Unstimulated saliva can be obtained by tilting the head forward and dribbling saliva from the lower lip into a graduated test tube fitted with a funnel. After five minutes, the subject expectorates any remaining saliva from the mouth. Dribbled saliva has a stability of five days at room temperature, but can be stored for longer times at 4°C to −20°C. Mechanical stimuli such as parafilm, paraffin wax, neutral gum base, or rubber bands can be used to collect stimulated saliva. Saliva from the parotid, submandibular, and sublingual glands can be obtained directly from the glandular ducts using specially designed collection systems. Several commercial devices have been developed for the collection of oral mucosal transudate specimens. The devices provide a homogeneous specimen rich in plasma-derived IgG and IgM that is passively transferred to the mouth across the mucosa and through the gingival crevices (for a detailed description of these devices; see Ref. 19). Oral mucosal transudate specimens may be stored for 21 days at temperatures of 4°C to 37°C or at −20°C for longer periods. Lastly, vitreous humor can be used for the detection of antibodies to HSV or VZV in individuals having eye infections with these agents (24,25).

Solid-Phase Immunoassays

Solid-phase immunoassays (SPIAs) have largely replaced other methods for the detection of antiviral antibodies because of their speed, convenience, ease of use, and excellent sensitivity and specificity (26–28). The assay format is quite versatile and is applicable to many viruses and large numbers of specimens at a relatively low cost. SPIAs require minimal training and

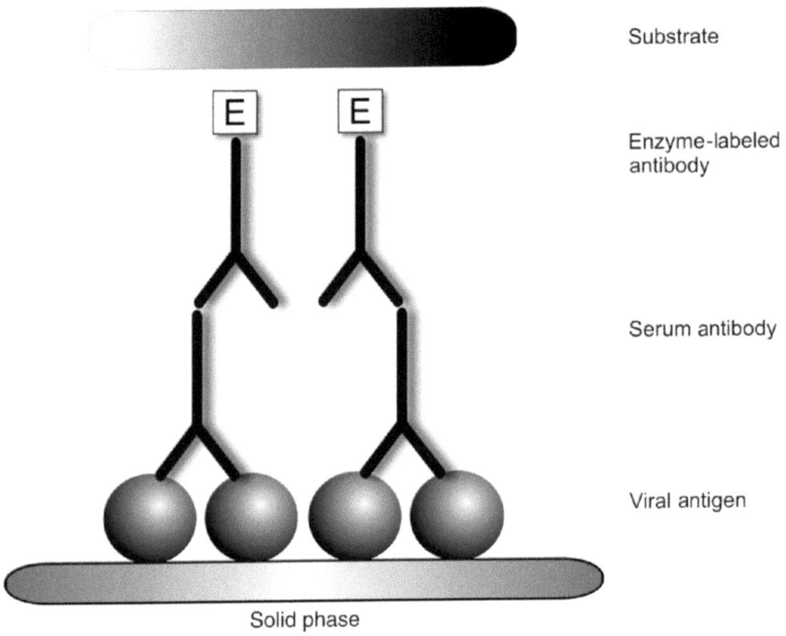

Figure 2 Schematic of a noncompetitive enzyme immunoassay for the detection of IgG antibody.

equipment and provide for objective results. A wide range of instruments are available to perform the assays and include simple, manually operated washers and readers to fully automated, high-capacity systems. Enzyme immunoassays (EIA) are the most popular SPIAs used in clinical virology laboratories, and offer the advantages of using highly standardized and stable immunoreagents with colorimetric measurements of captured antibodies. EIA kits that detect IgG or IgM antibodies to a number of different viruses are available from a variety of commercial sources. Both noncompetitive and competitive assays have been described, and results can be evaluated either qualitatively or quantitatively.

The noncompetitive EIA is one of the most frequently used antiviral antibody assays. The basic principle of the noncompetitive EIA is that viral antigen is immobilized on a solid phase and used to capture free virus-specific antibody from a clinical specimen (Fig. 2). Any unbound serum antibody is then washed away before the addition of an enzyme-labeled antihuman detector antibody. Following incubation and washing away of excess labeled antibody, a chromogenic substrate is added. Formation of specific antibody–antigen complexes results in binding of the enzyme-labeled secondary antibody and hydrolysis of the added colorless substrate to produce a color change. The intensity of the color generated is proportional to the amount of virus-specific antibody in the specimen. The results are measured in a spectrophotometer and compared with a set of positive and negative controls performed with each batch of specimens. The surface of microwell plates, polystyrene beads, test tubes, microparticles, or paramagnetic particles are normally used as the solid-phase carrier. Horseradish peroxidase and alkaline phosphatase are the most common enzyme labels. Fluorochrome, chemiluminescent, and electrochemiluminescent molecules, either as substrates for enzyme cleavage or directly conjugated to the detector antibody, have also been used to produce accurate signals that can be read in a fluorometer or luminometer (29,30).

In a competitive assay, enzyme-labeled antiviral antibody is mixed with test serum, and the presence of virus-specific antibodies in the specimen will compete with the labeled antibody for a limited number of antigen-binding sites on the solid phase (Fig. 3). Unbound antibody is washed away, and the amount of labeled antibody activity is measured as described above. If antibody is present in the clinical specimen, it will competitively inhibit the amount of labeled antibody that can react with the immobilized antigen. The decrease in detectable labeled antibody is inversely proportional to the quantity of antibody present in the sample. Competitive assays

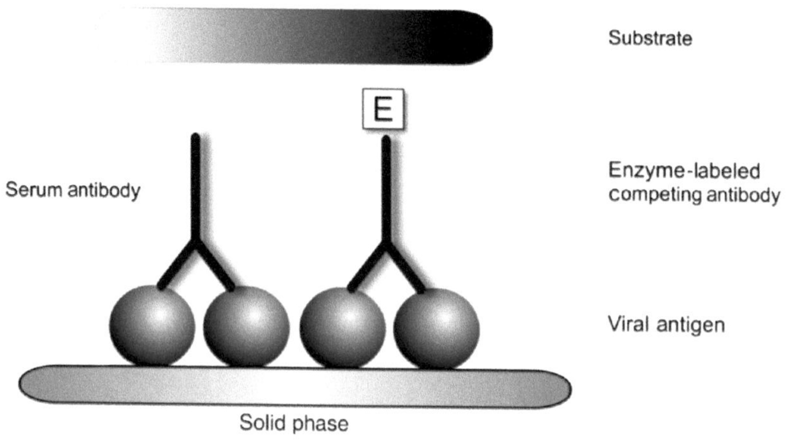

Figure 3 Schematic of a competitive enzyme immunoassay for the detection of IgG antibody.

are often used to provide greater specificity than noncompetitive assays; commercial tests with this format are available for the detection of total antibodies to hepatitis A (HAV) and HBV core antigen.

An antigen–antibody–antigen sandwich EIA is a variation of the more traditional EIAs that has been successfully applied to the detection of antibodies to HIV (for review, see Ref. 8). In this EIA, test serum is incubated with antigens bound to a solid phase and any virus-specific antibodies in the specimen will react with the antigens to form antigen–antibody complexes. A solution of the same antigens labeled with an enzyme is then added to form an antigen–antibody–antigen sandwich. The sandwich is then detected by adding a colorless substrate that is cleaved by the enzyme to give off a color reaction that is read in a spectrophotometer. This format has the distinct advantage of simultaneously detecting multiple classes of antibodies (e.g., IgG, IgM, and IgA), thereby providing a greater sensitivity over tests that only detect IgG antibody, and shortening the window period between infection with the virus and seroconversion.

More recently, rapid (~20 minutes) and simple SPIAs that use techniques involving membrane capture of antibody and lateral-flow immunochromatography have been applied to the detection of antiviral antibodies, including those for HIV (7,8,23,31,32), HBV (33), HSV (34–36), dengue virus (37,38), and others. The devices are self-contained and disposable and designed to be performed at the point of patient care or in the laboratory. In the membrane flow-through devices, antigens immobilized on a membrane will capture and concentrate virus-specific antibodies onto the surface of the device as the specimen flows through the membrane and is absorbed into an absorbent pad. Enzyme-labeled antihuman antibody and a colorless substrate are then sequentially added and enzymatic hydrolysis of the substrate leads to a colorimetric result that is read as a visually detectable symbol on the membrane. For lateral-flow immunochromatography, the specimen is applied to an absorbent pad and migrates by capillary action along a solid-phase strip, where it combines with viral antigens and detector reagents to produce a visible line on the strip when virus-specific antibodies are present. Procedural controls are included within both devices to verify the performance of the test. These assays offer the distinct advantages of requiring no specialized equipment and needing only limited technical expertise, and the reagents are stable for extended times at room temperature. As a general rule, these assays also have sensitivities and specificities that are comparable to the higher complexity laboratory-based assays.

The major disadvantage of the described SPIAs in this section is that the specificity of the reaction that has occurred cannot be directly evaluated. All that is known is that a color has developed following completion of testing on a serum specimen. With certain viruses, like HIV, additional testing may be required for specific confirmation of the results.

Immunofluorescence Assays

Immunofluorescence assays (IFA) are very useful and inexpensive methods that offer the advantages of speed and simplicity for the qualitative and quantitative detection of either IgM or IgG antiviral antibodies from clinical specimens (39,40). Commercial kits are readily available for many of the common viruses or antigen-coated slides and labeled secondary antibodies can be purchased separately for use in laboratory-developed tests. The kits contain all of the necessary reagents, including substrate slides, labeled secondary antibody, wash buffer, mounting fluid, and positive and negative control sera. IFAs are simple, well standardized, and highly reproducible and are nicely suited for low-volume testing. The major disadvantages of immunofluorescence assays are that they require a fluorescence microscope and dark room for examining slides and extensive training is needed to read and interpret the test results.

Antibody is usually detected either by an indirect immunofluorescence assay [Fig. 4(A)] or by using anticomplement immunofluorescence [Fig. 4(B)]. In the indirect IFA, dilutions of test serum are incubated with virus-infected cells that have been fixed to a glass microscope slide. Following incubation, the slide is washed to remove the excess serum and specific antibody–antigen complexes that form are detected using an antihuman antibody conjugated with a fluorochrome. Fluorescein isothiocyanate is the most commonly used fluorescent dye for IFA. The slides are then washed, dried, and examined using a fluorescence microscope. Many aspects of anticomplement immunofluorescence are similar to the indirect IFA. The method differs, however, in that the test serum is first heat-inactivated to remove endogenous complement

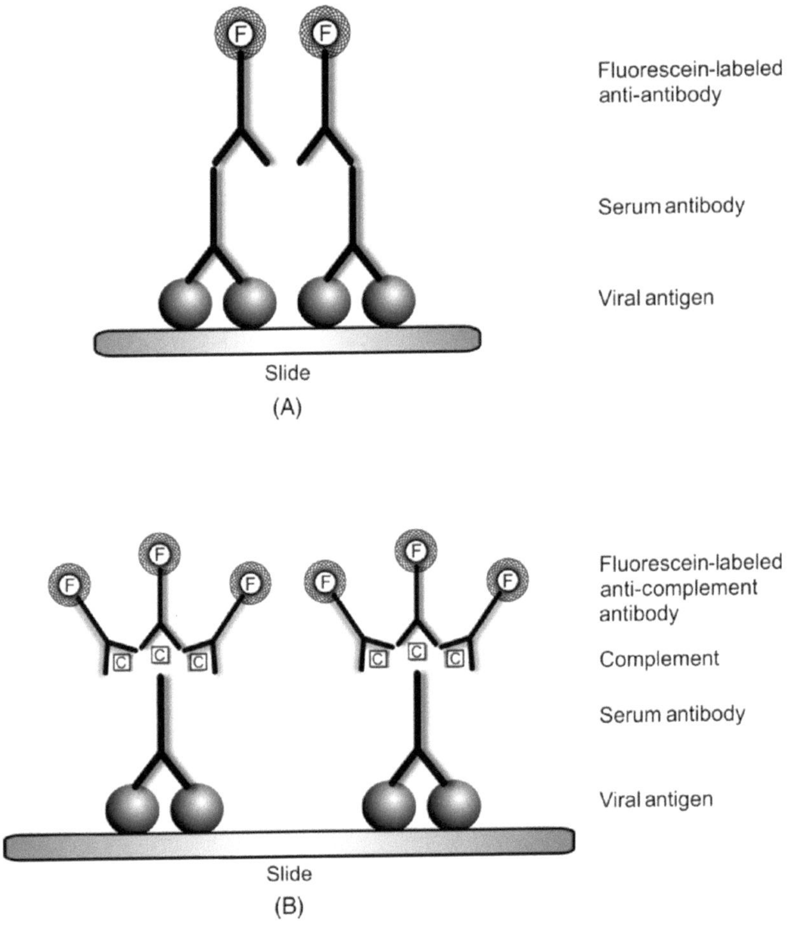

Figure 4 Diagrams of (**A**) an indirect immunofluorescence assay and (**B**) the anticomplement immunofluorescence assay for IgG antibody detection.

activity and then incubated with virus-infected cells on glass slides. After the excess serum is washed away, fresh complement is added and bound by any specific antigen–antibody complexes that have formed. A fluorescein-labeled anticomplement antibody is then added, and it binds to the C3 component of complement. The slides are then read using a fluorescence microscope. ACIF amplifies the fluorescence signal, allowing for the detection of small amounts of antibody or antibodies of low avidity. This method is routinely used to detect antibodies to the nuclear antigen of Epstein-Barr virus (EBV).

Immunofluorescence assays require critical evaluation to ensure reliable results. The number of positive-fluorescing cells as well as the quality and intensity of the fluorescence must be carefully examined and compared with that of cells reacted with positive and negative control sera. Most manufacturers of commercial kits provide antigen slides in which only 20% to 40% of the cells express viral antigens. Therefore, nonspecific binding of antibodies to the cells is easily discerned, since the staining pattern produced by this type of reaction normally involves all of the cells.

Agglutination Assays

Passive Latex Agglutination and Passive Hemagglutination

Passive latex agglutination (PLA) is currently the most commonly used agglutination assay available. It is uncomplicated, convenient, and inexpensive, and is best suited for testing low numbers of specimens. A suspension of latex particles coated with viral antigens is mixed with a clinical specimen and allowed to incubate with rotation for a short time (Fig. 5) (41). The antigen-coated particles will clump and produce agglutination in the presence of virus-specific antibody. The agglutination is visible to the naked eye. In the absence of specific antibody or in the presence of low concentrations of antibody, the latex particles will not agglutinate and will appear smooth and evenly dispersed. The test can be completed within 10 minutes and requires limited equipment and technical ability. Both IgG and IgM antibodies are detected without differentiation and the sensitivity and specificity of PLA is comparable to that observed for SPIAs and IFAs. Commercial PLA kits are currently available for the detection of heterophile antibodies in the diagnosis of EBV-associated infectious mononucleosis and for virus-specific antibodies to VZV and rubella virus. PLA is best suited for qualitative determinations of antiviral antibody, but quantitation also can be performed. The most important disadvantage of PLA is

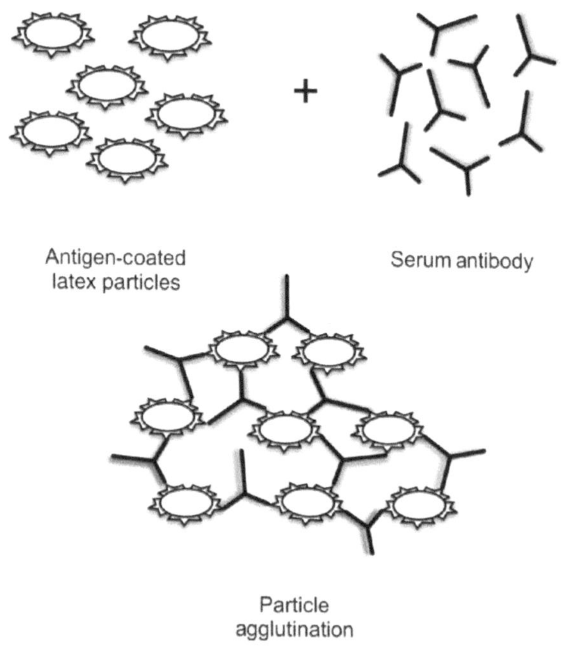

Figure 5 Illustration of a passive latex agglutination assay.

that the reading of an agglutination reaction can be subjective and the results may be difficult to interpret. Also, a prozone or reduction in the degree of agglutination can occur with sera that have high levels of specific antibody, resulting in the need to dilute negative specimens and repeat the assay. When erythrocytes are coated with viral antigen, the procedure described for PLA is termed *passive hemagglutination*.

Hemagglutination Inhibition

Hemagglutination inhibition assays (HIA) are used to detect antibodies to viruses that possess a hemagglutinin on their surface (42). This assay has been applied to the detection of antibodies to the arboviruses, influenza and parainfluenza viruses, measles, mumps, and rubella viruses, adenovirus, respiratory syncytial virus, and the polyomaviruses in seroepidemiological studies as well as to antiviral and vaccine trials for specific viruses. For performance of HIA, a known amount of viable virus is incubated with dilutions of the test serum. This is followed by the addition of the appropriate animal species of red blood cells (RBCs) to the mixture. If virus-specific antibodies are present, they will react with the viral hemagglutinin and prevent agglutination of the RBCs. In the absence of specific antibody, the added virus is capable of binding to the RBCs and causes a visible hemagglutination reaction. HIA is technically demanding and time consuming and requires appropriate quality control and adherence to procedures. The specific virus used as the source of hemagglutinin must be accurately titrated for best performance of the assay. Some serum specimens contain nonspecific inhibitors and natural agglutinins that must be removed before virus-specific antibodies can be detected. The inhibitors can be removed by pretreatment of the sera with receptor-destroying enzyme, heat inactivation or chemical treatment with potassium periodate or kaolin, while natural agglutinins are removed by absorption with the appropriate species of RBCs. The most common use for HIA is for the subtyping and antigenic characterization of influenza virus isolates, although PCR-based genotyping assays are rapidly displacing this methodology.

Complement Fixation

Complement fixation (CF) was one of the most widely used serologic assays in clinical virology for a number of years. The assay can measure antibodies against virtually any virus and has the distinct advantage of accurately detecting significant rises in IgG antibody levels during acute viral infections (43). To perform this assay, heat-inactivated test serum is mixed with specific viral antigens in the presence of a known amount of complement. If the serum contains virus-specific antibody, an antigen–antibody complex will form and complement will be activated (fixed or bound to the complex) and depleted from the mixture. If no specific antibody is present in the serum, then antigen–antibody complexes will not form and complement will remain unbound and free in the test system. A specific amount of antibody-coated (sensitized) sheep RBCs is then added to the reaction mixture. The sensitized sheep RBCs will activate any unbound complement that remains in the test system, causing lysis of the RBCs. The absence of hemolysis indicates that a specific antibody–antigen complex was formed, thereby binding and depleting the complement and preventing lysis of the sheep RBCs. The major disadvantages of the method include that it is technically demanding, requires rigid standardization and titration of reagents, and has a long turnaround time. The CF test is also less sensitive than other methods and anticomplement activity can occur due to nonspecific binding of serum components to the complement used in the assay. For these reasons, the assay has been largely replaced by the SPIAs, IFAs, and PLAs described above. The CF test currently has limited utility in clinical virology, being used only for less common viral agents (e.g., lymphocytic choriomeningitis virus) for which no other commercial reagents are available.

Neutralization

Virus-specific antibodies can be detected in serum by their ability to neutralize or block the infectivity and replication of a given virus within a cell culture system (44). In the neutralization (NT) assay, a defined quantity of viable virus is mixed with the test serum. Following incubation, dilutions of the mixture are prepared and inoculated into cultured cells that are normally susceptible to the virus used in the assay. The cells are incubated at a suitable temperature for viral growth and examined daily, usually for five to seven days, for the production of a virus-induced cytopathic effect (CPE) or some other indicator of viral growth. If the infectivity of the virus has been neutralized by specific antibody in the serum, CPE will not be observed. Conversely, CPE will be produced if no neutralizing antibody is present in the serum. By performing dilutions of the test serum, the quantity of neutralizing antibody can be determined. NT assays are cumbersome, expensive, and time-consuming. They also require that the quantity of virus used in the system be carefully titrated to obtain accurate results. The major advantage of the NT assay is that it can be performed on virtually all viruses that can be grown in cell culture. The information obtained also has biological relevance since the production of neutralizing antibodies in response to a viral infection is important in establishing protective immunity. The NT assay remains the method of choice for the detection of antibodies to the enteroviruses.

Immunoblotting

Immunoblotting is basically an SPIA that uses separated and immobilized viral antigens to detect antibodies to specific proteins (45–47). The technique is used mainly as a confirmatory or supplemental test to help verify the specificity of positive results obtained from other assays used to initially screen for virus-specific antibodies. Commercial kits are available for HIV-1 and HIV-2, hepatitis C virus (HCV), and human T-cell leukemia virus types I (HTLV-I) and II (HTLV-II). The major advantage of immunoblot assays is that the specific interaction of antibody and antigen can be directly visualized. These assays are highly sensitive and specific, but they are technically demanding, relatively expensive, and can be subject to interpretation.

The most commonly described immunoblot is the Western blot. In this assay, whole virus lysates of inactivated and disrupted viral proteins are separated by electrophoresis according to their molecular weight or relative mobility as they migrate through a polyacrylamide gel in the presence of sodium dodecyl sulfate (SDS) (Fig. 6). The resolved protein bands are then transblotted (transferred) to a sheet of nitrocellulose paper. The nitrocellulose paper is then cut into strips that are reacted with serum specimens. If virus-specific antibody is present in the serum, binding of antibody occurs in bands corresponding to the presence of the separated viral proteins. The bands are directly visualized by using an enzyme-labeled antihuman antibody followed by a chromogenic substrate. Immunoblots utilizing recombinantly derived proteins immobilized to nitrocellulose strips have been described for HCV (48,49), HSV-2 (35,36), hantavirus (50), and dengue virus (51,52).

IgG Avidity Assays

Assays for measurements of virus-specific IgG avidity have been developed and have proven useful for distinguishing primary from nonprimary infections, particularly in women suspected of having CMV or rubella virus during pregnancy (53–55) and in solid organ transplant recipients (56,57). Virus-specific IgG of low avidity is produced during the first weeks to months following primary infection, whereas IgG antibody of increasingly higher avidity is produced with past or nonprimary infections. Both commercial and user-developed avidity assays are available and the tests are accomplished by making simple modifications to the basic procedure of an SPIA. In one assay format (58), patient serum is added to viral antigens bound to a solid phase to allow virus-specific IgG antibody to bind and form antigen–antibody complexes. Any virus-specific antibody bound to the antigen source is then pretreated with a denaturing agent such as urea to determine the strength, or avidity, of the antibody binding. This is followed by the addition of an enzyme-labeled secondary antibody and chromogenic substrate to produce a color change that can be measured in a spectrophotometer. Virus-specific low-avidity IgG is detected indirectly since only high-avidity IgG remains bound to the solid phase and contributes to generation of the colorimetric signal. A second assay design (59), called avidity competition, involves the addition of soluble virus-specific antigen to a specimen before measuring its virus-specific IgG concentration. High- and low-avidity antibodies compete for binding sites on the soluble antigen, and virus-specific high-avidity IgG from the specimen is selectively captured. The pretreated sample is then combined with viral antigen bound to a solid phase to bind the virus-specific low-avidity IgG remaining in the specimen. An antihuman IgG antibody labeled with acridinium is then added and the resulting chemiluminescent reaction is measured in relative light units. For specimens that contain primarily low-avidity IgG, the virus-specific IgG concentration remains nearly the same in the presence or absence of soluble antigen, while it is

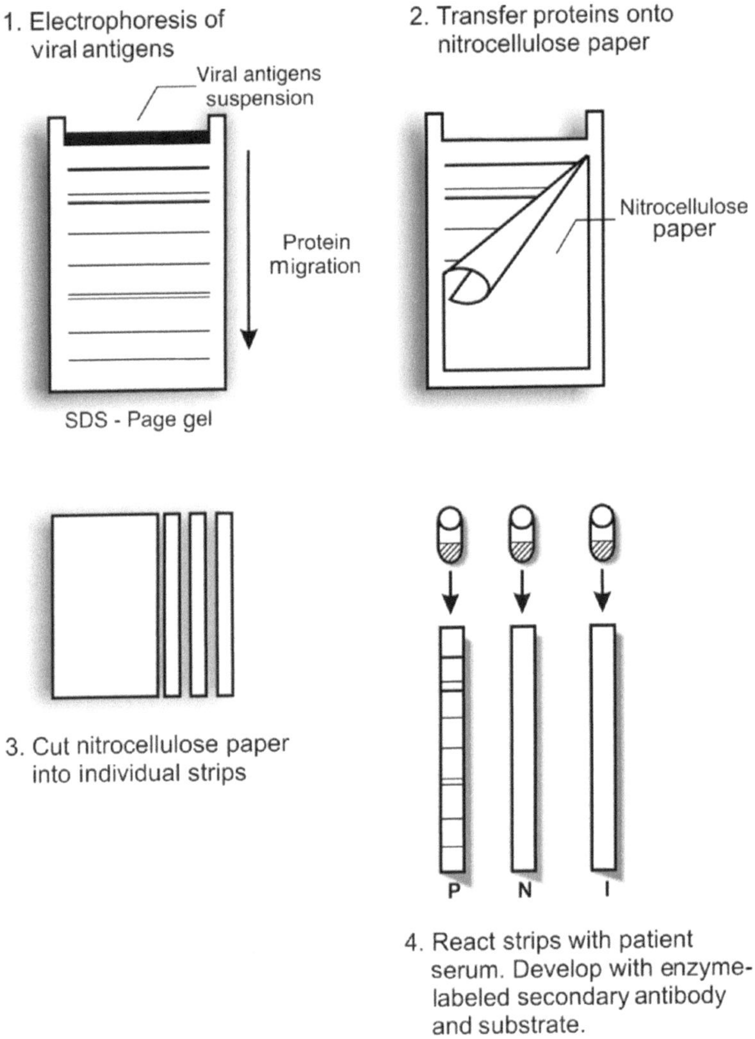

Figure 6 Illustration of the steps involved in the Western blot technique.

strongly reduced in specimens that contain mainly high-avidity virus-specific antibodies. For both assays, avidity levels are expressed as the percentage of virus-specific IgG bound to the antigen source when test results are generated and compared in the presence and absence of the pretreatment step.

Multiplexed Microsphere Immunoassays

Traditional methods for detection of viral antibodies normally require separate assays for each virus-specific antibody determination. More recently, a multiplexed technology has been developed that combines conventional immunoassay chemistry with flow cytometry to simultaneously measure antibody responses to multiple viruses or multiple serotypes of the same virus using polystyrene microspheres as the solid phase. They are internally dyed with red and infrared fluorescent compounds of different intensities and are given a unique number so that each microsphere can be differentiated from another based on a discrete color code. Up to 100 distinctly dyed microspheres can be classified and multiplexed together. The dyed microspheres can be covalently bound to different viral antigens and mixed in the same assay to capture multiple antiviral antibodies present in a given specimen. A fluorescent labeled secondary antibody is then added to detect the bound antigen–antibody complexes. When the assay is complete, the microspheres are read in single-file using a luminometer with dual lasers for classification and quantification of each antibody. The classification laser reads the internal dye composition of the microsphere and the reporter laser reads the relative fluorescence intensity of the external fluorescent label attached to the secondary detector antibody to quantify the antibody–antigen reaction that has occurred on the microsphere surface. This platform has been applied to the detection of antibodies to EBV (60–62), West Nile virus (63), different serotypes of human papillomavirus (64), human and avian influenza viruses (65,66), a panel of seven respiratory viruses (67), and HSV (unpublished commercial kit from Focus Technologies, Cyprus, CA).

IgM Antibody Determination

Virus-specific IgM antibodies are most commonly detected using IFAs and SPIAs (68), and commercial reagents and complete diagnostic kits are available for many viruses. The methods are similar to those used for detecting IgG antibodies, except that IgM bound to viral antigens on the solid phase is detected using secondary antihuman IgM antibodies labeled with suitable markers. Tests for virus-specific IgM antibody have been used as an aid in the diagnosis of infections with measles, mumps, and rubella viruses, hepatitis viruses A, B, D, and E, parvovirus B19, the herpesviruses, and arthropod- and rodent-borne viruses. Of major concern in measuring virus-specific IgM antibodies, however, is the occurrence of false-positive and false-negative reactions (68,69). False-negative reactions may occur as a result of high levels of specific IgG antibodies competitively blocking the binding of IgM to the viral antigen placed on the solid phase. False-positive reactions can occur when sera contain unusually high levels of rheumatoid factor. Rheumatoid factor is produced in some rheumatologic, vasculitic, and viral diseases, and is an IgM class immunoglobulin that reacts with the Fc portion of IgG. In the presence of virus-specific IgG antibodies, rheumatoid factor forms a complex with the IgG molecules. The IgG can then bind to the viral antigen on the solid phase, carrying nonviral IgM antibody with it and resulting in a false-positive result. The incidence of these false-negative and false-positive results can be minimized by separation of IgG and IgM from sera before testing.

A variety of methods have been developed for the removal of interfering rheumatoid factor and IgG molecules from serum, resulting in more reliable IgM tests (70). IgG and IgM antibodies can be physically separated using gel filtration, ion exchange chromatography, affinity chromatography, and sucrose density gradient centrifugation. Although such techniques are effective for separation of IgG and IgM, they are not very practical for clinical use. More rapid and simple procedures have been used for the selective absorption and removal of the IgG fraction from serum using hyperimmune antihuman IgG, staphylococcal protein A, or recombinant protein G from group G streptococci. These pretreatment methods are readily available and are now incorporated within most commercial IgM detection kits. More recently, reverse capture solid-phase IgM assays have been developed as an alternative to the physical fractionation of serum. In this method, the solid phase is coated with an antihuman IgM antibody that is used to capture the virus-specific IgM from the serum specimen (Fig. 7). This is followed by washing to remove competing IgG antibody and immune complexes that may interfere with the accuracy of the test. A specific viral antigen is then added and allowed to bind to the captured IgM. The antigen–antibody complexes are detected by adding an enzyme-labeled secondary antibody followed by a chromogenic substrate that produces a color when cleaved by the enzyme. IgM capture assays are considered to be more sensitive and specific than the more conventional IgM assay formats.

AUTOMATION

SPIAs have the greatest potential for automation, and a number of semiautomated and fully automated systems are now commercially available for the performance of many viral serological assays (Table 3) (71). Extensive test menus are available for HIV, hepatitis viruses, viruses associated with congenital infections, and other viruses of clinical importance. The product availability may vary by analyzer and from country to country and between the United States and international markets. Some manufacturers have developed multiple instrument models of various sizes, shapes, and complexities that are configured as bench top or free-standing units to accommodate large, moderate, or small volume laboratories. The automated systems provide walk-away simplicity to perform assays from sample processing through

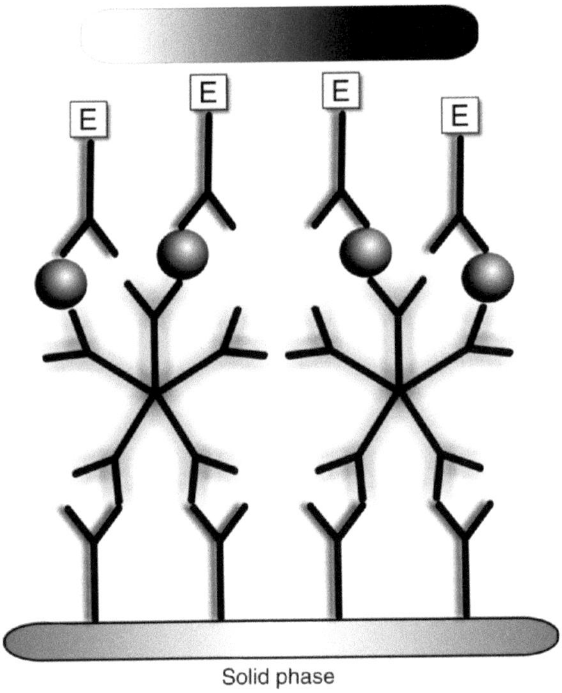

Figure 7 Schematic of a solid-phase capture enzyme immunoassay for the detection of IgM antibody.

interpretation of results. Instruments can automatically generate worklists of specimens to be tested, pipette and dilute the samples, dispense all reagents, time the incubations at a desired temperature, perform washes, and read and store the final results. Most manufacturers of automated instruments also provide software for the analysis and management of patient data and for monitoring the quality of the testing being performed. Many of the instruments can also interface with computer-based hospital laboratory information systems for seamless reporting of results. The quantity and choice of automated instruments used depends mainly on the volume of specimens for testing and the number of individual tests to be performed. Some instruments have been designed to run only those assays developed for the system by the manufacturer while other automated analyzers are open platforms that can be programmed to perform assays from a wide variety of manufacturers. Automation of viral serological assays can be advantageous to the laboratory that has a shortage of trained medical technologists or that needs to reduce costs or to improve the turnaround time for test results.

INTERPRETATION OF SEROLOGY RESULTS

Demonstration of seroconversion from a negative to a positive IgG antibody response between acute- and convalescent-phase sera or detecting the presence of virus-specific IgM in a single serum specimen can be diagnostic of primary viral infection. Fourfold or greater rises in IgG antibody titers in paired sera may support a recent viral infection due to reactivation or reinfection. Detection of virus-specific IgG in a single serum specimen or seeing no change in antibody levels between acute- and convalescent-phase sera indicates exposure to a virus some time in the past or a response to vaccination. Negative serum antibody titers may exclude viral infection.

The identification of intrathecal virus-specific antibody production in CSF can confirm the diagnosis of viral encephalitis (11). However, the appearance of virus-specific antibody in the CSF may be delayed for two to four weeks, and its presence may simply represent the passive transfer of serum antibodies across a damaged blood–brain barrier. Methods must be used to determine and compare the CSF/serum ratio of virus-specific antibody to the CSF/serum ratio of a defined marker such as albumin (11). Since albumin is not synthesized in the central

Table 3 Selected Automated Immunoassay Systems for Viral Serology

Manufacturer	Available instruments	Platform[a]	Website[b]
Abbott Laboratories	Axsym/Axsym Plus, Architect i2000, Architect i200SR, Commander, Prism, Quantum II	Closed	www.abbottdiagnostics.com
Adaltis	Eclectica, NexGen Four, Personal Lab	Open	www.adaltis.com
Beckman Coulter	Access 2, UniCel DxI 600 Access, UniCel DxI 800 Access, UniCel DxC 600i Synchron Access	Closed	www.beckman.com
BioMerieux-Vitek	Vidas, mini vidas, vidia	Closed	www.biomerieux-diagnostcs.com
Bio-Rad Laboratories	BioPlex 2200, Evolis, Elite	Closed (BioPlex) Open (Evolis, Elite)	www.bio-rad.com
Diamedix	MAGO Plus, DSX, DS2	Closed (MAGO) Open (DSX, DS2)	www.diamedix.com
Dynex Technologies	DSX, DS2	Open	www.dynextechnologies.com
Grifols-Quest	TRITURUS	Open	www.grifols.com
Luminex Corporation	Luminex 100 IS, Luminex 200	Open	www.luminexcorp.com
Ortho-Clinical Diagnostics	Vitros ECiQ, Vitros 3600, Vitros 5600	Closed	www.orthoclinical.com
Roche Diagnostics	ELECSYS 2010, Cobas e411, Cobas e601	Closed	www.roche-diagnostics.com
Siemens Healthcare Diagnostics	ADVIA Centaur XP and CP, Immulite 2500, 2000, and 1000	Closed	www.dadebehring.com
Zeuss Scientific	AtheNA Multi-Lyte	Open	www.zeusscientific.com

[a]Closed systems use only immunoassays produced by the manufacturer. Open systems can be programmed to process other manufacturer's immunoassays.
[b]The reader should contact the manufacturer for a comprehensive description and detailed test menu of a particular immunoassay analyzer.

nervous system, its presence in high concentrations within the CSF reflects the presence of contaminating serum proteins and an interruption of the blood–brain barrier. Demonstration of an intact blood–brain barrier in the presence of high levels of detectable virus-specific CSF antibody represents intrathecal production of antibody and is considered evidence of viral infection of the central nervous system.

In evaluating a fetus or newborn for congenital viral infections, the presence of virus-specific IgM strongly suggests infection since IgM antibodies do not cross the placenta. When testing for IgM in the fetus, blood should be collected after 22 weeks gestation since fetal synthesis of antibodies starts at 20 weeks gestation and may not reach detectable levels for one to two more weeks. Testing the fetus or newborn for virus-specific IgG is less helpful and rarely results in a definitive diagnosis since active transfer of maternal antibodies across the placenta begins at 18 weeks gestation. In the pregnant woman with symptoms of a viral disease, the presence of virus-specific IgM alone or a history of a positive seroconversion of IgG antibodies may be beneficial for the diagnosis of primary maternal infection and assessing fetal outcome. For pregnant women with preexisting virus-specific IgG and IgM antibodies, testing for IgG avidity may be more helpful in distinguishing primary from past infections and predicting fetal infection. Maternal testing for HSV glycoprotein G type-specific antibody may provide important information in pregnant women lacking symptoms but having a history of risk factors for genital herpes (35,36). Any viral agents for which sera obtained from the mother, fetus, or newborn are seronegative are very unlikely to have resulted in an infection. However, negative antibody levels in the mother and child may also be the result of the mother having a primary infection of recent onset without immediate production of virus-specific antibodies. When testing the newborn for virus-specific IgG antibody, single elevated IgG antibody titers to a specific viral agent are not useful and sera from both mother and newborn should be tested for accurate interpretation of results. If the amount of virus-specific IgG antibody in the newborn is lower than or the same as the corresponding IgG antibody in the mother, this may reflect passive antibody transfer. Having higher IgG antibody levels in the newborn than the mother may reflect active antibody production. If maternal and infant serum IgG antibody levels are the same, additional serum from the infant should be obtained one to two months later and periodically thereafter for six to nine months, to be tested and compared with the earlier antibody level. These sera should show a decrease in virus-specific antibody relative to the first specimen if the infant has not been congenitally infected with the agent tested. Testing in this manner is quite retrospective and of limited benefit to patient management, and maternal serum is seldom obtained and appropriate follow-up studies are infrequently performed. In general, serological diagnosis of congenital viral infections can be extremely difficult and is often not fully understood by those that request the tests. Accurate testing requires appropriate collection and timing of specimens from mother, fetus, or newborn, appropriate selection of assays, correct interpretation of results, and knowledge of the usefulness and limitations of the tests (for review, see Refs. 72 and 73).

The results of serologic tests for the detection of virus-specific antibodies must be interpreted with caution as measurements of an antibody response to viral infections can be complicated by a number of factors. There may be a lack of or delay in production of serum IgM or IgG antibodies, particularly in newborns, the elderly, immunocompromised hosts, and patients with agammaglobulinemia. IgM antibodies also may persist for extended periods after primary infection and can be present during reactivation of latent viral infections. Significant rises in IgG antibodies do not always occur as a result of recurrent infections or exogenous reinfection. Virus-specific IgG antibodies may be present in recipients of intravenous immunoglobulin, newborn infants possessing passively acquired maternal antibody, or patients who have received recent blood transfusions, making it difficult to interpret IgG tests. Rises in either IgM or IgG antibody to certain viruses also may be nonspecific and occur in response to recent infections with other viruses. This is especially true for the herpesviruses, since this group shares many common or cross-reactive epitopes to which antibodies can be produced. Because of the many caveats associated with serologic diagnoses of viral infections, isolation of the virus in culture or use of direct methods of detecting viral antigens or nucleic acids should also be considered whenever possible.

REFERENCES

1. Cowan KM. Antibody response to viral antigens. Adv Immunol 1973; 17:195–253.
2. Ogra PL, Morag A, Tiku ML. Humoral immune response to viral infections. In: Notkins AL, ed. Viral Immunology and Immunopathology. New York, NY: Academic Press, 1975:57.
3. James K. Immunoserology of infectious diseases. Clin Microbiol Rev 1990; 3:132–152.
4. Carpenter AB. Immunoassays for the diagnosis of infectious diseases. In: Murray PR, Baron EJ, Jorgensen JH, Landry ML, Pfaller MA, eds. Manual of Clinical Microbiology, Vol. 1. 9th ed. Washington, DC: ASM Press, 2007:257–270.
5. Leland DS. Concepts of immunoserological and molecular techniques. In: Leland DS, ed. Clinical Virology. Philadelphia, PA: WB Saunders Company, 1996:21–49.
6. Wassilak SG, Bernier RH, Herrmann KL, et al. Measles seroconfirmation using dried capillary blood specimens in filter paper. Pediatr Infect Dis 1984; 3:117–121.
7. Branson BM. State of the art for diagnosis of HIV infection. Clin Infect Dis 2007; 45:S221–S225.
8. Hodinka RL. Human immunodeficiency virus. In: Truant AL, ed. Manual of Commercial Methods in Clinical Microbiology. Washington, DC: ASM Press, 2002:100–127.

9. Varnier OE, Lillo FB, Reina S, et al. Whole blood collection on filter paper is an effective means of obtaining samples for human immunodeficiency virus antibody assay. AIDS Res Hum Retroviruses 1988; 4:131–136.
10. Sarge-Njie R, Schim van der Loeff M, Ceesay S, et al. Evaluation of the dried blood spot filter paper technology and five testing strategies of HIV-1 and HIV-2 infections in West Africa. Scand J Infect Dis 2006; 38:1050–1056.
11. Andiman WA. Organism-specific antibody indices, the cerebrospinal fluid-immunoglobulin index and other tools: A clinician's guide to the etiologic diagnosis of central nervous system infection. Pediatr Infect Dis J 1991; 10:490–495.
12. Jayashree S, Bhan MK, Kumar R, et al. Serum and salivary antibodies as indicators of rotavirus infection in neonates. J Infect Dis 1989; 158:1117–1120.
13. Parry JV, Perry KR, Panday S, et al. Diagnosis of hepatitis A and B by testing saliva. J Med Virol 1989; 28:255–260.
14. Perry KR, Brown DWG, Parry JV, et al. Detection of measles, mumps, and rubella antibodies in saliva using antibody capture radioimmunoassay. J Med Virol 1993; 40:235–240.
15. Adler SP, Wang JB. Salivary antibodies to cytomegalovirus (CMV) glycoprotein B accurately predict CMV infections among preschool children. J Clin Microbiol 1996; 34:2632–2634.
16. Elsana S, Sikuler E, Yaari A, et al. HCV antibodies in saliva and urine. J Med Virol 1998; 55:24–27.
17. Thieme T, Yoshihara P, Piacentini S, et al. Clinical evaluation of oral fluid samples for diagnosis of viral hepatitis. J Clin Microbiol 1992; 30:1076–1079.
18. Martinez PM, Torres AR, de Lejarazu RO, et al. Human immunodeficiency virus antibody testing by enzyme-linked fluorescent and Western blot assays using serum, gingival-crevicular transudate, and urine samples. J Clin Microbiol 1999; 37:1100–1106.
19. Hodinka RL, Nagashunmugam T, Malamud D. Detection of human immunodeficiency virus antibodies in oral fluids. Clin Diagn Lab Immunol 1998; 5:1–8.
20. Takahashi S, Machikawa F, Noda A, et al. Detection of immunoglobulin G and A antibodies to rubella virus in urine and antibody responses to vaccine-induced infection. Clin Diagn Lab Immunol 1998; 5:24–27.
21. Perry KR, Parry JV, Vandervelde EM, et al. The detection in urine specimens of IgG and IgM antibodies to hepatitis A and hepatitis B core antigens. J Med Virol 1992; 38:265–270.
22. Joshi MS, Walimbe AM, Arankalle VA, et al. Hepatitis E antibody profiles in serum and urine. J Clin Lab Anal 2002; 16:137–142.
23. Koopmans M, Sanchez-Martinez D, Patton J, et al. Evaluation of antigen and antibody detection in urine specimens from children with congenital human cytomegalovirus infection. J Med Virol 1995; 46:321–328.
24. Sarkies N, Gregor Z, Forsey T, et al. Antibodies to herpes simplex virus type 1 in intraocular fluids of patients with acute retinal necrosis. Br J Ophthalmol 1986; 70:81–84.
25. Suttorp-Schulten MSA, Zaal MJW, Luyendijk L, et al. Aqueous chamber tap and serology in acute retinal necrosis. Am J Ophthalmol 1989; 108:327–328.
26. Kemeny DM, Challacombe SJ. ELISA and Other Solid Phase Immunoassays. Theoretical and Practical Aspects. Chichester, UK: John Wiley & Sons, 1988.
27. Gosling JP. A decade of development in immunoassay methodology. Clin Chem 1990; 36: 1408–1427.
28. Leland DS. Enzyme immunoassay. In: Specter S, Hodinka RL, Young SA, eds. Clinical Virology Manual, 3rd ed. Washington, DC: ASM Press, 2000:93–104
29. Nakamura RM, Bylund DJ. Fluorescence immunoassays. In: Rose NR, Conway de Marcario E, Folds JD, Lane EC, Nakamura RM, eds. Manual of Clinical Laboratory Immunology, 5th ed. Washington, DC: ASM Press, 1997:39–48.
30. Kricka LJ. Chemiluminescence immunoassays. In: Rose NR, Conway de Marcario E, Folds JD, Lane EC, Nakamura RM, eds. Manual of Clinical Laboratory Immunology, 5th ed. Washington, DC: ASM Press, 1997:49–53.
31. Kassler WJ, Haley C, Jones WK, et al. Performance of a rapid, on-site human immunodeficiency virus antibody assay in a public health setting. J Clin Microbiol 1995; 33:2899–2902.
32. Stetler HC, Granade TC, Nunez CA, et al. Field evaluation of rapid HIV serologic tests for screening and confirming HIV-1 infection in Honduras. AIDS 1997; 3:369–375.
33. Sato K, Ichiyama S, Iinuma Y, et al. Evaluation of immunochromatographic assay systems for rapid detection of hepatitis B surface antigen and antibody, Dainacreen HbsAg and Dainascreen Ausab. J Clin Microbiol 1996; 34:1420–1422.
34. Ashley RL, Wald A, Eagleton M. Premarket Evaluation of the POCkit HSV-2 type-specific serologic test in culture-documented cases of genital herpes simplex virus type 2. Sex Transm Dis 2000; 27: 266–269.
35. Ashley RL. Performance and use of HSV type-specific serology test kits. Herpes 2002; 9:38–45.

36. Wald A, Ashley-Morrow R. Serological testing for herpes simplex virus (HSV)-1 and HSV-2 infection. Clin Infect Dis 2002; 35:S173–S182.
37. Vaughn DW, Nisalak A, Kalayanarooj S, et al. Evaluation of a rapid immunochromatographic test for diagnosis of dengue virus infection. J Clin Microbiol 1998; 36:234–238.
38. Wu S-J L, Paxton H, Hanson B, et al. Comparison of two rapid diagnostic assays for detection of immunoglobulin M antibodies to dengue virus. Clin Diagn Lab Immunol 2000; 7:106–110.
39. Rossier E, Miller HR, Phipps PH. Rapid Viral Diagnosis by Immunofluorescence: An Atlas and Practical Guide. Ottawa, Canada: University of Ottawa Press, 1989.
40. Schutzbank TE, McGuire R. Immunofluorescence. In: Specter S, Hodinka RL, Young SA, eds. Clinical Virology Manual, 3rd ed. Washington, DC: ASM Press, 2000:69–78.
41. Kasahara Y. Agglutination immunoassays. In: Rose NR, Conway de Marcario E, Folds JD, Lane EC, Nakamura RM, eds. Manual of Clinical Laboratory Immunology, 5th ed. Washington, D.C.: ASM Press, 1997:7–12.
42. Young SA, McLaren LC. Hemadsorption and hemagglutination inhibition. In: Specter S, Hodinka RL, Young SA, eds. Clinical Virology Manual, 3rd ed. Washington, DC: ASM Press, 2000:135–139.
43. Stark LM, Lewis AL. Complement fixation test. In: In: Specter S, Hodinka RL, Young SA, eds. Clinical Virology Manual, 3rd ed. Washington, DC: ASM Press, 2000:112–126.
44. Ballew HC. Neutralization. In: Specter S, Hodinka RL, Young SA, eds. Clinical Virology Manual, 3rd ed. Washington, DC: ASM Press, 2000:127–134.
45. Ledue TB, Garfin DE. Immunofixation and immunoblotting. In: Rose NR, Conway de Marcario E, Folds JD, Lane EC, Nakamura RM, eds. Manual of Clinical Laboratory Immunology, 5th ed. Washington, DC: ASM Press, 1997:54–64.
46. Timmons TM, Dunbar BS. Protein blotting and immunodetection. Methods Enzymol 1990; 182: 679–688.
47. Meads MB, Medveczky PG. Application of Western blotting to diagnosis of viral infections. In: Specter S, Hodinka RL, Young SA, eds. Clinical Virology Manual, 3rd ed. Washington, DC: ASM Press, 2000:182–187.
48. van der Poel CL, Cuypers HT, Reesink HW, et al. Confirmation of hepatitis C virus infection by new four-antigen recombinant immunoblot assay. Lancet 1991; 337:317–319.
49. Chaudhary RK, MacLean C. Evaluation of first- and second-generation RIBA kits for detection of antibody to hepatitis C virus. J Clin Microbiol 1991; 29:2329–2330.
50. Hjelle B, Jenison S, Torrez-Martinez N, et al. Rapid and specific detection of Sin Nombre virus antibodies in patients with hantavirus pulmonary syndrome by a strip immunoblot assay suitable for field diagnosis. J Clin Microbiol 1997; 35:600–608.
51. Kuno G, Cropp CB, Wong-Lee J, et al. Evaluation of an IgM immunoblot kit for dengue diagnosis. Am J Trop Med Hyg 1998; 59:757–762.
52. Ludolfs D, Schilling S, Altenschmidt J, et al. Serological differentiation of infections with dengue virus serotypes 1 to 4 by using recombinant antigens. J Clin Microbiol 2002; 40:4317–4320.
53. Bodeus M, Feyder S, Goubau P. Avidity of IgG antibodies distinguishes primary from nonprimary cytomegalovirus infection in pregnant women. Clin Diagn Virol 1998; 9:9–16.
54. Grangeot-Keros L, Mayaux MJ, Lebon P, et al. Value of cytomegalovirus (CMV) IgG avidity index for the diagnosis of primary CMV infection in pregnant women. J Infect Dis 1997; 175: 944–946.
55. Hofman J, Liebert UG. Significance of avidity and immunoblot analysis for rubella IgM-positive serum samples in pregnant women. J Virol Methods 2005; 130:66–71.
56. Lazzarotto T, Spezzacatena P, Pradelli P, et al. Avidity of immunoglobulin G directed against human cytomegalovirus during primary and secondary infections in immunocompetent and immunocompromised subjects. Clin Diagn Lab Immunol 1997; 4:469–473.
57. Lazzarotto T, Varani S, Spezzacatena P, et al. Delayed acquisition of high-avidity anti-cytomegalovirus antibody is correlated with prolonged antigenemia in solid organ transplant recipients. J Infect Dis 1998; 178:1145–1149.
58. Prince HE, Leber AL. Validation of an in-house assay for cytomegalovirus immunoglobulin G (CMV IgG) avidity and relationship of avidity to CMV IgM levels. Clin Diagn Lab Immunol 2002; 9:824–827.
59. Curdt I, Praast G, Sickinger E, et al. Development of fully automated determination of marker-specific immunoglobulin G (IgG) avidity based on the avidity competition assay format: application for Abbott Architect cytomegalovirus and Toxo IgG Avidity assays. J Clin Microbiol 2009; 47:603–613. doi:10.1128/JCM.01076-08.
60. Martins TB, Litwin CM, Hill HR. Evaluation of a multiplex fluorescent microsphere immunoassay for the determination of Epstein-Barr virus serologic status. Am J Clin Pathol 2008; 129:34–41.
61. Binnicker MJ, Jespersen DJ, Harring JA, et al. Evaluation of a multiplex flow immunoassay for detection of Epstein-Barr virus-specific antibodies. Clin Vaccine Immunol 2008; 15:1410–1413.

62. Gu A-D, Mo H-Y, Xie Y-B, et al. Evaluation of a multianalyte profiling assay and an enzyme-linked immunosorbent assay for serological examination of Epstein-Barr virus-specific antibody responses in diagnosis of nasopharyngeal carcinoma. Clin Vaccine Immunol 2008; 15:1684–1688.
63. Wong SJ, Demarest VL, Boyle RH, et al. Detection of human anti-flavivirus antibodies with a West Nile virus recombinant antigen microsphere immunoassay. J Clin Microbiol 2004; 42:65–72.
64. Opalka D, Lachman CE, MacMullen SA, et al. Simultaneous quantitation of antibodies to neutralizing epitopes on virus-like particles for human papillomavirus types 6, 11, 16, and 18 by a multiplexed luminex assay. Clin Diagn Lab Immunol 2003; 10:108–115.
65. Drummond JE, Shaw EE, Antonello JM, et al. Design and optimization of a multiplex anti-influenza peptide immunoassay. J Immunol Methods 2008; 334:11–20.
66. Watson DS, Reddy SM, Brahmakshatriya V, et al. A multiplexed immunoassay for detection of antibodies against avian influenza virus. J Immunol Methods 2009; 340:123–131.
67. Martins TB. Development of internal controls for the Luminex instrument as part of a multiplex seven-analyte viral respiratory antibody profile. Clin Diagn Lab Immunol 2002; 9:41–45.
68. Erdman DD. Immunoglobulin M determinations. In: Specter S, Hodinka RL, Young SA, eds. Clinical Virology Manual, 3rd ed. Washington, DC: ASM Press, 2000:146–153.
69. Meurman O. Detection of antiviral IgM antibodies and its problems – a review. Curr Top Microbiol Immunol 1983; 104:101–131.
70. Wiedbrauk DL, Chuang HYK. Managing IgM serologies: Getting the IgG out. Lab Manage 1988; 26:24–27.
71. Hodinka RL. Automated immunoassay analyzers. In: Truant AL, ed. Manual of Commercial Methods in Clinical Microbiology. Washington, DC: ASM Press, 2002:324–335.
72. Andrews JI Diagnosis of fetal infections. Curr Opin Obstet Gynecol 2004; 16:163–166.
73. Mendelson E, Aboudy Y, Smetana Z, et al. Laboratory assessment and diagnosis of congenital viral infections: Rubella cytomegalovirus (CMV), varicella-zoster virus (VZV), herpes simplex virus (HSV), parvovirus B19 and human immunodeficiency virus (HIV). Reprod Toxicol 2006; 21:350–382.

10 Surgical Pathology and Diagnostic Cytology of Viral Infections

James L. Caruso and John M. Childs
United States Navy and Duke University Medical Center, Durham, North Carolina, U.S.A.

David N. Howell
Duke University Medical Center and Veterans Affairs Medical Center, Durham, North Carolina, U.S.A.

INTRODUCTION

Light microscopic examination of histologic sections and cytologic preparations occupies an unusual place in the viral diagnostic armamentarium. As a tool for detecting and identifying viruses, conventional light microscopy is inferior in both the sensitivity and specificity to a number of other methods, including immunohistochemistry, in situ hybridization, and molecular diagnostic methods such as the polymerase chain reaction. Balanced against this deficiency, however, are several advantages.

First and foremost is the ability of conventional microscopy to detect a wide range of pathologic processes. Biopsies and cytology specimens are frequently obtained from patients with incompletely defined clinical problems, for which the differential diagnosis may include a variety of viral infections, infections with organisms other than viruses, and noninfectious disorders. In some cases, tissues are affected by multiple pathologic processes—in immunocompromised patients, simultaneous infection with more than one pathogen is not uncommon. Light microscopy is currently the best technique for sorting out such complex diagnostic problems.

In addition to detecting infections, light microscopy provides information regarding their severity and clinical relevance. It often allows differentiation between latent or innocuous infections and clinically significant ones. In tissues harboring more than one pathologic process, histologic examination can yield information about the relative contributions of the various processes to tissue damage.

Conventional histology and cytology can also serve as valuable adjuncts to other viral diagnostic methods. Embedded tissues and cytologic preparations can be probed for viruses by both immunochemistry and in situ hybridization; antibodies specific for most of the DNA viruses and many of the RNA viruses discussed in this chapter, developed for use with formalin-fixed, paraffin-embedded tissues, are commercially available. Initial light microscopic examination can be used to guide the selection of antibody or nucleic acid probes for specific viruses and to select focal areas of tissue damage for study by the ancillary methods. Light microscopic survey sections are also widely employed to choose tissue samples for subsequent analysis by electron microscopy (EM) (1).

In some instances, a formalin-fixed, paraffin-embedded tissue block or cytologic preparation may simply be the only sample available for study. Examples include biopsies taken from patients not suspected clinically of having a viral infection and cervical exfoliative cytology specimens obtained as a screening test. Histopathologic analysis for viral infection can also be applied to archival paraffin blocks maintained by most pathology practices and medical centers.

For these reasons, it is crucial for histopathologists, cytopathologists, and viral diagnosticians in general to be familiar with the patterns of tissue injury associated with viral infection. This chapter provides an overview of these patterns of injury; additional information can be obtained in the references provided, in the chapters on individual viruses in the second portion of this book, and in a comprehensive text on infectious disease pathology (2). Though no attempt has been made to organize the following discussion with taxonomic rigor, a majority of the pathogens have been grouped into sections on DNA and RNA viruses; within each section, the viruses are listed alphabetically. In a few instances, viruses from diverse taxa are known to elicit similar forms of systemic pathology; these are grouped together in a separate section at the end of the chapter.

SPECIMEN PREPARATION AND DIVISION

Histopathology Vs. Cytopathology

Pathologists utilize two major forms of specimen preparation for light microscopy. For histopathology, a solid tissue sample is sliced into thin sections and applied to glass slides. Prior to sectioning, the specimen must be rendered nonpliable, either by freezing or impregnation with substances such as paraffin or acrylic or epoxy resins. For cytopathology, the specimen consists of single cells and small cell clusters exfoliated from tissue surfaces or obtained from within tissues by aspiration with a needle. These can be spread directly onto slides ("smear" preparation), applied to slides using centrifugal force ("cytocentrifugation"), or aspirated onto filters and transferred to slides. As a final step, both histologic and cytologic preparations are treated with stains that render the specimens visible and allow differentiation of various cell and tissue components. Methodological details are provided by several textbooks (3–5).

Though both histopathology and cytopathology can be used to identify changes caused by viruses at the cellular level, the two methods have various strengths and weaknesses as applied to viral diagnosis. Examination of tissue sections allows a more accurate assessment of host responses to infection (e.g., inflammation, necrosis) and severity of infection than is afforded by cytologic preparations. Serial sections from tissues suspected of harboring viral infections are a convenient substrate for immunohistochemistry and molecular diagnostic studies. On the negative side, most histologic preparative methods are relatively time consuming, requiring several hours to days for completion. (Frozen sectioning can be accomplished more rapidly, but generally yields sections with more artifactual distortion than impregnation methods.) Focal viral infections can also go undetected in small tissue biopsies as a result of sampling error.

Cytologic studies, in contrast, can be performed quite rapidly, often at the patient's bedside. (The Tzanck preparation, a cytologic smear of material scraped from skin lesions, can be produced and examined in a matter of minutes.) Cytologic sampling can also cover larger areas than tissue biopsies; a cervical Papanicolaou smear, for example, samples the entire circumference of the cervix, while individual biopsies are limited to small sectors. Balanced against these advantages is the inability of cytopathology to detect and quantify many forms of virus-associated tissue damage, as well as other processes (e.g., transplant rejection) that may be present in some specimens. Exfoliative cytology may also fail to detect infections below the tissue surface (e.g., cytomegalovirus infection of vascular endothelium).

Given these considerations, selection of cytologic or histologic methods must be tailored to each clinical situation. In some circumstances (e.g., histologic examination of bronchoscopic biopsies and cytologic analysis of bronchoalveolar lavage fluids), application of the two techniques in tandem may improve diagnostic yield for viral infections. In other settings, virus-induced changes detected by screening cytologic examinations (e.g., human papillomavirus-induced dysplasia in cervical Pap smears) may serve as an indication for subsequent tissue biopsy.

Subdivision of Specimens

After a biopsy or cytologic specimen has been obtained, it must be apportioned for various diagnostic studies. If cultures or other special studies (e.g., negative staining and EM of liquid specimens, molecular diagnostic studies) are desired, it is generally advisable to reserve a portion of the specimen for them immediately and transport it promptly to the relevant laboratory; this minimizes the risk of contamination/degradation and maximizes the chance of recovering fastidious organisms.

Cytologic specimens are applied to slides using one or more of the methods described above and stained with a variety of reagents (Papanicolaou and Romanovsky stains are common choices). If sufficient material is available and immunocytochemistry or in situ hybridization are contemplated, additional unstained slides can be prepared; cytocentrifuge preparations are particularly good for this purpose. Following air-drying, such slides can be stored at 25°C for several days; for longer storage periods, slides can be wrapped in aluminum foil, sealed in a plastic bag, and stored at −70°C. Many cell suspensions (e.g., lavage fluids, body cavity effusions) can be stored for several days at 4°C before processing; this is not advisable for urine specimens, however, since storage for even short periods often allows cell degradation. Excess material can also be pelleted and embedded for histologic and/or electron microscopic examination.

Division of solid tissue specimens depends on the size of the specimen, the urgency of the diagnostic situation, and the necessity for special staining or microscopic procedures. Histologic sections of frozen tissue or cytologic preparations made by pressing the tissue lightly against slides ("touch preparations") can be produced in a matter of minutes, but identification of viral infection in such preparations is frequently difficult. Remnants of frozen tissue blocks can be thawed and processed for paraffin sections or stored frozen (preferably at −70°C) for subsequent studies.

For virtually all biopsy specimens, a portion (usually a majority of the specimen) should be fixed in neutral buffered formalin, embedded (generally in paraffin), sectioned, and stained with hematoxylin and eosin (H&E); with rapid processing methods, this can be accomplished within less than six hours for small biopsy specimens (6). A portion of the specimen can also be reserved in fixative for possible EM (glutaraldehyde is best, but formalin and glutaraldehyde/formalin mixtures can also be used); such tissues can be processed subsequently for EM or redirected for paraffin embedment and routine light microscopy if the initial histologic sections do not provide a diagnosis.

GENERAL KEYS TO THE DETECTION OF VIRUSES IN TISSUES

Though identification of specific viruses in tissues often depends on subtle details, recognition of a few general principles provides a valuable guide for initial screening for viral infection. Such screening can be used as a basis for more thorough histologic study, or can be used to direct selection of other, more specific diagnostic tests.

Viral infection causes a number of *changes in cell morphology* ("cytopathic effects") that can be detected at the light microscopic level (7–9). In some instances, clusters of replicating virus particles become large enough to be detected by light microscopy as inclusion bodies (8). DNA viral inclusions are generally intranuclear, while those of RNA viruses usually reside in the cytoplasm; exceptions to this rule are noted below. Large intranuclear inclusions associated with peripheral margination of chromatin are termed "Cowdry A" inclusions. Such inclusions are generally described as eosinophilic, but in practice, the tinctorial properties of intranuclear viral inclusions vary considerably in response to a number of factors, including stage of infection, tissue preparative method, and vagaries of the staining procedure.

Smaller, less distinctive clumps of intranuclear or intracytoplasmic material, classically described as "Cowdry B" inclusions, are often not associated with viral infection. Within the nucleus, nucleoli and chromatin clumps may masquerade as viral inclusion bodies. These items are frequently more widely distributed than true viral inclusions (see below).

Viral replication can also lead to alteration or destruction of a normal cytologic feature. Intranuclear replication of DNA viruses, for instance, frequently induces a loss of nuclear detail; nuclei in affected cells may have a translucent "ground glass" appearance or exhibit dense hyperchromasia. Care must be taken to distinguish such cells from naturally occurring cells with similar cytologic features, such as megakaryocytes [Fig. 1(A)]. Other, less specific forms of virus-induced injury include hemorrhage, necrosis, and cell fusion with resultant multinucleation; viral infection should be suspected when these features occur in the absence of another clear underlying cause.

As a result of their *tissue tropisms*, viruses generally infect only a subset of organs and their constituent tissues. This principle allows generation of lists of potential viral infectious agents for individual organs. Examples for several sites frequently examined by biopsy or cytodiagnosis are provided in Table 1. For most viruses, organ-specific distributions are maintained in both immunocompetent and immunocompromised individuals. Several viruses, however, including cytomegalovirus, herpes simplex virus, varicella zoster virus, and adenovirus, can cause diffuse, multisystemic infections in immunocompromised hosts.

Within a given organ, tissue tropisms dictate the cell type(s) subject to infection. This principle can direct the pathologist to sites within an organ worthy of particularly close scrutiny

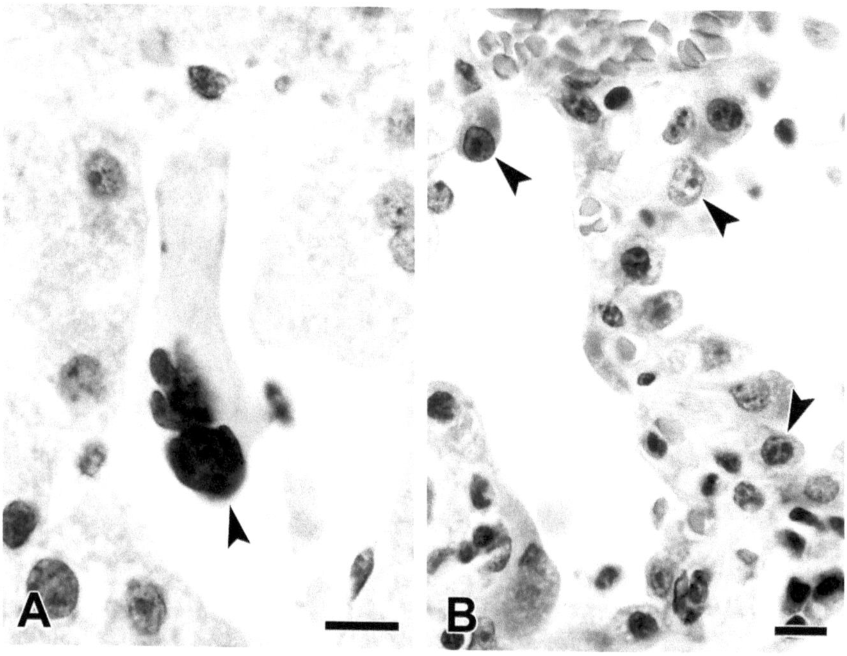

Figure 1 Possible confusing elements in viral diagnosis. (**A**) Megakaryocyte within hepatic sinusoid. The large, irregular, hyperchromatic nucleus (*arrowhead*) can be easily mistaken for viral cytopathic effect. (**B**) Diffuse alveolar damage in transbronchial biopsy from a lung transplant recipient. All of the alveolar lining cells are reactive, and many contain prominent nucleoli (*arrowheads*) that can be mistaken for viral inclusions (bars = 10 μm).

for viral cytopathic effects. It can help in the distinction of viral infections from other processes, which tend to be less selective in their tissue involvement (e.g., ischemic necrosis). It can also be useful in distinguishing viral infections from processes that selectively affect *other* tissues in a given organ. An illustration is provided by the histopathology of liver transplants, where acute rejection generally targets vascular endothelium and bile duct epithelium, while hepatitis viruses have a tropism for hepatocytes. Identification of tissue tropism is easiest in the early stages of infection, before collateral damage to adjacent tissues has occurred.

Within the subset of affected tissues dictated by tissue tropism, viral infection is generally *patchy* and *random*. In most cases, only a portion (often a minority) of a particular tissue will be involved. In contrast, many other forms of tissue damage (e.g., toxic or ischemic injury) tend to be more diffuse. For example, pulmonary viral infections are typically patchy, while other forms of alveolar damage, such as that encountered in acute respiratory distress syndrome [Fig. 1(B)], are often diffuse.

Viral infections frequently induce an influx of *inflammatory cells*, the presence of which can provide useful diagnostic clues. Careful inspection of cells and tissues in and adjacent to inflammatory foci can reveal specific viral cytopathic changes. Certain patterns of inflammation, though not virus specific, may suggest the presence of a particular viral pathogen (e.g., portal lymphoid aggregates in hepatitis C virus infection, microglial nodules in viral encephalitis, atypical lymphoid infiltrates in Epstein–Barr virus infection). Viral infection should also be considered when histologic examination reveals an inflammatory pattern that seems inconsistent with the patient's clinical course. Though a majority of inflammatory processes with acute clinical tempos induce tissue infiltrates of neutrophils, acute viral infections may elicit a predominantly mononuclear inflammatory response or, in immunocompromised individuals, little or no inflammation.

In addition to the changes described above, many viruses have a recognized or suspected role in tumorigenesis. Thus, identification of a virus-associated preneoplastic condition or neoplasm can be a clue to the presence of the virus itself. Several viruses, including Epstein–Barr virus (EBV), human herpesvirus-8 (HHV-8), some strains of human papillomavirus (HPV), and

Table 1 Tissue/Organ Tropisms of Pathogenic Viruses

Liver
Hepatitis viruses
Yellow fever
Lungs/upper respiratory tract
Adenovirus
Measles virus
Influenza virus
Parainfluenza virus
Respiratory syncytial virus
Central nervous system
Herpes simplex virus
Polyomavirus (JC)
Rabies virus
Measles virus (subacute sclerosing panencephalitis)
Urinary tract
Polyomavirus (BK)
Hematolymphatic tissue
Parvovirus B19
Epstein–Barr virus
Skin/mucosal surfaces
Poxvirus
Human papillomavirus
Herpes simplex virus
Varicella zoster virus
Widely distributed[a] (especially in immunocompromised patients)
Adenovirus
Cytomegalovirus
Herpes simplex virus
Varicella zoster virus

[a]Including, but not limited to, lungs, liver, GI tract, and urinary tract.

human T-lymphotropic virus-1 (HTLV-1), directly induce the genetic alterations responsible for the neoplastic transformation of target cells. In contrast, human immunodeficiency virus (HIV) causes a state of profound T-cell immunodeficiency that facilitates infection with other tumorigenic viruses and may also hamper immune surveillance and destruction of nascent tumors. Finally, some viruses, especially the hepatitis viruses, act as tumor promoters by establishing a chronic inflammatory state. A list of tumors with well-established virus associations is provided in Table 2. For a more exhaustive review, numerous current review articles are available (10–14).

Table 2 Virus–Tumor Associations

Virus	Tumor
Human papillomavirus	Squamous cell carcinomas of cervix, skin, and other squamous-lined mucosal sites
Human herpesvirus-8	Kaposi sarcoma
	Primary effusion lymphoma
	Multicentric Castleman disease
Epstein–Barr virus	Posttransplant lymphoproliferative disorder
	Diffuse large B-cell lymphoma
	Burkitt lymphoma
	Hodgkin lymphoma
	Nasopharyngeal carcinoma
	Transplant-associated smooth muscle neoplasms
Human T-lymphotrophic virus	Adult T-cell leukemia/lymphoma
Human immunodeficiency virus	HPV, HHV-8, EBV-driven tumors
Hepatitis B and C viruses	Hepatocellular carcinoma

HISTOLOGIC AND CYTOLOGIC FEATURES OF INFECTION WITH SPECIFIC VIRUSES

DNA Viruses

Adenovirus

Adenoviruses typically cause self-limited community-acquired respiratory infections in immunocompetent individuals, especially children (15). Infections causing acute respiratory failure have been reported in otherwise healthy adults (16). In immunocompromised hosts, adenoviruses can infect a wide variety of sites, including the lungs, liver, and gastrointestinal and urinary tracts. Patients with severe immunodeficiency occasionally develop devastating, multisystemic infections.

Intranuclear replication of adenoviruses produces collections of particles that can often be detected by light microscopy as inclusion bodies. At early stages of viral replication, the inclusions are small and tend to be eosinophilic or amphophilic on H&E-stained sections. Peripheralization of nuclear chromatin may produce a zone of clearing around some inclusions. As the inclusions enlarge, they become more basophilic, with associated degeneration of the nuclear membrane. The resulting "smudge cells," with indistinct nuclear-cytoplasmic borders, can be seen in both histologic (Fig. 2) and cytologic preparations. Multinucleation is uncommon, and intracytoplasmic inclusions are not seen. In cytologic preparations, detached apical portions of bronchial epithelial cells bearing cilia are sometimes detected ("ciliocytophthoria"); this finding, though diagnostically useful, can be encountered in many other forms of epithelial injury.

In the lung, adenoviruses infect epithelial cells lining respiratory passages (17); infection of respiratory bronchioles may lead to a form of inflammatory destruction referred to as bronchiolitis obliterans. In the liver, random foci of infection are often scattered throughout the parenchyma. In severe infections, foci of tissue necrosis are frequently seen. Inclusion bodies are usually easiest to identify at the borders of such necrotic foci. Adenovirus infection in the urinary tract usually takes the form of a hemorrhagic cystitis, but serious renal parenchymal infections have been documented in occasional transplant recipients (18).

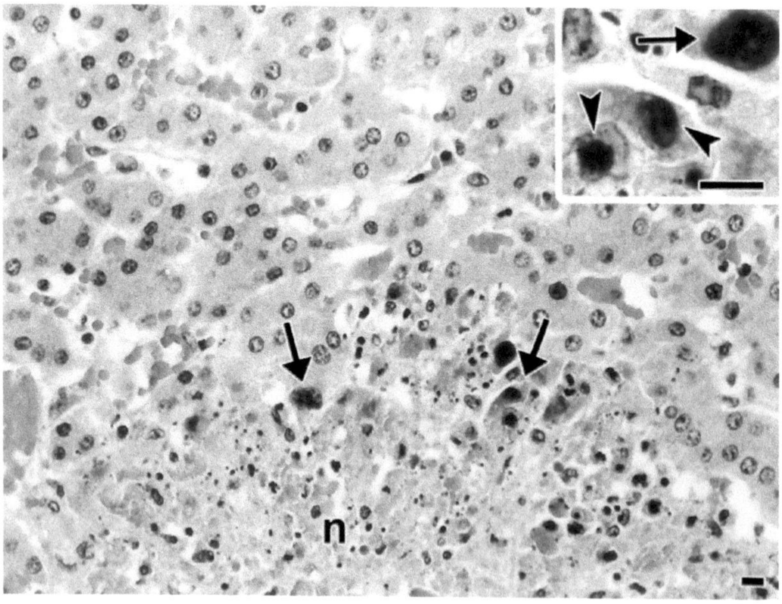

Figure 2 Adenovirus infection in liver tissue. Several virus-infected cells (*arrows*) are visible adjacent to a zone of necrosis (n). In the inset, nuclear inclusions are seen in two cells (*arrowheads*) adjacent to a characteristic "smudge cell" (*arrow*) (bars = 10 μm).

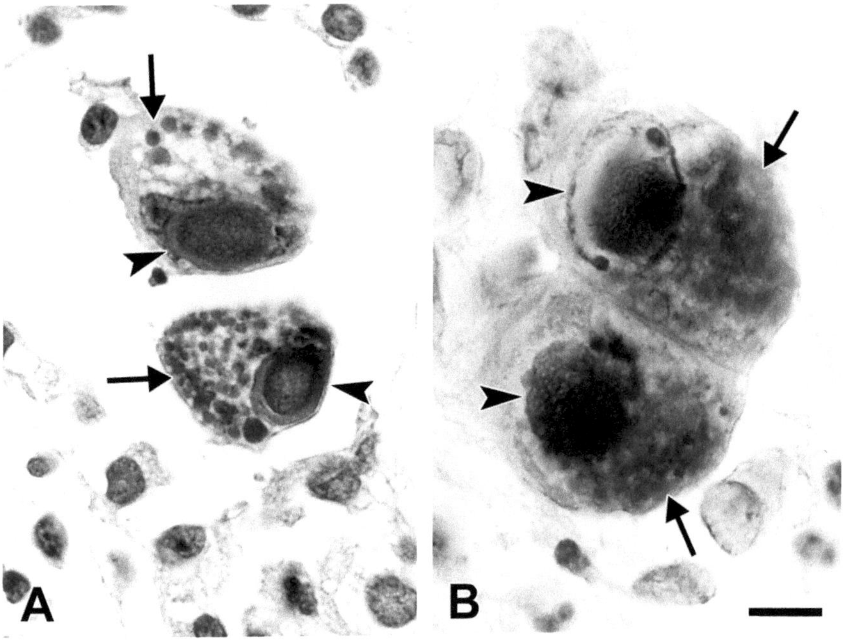

Figure 3 Cytomegalovirus infection in biopsy of a renal transplant (**A**) and cervical Papanicolaou smear (**B**). The infected cells are enlarged compared with surrounding cells, and contain both haloed intranuclear inclusions (*arrowheads*) and cytoplasmic inclusions (*arrows*). Bar = 10 μm, with same magnification in both panels. *Source*: Panel **B** courtesy of Ms. Rosiland Wallace, Duke University Medical Center.

Cytomegalovirus

Clinically apparent infections with cytomegalovirus (CMV) are usually encountered in immunocompromised hosts, though the virus can also cause a mononucleosis-like syndrome in individuals with normal immune function (19). Common sites of infection include the adrenals, lungs, gastrointestinal (GI) tract, central nervous system, and retina, but virtually any organ can be involved. Small numbers of CMV-infected cells are also an occasional incidental finding in biopsies from patients with noninfectious primary illnesses.

Cells infected with CMV exhibit both nuclear and cytoplasmic enlargement: the "cytomegaly" from which the virus draws its name. Intranuclear inclusion bodies with varying tinctorial properties are usually present. At early stages, these are easily confused with inclusions of other herpesviruses. As the inclusions mature, they become round to oval, with smooth borders. Chromatin margination and a peri-inclusional clear zone are often present, producing a pattern referred to as an "owl's eye" in both histologic [Fig. 3(A)] and cytologic [Fig. 3(B)] specimens. Clusters of complete virions that have budded through the nuclear membrane are often visible as multiple, basophilic, granular cytoplasmic inclusion bodies, a finding unique to CMV among herpesviruses (Fig. 3).

Cytomegalovirus can infect a wide variety of cell types, including vascular endothelium, glandular epithelium, histiocytes, neurons, and fibroblasts. Patterns of infection vary depending on the organ examined. In GI biopsies, endothelium of vessels within the lamina propria is the most common site of infection. In the liver, hepatocytes are often infected, though other cell types can also be involved. Infected hepatocytes are frequently surrounded by clusters of neutrophils. Respiratory epithelium is a common target in lung tissue; stromal and endothelial cells within bronchial lamina propria can also serve as targets.

Though the classic cytopathic changes associated with CMV are quite distinctive, characteristic inclusion bodies may be absent in some specimens, particularly those taken at very early or late points in the viral replicative cycle or from individuals with only moderate immunosuppression (e.g., solid-organ transplant recipients). For this reason, liberal use of ancillary techniques such as immunoperoxidase staining is recommended for specimens from patients with a high clinical index of suspicion for CMV infection.

Epstein–Barr Virus

In immunocompetent hosts, Epstein–Barr virus (EBV) causes infectious mononucleosis, a generally benign and self-limited systemic infection, in addition to nasopharyngeal carcinomas, the endemic (African) form of Burkitt lymphoma, and a significant number of Hodgkin lymphomas. In recipients of solid organ and bone marrow transplant, EBV is associated with posttransplant lymphoproliferative disorder (PTLD) and mesenchymal neoplasms, notably leiomyosarcoma (20). Patients with HIV infection are also at increased risk for developing several EBV-associated disorders, including oral hairy leukoplakia (21), a nonneoplastic oral lesion, and a variety of B-cell lymphomas, including primary CNS lymphoma, diffuse large B-cell lymphoma, and primary effusion lymphoma (22).

In infectious mononucleosis, EBV infects salivary gland epithelium and, subsequently, B-lymphocytes. Infected B-lymphocytes elicit proliferation and differentiation of T-lymphocytes, which become enlarged, with vacuolated cytoplasm and irregular, variable nuclear contours. These "atypical" T-lymphocytes are most often encountered in peripheral blood smears, but can also be seen in biopsies of lymphoid organs and other sites; they must be rigorously distinguished from neoplastic cells, which they resemble in many respects.

In contrast, the proliferating cells in PTLD are EBV-transformed B-lymphocytes. A spectrum of disorders, ranging from benign polyclonal proliferations to frank B-cell lymphomas, can be seen. The histologic hallmark is an infiltrate of lymphoid cells with varying degrees of cellular atypia; immunoperoxidase staining for B- and T-lymphocyte markers is useful in distinguishing PTLD (B-lymphocyte predominant) from transplant rejection (T-lymphocyte predominant). Affected sites include the transplants themselves (Fig. 4) and other organs, including the brain; extranodal involvement is frequent.

Oral hairy leukoplakia is unique among EBV-associated disorders in that lytic infection plays a major role in its pathogenesis. Infected oral epithelium is hyperplastic, and koilocyte-like cells with Cowdry type-A intranuclear inclusions are seen in the upper epithelial layers (21). In other EBV-associated conditions, latently infected or transformed cells lack specific inclusions or cytopathic effect, but can often be detected using immunohistochemical or chromogenic in situ hybridization techniques.

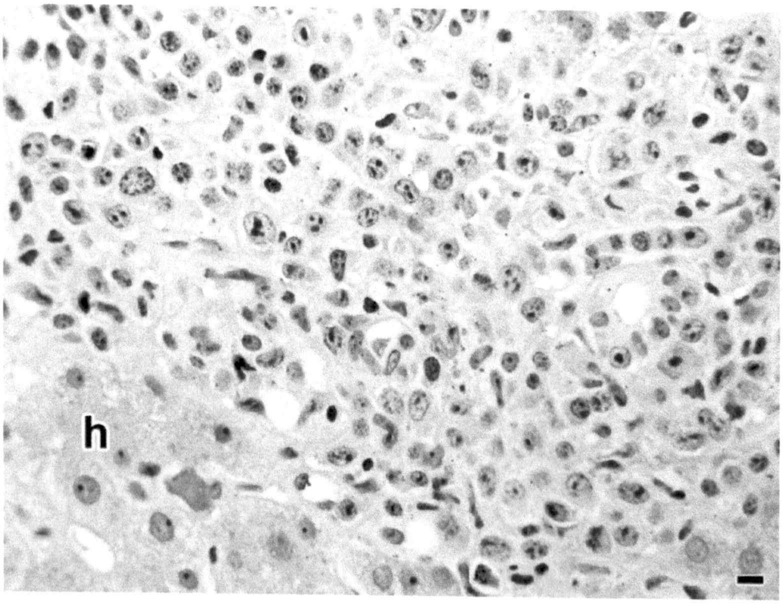

Figure 4 Posttransplant lymphoproliferative disorder in a liver transplant. Proliferation of atypical lymphocytes expands a portal tract; the proliferating cells have irregular nuclei, many of which contain conspicuous nucleoli. Adjacent hepatocytes (h) are normal in appearance (bar = 10 μm).

Herpes Simplex Virus and Varicella Zoster Virus

Herpes simplex virus (HSV) is a common cause of blistering infections of skin and mucous membranes, and can also cause pharyngitis, esophagitis, and encephalitis (most frequently affecting the temporal lobes). Varicella zoster virus (VZV) is the causative agent of chicken pox and herpes zoster ("shingles") (23). In immunocompromised hosts and infants, both HSV and VZV are occasionally associated with severe pneumonitis, hepatitis, or disseminated infection. Infection can either be primary or develop as a result of reactivation of latent virus in nerve roots and ganglia that innervate mucosal or cutaneous surfaces.

Cells infected with HSV-1, HSV-2, or VZV exhibit cytopathic changes that are indistinguishable by light microscopy. With any of the viruses, rounded, eosinophilic intranuclear inclusions of the Cowdry A type can be seen. Often, however, the nucleoplasm develops a milky, "ground glass" appearance without distinct inclusions. Some investigators have suggested that Cowdry A inclusions are encountered with greater frequency in secondary infections with HSV than in primary infections, but others have disputed this assertion (7). Cytomegaly of uninucleate cells is not generally seen. In many cases, however, multinucleated syncytial giant cells are formed by fusion of several virus-infected cells (Fig. 5).

The most commonly affected tissues are stratified squamous epithelia. In addition to the cytopathic effects noted above, infected epithelia exhibit a variety of nonspecific changes, including hyperplasia, intercellular edema, ballooning degeneration of the basal layer, and vesicle formation. In severe infections, ulceration and subepithelial inflammatory cell infiltrates can be seen. Herpes viruses have also been identified as a precipitating agent of erythema multiforme, an interface dermatitis with targetoid lesions; herpes associated erythema multiforme (HAEM) is now considered the most common form of the disorder (24).

Inflammation and viral inclusions can occasionally be seen in ganglia associated with areas of active herpes zoster. HSV hepatitis in immunocompromised hosts frequently causes large, geographic areas of necrosis; viral cytopathic changes are most easily detected at the edges of such lesions, and immunohistochemical stains are available for confirmation. Latent infection with HSV or VZV is not detectable by routine light microscopic methods.

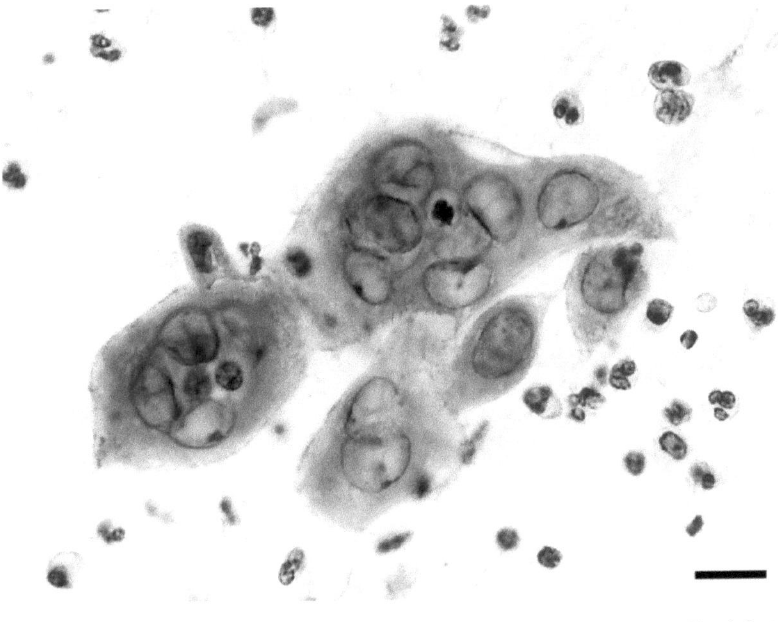

Figure 5 Herpes simplex virus infection in cervical Papanicolaou smear. The infected cells are multinucleated, and the nuclei have a "ground glass" chromatin pattern. Some of the nuclei contain small clumps of darkly stained material, but distinct viral inclusions are not present (bar = 10 μm). *Source*: Courtesy of Ms. Rosiland Wallace, Duke University Medical Center.

Human Herpesviruses 6, 7, and 8

Human herpesvirus (HHV)-6 causes exanthem subitum (roseola infantum, sixth disease), a skin rash of childhood, and has been linked to encephalitis, febrile illnesses, and bone marrow suppression in immunosuppressed adults (25). A possible role for HHV-7 in some cases of exanthem subitum has also been postulated. Both HHV-6 and HHV-7 have been suggested as potential inciting agents in pityriasis rosea, another benign and self-limited exanthem, although a definitive link has not yet been established (26). Genomic material from HHV-8 was first isolated from lesions of Kaposi's sarcoma, a malignant neoplasm of vascular origin, and a causal role for the virus in the pathogenesis of this tumor is well established. HHV-8 is also the tumorigenic virus responsible for primary effusion lymphoma and multicentric Castleman disease, particularly in immunosuppressed patients (27,28).

Biopsy is rarely employed in the evaluation of exanthem subitum or pityriasis rosea, but is of central importance to the diagnosis of HHV-8-driven tumors, each of which has distinctive histologic and immunohistochemical features. HHV-6, -7, and -8 are not associated with specific cytopathic effects in biopsy or cytologic specimens, but immunoperoxidase stains are available for the detection of HHV-8 in biopsy tissue.

Human Papillomavirus

Human papillomavirus (HPV) infects squamous epithelial cells lining a variety of mucosal and cutaneous surfaces, including the skin, oral cavity, larynx, and anogenital tract. The resulting lesions range from benign proliferations (warts) to malignant neoplasms. Cells infected with HPV frequently have darkly staining, basophilic nucleoplasm, though discrete inclusion bodies are seldom encountered. In both histologic and cytologic preparations, the nuclei of HPV-infected cells often appear shrunken and wrinkled, and are surrounded by a perinuclear clear zone. This phenomenon, termed "koilocytosis," is particularly common in genital warts (condyloma acuminatum). Binucleate cells are encountered in many specimens (Fig. 6).

There is frequently florid proliferation of epithelial cells with an exaggerated papillary configuration; in cutaneous warts (verrucae), marked hyperkeratosis is present. Human papillomavirus-infected cells can undergo a continuum of neoplastic changes, including varying

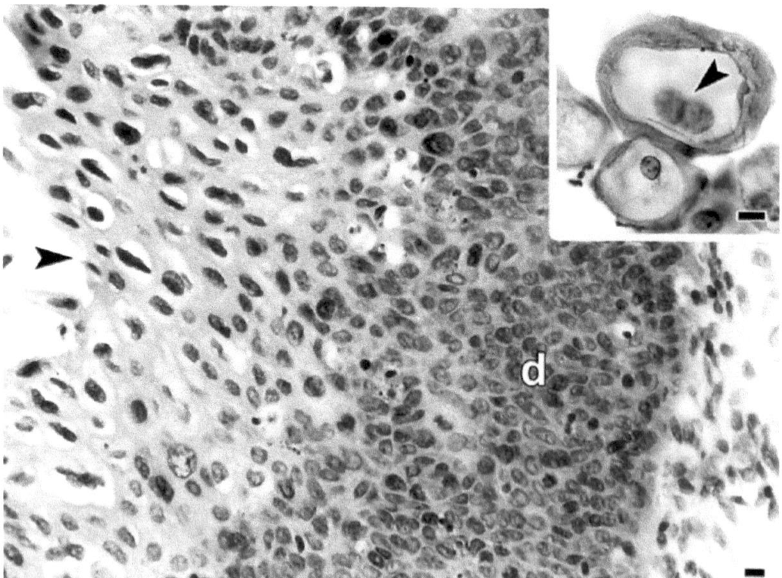

Figure 6 Human papillomavirus infection in cervical biopsy and Papanicolaou smear. Numerous koilocytes, with shrunken nuclei and perinuclear clear spaces, are present in the superficial epithelial layers of the biopsy tissue (arrowhead); dysplastic features (d) are present in the deeper layers. Two koilocytes are also present in the Papanicolaou smear (inset); one is binucleate (arrowhead) (bars = 10 μm). *Source*: Courtesy of Dr. Stanley Robboy (biopsy) and Ms. Rosiland Wallace (Papanicolaou smear), Duke University Medical Center.

degrees of dysplasia (Fig. 6), squamous cell carcinoma in situ, and invasive squamous cell carcinoma (29). In addition to direct tissue examination, molecular tests for high-risk HPV are routinely employed for detecting cervical infection (30). Immunohistochemical stains for HPV, including surrogate markers of infection such as p16 overexpression, are commercially available.

Parvovirus

Parvovirus B19 has been associated with several human disorders, including skin conditions (erythema infectiosum or "fifth disease" and a purpuric eruption known as "glove and socks" syndrome), arthropathies, red cell aplasia, and intrauterine infections with associated fetal hydrops (31). Infection of target cells requires the presence of the blood group P antigen, a globoside expressed primarily by cells of the erythroid lineage, but also to varying degrees by platelets and tissues from nonhematopoietic organs (32). Parvovirus infection of the bone marrow or sites of extramedullary hematopoiesis causes erythroid hypoplasia with giant pronormoblasts. The nuclei of affected erythroblasts often have a glassy appearance with clumped peripheral chromatin and eosinophilic inclusions of varying size (Fig. 7). Myeloid maturation is typically normal, though mild peripheral neutropenia and thrombocytopenia may be present (32).

Another parvovirus, human bocavirus, has been identified recently as a potential cause of acute respiratory disorders, particularly in children (33). The histologic features of infection with this virus are currently undefined.

Polyomavirus

Two well-characterized human polyomaviruses are associated with infections detectable by light microscopy. JC virus causes an ongoing, often devastating infection of the central nervous system referred to as progressive multifocal leukoencephalopathy (PML) (34), while BK virus infects the urinary tract (35). Clinically significant infections with both viruses are encountered almost exclusively in immunocompromised patients. PML is a significant cause of morbidity and mortality in patients with acquired immunodeficiency syndrome (AIDS) and hematopoietic malignancies, and has recently been reported in patients treated for autoimmune disorders with an antiintegrin monoclonal antibody. BK virus infections are fairly common in individuals with AIDS or undergoing cancer chemotherapy, and have emerged as a significant problem for renal transplant recipients in the past decade. An unusual skin disorder of immunocompromised

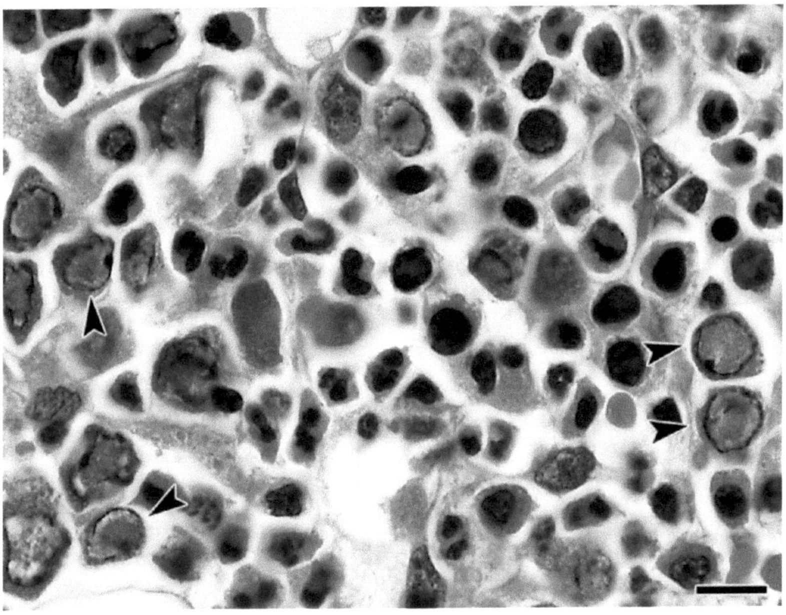

Figure 7 Parvovirus B19 infection in bone marrow. Numerous giant pronormoblasts with intranuclear inclusions are present (*arrowheads*) (bar = 10 μm).

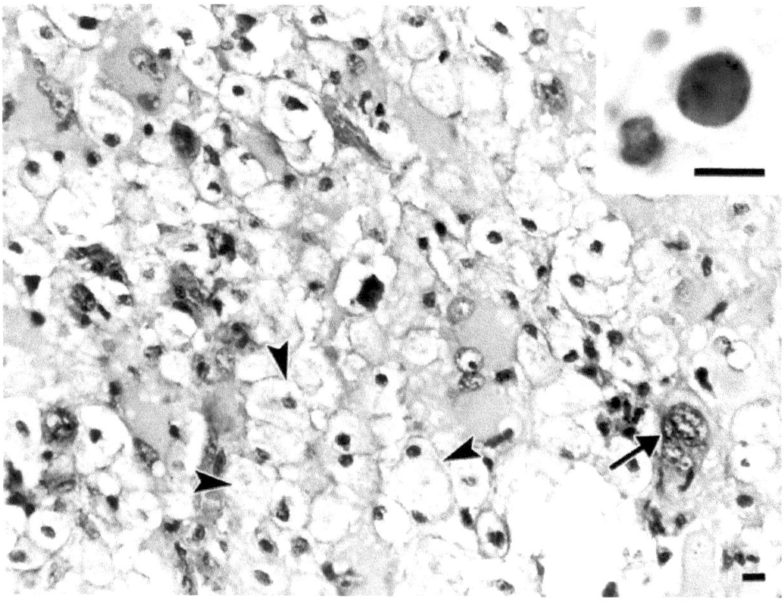

Figure 8 Progressive multifocal leukoencephalopathy. Large clusters of lipid-laden "gitter cells" (*arrowheads*) are present, as are astrocytes with large, bizarre nuclei (*arrow*). Occasional JC virus infected cells with a glassy, homogeneous chromatin pattern are visible at higher magnification (*inset*) (bars = 10 μm).

patients, trichodysplasia spinulosa, has also been attributed to a polyomavirus distinct from the JC and BK viruses (36).

Polyomavirus-infected cells exhibit a variety of nuclear changes, including nuclear enlargement, "ground glass" nucleoplasm, amphophilic intranuclear inclusions of varying sizes, and degeneration of the nuclear membrane with smudge cell formation. Cytoplasmic inclusions are not detected by light microscopy, and syncytial giant cells are absent.

The target cells for JC virus are oligodendrocytes within the white matter (Fig. 8). There is progressive demyelination in affected areas; lipid-laden phagocytic cells referred to as "gitter cells" generally abound. Also present in many cases are reactive astrocytes, which may have bizarre nuclear features. A sparse perivascular mononuclear inflammatory infiltrate may be present, but inflammation is frequently minimal except in the setting of immune reconstitution inflammatory syndrome (34).

In the native kidney, the primary target of BK virus is the transitional epithelium lining the urinary bladder, ureters, and renal pelvis; the resulting infection is typically a hemorrhagic cystitis. In renal transplant recipients and rare nontransplant patients, tubular epithelium may also be infected, leading in some cases to an intense tubulointerstitial nephritis that can culminate in graft loss (Fig. 9). This form of infection has some features, including mononuclear inflammation and lymphocytic infiltration of tubules, that overlap with those of cellular allograft rejection; useful distinguishing features of BK virus nephritis include nuclear viral cytopathic changes in infected cells, patchy distribution, and a prominent plasmacytic component in the inflammatory infiltrate. Exfoliated BK virus-infected cells can be detected in urine by cytologic examination, and are referred to as "decoy" cells.

Poxvirus

With the eradication of smallpox, the most prevalent human poxvirus infection encountered by anatomic pathologists is molluscum contagiosum, a generally innocuous infection of epidermal cells (37). Human cutaneous infections with various animal poxviruses also occur occasionally (38); one such virus, monkeypox, was introduced in the United States in 2003 via a chain of infections that included African rodents imported as pets and prairie dogs (39).

Unlike other pathogenic DNA viruses, poxviruses replicate in the cytoplasm, producing inclusion bodies referred to as "Guarneri bodies" in smallpox and "molluscum bodies" in

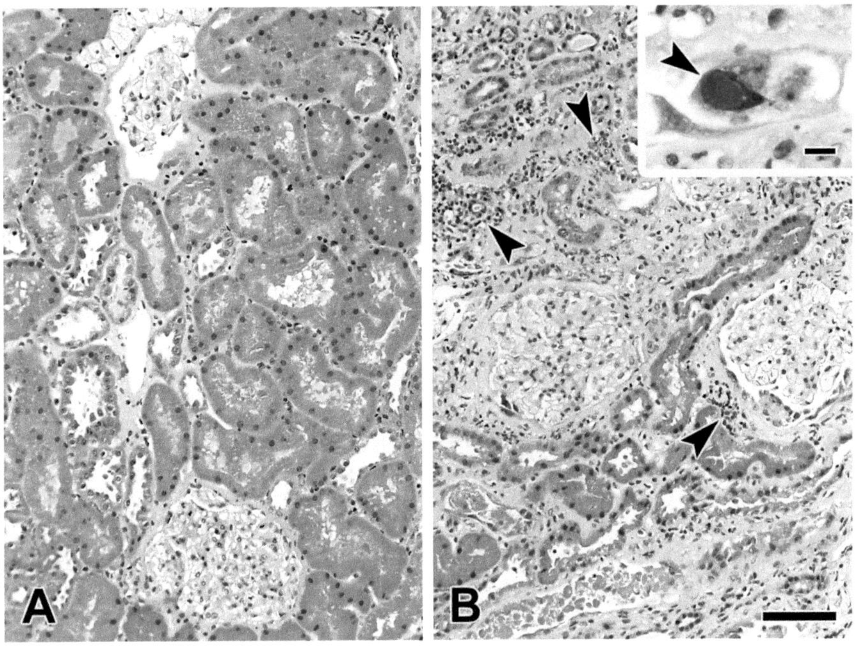

Figure 9 Polyomavirus infection in renal transplant. Two fields from the same biopsy show renal cortex with minimal abnormality (**A**) and extensive interstitial inflammation (*arrowheads*, **B**) (bar = 100 μm, with same magnification in both panels). A virus-infected tubular epithelial cell containing an intranuclear inclusion (*arrowhead*) is seen in the inset (bar = 10 μm).

molluscum contagiosum. In the latter disorder, the inclusions are initially visible as eosinophilic intracytoplasmic bodies in the stratum malpighii, above the basal layer of keratinocytes. As the infected cells progress to the epithelial surface, the inclusions enlarge, displacing the nucleus, and eventually become basophilic. The proliferating epidermis is displaced downward into the dermis in a lobular pattern, generally with minimal inflammatory response (Fig. 10).

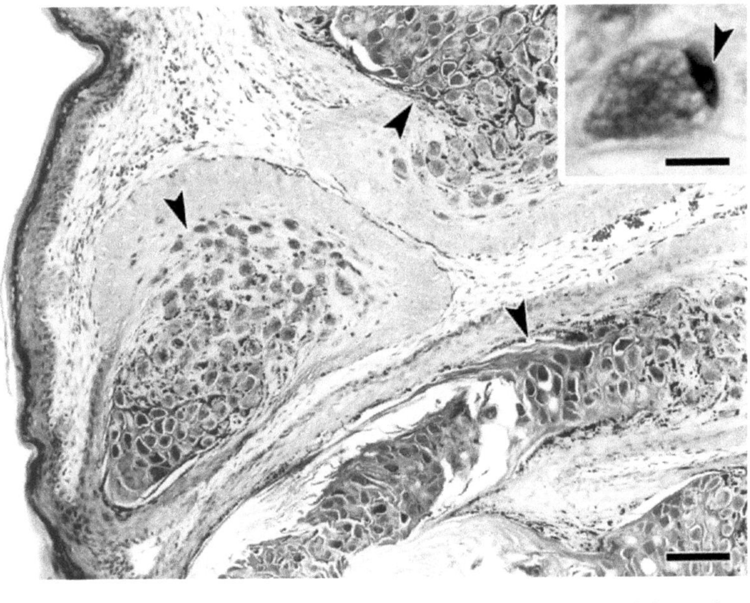

Figure 10 Molluscum contagiosum in skin biopsy. Nest of cells containing molluscum bodies (*arrowheads*) lined by an epidermal layer are displaced into the dermis. In a single infected cell viewed at higher magnification (*inset*), displacement of the nucleus (*arrowhead*) by a large molluscum body is seen (bar in main panel = 100 μm, bar in inset = 10 μm). *Source*: Courtesy of Dr. Victor Prieto, Duke University Medical Center.

In many of the other orthopox virus infections, including monkeypox and smallpox, the lesions take the form of vesicles/bullae that progress to pustules and heal with varying degrees of residual scarring. Mixed inflammation, necrosis, edema, and keratinocytes containing cytoplasmic inclusions are seen at the periphery of the lesions. Multinucleated keratinocytes have been reported in monkeypox (39), a feature that could conceivably lead to confusion with herpesvirus infection.

RNA Viruses

Enteroviruses

The enteroviruses are a diverse group of small RNA viruses that are associated with clinical syndromes ranging from a mild, acute, self-limited gastroenteritis to life-threatening infections of the heart, liver, and central nervous system. A benign, self-limited viral exanthem, hand, foot, and mouth disease, is also caused by enteroviruses. Though light microscopy plays a limited role in the diagnosis of enterovirus infections, several are of sufficient current or historical importance to merit a brief discussion.

During the first half of the past century, poliovirus was the cause of seasonal epidemics resulting in paralytic illness in a small percentage of the individuals afflicted. Poliovirus infection in both the central and peripheral nervous system yields a combination of histologic findings that includes neuronal injury/necrosis, reactive gliosis, and inflammatory cell infiltration. Early accumulation of neutrophils is supplanted by mononuclear inflammation, often with the formation of aggregates of glial cells and lymphocytes (microglial nodules). Phagocytosis of injured nerve cells (neuronophagia) may occur (40,41).

Other members in the enterovirus group, including coxsackievirus group A, coxsackievirus group B, and echoviruses, have also been linked etiologically with severe illnesses such as meningitis, encephalitis, myopericarditis, ophthalmic infections, and systemic infections (42). Though the histologic findings in these disorders are by no means specific, tissue biopsy is sometimes useful in distinguishing them from other pathologic processes and in assessing disease progression. Endomyocardial biopsy, for example, is sometimes of value in the diagnosis and staging of myocarditis associated with coxsackieviruses and other viruses, particularly when coupled with immunohistochemistry and molecular diagnostic techniques (43).

Human Retroviruses

Discovered in 1979, human T-lymphotropic virus (HTLV)-1 was the first human retrovirus described and the first infectious cause identified for a human leukemia, adult T-cell leukemia/lymphoma (ATLL). ATLL is prevalent in Japan, the Caribbean, and some parts of South America and Africa, locations where HTLV-1 infection is endemic. The disease is characterized by leukocytosis, lymphadenopathy, skin involvement, organomegaly, and hypercalcemia. The leukemic cells in ATLL have convoluted and polylobated nuclei, and have been dubbed "flower cells." Occasionally, the malignant cells have a Hodgkin-like cellular morphology. No specific viral inclusion can be identified, and the diagnosis rests on morphologic tissue patterns and other ancillary tests (44,45). Infection with HTLV-1 (and possibly with a related virus, HTLV-2) can also cause a chronic encephalomyelopathy called HTLV-1 associated myelopathy or tropical spastic paraparesis (HAM/TSP) (46).

Human immunodeficiency virus (HIV)-1 is the causative agent of AIDS. As noted elsewhere in this chapter, opportunistic viral infections (and in some cases, tumors promoted by them) are a source of extensive morbidity and mortality in patients with AIDS. Several pathologic alterations directly attributable to HIV-1 have also been described. Lymph nodes in HIV-1-infected individuals undergo a series of alterations beginning with florid follicular hyperplasia and culminating in involution, often with intervening mixed patterns. Infection of macrophages/microglia within the nervous system by HIV-1 plays a role in the pathogenesis of several complications of AIDS, including cognitive disorders, vacuolar myelopathy (a form of spinal cord degeneration), and sensory neuropathy (47). Direct infection of renal tubular and glomerular cells by HIV-1 is also responsible for HIV-associated nephropathy, a renal disorder involving both collapsing glomerular sclerosis and tubulointerstitial degeneration (48).

Influenza and Parainfluenza Virus

Influenza viruses replicate in epithelia spanning the entire respiratory tree, causing febrile illnesses ranging from tracheobronchitis to pneumonia. In upper respiratory infections, damage to and sloughing of tracheal and bronchial epithelium are accompanied by varying degrees of submucosal inflammation and accumulation of luminal secretions. Epithelial regeneration may be accompanied by squamous metaplasia. Reported features of influenza pneumonia include microvascular thrombosis, interstitial edema and hemorrhage, and diffuse alveolar damage with hyaline membrane formation. Neutrophil infiltration is frequently absent or modest, particularly at early stages of infection; large numbers of neutrophils often indicate bacterial superinfection (49).

Parainfluenza viruses also cause diverse respiratory syndromes, including croup (acute laryngotracheobronchitis), bronchiolitis, and pneumonia. Like influenza viruses, parainfluenza viruses have respiratory epithelia as their target cells. Infected epithelial cells may form syncytial giant cells, and a proteinaceous exudate is often present (50).

Measles Virus

Measles virus (rubeola virus) causes measles, a febrile illness characterized by a maculopapular rash, upper respiratory symptoms, and conjunctivitis. The virus uses the respiratory tract as a portal of entry, then spreads via hematopoietic cells to a variety of lymphoid and other organs. Though measles itself is generally benign and self-limited, rare children develop central nervous system sequelae, including acute demyelinating encephalomyelitis, measles inclusion body encephalitis, and subacute sclerosing panencephalitis, which are often fatal (51).

Measles virus nucleocapsids, unlike those of most RNA viruses, can be found both in the nucleus and the cytoplasm. As a result, light microscopy of infected cells may show both intranuclear and cytoplasmic inclusions. The intranuclear inclusions are generally eosinophilic and are often surrounded by a clear halo; those in the cytoplasm are usually smaller and less conspicuous. Infection with measles virus also frequently leads to the formation of syncytial giant cells. During the incubation period of measles, such cells, termed Warthin–Finkeldey cells, may be present in lymphoid tissue (Fig. 11), and occasionally allow a presumptive diagnosis to

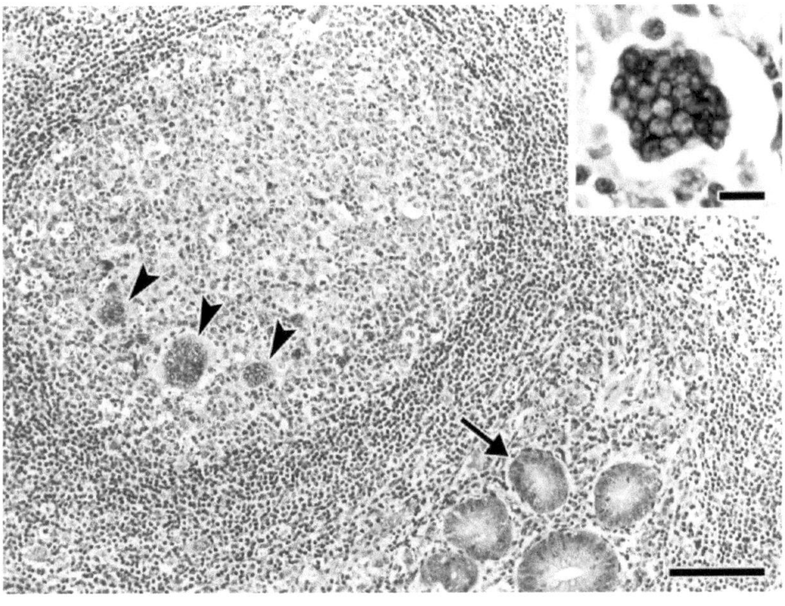

Figure 11 Vermiform appendix from a child with measles. Several Warthin–Finkeldey cells (*arrowheads*) are present in a lymphoid follicle; a cluster of mucosal glands (*arrow*) is adjacent to the follicle. At higher magnification (*inset*), the multinucleation of a Warthin–Finkeldey cell is apparent (bar in main panel = 100 μm, bar in inset = 10 μm). *Source*: Courtesy of Dr. Jimmy Green, Naval Medical Center, Portsmouth, Virginia.

be made. In the lung, measles virus may cause an interstitial pneumonitis with inclusion-bearing giant cells.

Mumps Virus

Mumps virus is an enveloped RNA virus that causes a characteristic swelling in the parotid gland. In addition to the salivary glands, affected sites may include the gonads, pancreas, heart, kidney, respiratory tract, and central nervous system. Severe, multisystemic infections occur occasionally, particularly in adults. There is no characteristic viral cytopathology; affected organs usually display interstitial edema and a lymphocytic infiltrate with occasional hemorrhage, infarction, and necrosis. Vacuolization and desquamation of ductal epithelium with duct ectasia and obstruction have also been reported (52). Multinucleated giant cells and eosinophilic cytoplasmic inclusions have been seen in tissue culture, but not in histologic sections.

Rabies Virus

Rabies virus and other members of the Lyssavirus genus cause rabies, an almost invariably fatal encephalomyelitis. In countries with effective vaccination programs for domestic animals, rabies is largely limited to wild and feral animals, but the disease is an important (and probably under-recognized) public and veterinary health problem in the developing world (53). In the past, the diagnosis of rabies depended largely on histologic identification of viral inclusions referred to as Negri bodies: eosinophilic, intracytoplasmic inclusions most often found in the large neurons of Ammon's horn in the hippocampus and in the Purkinje cells of the cerebellum of infected hosts (Fig. 12) (54). Because of increased sensitivity, direct fluorescent antibody methodology has replaced histologic examination for Negri bodies as the primary diagnostic modality for detecting rabies virus in tissues (53,54).

Respiratory Syncytial Virus

Respiratory syncytial virus (RSV) is a large RNA virus that causes upper and lower respiratory tract infections, most commonly in infants and young children. Younger infants and those with congenital cardiopulmonary abnormalities as well as the immunocompromised are at greater

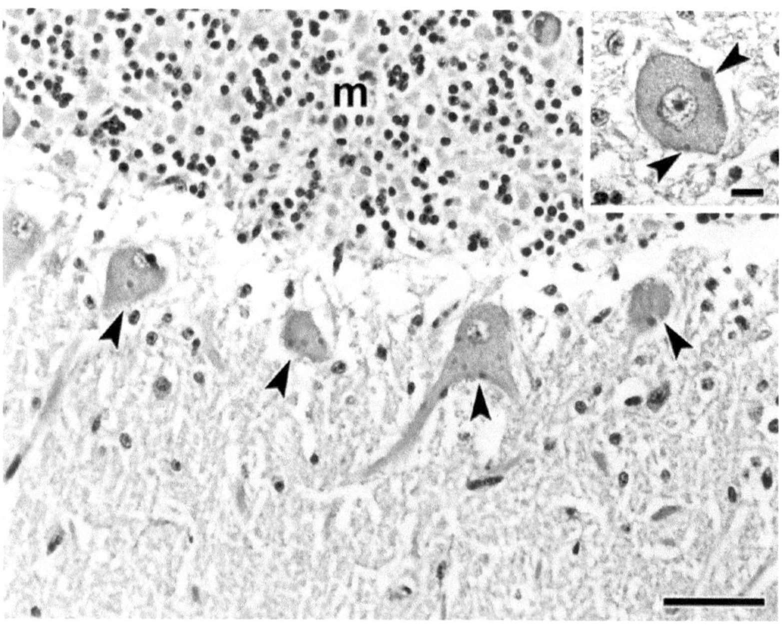

Figure 12 Rabies encephalitis in horse cerebellum. Large pyramidal cells (*arrowheads*) adjacent to the molecular layer (m) contain numerous cytoplasmic inclusions (Negri bodies). A single infected cell viewed at higher magnification (*inset*) contains several Negri bodies (*arrowheads*) (bar in main panel = 100 μm, bar in inset = 10 μm). *Source*: Courtesy of Drs. Glenn Sandberg and Kymberly Gyure, Armed Forces Institute of Pathology.

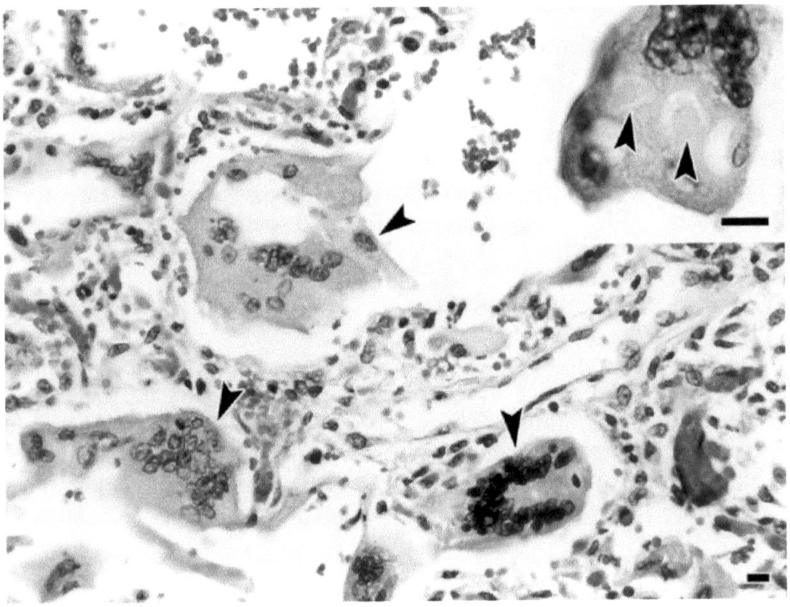

Figure 13 Respiratory syncytial virus infection in lung tissue. Several multinucleated giant cells are present (*arrowheads*). At higher magnification (*inset*), pale cytoplasmic inclusions surrounded by haloes are seen in a giant cell (bars = 10 μm). *Source*: Courtesy of Dr. Victor Roggli, Duke University Medical Center.

risk for severe infections. In such cases, microscopic examination of the lung reveals sloughed epithelium, mucus, and debris within the small airways with consequent hyperinflation of the distal lung segments. As the name implies, infection results in the presence of large syncytial giant cells in alveolar spaces (55). Epithelial cells may contain intracytoplasmic inclusions surrounded by a thin, peripheral halo (Fig. 13). In milder infections, including a majority of those encountered in lung transplant biopsies, the histologic findings are generally nonspecific.

Rubella Virus

Postnatal exposure to the rubella virus causes a benign childhood exanthem, but in utero infection can cause severe congenital abnormalities (56). Gross pathologic examination of an infant or fetus exposed to rubella virus during the first trimester in utero may reveal cardiac defects, growth retardation, or ocular abnormalities; there is also an increased risk of spontaneous abortion. Microscopic findings may include interstitial pneumonitis, hepatitis, mineralization of the cerebral arterioles, and chronic inflammation of inner ear structures. A lymphohistiocytic villitis may be present in the placenta, and eosinophilic cytoplasmic inclusions have been noted in trophoblast and endothelial cells. Distinct viral cytopathology is absent in most cases.

Systemic Pathologies Associated with Diverse Viral Groups

Arbovirus Encephalitis

A diverse group of viruses transmitted by arthropod vectors is capable of causing encephalitis. Included within the arbovirus group are members of the Flaviviridae, Alphaviridae, Bunyaviridae, and Reoviridae. Most elicit a similar histologic pattern, which includes perivascular and meningeal infiltration of predominantly mononuclear inflammatory cells. Variable necrosis of neurons may be encountered, often associated with inflammatory cell clusters. Microglial nodules have been linked to viral encephalitis, but the specificity of this finding for viral infection is unclear. Eastern equine encephalitis is frequently associated with a more severe, acute form of tissue damage, including widespread necrosis and neutrophil infiltrates (57–59). West Nile virus, a flavivirus, has reached epidemic proportions in many parts of the western hemisphere in the past decade (see below).

Gastroenteritis Viruses

A wide variety of viruses, including picornaviruses, caliciviruses, rotavirus, and coronavirus, have been identified as causative agents of gastroenteritis. Light microscopic changes associated with infection by these pathogens are nonspecific, and include distortion of the intestinal villi, increased inflammatory cells in the lamina propria, and degenerative changes in the intestinal epithelial cells (60). In severe infections, more extensive inflammation and associated tissue necrosis may be seen. Biopsies are rarely performed in suspected cases of viral gastroenteritis, as other, less invasive diagnostic modalities are more sensitive and specific. The histopathologic features of enteric viral infection overlap somewhat with those of small bowel allograft rejection, however, a potential source of confusion in transplant biopsies (61).

Hepatitis Viruses

The hepatitis viruses are a diverse group of DNA and RNA viruses with specific tropisms for liver tissue. Infections with these pathogens lead to a broad spectrum of clinical presentations, ranging from acute, fulminant hepatic failure to chronic, progressive hepatic dysfunction and cirrhosis. Chronic infections with hepatitis B and C viruses (HBV and HCV) also potentiate the development of hepatocellular carcinoma (10,11,14). Detection and identification of specific hepatitis viruses is accomplished primarily by serologic or molecular diagnostic methods. Histopathology, however, plays a vital role in assessing the course and progression of infections with these agents.

All of the hepatitis viruses infect hepatocytes. In acute infections, the histologic changes are primarily in the hepatic lobules, and include hepatocyte death (with the ultimate formation of anucleate eosinophilic remnants termed "acidophil" or "Councilman" bodies), reparative activity (mitoses, multinucleation, variability in nuclear size), and infiltration by inflammatory cells of various lineages. The histologic pattern in chronic hepatitis is usually predominated by secondary changes in and around portal triads, including accumulation of mononuclear inflammatory cells, fibrosis, and injury to periportal hepatocytes ("piecemeal necrosis") [Fig. 14(A)] (62). In addition to its effects on the liver, HCV has been linked to a variety of systemic

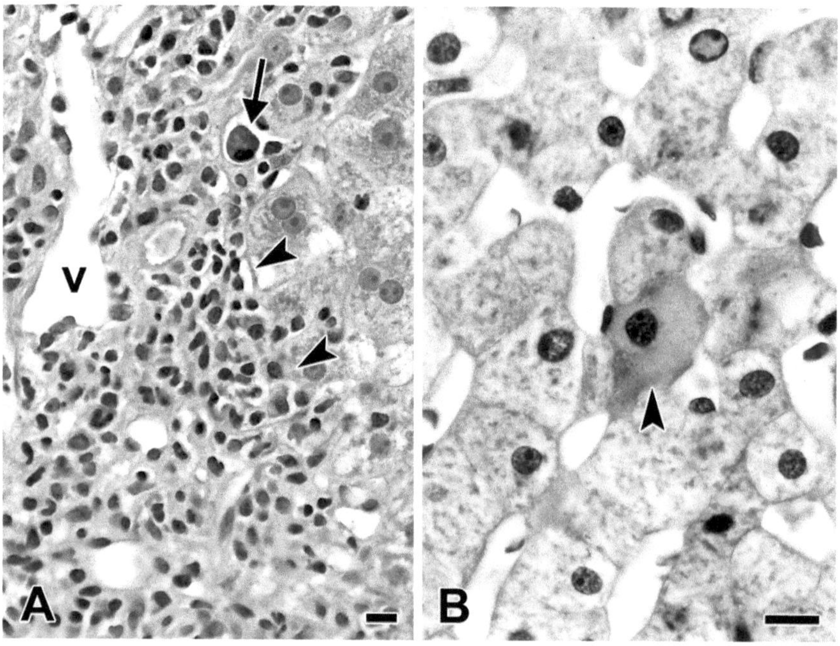

Figure 14 Viral hepatitis. (**A**) Liver with hepatitis C virus infection. A portal tract containing a venule (v) is expanded by an inflammatory infiltrate that involves the portal–hepatocyte interface (*arrowheads*). An apoptotic hepatocyte, or acidophil body (*arrow*), is present near the interface. (**B**) Liver with hepatitis B virus infection. An infected hepatocyte with "ground glass" cytoplasm (*arrowhead*) is seen (bars = 10 μm). *Source*: Panel B courtesy of Dr. Paul Killenberg, Duke University Medical Center.

manifestations, including type II mixed cryoglobulinemia and several renal glomerular disorders (63).

Distinction of the viral hepatitides from one another and from hepatitis due to other etiologies (e.g., drug toxicity, autoimmune diseases, and metabolic disorders) is often impossible by histology alone. Infections with HBV and HCV, the two most common types encountered in biopsy material, can occasionally be distinguished on histologic grounds, however. Several features, including the presence of portal lymphoid aggregates, inflammatory injury to bile duct epithelium, and lobular steatosis, are more common in chronic HCV infections than those with HBV (64). In chronic HBV infection, accumulation of hepatitis B surface antigen (HBsAg) in hepatocyte cytoplasm will occasionally impart a "ground glass" appearance on H&E-stained sections [Fig. 14(B)]. This finding should be confirmed by immunoperoxidase staining for HBsAg, as glycogen-rich cytoplasm can have a similar appearance.

Viral Hemorrhagic Fevers

The viral hemorrhagic fevers are caused by members of four viral families: the Flaviviridae, Arenaviridae, Bunyaviridae, and Filoviridae. Arthropod vectors and rodent reservoirs are frequently involved in the transmission of these diseases, and some of the viruses have a sylvatic or jungle cycle that involves mosquitoes and nonhuman primates. Hemorrhage fever viruses can infect a broad range of cells, including components of the immune system (macrophages, monocytes, dendritic cells), endothelial cells, hepatocytes, and adrenal cortical cells. Disruption of the immune system engenders unchecked viral replication, inappropriate elaboration of inflammatory mediators with development of a shock-like syndrome, and defects in the clotting cascade. These factors, coupled in some cases with direct endothelial viral replication, lead to vascular damage, coagulation disorders, and hemorrhage (65,66).

There are no definitive diagnostic light microscopic features in any of the viral hemorrhagic fevers, though some have a propensity to cause more extensive damage in a particular organ. Most induce varying degrees of multisystemic vascular thrombosis, hemorrhage, edema, and tissue necrosis, often with minimal associated inflammation. Hepatocyte necrosis, frequently with a midzonal pattern, is particularly prevalent in yellow fever, but has also been reported in other hemorrhagic fevers, including dengue hemorrhagic fever (67). Two forms of hantavirus infection that target the kidneys and lungs are described below.

Emerging Viral Infections

Though most of the viruses described in this chapter have been well characterized for many years, several novel agents that can cause life-threatening infections have been described recently. Many of them are zoonotic pathogens that have developed the ability to infect human hosts. Few anatomic pathologists have had direct experience with these viruses. Some are currently rare in humans and/or limited in geographic distribution; for others, serologic or nucleic-acid-based tests are the usual standard for diagnosis rather than tissue biopsy. The following is a brief introduction to some of the emerging viruses, with references to recent literature reviews for additional information.

Hantaviruses, members of the family Bunyaviridae, are rodent-borne pathogens that cause two major human syndromes, both of which involve injury to blood vessels and vascular leakage (68,69). Hemorrhagic fever with renal syndrome (HFRS), caused by a group of Old-World hantaviruses, has been recognized for decades (with reports of similar outbreaks dating back centuries), and is characterized by a febrile illness with sequential phases of hypotension, oliguria, and diuresis. Renal biopsies in patients with HFRS show acute hemorrhagic interstitial nephritis primarily affecting the medulla (68). Hantavirus pulmonary syndrome (HPS), for which the pathogenic agents are New-World hantaviruses, was first recognized in 1993, though earlier infections have been documented by retrospective analysis of archival tissue. Patients with HPS experience pulmonary vascular leak with associated, often fatal pulmonary edema.

West Nile virus, a mosquito-borne flavivirus, causes a range of central nervous system disorders, including meningitis, encephalitis, and poliomyelitis (70,71). First identified in 1937, this virus has spread rapidly in the western hemisphere since 1999. Its histologic manifestations include perivascular inflammation, microglial nodules, necrosis, and neuronal loss. Injury to anterior horn motor neurons in the spinal cord appears to underlie the flaccid paralysis seen in some patients (70,71).

Severe acute respiratory syndrome (SARS), a coronavirus infection first documented in 2002, and avian influenza A subtype H5N1, which emerged as a significant clinical threat in the late 1990s, both cause severe lower respiratory tract infections with a pattern of diffuse alveolar damage (72–74). In SARS, alveolar injury and hyaline membranes are accompanied by a variety of histologic stigmata, including multinucleate giant cells and vascular injury; a healing phase with airspace fibrosis similar to bronchiolitis obliterans organizing pneumonia (BOOP) has been reported (72,74). A more aggressive, necrotizing pattern of alveolar damage has been described for avian influenza, with subsequent fibrosis occurring in an interstitial, non-BOOP-like pattern (73,74). Multisystemic involvement has been reported for both viruses (72–74).

Nipah (75) and Hendra (75,76) viruses are paramyxoviruses that can infect a variety of organ systems. Both have a predilection for the central nervous system, but can also cause pneumonitis in humans and/or animal hosts. Both viruses have a tropism for vascular endothelium, where they cause several forms of cellular and tissue disruption, including multinucleation and necrosis. Intracytoplasmic and occasionally intranuclear viral inclusions have been reported in cells infected with Nipah virus (76). Nipah virus can also infect pulmonary epithelial cells in pigs, facilitating zoonotic transmission (76).

ACKNOWLEDGMENTS

The authors are indebted to Ms. Susan Reeves and Dr. Sara Miller for assistance with photomicroscopy and figure preparation. Drs. Caruso and Childs are active duty physicians in the U. S. Navy. The opinions expressed herein are solely the authors' and do not represent official policy of the U.S. Department of the Navy, the U.S. Department of Defense, or the U.S. Government.

REFERENCES

1. Miller SE, Levenson RM, Aldridge C, et al. Identification of focal viral infections by confocal microscopy for subsequent ultrastructural analysis. Ultrastruct Pathol 1997; 21:183–193.
2. Connor DH, Chandler FW, Schwartz DA, et al. Pathology of Infectious Diseases. Stamford, CT: Appleton & Lange, 1997.
3. Bancroft JD, Gamble M. Theory and Practice of Histological Techniques, 5th ed. London, U.K.: Churchill Livingstone, 2002.
4. DeMay RM. The Art & Science of Cytopathology. Chicago, IL: ASCP Press, 1996.
5. Keebler CM. Cytopreparatory techniques. In: Bibbo M, ed. Comprehensive Cytopathology. Philadelphia, PA: W. B. Saunders Co., 1997:889–917.
6. Kok LP, Boon M. Microwaves for the Art of Microscopy. Leiden: Coulomb Press, 2003.
7. Gupta PK. Microbiology, inflammation, and viral infections. In: Bibbo M, ed. Comprehensive Cytopathology. Philadelphia, PA: W. B. Saunders Company, 1997:125–160.
8. Netherton C, Moffat K, Brooks E, et al. A guide to viral inclusions, membrane rearrangements, factories, and viroplasm produced during virus replication. Adv Virus Res 2007; 70:101–182.
9. Nuovo GJ. The surgical and cytopathology of viral infections: Utility of immunohistochemistry, in situ hybridization, and in situ polymerase chain reaction amplification. Ann Diagn Pathol 2006; 10 (2):117–131.
10. Boccardo E, Villa LL. Viral origins of human cancer. Curr Med Chem 2007; 14(24):2526–2539.
11. Butel JS. Viral carcinogenesis: Revelation of molecular mechanisms and etiology of human disease. Carcinogenesis 2000; 21(3):405–426.
12. Damania B. DNA tumor viruses and human cancer. Trends Microbiol 2007; 15 (1):38–44.
13. Elgui de Oliveira D. DNA viruses in human cancer: An integrated overview on fundamental mechanisms of viral carcinogenesis. Cancer Lett 2007; 247(2):182–196.
14. McLaughlin-Drubin ME, Munger K. Viruses associated with human cancer. Biochim Biophys Acta 2008; 1782(3):127–150.
15. Langley JM. Adenoviruses. Pediatr Rev 2005; 26(7):244–249.
16. Hakim FA, Tleyjeh IM. Severe adenovirus pneumonia in immunocompetent adults: A case report and review of the literature. Eur J Clin Microbiol Infect Dis 2008; 27(2):153–158.
17. Hogg JC, Hegele RG. Adenovirus and Epstein-Barr virus in lung disease. Semin Respir Infect 1995; 10:244–253.

18. Emovon OE, Lin A, Howell DN, et al. Refractory adenovirus infection after simultaneous kidney-pancreas transplantation: Successful treatment with intravenous ribavirin and pooled human intravenous immunoglobulin. Nephrol Dial Transplant 2003; 18(11):2436–2438.
19. Taylor GH. Cytomegalovirus. Am Fam Physician 2003; 67(3):519–524.
20. Kutok JL, Wang F. Spectrum of Epstein-Barr virus-associated diseases. Annu Rev Pathol 2006; 1:375–404.
21. Fowler CB, Reed KD, Brannon RB. Intranuclear inclusions correlate with the ultrastructural detection of herpes-type virions in oral hairy leukoplakia. Am J Surg Pathol 1989; 13(2):114–119.
22. Grogg KL, Miller RF, Dogan A. HIV infection and lymphoma. J Clin Pathol 2007; 60(12): 1365–1372.
23. Dwyer DE, Cunningham AL. 10: Herpes simplex and varicella-zoster virus infections. Med J Aust 2002; 177(5):267–273.
24. Weston WL. Herpes-associated erythema multiforme. J Invest Dermatol 2005; 124(6):xv–xvi.
25. De Bolle L, Naesens L, De Clercq E. Update on human herpesvirus 6 biology, clinical features, and therapy. Clin Microbiol Rev 2005; 18(1):217–245.
26. Broccolo F, Drago F, Careddu AM, et al. Additional evidence that pityriasis rosea is associated with reactivation of human herpesvirus-6 and -7. J Invest Dermatol 2005; 124(6):1234–1240.
27. Ablashi DV, Chatlynne LG, Whitman JE Jr., et al. Spectrum of Kaposi's sarcoma-associated herpesvirus, or human herpesvirus 8, diseases. Clin Microbiol Rev 2002; 15(3):439–464.
28. Du MQ, Bacon CM, Isaacson PG. Kaposi sarcoma-associated herpesvirus/human herpesvirus 8 and lymphoproliferative disorders. J Clin Pathol 2007; 60(12):1350–1357.
29. Wright TC, Kurman RJ, Ferenczy A. Precancerous lesions of the cervix. In: Kurman RJ, ed. Blaustein's Pathology of the Female Genital Tract. New York, NY: Springer, 2002:253–324.
30. Dehn D, Torkko KC, Shroyer KR. Human papillomavirus testing and molecular markers of cervical dysplasia and carcinoma. Cancer 2007; 111(1):1–14.
31. Vafaie J, Schwartz RA. Parvovirus B19 infections. Int J Dermatol 2004; 43(10):747–749.
32. Florea AV, Ionescu DN, Melhem MF. Parvovirus B19 infection in the immunocompromised host. Arch Pathol Lab Med 2007; 131(5):799–804.
33. Schildgen O, Muller A, Allander T, et al. Human bocavirus: Passenger or pathogen in acute respiratory tract infections? Clin Microbiol Rev 2008; 21(2):291–304, table of contents.
34. Koralnik IJ. Progressive multifocal leukoencephalopathy revisited: Has the disease outgrown its name? Ann Neurol 2006; 60(2):162–173.
35. Randhawa P, Brennan DC. BK virus infection in transplant recipients: An overview and update. Am J Transplant 2006; 6 (9):2000–2005.
36. Osswald SS, Kulick KB, Tomaszewski MM, et al. Viral-associated trichodysplasia in a patient with lymphoma: A case report and review. J Cutan Pathol 2007; 34(9):721–725.
37. Smith KJ, Skelton H. Molluscum contagiosum: Recent advances in pathogenic mechanisms, and new therapies. Am J Clin Dermatol 2002; 3(8):535–545.
38. Xu X, Erickson LA, Elder DE. Diseases caused by viruses. In: Elder DE, Elenitsas R, Johnson BL Jr., Murphy GF, eds. Lever's Histopathology of the Skin. Philadelphia, PA: Lippincott Williams & Wilkins, 2005:651–679.
39. Bayer-Garner IB. Monkeypox virus: Histologic, immunohistochemical and electron-microscopic findings. J Cutan Pathol 2005; 32(1):28–34.
40. Love S, Wiley CA. Viral diseases. In: Graham DI, Lantos PL, eds. Greenfield's Neuropathology, Vol. 2. London, U.K.: Arnold, 2002:1–105.
41. Steiner I, Budka H, Chaudhuri A, et al. Viral encephalitis: A review of diagnostic methods and guidelines for management. Eur J Neurol 2005; 12(5):331–343.
42. Zeichhardt H, Grunert H-P. Enteroviruses: Polioviruses, Coxsackie viruses, echoviruses and enteroviruses 68–71. In: Cohen J, Powderly WG, eds. Infectious Diseases, Vol 2. Edinburgh, UK: Mosby, 2004:1993–2005.
43. Esfandiarei M, McManus BM. Molecular biology and pathogenesis of viral myocarditis. Annu Rev Pathol 2008; 3:127–155.
44. Matutes E. Adult T-cell leukaemia/lymphoma. J Clin Pathol 2007; 60(12):1373–1377.
45. Ohshima K. Pathological features of diseases associated with human T-cell leukemia virus type I. Cancer Sci 2007; 98(6):772–778.
46. Araujo A, Hall WW. Human T-lymphotropic virus type II and neurological disease. Ann Neurol 2004; 56(1):10–19.
47. McArthur JC, Brew BJ, Nath A. Neurological complications of HIV infection. Lancet Neurol 2005; 4(9):543–555.
48. Wyatt CM, Rosenstiel PE, Klotman PE. HIV-associated nephropathy. Contrib Nephrol 2008; 159: 151–161.

49. Taubenberger JK, Morens DM. The pathology of influenza virus infections. Annu Rev Pathol 2008; 3:499–522.
50. Henrickson KJ. Parainfluenza viruses. Clin Microbiol Rev 2003; 16(2):242–264.
51. Rima BK, Duprex WP. Morbilliviruses and human disease. J Pathol 2006; 208(2):199–214.
52. Henson D, Siegel S, Strano AJ, et al. Mumps virus sialoadenitis. An autopsy report. Arch Pathol 1971; 92:469–474.
53. Nel LH, Markotter W. Lyssaviruses. Crit Rev Microbiol 2007; 33(4):301–324.
54. Iwasaki Y, Tobita M. Pathology. In: Jackson AC, Wunner WH, eds. Rabies. Amsterdam, The Netherlands: Academic Press, 2002:283–306.
55. Neilson KA, Yunis EJ. Demonstration of respiratory syncytial virus in an autopsy series. Pediatr Pathol 1990; 10:491–502.
56. Dwyer DE, Robertson PW, Field PR. Broadsheet: Clinical and laboratory features of rubella. Pathology 2001; 33(3):322–328.
57. Esiri MM, Kennedy PGE. Viral diseases. In: Graham DI, Lantos PL, eds. Greenfield's Neuropathology, Vol. 2. London, U.K.: Arnold, 1997:3–63.
58. Rennels MB. Arthropod-borne virus infections of the central nervous system. Neurol Clin 1984; 2:241–254.
59. Reyes MG, Gardner JJ, Poland JD, et al. St. Louis encephalitis. Quantitative histologic and immunofluorescent studies. Arch Neurol 1981; 38:329–334.
60. Wilhelmi I, Roman E, Sanchez-Fauquier A. Viruses causing gastroenteritis. Clin Microbiol Infect 2003; 9(4):247–262.
61. Kaufman SS, Chatterjee NK, Fuschino ME, et al. Characteristics of human calicivirus enteritis in intestinal transplant recipients. J Pediatr Gastroenterol Nutr 2005; 40(3):328–333.
62. Batts KP, Ludwig J. Chronic hepatitis. An update on terminology and reporting. Am J Surg Pathol 1995; 19:1409–1417.
63. Kamar N, Izopet J, Alric L, et al. Hepatitis C virus-related kidney disease: An overview. Clin Nephrol 2008; 69(3):149–160.
64. Lefkowitch JH, Schiff ER, Davis GL, et al. Pathological diagnosis of chronic hepatitis C: A multicenter comparative study with chronic hepatitis B. Gastroenterology 1993; 104:595–603.
65. Lacy MD, Smego RA. Viral hemorrhagic fevers. In: Aronoff SC, Hughes WT, Kohl S, Wald ER, eds. Advances in Pediatric Infectious Diseases, Vol. 12. St. Louis, MO: Mosby, 1997:21–53.
66. Marty AM, Jahrling PB, Geisbert TW. Viral hemorrhagic fevers. Clin Lab Med 2006; 26(2):345–386, viii.
67. Leong AS, Wong KT, Leong TY, et al. The pathology of dengue hemorrhagic fever. Semin Diagn Pathol 2007; 24(4):227–236.
68. Ferluga D, Vizjak A. Hantavirus nephropathy. J Am Soc Nephrol 2008; 19(9):1653–1658.
69. Simmons JH, Riley LK. Hantaviruses: An overview. Comp Med 2002; 52(2):97–110.
70. Hayes EB, Sejvar JJ, Zaki SR, et al. Virology, pathology, and clinical manifestations of West Nile virus disease. Emerg Infect Dis 2005; 11(8):1174–1179.
71. Kramer LD, Li J, Shi PY. West Nile virus. Lancet Neurol 2007; 6(2):171–181.
72. Gu J, Korteweg C. Pathology and pathogenesis of severe acute respiratory syndrome. Am J Pathol 2007; 170(4):1136–1147.
73. Korteweg C, Gu J. Pathology, molecular biology, and pathogenesis of avian influenza A (H5N1) infection in humans. Am J Pathol 2008; 172(5):1155–1170.
74. Ng WF, To KF, Lam WW, et al. The comparative pathology of severe acute respiratory syndrome and avian influenza A subtype H5N1—A review. Hum Pathol 2006; 37(4):381–390.
75. Wong KT, Shieh WJ, Zaki SR, et al. Nipah virus infection, an emerging paramyxoviral zoonosis. Springer Semin Immunopathol 2002; 24(2):215–228.
76. Hooper P, Zaki S, Daniels P, et al. Comparative pathology of the diseases caused by Hendra and Nipah viruses. Microbes Infect 2001; 3(4):315–322.

11 Electron Microscopy of Viral Infections

Sara E. Miller
Department of Pathology, Duke University Medical Center, Durham, North Carolina, U.S.A.

INTRODUCTION

Electron microscopy (EM) is an important adjunct to other diagnostic virology tests. In some cases, it may be the only procedure for detecting viruses because they are fastidious or not cultivable in the routine culture laboratory or because biochemical probes are not readily available. Even when biochemical tests for a virus exist, their selection and application require a preconceived notion of which viral agent(s) is present. Additionally, cross-reactions may give false-positive results.

Negative staining of liquid samples is the most widely used technique for virus detection by EM because body fluids are more readily obtainable, and also because the technique is rapid. Further, no specific reagents, such as antibodies, protein standards, or nucleic acid probes, are required. Even in instances where the sample must be concentrated, the whole procedure takes only one to two hours. Thin sectioning, used for examining cells and tissues, is a more lengthy procedure, but can also be accomplished routinely in a couple of days and for rush cases, in three to four hours.

Virus identification and methodology have been extensively described (1–11) and are not detailed here. Rather, general specimen handling procedures, the samples most likely to be received in the diagnostic EM laboratory, and the types of viruses that are most often visualized in those specimens are illustrated. Note that all low-magnification micrographs of thin sections are printed at 20,000× to show the relationship of viruses to their host cells. All insets and micrographs of negative stains are at 100,000× to show virus detail and for direct size comparison of one to another. Additional information on virus identification based on morphological grouping and specimen affinity has been published (5,7,11).

METHODS

Negative Staining

Stains and Staining Characteristics

Particulate matter, such as viruses in suspension, can be contrasted by heavy metal salts [reviewed by Hayat and Miller (4)]. Those most commonly used in diagnostic virology are 0.5% to 2% and up to saturated uranyl acetate, 1% to 2% phosphotungstic acid (PTA), or 0.5% to 5% ammonium molybdate, all in water. Uranyl acetate gives a finer grain than PTA and also acts as a fixative to preserve viral structure. PTA can actively degrade some viruses; immediate visualization is possible, but virions may fall apart after storage of PTA-stained grids for several hours unless they are first fixed (1% glutaraldehyde) and washed before staining. PTA has the advantage of being nonradioactive and, in some cases, better delineates surface projections of enveloped viruses. Ammonium molybdate produces a finer-grained background than uranyl acetate and gives less contrast; it is sometimes used with 1% carbohydrate (glucose or trehalose) to support structures while drying. It is good for particles in clumps, does not precipitate with salts, and does not cause virus shrinkage. It is prudent to keep all these stains available.

Negative staining relies on the pooling of the stain around particulate specimens and in crevices on their surfaces. If the stain also binds directly to a component of the specimen (positive staining), sometimes due to electrostatic interactions, the appearance may be confusing. PTA can be adjusted to pH 7, which is above the isoelectric points of most proteins where both the stain, an anion, and the proteins are negatively charged; positive staining is thus minimized. Uranyl acetate, a cation, precipitates at pH 7 and is usually used at ~4.5 to 5 where it and most proteins are positively charged, and stain–protein electrostatic interactions are not prevalent. However, uranyl acetate has a strong affinity for phosphate and carboxyl groups in DNA, lipid, and some proteins, which may result in positive staining. Positively stained virions appear darker and smaller than negatively stained ones, and since, in many cases, virus identification is dependent on size measurements, only negatively stained ones should be measured. Sometimes uranyl acetate produces both positive and negative staining on the same grid.

The appearance of the grid depends on the spreading characteristics of the sample and the stain, and these, in turn, depend on the hydrophobic or hydrophilic nature of the sample and support film. Freshly carbon-coated films generally work well, but older ones may need to be glow-discharged in a vacuum evaporator for 30 to 60 seconds to add charges and produce an even spread. Alternatively, a five-minute pretreatment with 1% alcian blue, 0.01% to 0.25% poly-L-lysine, or 0.05% bacitracin, followed by washing with water, will aid spreading. The appearance of the stain should be uniform and finely granular. If it piles up heavily in some areas and does not stain others, or if it has a slick, glassy, or amorphous appearance, the grid would benefit from glow discharge or spreading treatment prior to applying the sample.

Support Films

Particulate material from suspensions must be supported on a thin plastic film (e.g., Formvar or collodion) coated in the vacuum evaporator with a fine layer of carbon for conductivity and stabilization in the beam [described in detail by Hayat and Miller (4)]. Grids with support films may also be purchased from an EM supplier, but their age will be unknown, and they may or may not be hydrophilic.

Virus Concentration

Viruses must be present at $\sim 10^5$ to 10^6 particles/mL for detection. Gastroenteritis viruses are frequently shed in numbers sufficient to be seen without concentrating; however, viruses in other samples such as urine, cerebrospinal fluid (CSF), lavages, and aspirates often must be concentrated. Ultracentrifugation ($\sim$50,000 $\times$ *g* for 45 minutes) or in an Airfuge (30 lb pressure/square inch for 30 minutes; Beckman, Palo Alto, CA) will pellet viruses. The latter instrument can be equipped with the EM-90 rotor, which accepts grids for depositing viruses directly onto the film (12). Prior to ultracentrifugation, large cell debris must be eliminated by low speed clarification (1000 $\times$ *g*). For EM-90 centrifugation, a higher clarification speed (e.g., in a microfuge at 10,000 $\times$ *g* for one to five minutes) is necessary; however, this may pellet out clumped viruses and reduce their concentration on the grid, making them harder to find. Other possible concentration procedures include agar diffusion (13) and pseudoreplica (14).

Immunoaggregation

If appropriate antiserum is available, immunoaggregation can be used both to concentrate viruses in dilute suspensions and to distinguish morphologically similar viruses. They can be clumped and pelleted at lower *g*-force (15), or attracted onto the grid by treating it with antiserum by the Serum-in-Agar or Solid-Phase Immunoelectron Microscopy (IEM) techniques (1).

Use of antibodies (or any other biochemical probes) necessitates that one know the kind of virus present, or at least be able to narrow the possibilities, to select the correct reagent. These procedures have been used to document viral infections with convalescent serum, follow epidemics, serotype viruses with known antiserum, and test antiserum for reactivity with known viruses. If the sample may contain an unknown virus, one can attempt to concentrate it by using pooled gamma globulin at about 1:100 final concentration. Success, of course, necessitates that the serum contain some specific antibodies against the virus in question.

Sample Storage

If examination is not immediate, fluid samples can be sealed to prevent drying and stored at 4°C. Long-term storage can be done at –70°C or in liquid nitrogen, but repeated freezing and thawing, such as in a self-defrosting freezer, is not recommended (Centers for Disease Control and Prevention Conference, 1989).

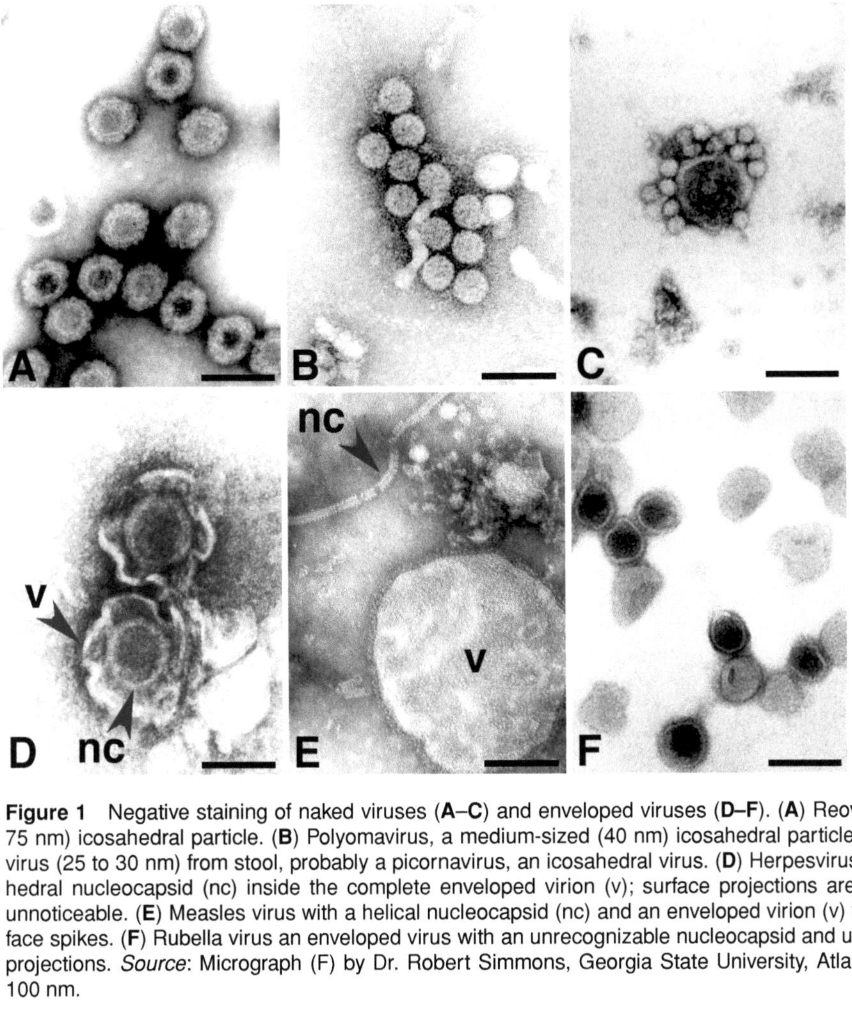

Figure 1 Negative staining of naked viruses (**A–C**) and enveloped viruses (**D–F**). (**A**) Reovirus, a large (70 to 75 nm) icosahedral particle. (**B**) Polyomavirus, a medium-sized (40 nm) icosahedral particle. (**C**) A small round virus (25 to 30 nm) from stool, probably a picornavirus, an icosahedral virus. (**D**) Herpesvirus showing an icosahedral nucleocapsid (nc) inside the complete enveloped virion (v); surface projections are so short as to be unnoticeable. (**E**) Measles virus with a helical nucleocapsid (nc) and an enveloped virion (v) with 8- to 9-nm surface spikes. (**F**) Rubella virus an enveloped virus with an unrecognizable nucleocapsid and unnoticeable surface projections. *Source*: Micrograph (F) by Dr. Robert Simmons, Georgia State University, Atlanta. Bars represent 100 nm.

Virus Morphology in Negative Stains

Morphologically, there are two main categories of viruses: naked ones and enveloped ones. Naked human viruses are all icosahedral (roughly spherical in negative stains) and can be grouped into three size categories (65 to 90 nm, 40 to 55 nm, and 22 to 35 nm) [Figs. 1(A) to 1(C)]. Spherical particles falling outside these ranges are not human pathogens. Thus, the importance of proper microscope magnification calibration is evident.

The other large morphological grouping is enveloped viruses [Figs. 1(D) to 1(F)]; these have a lipoprotein membrane around the nucleocapsid that is usually derived by budding through host membranes (exception: poxviruses). Nucleocapsids inside enveloped viruses may be spherical [Fig. 1(D), compare to Fig. 1(A)], helical like a "Slinky" [Fig. 1(E)], or morphologically nondescript [Fig. 1(F)]; even some of the enveloped viruses with icosahedral nucleocapsids have unrecognizable nucleocapsid morphology in negative stains (e.g., togaviruses).

Enveloped viruses can be very small (e.g., flaviviruses; ~40 to 60 nm) or large (e.g., paramyxoviruses; 200 to 400 nm). They may or may not have visible projections on their surfaces. Some, e.g., herpesviruses and togaviruses, have very short projections that are not usually visible in negative stains of clinical material. Others (e.g., paramyxoviruses and coronaviruses) have long, clearly visible spikes (9 nm and 20 nm, respectively). A virus is difficult to recognize amongst cell debris if the particle is enveloped, the spikes are short, and the stain has not penetrated the membrane so that a recognizable nucleocapsid is visible.

Thin Sectioning

Procedures

Cells and solid tissues should be thin sectioned for EM examination. While viruses have been reported in aqueous extracts of infected homogenized tissue by negative staining, the yield is usually low, and diagnosis is unreliable.

Any routine fixation procedure using glutaraldehyde, osmium, and, frequently, uranyl acetate is sufficient for samples suspected of harboring viruses [see Miller (7) for details]. Ultrastructural preservation of material stored in formalin is also generally acceptable; it should simply be transferred into glutaraldehyde and then processed routinely. Additionally, some viruses, such as the large icosahedral viruses, survive retrieval from paraffin-embedded blocks or wax sections on slides. Ultrastructure of the tissue is poorly preserved in these samples, and smaller viruses (e.g., picornaviruses) and many enveloped viruses with nondistinct nucleocapsids (e.g., flaviviruses) cannot be distinguished from clumped and degraded cell structures.

Epoxy resins are usually used for routine embedment, the most common being Epon substitutes and Spurr, available from any EM supplier. Rapid methods have been described (16–18) and consist of the use of very thin slivers of tissue, decreased processing times, increased catalyst (e.g., double), and shortened baking times at hotter temperatures (e.g., 25 minutes at 95°C). Alternatively, microwave processing can also decrease the time required (19). Sections are finally poststained with uranyl acetate and lead (20).

Immunostaining and in situ hybridization, techniques widely used in light microscopy to detect and identify viruses, have been applied with some success at the EM level. However, they are time-consuming, limited by reagent availability, and have not proven useful in the diagnostic setting.

Several techniques have been devised to circumvent the sample size limitations imposed by thin sectioning [referenced in Miller et al. (8)]. These generally employ some form of optical microscopy as a survey tool to select promising areas of tissue injury for subsequent processing and EM examination. In one such technique, laser scanning confocal microscopy is used to examine vibrating microtome sections of wet tissue. Areas of unusual architecture, such as inflammation, necrosis, hemorrhage, enlarged cells, multinucleate cells, or enlarged nuclei, can be identified and cut out for embedment and ultramicrotomy (8).

Virus Morphology in Thin Sections

Virus location within cells is a clue to their identity. Most DNA viruses are constructed in the nucleus [Figs. 2(A) to 2(C)]; an exception is poxviruses [Fig. 2(D)]. Figures 2(A) and 2(B) show naked DNA viruses; Figure 2(A) demonstrates the largest and Figure 2(B) the smallest. Naked viruses get out of cells by lysing them, and late in infection, DNA viruses may be seen throughout the cell after the nuclear membrane has broken down. Nucleocapsids of enveloped DNA viruses (herpesviruses) are found mostly in the nucleus, though some can escape through the nuclear pores into the cytoplasm. Complete (enveloped) virions are not found in the nucleus; instead, they are seen in the cytoplasm or budding into the extracellular space [Fig. 2(C)]. Poxvirus, a DNA virus, is an exception; it develops in the cytoplasm [Fig. 2(D)]. Its envelope is synthesized *de novo*, rather than by budding through cellular membranes, although the virion can sometimes obtain an extra membrane layer of cell origin.

Most RNA viruses are constructed in the cytoplasm [Figs. 3(A) to 3(F)]. An exception is that measles virus nucleocapsids (but not complete enveloped particles) can occasionally be found in nuclei. Some processes of orthomyxovirus replication occur in nuclei, and light microscopic immunohistochemistry may show nuclear staining; however, ultrastructural evidence of infection is confined to the cytoplasm. Figures 3(A) and 3(B) show naked RNA viruses in the cytoplasm of infected cells. Enveloped RNA viruses may get their outer covering by budding into internal vesicles [Fig. 3(C)] or out through the cytoplasmic membrane [Figs. 3(D) to 3(F)].

In enveloped viruses, nucleocapsid size and shape are important characteristics for identification. Like icosahedral naked viruses [Figs. 2(A) and 2(B)], icosahedral nucleocapsids of enveloped viruses [Fig. 2(C)] are spherical in sections. Helical nucleocapsids appear like "worms" going in and out of the plane of section [Fig. 3(E), top inset]. Some RNA nucleocapsids appear simply as electron dense material without characteristic shape [Fig. 3(F)].

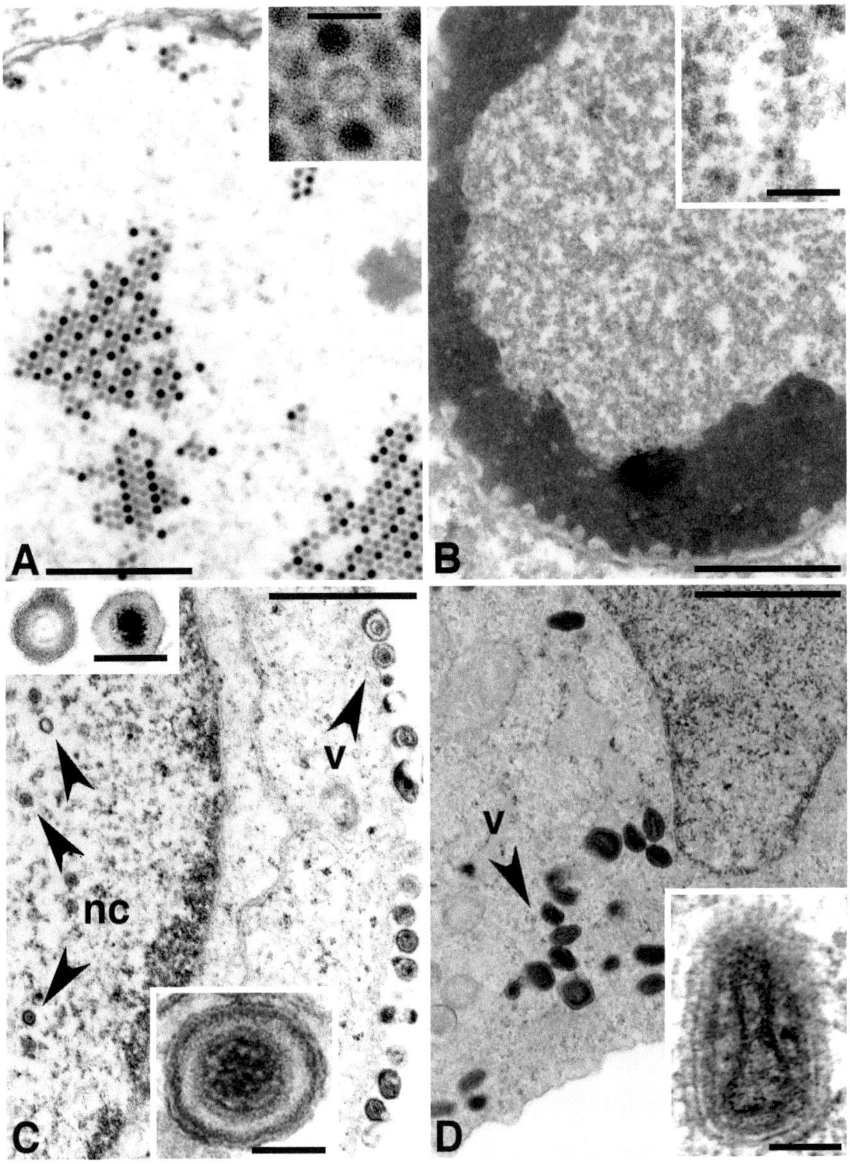

Figure 2 DNA viruses shown in thin sections of infected cells. (**A**) Adenovirus, a large (70 to 90 nm) naked icosahedral virus; note the paracrystalline arrays of particles in the nucleus. The inset shows a high magnification of particles cut through different planes. (**B**) Parvovirus, a very small naked virus not visible in the low magnification; the center of the nucleus is filled with viral material pushing the chromatin to the margin and producing a ring-shape appearance. The high magnification inset shows faint 22-nm spheres, barely recognizable in clumped nuclear material. (**C**). Herpesvirus with ~100-nm icosahedral nucleocapsids (nc) in the nucleus and enveloped virions (v) at the cell surface. Complete virus can also be seen in the cytoplasm (not pictured here). Nucleocapsids (upper inset) and a complete, enveloped virion (lower inset) are shown at high magnification. (**D**) Cowpox virus (v), an enveloped DNA virus (~200 × 200 × 250 nm) with a complex nucleocapsid. The envelope is constructed *de novo* in the cytoplasm, i.e., it is not added by budding through cellular membranes as in (**C**), although, sometimes virions can take on an extra membranous cell-derived layer. The mature core is a dumbbell-shaped structure (see inset). Bars in low magnifications (20,000 ×) represent 1 μm; those in high-magnification insets (100,000 ×) represent 100 nm.

The kind of cell membranes [nuclear (e.g., herpesvirus), vacuolar [Fig. 3(C)], or cytoplasmic [Figs. 3(E) and 3(F)] with which enveloped viruses are associated can also be a clue to identification. Membranes containing viral projection proteins may appear to have fuzz or be denser than cell membranes not containing viral proteins. [see budding particles Fig. 3(E)].

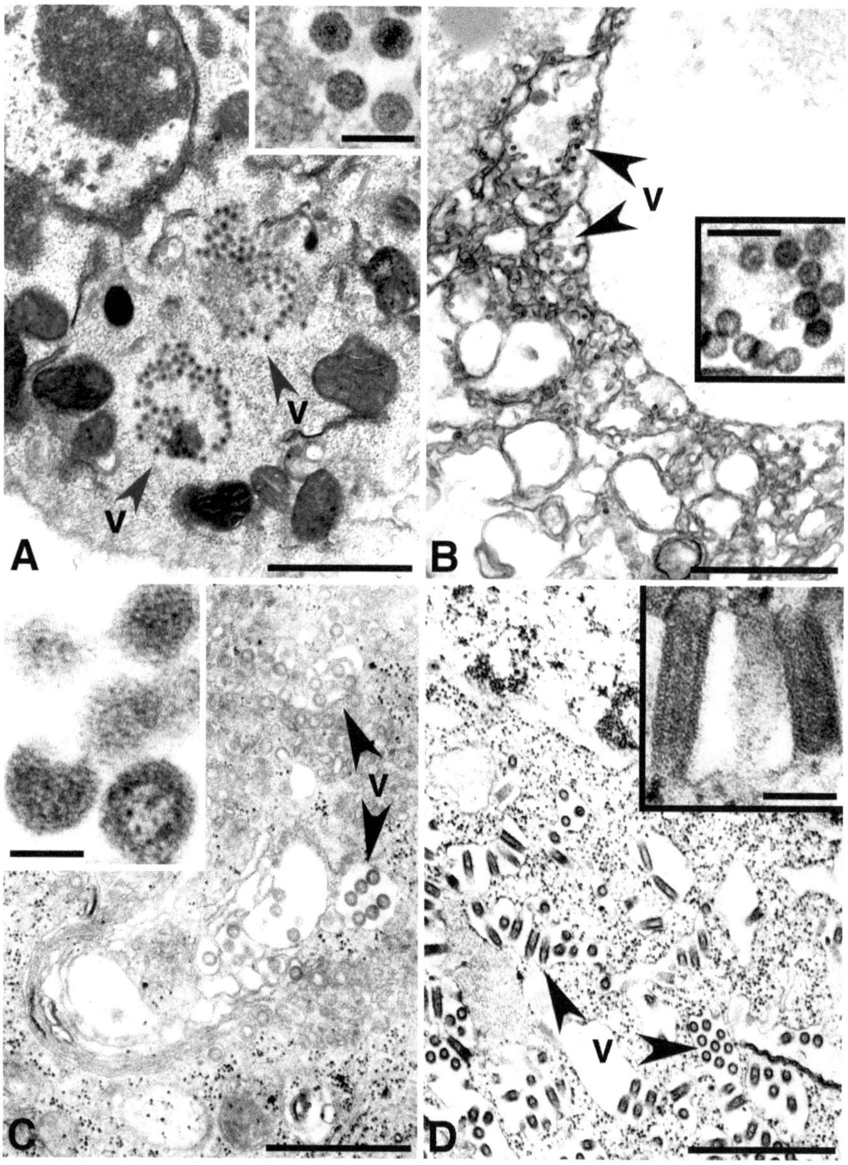

Figure 3 RNA viruses in thin sections of infected cells. (**A**) and (**B**), naked viruses; (**C**)–(**F**), enveloped viruses. (**A**) Reovirus produced in a dense protenaceous matrix in the cytoplasm. The inset shows a high magnification of ~75-nm particles on the right and single-shelled ~55-nm particles on the left. (**B**) Poliovirus, an enterovirus. Though seen here associated with membranes, the virions (v) are not budding. The inset is a high magnification of the ~27-nm particles. (**C**) Rift Valley fever virus, an arenavirus. Virions (v) bud into smooth cytoplasmic vesicles associated with the Golgi apparatus. The high magnification inset shows roughly spherical virions containing ribosomes (sometimes called "grains of sand") inside. (**D**) Rabies, a rhabdovirus. Complete virions (v) bud from the cytoplasmic membrane; cross sections show 80-nm circles, while longitudinal sections show bullet-shaped particles 200 to 300 nm long. The high magnification inset shows spikes on the outside and hints at the helical nucleocapsid inside. (Micrographs (**C**) and (**D**) by Ms. Alyne Harrison, Centers for Disease Control, Atlanta.)

CLINICAL SPECIMENS

Stool

The most advantageous use of EM in diagnostic virology is in the detection of viruses in fecal specimens from patients with gastroenteritis. All of these viruses are either fastidious or noncultivable in the routine culture laboratory, though specialized culture techniques have

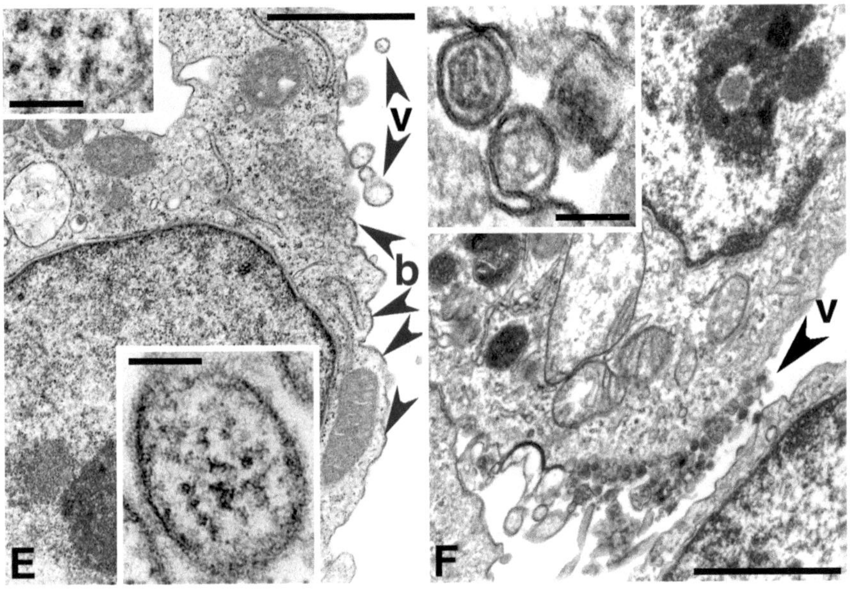

Figure 3 ***(Continued)*** (**E**) Measles virus. Complete virions (v) are seen in the extracellular space; budding particles (b) appear as thickened areas on the cytoplasmic membrane. The upper high magnification inset shows ~22-nm-diameter helical nucleocapsids cut in cross section on the left and one cut in longitudinal section on the right. The lower inset shows a complete virion with the nucleocapsids inside cut in cross section (left side) and fuzz (spikes) on the outside of the envelope (right side). (**F**) HTLV, a retrovirus. Virions (v) bud from the cytoplasmic membrane into the extracellular space; nucleocapsids are nondescript. Sometimes the particles are roughly hexagonal or angular with one or more flat sides (center particle in high magnification inset). (HTLV-infected cells provided by Dr. Barton Haynes, Duke University Medical Center.) Bars in low magnifications (20,000 ×) represent 1 μm; those in high-magnification insets (100,000 ×) represent 100 nm.

been described for some agents. Furthermore, specific biochemical reagents are not available for all, and if the wrong probe is used, results will be negative. Finally, viruses in stool are usually present in large quantities, facilitating detection by EM. A caveat is that in some illnesses, viruses may be shed for only a short window of time after symptoms begin; thus, prompt specimen collection shortly after onset is recommended. Sometimes collection at several different times after onset can enhance the likelihood of detecting viruses.

For examination, a suspension of approximately 10% to 20% (w/v) stool in water or volatile buffer, such as ammonium acetate, is made, and the solid material is pelleted at low speed (1000–2000 × *g*). The supernatant is placed onto a grid and negatively stained as described above. If this direct observation is negative, concentration by ultracentrifugation is warranted. Additional EM and IEM methods for virus detection have been published (15,21,22). Figure 4 demonstrates viruses found in stool samples.

Rotavirus

Rotavirus [Fig. 4(A)], a large, naked icosahedral virus, 65 to 70 nm in diameter, is characterized by tubular capsomers arranged like the spokes of a wheel (hence its name "rota"). The intact virion has a double shell, and sometimes both double- and single-shelled (~55 nm) particles may be present. Occasionally, sheets or tubular forms of capsid material resembling "chicken wire" may be seen.

Rotavirus infection in the USA is seasonal, usually appearing in the west in late fall and in the east in winter or early spring. It is rarely diagnosed in the summer months, except in immunosuppressed patients. It can be detected as early as one to three days before onset of diarrhea and up to nine days after onset. Asymptomatic shedding has been observed in neonates and adults, and the number of infected children exhibiting symptoms increases with age between 1 and 24 months (23).

There are seven rotavirus serogroups A through G (24); groups a through c are found in humans, while agents from all groups are seen in animals. Group A, the most common type in the USA and worldwide, is usually seen in children under one or two years of age and elderly

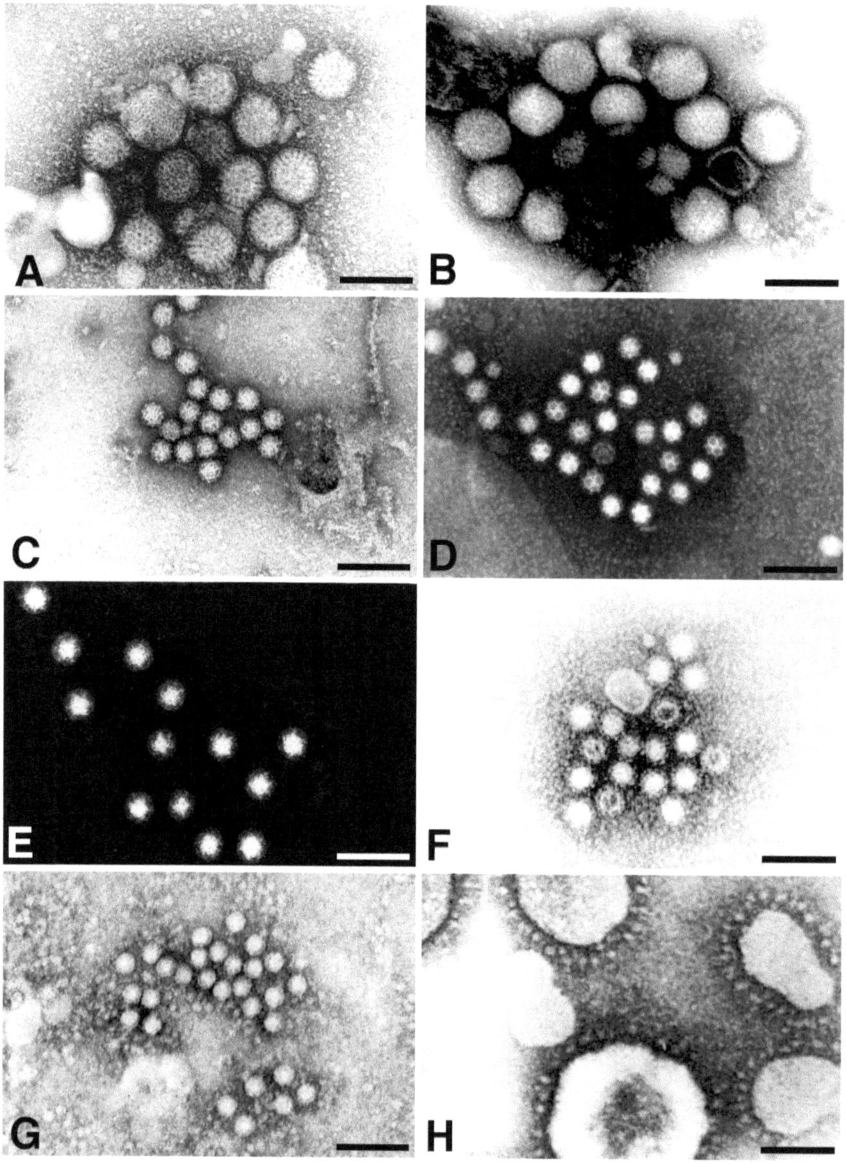

Figure 4 Viruses seen in stool by negative staining. (**A**) Rotavirus, a large (70 to 75 nm) naked icosahedral virus with tube-shaped capsomers that radiate out from the center like spokes of a wheel and a thin outer rim. (**B**) Adenovirus, a large (75 to 90 nm) naked icosahedral virus with bead-shaped capsomers in flat triangular facets sometimes causing the particles to appear hexagonal. (**C**) Norwalk virus, a 25- to 27-nm naked icosahedral calicivirus that appears rough but does not have distinctive capsomeric organization. (Virus preparation provided by Dr. Christine Moe, Emory University.) (**D**) Calicivirus, a 30- to 35-nm naked icosahedral virus with capsomers that form cup-shaped indentations on the surface. (**E**) Astrovirus, a 27- to 32-nm naked icosahedral virus with a surface organization that sometimes appears in a star pattern, particularly if the micrograph is overexposed (dark). (**F**) Minireovirus, a calicivirus that looks like a small reovirus. (Micrographs (**D**), (**E**), and (**F**) courtesy of Mrs. Maria Szymanski, Hospital for Sick Children, Toronto.) (**G**) Enterovirus, a small (27 to 30 nm) naked icosahedral virus without characteristic capsomeric appearance. (**H**) Coronavirus, the only enveloped virus sometimes associated with gastroenteritis. Its size ranges from 75 to 160 nm, and it has 20-nm club-shaped peplomers on the surface. (Micrograph by Dr. E. O. Caul, Bristol Public Health Laboratory, U.K.) Bars represent 100 nm.

people (25). In immunocompetent adults, it usually causes a short illness of two to eight days. Group B, found mostly in China, is more virulent and infects adults as well as children, causing serious dehydration (26). In underdeveloped countries, gastroenteritis is the number one killer of infants and children, and rotavirus tops the list of viral infections causing over 500,000 deaths per year (27).

Adenovirus

Adenovirus [Fig. 4(B)] is a large (70 to 90 nm) naked icosahedral virus with bead-like capsomers arranged in flat triangular facets on the surface. Sometimes the overall shape may appear hexagonal, depending on how the virion lands on the grid. Adenovirus and rotavirus are the only two nonenveloped viruses in this size range, and their morphologies are easily distinguished [compare Fig. 4(A) with Fig. 4(B)].

There are numerous adenovirus strains (24); serotypes 40 and 41 (group F) are recognized as etiologic agents of gastroenteritis, though the percentage of cases caused by adenovirus is fairly low (1–10%) (28–30). Other strains can be seen in stool, sometimes swallowed from respiratory infections; thus, an occasional virion in stool may not be indicative of intestinal infection. However, some of these other strains can actually multiply in the gut and cause diarrhea. One guide for the electron microscopist is to report viral load. If large numbers of virions are present, along with diarrhea, there is a good chance of significant relationship. Only IEM (or other immunological tests) can specifically distinguish serotypes, since they are morphologically identical.

Norwalk Virus and Other Caliciviruses

The Caliciviridae family has two genera pathogenic for humans (24). Norovirus [Fig. 4(C)] contains Norwalk virus (named after the town in Connecticut where it was first identified) and others (Desert Shield, Hawaii, Lordsdale, Mexico, Snow Mountain and Southampton viruses). Minireovirus, originally named for its appearance like a small reovirus, is 52% homologous to Norwalk virus (28). The Sapovirus genus contains Sapporo virus (originally described in Japan) and other strains isolated in the United States (Houston 86 and 90, Parkville) and the United Kingdom (London, Manchester).

Infection with Noroviruses has been called winter vomiting disease. About 70% of patients have vomiting, cramps, diarrhea, or fever, but not all of these symptoms may be present in each patient; infections can also be subclinical. This virus is frequently seen in epidemics; common settings include college campuses and cruise ships. Illness lasts three to five days with peak shedding at onset, and after three days, the virus may be undetectable by EM. Disease caused by these agents is sometimes referred to as acute infectious nonbacterial gastroenteritis.

Norwalk virus is a small (25 to 27 nm) round structured virus (SRSV) (31). Its morphology by negative stain does not usually suggest the characteristic cup-like capsomeric arrangement of some species in the family. Other caliciviruses (30 to 35 nm) may exhibit a more pronounced cup-shaped capsomeric pattern, which can appear as a 4-, 6-, or 10-pointed star pattern on their surface [Fig. 4(D)] and occasionally resemble astrovirus. This pattern may not be visible on all particles in a sample; many virions can appear simply as small round viruses (SRVs) or SRSVs (see below).

Astrovirus

Astroviruses [Fig. 4(E)] are small (27 to 32 nm) naked icosahedral particles that occasionally may have a five- or six-pointed star pattern on their surface. Not all particles have this appearance, and the star may not be evident on the microscope screen, but only in the micrograph, particularly if overdeveloped (dark). Thus, differentiation from other SRVs or SRSVs may not be possible. Astroviruses are classified in the Astroviridae family (24) but may sometimes be confused in electron micrographs with caliciviruses; however, they do not have the distinct cup-like indentations [compare Fig. 4(D) with 4(E)]. Marin County virus, now classified as astrovirus type 5, was once thought to be a calicivirus.

Astroviruses have been seen in both children and adults with gastroenteritis. The disease is usually milder than that with rotavirus. There may be a variety of constitutional symptoms lasting from two to four days (32).

Small Round Viruses and Small Round Structured Viruses

Many times, small enteric viruses do not have identifying characteristics and appear simply as rough or smooth. Caul and Appleton (33,34) proposed a morphological classification scheme for these agents. Those with rough surfaces, SRSVs measure 27 to 40 nm and include astro-, calici-, Norwalk, Hawaii, Tauton, and Snow Mountain viruses, as well as the 35- to 40-nm agents such as minireo- [Fig. 4(F)], Sapporo, and Otofuke viruses.

They referred to the 22- to 26-nm featureless particles [Fig. 4(G)] as SRVs and include Cockle, W-Ditchling, and Parametta viruses. Some investigators feel that SRVs are not pathogenic, and it is difficult to distinguish some SRVs from nonpathogenic parvovirus (see below). However, as can be appreciated from the micrographs of calicivirus and astrovirus, a rough pattern may not always be visible due to variations in the staining pattern. Also, antibodies can coat viruses, obscuring the surface pattern. Some SRVs in feces may be enteroviruses or Coxsackie viruses (picornavirus family), and enterovirus infections of other organ systems often produce virus in stool. For the EM diagnosis, numerous small viruses in the presence of obvious diarrhea should be reported as positive (calicivirus or astrovirus, if evident; or SRSV or SRV).

The very small 18- to 22-nm featureless parvoviruses seen in stool with adenoviruses (adenovirus-associated virus, AAV) have not been shown to be pathogenic. Parvovirus seen at other sites such as in blood or amniotic or joint fluid (parvovirus B19) is pathogenic.

A problem in identification of enteric viruses is that some tailless bacteriophages may resemble SRVs. In this case, one should note whether the small round particles are closely aligned with bacterial debris, such as cell wall fragments or flagella and pili. Another drawback is that lipids, proteins, and lipoprotein molecules can form small droplets that may resemble SRVs/SRSVs (see Cerebrospinal fluid below). Large numbers of particles, virus clumps, and uniformity of shape and size can be clues to viral identity.

Enteroviruses and Other Picornaviruses

Entero-, polio-, coxsackie-, echo-, and hepatitis A viruses belong to the Picornaviridae family (24); all are excreted in stool (35). Morphologically, they all appear as SRVs [Fig. 4(G)] and cannot be differentiated by EM without specific antiserum. Though transmitted by the fecal-oral route, some may have an affinity for other organs. Polioviruses home to the motor neurons of the central nervous system, causing paralysis; coxsackieviruses are found in heart disease and diabetes; and echoviruses cause respiratory illness. These naked icosahedral viruses are particularly insensitive to methods of disinfection such as UV light and chlorination.

Hepatitis Viruses

Hepatitis A virus (HAV), or enterovirus 72, is shed in stool early in the course of illness—from a week or so after infection, continuing for about a month to a couple of weeks prior to jaundice. HAV is a picornavirus (24) appearing in negative stains as an SRV (36). Hepatitis E virus (HEV) is a calicivirus (24) and appears as an SRSV in negative stains (37). Four human strains have been described [HEV-1 (Burma), HEV-2 (Mexico), HEV-3 (Meng), HEV-4]. The incubation period for HEV is usually longer than for HAV, and the period of shedding in stool begins about the second week and lasts four to five weeks. Specific identification of either virus by EM requires reaction with antiserum (see Immunoaggregation); infection is usually diagnosed serologically. Transfusion transmitted virus (TT virus or TTV) is a tiny (17 to 20 nm) virus in the Circoviridae family (24) that has been detected in stool as well as saliva and blood (38). By EM it would simply appear as a very small SRV without distinguishing characteristics.

Coronavirus and Coronavirus-Like Particles (CVLP)

Coronaviruses [Fig. 4(H)] are classified into three groups, two of which are pathogenic to humans (24). They are 75 to 160 nm in size, and their enveloped surfaces are studded with 20-nm club-shaped projections called "peplomers." The role of coronaviruses in gastroenteritis has been controversial, as they have been reported from both asymptomatic individuals and patients with diarrhea. However, the SARS (Severe Adult Respiratory Syndrome) virus (Group 2) does cause diarrhea in some of the patients. It should be pointed out that EM played a large role in the classification of this agent as a coronavirus (39,40). Additionally, other coronaviruses do cause respiratory symptoms occasionally accompanied by abdominal pain and diarrhea. Abdominal symptoms are seen more frequently in children. Coronaviruses have been associated with neonatal necrotizing enterocolitis (37).

Torovirus, a genus in the Coronavirus family, is genetically similar to coronavirus (41), but differs morphologically in that it appears in negative stains as a torus (doughnut-shaped) or a kidney- or rod-shaped particle. These viruses are 120 to 140 nm in diameter with club-shaped peplomers that are not as distinct as those on other coronaviruses. The 35- to 170-nm nucleocapsid is tubular and may be rod-shaped or may be curved inside the virion to form an open torus (24). They are primarily seen in livestock and cats, but have been reported in humans (42). Breda virus, first described in Iowa, is known to cause gastroenteritis and respiratory disease in cattle and pigs, and antibody has been detected in humans. Berne virus is a similar agent reported from a horse in Switzerland. Similar particles were described in human stools from Birmingham, England, and Bordeaux, France. We identified by EM a similar virus in a neonate with diarrhea, as well as in his mother and one of his nurses with milder diarrhea (unpublished observations).

Other Viruses Reported in Stool

Two additional viruses have been reported in stools of individuals with gastroenteritis. One was of a 35- to 37-nm icosahedral virus with double-stranded RNA resembling a small birnavirus (43,44). Another report was of antigens of a pestivirus detected in stools of children with gastroenteritis (45). In the latter case, EM would not be an appropriate method of searching for these flaviviruses because of the lack of distinctive morphology.

Urine

The most commonly observed viruses in urine are grouped in Figure 5. The differential diagnosis of viruses in urine is usually between BK polyomavirus [Fig. 5(A)], which does not readily grow in culture, and adenovirus [Fig. 5(B)]. The distinction can easily be made based on virion size. Differentiation is important because therapies for the two viruses differ: adenovirus infection in bone marrow recipients might be controlled with ribavirin until marrow engraftment (46), while polyomavirus in renal transplant patients may be controlled by modification of the immunosuppressants and possibly cidofovir therapy (47). Polyomaviruses can cause gross hematuria, but have also been seen in urine from asymptomatic pregnant women and immunosuppressed individuals without overt urinary tract disease. They have been the cause of significant renal transplant loss; however, with early detection and modification of immunosuppression, graft loss has been reduced (47).

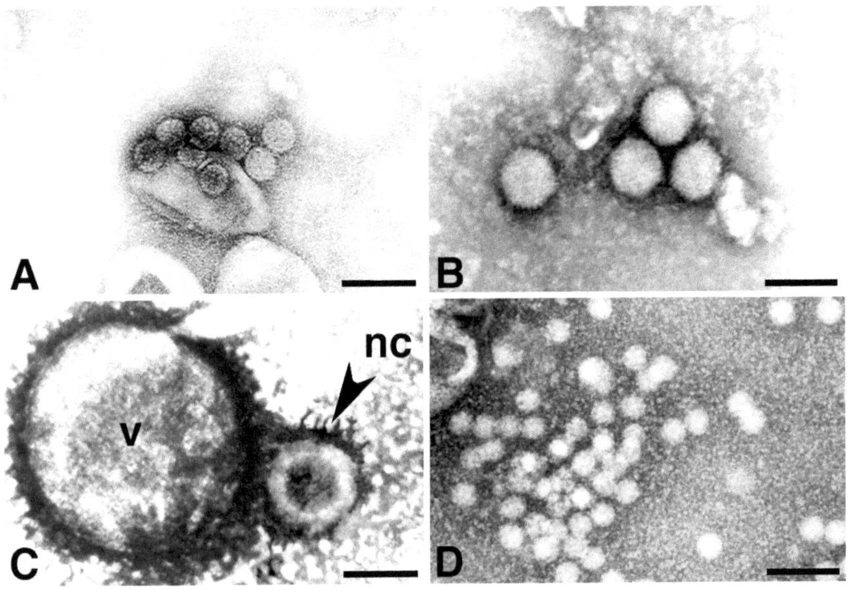

Figure 5 Viruses observed by negative staining of urine. (**A**) Polyomavirus, a medium-sized (~40 nm) naked icosahedral particle with bumps on its surface. (**B**) Adenovirus, a large (70 to 90 nm) naked icosahedral virus with flat triangular facets. (**C**) Cytomegalovirus, a large (120- to 200-nm) enveloped virus (v) with a 90- to 100-nm spherical nucleocapsid (nc). (**D**) Enterovirus, a small (25 to 30 nm) round virus in the picornavirus family with unidentifiable surface features. Bars represent 100 nm.

EM is a method of choice for detection of viruses in urine samples because it is a rapid and noninvasive method of detection, and the testing sensitivity is significant. EM is less sensitive than polymerase chain reaction (PCR), and a detectable number in urine suggests a significant rise in titer. Analysis of urine by PCR can be problematic because the majority of adults (80%) have been exposed; it therefore can yield a positive result when the actual amount of virus present is insignificant to disease. PCR is better used on blood to detect viremia.

CMV [Fig. 5(C)], a herpesvirus, and rarely, rubella virus [Fig 1(F)] can also be present in urine, particularly in congenital infections; rubella virus is difficult to detect by negative staining in clinical specimens because of its lack of distinctive surface morphology. Mumps and measles virus [Fig. 1(E)], both paramyxoviruses, may be present in urine during viruria in these infections. A hemolytic-uremic syndrome (HUS) can be caused by enteroviruses and coxsackieviruses [SRVs; Fig. 5(D)], which can be seen as SRVs but not differentiated further without specific antiserum.

Blister Fluid

Viral skin lesions are another type of specimen that can be easily and rapidly examined by EM. Aqueous suspensions, made from skin lesions, or vesicular fluid can be negatively stained (48). Biopsies can be examined by thin sectioning (see Tissue below). The differential, for the most part, is between herpes viruses [Fig. 6(A)] and poxviruses [Figs. 6(B) to 6(D)]. In addition, EM labs need to be able to differentiate orthopoxviruses [Fig 6(C)] from parapoxviruses [Fig. 6(D)]. EM, for its ability to identify poxviruses rapidly, is on the front line of bioterrorism surveillance (49). Some picornaviruses (e.g., coxsackie- and enteroviruses) can cause vesicular rashes. These agents would appear by EM simply as SRVs [Figs. 1(C), 4(G), 5(D), 7(B), and 8(C)]. Other viruses that have occasionally been seen in sections of skin lesions are measles-, papilloma-, and parvoviruses, but these agents are not readily demonstrated by tissue homogenization and negative staining, and lesions are not usually vesicular.

Blood/Serum

Hepatitis B virus (HBV) [Fig. 9(A)] can be seen in serum of infected patients early in the infection (50), but diagnosis is usually made by serology. Hepatitis delta is an RNA agent inside the HBV surface coat. It is heterogeneous in size, ranging from 28 to 39 nm, but it is not diagnosed by EM.

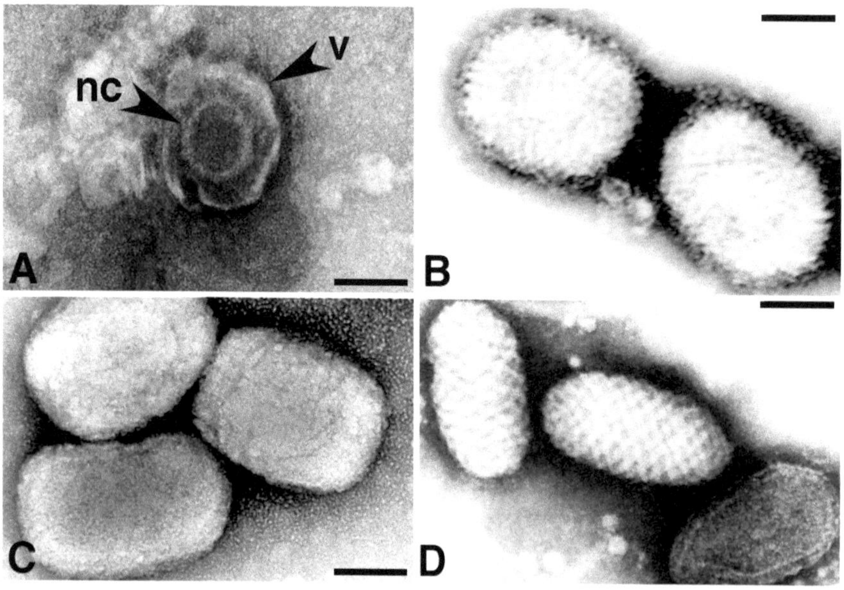

Figure 6 Negative stains of viruses found in skin lesions. (**A**) Herpes simplex virus, a 120- to 200-nm enveloped virus (v) with a spherical 90- to 100-nm nucleocapsid (nc). (**B**)–(**D**) Poxviruses. (**B**) Molluscum contageosum (~320 × 250 × 200 nm). (**C**) Cowpox virus (~200 × 200 × 250 nm). (Culture provided by Dr. David Pickup.) (**D**) Orf virus, a parapox virus (~220 to 300 nm × 140 to 170 nm). Bars represent 100 nm.

Parvovirus B19 [Fig. 9(B)] causes erythema infectiosum (fifth disease), an illness of rash and fever in children. It has been seen by EM in blood (51,52) and in amniotic and joint fluids. SRVs (entero-, coxsackieviruses) [Figs. 1(C), 4(G), 5(D), 7(B), 8(C), and 9(C)] may occasionally be found in blood, and great care must be exercised to differentiate them from lipoproteins [Fig. 8(F)] based on correct size and uniformity of size and shape (see Cerebrospinal fluid below).

Human herpes virus 6 (HHV 6) can be isolated from peripheral blood monocytes of patients with exanthem subitum, and herpesviruses are recognizable by EM; however, culture is a preferable diagnostic modality in this case. Arboviruses and arenaviruses circulate in blood, but cannot easily be discerned by morphology in negative stains. Retroviruses such as human immunodeficiency virus (HIV) and human T cell leukemia/lymphoma virus (HTLV) circulate in blood, but are not of high titer. Further, they are not recognizable by EM, since envelope spikes are short and the nucleocapsid is not distinctive. Filoviruses are found in blood, but EM in not required for diagnosis.

Nasopharyngeal Fluids, Lung Lavages, Pleural Effusions

Numerous viruses cause respiratory tract infections (53,54). Those most likely to be observed by negative staining of respiratory samples are shown in Figure 7. Adenoviruses [Fig. 7(A)], SRVs, e.g., enteroviruses, rhinoviruses, coxsackieviruses [Fig. 7(B)]; ortho- and paramyxoviruses, e.g., influenza and parainfluenza viruses, measles, mumps, respiratory syncytial virus (RSV) [Figs. 7(C) and 7(D)]; and coronavirus [Fig. 7(E)] can be visualized in respiratory tract secretions.

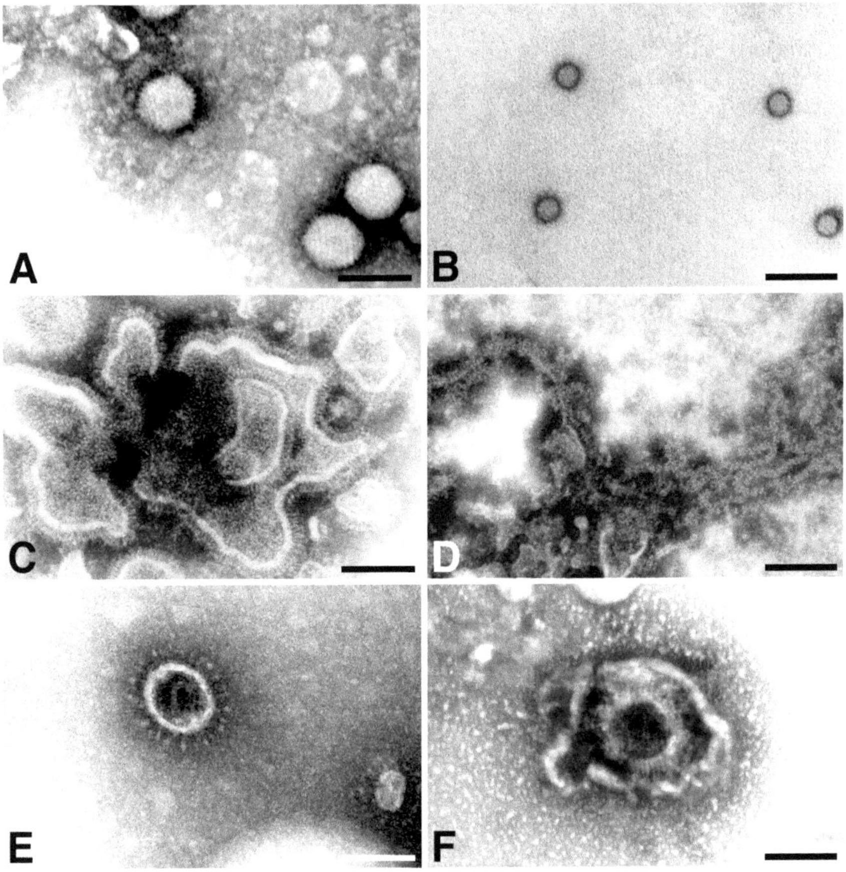

Figure 7 Respiratory viruses viewed by negative staining. (**A**) Adenovirus, a large (70 to 90 nm) naked icosahedral virus. (**B**) Coxsackievirus, a picornavirus (~27–30 nm). (**C**) Respiratory syncytial virus, an enveloped virus in the paramyxovirus family; note the fuzz around the outside of virions. (**D**) Helical nucleocapsids of respiratory syncytial virus. (Micrographs (C) and (D) supplied by Mrs. Maria Szymanski, Hospital for Sick Children, Toronto.) (**E**) Coronavirus, an enveloped virus with long 20-nm spikes on the surface. (**F**) Cytomegalovirus, an enveloped virus with short surface projections so short as not usually to be visualized in clinical material; the icosahedral nucleocapsid inside is 90 to 100 nm. Bars represent 100 nm.

Ortho- and paramyxoviruses must be differentiated from cell debris such as mitochondria, but are difficult to differentiate from each other. Orthomyxoviruses can be ovoid (~100 nm) or long and filamentous (~100 × 1500 nm), while paramyxoviruses vary in size (~100 nm to 300 nm) but usually do not form long particles. Coronaviruses have longer projections on the envelope surface. Herpesviruses, particularly CMV [Fig. 7(F)] and HSV, can be found in immunosuppressed patients. Rarely, polyomaviruses [Figs. 1(B) and 5(A)] may be seen.

Many times these samples can be thick with mucus that would obscure the field of view; they can be thinned with Sputolysin (Behring Diagnostics, Inc., Somerville, NJ) and then ultracentrifuged to concentrate viruses. The stock Sputolysin (1% dithiothreitol) is diluted 1:10 and added one part to one part specimen. The mixture is incubated 30 to 60 minutes with shaking and then ultracentrifuged to pellet viruses.

Cerebrospinal Fluid

Viruses in central nervous system infections (55) can sometimes be seen in CSF. All CSF samples should be ultracentrifuged, and the larger the volume, the more likely it is that viruses may be observed in infected samples. Mumps [Figs. 8(A) and 8(B)] and enteroviruses [Figs. 1(C), 4(G), 5(D), 7(B), 8(C), and 9(C)] are the most common isolates from CSF. Herpesviruses [VZV, CMV (Fig. 8(D)), HSV] and adenoviruses [Figs. 4(B), 5(B), 7(A), and 8(E)] may be found in CSF from immunosuppressed patients. The small, enveloped viruses such as toga- and flaviviruses are not detected by EM in CSF, as neither the envelope nor the nucleocapsid has distinctive morphology.

Other Fluids

Negative staining and EM can be used for other liquid samples, including tears and aspirates such as pleural, pericardial, and ascites fluids. In these cases, viral concentration is usually low, and the sample should be ultracentrifuged if enough volume is supplied. Some of it should also be sent to the culture laboratory for virus amplification. Adenoviruses [Figs. 4(B), 5(B), 7(A), and 8(E)] and SRVs [Figs. 1(C), 4(G), 5(D), 7(B), 8(C), and 9(C) are the most frequently seen viruses in these samples. SRVs may be present in pericardial fluid (e.g., coxsackieviruses). Tears may have SRVs (e.g., enteroviruses) or adenoviruses; viral eye diseases have been discussed in detail (56). Herpesviruses [Figs. 1(D), 5(C), 6(A), 7(F), and 8(D) may be present in saliva, tears, and lung lavages. Rabies and HIV have been demonstrated in saliva, but neither is diagnosed by EM.

Tissues

Solid tissues may be submitted for EM, particularly specimens from immunologically suppressed patients, such as children born with immunodeficiency, AIDS patients, solid organ and bone marrow transplant recipients, and patients on anticancer therapy. While light microscopic immunohistochemistry permits a larger area of examination, and molecular diagnostic tests are more sensitive, EM can be particularly useful where specific probes do not detect viruses. The most frequent tissues received by the EM lab for viral examination are liver, lung, brain, intestine, skin, and kidney.

Useful information to remember when looking for human viruses in thin sections is that, usually, DNA viruses are assembled in the nucleus and RNA viruses are assembled in the cytoplasm. Naked icosahedral DNA viruses start out there and get out of the cell by lysis; late in infection, they may be seen in the cytoplasm after the nuclear membrane has been broken. Nucleocapsids of enveloped DNA viruses are constructed in the nucleus and may bud through the nuclear membrane to obtain their envelopes or may get through the nuclear membrane unenveloped and then bud through cytoplasmic membranes to obtain their envelopes (herpesviruses). Poxviruses are an exception in that they are DNA viruses that are assembled in the cytoplasm only. An exception to the RNA virus construction in the cytoplasm is that, occasionally, helical nucleocapsids, but not the whole enveloped virion, of some paramyxoviruses can be seen in the nucleus. Therefore, if virus-like structures are seen in the cytoplasm, one should check to see if they are also found in the nucleus; if so, then they are DNA viruses; if not, they are probably RNA viruses. Naked RNA viruses can get out of cells by lysis. Enveloped ones may bud through the cytoplasmic membrane into the extracellular space or through internal membranes to obtain their outer covering.

In searching for an unknown virus in any tissue, especially brain, one should look at low magnification for unusual features, such as inflammation or enlarged cells and then go to

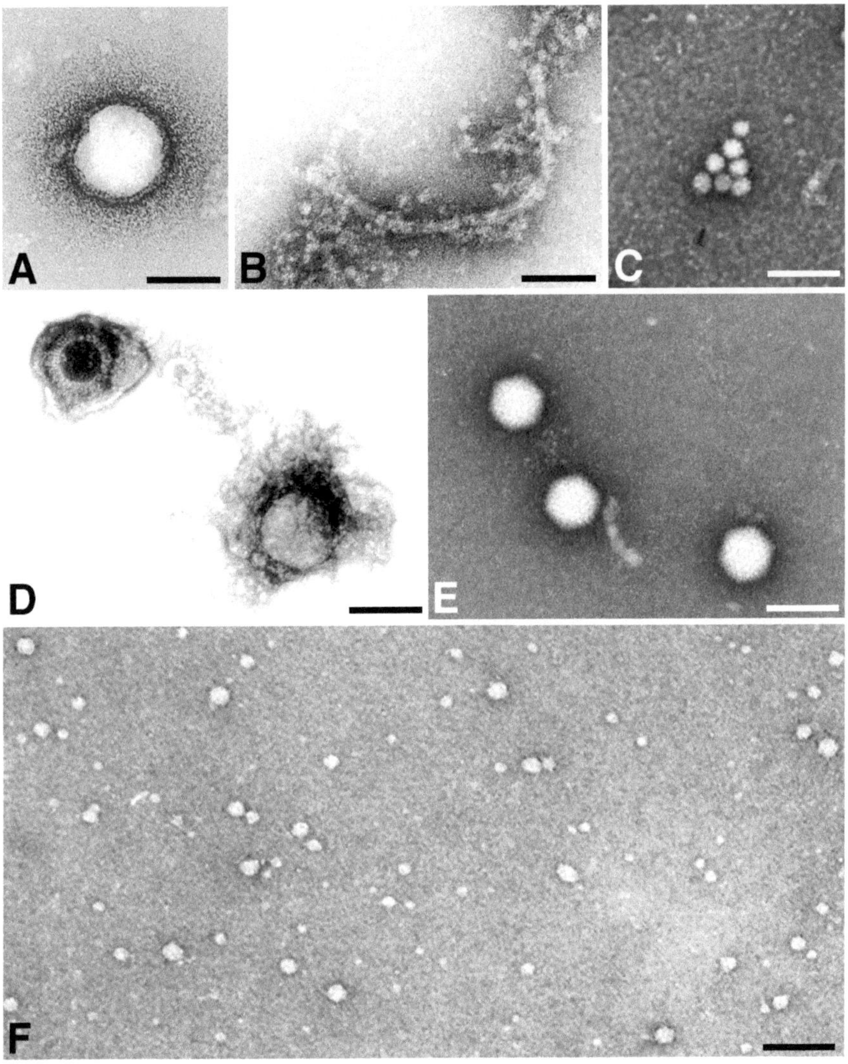

Figure 8 Viruses found in CSF by negative staining. (**A**) Mumps virus, a paramyxovirus with spikes on the surface. (**B**) Helical nucleocapsid of mumps virus. Compare (**A**) and (**B**) to other paramyxoviruses in Figures 1(**E**) and 7(**C**) and (**D**). (**C**) Enterovirus, a small (27 to 30 nm) round virus in the picornavirus family; its surface is nondescript. (**D**) Cytomegalovirus, a large (120 to 200 nm) enveloped virus. The stain has penetrated the envelope of the top left particle and outlines the 100-nm nucleocapsid. The bottom right particle is not penetrated by stain and would be unidentifiable if it were alone. (**E**) Adenovirus, a large (70 to 90 nm) naked icosahedral particle. (**F**) Apolipoprotein enriched by ultracentrifugation (droplets come to the top) from CSF; this micrograph demonstrates confusion created by lipoprotein molecules in this and other samples, such as stool and blood. Note the irregularity of size. Bars represent 100 nm. (CSF preparation provided by Dr. John Guyton, Duke University Medical Center.) Bars represent 100 nm.

high magnification for virus identification. Techniques for locating focal infections have been described (8,57). Great care must be taken to distinguish viruses from normal cellular structures (see Confusing Structures below).

Liver

Some viruses that cause hepatitis (58), such as hepatitis A virus [an SRV; see enterovirus, Fig. 3(B)], HBV [Figs. 10(A), 10(a), 10(a′), and 10(a″)], and CMV [Fig. 10(B)], have been demonstrated in liver by EM but are most readily and efficiently diagnosed by serology. Adenovirus [Fig. 10(C)] can also been seen in liver. Ebola virus can be seen in liver, but EM is not required for diagnosis, and Ebola virus infection is rare. EM can demonstrate defective or incomplete

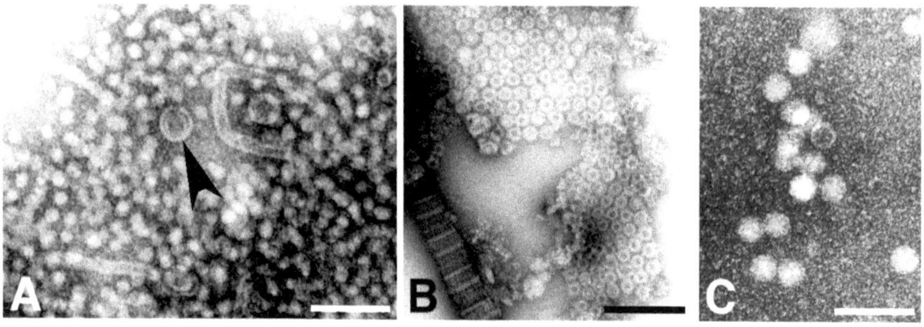

Figure 9 Viruses seen in blood by negative staining. (**A**) Hepatitis B virus. The complete 42-nm virus is called a Dane particle (arrow); the filaments and small (~22 nm), round particles are noninfectious viral surface antigen. (Micrograph provided by the Centers for Disease Control and Prevention, Atlanta.) (**B**) Parvovirus B19, a very small (22 nm) particle obtained, in this case, by homogenizing infected liver tissue and extracting with water. (**C**). Enterovirus, a small (~27 nm) icosahedral particle in the picornavirus family. Bars represent 100 nm.

virus particles, a useful means of assessing response to antiviral agents [Fig. 3.42 in Miller (7)]. Hepatitis C virus (a flavivirus) causes a large percentage of the hepatitis in the United States, but has not been visualized by EM in liver tissue.

Lung and Bronchoalveolar Cells

The most frequent viruses seen in the respiratory tract include adenovirus [Fig. 2(A)], paramyxoviruses, e.g., respiratory syncytial virus (RSV), parainfluenza virus, occasionally measles virus [Fig. 3(E)], orthomyxoviruses (influenza viruses), CMV [Fig. 10(B)], coronavirus (Fig. 11), and SRVs [enteroviruses (Fig. 3(B)) and rhinoviruses] (36). Adenovirus [Figs. 2(A) and 10(C)] and CMV [Fig. 10(B)] are easily identified by EM, and polyomavirus has been seen in cells from bronchoalveolar lavage. The enveloped viruses with helical nucleocapsids (ortho- and paramyxoviruses) require an experienced eye, particularly since some cells have projections on the surface that resemble the surface spikes of the myxoviruses. SRVs can be easily confused with ribosomes and glycogen.

Central Nervous Tissue (Brain, Meninges)

Central nervous system viral infections, viral neurotropism, and differential diagnosis have been discussed (55,59). Adenoviruses [Figs. 2(A) and 10(C)], herpesviruses [Figs. 2(C), 10(B), and 13(A)], and SRVs [Fig. 3(B)] are the viruses most likely to be recognized by EM in brain. HSV has a predilection for temporal lobe. The morphologically undifferentiated SRVs include entero-, coxsackie-, and echoviruses. EM is particularly useful in diagnosing progressive multifocal leukoencephalopathy, which is caused by a polyomavirus (JC virus) [Figs. 12(A), a]; by light microscopy, nuclear inclusions can resemble adenovirus infection.

Arboviruses such as alphaviruses, e.g., St. Louis encephalitis virus; flaviviruses [Figs. 12(B) and (b)], e.g., eastern equine encephalitis virus; and enteroviruses [Fig. 3(B)], e.g., poliovirus have occasionally been seen in brain. Bunyaviruses, e.g., the California encephalitis group of viruses; and an orbivirus [see reovirus, Fig. 3(A)], Colorado tick fever virus, infect brain. Alphaviruses are enveloped 60- to 70-nm particles and bud from the cytoplasmic membrane; flaviviruses [Fig. 12(B)] are 40- to 50-nm enveloped particles and do not bud from the cytoplasmic membrane but mature in association with endoplasmic reticulum. Bunyaviruses are 90- to 100-nm enveloped particles; they bud into smooth vesicles associated with the Golgi apparatus. Orbiviruses are 65- to 80-nm nonenveloped particles [see reovirus, Fig. 3(A)]. Rabies [Fig. 3(D)], a rhabdovirus, infects brain but is usually diagnosed by fluorescence microscopy following a bite from a rabid animal.

Measles virus [Fig. 3(E)] can be seen in brain in the chronic demyelinating disease subacute sclerosing panencephalitis (SSPE). In SSPE, the helical nucleocapsids may be found by EM in nuclei, an unusual situation because they contain RNA. Rubella can occasionally cause a demyelinating disease. Fortunately, both are rare today because of vaccines.

Chronic degenerative CNS diseases or spongioform encephalopathies are caused by subviral particles (prions), but EM is of limited utility in diagnosis because the altered prion proteins are morphologically similar to their normal protein counterparts. Some viruses, e.g., measles,

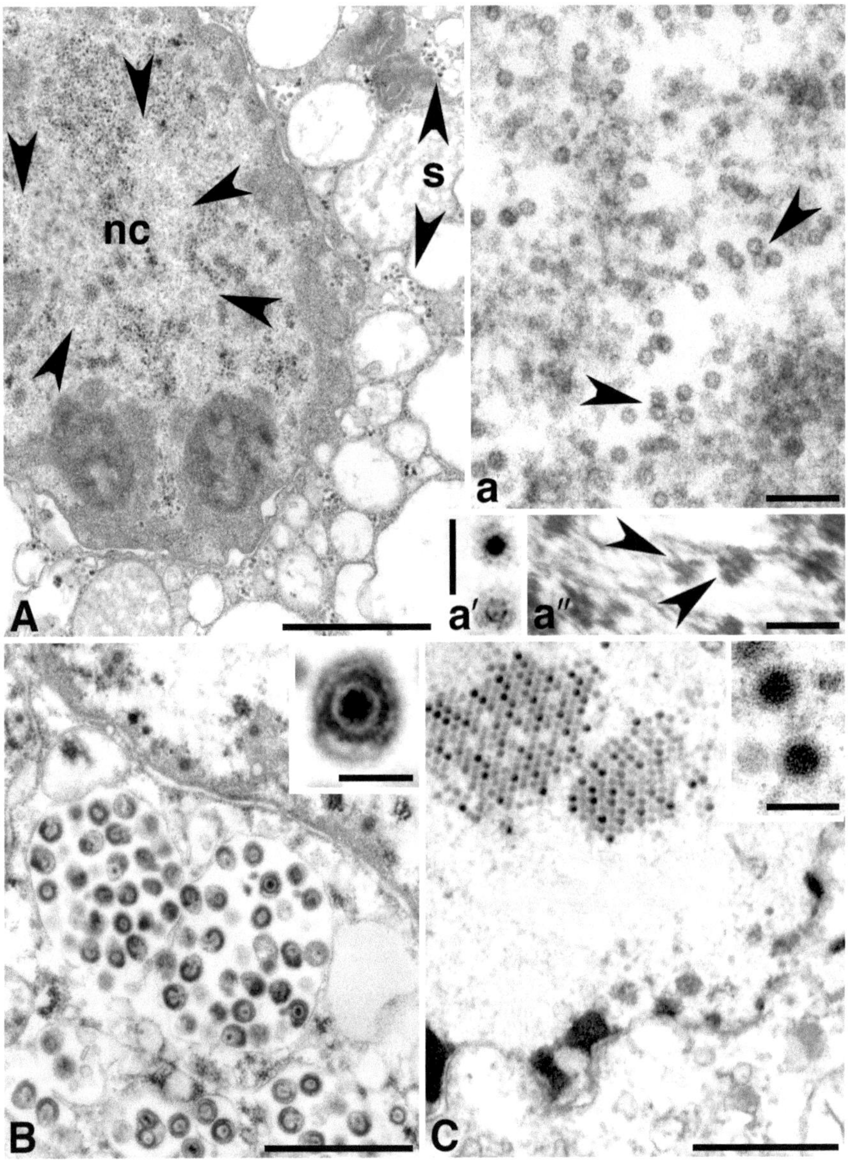

Figure 10 Hepatitis viruses that may be seen in thin sections of liver. (**A**) Hepatitis B virus; the 28-nm nucleocapsids (nc) in the nucleus are barely visible at this low magnification (20,000 ×). Collections of surface antigen(s) can be seen associated with the endoplasmic reticulum. (**a**) High magnification of small nucleocapsids in the nucleus. (**a′**) High magnification of 42-nm complete (Dane) particles. (**a″**) High magnification of surface antigen (irregular densities at arrows) associated with the endoplasmic reticulum. (Micrograph a″ courtesy of Dr. Joseph Harb, Medical College of Wisconsin, Milwaukee.) (**B**) Cytomegalovirus complete virions inside vacuoles in the cytoplasm; note the large numbers of particles together in vesicles. Compare this to herpes simplex [Fig. 2(C)] where the virus particles do not collect in large aggregates in the cytoplasm. (**Inset**) High magnification of complete virion. (**C**) Adenovirus paracrystalline arrays in the nucleus. (**Inset**) High magnification of 2 viruses. Bars in (**A**), (**B**), and (**C**) represent 1 μm; those in (**a**), (**a′**), (**a″**), (**B**) inset, and (**C**) inset represent 100 nm.

varicella, rubella, mumps, Epstein–Barr virus (EBV), CMV, VZV, and influenza, may cause a postinfectious encephalomyelitis, which appears to have an autoimmune component, and little or no virus is seen.

Meningitis can be caused by SRVs (entero-, coxsackieviruses) [Fig. 3(B)], mumps [see measles; Fig. 3(E)], herpesviruses [Figs. 2(C); 13(A), (a), (a′)], and lymphocytic choriomeningitis virus (a rare arenavirus contracted from mice and hamsters).

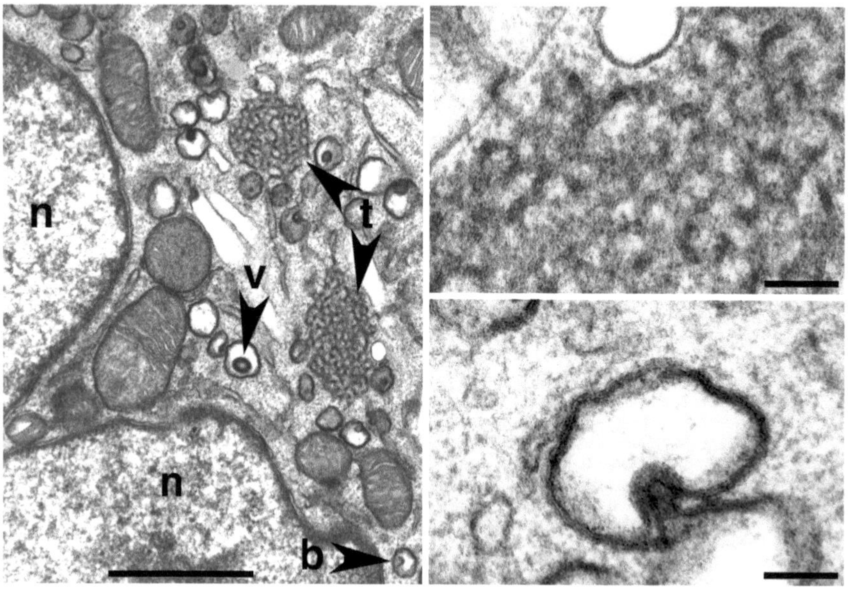

Figure 11 Thin section of coronavirus; occasionally seen in lung. Virions bud into cytoplasmic vesicles (v); tubular inclusions (t) are present in the cytoplasm, and parts of two nuclei (n) from this syncytial cell are visible at the left. Later infections have numerous viral particles within cytoplasmic vesicles. Upper right is a high magnification (100,000 ×) of a tubular inclusion; lower right is a high magnification of a virus budding into a cytoplasmic vesicle. (Infected cells provided by Dr. Ralph Baric, University of North Carolina, Chapel Hill.) Other viruses seen in lung include adenovirus [Fig. 2 (A)], herpesviruses, e.g., HSV [Fig. 2(C)] and CMV [Fig. 10(B)], reovirus (rarely) [Fig. 3(A)], picornaviruses, e.g., entero- and rhinoviruses [Fig. 3(B)], measles virus [Fig. 3(E)], and polyomavirus (see Fig. 12 below). Bar in low magnification at the left represents 1 μm; bars in high magnifications on the right represent 100 nm.

Intestine, Stomach, and Esophagus

Most gastroenteritis viruses (25,60) are shed in numbers sufficient to be detected by negative staining of fecal material, but adenoviruses [Figs. 2(A) and 10(C)] and rotaviruses [see reovirus; Fig. 3(A)] may be seen in thin sections of intestine. Other viruses such as CMV [Fig. 10(B)] and HSV [Fig. 2(C)] have been isolated from esophagus and intestine from immunosuppressed patients. Enteroviruses multiply in the intestinal tract, and may or may not be associated with diarrhea.

Skin

Viral skin lesions may be a result of cutaneous infection or a secondary reaction to virus infection elsewhere in the body (61). Viruses that cause vesicular rashes can frequently be demonstrated in the blister fluid by negative staining (Fig. 6). Thin sections of skin biopsies may also show herpesviruses [Figs. 2(C); 13(A), (a), (a′)] and poxviruses [Figs. 13(B), (b), and (b′), e.g., molluscum contagiosum and the parapox of milker's nodes. The outbreak of monkeypox was demonstrated first by EM of negative stains and thin sections to be a poxvirus (62).

Polyomaviruses [Figs. 12(A), a] have been seen by EM in skin of immunosuppressed patients (63). Papillomaviruses (morphologically similar to polyomaviruses) have been demonstrated in skin, but diagnosis usually rests with histology rather than EM. Further, the papilloma viral genome can exist in cells without producing complete virions so that it may be hard to detect morphologically. A number of other viruses, e.g., measles, rubella, parvovirus, some echoviruses, some coxsackieviruses, EBV, CMV, and some hemorrhagic fever viruses, cause maculopapular rashes and have been seen in thin sections, but diagnosis does not depend on EM.

Kidney

Viruses found in the urinary tract can frequently be detected by negative staining of urine, a noninvasive procedure. Nonetheless, since surgical pathology EM laboratories frequently get kidney tissue for examination, it is necessary to be aware of viruses that might be found there.

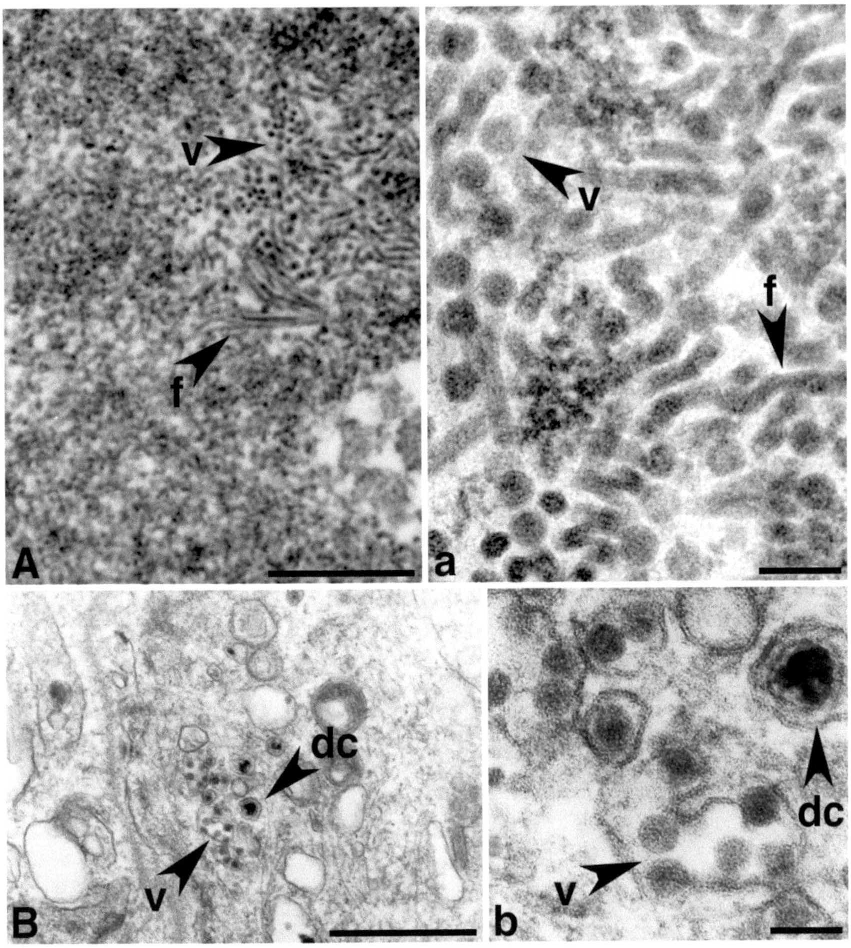

Figure 12 A naked DNA and an enveloped RNA virus seen in thin sections of brain. (**A**) Polyomavirus from a case of progressive multifocal leukoencephalopathy; low magnification of a nucleus filled with viral particles (v) and filaments (f). Sometimes viruses are in a paracrystalline array. a: High magnification of filaments and spherical virions (v) and filaments (f). (**B**) A flavivirus seen in a brain biopsy of a patient with encephalitis. Virions (v) are associated with endoplasmic reticulum. Note the neurosecretory or dense core granule (dc), a normal organelle in brain. b: High magnification of viruses (v) and a dense core granule (dc). The bars in (**A**) and (**B**) represent 1 μm; those in (**a**) and (**b**) represent 100 nm.

Polyomavirus [Figs. 12(A), (a)], adenovirus [Figs. 2(A) and 10(C)], CMV and HSV [Figs. 2(C) and 10(B)], and enteroviruses [Fig. 3(B)] are potential pathogens in that site.

Other Tissues

Any tissue submitted can be examined by EM for viruses, but those mentioned above are the most likely to be received. Other rare samples include lymph node, salivary gland, pericardial tissue, cells pelleted from pericardial fluid, joint fluid, or lung lavage. Likely viruses include adenovirus, herpesviruses, polyomaviruses, paramyxoviruses, and SRVs. Hematolymphatic and joint tissues may support parvovirus B19, measles, HHV-6, and retroviruses, though visualization of retroviruses is rare unless the specimen is co-cultivated with susceptible cells (e.g., placental cord blood cells).

EM IN EMERGING VIRAL DISEASES AND BIOTERRORISM SURVEILLANCE

Infectious organisms continue to emerge and reemerge (64), and many reagents are in development for the surveillance of these agents (65). However, in the event of a new or unusual presentation of disease with unknown origin, choosing the correct molecular probe is tricky

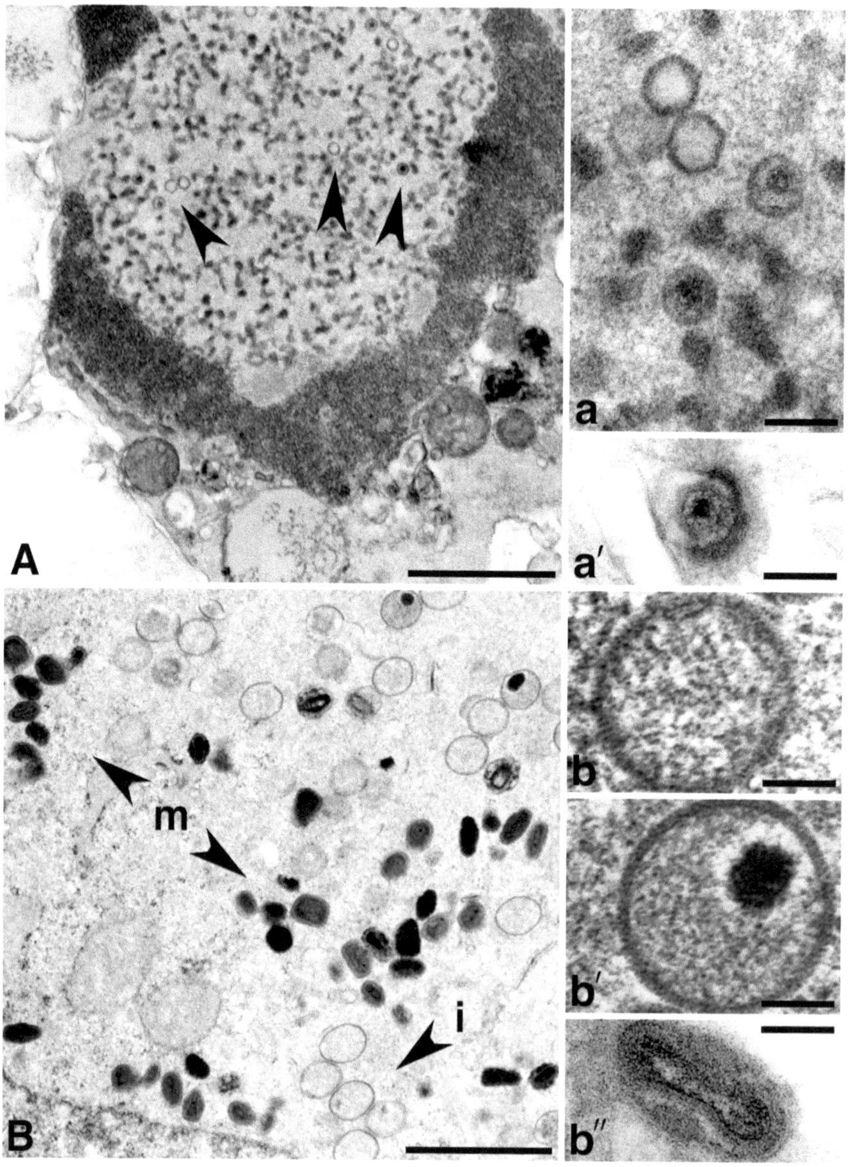

Figure 13 Thin sections of viruses seen in skin lesions. (**A**) Varicella zoster virus nucleocapsids (arrows) in the nucleus of a cell. (**a**): High magnification of nucleocapsids. (**a**′): High magnification of a complete virion in the cytoplasm. (Tissue provided by Dr. James Caruso, Duke University Medical Center, Durham, NC). (**B**) Cowpox from tissue culture. Immature (i) oval particles are forming in the cytoplasm along with dense brick shaped mature (m) virions. A small portion of the nucleus is seen at the bottom left. (**b**): High magnification of an immature particle. (**b**′): High magnification of an immature particle with a condensing nucleocapsid. (**b**″): High magnification of a mature particle showing a dumbbell-shaped core. The bars in (**A**) and (**B**) represent 1 μm; those in (**a**), (**a**′), (**b**), (**b**′), and (**b**″) represent 100 nm.

and time consuming. Owing to the facts that EM can be a rapid technique (4,66,67) and that it does not require specific reagents (antibodies, antigens, PCR probes, nucleic acid standards), it is on the forefront in the identification of unknown agents (68,69). A few examples include the following.

The initial Ebola outbreak in Zaire was diagnosed by EM of the virus isolated from human specimens (70). Polyomavirus infection of skin (tricodysplasia spinulosa) was first discovered by EM (63). The henipavirus (Hendra, Nipah) outbreaks in Australia and Asia benefited from EM (71). Some hantaviruses (bunyaviruses) cause no detectable cytopathology in vertebrate cell cultures (24) and cause persistent, nonpathogenic infections of rodents that can be transmitted to humans through urine and feces. EM complemented the characterization of the agent responsible for the hantavirus pulmonary syndrome in the southwestern United States (72). The identification of both monkeypox (62,73) and SARS coronavirus (39,40) was speeded by ultrastructural studies to determine virus family.

Finally, EM is critical in the surveillance of viral agents that could be distributed by terrorists. Rapid response laboratories in the Laboratory Response Network (LRN) are paired with EM facilities, and procedures for dealing with potential viral agents have been published (48,49). Thus, the future of EM in the study of viral and bacterial diseases (69,74,75) is secure in that no other single technique provides such a wide view of potential pathogenic agents in such a short time.

CELL CULTURES

Monitoring

Cell lines used for isolation of viruses in clinical material may contain endogenous agents or can become contaminated with persistent viruses; EM can be used to monitor them (1,3,5). Examples are retroviruses [Fig. 3(F)], SV 40 (a polyomavavirus), SV 5 [a paramyxovirus; Fig. 3(E)], and foamy viruses (retroviruses that bud into endoplasmic reticulum) (1). These contaminants may impede culture diagnosis of viruses by obscuring or altering virus-induced cytopathic effect (CPE) or by causing their own cytopathic alterations. They can prevent inoculated pathogens from growing, cause unusual cytopathology themselves, or confuse diagnosis by simply being visible in EM and mistaken for a pathogen. Mycoplasmas and other bacteria can also contaminate tissue cultures and hinder viral diagnosis.

Procedures

Some viral agents can be identified by negative staining of cell media or aqueous extracts of frozen-thawed cultures. Some cell-associated viruses, mycoplasmas, and other bacteria are best demonstrated by thin sectioning.

For viral examination by negative staining, cells in media can be frozen in an acetone/dry ice bath or liquid nitrogen and thawed in a 37°C bath two to three times. The cell debris is pelleted at low speed, and the supernatant is placed onto a filmed grid and negatively stained as described earlier. The clarified cell homogenate may need to be ultracentrifuged to concentrate virions.

For sectioning, monolayer cells should be briefly rinsed in serum-free medium or buffer, fixed for five minutes in buffered glutaraldehyde, scraped off the substrate, and pelleted in a microfuge. If the culture appears cloudy, it may be a sign that bacterial contamination has occurred or that cells have become detached from the substrate. In this case, one should be sure to collect the floating cells as well as the adherent ones by centrifugation for fixation. The pellet should be allowed to fix further in glutaraldehyde for at least 30 minutes. If the pellet sticks together well, it can be cut into millimeter-sized pieces and processed gently as tissue. If not, it can be embedded in 1% molten but cooled agar to keep it together (5,17). Nonadherent culture cells can be pelleted and treated in the same manner.

For agar embedment, the pellet should be scraped out of the tube, or if the tube is small and plastic, the end can be cut off with a razor blade. The pellet is then placed onto Parafilm, drained with the tip of a wedge of filter paper to a "cooked oatmeal" consistency, and surrounded by agar; cells should not be resuspended in the agar or they will be too dispersed to be found easily in the microscope. Cells can be embedded in agar directly after pelleting from glutaraldehyde fixation without a buffer wash, but the agar itself should not be glutaraldehyde fixed, as this will decrease penetration of other solutions. Further processing as for tissue should follow.

CONFUSING STRUCTURES

The most difficult task in diagnosing viruses by EM is to determine whether unusual structures are indeed viruses rather than spherical structures and membrane debris in negative stains and normal cellular organelles in thin sections. Numerous confusing structures have been described (1,5,9,76).

Lipids, lipoproteins, and tailless bacteriophages can closely resemble small icosahedral viruses in negative stains [Fig. 7(F)]. Cellular debris and mitochondrial fragments can masquerade as enveloped viruses with spikes.

In thin sections, clathrin-coated vesicles, synaptic vesicles, dense core granules, and caveolae can resemble enveloped viruses. Nuclear granules, ribosomes, and glycogen can be mistaken for small icosahedral viruses. Nuclear pores, clumped chromatin, and mitochondrial granules can resemble large icosahedral viruses. Microtubules and intermediate filaments may resemble helical nucleocapsids.

Once it has been determined that an unusual structure is probably not a normal cellular component, a viral atlas can be consulted. Several excellent ones are available (1,3,10,24,76–77).

SAFETY

All specimens should be handled under universal precautions. Negative stains cannot be assumed to kill viruses; aldehydes do not inactivate prions. EM fixatives and resins may be skin irritants or carcinogens (7,78).

CONCLUSION

EM can be an important adjunct to other methods for virus identification. Its best selling points are its rapidity and its lack of requirement for specific reagents. The down sides are that high viral density in liquids is required and that only small tissues areas can be examined by thin sectioning; however, techniques have been described for enhancing chances for virus detection. The most frequent samples examined by EM are liquid specimens, as they are easily obtained, rapidly observed, and frequently contain viruses not readily detectable by other means. For this reason, EM is an important modality in the surveillance of emerging diseases and bioterrorism events.

REFERENCES

1. Doane FW, Anderson N. Electron microscopy in diagnostic virology. A Practical Guide and Atlas. New York, NY: Cambridge University Press, 1987.
2. Howell DN, Payne CM, Miller SE, et al. Special techniques in diagnostic electron microscopy. Hum Pathol 1998; 29(12):1339–1346.
3. Hsiung GD, Fong CKY, Landry ML. Hsiung's Diagnostic Virology: As Illustrated by Light and Electron Microscopy. New Haven, CT: Yale University Press, 1994.
4. Hayat MA, Miller SE. Negative Staining: Applications and Methods. New York, NY: McGraw-Hill, 1990.
5. Miller SE. Detection and identification of viruses by electron microscopy. J Electron Microsc Tech 1986; 4:265–301.
6. Miller SE. Evaluation of electron microscopic information available from clinical samples. In: de la Maza LM, Peterson EM, eds. Medical Virology, Vol 10. New York, NY: Plenum Press, 1991:21–54.
7. Miller SE. Diagnosis of viral infection by electron microscopy. In: Lennette EH, Lennette DA, Lennette ET, eds. Diagnostic Procedures for Viral, Rickettsial, and Chlamydial Infections. Washington, DC: American Public Health Association, 1995:37–78.
8. Miller SE, Levenson RM, Aldridge C, et al. Identification of focal viral infections by confocal microscopy for subsequent ultrastructural analysis. Ultrastruct Pathol 1997; 21(2):183–193.
9. Oshiro LS, Miller SE. Application of electron microscopy to the diagnosis of viral infections. In: Lennette EH, ed. Laboratory Diagnosis of Viral Infections. New York, NY: Marcel Dekker, 1992:45–68.
10. Palmer EL, Martin ML. Electron Microscopy in Viral Diagnosis. Boca Raton, FL: CRC Press, 1988.
11. Miller SE. Use of electron microscopy to diagnose viral illnesses. Ann Saudi Med 1997; 17:66–76.
12. Hammond GW, Hazelton PR, Chuang I, et al. Improved detection of viruses by electron microscopy after direct ultracentrifuge preparation of specimens. J Clin Microbiol 1981; 14(2):210–221.
13. Anderson N, Doane FW. Agar diffusion method for negative staining of microbial suspensions in salt solutions. Appl Microbiol 1972; 24(3):495–496.
14. Lee FK, Takei Y, Dannenbarger J, et al. Rapid detection of viruses in biopsy or autopsy specimens by the pseudoreplica method of electron microscopy. Am J Surg Pathol 1981; 5(6):565–572.
15. Kapikian AZ, Dienstag JL, Purcell RH. Immune electron microscopy as a method for the detection, identification, and characterization of agents not cultivable in an *in vitro* system. In: Rose NL, Friedman H, eds. Manual of Clinical Immunology. Washington, DC: American Society for Microbiology, 1976:467–480.
16. Doane FW, Anderson N, Chao J, et al. Two-hour embedding procedure for intracellular detection of viruses by electron microscopy. Appl Microbiol 1974; 27(2):407–410.

17. Miller SE. Rapid processing of tissues for diagnostic electron microscopy. In: Jones BR, ed. Electron Microscopy: 41 Exercises by 17 Scientists. Monroe, NY: Library Research Associates, 1985:316–320.
18. Miller SE, Lang DJ. Rapid diagnosis of herpes simplex infection: amplification for electron-microscopy by short-term in vitro replication. J Infect 1982; 4(1):37–41.
19. Kok LP, Boon ME. Microwave Cookbook for Microscopists. Art and Science of Visualization, 3rd ed. Leiden: Coulomb Press Leyden, 1992.
20. Bozzola JJ, Russell LD. Electron Microscopy. Boston, MA: Jones and Bartlett Publishers, Inc., 1991.
21. Davies HA. Electron Microscopy and Immune Electron Microscopy for Detection of Gastroenteritis Viruses. New York, NY: Marcell Dekker, 1982.
22. Kapikian AZ. Viral Infections of the Gastrointestinal Tract, 2nd ed. New York, NY: Marcel Dekker, 1994.
23. Champsaur H, Questiaux E, Prevot J, et al. Rotavirus carriage, asymptomatic infection, and disease in the first two years of life. I. Virus shedding. J Infect Dis 1984; 149(5):667–674.
24. Fauquet CM, Mayo MA, Maniloff J, et al. Virus Taxonomy: Classification and Nomenclature of Viruses. Eighth Report of the International Committee on the Taxonomy of Viruses. New York: Elsevier Academic Press, 2005:1259.
25. Kapikian AZ, Hoshino Y, Chanock RM. Rotaviruses. In: Knipe DM, Howley PM, Griffin DE, et al., eds. Fields Virology, Vol 2. New York, NY: Lippincott Williams & Wiilkins, 2001:1787–1833.
26. Kelkar SD, Zade JK. Group B rotaviruses similar to strain CAL-1, have been circulating in Western India since 1993. Epidemiol Infect 2004; 132(4):745–749.
27. Rotavirus vaccines. Wkly Epidemiol Rec 2007; 82(32):285–295.
28. Lew JF, Petric M, Kapikian AZ, et al. Identification of minireovirus as a Norwalk-like virus in pediatric patients with gastroenteritis. J Virol 1994; 68(5):3391–3396.
29. Uhnoo I, Wadell G, Svensson L, et al. Importance of enteric adenoviruses 40 and 41 in acute gastroenteritis in infants and young children. J Clin Microbiol 1984; 20(3):365–372.
30. Wadell G, Allard A, Johansson M, et al. Enteric adenoviruses. In: Bock G, Whelan J, eds. Novel Diarrhoea Viruses. New York, NY: John Wiley & Sons, Inc., 1987:63–91.
31. Atmar RL, Estes MK. Norwalk and related diarrhea viruses. In: Richman DD, Whitley RJ, Hayden FG, eds. Clinical Virology. Washington, DC: ASM Press, 2002:1041–1060.
32. Matsui SM. Astrovirus. In: Richman DD, Whitley RJ, G. HF, eds. Clinical Virology. Washington, DC: ASM Press, 2002:1075–1086.
33. Appleton H. Small round viruses: Classification and role in food-borne infections. Ciba Found Symp 1987; 128:108–125.
34. Caul EO, Appleton H. The electron microscopical and physical characteristics of small round human fecal viruses: An interim scheme for classification. J Med Virol 1982; 9(4):257–265.
35. Fong TT, Lipp EK. Enteric viruses of humans and animals in aquatic environments: health risks, detection, and potential water quality assessment tools. Microbiol Mol Biol Rev 2005; 69(2):357–371.
36. Feinstone SM, Kapikian AZ, Purceli RH. Hepatitis A: Detection by immune electron microscopy of a viruslike antigen associated with acute illness. Science 1973; 182(116):1026–1028.
37. Balayan MS, Andjaparidze AG, Savinskaya SS, et al. Evidence for a virus in non-A, non-B hepatitis transmitted via the fecal-oral route. Intervirology 1983; 20(1):23–31.
38. Miyata H, Tsunoda H, Kazi A, et al. Identification of a novel GC-rich 113-nucleotide region to complete the circular, single-stranded DNA genome of TT virus, the first human circovirus. J Virol 1999; 73(5):3582–3586.
39. Drosten C, Gunther S, Preiser W, et al. Identification of a novel coronavirus in patients with severe acute respiratory syndrome. N Engl J Med 2003; 348(20):1967–1976.
40. Ksiazek TG, Erdman D, Goldsmith CS, et al. A novel coronavirus associated with severe acute respiratory syndrome. N Engl J Med 2003; 348(20):1953–1966.
41. Smits SL, Lavazza A, Matiz K, et al. Phylogenetic and evolutionary relationships among torovirus field variants: evidence for multiple intertypic recombination events. J Virol 2003; 77(17):9567–9577.
42. Jamieson FB, Wang EE, Bain C, et al. Human torovirus: a new nosocomial gastrointestinal pathogen. J Infect Dis 1998; 178(5):1263–1269.
43. Pereira HG, Fialho AM, Flewett TH, et al. Novel viruses in human faeces. Lancet 1988; 2(8602):103–104.
44. Taniguchi K, Wakuda M. [Picobirnavirus]. Uirusu 2005; 55(2):297–302.
45. Yolken R, Dubovi E, Leister F, et al. Infantile gastroenteritis associated with excretion of pestivirus antigens. Lancet 1989; 1(8637):517–520.
46. Abe S, Miyamura K, Oba T, et al. Oral ribavirin for severe adenovirus infection after allogeneic marrow transplantation. Bone Marrow Transplant 2003; 32(11):1107–1108.
47. Hirsch HH, Brennan DC, Drachenberg CB, et al. Polyomavirus-associated nephropathy in renal transplantation: Interdisciplinary analyses and recommendations. Transplantation 2005; 79(10):1277–1286.
48. CDC. Negative staining electron microscopic protocol for rash illness, 2006. (http://www.bt.cdc.gov/agent/smallpox/lab-testing/pdf/em-rash-protocol.pdf)

49. Miller SE. Bioterrorism and electron microscopic differentiation of poxviruses from herpesviruses: dos and don'ts. Ultrastruct Pathol 2003; 27(3):133–140.
50. Dane DS, Cameron CH, Briggs M. Virus-like particles in serum of patients with Australia-antigen-associated hepatitis. Lancet 1970; 1(7649):695–698.
51. Cossart YE, Field AM, Cant B, et al. Parvovirus-like particles in human sera. Lancet 1975; 1(7898):72–73.
52. Brown KE. Detection and quantitation of parvovirus B19. J Clin Virol 2004; 31(1):1–4.
53. Treanor J. Respiratory infections. In: Richman DD, Whitley RJ, Hayden FG, eds. Clinical Virology. New York, NY: Churchill Livingstone, 1997:5–33.
54. Gillim-Ross L, Subbarao K. Emerging respiratory viruses: Challenges and vaccine strategies. Clin Microbiol Rev 2006; 19(4):614–636.
55. Booss J, Esiri MM. Viral Encephalitis in Humans. Washington, DC: ASM Press, 2003.
56. Miller SE, Howell DN, Proia AD. Viral disease. In: Garner A, Klintworth G, eds. Garner and Klintworth's Pathobiology of Ocular Disease, Part A. New York, NY: Informa Health Healthcare, 2008:155–211.
57. Chu CT, Howell DN, Morgenlander JC, et al. Electron microscopic diagnosis of human flavivirus encephalitis: use of confocal microscopy as an aid. Am J Surg Pathol 1999; 23(10):1217–1226.
58. Thomas HC, Lemon S, Zuckerman AJ. Viral Hepatitis, 3rd ed. Malden, MA: Blackwell Publishing, 2005.
59. Cassady KA, Whitley RJ. Central nervous system infections. In: Richman DD, Whitley RJ, Hayden FG, eds. Clinical Virology. New York, NY: Churchill Livingstone, 1997:35–53.
60. Bock G, Whelan JE. Novel Diarrhoea Viruses–Ciba Foundation Symposium 128. New York, NY: Wiley, 1987.
61. Vander Straten M, Carrasco D, Tyring SK. Infections involving the skin. In: Richman DD, Whitley RJ, Hayden FG, eds. Clinical Virology. Washington, DC: ASM Press, 2002:117–134.
62. Bayer-Garner IB. Monkeypox virus: Histologic, immunohistochemical and electron-microscopic findings. J Cutan Pathol 2005; 32(1):28–34.
63. Haycox CL, Kim S, Fleckman P, et al. Trichodysplasia spinulosa–A newly described folliculocentric viral infection in an immunocompromised host. J Investig Dermatol Symp Proc 1999; 4(3):268–271.
64. Murphy FA. Emerging zoonoses: The challenge for public health and biodefense. Prev Vet Med 2008; 86(3–4):216–223.
65. Petrenko VA, Sorokulova IB. Detection of biological threats. A challenge for directed molecular evolution. J Microbiol Methods 2004, 58(2):147–168.
66. Schroeder JA, Gelderblom HR, Hauroeder B, et al. Microwave-assisted tissue processing for same-day EM-diagnosis of potential bioterrorism and clinical samples. Micron 2006; 37(6):577–590.
67. Giberson RT, Austin RL, Charlesworth J, et al. Microwave and digital imaging technology reduce turnaround times for diagnostic electron microscopy. Ultrastruct Pathol 2003; 27(3):187–196.
68. Gentile M, Gelderblom HR. Rapid viral diagnosis: Role of electron microscopy. New Microbiol 2005; 28(1):1–12.
69. Curry A, Appleton H, Dowsett B. Application of transmission electron microscopy to the clinical study of viral and bacterial infections: Present and future. Micron 2006; 37(2):91–106.
70. Johnson KM, Lange JV, Webb PA, et al. Isolation and partial characterisation of a new virus causing acute haemorrhagic fever in Zaire. Lancet 1977; 1(8011):569–571.
71. Chua KB, Wong EM, Cropp BC, et al. Role of electron microscopy in Nipah virus outbreak investigation and control. Med J Malaysia 2007; 62(2):139–142.
72. Zaki SR, Greer PW, Coffield LM, et al. Hantavirus pulmonary syndrome. Pathogenesis of an emerging infectious disease. Am J Pathol 1995; 146(3):552–579.
73. Guarner J, Johnson BJ, Paddock CD, et al. Monkeypox transmission and pathogenesis in prairie dogs. Emerg Infect Dis 2004; 10(3):426–431.
74. Hazelton PR, Gelderblom HR. Electron microscopy for rapid diagnosis of infectious agents in emergent situations. Emerg Infect Dis 2003; 9(3):294–303.
75. Biel SS, Gelderblom HR. Diagnostic electron microscopy is still a timely and rewarding method. J Clin Virol 1999; 13(1–2):105–119.
76. Dalton AJ, Haguenau F. Ultrastructure of Animal Viruses and Bacteriophages: An Atlas. New York, NY: Academic Press, 1973.
77. Palmer EL, Martin ML. An Atlas of Mammalian Viruses. Boca Raton, FL: CRC Press, 1982.
78. Fleming DO, Hunt DL. Biological Safety: Principles and Practices, 4th ed. Washington, DC: American Society for Microbiology Press, 2006.

12 Laboratory Diagnosis of Viruses in Resource-Limited Settings

Rosemary C. She
Department of Pathology, University of Utah School of Medicine, Salt Lake City, Utah, U.S.A.

Jeffrey M. Bender
Department of Pediatrics, University of Utah School of Medicine, Salt Lake City, Utah, U.S.A.

Cathy A. Petti
Department of Pathology, University of Utah School of Medicine, Associated Regional University Pathologists (ARUP) Laboratories, and Department of Medicine, University of Utah School of Medicine, Salt Lake City, Utah, U.S.A.

A strong laboratory infrastructure to perform diagnostic testing is necessary to make and confirm clinical diagnoses as well as conduct disease surveillance to stem potential epidemics. Barriers to defining illness in resource-limited regions are vast, and patients are often given a presumptive diagnosis of infection based solely on clinical findings for well-recognized syndromes. This practice frequently leads to misdiagnosis with failure to treat alternative diagnoses, causing worse outcomes and increased mortality (1–5). In particular, syndromes with a viral etiology are rarely diagnosed because in the absence of laboratory confirmation, few epidemiologic data exist to define their disease prevalence that could support the development of clinical algorithms based on syndromic diagnosis. For this reason, building laboratory capacity in resource-limited regions has become a focal point in the fight to prevent and reduce infectious diseases (5–9).

Despite numerous challenges in resource-limited settings, innovative, low-complexity, culture-independent methods exist to diagnose viral infections such as antigen-based point-of-care testing. The advantages of rapid tests are that most do not require refrigeration or the use of complex machinery, they are easy to perform, they potentially offer a cost-effective option, and many have shelf lives of up to one year or more. Of course, many considerations are warranted in determining whether or not a specific test should be performed, which are based on local and regional likelihoods of viral-associated infections, expected test volume at a particular site, skill level of laboratory personnel, and ability to implement quality assurance measures. This chapter discusses the indications for virological testing for common clinical settings and provides specific recommendations on test menus for resource-limited laboratories. Recommendation criteria are outlined in Table 1.

BLOOD-BORNE VIRUSES

HIV

Much of the disease burden of HIV falls upon regions in which resources are limited, such as Africa and Southeast Asia. Conventionally, the diagnosis of HIV is made by screening for anti-HIV antibodies with an enzyme immunoassay (EIA) followed by confirmation of positive results with a second assay, e.g., western blot or immunofluorescence antibody testing. Confirmatory tests are relatively expensive and require technical expertise often not available in resource-limited settings. Thus, rapid antigen assays are often more practical for HIV testing in these regions. Specifically, available methods include immunochromatography, microfiltration-bound EIA, and latex bead agglutination. These assays have sensitivity and specificity comparable to conventional testing and are simple to perform (7,10). They are convenient for individual use but for high sample volumes, EIA is more convenient if equipment and personnel are available.

Table 1 Levels of Recommendation: Scoring and Definitions[a]

Criteria	Score
Clinical applicability	2 = Recommended for diagnosis
	1 = Helpful in select circumstances
	0 = Not recommended
Cost	2 = Low
	1 = Moderate
	0 = Prohibitive
Ease of performance	2 = Low complexity, minimal technical expertise required
	1 = Moderate complexity, some technical expertise required
	0 = High complexity, extensive training and expertise required
Total score	5–6 = Highly recommended (I)
	3–4 = Moderately recommended (II)
	0–2 = Not recommended (III)

[a]These recommendations are based on personal field experience and expert opinion.

Numerous documents pertaining to rapid HIV test kit performance, appropriate kit evaluation, and quality assurance in resource-limited countries have been published by the World Health Organization (WHO) (6,11,12). Importantly, the WHO recommends that countries validate the test kit being used, as performance will vary by regional differences in HIV strain and prevalence. Depending on the prevalence of HIV in a given region, the WHO provides several different testing algorithms (6). In general, algorithms rely on the serial or parallel use of two to three different rapid kits for detection of anti-HIV antibodies. Patients with an indeterminate result should have testing repeated two weeks after the initial testing, and if results remain indeterminate, repeat testing should be performed again at four weeks, three months, six months, and one year as appropriate.

Numerous kits are available and they cost between approximately $1.50 and $5.00 USD per test. Most use serum or plasma, but others have been validated for finger stick whole blood (OraQuick HIV–1/2, Orasure Technologies, Inc., Bethlehem, PA; UniGold HIV; Trinity Biotech, Plc, Bray, Co Wicklow, Ireland; Determine HIV–1/2, Abbott Laboratories, Abbott Park, IL), or saliva (OraQuick HIV–1/2; Saliva-Strip HIV–1/2, Saliva Diagnostic Systems, Ltd., Framingham, MA). Studies have shown improved rates of confirmed diagnoses and treatment due to the rapid diagnostic capabilities of these assays, with turnaround times as low as 20 minutes per test (13,14). A list of rapid HIV assays is available at http://www.rapid-diagnostics.org.

HIV viral load testing is limited by the issue of RNA stability and high costs of commercial assays. Routine viral load testing is not currently recommended by the WHO in resource-poor settings due to its complexities (6). However, with improvement of technologies and decrease in test costs, viral load testing may become increasingly available in the future as it remains the cornerstone of monitoring antiretroviral treatment. When viral load testing is possible, RNA stability needs to be ensured and plasma should be immediately separated after specimen collection. Hand-cranked centrifuges are often used in the field for this purpose. When testing is not performed immediately, refrigeration (e.g., large coolers can be used to transport specimens from field-based sites to the laboratory) or freezing is necessary to ensure specimen stability. This obstacle can be circumvented by using dried blood spots on filter paper for testing, as RNA remains stable in the filter paper. This specimen type has been validated for use by several commercial RT-PCR kits (Amplicor HIV-1 Monitor version 1.5, Roche, Indianapolis, IN; Organon Teknika Nuclisens QT, bioMérieux, Marcy l'Etoile, France; Primagen, Amsterdam, The Netherlands Retina Rainbow/NucliSens EasyQ, Marcy l'Etoile, France) (15–17). If laboratory resources are available for determination of viral load by RT-PCR, the circulating HIV-1 subtypes in a given area must also be taken into consideration. Commercially available assays may not be as sensitive in detecting clade O or non-B subtypes of HIV-1 (13,14).

Cost-effective options for viral load testing currently include use of home-brew RT-PCR assays, which save on reagent costs but still require substantial start-up costs for required equipment and technical expertise (18). Ultrasensitive p24 antigen testing as performed in an EIA format has been studied for use as a surrogate for nucleic acid quantitation, with mixed results (19,20). It has also been studied as an option for use with dried blood spots as an approach to diagnose HIV infection in infants (19,21).

For infant testing, clinical data are used in conjunction with laboratory testing. WHO guidelines are available for infant testing and diagnosis (22). Nucleic acid testing (NAT), where available, is strongly recommended at four to six weeks of age for infants known to have exposure to HIV. Dried whole blood spots may be used if testing cannot be performed by the local laboratory. Serologic testing may be helpful in infants of unknown HIV exposure to indicate the possibility of exposure or rule out current infection (22,23).

Hepatitis B Virus and Hepatitis C Virus

Hepatitis B virus (HBV) and hepatitis C virus (HCV) infections are seen frequently in HIV positive individuals and are endemic in many areas of Asia and Africa (24,25). HBV and, to a lesser extent, HCV are responsible for most cases of hepatocellular carcinoma worldwide (25). Viral testing is necessary to confirm infection. However, as with other infections discussed in this chapter, diagnosis is often made clinically without laboratory confirmation as testing is not widely available in many resource-limited areas. Intense vaccination efforts for HBV are ongoing.

Detection of hepatitis B surface antigen (HBsAg) is the mainstay of diagnosing HBV infection. As in developed countries, this can be accomplished by EIA-based formats where resources are available and testing volume is sufficient. Positive results should be confirmed by a neutralization assay. In developing countries, rapid tests for detection of HBsAg may be more practical and many have been evaluated by the WHO (26,27) and other investigators (28,29) with generally excellent concordance with EIA testing. The formats include immunochromatography, agglutination, and immunofiltration. When these tests are used and confirmation testing by a neutralization assay is not possible, diagnosis of HBV infection can be achieved by the combination of a positive rapid test result with clinical findings or results of other HBV serologic markers for monitoring the disease (27). A list of rapid HBV assays is available at http://www.rapid-diagnostics.org.

HCV infection is also most conveniently established serologically. EIA and the recombinant immunoblot assay, as currently included in the diagnostic algorithm recommended by the Centers for Disease Control and Prevention (CDC), may not be easily performed in laboratories in resource-limited settings. Rapid formats such as for detection of anti-HCV antibodies are also available and many kits have been evaluated (30,31). Genotyping is not routinely done due to its complexity. As treatment of chronic HCV infection becomes more commonplace, the knowledge of genotype will be important in determining treatment and prognosis. HCV genotypes 4, 5, and 6, commonly found outside of North America and Europe, are less well studied but importantly may have responses to treatment and prognoses distinct from the better studied genotypes 1 and 2 (32). Our recommendations for HIV, HBV, and HCV testing are summarized in Table 2.

Blood Bank Screening for Viruses

Unfortunately, there are currently no standard procedures for blood bank screening of infectious agents in developing regions. In fact, blood may be transfused without such testing (33) even though HIV, HBV, and HCV are endemic in most resource-limited countries. However, organizations such as the WHO are working to ensure standard testing of all blood products in developing regions to include screening for blood-borne pathogens. Reports of rapid antigen kits for HIV, HBV, and HCV testing for blood bank use in developing countries are available through the WHO to which the reader is referred (26,27,30,31). These kits can be useful in blood banks where a smaller number of units is tested. Nucleic acid testing for HIV and HCV provides the most optimal sensitivity. However, the decision for testing is based on regional resources and healthcare priorities.

RESPIRATORY VIRUSES

Viral respiratory tract infections are the most common infections in children in resource-limited regions. Despite this, laboratory testing for common respiratory viruses such as influenza virus and respiratory syncytial virus (RSV) is limited. Furthermore, there is a dearth of data on the

Table 2 Recommendations for Use of Commercially Available Diagnostic Assays for Blood-Borne Viruses

	Clinical applicability	Cost	Ease of performance	Total score/ recommendation level
HIV				
Rapid kit	2	2	2	6 (I)
Serology (EIA)	2	2	1	5 (I)
NAT	2	0	1	3 (II)
Hepatitis B				
Rapid kit	2	1	2	5 (I)
Serology (EIA)	2	1	1	4 (II)
NAT	1	0	1	2 (III)
Hepatitis C				
Rapid kit	2	1	2	5 (I)
Serology (EIA)	2	1	1	4 (II)
NAT	2	0	1	3 (II)

epidemiology and mortality associated with these respiratory viruses outside of the developed world. Thus, the utility of diagnosing influenza and other viruses in resource-limited countries remains unclear. Through programs established by international health organizations such as the CDC and the WHO, new information continues to be incorporated to improve the accuracy of syndromic diagnostic algorithms related to respiratory viruses.

As respiratory viruses present clinically with vague symptoms easily mimicked by a myriad of other pathogens, establishing a diagnosis based on clinical findings alone is problematic (34,35). Laboratory diagnosis of respiratory viruses relies primarily on detection of either the virus itself or the patient's immune response to the virus. Although many viral testing procedures have been simplified in recent years, they remain challenging in the significantly resource-limited setting.

EIA kits for the rapid detection of respiratory viruses remain the primary source of testing in resource-limited countries. Although the sensitivity and specificity of these tests vary significantly based on patient population, disease prevalence, length of illness prior to testing, and patient age, these kits are easily stored and used with little resources. The WHO established guidelines for influenza testing in resource-limited settings based on EIA in 2005 (36). These guidelines are based on an understanding of local influenza epidemiology. Unfortunately, respiratory virus surveillance and activity data are frequently unknown in the setting of laboratories unable to confirm with immunofluorescence microscopy, cell culture, or PCR. More recent studies from rural Southeast Asia have shown improved surveillance and decreased inappropriate antibiotic use with rapid EIA influenza testing (37,38). Rapid RSV tests are now being used in resource-limited settings and have begun to further our understanding of the burden of RSV disease in Africa (39,40).

Immunofluorescence microscopy remains a primary method of detecting most respiratory virus infections in microbiology laboratories worldwide. Sensitivity and specificity for direct fluorescent antibody testing (DFA) have been well established and on the whole are relatively good. Recently, studies have begun to look at the burden of disease in resource-limited settings using this method in select locations (39). The challenges of performing DFA in resource-limited areas are that it requires access to a fluorescent microscope with a reliable power supply and trained personnel for results interpretation. Light-emitting diode (LED) light sources, powered by rechargeable batteries, can provide five or more hours of use with a single charge, and may prove to be a viable option for fluorescence microscopy in the future.

Other methods of virus detection such as culture, nucleic acid amplification, and serological response assays pose significant challenges in resource-limited settings. The requirement for maintenance of multiple cell lines for conventional or shell vial culture precludes their use in many of these microbiology laboratories. Though new rapid multiplex PCR techniques are being developed to identify more than 20 different respiratory viruses, including influenza H5N1 and severe acute respiratory syndrome (SARS) coronavirus, current methods remain labor

Table 3 Recommendations for Use of Commercially Available Diagnostic Assays for Respiratory Viruses

	Clinical applicability	Cost	Ease of performance	Total score/ recommendation level
Influenza A and B				
Rapid EIA	2	1	2	5 (I)
DFA	2	1	1	4 (II)
NAT (including highly pathogenic strains)	1	0	0	1 (III)[a]
RSV				
Rapid EIA	2	1	2	5 (I)
DFA	2	1	1	4 (II)
NAT	1	0	0	1 (III)
Parainfluenza, adenovirus, hMPV, SARS coronovirus, and others				
DFA	1	1	1	3 (II)
NAT	1	0	0	1 (III)

[a]While not recommended in field laboratories, this test is important for monitoring vaccine efficacy and should be performed by reference laboratories.

intensive and require significant technical expertise and resources. Serologic tests in resource-limited laboratories may be helpful in quantifying responses to vaccination (41) and have been used on occasion to help confirm rapid EIA assays (40,42). Recommendations for viral respiratory testing are listed in Table 3.

Influenza Virus

Influenza virus infection remains a significant source of morbidity and mortality worldwide, though its specific impact on resource-limited regions remains unclear. The diagnosis of influenza has become a key factor in dealing with seasonal influenza epidemics. The emergence of novel and highly pathogenic influenza viruses such as H5N1 coupled with the looming threat of a future global influenza pandemic has pushed for increased surveillance of respiratory viruses worldwide. Programs such as the International Emerging Infections Program at the CDC have made tremendous efforts to help improve respiratory virus testing worldwide. In conjunction with national ministries of health, these programs have helped to establish National Influenza Centers to act as state-of-the-art regional reference laboratories. Although studies specifically focused on resource-limited countries remain ongoing, numerous studies in the United States have demonstrated a decrease in antibacterial use, ancillary testing, hospital stays, and health care costs as a direct result of improved point-of-care influenza testing (43–45). Despite these efforts, reliable influenza virus testing in much of the developing world remains scarce.

GASTROINTESTINAL-ASSOCIATED VIRUSES

Viruses Associated with Gastroenteritis

Viruses cause most cases of gastroenteritis in children in developing countries, with rotavirus, adenovirus, norovirus, and astrovirus playing major roles in morbidity and mortality (46,47). It is estimated that rotavirus alone accounts for approximately half a million deaths in low-income countries (48). Interestingly, some studies have found that up to 53% of control patients in developing regions have detectable enteric viral pathogens. While an even higher percentage of patients suffering from gastroenteritis have detectable viral pathogens, these data emphasize that interpretation of a positive microbiologic result can be challenging (47,49). Nonetheless, rapid methods of diagnosis are available and can help prevent the use of costly antibiotics, and monitor the efficacy of vaccines.

For detection of adenovirus, rotavirus, and norovirus antigens in stool, immunochromatographic methods have been developed and function as a quick dipstick test with results in as few as five minutes. These employ latex agglutination or immunochromatography testing methods. Performance varies depending on the study and the kit used. In general, immunochromatography offers superior performance to most latex agglutination-based assays. Combination strips are also available for concomitant testing for multiple pathogens.

Many rotavirus studies report sensitivities of greater than 90% and specificities close to 100% when the quick dipstick tests are compared to enzyme immunoassay or RT-PCR (50–53). Few studies have examined the utility of rapid tests in resource-limited regions and thus the reported sensitivities and specificities may not be reflective of actual performance in the field (54). For adenovirus, even fewer data on rapid antigen assays have been published. One study demonstrated poor performance compared to PCR, with a sensitivity of 22% and specificity of 84% (54).

Hepatitis A Virus and Hepatitis E Virus

In developing areas with poor sanitation, most children have been infected with hepatitis A virus (HAV) by the age of five and experience self-limited disease. Laboratory diagnosis in such areas, such as Africa, Southeast Asia, and parts of South America, is not a necessity. In areas with improving hygiene, relatively fewer people have been exposed therefore outbreaks are more apt to occur (24). Laboratory diagnosis may be helpful to confirm outbreaks with detection of anti-HAV IgM antibody. The ELISA assay for HAV has been modified for testing on urine in which it has comparable sensitivity and specificity (55,56).

Hepatitis E virus (HEV) is endemic in Western and Northern Africa, Mexico, and Southeast and Central Asia and causes periodic outbreaks. It is important to diagnose HEV infections for both surveillance and in the case of outbreaks from contaminated water or foods in the interest of public health. A rapid immunochromatographic test (Genelab Diagnostics, Singapore) has been developed and evaluated with good correlation with ELISA-based serology (57,58).

Poliovirus

In 1988, the World Health Assembly resolved to eradicate poliomyelitis. Today, as national ministries of health and global networks continue to improve immunization practices and reporting of disease, the goal of poliovirus eradication is coming closer (59). Unfortunately, testing for poliovirus has not gone through similar growth. Centralized laboratories that participate in poliovirus surveillance continue to rely upon viral culture of stool and PCR confirmation that may take as long 21 days to isolate and identify wild or vaccine-like poliovirus (60). For a summary of recommendations on testing for gastrointestinal viruses, see Table 4.

MUCOCUTANEOUS VIRUSES

Herpes Simplex Virus

The seroprevalance of herpes simplex virus (HSV), particularly HSV-2, is high in many developing regions, with reported prevalence of 87% in HIV-1 positive pregnant women in Kenya (61), 15% in school children in Tanzania up to eight years of age (62) and 64% in a coastal Kenyan population (63). The importance of HSV in these settings is further highlighted by several reports of its association with increased acquisition of HIV. In resource-limited settings, the diagnosis of genital ulcer disease relies mainly on clinical judgment without laboratory confirmation. The most common pathogens, HSV, *Treponema pallidum*, and *Haemophilus ducreyi*, can be hard to differentiate clinically and thus syndromic diagnosis remains unreliable (64–66). On the other hand, waiting for a laboratory diagnosis delays treatment and increases costs. Therefore, syndromic treatment rather than treatment based on a specific pathogen has been advocated for genital ulcer disease (11,66) and recommended by the WHO for resource-poor settings, even where sophisticated laboratory capacities are available (12).

In the case where a laboratory diagnosis is required (e.g., persistent genital ulcer disease despite appropriate therapy, or skin lesions in a newborn), a Tzanck smear is inexpensive and can provide a quick diagnosis. The Tzanck smear, however, cannot differentiate between HSV or VZV cytopathic effects. DFA, where equipment is available, is highly sensitive in early lesions and highly specific when compared to viral culture (67,68). Serology may also aid in

Table 4 Recommendations for Use of Commercially Available Diagnostic Assays for Gastrointestinal Viruses

	Clinical applicability	Cost	Ease of performance	Total score/ recommendation level
Adenovirus 40/41, norovirus				
Rapid kit	0	1	2	3 (II)
NAT	0	0	0	0 (III)
Rotavirus				
Rapid kit	1	1	2	4 (II)[a]
NAT	2	0	0	2 (III)[a]
HAV, HEV				
Rapid EIA (HEV only)	1	1	2	4 (II)[b]
Serology (EIA)	1	1	1	3 (II)[b]
Poliovirus				
Serology (serum neutralization)	1	0	0	1 (III)[c]
Culture	2	0	0	2 (III)[c]
NAT	2	0	0	2 (III)[c]

[a]This test is helpful in determining vaccine efficacy.
[b]This test is useful in public health surveillance and in cases of outbreaks.
[c]While not recommended for use by field laboratories, this test is important for monitoring vaccine efficacy and should be performed by reference laboratories.

the diagnosis but as with latent viruses, interpretation is difficult unless paired serology or IgM testing is available.

Varicella Zoster Virus

The diagnosis of varicella or herpes zoster can usually be made clinically. Varicella zoster virus (VZV) typically has a more severe course in patients who are HIV-positive and may cause atypical, chronic skin lesions (69,70). In such cases with unusual clinical manifestation, the Tzanck smear or DFA (where equipment is available) of skin lesions can help establish the diagnosis. DFA has been shown to have superior sensitivity to culture (71).

ONCOGENIC VIRUSES

Epstein–Barr Virus

Most individuals acquire Epstein–Barr virus (EBV) by early childhood in developing countries. Infectious mononucleosis is primarily a clinical diagnosis. In northern Africa, China, and Southeast Asia, there is a relatively high incidence of nasopharyngeal carcinoma, which is caused by EBV. While histopathology provides definitive diagnosis, diagnosis of nasopharyngeal carcinoma is facilitated by detection of EBV antibodies by immunofluorescence antibody testing or EIA (72). In addition to IgG, IgA antibodies directed against viral capsid antigen, Epstein–Barr nuclear antigen, and early antigen have been used. Screening studies in areas of high prevalence of nasopharyngeal carcinoma may be facilitated by collection of patient blood on filter paper (73). Amidst the HIV epidemic, EBV has become an important cause of morbidity and mortality. Laboratory determination of viral load is usually carried out in central laboratories.

Human Papilloma Virus

Cervical cancer caused by human papilloma virus (HPV) is the second leading cause of cancer deaths worldwide, with 80% of new cases occurring in developing countries. Its significance in developing countries is magnified by the association between HIV infection and increased rates of cervical cancer (74,75). Screening for cervical cancer in low-income countries is hampered by lack of an adequate healthcare infrastructure and lack of public education. Diagnosis, where available, is based primarily on colposcopic examination with acetic acid treatment of the cervix. While cost-effective, this method is relatively crude and insensitive compared to Papanicolaou-stained smears of cervical cells and HPV molecular testing (12). There are ongoing efforts to provide resource-limited areas with affordable HPV assays that require minimal technical expertise and no special equipment such as the *care*HPV assay (QIAGEN GmbH, Germany), developed in partnership with PATH (www.path.org). Along with vaccination efforts against HPV, such new strategies may prove a breakthrough in the reduction of cervical cancer deaths worldwide.

CHILDHOOD VIRAL ILLNESSES: MEASLES, MUMPS, AND RUBELLA

Vaccine preventable childhood illnesses remain a significant source of morbidity and mortality worldwide. Childhood illness eradication campaigns by the WHO, United Nations Children's Fund (UNICEF), and national ministries of health have sparked a significant increase in childhood vaccination rates worldwide. Although this campaign continues to be an international success, the challenge remains for investigators to track the progress of these programs with limited diagnostic laboratory resources.

Due to the highly communicable and pathogenic nature of measles virus, in 2000 the WHO established the Global Measles Laboratory Network to assist in the eradication of the disease (76). Standardized testing procedures have been established and effectively implemented across much of the world. As they stand currently, measles case identification relies on the detection of measles-specific IgM from a single serum sample collected from a suspected case at first contact with a health facility, anytime within 28 days of rash onset (77). Typically, this test requires a centralized laboratory employing a validated EIA IgM assay, which is relatively simple and rapid to perform. Measles virus detection using specific antigens or nucleic acid amplification is not practically employed as a diagnostic tool. This is due to the relatively short period of measles viremia following rash onset as well as the difficulties faced by laboratories in resource-limited settings (76). Nonetheless, when possible, the virus is cultured and sequenced at reference laboratories to assist investigators in identifying the source and transmission routes of the virus (78).

Testing for mumps remains problematic even in reference laboratories. The current WHO definition for laboratory confirmed mumps infection requires identification of mumps-specific IgM by EIA, a fourfold increase in IgG titers, or identification of the virus by antigen detection or nucleic acid amplification (79). Unfortunately, the mumps IgM assays are particularly susceptible to cross reaction with other viruses (80,81). Viral culture and detection assays can be performed in centralized laboratories to confirm the diagnosis. Few studies have been done to address the importance of mumps virus in resource-limited settings.

The WHO clinical case definition for measles also captures rubella cases. Confirmation of rubella cases is similar to that of measles. Standard protocols for EIA IgM assays have been established, primarily to help eliminate false positive reporting of measles cases (77). Rubella virus culture as well as antigen and nucleic acid detection remain more difficult, but also can be helpful in identifying the epidemiology of an outbreak. Again, confirmatory testing is primarily done at reference laboratories, leaving clinical diagnosis the mainstay in resource-limited settings.

ARTHROPOD-BORNE VIRUSES

It is important to diagnose arthropod-borne viruses for surveillance purposes and identification of outbreaks. While rapid testing is not available for most of these viruses, the WHO supports a laboratory network for surveillance of yellow fever and Japanese encephalitis, diseases which are vaccine preventable. Specimens from peripheral labs may be sent to a network laboratory for the serological testing for antibodies to these pathogens, of which capture IgM EIA is the test of choice. For yellow fever, serum is collected, but for Japanese encephalitis, CSF is the preferred specimen source.

For Dengue virus, the most thoroughly evaluated rapid kit is an immunochromatographic test strip from PanBio (Brisbane, Australia) for detection of anti-Dengue virus antibodies in serum (82–85). Results of rapid kits for Dengue virus should be interpreted with caution, as prospective field testing has shown that performance of these assays may not be as good as claimed by the manufacturer, with some kits showing sensitivities of <20% (86). Although point-of-care testing does not currently exist for the diagnosis of other arboviruses or agents of hemorrhagic fever, many promising rapid platforms are in development. For further information on laboratory diagnosis, the reader is referred to other chapters in this textbook.

GENERAL CONSIDERATIONS OF LABORATORY PRACTICE

With the involvement of international agencies and other organizations, laboratories in developing regions are increasing their diagnostic capacity. Nonetheless, it is important to recognize the obstacles that these laboratories face due to continued lack of resources and infrastructure. Even in the most basic of laboratory operations, good laboratory practices must be maintained. To this end, written standard operating procedures, an external assessment system, quality compliance, and quality assurance are necessary measures. Even though these measures may be burdensome to resource-poor laboratories, they ensure that reliable results are reported. Ways to conserve resources while maintaining quality are possible. For example, in the absence of available specimens for validation or proficiency testing, sharing of specimens between "sister" laboratories using split samples can be helpful. Laboratory personnel must be trained and proficient with the assays they are expected to perform. This can be facilitated by using standardized training modules with an up-to-date curriculum.

Other elements to consider in developing laboratory tests are an organized record keeping and test reporting system, laboratory workflow, and biosafety precautions. Computerized records are rare in the majority of these laboratories making evaluation of laboratory results difficult. If a test must be referred to another laboratory, there must exist a reliable transport system for both the specimen and the return of results. Unfortunately, reliable transportation of specimens and maintaining specimen integrity are often beyond the capabilities of the existing healthcare infrastructure.

Distance-based learning tools can be extremely valuable to laboratories in resource-limited areas. Such tools include teleconferencing, web-based access to educational resources, and image transmittal for remote consultation, for example, histopathology images or DFA images. Continuing education can also be facilitated via broadcasts of presentations between institutions. Indeed, educational partnerships with global centers, whether on the same continent or abroad, serve an important role in improving the quality of laboratory testing and providing technical assistance where resources are limited.

As mentioned throughout this chapter, there are ways in which seemingly complicated testing can be performed in resource-limited areas. Battery-operated equipment such as microscopes makes access to a reliable electricity source less of an issue. LED light bulbs provide numerous hours of use with minimal energy consumption. Dried blood spots are an ingenious way of transporting specimens to reference laboratories for molecular and serologic testing. Furthermore, new automated molecular testing techniques requiring less resources and expertise continue to be developed.

In considering the addition of new tests, individual laboratories must prioritize based upon endemicity of disease and feasibility of implementing the test. Regulatory agencies overseeing the operations of laboratories are often nonexistent. The selection of diagnostic test kit should be made by laboratory personnel along with their Ministries of Health since human and capital resource constraints, cultural practices, and healthcare policies vary with each region and may impact assay selection. Hence, laboratory decision making should not be based solely upon published reports, manufacturer's claims, or recommendations of international organizations. Additionally, local evaluation of a kit is necessary to truly determine its performance characteristics as few commercial kits have been field tested in rural settings.

Invaluable resources on the topic of viral testing in the developing world include the WHO and the CDC. For example, the WHO has implemented a global laboratory network to support the laboratory diagnosis of measles, rubella, poliomyelitis, yellow fever, and HPV. Through this network the WHO is working toward improving quality assurance, proficiency testing, personnel training, laboratory equipment, the communication of data for global surveillance, and self-sufficiency. The CDC has similar networks for influenza and other respiratory viruses.

CONCLUSIONS

Here we briefly outlined the various testing options for the diagnosis of viral disease in the peripheral laboratories of developing countries. Regional reference laboratories and government agencies such as local ministries of health should be sought as a resource for confirmation testing of various agents important to the public health. They are also vital in performing the more complex assays, such as nucleic acid amplification and sequencing. The assays discussed in this chapter are those available at the current time, but certainly, as technology advances, rapid nucleic acid-based tests and other innovative platforms for viral detection will become more readily available to resource-limited areas.

REFERENCES

1. Berkley JA, Lowe BS, Mwangi I, et al. Bacteremia among children admitted to a rural hospital in Kenya. N Engl J Med 2005; 352(1):39–47.
2. Berkley JA, Versteeg AC, Mwangi I, et al. Indicators of acute bacterial meningitis in children at a rural Kenyan district hospital. Pediatrics 2004; 114(6):e713–e719.
3. Evans JA, Adusei A, Timmann C, et al. High mortality of infant bacteraemia clinically indistinguishable from severe malaria. QJM 2004; 97(9):591–597.
4. Reyburn H, Mbatia R, Drakeley C, et al. Overdiagnosis of malaria in patients with severe febrile illness in Tanzania: a prospective study. BMJ 2004; 329(7476):1212.
5. Petti CA, Polage CR, Quinn TC, et al. Laboratory medicine in Africa: A barrier to effective health care. Clin Infect Dis 2006; 42(3):377–382.
6. Guidelines for HIV Diagnosis and Monitoring of Antiretroviral Therapy, WHO Project: ICP BCT 001. World Health Organization; 2005. http://www.searo.who.int/LinkFiles/Publications_HLM-382Rev1.pdf. Accessed June 15, 2008.
7. Tegbaru B, Wolday D, Messele T, et al. Assessment of the implementation of HIV-rapid test kits at different levels of health institutions in Ethiopia. Ethiop Med J 2007; 45(3):293–299.
8. Berkelman R, Cassell G, Specter S, et al. The "Achilles heel" of global efforts to combat infectious diseases. Clin Infect Dis 2006; 42(10):1503–1504.
9. Muula AS, Maseko FC. Medical laboratory services in Africa deserve more. Clin Infect Dis 2006; 42(10):1503.
10. Plate DK. Evaluation and implementation of rapid HIV tests: the experience in 11 African countries. AIDS Res Hum Retroviruses 2007; 23 (12):1491–1498.
11. Htun Y, Morse SA, Dangor Y, et al. Comparison of clinically directed, disease specific, and syndromic protocols for the management of genital ulcer disease in Lesotho. Sex Transm Infect 1998; 74(suppl 1):S23–S28.
12. World Health Organization, Department of Reproductive Health and Research. Management of symptomatic STIs/RTIs. In: Sexually Transmitted and Other Reproductive Tract Infections – A guide to essential practice. Geneva: World Health Organization, 2005:109–113.
13. Swanson P, Soriano V, Devare SG, et al. Comparative performance of three viral load assays on human immunodeficiency virus type 1 (HIV-1) isolates representing group M (subtypes A to G) and group O: LCx HIV RNA quantitative, AMPLICOR HIV-1 MONITOR version 1.5, and Quantiplex HIV-1 RNA version 3.0. J Clin Microbiol 2001; 39(3):862–870.
14. Swanson P, de Mendoza C, Joshi Y, et al. Impact of human immunodeficiency virus type 1 (HIV-1) genetic diversity on performance of four commercial viral load assays: LCx HIV RNA Quantitative, AMPLICOR HIV-1 MONITOR v1.5, VERSANT HIV-1 RNA 3.0, and NucliSens HIV-1 QT. J Clin Microbiol 2005; 43(8):3860–3868.
15. Ayele W, Schuurman R, Messele T, et al. Use of dried spots of whole blood, plasma, and mother's milk collected on filter paper for measurement of human immunodeficiency virus type 1 burden. J Clin Microbiol 2007; 45(3):891–896.
16. Fiscus SA, Cheng B, Crowe SM, et al. HIV-1 viral load assays for resource-limited settings. PLoS Med 2006; 3(10):e417.
17. Brambilla D, Jennings C, Aldrovandi G, et al. Multicenter evaluation of use of dried blood and plasma spot specimens in quantitative assays for human immunodeficiency virus RNA: Measurement, precision, and RNA stability. J Clin Microbiol 2003; 41(5):1888–1893.
18. Rouet F, Ekouevi DK, Chaix ML, et al. Transfer and evaluation of an automated, low-cost real-time reverse transcription-PCR test for diagnosis and monitoring of human immunodeficiency virus type 1 infection in a West African resource-limited setting. J Clin Microbiol 2005; 43(6):2709–2717.
19. Stevens G, Rekhviashvili N, Scott LE, et al. Evaluation of two commercially available, inexpensive alternative assays used for assessing viral load in a cohort of human immunodeficiency virus type 1 subtype C-infected patients from South Africa. J Clin Microbiol 2005; 43(2):857–861.

20. Prado JG, Shintani A, Bofill M, et al. Lack of longitudinal intrapatient correlation between p24 antigenemia and levels of human immunodeficiency virus (HIV) type 1 RNA in patients with chronic hiv infection during structured treatment interruptions. J Clin Microbiol 2004; 42(4):1620–1625.
21. Patton JC, Sherman GG, Coovadia AH, et al. Ultrasensitive human immunodeficiency virus type 1 p24 antigen assay modified for use on dried whole-blood spots as a reliable, affordable test for infant diagnosis. Clin Vaccine Immunol 2006; 13(1):152–155.
22. World Health Organization. Report of the WHO Technical Reference Group, Paediatric HIV/ART Care Guideline Group Meeting. Geneva: World Health Organization, 2008.
23. Creek TL, Sherman GG, Nkengasong J, et al. Infant human immunodeficiency virus diagnosis in resource-limited settings: issues, technologies, and country experiences. Am J Obstet Gynecol 2007; 197(suppl 3):S64–S71.
24. Bojuwoye BJ. The burden of viral hepatitis in Africa. West Afr J Med 1998; 16(4):198–203.
25. Parkin DM. The global health burden of infection-associated cancers in the year 2002. Int J Cancer 2006; 118(12):3030–3044.
26. World Health Organization. Hepatitis B Surface Antigen Assays: Operational Characteristics (Phase 1) Report 1. Geneva: World Health Organization, 2001.
27. World Health Organization. Hepatitis B Surface Antigen Assays: Operational Characteristics (Phase 1) Report 2. Geneva: World Health Organization, 2004.
28. Lau DT, Ma H, Lemon SM, et al. A rapid immunochromatographic assay for hepatitis B virus screening. J Viral Hepat 2003; 10(4):331–334.
29. Randrianirina F, Carod JF, Ratsima E, et al. Evaluation of the performance of four rapid tests for detection of hepatitis B surface antigen in Antananarivo, Madagascar. J Virol Methods 2008; 151(2): 294–297.
30. World Health Organization. Hepatitis C Assays: Operational Characteristics (Phase I) Report 1. Geneva: World Health Organization, 2001.
31. World Health Organization. Hepatitis C Assays: Operational Characteristics (Phase I) Report 2. Geneva: World Helath Organization, 2001.
32. Nguyen MH, Keeffe EB. Prevalence and treatment of hepatitis C virus genotypes 4, 5, and 6. Clin Gastroenterol Hepatol 2005; 3(10 suppl 2):S97–S101.
33. Owusu-Ofori S, Temple J, Sarkodie F, et al. Predonation screening of blood donors with rapid tests: Implementation and efficacy of a novel approach to blood safety in resource-poor settings. Transfusion 2005; 45(2):133–140.
34. Peltola V, Reunanen T, Ziegler T, et al. Accuracy of clinical diagnosis of influenza in outpatient children. Clin Infect Dis 2005; 41(8):1198–1200.
35. Pachucki CT. Rapid Tests for Influenza. Curr Infect Dis Rep 2005; 7(3):187–192.
36. World Health Organization. WHO recommendations on the use of rapid testing for influenza diagnosis. Geneva: World Health Organization, 2005. http://www.who.int/csr/disease/avian_influenza/guidelines/RapidTestInfluenza_web.pdf. Accessed June 22, 2008.
37. Bhavnani D, Phatinawin L, Chantra S, et al. The influence of rapid influenza diagnostic testing on antibiotic prescribing patterns in rural Thailand. Int J Infect Dis 2007; 11(4);355–359.
38. Simmerman JM, Chittaganpitch M, Erdman D, et al. Field performance and new uses of rapid influenza testing in Thailand. Int J Infect Dis 2007; 11(2):166–171.
39. Nokes DJ, Okiro EA, Ngama M, et al. Respiratory syncytial virus infection and disease in infants and young children observed from birth in Kilifi District, Kenya. Clin Infect Dis 2008; 46(1):50–57.
40. Poulsen A, Stensballe LG, Nielsen J, et al. Long-term consequences of respiratory syncytial virus acute lower respiratory tract infection in early childhood in Guinea-bissau. Pediatr Infect Dis J 2006; 25(11):1025–1031.
41. Prince HE, Leber AL. Comparison of complement fixation and hemagglutination inhibition assays for detecting antibody responses following influenza virus vaccination. Clin Diagn Lab Immunol 2003; 10(3):481–482.
42. Okiro EA, Sande C, Mutunga M, et al. Identifying infections with respiratory syncytial virus by using specific immunoglobulin G (IgG) and IgA enzyme-linked immunosorbent assays with oral-fluid samples. J Clin Microbiol 2008; 46(5):1659–1662.
43. Bonner AB, Monroe KW, Talley LI, et al. Impact of the rapid diagnosis of influenza on physician decision-making and patient management in the pediatric emergency department: Results of a randomized, prospective, controlled trial. Pediatrics 2003; 112(2):363–367.
44. Benito-Fernandez J, Vazquez-Ronco MA, Morteruel-Aizkuren E, et al. Impact of rapid viral testing for influenza A and B viruses on management of febrile infants without signs of focal infection. Pediatr Infect Dis J 2006; 25(12):1153–1157.
45. Barenfanger J, Drake C, Leon N, et al. Clinical and financial benefits of rapid detection of respiratory viruses: an outcomes study. J Clin Microbiol 2000; 38(8):2824–2828.

46. Sdiri-Loulizi K, Gharbi-Khelifi H, de Rougemont A, et al. Acute infantile gastroenteritis associated with human enteric viruses in Tunisia. J Clin Microbiol 2008; 46(4):1349–1355.
47. Reither K, Ignatius R, Weitzel T, et al. Acute childhood diarrhoea in northern Ghana: epidemiological, clinical and microbiological characteristics. BMC Infect Dis 2007; 7(104):104.
48. Miller MA, McCann L. Policy analysis of the use of hepatitis B, Haemophilus influenzae type b-, Streptococcus pneumoniae-conjugate and rotavirus vaccines in national immunization schedules. Health Econ 2000; 9(1):19–35.
49. Mertens TE, Wijenayake R, Pinto MR, et al. Microbiological agents associated with childhood diarrhoea in the dry zone of Sri Lanka. Trop Med Parasitol 1990; 41(1):115–120.
50. Bon F, Kaplon J, Metzger MH, et al. [Evaluation of seven immunochromatographic assays for the rapid detection of human rotaviruses in fecal specimens]. Pathol Biol (Paris) 2007; 55(3–4):149–153.
51. Wilhelmi I, Colomina J, Martin-Rodrigo D, et al. New immunochromatographic method for rapid detection of rotaviruses in stool samples compared with standard enzyme immunoassay and latex agglutination techniques. Eur J Clin Microbiol Infect Dis 2001; 20(10):741–743.
52. Dennehy PH, Hartin M, Nelson SM, et al. Evaluation of the ImmunoCardSTAT! rotavirus assay for detection of group A rotavirus in fecal specimens. J Clin Microbiol 1999; 37(6):1977–1979.
53. Gilchrist MJ, Bretl TS, Moultney K, et al. Comparison of seven kits for detection of rotavirus in fecal specimens with a sensitive, specific enzyme immunoassay. Diagn Microbiol Infect Dis 1987; 8(4):221–228.
54. Weitzel T, Reither K, Mockenhaupt FP, et al. Field evaluation of a rota- and adenovirus immunochromatographic assay using stool samples from children with acute diarrhea in Ghana. J Clin Microbiol 2007; 45(8):2695–2697.
55. Joshi MS, Chitambar SD, Arankalle VA, et al. Evaluation of urine as a clinical specimen for diagnosis of hepatitis a. Clin Diagn Lab Immunol 2002; 9(4):840–845.
56. Rodriguez Lay Lde L, Larralde Diaz O, Martinez Casanueva R, et al. Anti-hepatitis A virus immunoglobulin M antibodies in urine samples for rapid diagnosis of outbreaks. Clin Diagn Lab Immunol 2003; 10(3):492–494.
57. Myint KS, Guan M, Chen HY, et al. Evaluation of a new rapid immunochromatographic assay for serodiagnosis of acute hepatitis E infection. Am J Trop Med Hyg 2005; 73(5):942–946.
58. Chen HY, Lu Y, Howard T, et al. Comparison of a new immunochromatographic test to enzyme-linked immunosorbent assay for rapid detection of immunoglobulin m antibodies to hepatitis e virus in human sera. Clin Diagn Lab Immunol 2005; 12(5):593–598.
59. Centers for Disease Control and Prevention (CDC) Progress toward interruption of wild poliovirus transmission–worldwide, January 2007-April 2008. MMWR Morb Mortal Wkly Rep 2008; 57(18):489–494.
60. World Health Organization. Supplement to the WHO Polio Laboratory Manual: An alternative test algorithm for poliovirus isolation and characterization. Geneva: World Health Organization, 2007. http://www.who.int/immunization_monitoring/Supplement_polio_lab_manual.pdf. Accessed July 1, 2008.
61. Drake AL, John-Stewart GC, Wald A, et al. Herpes simplex virus type 2 and risk of intrapartum human immunodeficiency virus transmission. Obstet Gynecol 2007; 109(2 Pt 1):403–409.
62. Kasubi MJ, Nilsen A, Marsden HS, et al. Prevalence of antibodies against herpes simplex virus types 1 and 2 in children and young people in an urban region in Tanzania. J Clin Microbiol 2006; 44(8):2801–2807.
63. O Ng'ayo M, Bukusi E, A Morrow R, et al. Sexual and demographic determinants for herpes simplex virus type 2 among fishermen along Lake Victoria, Kenya. Sex Transm Infect 2008; 84(2):140–142.
64. Beyrer C, Jitwatcharanan K, Natpratan C, et al. Molecular methods for the diagnosis of genital ulcer disease in a sexually transmitted disease clinic population in northern Thailand: Predominance of herpes simplex virus infection. J Infect Dis 1998; 178(1):243–246.
65. Ndinya-Achola JO, Kihara AN, Fisher LD, et al. Presumptive specific clinical diagnosis of genital ulcer disease (GUD) in a primary health care setting in Nairobi. Int J STD AIDS 1996; 7 (3):201–205.
66. Behets FM, Andriamiadana J, Randrianasolo D, et al. Chancroid, primary syphilis, genital herpes, and lymphogranuloma venereum in Antananarivo, Madagascar. J Infect Dis 1999; 180(4):1382–1385.
67. Moseley RC, Corey L, Benjamin D, et al. Comparison of viral isolation, direct immunofluorescence, and indirect immunoperoxidase techniques for detection of genital herpes simplex virus infection. J Clin Microbiol 1981; 13(5):913–918.
68. Reina J, Saurina J, Fernandez-Baca V, et al. Evaluation of a direct immunofluorescence cytospin assay for the detection of herpes simplex virus in clinical samples. Eur J Clin Microbiol Infect Dis 1997; 16(11):851–854.
69. Glesby MJ, Moore RD, Chaisson RE. Clinical spectrum of herpes zoster in adults infected with human immunodeficiency virus. Clin Infect Dis 1995; 21(2):370–375.

70. Tyndall MW, Nasio J, Agoki E, et al. Herpes zoster as the initial presentation of human immunodeficiency virus type 1 infection in Kenya. Clin Infect Dis 1995; 21(4):1035–1037.
71. Dahl H, Marcoccia J, Linde A. Antigen detection: The method of choice in comparison with virus isolation and serology for laboratory diagnosis of herpes zoster in human immunodeficiency virus-infected patients. J Clin Microbiol 1997; 35(2):347–349.
72. Fachiroh J, Paramita DK, Hariwiyanto B, et al. Single-assay combination of Epstein-Barr Virus (EBV) EBNA1- and viral capsid antigen-p18-derived synthetic peptides for measuring anti-EBV immunoglobulin G (IgG) and IgA antibody levels in sera from nasopharyngeal carcinoma patients: Options for field screening. J Clin Microbiol 2006; 44 (4):1459–1467.
73. Fachiroh J, Prasetyanti PR, Paramita DK, et al. Dried-blood sampling for epstein-barr virus immunoglobulin G (IgG) and IgA serology in nasopharyngeal carcinoma screening. J Clin Microbiol 2008; 46(4):1374–1380.
74. Hawes SE, Critchlow CW, Faye Niang MA, et al. Increased risk of high-grade cervical squamous intraepithelial lesions and invasive cervical cancer among African women with human immunodeficiency virus type 1 and 2 infections. J Infect Dis 2003; 188(4):555–563.
75. Hawes SE, Critchlow CW, Sow PS, et al. Incident high-grade squamous intraepithelial lesions in Senegalese women with and without human immunodeficiency virus type 1 (HIV-1) and HIV-2. J Natl Cancer Inst 2006; 98(2):100–109.
76. Featherstone D, Brown D, Sanders R. Development of the Global Measles Laboratory Network. J Infect Dis 2003, 187(suppl 1):S264–S269.
77. World Health Organization, Department of Immunization, Vaccines, and Biologicals. Manual for the laboratory diagnosis of measles and rubella virus infection. In: World Health Organization, ed. Expanded Programme on Immunization (EPI) team of the Department of Immunization, Vaccines and Biologicals, 2nd ed. World Health Organization, 2007.
78. Expanded Programme on Immunization (EPI). Standardization of the nomenclature for describing the genetic characteristics of wild-type measles viruses. Wkly Epidemiol Rec 1998; 73(35):265–269.
79. Global status of mumps immunization and surveillance. Wkly Epidemiol Rec 2005; 80(48):418–424.
80. Tsurudome M, Yamada A, Hishiyama M, et al. Monoclonal antibodies against the nucleoprotein of mumps virus: their binding characteristics and cross-reactivity with other paramyxoviruses. Acta Virol 1990; 34(3):220–227.
81. Harmsen T, Jongerius MC, van der Zwan CW, et al. Comparison of a neutralization enzyme immunoassay and an enzyme-linked immunosorbent assay for evaluation of immune status of children vaccinated for mumps. J Clin Microbiol 1992; 30(8):2139–2144.
82. Wu SJ, Paxton H, Hanson B, et al. Comparison of two rapid diagnostic assays for detection of immunoglobulin M antibodies to dengue virus. Clin Diagn Lab Immunol 2000; 7(1):106–110.
83. Blacksell SD, Newton PN, Bell D, et al. The comparative accuracy of 8 commercial rapid immunochromatographic assays for the diagnosis of acute dengue virus infection. Clin Infect Dis 2006; 42(8):1127–1134.
84. Vaughn DW, Nisalak A, Kalayanarooj S, et al. Evaluation of a rapid immunochromatographic test for diagnosis of dengue virus infection. J Clin Microbiol 1998; 36(1):234–238.
85. Palmer CJ, King SD, Cuadrado RR, et al. Evaluation of the MRL diagnostics dengue fever virus IgM capture ELISA and the PanBio Rapid Immunochromatographic Test for diagnosis of dengue fever in Jamaica. J Clin Microbiol 1999; 37(5):1600–1601.
86. Blacksell SD, Bell D, Kelley J, et al. Prospective study to determine accuracy of rapid serological assays for diagnosis of acute dengue virus infection in Laos. Clin Vaccine Immunol 2007; 14(11):1458–1464.

13 Antiviral Therapy

Brian Rha, David Kimberlin, and Richard Whitley
Department of Pediatrics, University of Alabama at Birmingham, Birmingham, Alabama, U.S.A.

INTRODUCTION

Recent advances in the development of antiviral therapy have made prompt and accurate diagnosis of viral infections essential for optimal patient care. The proliferation of antiviral agents over the past two decades has revolutionized therapy. While the first antiviral agents were approved for use in the United States in the 1960s, the introduction of acyclovir in the 1980s resulted in rapid development of new antiviral drugs. Currently, licensed antiviral drugs include those active against herpes viruses [herpes simplex viruses (HSV), varicella zoster virus (VZV), and cytomegalovirus (CMV)], respiratory viruses [influenza A virus, influenza B virus, and respiratory syncytial virus (RSV)], hepatitis B and C viruses, human papillomaviruses (HPV), and human immunodeficiency virus (HIV). The development of new antiviral agents continues to be a constant and active area of research and development.

Like most of the earliest drug discoveries, the first compounds with antiviral properties were found serendipitously. Examples include the earliest nucleoside analogues that were initially developed to target DNA replication of rapidly dividing tumor cells as part of cancer treatment but subsequently became mainstays of antiviral therapy (1,2). Further advances in molecular biology led to more target-directed development of antiviral agents. More specifically, knowledge of each step of viral replication has led to the development of compounds that target individual steps within the viral life cycle. Currently, licensed therapeutic agents can be categorized into broad groups according to their mechanism of action: those designed to prevent entry of viruses into host cell, inhibit transcription or replication of the viral genome, interfere with viral protein synthesis, alter cell fusion, or disrupt viral assembly and release.

One of the major challenges in drug development has been in designing therapy specific enough to avoid toxicity to normal host cells. Because viruses require host cell machinery to replicate, it was originally thought that any action interfering with viral replication would also necessarily kill host cells. However, the strategy of targeting enzymes unique to viruses has since yielded several safe and effective therapies. Today, the indications for the use of antiviral drugs include the treatment of active viral disease, as well as for prophylactic (for uninfected but at-risk individuals) and preemptive (infected but asymptomatic) therapies.

Despite these successes, cellular toxicity remains an important therapeutic consideration, as adverse effects limit the successful use of many antiviral drugs. In addition, the expanded use of antiviral drugs has led to drug-resistant strains that further limit effectiveness of therapy as best illustrated by the rapidly evolving nature of HIV therapy. Although the devastating consequences of untreated disease allow a higher threshold for acceptable side effects, poorly tolerated agents still lead to higher rates of noncompliance and inconsistent drug exposure that, in turn, lead to further development of resistant strains. This chapter will examine currently available therapies for viral infections.

THERAPY FOR RESPIRATORY VIRUS INFECTIONS

Influenza

There are three influenza viruses (A, B, and C) that are members of the orthomyxovirus family. These viruses have segmented negative-sense RNA genomes, an envelope derived from the host cell, and characteristic surface glycoproteins that are involved in the entry and release of the virus from host cells. Influenza C causes only minor illness that does not usually require therapy. Influenza A and B, however, can both cause seasonal epidemics with significant morbidity and mortality. Influenza A is also the source of occasional pandemics.

Recognizing that vaccination against the influenza virus is a more effective measure in reducing the burden of disease (3), specific antiviral agents are useful in the prophylaxis and treatment of infection. Two specific viral proteins are targets for current therapies: the matrix 2 (M2) protein, which is an ion channel in the viral membrane of influenza A, and neuraminidase, which is a surface glycoprotein common to both influenza A and B.

Amantadine, Rimantadine: The Adamantanes

Amantadine and rimantadine are structurally related tricyclic amines that bind to the M2 protein found in the nucleocapsid membrane of influenza A. This protein is an ion channel that allows protons to cross the membrane barrier, thereby acidifying the cytoplasm. This drop in pH enables viral uncoating, a step necessary to initiate viral replication (4). The M2 inhibitors work by blocking the acidification step, thereby, preventing replication.

Amantadine and rimantadine are useful for the treatment and prophylaxis of influenza A infections. Treatment is effective in reducing the duration of illness by about one day if given within two days of the onset of symptoms (5,6). Prophylaxis for high-risk individuals is indicated for those who cannot tolerate the influenza vaccine either due to allergic reaction or immunosuppressed status, and is also indicated for two weeks following vaccination if virus is already circulating in the community (7).

Both medications are given orally. Amantadine is not metabolized systemically and is excreted by the kidneys largely unchanged; rimantadine is metabolized extensively by the liver prior to renal clearance. As a result, the dose of amantadine must be reduced in patients with renal insufficiency; rimantadine dosing should be adjusted in patients with liver failure.

Side effects are similar with both drugs, but are typically less severe with rimantadine. Most commonly, these include gastrointestinal upset including nausea and vomiting, and central nervous system (CNS) symptoms such as anxiety, depression, confusion, insomnia, and difficulty concentrating (8).

A major area of concern with the use of the adamantanes has been the rise of viral resistance to these drugs. Resistance arises from single point mutations in the viral RNA that encodes the M2 protein transmembrane domain, and typically appears within two to three days of initiating therapy (9). Up to a third of treated patients develop resistant strains by the fifth day of treatment (10). While infection of the treated index case typically resolves despite the development of resistant strains, transmission of the strain to others often results in failure of drug prophylaxis for household contacts (9). Indeed, surveillance of influenza strains in the United States from 2005 to 2006 revealed over 90% had developed adamantane resistance (11).

Zanamivir, Oseltamivir: The Neuraminidase Inhibitors

Zanamivir and oseltamivir are structurally similar compounds that work by competitively binding neuraminidase, a surface glycoprotein that is common to both influenza A and B. Neuraminidase is essential for the release and spread of newly formed virus, making this enzyme an attractive target for inhibiting viral replication. Specifically, neuraminidase enables the release of the new viral particle from the host cell by cleaving the terminal sialic acid from glycoproteins on the cell surface. It also facilitates the migration of virus through mucous, allowing spread through the respiratory tract. Zanamivir is a synthetic competitive inhibitor, while oseltamivir is an ethyl ester prodrug that is converted to its active form by hepatic esterases. Peramivir, another similar compound, is an additional promising drug currently undergoing clinical trials.

Zanamivir and oseltamivir have both been shown to be effective in the prophylaxis and treatment of influenza A and B. Prophylaxis with zanamivir or oseltamivir reduces the rate of infection by up to 79% and 75–85%, respectively (12–18). Treatment with zanamivir within two days of the onset of symptoms lessens the severity of disease and shortens the duration of symptoms by an average of one day (19). Similarly, oseltamivir treatment started within one to two days of disease onset ameliorates symptoms and reduces duration by 1 to 1.5 days (15,20–21). In a retrospective evaluation of managed care databases, patients treated with oseltamavir had decreased hospitalization rates and respiratory complications (22).

Zanamivir requires administration by oral inhalation due to its poor oral bioavailability. The amount of drug reaching airway and lung mucosa is adequate to inhibit viral replication

(22). Zanamivir is generally well tolerated, but some patients experience exacerbation of reactive airways disease with treatment (23,24). Systemic absorption is approximately 15% (22), which likely explains the drug's benign side effect profile. Renal dose adjustment is rarely needed despite the fact that the unchanged compound is excreted via the kidneys.

In contrast, oseltamivir is well tolerated orally, with over 90% of the drug converted to its active metabolite (21). Dose adjustment is required in patients with renal insufficiency, as the active form is excreted by the kidney. Side effects are rare and mild, and typically consist of nausea and vomiting (25). Rare case reports of delirium and abnormal behavior in children taking oseltamivir, mostly in Japan, have prompted revision in the warning label of the drug (26). However, due to its excellent oral availability and minimal side effect profile, in light of widespread resistance to the adamantanes, oseltamivir has become the most widely used drug for the treatment of influenza.

The development of viral resistance to zanamivir is rare; however, widespread resistance of the H1N1 influenza strain to oseltamavir has been detected worldwide, with over 7% in the United States by March of 2008 (27). More recent data show that virtually all H1N1 strains of influenza in the United States for the 2008–2009 season are resistant to oseltamavir (28). Point mutations in the viral genome (H274Y) alter the active site of neuraminidase and block binding of the drug (27).

Respiratory Syncytial Virus

Respiratory syncytial virus (RSV) is a nonsegmented, single-stranded negative sense RNA virus that is a member of the paramyxoviridae family. RSV is a major cause of lower-respiratory infection in children and is also associated with significant morbidity and mortality in immunocompromised hosts.

Ribavirin

Ribavirin is a synthetic nucleoside analogue structurally similar to guanosine that inhibits viral RNA replication by interfering with messenger RNA synthesis (29). Ribavirin has activity against a broad spectrum of viruses, including influenza A and B, hantaviruses, herpes viruses, measles, and Lassa fever. Activity is greater against RNA viruses than DNA viruses. Historically, inhaled ribavirin was used to treat RSV respiratory infections. Ribavirin is no longer used because of the lack of improvement in clinical end points such as duration of hospitalization or required oxygen therapy (30–32). Intravenous ribavirin is used in the treatment of Lassa fever, while an oral formulation is used to treat hepatitis C in conjunction with pegylated interferon-α. Ribavirin's role in the treatment of hepatitis C is discussed later in the chapter.

Ribavirin can be administered in aerosolized, oral, and intravenous forms. Inhaled drug reaches therapeutic levels in the respiratory mucosa with only small amounts absorbed systemically. The oral bioavailability of ribavirin is about 40% (33), whereas intravenous therapy achieves approximately ten times higher peak concentrations. Less than a third of systemically administered drug passes through the urine unchanged, with an additional one-third of the drug excreted as metabolites (33).

Side effects of the inhaled form include mild conjunctivitis, rash, and bronchospasm. Special precautions must be taken due to ribavirin's mutagenic and teratogenic properties. Systemic administration has been associated with reversible anemia (33). To date, RSV resistance to ribavirin has not been observed.

THERAPY OF HERPES VIRUSES

The pathology caused by herpes viruses is as diverse as the viruses themselves. Fortunately, over the past 30 years several drug therapies have emerged that have proven to be safe and effective. As a group, the human herpes viruses consist of eight large enveloped viruses with double-stranded DNA genomes. Effective antiviral therapies include those against HSV-1, HSV-2, CMV, and VZV. Efficacious therapies that target Epstein–Barr virus (EBV) and human herpes viruses 6, 7, and 8, have yet to be established.

Acyclovir, Valacyclovir

Acyclovir is a synthetic deoxyguanosine analogue that prevents viral replication by competitively inhibiting viral DNA polymerase. Its active form, acyclovir triphosphate, is a specific substrate for the polymerase binding site, functioning as a competitive inhibitor of the DNA polymerase and a chain terminator. The therapeutic advantage of acyclovir lies in its specificity of action to viral-infected cells. Acyclovir is taken up preferentially by infected cells and is initially phosphorylated by the viral enzyme thymidine kinase. This monophosphate form is then diphosphorylated by host cellular enzymes to the active triphosphate form. Finally, the active form targets the viral DNA polymerase preferentially over host polymerases. The end result is effective drug therapy with minimal cellular toxicity.

Acyclovir is useful for the treatment of active HSV-1, HSV-2, and VZV disease, as well as for prophylactic and suppressive therapy of both immunocompetent and immunocompromised patients. Acyclovir is effective in the treatment of life-threatening infections including HSV encephalitis, neonatal HSV infections, and VZV infections in immunocompromised hosts. It is also indicated for the treatment of disseminated HSV and VZV infections, as well as for mucocutaneous HSV infections in immunocompromised hosts (34). Acyclovir has been used for the treatment of primary and recurrent genital HSV infections (35). Prophylactic or suppressive therapy can be used in patients with recurrent genital infections as well as to prevent reactivation of herpes labialis (35–41). It can be administered to prevent reactivation in HSV-seropositive transplant patients undergoing immunosuppression (39).

Acyclovir can be given intravenously, topically, or orally with less than 30% bioavailability (42). Drug distribution is good with penetration to kidney, lung, liver, and cardiac tissue as well as to skin lesions. Drug concentrations in cerebrospinal fluid are about half that of plasma levels (42). Acyclovir is primarily excreted by the kidney, necessitating dose adjustment in patients with renal insufficiency.

Acyclovir has shown to be a safe drug with minimal side effects relative to other antiviral drugs. The oral form is generally well tolerated with only mild associated gastrointestinal upset. Extravasation of the intravenous form can cause tissue inflammation and necrosis. Standard courses of therapy have not been shown to cause bone marrow suppression in adults, although neutropenia has been observed in neonates undergoing high-dose therapy (43,44). Acyclovir is expressed in breast milk and crosses the placenta, but association with congenital defects has not been demonstrated. The most serious adverse effects of acyclovir are nephrotoxicity and neurotoxicity. Renal dysfunction is reversible and is typically worse in settings of dehydration. Neurotoxicity is also worse in dehydrated patients, and can manifest as seizures, lethargy, confusion, hallucinations, delirium, and extrapyramidal signs. These too typically resolve after withdrawal of therapy.

Acyclovir-resistant HSV and VZV can be problematic for immunocompromised patients receiving chronic therapy (45), as rates of resistance range from 6% to 12% (46). Rarely, resistant strains of HSV have been observed in normal hosts, including patients with recurrent genital infections undergoing long-term suppressive therapy. Resistance is most commonly conferred by mutations in the viral thymidine kinase gene, and more rarely from mutations in the viral DNA polymerase gene (47).

The poor oral bioavailability of acyclovir led to the development of valacyclovir, which is its L-valine ester prodrug. Valacyclovir is completely converted to acyclovir by first-pass hepatic metabolism, which increases bioavailability to over 50% (48). Its indications are the same as those for acyclovir, although it should not be used for life-threatening conditions where accurate monitoring of levels is necessary. A pediatric formulation has not yet been licensed.

Penciclovir, Famciclovir

Like acyclovir, penciclovir is a guanosine analogue that has activity against HSV-1, HSV-2, and VZV in vitro. It is similarly phosphorylated by viral thymidine kinase and subsequently converted to its active form, penciclovir triphosphate. However, its mechanism of action differs from acyclovir in that, while a competitive inhibitor of DNA polymerase, it does not cause chain termination. Penciclovir's clinical utility is limited to topical treatment for herpes labialis due to its minimal bioavailability. Famciclovir is the diacetyl ester prodrug of penciclovir and confers 70% bioavailability (49). It is indicated for treatment of herpes zoster infections as well as genital herpes, and has similar efficacy to valacyclovir. It is excreted by the kidney and thus requires dose adjustment in patients with renal insufficiency. Famciclovir is tolerated well with minimal side effects with headache and gastrointestinal upset being most common. The resistance profile is similar to that of acyclovir with mutation of the viral thymidine kinase being the most common (47).

Because of improved pharmacokinetics and pharmacodynamics, valacyclovir and famciclovir are the recommended therapies for HSV and VZV infections.

Ganciclovir, Valganciclovir

Ganciclovir is a synthetic analogue of 2′-deoxyguanosine structurally similar to acyclovir. Its main therapeutic use is for the treatment of CMV infections. In CMV-infected cells, viral phosphotransferase phosphorylates the drug initially. Cellular enzymes subsequently phosphorylate the monophosphate derivate to yield the active triphosphate compound. It also has activity against HSV and VZV in vitro by inhibition of viral DNA polymerase and subsequent chain termination.

Ganciclovir is indicated for use in the treatment and prevention of CMV disease, the most significant of which occurs in immunocompromised hosts. The morbidity and mortality associated with CMV retinitis and pneumonitis are significant in patients with HIV and after bone marrow or solid organ transplants; in these settings ganciclovir or its prodrug, valganciclovir, is the mainstay of treatment. Ganciclovir is also used for the treatment of neonates with congenital CMV infections, and has shown to decrease hearing deterioration in these patients (50). Prophylaxis with ganciclovir is used in transplant recipients, and preemptive therapy is used routinely in seropositive posttransplant patients (51–54).

Ganciclovir is available in intravenous, intraocular, and oral forms. The oral form has poor bioavailability, with less than 10% drug absorption (55,56). The drug is excreted by the kidneys, necessitating drug dose adjustment in patients with renal insufficiency.

Myelosuppresion and neurotoxicity are the most significant adverse effects of ganciclovir. Hematologic toxicity occurs in up to a third of recipients and most commonly includes neutropenia, although thrombocytopenia and anemia can also be observed. For this reason, close monitoring of the complete blood count is necessary to detect early bone marrow suppression. Discontinuation of the drug results in resolution of cytopenias. CNS side effects occur in up to 5% of recipients and can include headache, altered mentation, confusion, anxiety, hallucinations, seizures, and tremors (55). Fever, liver function abnormalities, and rash are less likely but have also been observed. The seriousness of the adverse effects of ganciclovir makes either valacyclovir or famciclovir the preferred agent in treatment of HSV and VZV infections, even though both drugs have similar efficacy.

Resistance to ganciclovir is seen most often in patients receiving chronic therapy, and usually results from mutations in the UL97 gene that encodes the viral phosphotransferase. Resistance via mutations in the viral DNA polymerase, encoded by the UL54 gene, is less common.

Valganciclovir is the L-valine ester prodrug of ganciclovir that is rapidly metabolized after oral administration. Its oral bioavailability is improved to 60%, and is further increased by administration with food (57). It has similar indications and resistance mechanisms as ganciclovir and offers an effective alternative to intravenous ganciclovir. As with ganciclovir, patients with renal insufficiency should have doses adjusted accordingly. Neutropenia, anemia, and headache are seen in some recipients, but nausea and diarrhea are more common adverse effects (58).

Cidofovir

Cidofovir is a synthetic acyclic cytosine nucleotide that has activity against a broad variety of DNA and RNA viruses, but whose main indication is the treatment of CMV retinitis in patients with AIDS. Like the nucleoside analogues, cidofovir competitively inhibits viral DNA polymerase. Initial phosphorylation by a viral enzyme is not necessary as the compound already has a monophosphate group. Host cellular enzymes subsequently phosphorylate the drug, resulting in the active form. Specificity to viral-infected cells results from cidofovir's 25- to 50-fold higher affinity for viral DNA polymerase over cellular DNA polymerase.

In addition to its approved use of treatment of AIDS-associated CMV retinitis, cidofovir has a broad spectrum of activity, and can be used against acyclovir- and ganciclovir-resistant strains of HSV and CMV, respectively. Its potential therapeutic role has been tested in other clinical situations as well. A topical form was used for refractory HSV lesions in an AIDS patient (59). Case reports also suggest that cidofovir may be beneficial in the treatment of BK virus nephritis in renal transplant patients (60,61). Cidofovir was used to treat progressive multifocal leukoencephalopathy caused by JC virus in patients with AIDS, but failed to prove efficacious (62,63). Activity has also been reported for orthopoxviruses, adenovirus, hepatitis B virus, human papilomavirus, and EBV.

Because of its poor oral bioavailability, cidofovir is used primarily in its intravenous formulation. Weekly maintenance dosing is possible due to its long half-life. Ninety percent of the drug is excreted by the kidneys, thus necessitating dose adjustment in patients with renal insufficiency (64).

Despite its broad activity, the clinical utility of cidofovir is limited by its potential for severe renal toxicity. Aggressive intravenous hydration, co-administration of probenecid, and avoidance of other nephrotoxic drugs minimizes the risk, yet nephrotoxicity still causes the discontinuation of cidofovir in 25% of patients. Other side effects include neutropenia, fever, myalgias, nausea, and hair loss, but it is its renal toxicity that renders the drug a therapy of last resort.

Although cidofovir-resistant strains of CMV due to mutations in the viral DNA polymerase gene have been isolated (65,66), resistance is typically not a clinical concern.

Foscarnet

Foscarnet is a pyrophosphate analogue and the only antiherpes drug that is not a nucleoside or nucleotide analogue. It has activity against all of the herpes viruses, and directly inhibits the viral DNA polymerase by blocking the pyrophosphate-binding site (67) and terminating chain elongation. Foscarnet does not require phosphorylation by viral or cellular kinases.

Foscarnet is indicated for the treatment of CMV retinitis and is also effective in the treatment of resistant HSV, VZV, and CMV infections (68–70), an important problem in immunocompromised hosts. Foscarnet also has activity against influenza A and retroviruses including HIV, although it is not used clinically due to its toxicity. Because of the severity of its associated adverse effects, foscarnet is typically reserved as a potent therapeutic option in select situations.

Foscarnet is available only as an intravenous formulation due to its poor oral bioavailability (20%). It is cleared renally, and thus requires dose adjustment in patients with renal insufficiency.

Nephrotoxicity and electrolyte disturbances are the major side effects associated with foscarnet. Serum creatinine elevations of up to threefold are observed in about half of the recipients. Risk factors for renal dysfunction include preexisting renal disease and concurrent use of other nephrotoxic drugs. In addition, factors such as hydration status and manner of infusion also affect nephrotoxicity (71). Renal toxic effects are typically reversible within two to four weeks of discontinuing therapy. Foscarnet is a chelating agent that can cause significant electrolyte abnormalities, including hypo- and hypercalcemia and hypo- and hyperphosphatemia (72). Hypocalcemia is seen in up to a third of patients, and can result in seizures, tetany, and arrhythmias. Hypomagnesemia and hypokalemia can also occur in some patients. CNS side effects include headache, seizures, hallucinations, tremors, and neuropathies. Other adverse effects include fever, nausea, vomiting, hepatic dysfunction, and cytopenias.

Resistance to foscarnet occurs through mutations in the viral DNA polymerase, and has been observed in some strains of CMV, HSV, and VZV (73,74).

Trifluridine

Trifluridine is an ophthalmic agent used specifically for HSV keratitis. Trifluridine and its predecessor, idoxuridine, are thymidine analogs that inhibit viral DNA polymerase. Like many of the other antiherpetic drugs, trifluridine's action depends on phosphorylation of the compound by viral and cellular thymidine kinases. But because the drug undergoes phosphorylation by cellular kinases, there is significant toxicity with systemic administration.

Trifluridine is therefore limited to topical ophthalmic use for herpes keratitis. Patients should be monitored for side effects such as occlusion of the puncta and keratinization of lid margins.

THERAPY FOR HEPATITIS VIRUSES

Infections with hepatitis B and C viruses cause both acute and chronic liver disease, with serious morbidity and mortality worldwide. Chronic infections often lead to cirrhosis and hepatocellular carcinoma, and are a leading cause for liver transplantations in the United States. Hepatitis B virus (HBV) has a circular double-stranded DNA genome enclosed in an icosahedral envelope. Viral replication occurs through a reverse transcriptase. Hepatitis C virus (HCV), a member of the Flaviviridae family, is an enveloped virus with a positive-sense RNA genome. Current therapeutic regimens designed to clear infection and prevent late sequelae involve the use of nucleoside and nucleotide analogues, as well as immune modulators.

Lamivudine

Lamivudine is a nucleoside analogue that was initially designed as a reverse transcriptase inhibitor for the treatment of HIV infection, but was later shown to also inhibit HBV reverse transcriptase. It is phosphorylated by cellular kinases to its active form, lamivudine triphosphate, which is then incorporated to the growing DNA chain and subsequently terminates elongation. This drug is indicated for treatment of HIV as well as chronic HBV infection, although at different doses. Its role in the treatment of HIV will be discussed later in the chapter.

Lamivudine can be used alone or in combination with other medications such as interferon-α for the treatment of chronic hepatitis B infection. In patients with chronic HBV infection undergoing one year of treatment, lamivudine monotherapy improves clinical outcome, with normalization of ALT (72%), HBeAg seroconversion (16%), and improved histological inflammatory score (56%) (75).

Treatment with lamivudine requires long-term oral administration, and is generally well tolerated. Adverse reactions include headache, fatigue, nausea, vomiting, diarrhea, peripheral neuropathy, and hair loss. More serious but rarer side effects include pancreatitis and lactic acidosis.

The main limitation of lamivudine is the development of relapse following treatment with reappearance of HBV DNA in serum after its initial clearance. This is due to the rapid development of drug resistance that arises from mutations in the catalytic domain of HBV reverse transcriptase. Lamivudine-resistant strains of HBV have been observed in one-third of patients by the end of one year of treatment, and in up to two-thirds after four years of therapy (76).

Adefovir

Adefovir is a nucleotide analogue of adenosine monophosphate that is administered orally as its prodrug, adefovir dipivoxil. It was initially designed for HIV therapy, and although it has been shown to inhibit HIV in vitro, an efficacious dose with a margin of safety could not be achieved in human studies. The prodrug is metabolized to adefovir, which is phosphorylated by cellular kinases to adefovir diphosphate, which competitively inhibits HBV reverse transcriptase and terminates DNA synthesis upon incorporation into the growing chain.

Adefovir is indicated for the treatment of chronic hepatitis B in adults with evidence of active disease, which include elevations of serum HBV DNA, serum aminotransferases, or worsening histologic findings. In HBeAg-positive patients, 48 weeks of treatment with adefovir monotherapy normalized ALT in 48%, caused HBeAg seroconversion in 12%, and resulted in histologic improvement in 53% (77).

Adefovir has approximately 60% oral bioavailability and is excreted by the kidneys and therefore requires dose adjustment in patients with impaired renal function.

Adefovir is generally well tolerated, with headache, pharyngitis, abdominal pain, and peripheral neuropathy being the most commonly reported side effects. Nephrotoxicity has also been observed in some patients, with those receiving higher doses and longer courses of therapy at greater risk (78). Exacerbation of hepatitis has been reported in patients immediately following discontinuation of adefovir. Most of these exacerbations occur within 12 weeks of stopping therapy, and elevations of ALT up to 10 times the upper limit of normal can be observed in over a quarter of patients (78).

Adefovir has a lower propensity to induce drug resistance than lamivudine, making it a preferable therapeutic choice. Clinical trials of patients receiving 48 weeks of therapy did not identify any cases of resistance (79). Longer courses of treatment yield resistant strains of HBV with mutations in the DNA polymerase gene; other rare variants of resistant strains have been identified (80,81). Regardless, adefovir's utility is underscored by the fact that lamivudine-resistant strains of HBV have been shown to retain susceptibility to adefovir.

Tenofovir

Tenofovir is a nucleotide analog structurally similar to adefovir, and has been licensed for use in the treatment of HIV infection as a reverse transcriptase inhibitor. Recently, tenofovir was approved for the treatment of chronic hepatitis B infections based on data from ongoing clinical trials demonstrating its efficacy. Previous studies had suggested a role for tenofovir in patients with lamivudine-resistant strains of virus, including one study which demonstrated decline in HBV DNA levels below 10^5 copies/mL at 48 weeks of tenofovir therapy in 100% of patients versus 44% of patients on adefovir therapy (82). Case reports of patients with primary resistance to adefovir responding to tenofovir have also been documented (80). Tenofovir also has great potential in the treatment of those with chronic hepatitis B and HIV infections. One prospective randomized placebo-controlled trial with 52 coinfected patients demonstrated tenofovir's activity against HBV to be noninferior to that of adefovir at 48 weeks (83). The most current NIH AIDS treatment guidelines include tenofovir as an agent to be used as part of an antiretroviral regimen in patients coinfected with HBV (84). Although tenofovir is generally well tolerated, the most common side effects noted in clinical trials in patients with chronic HBV included nausea and gastrointestinal upset, headache, dizziness, fatigue, and rash (85).

Entecavir

Entecavir is an analogue of guanosine and a more recent addition to the nucleotide analogues targeting HBV DNA polymerase. A study comparing entecavir monotherapy versus lamivudine in HBeAg-positive patients with chronic hepatitis B demonstrated better outcomes in those receiving entecavir, with normalization of ALT in 68%, HBeAg seroversion in 21%, and histologic improvement in 72% (86). Severity of adverse reactions was comparable to that of lamivudine, with headache, fatigue, upper respiratory infections, and abdominal pain being most common. Lactic acidosis and hepatic steatosis were rarely observed. Furthermore, emergence of resistant strains was not demonstrated over the 48-week course of therapy (86). A subset of subjects in this study was followed up to 96 weeks, and a greater proportion of those in the entecavir group demonstrated undetectable HBV DNA levels and normalization of ALT (87). Improved efficacy over lamivudine was also demonstrated in patients with chronic hepatitis B that were HBeAg-negative (88), and those with lamivudine-resistant strains (89).

Telbivudine

Telbivudine is a synthetic thymidine nucleoside analogue approved for the treatment of chronic hepatitis B infection. Its active phosphorylated form competitively inhibits HBV DNA polymerase. Initial clinical studies have demonstrated that a greater proportion of HBeAg-positive subjects receiving telbivudine had suppression of HBV DNA than those receiving lamivudine (90,91). In one phase III trial, a significantly higher proportion of patients receiving telbivudine (75%) reached the primary end point of suppression of HBV DNA and loss of serum HBeAg than those receiving lamivudine (67%) over 52 weeks (90). Noninferiority of telbivudine was also demonstrated for HBeAg-negative patients (90).

Telbivudine is taken orally and cleared renally, necessitating dose adjustment in patients with renal insufficiency. Adverse effects are similar to those of lamivudine and include upper respiratory tract infection, headache, fatigue, and gastrointestinal upset (92). Myopathy is a rare side effect but has been observed in some patients several weeks into the course with associated rise in serum creatine kinase levels (92). Acute exacerbations of hepatitis have also been observed upon discontinuation of therapy (92). The rate of resistance was shown to be less in those receiving telbivudine versus those receiving lamivudine (90).

Interferons

Interferons are a group of naturally occurring cytokine proteins that help mediate a variety of physiologic functions including immunomodulation and antiproliferation. There are three classes of interferons based on whether they are produced by leukocytes (α), fibroblasts (β), or lymphocytes (γ). Recombinant technology allows their production for therapeutic purposes. Interferons effect their various actions by binding specific cell receptors and inducing cell responses.

The antiviral properties of interferons are based on their ability to act on infected cells as well as modulate innate host immune responses. Interferon α and β are secreted in response to viral infection, while interferon γ is secreted by activated lymphocytes. These in turn act on infected cells to inhibit multiple steps in the viral life cycle including viral penetration and uncoating, mRNA synthesis and protein translation, and viral assembly and release (93). Interferons can also act on host cells directly involved in the immune response to infection, including increasing activity of NK cells and expression of MHC class I on infected cells.

Interferon-α (IFN-α) is indicated for the treatment of HBV and HCV hepatitis, as well as for lesions caused by human papillomaviruses. Treatment with IFN-α alone has shown to be effective therapy for chronic hepatitis caused by HBV and HCV (94). Improvement in serum aminotransferase levels and histologic abnormalities in up to 40% of patients receiving IFN-α was observed, with up to 20% of patients clearing hepatitis B surface antigen (95). Less than half of subjects with chronic hepatitis C showed similar biochemical and histologic improvements, usually accompanied by the loss of detectable serum viral RNA, and relapse occurred in about half (96,97).

Combining IFN-α with polyethylene glycol (PEG) improves its efficacy over IFN-α alone. Pegylation has the effect of increasing the half-life of IFN by slowing absorption, decreasing renal clearance, and reducing immunogenicity. The end result is a drug that has demonstrated greater efficacy in patients with chronic hepatitis C than IFN-α both as monotherapy (98,99) and as combined therapy with ribavirin (100–102). Pegylated interferon shows only minimal improvement in efficacy in patients with chronic hepatitis B infections, but has the advantage of requiring weekly dosing as compared to three times a week for standard interferon (103). In patients with HBeAg-positive chronic hepatitis B infection, treatment with pegylated interferon alone caused HBeAg seroconversion in 32% of recipients, normalization of ALT in 41%, and suppression of HBV DNA in 32%, making it more efficacious than lamivudine monotherapy (104).

Finally, as alluded to above, combining oral ribavirin with interferon-α improves outcomes over treatment with interferon-α alone in patients with chronic hepatitis C infections. Forty-one percent of patients treated with combination therapy for 48 weeks had cleared viral disease, in contrast to 16% of patients treated with interferon alone (105,106). The best response rates are achieved for patients with nonserotype-1 infections. Furthermore, combination therapy has been shown to be effective in the treatment of some patients who relapsed after a previous course of interferon alone (107). The development of pegylated IFN-α has led to the current standard of care regimen of pegylated IFN-α and ribavirin. Over half of patients successfully completing a 48-week regimen have attained sustained virologic clearance (100–102).

Interferon-α is also indicated for the treatment of lesions caused by human papillomaviruses (HPV) such as condylomata acuminata. Therapy can be given by direct injection into the lesions or systemically for more extensive disease. Clearance of injected warts has been observed in up to 60% of patients (108,109). It is also worth noting here that another licensed therapy for HPV lesions that modulates the immune response is imiquimod, which is a topical agent that has toll-like receptor activity.

Interferon-α is given intramuscularly or subcutaneously with over 80% absorption (110). As discussed above, pegylation increases the half-life of the drug and allows for higher steady-state concentrations. Interferon is found in only small amounts in the body tissues and fluid, with only minimal excretion in the urine (110).

The side effects of interferon therapy render it poorly tolerated in many patients. Flu-like symptoms of fever, chills, headache, myalgias, malaise, and gastrointestinal upset are commonly seen, especially early in treatment. Up to half of patients also experience significant increases in serum ALT levels, presumably secondary to lysis of infected hepatocytes. At higher doses, neurotoxicity can occur including behavioral disturbances, depression, somnolence, confusion, and occasionally seizures and coma. Neutropenia and thrombocytopenia can also occur (111). These side effects necessitate close monitoring of the patient, especially during prolonged courses.

Viral resistance to interferon has not been observed, which is one advantage that interferon has over the nucleoside analogues. Other advantages include the fact that interferon therapy has clearly defined durations. For example, in patients with active hepatitis B disease, the current recommended duration for pegylated interferon-α is 48 weeks, while treatment with nucleoside or nucleotide analogues may take up to years. Finally, clinical relapse occurs more commonly following courses of therapy with nucleoside or nucleotide analogues as compared with interferon therapy.

THERAPY FOR HUMAN IMMUNODEFICIENCY VIRUS

Antiviral agents for human immunodeficiency virus (HIV) have revolutionized the care of patients with HIV and its associated syndrome, AIDS, and therapy is an ever-changing and rapidly developing field. Current therapeutic strategy targets minimizing viral load in infected patients, and the successes over the past 30 years have led to drastic improvements in quality of life and reduced spread of disease where treatment is available.

HIV-1 and the less common type HIV-2 are enveloped viruses with single-stranded RNA genomes that are dependent on the action of a reverse transcriptase for replication. Antiviral agents can be categorized by the step of the viral life cycle they target. Fusion inhibitors are designed to prevent the virion's attachment and entry into prospective host cells. Reverse transcriptase inhibitors target viral replication, integrase inhibitors target the integration of proviral DNA into the host DNA, and protease inhibitors impede the conversion of the new virion to its infectious form prior to release.

One of the greatest challenges in treating HIV is in counteracting the virus' ability to mutate and develop drug-resistant strains. The rapid rate at which resistance arises under monotherapy mandates multidrug therapy and strict adherence to treatment regimens. This in turn necessitates that drug regimens are well tolerated and easy to follow in order to ensure patient compliance. Without a cure or vaccine on the horizon, new drugs are constantly being introduced and tested to provide a more effective therapy. The rapidly changing nature of this field precludes comprehensive discussion of all the agents available for use today and is beyond the scope of this chapter. Examples of drugs in each category of therapy will be highlighted.

Nucleoside/Nucleotide Reverse Transcriptase Inhibitors: Zidovudine, Didanosine, Stavudine, Lamivudine, Abacavir, Tenofovir, Emtrictabine

Nuclesoside/nucleotide reverse transcriptase inhibitors (NRTIs) were the first class of drugs used in HIV patients and are the cornerstone of highly active antiretroviral therapy (HAART) regimens. NRTIs are 3′-modified deoxynucleosides that require phosphorylation by host cell enzymes. Once activated, NRTIs inhibit reverse transcriptase activity by competing with dGTP for incorporation into the growing viral DNA chain, which subsequently terminates DNA chain elongation.

In general, NRTIs are well absorbed orally with over 80% bioavailability for some agents (84). Except for abacavir, the NRTIs are excreted renally with minimal hepatic metabolism and require dose adjustment in cases of renal insufficiency (84). Side effects often include gastrointestinal upset such as nausea and vomiting, and rarely but significantly, lactic acidosis and hepatic steatosis. Pancreatitis, peripheral neuropathy, myopathy, and lipodystrophy also complicate the administration of some agents. Zidovudine, the first compound licensed for AIDS therapy, is a thymidine analogue that can cause anemia and neutropenia as an adverse effect (84). Abacavir, one of the newest NRTIs, has been associated with a potentially fatal hypersensitivity reaction in 5–10% of recipients, with life-threatening hypotension making close monitoring during administration necessary (84).

As mentioned previously, the rapid development of resistance in HIV renders monotherapy with NRTIs virtually useless. Combination with at least one other NRTI in addition to another class of drug is required for successful therapy. To this end, several combination drugs have appeared on the market including Combivir (zidovudine/lamivudine), and Truvada (tenofovir/emtricitabine), which have made HAART therapy more convenient for patients.

Nonnucleoside Reverse Transcriptase Inhibitors: Nevirapine, Delvirdine, Efavirenz, Etravirine

Nonnucleoside reverse transcriptase inhibitors (NNRTIs) are an important addition to the NRTIs in inhibiting HIV reverse transcriptase while minimizing the induction of resistance. NNRTIs do not require intracellular phosphorylation, and terminate DNA chain elongation by directly binding reverse transcriptase.

NNRTIs as a class are also generally well absorbed orally with up to greater than 85% bioavailability for delavirdine and nevirapine (84). Unlike NRTIs, the NNRTIs are primarily metabolized in the liver by the CYP enzyme system. This is the basis of many drug interactions that can occur and is of particular importance in patients with comorbidities. Other side effects include rash, which can progress to Stevens–Johnson syndrome in patients taking nevirapine (84). Nevirapine also has associated hepatic toxicity (84). Efavirenz results in birth defects in animal models and therefore should be avoided in pregnant women during the first trimester and in women with child-bearing potential (84). CNS side effects such as dizziness, poor concentration, anxiety, hallucinations, and insomnia are also observed in some patients who take efavirenz.

NNRTIs should not be used as monotherapy due to rapid development of resistance. Furthermore, NNRTIs should not be used together as drug resistance can be conferred to other members of the same class. Despite side effects, NNRTIs are useful in combination therapy with NRTIs and have the added convenience of less-frequent dosing due to relatively long half-lives (84).

Protease Inhibitors: Saquinivir, Ritonavir, Indinavir, Nelfinavir, Lopinavir/Ritonavir, Atazanivir, Fosamprenavir, Tipranavir, Darunavir

Protease inhibitors (PIs) offer another class of potent agents to use in combination with NRTIs. Protease is a viral enzyme that cleaves viral protein precursors into their smaller, functional end products. PIs inhibit this cleavage step, preventing effective protein function and rendering the virion immature and noninfectious (112,113). Several drugs have been developed in this class over the past 20 years and offer a variety of therapeutic choices. Most are peptide-like compounds that bind the viral protease.

PIs are administered orally and have severe gastrointestinal side effects that include nausea, vomiting, diarrhea, and abdominal pain. Administration with food is often recommended, which can affect drug levels. To achieve sufficient levels, many are boosted with ritonavir, an effect that was discovered serendipitously but is now its main indicated use. The combination of lopinavir/ritonavir, or Kaletra, is one example. Hepatic toxicity and other metabolic complications such as dyslipidemia, lipodystrophy, and insulin resistance have been observed in some agents (84).

PIs are metabolized by the CYP hepatic enzyme system, which results in the other major limitation of significant drug–drug interactions (84). As always, combination therapy with agents in other classes is recommended to avoid development of resistance.

Atazanavir is among the more popular and well-tolerated agents, but still highlights some of the considerations when using a PI. It is taken orally and absorbed best with food and in an acidic environment, which limits the use of acid-reducing agents like proton pump inhibitors. Dosing must be adjusted in patients with hepatic impairment (84). It has less metabolic disturbances than other PIs, but can cause an asymptomatic hyperbilirubinemia. A few side effects unique to this drug are associations with prolonged PR interval and slow cardiac conduction, as well as with nephrolithiasis (84).

New Classes: Enfuviritide, Maraviroc, Raltegravir

These drugs represent some of the newest agents available for the treatment of HIV infection, and at this time are generally reserved for multidrug resistant strains. Each drug's mechanism of action targets stages of the viral life cycle different than that of previous classes. Enfuviritide is a fusion inhibitor that interferes with the entry of HIV into cells by preventing fusion of the viral membrane with the cell membrane. It has shown to be effective in reducing HIV RNA levels in patients with multiresistant strains (114,115). Its convenience of use is limited by the fact that is available only in injectable form. Maraviroc, a novel agent, also blocks viral entry by specifically binding the CCR5 receptor of CD4 T-cells, which is a critical co-receptor for cellular entry of select viral strains. It is considered a CCR5 antagonist, only has activity against HIV strains that are tropic for the CCR5 receptor (R5 strains), and is therefore not indicated against other strains. Raltegravir represents the newest addition to the HIV armamentarium and is an integrase inhibitor. This compound inhibits viral integrase from inserting HIV DNA into the host genome, thus preventing viral replication. Based on data from separate phase III clinical studies that documented improved virologic response to maraviroc and raltegravir over placebo in patients who failed a prior optimized antiretroviral therapy regimen, current guidelines recommend these two novel drugs as options for patients with treatment failure due to resistance (84).

CONCLUSION

The past three decades have witnessed a robust pipeline of drugs that have been licensed for the treatment of human viral infections. The remarkable human benefit from these advances is translated into saved lives and decreased morbidity. Advances have been stunning in the management of HIV/AIDS as witnessed by prolonged survival and quality of life. Treatment of HSV infections of the CNS has similarly improved the quality of life for afflicted patients.

In spite of the recognized advances, there are still unmet and under-addressed medical needs. Even with successful therapy of HSV infections of the brain, mortality and morbidity remain too high, indicating an obvious need for improved therapies or combination therapies of drugs with different mechanisms of action. Similarly in the management of chronic hepatitis C infection, the obvious need for small molecules that can be used in combination, and that are active against serotype 1, would avoid the use of pegylated interferon and ribavirin with their associated toxicities. Equally importantly, with broader use of antiviral drugs, resistance to licensed medications becomes an increasing problem. Resistance has been well established for the drugs used to treat HIV/AIDS and has resulted in an ever-expanding arsenal of medications with different mechanisms of action. More recently, the resistance of H1N1 influenza to oseltamivir has become a significant problem in South Africa and Scandinavian countries. In the end, the requisite need to replenish the pipeline of new medications is essential.

REFERENCES

1. Darby G. A History of Antiherpes Research. Antivir Chem Chemother 1994; 5(suppl 1):54–63.
2. Field H, Whitley RJ. Antiviral chemotherapy. In: Mahy BWJ, Meulen Vt, eds. Topley Wilson's Microbiology and Microbial Infections. London, UK: Hodder Arnold, 2005:1605–1645.
3. Centers for Disease Control and Prevention. Prevention and Control of Influenza: Recommendations of the Advisory Committee on Immunization Practices (ACIP), 2008. MMWR Recomm Rep 2008; 57:1–60.
4. Bui M. Whittaker G, Helenius A. Effect of M1 protein and low pH on nuclear transport of influenza virus ribonucleoproteins. J Virol 1996; 70(12):8391–8401.
5. Hall CB, Dolin R, Gala CL, et al. Children with influenza A infection: treatment with rimantadine. Pediatrics 1987; 80(2):275–282.
6. Van Voris LP, Betts RF, Hayden FG, et al. Successful treatment of naturally occurring influenza A/USSR/77 H1N1. JAMA 1981; 245(11):1128–1131.
7. Anonymous, Immunization Practices Advisory Committee. Prevention and control of influenza. MMWR Morb Mortal Wkly Rep 1992; 41:1–14.
8. Dolin R, Reichman RC, Madore HP, et al. A controlled trial of amantadine and rimantadine in the prophylaxis of influenza A infection. N Engl J Med 1982; 307(10):580–584.
9. Hayden FG, Couch RB. Clinical and epidemiologic importance of influenza A viruses resistant to amantadine and rimantadine. Rev Med Virol 1992; 2:89–96.
10. Hayden FG, Belshe RB, Lover RDC, et al. Emergence and apparent transmission of rimantadine-resistant influenza A virus in families. N Engl J Med 1989; 321(25):1696–1702.
11. Bright RA, Shay DK, Shu B, et al. Adamantane resistance among influenza A viruses isolated early during the 2005–2006 influenza season in the United States. JAMA 2006; 295(8):891–894.
12. Hayden FG, Gubareva LV, Monto AS, et al. Inhaled zanamivir for the prevention of influenza in families. N Engl J Med 2000; 343:1282–1289.
13. Monto AS, Robinson DP, Herlocher ML, et al. Zanamivir in the prevention of influenza among healthy adults: a randomized controlled trial. JAMA 1999; 282:31–35.

14. Hayden FG, Atmar RL, Schilling M, et al. Use of the selective oral neuraminidase inhibitor oseltamivir to prevent influenza. N Engl J Med 1999; 341:1336–1343.
15. Treanor JJ, Hayden FG, Vrooman PS, et al; US Oral Neuraminidase Study Group. Efficacy and safety of the oral neuraminidase inhibitor oseltamivir in treating acute influenza: A randomized controlled trial. JAMA 2000; 283:1016–1024.
16. Whitley RJ, Hayden FG, Reisinger KS, et al. Oral oseltamivir treatment of influenza in children. Pediatr Infect Dis J 2001; 20:127–133.
17. Nicholson KG, Aoki FY, Osterhause A, et al. Treatment of acute influenza: efficacy and safety of the oral neuraminidase inhibitor oseltamivir. Lancet 2000; 355:1845–1850.
18. Welliver R, Monto AS, Carewicz O, et al. Effectiveness of oseltamivir in preventing influenza in household contacts: a randomized controlled trial. JAMA 2001; 285:748–754.
19. Hayden FG, Osterhaus AD, Treanor JJ, et al. Efficacy and safety of the neuraminidase inhibitor zanamivir in the treatment of influenza virus. N Engl J Med 1997; 337:874–880.
20. Nicholson KG, Aoki FY, Osterhaus AD, et al. Efficacy and safety of oseltamivir in treatment of acute influenza: A randomised controlled trial. Neuraminidase Inhibitor Flu Treatment Investigator Group. Lancet 2000; 355(9218):1845–1850.
21. Wood ND, Aitken M, Sharp S, et al. Tolerability and pharmacokinetics of the influenza neuraminidase inhibitor Ro-64–0802 (GS4071) following oral administration of the prodrug Ro-64–0796 (GS4104) to healthy male volunteers. In: 37th Interscience Conference on Antimicrobial Agents and Chemotherapy (ICAAC), Toronto, 1997.
22. Cass LM, Brown J, Pickford M, et al. Pharmacoscintigraphic evaluation of lung deposition of inhaled zanamivir in healthy volunteers. Clin Pharmacokinet 1999; 1:21–31.
23. Cheer SM, Wagstaff AJ. Zanamivir: An update of its use in influenza. Drugs 2002; 62:71–106.
24. Cass LM, Gunawardena KA, Macmahon MM, et al. Pulmonary function and airway responsiveness in mild to moderate asthmatics given repeated inhaled doses of zanamivir. Respir Med 2000; 94(2): 166–173.
25. Hayden FG, Treanor JJ, Fritz RS, et al. Use of the oral neuraminidase inhibitor oseltamivir in experimental human influenza: Randomized controlled trials for prevention and treatment. JAMA 1999; 282(13):1240–1246.
26. U.S. Food and Drug Administration. 2008 Safety Alerts for Human Medical Products (Drugs, Biologics, Medical Devices, Special Nutritionals, and Cosmetics). Tamiflu (oseltamivir phosphate). Silver Spring, MD: U.S.F.D.A. 2008. www.fda.gov/medwatch/safety/2008/safety08.htm#Tamiflu. Accessed September 25, 2008.
27. Sheu TG, Deyde VM, Okomo-Adhiambo M, et al. Surveillance for neuraminidase inhibitor resistance among human influenza A and B viruses circulating worldwide from 2004 to 2008. Antimicrob Agents Chemother 2008; 52(9):3284–3292.
28. Centers for Disease Control and Prevention. Additional Considerations to the Interim Recommendations for the Use of Influenza Antiviral Medications in the Setting of Oseltamivir Resistance among Circulating Influenza A (H1N1) Viruses, 2008–09 Influenza Season. Atlanta, GA: CDC. http://www.cdc.gov/flu/professionals/antivirals/additional_considerations.htm. Accessed February 17, 2009.
29. Sidwell RW, Robins RK, Hillyard IW. Ribavirin: An antiviral agent. Pharmacol Ther 1979; 6(1): 123–146.
30. Hall CB, McBride JT, Walsh EE, et al. Aerosolized ribavirin treatment of infants with respiratory syncytial viral infection. A randomized double-blind study. N Engl J Med 1983; 308(24):1443–1447.
31. Groothuis JR, Woodin KA, Katz R, et al. Early ribavirin treatment of respiratory syncytial viral infection in high-risk children. J Pediatr 1990; 117(5):792–798.
32. Conrad DA, Christenson JC, Waner JL, et al. Aerosolized ribavirin treatment of respiratory syncytial virus infection in infants hospitalized during an epidemic. Pediatr Infect Dis J 1987; 6(2):152–158.
33. Laskin OL, Longstreth JA, Hart CC, et al. Ribavirin disposition in high-risk patients for acquired immunodeficiency syndrome. Clin Pharmacol Therap 1987; 41(5):546–555.
34. Kimberlin DW, Prober CG. Antiviral agents. In: Long SS, Pickering LK, Prober CG, eds. Principles and Practice of Pediatric Infectious Diseases. Philadelphia, PA: Churchill Livingstone, 2003:1527–1547.
35. Whatley JD, Thin RN. Episodic acyclovir therapy to abort recurrent attacks of genital herpes simplex infection. J Antimicrob Chemother 1991; 27(5):677–681.
36. Reichman RC, Badger GJ, Mertz GJ, et al. Treatment of recurrent genital herpes simplex infections with oral acyclovir. A controlled trial. JAMA 1984; 251(16):2103–2107.
37. Wald A, Carrell D, Remington M, et al. Two-day regimen of acyclovir for treatment of recurrent genital herpes simplex virus type 2 infection. Clin Infect Dis 2002; 34(7):944–948.
38. Goldberg LH, Kaufman R, Kurtz TO, et al; Acyclovir Study Group. Long-term suppression of recurrent genital herpes with acyclovir. A 5-year benchmark. Arch Dermatol 1993; 129(5):582–587.

39. Gold D, Corey L. Acyclovir prophylaxis for herpes simplex virus infection. Antimicrob Agents Chemother 1987; 31(3):361–367.
40. Mindel A, Faherty A, Carney O, et al. Dosage and safety of long-term suppressive acyclovir therapy for recurrent genital herpes. Lancet 1988; 1(8591):926–928.
41. Rooney JF, Straus SE, Mannix ML, et al. Oral acyclovir to suppress frequently recurrent herpes labialis. A double-blind, placebo-controlled trial. Ann Intern Med 1993; 118(4):268–272.
42. Wagstaff AJ, Faulds D, Goa KL. Aciclovir. A reappraisal of its antiviral activity, pharmacokinetic properties and therapeutic efficacy. Drugs 1994; 47(1):153–205.
43. Kimberlin DW, Lin CY, Jacobs RF, et al. Safety and efficacy of high-dose intravenous acyclovir in the management of neonatal herpes simplex virus infections. Pediatrics 2001; 108(2):230–238.
44. Kimberlin D, Powell D, Gruber W, et al. Administration of oral acyclovir suppressive therapy after neonatal herpes simplex virus disease limited to the skin, eyes and mouth: results of a phase I/II trial. Pediatr Infect Dis J 1996; 15(3):247–254.
45. Englund JA, Zimmerman ME, Swierkosz EM, et al. Herpes simplex virus resistant to acyclovir. A study in a tertiary care center. Ann Intern Med 1990; 112(6):416–422.
46. Whitley RJ, Kimberlin D. Antiviral Agents. In: Burg FD, Ingelfinger JR, Polin RA, et al., eds. Gellis and Kagan's Current Pediatric Therapy, Vol. 17. Philadelphia, PA: W. B. Saunders Co, 2002:244–256.
47. Kimberlin DW, Coen DM, Biron KK, et al. Molecular mechanisms of antiviral resistance. Antiviral Res 1995; 26(4):369–401.
48. Soul-Lawton J, Seaber E, On N, et al. Absolute bioavailability and metabolic disposition of valaciclovir, the L-valyl ester of acyclovir, following oral administration to humans. Antimicrob Agents Chemother 1995; 39(12):2759–2764.
49. Boike SC, Pue MA, Freed MI, et al. Pharmacokinetics of famciclovir in subjects with varying degrees of renal impairment. Clin Pharmacol Therap 1994; 55(4):418–426.
50. Kimberlin DW, Lin C, Sanchez P, et al. Ganciclovir (GCV) treatment of symptomatic congenital cytomegalovirus (CMV) infections: Results of a phase III randomized trial. In: 40th Interscience Conference on Antimicrobial Agents and Chemotherapy (ICAAC), Toronto, Ontario, Canada, 2000.
51. Rayes N, Oettle H, Schmidt CA, et al. Preemptive therapy in CMV-antigen positive patients after liver transplantation–a prospective trial. Ann Transplant 1999; 4(2):12–17.
52. Kelly J, Hurley D, Raghu G. Comparison of the efficacy and cost effectiveness of pre-emptive therapy as directed by CMV antigenemia and prophylaxis with ganciclovir in lung transplant recipients. J Heart Lung Transplant 2000; 19(4):355–359.
53. Mori T, Okamoto S, Matsuoka S, et al. Risk-adapted pre-emptive therapy for cytomegalovirus disease in patients undergoing allogeneic bone marrow transplantation. Bone Marrow Transplant 2000; 25(7):765–769.
54. Reddy V, Hao Y, Lipton J. et al. Management of allogeneic bone marrow transplant recipients at risk for cytomegalovirus disease using a surveillance bronchoscopy and prolonged pre-emptive ganciclovir therapy. J Clin Virol 1999; 13(3):149–159.
55. Markham A, Faulds D. Ganciclovir. An update of its therapeutic use in cytomegalovirus infection. Drugs 1994; 48(3):455–484.
56. Frenkel LM, Capparelli, EV. Dankner WM,et al. Oral ganciclovir in children: Pharmacokinetics, safety, tolerance, and antiviral effects. J Infect Dis 2000; 182(6):1616–1624.
57. Brown F, Banken L, Saywell K, et al. Pharmacokinetics of valganciclovir and ganciclovir following multiple oral dosages of valganciclovir in HIV- and CMV-seropositive volunteers. Clin Pharmacokinet 1999; 37(2):167–176.
58. Cocohoba JM, McNicholl IR. Valganciclovir: An advance in cytomegalovirus therapeutics. Ann Pharmacother 2002; 36(6):1075–1079.
59. Kopp T, Geusau A, Rieger A, et al. Successful treatment of an aciclovir-resistant herpes simplex type 2 infection with cidofovir in an AIDS patient. Br J Dermatol 2002; 147(1):134–138.
60. Kadambi PV, Josephson MA, Williams J, et al. Treatment of refractory BK virus-associated nephropathy with cidofovir. Am J Transplant 2003; 3(2):186–191.
61. Vats A, Shapiro R, Singh Randhawa P, et al. Quantitative viral load monitoring and cidofovir therapy for the management of BK virus-associated nephropathy in children and adults. Transplantation 2003; 75(1):105–112.
62. Marra CM, Rajicic N, Barker DE, et al. A pilot study of cidofovir for progressive multifocal leukoencephalopathy in AIDS. AIDS 2002; 16(13):1791–1797.
63. Center for Disease Control and Prevention. Treating opportunistic infections among HIV-infected adults and adolescents: recommendations from the CDC, the National Institutes of Health, and the HIV Medicine Association/Infectious Diseases Society of America. MMWR Morb Mortal Wkly Rep 2004; 53(No. RR-15).

64. Lalezari JP, Drew WL, Glutzer E, et al. (S)-1-[3-hydroxy-2-(phosphonylmethoxy)propyl]cytosine (cidofovir): results of a phase I/II study of a novel antiviral nucleotide analogue. J Infect Dis 1995; 171(4):788–796.
65. Lurain NS, Spafford LE, Thompson KD. Mutation in the UL97 open reading frame of human cytomegalovirus strains resistant to ganciclovir. J Virol 1994; 68(7):4427–4431.
66. Cherrington JM, Miner R, Allen SJW, et al. Sensitivities of human cytomegalovirus (HCMV) clinical isolates to cidofovir. In: 8th International Conference on Antiviral Research, 1995. Sante Fe, N.M.: Antiviral Research.
67. Wagstaff AJ, Bryson HM. Foscarnet. A reappraisal of its antiviral activity, pharmacokinetic properties and therapeutic use in immunocompromised patients with viral infections. Drugs 1994; 48(2): 199–226.
68. Jacobson MA, Drew WL, Feinberg J, et al. Foscarnet therapy for ganciclovir-resistant cytomegalovirus retinitis in patients with AIDS. J Infect Dis 1991; 163(6):1348–1351.
69. Safrin S, Berger TG, Gilson I, et al. Foscarnet therapy in five patients with AIDS and acyclovir-resistant varicella-zoster virus infection. Ann Intern Med 1991; 115(1):19–21.
70. Safrin S, Crumpacker C, Chatis P, et al.; The AIDS Clinical Trials Group. A controlled trial comparing foscarnet with vidarabine for acyclovir-resistant mucocutaneous herpes simplex in the acquired immunodeficiency syndrome. N Engl J Med 1991; 325(8):551–555.
71. Deray G, Martinez F, Katlama C, et al. Foscarnet nephrotoxicity: mechanism, incidence and prevention. American J Nephrol 1989; 9(4):316–321.
72. Jacobson MA, Gambertoglio JG, Aweeka FT, et al. Foscarnet-induced hypocalcemia and effects of foscarnet on calcium metabolism. J Clin Endocrinol Metab 1991; 72(5):1130–1135.
73. Safrin S, Kemmerly S, Plotkin B, et al. Foscarnet-resistant herpes simplex virus infection in patients with AIDS. J Infect Dis 1994; 169(1):193–196.
74. Snoeck R, Andrei G, Gerard M, et al. Successful treatment of progressive mucocutaneous infection due to acyclovir- and foscarnet-resistant herpes simplex virus with (S)-1-(3-hydroxy-2-phosphonylmethoxypropyl)cytosine (HPMPC). Clin Infect Dis 1994; 18(4):570–578.
75. Lai CL, Chien RN, Leung NW, et al; Asia Hepatitis Lamivudine Study Group. A one-year trial of lamivudine for chronic hepatitis B. N Engl J Med 1998; 339:61–68.
76. Dienstag JL, Schiff ER, Wright TL, et al. Lamivudine as initial treatment for chronic hepatitis B in the United States. N Engl J Med 1999; 341(17):1256–1263.
77. Marcellin P, Chang TT, Lim SG, et al. Adefovir dipivoxil for the treatment of hepatitis B e antigen-positive chronic hepatitis B. N Engl J Med 2003; 348(9):808–816.
78. Hadziyannis SJ, Tassopoulos NC, Heathcote EJ, et al. Long-term therapy with adefovir dipivoxil for HBeAg-negative chronic hepatitis B. N Engl J Med 2005; 352(26):2673–2681.
79. Westland CE, Yang H, Delaney WEt, et al. Week 48 resistance surveillance in two phase 3 clinical studies of adefovir dipivoxil for chronic hepatitis B. Hepatology 2003; 38(1):96–103.
80. Schildgen O, Sirma H, Funk A, et al. Variant of hepatitis B virus with primary resistance to adefovir. N Engl J Med 2006; 354(17):1807–1812.
81. Chang TT, Lai CL. Hepatitis B virus with primary resistance to adefovir. N Engl J Med 2006; 355(3):322–323; author reply 323.
82. van Bommel F, Wunsche T, Mauss S, et al. Comparison of adefovir and tenofovir in the treatment of lamivudine-resistant hepatitis B virus infection. Hepatology 2004; 40(6):1421–1425.
83. Peters MG, Andersen J, Lynch P, et al. Randomized controlled study of tenofovir and adefovir in chronic hepatitis B virus and HIV infection: ACTG A5127. Hepatology 2006; 44(5):1110–1116.
84. Department of Health and Human Services. Panel on Antiretroviral Guidelines for Adults and Adolescents, Guidelines for the use of antiretroviral agents in HIV-1-infected adults and adolescents. Department of Health and Human Services. Rockville, MD. January 29, 2008:1–128. http://www.aidsinfo.nih.gov/ContentFiles/AdultandAdolescentGL.pdf. Accessed September 10, 2008.
85. Tenofovir (Viread) [package insert]. Foster City, CA: Gilead Sciences; 2008.
86. Chang TT, Gish RG, de Man R, et al., A comparison of entecavir and lamivudine for HBeAg-positive chronic hepatitis B. N Engl J Med 2006; 354(10):1001–1010.
87. Gish RG, Lok AS, Chang TT, et al. Entecavir therapy for up to 96 weeks in patients with HBeAg-positive chronic hepatitis B. Gastroenterology 2007; 133(5):1437–1444.
88. Lai CL, Shouval D, Lok AS, et al. Entecavir versus lamivudine for patients with HBeAg-negative chronic hepatitis B. N Engl J Med 2006; 354(10):1011–1020.
89. Sherman M, Yurdaydin C, Sollano J, et al. Entecavir for treatment of lamivudine-refractory, HBeAg-positive chronic hepatitis B. Gastroenterology 2006; 130(7):2039–2049.
90. Lai CL, Gane E, Liaw YF, et al. Telbivudine versus lamivudine in patients with chronic hepatitis B. N Engl J Med 2007; 357(25):2576–2588.

91. Hou J, Yin YK, Xu D,et al. Telbivudine versus lamivudine in Chinese patients with chronic hepatitis B: Results at 1 year of a randomized, double-blind trial. Hepatology 2008; 47(2):447–454.
92. Telbivudine (Tyzeka) [package insert]. East Hanover, NJ: Novartis; 2006.
93. Gen GC, Ransohoff RM. Interferon-induced antiviral actions and their regulation. Adv Virus Res 1993; 42:57–102.
94. Wong DK, Cheung AM, O'Rourke K, et al. Effect of alpha-interferon treatment in patients with hepatitis B e antigen-positive chronic hepatitis B. A meta-analysis.[see comment]. Ann Intern Med 1993; 119(4):312–323.
95. Hirschman SZ. Current therapeutic approaches to viral hepatitis. Clin Infect Dis 1995; 20(4):741–743.
96. Davis GL, Balart LA, Schiff ER, et al. Treatment of chronic hepatitis C with recombinant interferon alfa. A multicenter randomized, controlled trial. Hepatitis Interventional Therapy Group. N Engl J Med 1989; 321(22):1501–1506.
97. Hagiwara H, Hayashi N, Mita E, et al. Detection of hepatitis C virus RNA in serum of patients with chronic hepatitis C treated with interferon-alpha. Hepatology 1992; 15(1):37–41.
98. Zeuzem S, Feinman SV, Rasenack J, et al. Peginterferon alfa-2a in patients with chronic hepatitis C. N Engl J Med 2000; 343(23):1666–1672.
99. Heathcote EJ, Shiffman ML, Cooksley WG, et al. Peginterferon alfa-2a in patients with chronic hepatitis C and cirrhosis. N Engl J Med 2000; 343(23):1673–1680.
100. Manns MP, McHutchison JG, Gordon SC, et al. Peginterferon alfa-2b plus ribavirin compared with interferon alfa-2b plus ribavirin for initial treatment of chronic hepatitis C: A randomised trial. Lancet 2001; 358(9286):958–965.
101. Fried MW, Shiffman ML, Reddy KR, et al. Peginterferon alfa-2a plus ribavirin for chronic hepatitis C virus infection. N Engl J Med 2002; 347(13):975–982.
102. Hadziyannis SJ, Sette H Jr., Morgan TR,et al. Peginterferon-alpha2a and ribavirin combination therapy in chronic hepatitis C: A randomized study of treatment duration and ribavirin dose. Ann Intern Med 2004; 140(5):346–355.
103. Cooksley WG, Piratvisuth T, Lee SD, et al. Peginterferon alpha-2a (40 kDa): an advance in the treatment of hepatitis B e antigen-positive chronic hepatitis B. J Viral Hepat 2003; 10(4):298–305.
104. Lau GK, Piratvisuth T, Luo KX, et al. Peginterferon Alfa-2a, lamivudine, and the combination for HBeAg-positive chronic hepatitis B. N Engl J Med 2005; 352(26):2682–2695.
105. McHutchison JG, Gordon SC, Schiff ER, et al. Interferon alfa-2b alone or in combination with ribavirin as initial treatment for chronic hepatitis C. Hepatitis Interventional Therapy Group. N Engl J Med 1998; 339(21):1485–1492.
106. Poynard T, Marcellin P, Lee SS, et al; International Hepatitis Interventional Therapy Group (IHIT). Randomised trial of interferon alpha2b plus ribavirin for 48 weeks or for 24 weeks versus interferon alpha2b plus placebo for 48 weeks for treatment of chronic infection with hepatitis C virus. Lancet 1998; 352(9138):1426–1432.
107. Davis GL, Esteban-Mur R, Rustgi V, et al; International Hepatitis Interventional Therapy Group. Interferon alfa-2b alone or in combination with ribavirin for the treatment of relapse of chronic hepatitis C. N E J Med 1998; 339(21):1493–1499.
108. Eron LJ, Judson F, Tucker S, et al. Interferon therapy for condylomata acuminata. N Engl J Med 1986; 315(17):1059–1064.
109. Friedman-Kien AE, Eron LJ, Conant M, et al. Natural interferon alfa for treatment of condylomata acuminata. JAMA 1988; 259(4):533–538.
110. Wills RJ. Clinical pharmacokinetics of interferons. Clin Pharmacokinet 1990; 19(5):390–399.
111. CASL Hepatitis Consensus Group. Treatment of chronic viral hepatitis with alpha interferon: A consensus conference report. Can J Infect Dis 1994; 5:107–112.
112. Debouck C. The HIV-1 protease as a therapeutic target for AIDS. AIDS Res Hum Retroviruses 1992; 8(2):153–164.
113. Kohl NE, Emini EA, Schleif WA, et al. Active human immunodeficiency virus protease is required for viral infectivity. Proc Natl Acad Sci U S A 1988; 85(13):4686–4690.
114. Lazzarin A, Clotet B, Cooper D, et al. Efficacy of enfuvirtide in patients infected with drug-resistant HIV-1 in Europe and Australia. N Engl J Med 2003; 348(22):2186–2195.
115. Lalezari JP, Henry K, O'Hearn M, et al. Enfuvirtide, an HIV-1 fusion inhibitor, for drug-resistant HIV infection in North and South America. N Engl J Med 2003; 348(22):2175–2185.

14 Common Viral Infections of Childhood

Danielle M. Zerr

Department of Pediatrics, University of Washington and Seattle Children's Hospital, Seattle, Washington, U.S.A.

INTRODUCTION

Viral infections are a primary cause of childhood morbidity and mortality globally. The World Health Organization estimated that between 2000 and 2003, 37% of deaths in children younger than five years were due to pneumonia or diarrhea (1). In addition, preterm delivery was estimated to account for 10% of deaths, of which a portion are likely associated with viral infections. Every child acquires certain common respiratory and gastrointestinal viruses in the first few years of life. These common viruses exact a large toll in terms of morbidity and healthcare resources even in developed countries. For example, prior to the availability of rotavirus vaccination, the United States annual rate of hospitalizations for diarrhea in children younger than five years was estimated to be 97 per 10,000 persons (average, 185,742 per year) (2). Rotavirus was estimated to account for 231 deaths, >87,000 hospitalizations, and almost 700,000 outpatient visits for children younger than five years of age in the European Union (3). Similarly, population-based rates of acute viral respiratory infection hospitalizations in children younger than five years in the United States have been estimated to be 180 per 10,000 children per year (4).

The relatively high morbidity and mortality associated with certain viral infections in early childhood is due to multiple factors, including lack of preexisting immunity. RSV is more likely to present with lower respiratory tract disease and rotavirus with dehydration in young children in the first few years of life when they are likely experiencing their first infection due to these pathogens. Similarly, neonates, especially premature neonates, show an extreme predisposition for severe disease manifestations as a result of certain infections such as herpes simplex virus and enterovirus. Limitations in both innate and adaptive immunity, especially that related to the cellular immune response, are at least partly responsible for this phenomenon (5).

Children also present certain challenges when attempting to diagnose viral infections. It is well recognized that influenza is under-recognized and under-diagnosed in young children due to its less than characteristic presentation in children versus adults (6). Similarly, in the neonate, enterovirus may present as a sepsis syndrome leading clinicians to consider only bacterial sources. Diagnostic tests do not always provide an easy means of clarifying the situation. During infancy, the presence of maternal antibody makes diagnosis using serological methods challenging. Thus, the importance of direct detection of the virus is emphasized in certain scenarios, especially when considering young infants.

This chapter discusses the clinical presentation and diagnosis of important viral infections occurring during childhood, including those viruses that cause congenital infection syndromes (rubella, herpes simplex virus, cytomegalovirus, varicella zoster virus, and parvovirus B19) as well as viruses causing disease in children and infants postnatally, including enteroviruses, parvovirus B19, herpes simplex virus, human herpes viruses 6 and 7, measles, mumps, and rubella. Aspects of several of these viruses will be discussed in detail in other chapters of this text. In addition, respiratory and gastrointestinal tract viral infections are discussed thoroughly in chapters 15 and 16, and therefore will not be specifically discussed in this chapter.

PATHOGENS AND SPECIFIC SYNDROMES

Congenital Viral Infections

Viral infections acquired during pregnancy have the potential to cause fetal and neonatal mortality and morbidity as well as late morbidity in older children. The acronym TORCH (toxoplasmosis, other—syphilis, rubella, cytomegalovirus, herpes simplex) has been used to signify those pathogens causing congenital infection with similar presentations. However, it is recognized that the presentation of infections caused by these organisms can be variable and that other pathogens, such as varicella zoster virus and parvovirus B19, may also cause a congenital syndrome and significant harm when congenitally acquired. This section will focus on the most common viral pathogens that cause congenital infection syndromes: rubella virus, herpes simplex virus (HSV), cytomegalovirus (CMV), varicella zoster virus (VZV), and parvovirus B19. Most of these viruses are covered in greater detail in other sections of this chapter or other chapters of this text; this discussion will focus on the clinical presentations and diagnosis of these agents in the setting of congenital infection.

Clinical Presentation and Viral Agents

Infants born with infections due to rubella, HSV, CMV, VZV, and parvovirus B19 are variably asymptomatic or affected with obvious signs of congenital infection, including fever, rash (maculopapular, petechial, or purpuric), hepatosplenomegaly, microcephaly, seizures, jaundice, musculoskeletal abnormalities, and thrombocytopenia (Table 1). In general, symptomatic infection carries higher morbidity and mortality. Long-term sequelae in survivors most commonly include developmental delay and deafness.

Rubella

Congenital rubella syndrome is a devastating disease. Since the advent of comprehensive rubella vaccination, it has become a rare entity in the United States, although it continues to occur among infants born to women who have emigrated from countries without rubella control programs, or from countries that have recently implemented them. The characteristic presentation includes sensorineural deafness, cataracts, cardiac malformation, and neurological findings. A purpuric rash, referred to as "blueberry muffin" rash, classically accompanies congenital rubella syndrome and is due to extramedullary hematopoiesis. Ten to twenty percent of children with congenital rubella syndrome will be developmentally delayed. Risk of infection and congenital anomalies is highest when infection is acquired by the mother during the first trimester of pregnancy.

HSV

HSV is more commonly recognized as a cause of neonatal disease acquired from birth. This form of HSV infection is referred to as "neonatal HSV" and is discussed below in the section "Herpes simplex virus." Congenital HSV is relatively rare (~5% of neonatal cases). The syndrome is characterized by skin vesicles or scarring, chorioretinitis, microphthalmia, microcephaly, and hydranencephaly and it can occur either as a consequence of primary or recurrent maternal infection (7). Hydrops fetalis due to HSV has also been reported.

CMV

CMV is the most common congenital pathogen in the United States, affecting ~1% of all live births each year. Approximately 90% of infants born with congenital CMV are asymptomatic. However, 10% to 15% of these children will develop sensorineural hearing loss or other neurological, ocular, or developmental problems over time. The 10% of neonates who are symptomatic at birth may present with isolated hepatosplenomegaly, jaundice, and rash (usually petechial), but approximately half will present with more fulminant disease including the above findings as well as chorioretinitis, cerebral calcifications, microcephaly, seizures, and respiratory distress (8,9). Mortality in these cases can be as high as 12% to 30% and neurodevelopmental problems are common in survivors.

VZV

Approximately 24% of episodes of primary VZV during pregnancy will result in intrauterine infection (10). Congenital infection may be asymptomatic, cause fetal loss, or result in congenital varicella syndrome. Congenital varicella syndrome was observed to occur in the infants of 9 (0.7%) of 1373 women who acquired varicella during their pregnancy (11). In this study, the highest risk period of pregnancy was between 13 and 20 weeks of gestation with seven

Table 1 Common Clinical Manifestations of Symptomatic Congenital Viral Infections and Their Diagnosis

	Rubella	HSV	VZV	CMV	Parvovirus
Findings at birth					
Growth retardation			x	x	
Rash					
Maculopapular/petechial/purpural	x			x	
Vesicles/scarring		x	x		
Thrombocytopenia	x			x	
Hydrops fetalis		x			x
Hepatosplenomegaly	x			x	
Jaundice				x	
Cardiovascular malformation	x				
Neurological findings					
Cerebral calcifications			x	x	
Microcephaly		x	x	x	
Meningoencephalitis			x		
Seizures			x	x	
Sensorineural hearing loss	x			x	
Ocular findings					
Micropthalmia		x	x		
Cataracts	x		x		
Chorioretinitis		x	x	x	
Keratoconjunctivitis		x			
Bone/skeletal abnormalities	x		x		
Standard diagnostic approaches					
Serology	Detection of IgM		Detection of IgM		
PCR		CSF, blood/serum	Blood/serum, CSF, lesion, amniotic fluid	Blood/serum, amniotic fluid	Blood/serum (maternal and neonatal), tissue
Culture	Nasopharyngeal secretions	Swabs of lesions and mucous membranes		Urine, saliva, amniotic fluid	
Direct detection of antigen		Swabs of lesions and mucous membranes	Swabs of lesions		

infants identified among 351 pregnancies (2%). Only two cases arose from 472 pregnancies (0.4%) in which maternal varicella occurred before 13 weeks gestation. No cases occurred in pregnancies where infection occurred after 20 weeks gestation. The characteristic findings of this syndrome include skin lesions or scars in dermatomal distribution, neurological problems, eye diseases, and skeletal anomalies, such as limb hypoplasia. About 30% of infants born with congenital varicella syndrome die in the first months of life and survivors often experience developmental abnormalities (12). VZV acquired near the end of pregnancy may result in infants with disseminated varicella complicated by hepatitis and/or pneumonitis.

Parvovirus B19

Parvovirus B19 has been estimated to infect 25% to 50% of fetuses of women who acquired the virus during pregnancy (13). In most studies of pregnancies complicated by parvovirus B19 infection, fetal loss attributed to parvovirus has been estimated to occur in less than 5%. Most newborns with congenital parvovirus infection are asymptomatic. Nonimmune hydrops fetalis is a rare outcome of congenital parvovirus B19 infection and parvovirus B19 accounts for a minority (~10%) of the cases of hydrops fetalis (13). CNS abnormalities have also been reported in association with congenital parvovirus (14).

Laboratory Testing

Rubella

Diagnosis of rubella should be pursued when infection has been documented in the mother or in a neonate with the clinical stigmata of congenital viral infection. Virus is typically shed for months from infants with congenital rubella syndrome and poses a potential risk to susceptible individuals. Diagnosis is confirmed by the isolation of rubella virus from nasal secretions. Rubella can also be recovered from throat swabs, blood, urine, or CSF. Additionally, rubella can be diagnosed by serological (detection of IgM) and molecular (15) methods.

HSV

The approach described for diagnosis of neonatal HSV (below) can be applied for diagnosis of congenital HSV.

CMV

Diagnosis of congenital CMV should be considered in cases where mothers were documented to acquire CMV during pregnancy and in neonates with clinical findings of congenital viral infection. Culture and PCR are preferred methods. In mothers with documented infection during pregnancy, prenatal assessment for congenital CMV can be accomplished by testing amniotic fluid after 21 to 23 weeks of gestation and at least 6 to 9 weeks past maternal infection (16,17). Postnatally, detection of the virus in the first two weeks of life is indicative of congenital infection versus infection acquired during or after delivery. High quantities of CMV are excreted in the urine and saliva of the congenitally infected neonate, making these fluids ideal specimens for culture. PCR testing of blood or respiratory secretions collected from the neonate in the first two weeks of life may also be used to diagnose congenital CMV. Detection of CMV-specific IgM is not a sensitive means of diagnosing congenital CMV.

VZV

Diagnosis of congenital varicella syndrome is based on detection of anti-VZV IgM, detection of viral nucleic acids by PCR, or direct detection of VZV antigen. Detection of the virus may not be possible because the period of viral replication is typically early in gestation. However, if vesicular lesions are present, they can be unroofed, scraped, and tested for VZV antigen by immunofluorescence or immunoperoxidase assays or VZV DNA by PCR. Other specimens, such as serum or plasma, CSF, or amniotic fluid may also be tested for VZV DNA by PCR. Culture is not used to diagnose VZV due to its low sensitivity.

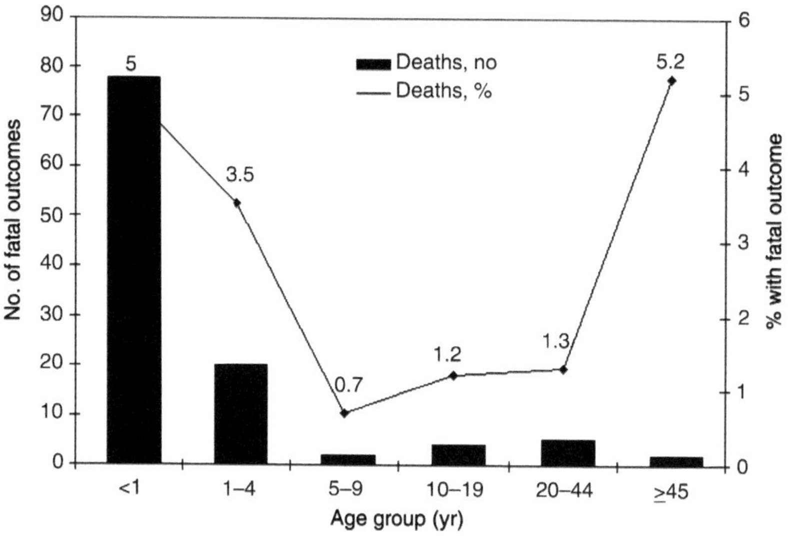

Figure 1 Number and percentage of enterovirus detection reports with fatal outcomes, by age group – National Enterovirus Surveillance System, United States, 1983–1998.
Source: From Ref. 21.

Parvovirus B19

Detection of parvovirus B19 nucleic acids is considered the superior method for diagnosing congenital infection. Please see the section below for more detailed discussion of diagnosis of parvovirus B19.

Enteroviruses (Nonpolio)

Enteroviruses are small, nonenveloped, single-strand RNA viruses belonging to the *Picornaviridae* family. Nonpolio enteroviruses are classically divided into four subgenera based on differences in pathogenicity in humans and experimental animals: group A and B coxsackieviruses, echoviruses, and enteroviruses (18). Each subgenus contains unique serotypes, which can be distinguished from one another on the basis of neutralization by specific antisera. Given the difficulty assigning enteroviruses to subgroups, more recently identified human enteroviruses have been numbered in their order of identification as serologically distinct new isolates (enteroviruses 68–71). Molecular techniques have also led to a new classification scheme that classifies nonpolio enteroviruses based on homology within the RNA region coding for the VP1 capsid protein (19). With new molecular tools, many new enterovirus serotypes have been characterized, bringing the number of known serotypes to over 90 (20).

Epidemiology

Enteroviruses are ubiquitous throughout the world, infecting individuals repeatedly throughout life, but causing more clinically significant disease in infants and young children (Fig. 1) (20). Enterovirus infections occur more frequently during the summer and fall in the United States. In a large U.S. surveillance study, spanning 1970–2005 and performed by the Centers for Disease Control and Prevention, 78% of enterovirus positive specimens were collected during the months of June through October (21). This and other studies have demonstrated that predominant serotypes change over time, and depending on the serotype, demonstrate epidemic or endemic patterns (Table 2) (21,22).

Clinical Manifestations

The clinical manifestations of infection are varied and include asymptomatic or subclinical illness, nonspecific febrile illness, rash, conjunctivitis, central nervous system infections, pleurodynia, and myopericarditis. Specific manifestations and severity vary by age and immune status of the host and the enterovirus subgroup or serotype.

Table 2 Distribution of the 15 Most Commonly Reported Nonpolio Enterovirus Serotypes, by Rank and Year—National Surveillance System, United States, 2002–2004

	2002 ($n = 710$)		**2003 ($n = 1811$)**		**2004 ($n = 1109$)**		**2002–2004 ($n = 3630$)**	
Rank	**Serotype**	**%**	**Serotype**	**%**	**Serotype**	**%**	**Serotype**	**%**
1	Echo 7	22.5	Echo 9	41.0	Echo 30	40.3	Echo 9	30.4
2	Echo 9	21.5	Echo 30	32.4	Echo 9	18.9	Echo 30	29.1
3	Coxsackie B1	10.8	Coxsackie B1	4.6	Coxsackie A9	6.9	Echo 7	5.0
4	Echo 11	6.8	Coxsackie B4	2.0	Coxsackie B5	5.2	Coxsackie B1	4.6
5	Coxsackie B5	5.0	Coxsackie A9	2.7	Coxsackie B4	4.7	Coxsackie A9	3.8
6	Coxsackie B3	4.1	Coxsackie A24	2.5	Echo 18	4.3	Coxsackie B5	3.4
7	Echo 4	4.1	Echo 11	1.8	Coxsackie B3	2.6	Coxsackie B4	3.1
8	Echo 6	3.4	Coxsackie B5	1.7	Coxsackie B2	2.3	Echo 11	2.0
9	Echo 30	3.3	Enterovirus 71	1.4	Echo 11	2.2	Coxsackie B3	2.2
10	Echo 18	2.8	Coxsackie B3	1.1	Echo 6	2.0	Echo 18	2.0
11	Coxsackie B2	2.7	Echo 5	0.9	Coxsackie A1	1.4	Coxsackie B2	1.7
12	Echo 13	2.7	Echo 7	0.9	Echo 13	1.3	Echo 6	1.4
13	Coxsackie A9	1.7	Coxsackie B2	0.9	Echo 5	1.1	Enterovirus 71	1.3
14	Enterovirus 71	1.6	Enterovirus 68	0.7	Enterovirus 71	1.0	Coxsackie A24	1.2
15	Echo 3	1.3	Echo 13	0.7	Coxsackie B1	0.7	Echo 13	1.2
Total (top 15)	94.3		96.2		94.9		93.3	

Source: From Ref. 22.

Table 3 Common Childhood Viral Infections with Typically Generalized Nonvesicular Rashes

Virus	Classic terms	Contemporary terms and descriptions of rashes
Measles	First disease	*Measles or rubeola*: generalized erythematous, maculopapular rash following prodrome of fever, coryza, and conjunctivitis. Rash primarily involves the head, neck, and shoulders first and then spreads down the body to involve the upper extremities and the trunk, and finally the lower extremities. Koplik's spots may be present on the buccal mucosa just before onset of the rash.
Rubella	Third disease	*Rubella or German measles*: generalized erythematous, maculopapular rash. Rash begins on the face, spreading down the body.
Enterovirus	[a]Fourth disease or Dukes' disease	*Generalized rashes*: Generalized maculopapular, petechial, or purpural rashes
		Hand foot and mouth syndrome: Small deep vesicles on the palms and soles, rash involving the genitalia, oral vesicles involving the tongue and buccal mucosa
		Herpangina: Vesicular enanthem of the posterior pharynx
Parvovirus B19	Fifth disease	*Erythema infectiosum*: erythematous malar rash with perioral sparing ("slapped cheek" rash) follows a prodrome of fever, coryza, headache, and diarrhea. The malar rash is typically accompanied by a generalized reticular rash.
HHV-6	Sixth disease	*Exanthem subitum/roseola*: generalized erythematous maculopapular rash that appears suddenly following resolution of fever of 1–3 days duration.
		Generalized rash: children may also develop generalized erythematous maculopapular rash concurrent with fever and other symptoms.

[a]The etiology of "fourth" or "Dukes" disease is unclear. It is now thought it may have been misdiagnosed measles, rubella, enterovirus, or possibly staphylococci toxin-mediated rash. Neither term is used currently. "Second disease" is due to *Streptococcus pyogenes*. The term is no longer in use.

Rash Illnesses

There are a variety of rashes that may accompany enterovirus infections; echoviruses and coxsackieviruses are the most commonly associated subgenera. Potential skin findings include nonspecific maculopapular rashes or petechial/purpural rashes (Table 3). Two specific rashes include herpangina, a vesicular enanthem of the posterior pharynx, and hand, foot, and mouth syndrome, which is characterized by small deep vesicles on the palms and soles, a rash involving the genitalia, and oral vesicles involving the tongue and buccal mucosa. Both herpangina and hand, foot, and mouth syndrome mainly affect children and both are caused primarily by coxsackie A viruses. Hand, foot, and mouth syndrome may also be associated with enterovirus 71, especially in the setting of central nervous system disease.

Central Nervous System Infections

Enterovirus-associated central nervous system infections occur primarily in young children. Meningitis is the more common CNS manifestation of enterovirus infection while encephalitis is relatively rare, except in neonates where CNS involvement is often manifested as encephalitis. Enterovirus is by far the most frequent cause of viral meningitis, accounting for up to 99% of cases of viral meningitis when an etiology is identified (23). In comparison, enteroviruses follow herpes simplex virus and arboviruses in frequency as etiologies of viral encephalitis. Chronic encephalitis/meningitis due to enteroviruses may occur in individuals with defects in B cell function, especially children with X-linked agammaglobulinemia. Enterovirus 71, like polioviruses, may cause acute paralysis by infecting the motor nuclei and anterior horn cells of the brain and spinal cord. This serotype has also been reported to cause large outbreaks with severe manifestations, including encephalitis and fulminant infection with resulting high case-fatality rates (24,25).

Neonatal Infection

Neonates are highly susceptible to severe, often fulminant, enterovirus infections characterized by a sepsis-like syndrome, hepatitis, myocarditis, or central nervous system infection. Coxsackieviruses and echoviruses are the commonly implicated subgroups (26). Enteroviruses may be acquired from the mother perinatally. Frequently these mothers will report a febrile illness during the last week of pregnancy. Outbreaks in neonatal units, which implicate lapses in basic infection control approaches, have also been described (27).

Laboratory Testing

Enterovirus infections may be diagnosed by viral isolation, detection of viral nucleic acids, or by seroconversion.

Specimen Types

A variety of specimens can be submitted to the laboratory for culture or direct detection of enterovirus. In general, the ideal specimen is taken from the site of disease or symptoms as soon as possible after onset of symptoms. For example, when attempting to diagnose central nervous system disease, CSF is the optimal specimen. If obtaining a specimen from the site of infection is not possible, stool specimens, rectal swabs, and throat swabs can be utilized. It should be recognized, however, that enteroviruses can shed in the stool of children for weeks after acute illness.

Viral Isolation

Cell culture is used for viral isolation. Cultures are routinely held up to two weeks, but enterovirus CPE is typically apparent after only two to six days.

Identification/Typing

Isolates may be identified as enterovirus through group-specific reagents, such as group-specific monoclonal antibodies, or by serotyping which can be accomplished by specific neutralization, complement fixation, hemagglutination inhibition, and type-specific monoclonal antibodies.

Nucleic Acid Detection

Direct detection of enterovirus by PCR has the advantages of improved sensitivity in most clinical scenarios and the potential of faster turnaround time, especially when PCR is available in the hospital laboratory. PCR is clearly more sensitive than culture in detecting enterovirus in CSF (28–30). PCR has also been shown to compare favorably to cell culture for detection of enterovirus in blood, throat swab, urine, and stool (31,32).

Serologic Testing

Serological diagnosis can either be accomplished via serotype- or group-specific testing. Serotype-specific testing is most commonly accomplished with the neutralization assay and is typically only practical and pursued when a specific serotype is suspected. Paired sera, one sample obtained as soon as possible after onset of illness and one sample obtained two to four weeks later, are required. A fourfold rise in IgG titers indicates recent infection. Assays detecting IgM antibodies have been developed; however, sensitivity of these assays has been variable and they are not widely available.

Human Parvovirus B19

Human parvovirus B19 (parvovirus B19), a small, nonenveloped, single-stranded DNA virus, was the first-described human pathogen of the *Parvoviridae* family (the others being adeno-associated virus and human bocavirus).

Clinical Presentation

The clinical presentation of parvovirus B19 depends on the age of the host as well as underlying medical conditions. Parvovirus B19 infection in otherwise healthy children is typically mild and classically presents as erythema infectiosum or "fifth disease" (Table 3). Erythema infectiosum is characterized by a febrile illness accompanied by nonspecific symptoms, such as coryza, headache, and diarrhea followed a few days later by an erythematous malar rash with perioral sparing ("slapped cheek" rash). The malar rash may be accompanied by a generalized reticular rash. The rash of erythema infectiosum may wax and wane for weeks, recrudescing with various stimuli. Parvovirus B19 infection can be complicated by nondestructive small joint arthralgias and arthritis, in adolescents and adults, especially females. Parvovirus B19 infection in individuals with increased red blood cell destruction (sickle cell anemia, thalassemia, etc.) or decreased red blood cell production (iron deficiency, anemia, etc.) may present with transient aplastic crisis, resulting in a significant drop in hematocrit due to a complete arrest of erythropoiesis. It is diagnosed by the inability to detect reticulocytes in the peripheral blood smear. Immunocompromised patients may have difficulty resolving parvovirus B19 infection. Infection in these cases can become chronic, complicated by chronic anemia, pancytopenia, hepatitis, myocarditis, or pneumonia (33). Reduction of immunosuppression and intravenous immune globulin are commonly used means of treating parvovirus B19 infections in immunocompromised patients.

Epidemiology

Most individuals become infected with parvovirus during their lifetime. About 50% of children by age 10 and at least 60% to 70% of adults have detectable antibodies (34).

Laboratory Testing

The choice of diagnostic approach depends on the clinical setting. In immunocompetent individuals, serological methods (detection of IgM and IgG) are usually preferred. IgG and IgM appear early after infection and IgG is thought to persist for life. In a study of patients with erythema infectiosum, IgM was present in 97% of cases but only 1% of controls (35). IgM persisted in 83% of cases four to six months after infection. In comparison, viral DNA was detected in 94% of acute cases and was absent in controls positive for both IgG and IgA antibodies. Unlike IgM, viral DNA was not present four to six months after infection. For diagnosis of congenital infection or infection in an immunocompromised individual, detection of nucleic acids is the superior method.

Specimen Types/Handling

A single serum specimen may be used for detection of IgM and IgG antibodies. Parvovirus DNA can be detected in a variety of specimen types; however, serum or diseased tissue is the preferred specimen.

Detection of Nucleic Acids

PCR is the most sensitive method available for detecting parvovirus B19. Low levels of viral DNA can be detected for weeks after infection, but typically not beyond four to six months (35). Detection of parvovirus B19 DNA in maternal serum has been shown to be useful in diagnosis of congenital infection at the time nonimmune fetal hydrops is diagnosed (36).

Serologic Testing

EIA assays are generally considered the most sensitive and specific means of detecting antiparvovirus B19 antibodies.

Human Herpesviruses

The human herpesvirus family includes herpes simplex virus (HSV) 1 and 2, varicella zoster virus (VZV), cytomegalovirus (CMV), Epstein–Barr virus (EBV), Human herpesvirus 6 (HHV-6), human herpesvirus 7 (HHV-7), and human herpesvirus 8 (HHV-8). Human herpesviruses are large, double-stranded DNA viruses and, in general, are ubiquitous in humans. This chapter covers those aspects of these viruses that are common in children and not covered by other chapters in this text.

Herpes Simplex Virus

By 12 to 19 years of age, approximately 44% of children will have acquired HSV-1 and 6% will have acquired HSV-2 (37). HSV-1 seroprevalence increases steadily throughout the lifespan to approximately 90% in those ≥70 years of age. HSV-2 seroprevalence increases during young adulthood to 25% to 28% after the third decade of life. This section focuses on the common clinical entities occurring in childhood and relevant diagnostic strategies. Other chapters in this book address HSV in detail. (Please see chap. 21 for discussion of herpes simplex encephalitis and chap. 22 for discussion of genital HSV.)

Clinical Presentations

Neonatal HSV

Though relatively rare, neonatal HSV is a life-threatening infection and often leaves survivors with permanent sequelae. HSV-2 is the more common cause, but HSV-1 also contributes. The majority of neonatal herpes cases are acquired during birth, while a much smaller proportion are acquired either congenitally or postnatally. Mothers who acquire HSV late in their pregnancy are at very high risk of delivering an infant who will be affected by neonatal herpes. Most cases of neonatal HSV present within the first three weeks of life. Clinical presentations are classified as:

1. disseminated disease, characterized by involvement of multiple organs including lung, liver, skin, and/or brain
2. central nervous system disease, characterized by encephalitis with or without involvement of the skin, or
3. skin/eye/mouth disease, characterized by vesicular rash, conjunctivitis, and/or excretion of virus from the oropharynx.

Virus type and disease classification both predict morbidity and mortality (38). While development of effective treatment regimens with acyclovir has greatly improved survival and functional outcome, mortality and morbidity are still high for disseminated and CNS disease. Disseminated disease carries the highest mortality, currently about 30%. Mortality of CNS disease is approximately 4%, but a high frequency (70%) of survivors experience neurological or developmental problems (39). Early initiation of antiviral therapy improves outcome of neonatal herpes (40). Thus, early recognition of the possibility of neonatal HSV and prompt initiation of the laboratory evaluation and institution of empiric acyclovir therapy is critical. This can be challenging as infants with neonatal HSV often present with nonspecific findings such as lethargy or poor feeding and the differential is broad.

Oral Herpes/Gingivostomatitis

Most HSV-1 infections acquired outside the neonatal period are asymptomatic or subclinical. Gingivostomatitis is the most common clinical syndrome accompanying symptomatic primary infection in young children. Gingivostomatitis is characterized by a painful vesiculo-ulcerative eruption on the palate, gingival surfaces, tongue, and lips. The discomfort is significant and often results in an inability to swallow and drooling. Patients may also have fever, malaise, irritability, and tender cervical lymphadenopathy. Hospitalization may be necessary to provide pain control and/or hydration. Older children and adults may experience pharyngitis with acquisition of HSV. Recurrences manifest as herpes labialis, usually one, but possibly more, painful lesions on the vermillion border of the lip.

Cutaneous HSV/Herpetic Whitlow

Herpetic whitlow is a less common presentation of HSV infection in children (41). Whitlows can result from autoinoculation from oral herpes or from transmission from another individual. Herpetic whitlow usually involves the fingers, but may also involve other areas of the body, including toes and face. The lesion is characterized by erythema and painful vesicles and is often misdiagnosed in childhood as a bacterial infection.

Laboratory Testing

HSV may be diagnosed through a variety of means including serology, direct fluorescent antigen detection, culture, and detection of nucleic acids. For neonatal herpes, it is useful to use more than one approach given the seriousness of the diagnosis and the challenges that are sometimes experienced in securing it. The classic gold standard for diagnosis of HSV infection is viral culture, but PCR has greatly enhanced diagnosis due to its increased sensitivity over other approaches and rapid turnaround times.

Specimen Types

To diagnose neonatal herpes, it is important to obtain CSF and serum or plasma for PCR. Viral culture on CSF is insensitive and should not be performed. In addition, swab specimens of the conjunctiva, nasopharynx, and rectum and scrapings of any suspicious skin or mucous membrane lesion, should be obtained and placed in viral transport media for direct fluorescent antibody testing and culture. HSV gingivostomatitis can be diagnosed on the basis of a scraping of the oral lesions placed in viral transport media and tested by direct fluorescent antibody testing and culture.

Direct Examination

Antigen detection is typically performed on samples from mucous membranes or lesions. There are several techniques available, including fluorescent antibody detection, immunoperoxidase detection, and enzyme immunoassays. The training and skill required varies by assay. In general these approaches are not as sensitive as culture, but have a fast turnaround time that can be helpful in the diagnostic evaluation of neonates.

Nucleic Acid Detection

Diagnosis of neonatal HSV has been facilitated by the application of PCR, which offers greater sensitivity for detection of HSV in both CSF and blood (42–44). Interpretation of results must be correlated with the patient's clinical presentation and course. It is important to note that a negative result does not necessarily rule out HSV. It has been shown that initial negative results may be obtained from as many as 24% of CSF specimens obtained before day 3 of disease in pediatric cases of herpes simples encephalitis (45). When neonatal HSV CNS disease is documented, an end of therapy CSF specimen should be analyzed to document clearance of viral DNA. If DNA is still present at that time, antiviral therapy should be continued until negativity is achieved (43).

Viral Isolation

Culture is typically used to isolate HSV from swabs of mucous membranes or lesions. HSV can be isolated from a variety of cell culture systems. Once cytopathic effect has been observed, it is critical to perform additional tests to definitively identify HSV and the subtype of HSV. Rapid culture methods have also been developed, which allow for detection of HSV prior to the ability to visualize cytopathic effect. This approach involves centrifugation of the sample onto a monolayer of cells and performance of antigen detection tests between 16 and 48 hours after inoculation.

Typing Systems

Subtype identification can be achieved with antigen detection, culture, and PCR approaches.

Serologic Testing

Serology, in particular that which accurately discriminates between HSV-1 and HSV-2 antibodies, can be a helpful adjunct to the other methods described above, when the diagnosis, especially of neonatal disease, remains in question. Type-specific HSV tests exploit the type-specific glycoprotein G (gG). Several different methodologies exist, including Western blot and enzyme immunoassay. Acute serum from a neonate with perinatal acquisition of HSV is expected to be negative or representative of maternal antibodies. Serology obtained 6 to 12 months after the illness represents the infant's own immune response.

Varicella Zoster Virus (VZV)
Please see chapter 17.

Epstein–Barr Virus (EBV)
Please see chapter 24.

Cytomegalovirus (CMV)
Please see the "Congenital Infection," above. Please also see chapter 24 for discussion of cytomegalovirus in the immunocompromised host.

Human Herpesvirus 6 (HHV-6)
Please see chapter 24 for discussion of HHV-6 in the immunocompromised host.

Human herpesvirus 6 (HHV-6) is a member of the Roseolovirus genus of the β-herpesvirus subfamily of human herpesviruses. Like other herpesviruses, it establishes latency after primary infection. There are two subtypes of HHV-6–type A and type B (HHV6-A and B, respectively)—which share certain biological properties and a high level of sequence homology, but differ in their epidemiology.

Occasionally, HHV-6 can be found integrated in the host chromosomes. This phenomenon has been estimated to occur in 0.2% to 0.8% of the population (46,47). Individuals with chromosomally integrated HHV-6 have viral DNA present in every cell in the body as a result of passage of viral DNA through the germ-line. High levels of viral DNA are detectable in serum, whole blood, and CSF from these individuals (48–50). Whether there is any clinical consequence from chromosomal integration of HHV-6 is unknown.

Epidemiology
HHV-6B is ubiquitous, infecting virtually all children within the first two to three years of life (51). The peak age of infection is between 9 and 21 months of age (52). The epidemiology and clinical importance of HHV-6 A remains largely undefined.

Clinical Presentation
Most children (94%) are symptomatic with primary HHV-6B infection. Common symptoms include fever (58%), fussiness (70%), and rhinorrhea (66%), while cough (34%), vomiting (8%), diarrhea (26%), and roseola (24%) occur less frequently (52). Compared with other illnesses commonly occurring during early childhood, HHV-6B is significantly more likely to be accompanied by fever, fussiness, diarrhea, rash, and roseola (high fever followed by a rash with defervescence) and result in physician visits (52). In the acute-care setting, primary HHV-6B infection is present in 10% to 20% of young children evaluated for fever. In this setting HHV-6B has been associated with fever, irritability, otitis media, roseola, and seizures (53,54). HHV-6 can also be a cause of encephalitis in young children (55). The vast majority of documented primary HHV-6 infections are due to HHV-6B (52–54). The epidemiology and clinical findings associated with acquisition of HHV-6 A remain unknown.

Laboratory Testing
Diagnosis of clinically relevant HHV-6 can be challenging, due to the high prevalence of infection and persistence of the virus. HHV-6 infections are most commonly diagnosed by detection of viral nucleic acids. Serology and viral isolation are other possible means.

Specimen Types
HHV-6 is detectable in multiple cell types and specimens including CSF, blood, and saliva. CSF, plasma or serum, whole blood, or peripheral blood mononuclear cells are the typical specimens used for diagnosis of HHV-6.

Nucleic Acid Detection
Detection of viral nucleic acids may indicate active or latent infection depending on the clinical setting and the specimen tested. Detection of viral DNA in white blood cell fractions by PCR can be difficult to interpret since the mononuclear cell is a site of latency. Quantitative PCR methods

improve interpretability as levels indicative of active infection can be established. Detection of HHV-6 DNA in plasma or serum correlates well with indicators of active replication and is therefore more directly interpretable (56–60). Reverse transcription PCR detects messenger RNA and thus indicates actively replicating virus even when peripheral blood mononuclear cells are assayed (61). Methods involving detection of antigenemia have also been described, (62) but further study is needed to understand their applicability, advantages, and limitations.

Individuals with chromosomal integration of HHV-6 will have high, persistent levels of HHV-6 DNA detected. Fluorescent in situ hybridization or FISH can be used to demonstrate the integrated HHV-6 DNA in the human chromosome (48). Integrated HHV-6 DNA can also be demonstrated in unusual tissues/samples such as hair follicles (49). If these studies are not easily available, the expected levels of HHV-6 DNA in integrated versus nonintegrated states may be helpful in distinguishing the two entities. HHV-6 DNA levels in patients with chromosomally integrated HHV-6 are much higher than what can be detected in nonintegrated latent infections, both in whole blood (>6 $\log_{10}$ copies/mL whole blood or >1 copy per leukocyte versus ~2 $\log_{10}$ copies/mL whole blood or 1 copy per 10^4 to 10^5 leukocytes) and serum (4.6–6.4 $\log_{10}$ copies/mL serum versus undetectable) (49). The HHV-6 DNA levels documented with chromosomal integration are also typically higher than what is documented with primary infection (50). In addition, the high levels of HHV-6 DNA observed with chromosomal integration are persistent over time and do not decrease with antiviral therapy. Distinguishing between chromosomal integration and active infection can be difficult and the possibility of chromosomal integration should be kept in mind when interpreting positive PCR results.

Viral Isolation

HHV-6 can be isolated from peripheral blood mononuclear cells, typically in co-culture with cord blood lymphocytes, during primary infection. Isolation of HHV-6 from the blood indicates active viral infection. However, this technique is labor intensive and takes up to three weeks and is therefore not used in many clinical laboratories. Techniques for rapid viral culture have also been developed, but are not widely available (63).

Typing Systems

Distinguishing between HHV-6 subtypes is mainly accomplished using PCR techniques, including real-time PCR assays based on melting curves or variant-specific primers (64,65).

Serologic Testing

Serological methods have many limitations, including complications posed by maternal antibodies in the setting of primary infection and the unreliability of antibody assays in severely immunocompromised patients. Serological methods using antibody avidity assays exploit the fact that, during primary infection, the first immunoglobulin G antibodies are low avidity, but with time and maturation of the immune response, higher avidity antibodies are produced (66). This allows the immune response to primary infection to be distinguished from either maternal antibodies or established infection. There are currently no type-specific antibody tests.

Antiviral Resistance Testing

Foscarnet, ganciclovir, and cidofovir have been shown to have in vitro inhibitory effects against HHV-6. Primary infection is typically self-limited and antivirals are not indicated. Severe disease, however, especially in immunocompromised populations, is treated, usually with foscarnet or ganciclovir. Recently, a mutant HHV-6 strain carrying an amino acid substitution in the ganciclovir phosphorylating pU69 kinase, the functional homologue of the cytomegalovirus UL97 gene product, has been isolated both from cell culture and from patients (67). PCR systems to detect such mutants can be designed and may play a role in monitoring for resistance in immunocompromised populations in the future.

Human Herpesvirus 7 (HHV-7)

Along with HHV-6, HHV-7 belongs to the *Roseolovirus* genus of the β-herpesvirus subfamily of human herpesviruses. HHV-7 shares homology with HHV-6 (68), but beyond having a distinct genome, it also has differences from HHV-6 in cell tropism, viral entry, effects on cells, and epidemiology.

Clinical Presentation

Similar to HHV-6, acquisition of HHV-7 has been associated with fever, rash, roseola (fever for one to three days, with sudden appearance of a generalized maculopapular rash with resolution of fever), upper respiratory tract symptoms, diarrhea, and seizures (69–71).

Epidemiology

Like HHV-6, HHV-7 is ubiquitous, infecting at least 95% of people (72,73). The peak age of infection is slightly later than HHV-6, between two and three years of age.

Laboratory Testing

Clinical testing for HHV-7 has not been standardized and is not routinely available in clinical laboratories. Like HHV-6, after primary infection, HHV-7 persists for life, establishing latent infection in lymphocytes. HHV-7 is also routinely detectable in saliva after infection and has been detected in CSF.

Nucleic Acid Detection

Various procedures for detection of HHV-7 DNA have been described. Further work is needed to determine the clinical significance of a positive PCR result given specific clinical scenarios and specimen types.

Viral Isolation

A number of techniques for isolation of HHV-7 have been described. An approach similar to that described for HHV-6 can be used (74,75).

Serologic Testing

Many early versions of serological tests for HHV-7 demonstrated cross-reactivity with HHV-6. Immunofluorescence, enzyme-linked immunosorbent, and Western blot assays have been developed that distinguish between HHV-6 and HHV-7 responses. As with HHV-6, serological methods have many limitations, including the ubiquitous and chronic nature of infection and the unreliability of antibody assays in severely immunocompromised patients. Approaches using antibody avidity have also been developed to distinguish primary HHV-7 infection from reactivation (76).

Measles

Measles is a vaccine-preventable disease caused by the measles virus, a member of the family *Paramyxoviridae*, genus *Morbillivirus*. The *Paramyxoviridae* family also includes human parainfluenza virus types 1–4 and mumps virus.

Clinical Presentation

Following exposure, the typical incubation period of classic measles is 8 to 12 days. Measles illness is characterized by a prodromal phase including fever, cough, coryza, and conjunctivitis. Symptoms intensify over three to four days and a generalized maculopapular rash appears on day 4–5 of illness. The rash primarily involves the head, neck, and shoulders first, followed by the upper extremities and the trunk, and finally the lower extremities. The rash may become confluent in the areas where it first develops. Koplik's spots, pathognomonic for measles, are small, erythematous lesions with raised whitish centers appearing on the buccal mucosa, often across from the molars, just before onset of the rash. Measles is often complicated by upper and lower respiratory tract complications including laryngotracheitis, bronchitis, pneumonitis, and secondary bacterial infection. In addition to the aforementioned complications, hepatitis, premature labor, and spontaneous abortion have been reported in pregnant women with measles (77).

Modified measles and atypical measles are two potential manifestations of measles infection in recipients of measles vaccine. Modified measles is a mild form of the disease characterized by a relatively mild rash of short duration. It may occur in individuals who failed to make a full immunological response to the vaccine, those who received immune globulin as post exposure prophylaxis, or in young infants who have residual maternal antibodies. Atypical measles occurs in individuals who received the killed virus vaccine (distributed in the United States between 1963 and 1967). This vaccine sensitized recipients to the measles virus without providing protection. The illness is characterized by fever, pneumonia, pleural effusions, and edema. The rash can be maculopapular, petechial, purpuric, or urticarial, and unlike typical measles, it starts on the extremities and spreads to the trunk. The rash may involve the palms and soles and spare the head, neck, and upper chest. Atypical measles is usually self-limited; however, complications can include organ dysfunction/failure.

Neurological complications of measles are rare and include acute encephalitis, acute disseminated encephalomyelitis (ADEM), and subacute sclerosing panencephalitis (SSPE). SSPE is an extremely rare, degenerative central nervous system disease, which is believed to be due to persistent infection of the CNS and is usually fatal. As opposed to classic measles, which has a typical incubation period of 8 to 12 days, SSPE has an average incubation period of 10 years.

Case fatality rates of classic measles range between 1 and 3/1000. Risk of death is higher in younger children, malnourished or immunocompromised individuals, and pregnant women.

Epidemiology

Prior to availability of measles vaccine, measles was endemic throughout the world. Epidemics occurred approximately every two years in the United States and resulted in over 500,000 cases per year. Current high vaccination rates have resulted in annual rates of <1 case per million population in the United States since 1997 (78). Most of these cases are imported or arise from sporadic outbreaks linked to imported cases (79). Measles remains an important cause of childhood morbidity and mortality worldwide. The World Health Organization estimated it accounted for 5% of all deaths in children younger than five years in 2002.

Laboratory Testing

Laboratory testing should be considered in persons with clinical findings compatible with acute measles, who are nonvaccinated or who have had a suspected exposure. Testing of possible cases should be accomplished rapidly so that control measures may be taken in a timely manner to prevent further spread. The simplest approach for establishing the diagnosis of classic measles is testing for IgM antibody on a single serum specimen obtained during the rash phase of the illness. The sensitivity of most IgM assays is less than 100% during the first 72 hours of the rash, but increases to 100% during days 4–10 of the rash (80). If the initial sample is obtained during the first 72 hours of rash and the result is negative, and the patient has a rash lasting for >72 hours, a repeat sample should be obtained. Confusion can arise when interpreting test results in individuals who have been recently vaccinated. Vaccination results in an IgM response detectable between one and eight weeks after immunization (81). Measles can also be diagnosed using acute and convalescent serology or by isolation of the virus from clinical specimens. Regardless of the method used for diagnosis, suspected cases should be reported to public health while awaiting results.

Specimen Types/Handling

A single serum specimen obtained during the rash phase of the illness and tested for IgM is the preferred approach for diagnosis of classic measles. Antibody testing in both serum and CSF can be used in the case of possible SSPE (82). Urine, blood, throat, or respiratory secretions, CSF, or tissue can be used for isolation of the virus.

Serologic Testing

The most sensitive serological method is the EIA and there are a number of sensitive and specific commercial kits with rapid turnaround time available. A fourfold rise in measles IgG may also be used for diagnosis of measles; however, acute and convalescent serum specimens are required, the convalescent specimen being obtained 10 to 14 days after the onset of the rash.

Direct Detection

Detection of viral antigens (using IFA) or nucleic acids (using PCR) are also possible methods for diagnosing measles; however, these methods do not offer greater sensitivity over IgM for diagnosis of classic measles and they are not widely available.

Virus Isolation

Virus is most easily isolated from blood (leukocytes) or respiratory secretions (nasal wash) during the prodromal phase until the first or second day of the rash. The virus can be isolated in multiple different cell culture types and lines. The laboratory should be notified that measles is included in the differential.

Mumps

Mumps virus, a single-strand RNA virus, is a member of the *Paramyxoviridae* family, which also includes human parainfluenza virus and measles.

Clinical Presentation

Mumps infection is typically subclinical or mild. Symptoms and signs, when they occur, include low-grade fever lasting for three to four days and enlargement of one or both parotid glands lasting for seven to ten days. The most common serious complications of mumps infection are encephalitis, deafness, and orchitis. Up to 15% of mumps cases will have signs of meningeal inflammation. Reported rates of mumps encephalitis range as high as five cases per 1000 reported mumps cases (83). Permanent sequelae are rare and the reported encephalitis case-fatality rate is approximately 1.4%. Sensorineural deafness is one of the most serious of the rare complications involving the central nervous system (CNS). It occurs with an estimated frequency of 0.5–5.0 per 100,000 reported mumps cases. Orchitis (usually unilateral) has been reported as a complication in 20% to 30% of clinical mumps cases in postpubertal males. Some degree of testicular atrophy occurs in about a third of cases of mumps orchitis, but sterility rarely occurs. Mumps involvement of other organs has been observed less frequently.

Epidemiology

Following the introduction of the live mumps virus vaccine in 1967 and recommendation of its routine use in 1977, the incidence rate of reported mumps cases decreased steadily in the United States. Despite widespread use of the vaccine, there have been resurgences in mumps activity in the United States. Most recently, in 2006 there was a large outbreak involving 6584 cases focused in Midwest college–aged students, but involving 45 states and a wide range of ages (84,85). This outbreak occurred in a highly vaccinated population, and may have occurred due to waning immunity. It is thought, however, that the high vaccination level prevented a much larger outbreak and a higher rate of complications (85).

Laboratory Testing

Laboratory testing should be considered in any patient with parotid gland swelling or other manifestations of mumps, especially if there is an epidemiological link to other cases. Mumps can be diagnosed by isolating the virus in cell culture, by detection of the viral RNA by reverse transcriptase-PCR, or by serological methods. Serological methods are the most widely available and least expensive and are considered the method of choice for identifying mumps infection. Tests are often negative when mumps occurs in previously vaccinated individuals; therefore, mumps testing can confirm but not rule out infection in this setting (85). Public Health should be contacted when mumps is thought to be the likely diagnosis even when testing is negative.

Specimen Types

Single specimens or acute and convalescent serum samples may be used for serologic studies. Saliva, throat swab, urine, or CSF can be used to inoculate cell cultures for viral isolation. The virus is present in saliva during the first few days of illness and is present in urine for as long as two weeks after onset of symptoms. The laboratory should be informed that mumps is on the differential so that appropriate steps may be taken with the viral culture.

PCR

Detection of viral RNA by PCR is a sensitive means of identifying mumps virus and should be considered for detection of mumps in CSF specimens.

Viral Isolation

Various cell culture approaches can be used to isolate mumps virus. Cultures are typically held for 14 days. Cells infected with mumps virus will demonstrate a nonspecific CPE and, as with other paramyxoviruses, will demonstrate hemadsorption. Confirmation of mumps virus can be achieved with a hemadsorption inhibition test (cell cultures are pretreated with a mumps-specific antibody which will block hemadsorption if cells are infected with the mumps virus), indirect immunofluorescence staining, or neutralization testing. Of the three, indirect immunofluorescence staining is most commonly used due to its sensitivity and ease.

Serologic Testing

Multiple methods are available for detecting both IgG and IgM against mumps. Depending on the method, detection of mumps IgG can be complicated by cross reaction with parainfluenza virus-specific antibody (86). Cross reaction is less of an issue with IgM assays; however, many previously vaccinated individuals who are infected with mumps virus will not have detectable IgM. IgG testing should be considered for these individuals (85).

Rubella

Rubella, a small, enveloped, RNA virus, is a member of the *Togaviridae* family. Infection during early pregnancy carries serious consequences for the fetus (see "Congenital Viral Infections, above), while infections outside the fetal period are typically very mild or asymptomatic. When postnatal infection is apparent it is characterized by fever, lymphadenopathy, and rash. The rash is an erythematous, maculopapular exanthema that classically begins on the face and spreads down the body. The appearance of the rash is similar to that of measles except that the rash in patients with rubella generally does not darken as in measles. For further discussion of rubella, see "Congenital Viral Infections," above.

REFERENCES

1. Bryce J, Boschi-Pinto C, Shibuya K, et al. WHO estimates of the causes of death in children. Lancet 2005; 365:1147–1152.
2. Jin S, Kilgore PE, Holman RC, et al. Trends in hospitalizations for diarrhea in United States children from 1979 through 1992: estimates of the morbidity associated with rotavirus. Pediatr Infect Dis J 1996; 15:397–404.
3. Soriano-Gabarro M, Mrukowicz J, Vesikari T, et al. Rotavirus Infection in Europe: Time for Effective Prevention? Pediatr Infect Dis J 2006; 25:S7–S11.
4. Iwane MK, Edwards KM, Szilagyi PG, et al. Population-based surveillance for hospitalizations associated with respiratory syncytial virus, influenza virus, and parainfluenza viruses among young children. Pediatrics 2004; 113:1758–1764.
5. Lewis DB, Wilson CB. Developmental immunology and role of host defenses in fetal and neonatal susceptibility to infection. In: Remington JS, Klein JO, Wilson CB, Baker CJ, eds. Infectious Diseases of the Fetus and Newborn Infant. Philadelphia, PA: Elsevier Saunders, 2006:87–210.
6. Peltola V, Ziegler T, Ruuskanen O. Influenza A and B Virus Infections in Children. Clin Infect Dis 2003; 36:299–305.
7. Hutto C, Arvin A, Jacobs R, et al. Intrauterine herpes simplex virus infections. J Pediatr 1987; 110: 97–101.
8. Boppana SB, Pass RF, Britt WJ, et al. Symptomatic congenital cytomegalovirus infection: Neonatal morbidity and mortality. Pediatr Infect Dis J 1992; 11:93–99.
9. Kylat RI, Kelly EN, Ford-Jones EL. Clinical findings and adverse outcome in neonates with symptomatic congenital cytomegalovirus (SCCMV) infection. Eur J Pediatr 2006; 165:773–778.
10. Paryani SG, Arvin AM. Intrauterine infection with varicella-zoster virus after maternal varicella. N Engl J Med 1986; 314:1542–1546.
11. Enders G, Miller E. Consequences of varicella and herpes zoster in pregnancy: Prospective study of 1739 cases. Lancet; 1994; 343:1548–1551.
12. Sauerbrei A, Wutzler P. The congenital varicella syndrome. J Perinatol 2000; 20:548–554.
13. Koch WC, Harger JH, Barnstein B, et al. Serologic and virologic evidence for frequent intrauterine transmission of human parvovirus B19 with a primary maternal infection during pregnancy. Pediatr Infect Dis J 1998; 17:489–494.
14. Katz VL, McCoy MC, Kuller JA, et al. An association between fetal parvovirus B19 infection and fetal anomalies: a report of two cases. Am J Perinatol 1996; 13:43–45.

15. Bosma TJ, Corbett KM, Eckstein MB, et al. Use of PCR for prenatal and postnatal diagnosis of congenital rubella. J Clin Microbiol 1995; 33:2881–2887.
16. Enders G, Bäder U, Lindemann L, et al. Prenatal diagnosis of congenital cytomegalovirus infection in 189 pregnancies with known outcome. Prenat Diagn 2001; 21:362–377.
17. Gouarin S, Palmer P, Cointe D, et al. Congenital HCMV infection: a collaborative and comparative study of virus detection in amniotic fluid by culture and by PCR. J Clin Virol 2001; 21:47–55.
18. Hyypia T, Hovi T, Knowles NJ, et al. Classification of enteroviruses based on molecular and biological properties. J Gen Virol 1997; 78:1–11.
19. Oberste MS, Maher K, Kilpatrick DR, et al. Typing of human enteroviruses by partial sequencing of VP1. J Clin Microbiol 1999; 37:1288–1293.
20. Oberste MS, Maher K, Michele SM, et al. Enteroviruses 76, 89, 90 and 91 represent a novel group within the species Human enterovirus A. J Gen Virol 2005; 86:445–451.
21. Khetsuriani N, LaMonte-Fowlkes A, Oberste MS, et al. Enterovirus Surveillance — United States, 1970–2005. Morb Mortal Wkly Rep Surveill Summ 2006; 55(SS08):1–20.
22. CDC. Enterovirus Surveillance — United States,2002–2004. MMWR Morb Mortal Wkly Rep 2006; 55:153–156.
23. Berlin LE, Rorabaugh M, Heldrich F, et al. Aseptic meningitis in infants <2 years of age: diagnosis and etiology. J Infect Dis 1993; 168:888–892.
24. Chan LG, Parashar UD, Lye MS, et al. Deaths of children during an outbreak of hand, foot, and mouth disease in sarawak, malaysia: clinical and pathological characteristics of the disease. For the Outbreak Study Group. Clin Infect Dis 2000; 31:678–683.
25. Lin TY, Chang LY, Hsia SH, et al. The 1998 Enterovirus 71 Outbreak in Taiwan: Pathogenesis and Management. Clin Infect Dis 2002; 34:S52–S57.
26. Khetsuriani N, Lamonte A, Oberste MS, et al. Neonatal enterovirus infections reported to the national enterovirus surveillance system in the United States, 1983–2003. Pediatr Infect Dis J 2006; 25:889–893.
27. Syriopoulou VP, Hadjichristodoulou C, Daikos GL, et al. Clinical and epidemiological aspects of an enterovirus outbreak in a neonatal unit. J Hosp Infect 2002; 51:275–280.
28. Glimaker M, Johansson B, Olcen P, et al. Detection of enteroviral RNA by polymerase chain reaction in cerebrospinal fluid from patients with aseptic meningitis. Scand J Infect Dis 1993; 25:547–557.
29. Sawyer MH, Holland D, Aintablian N, et al. Diagnosis of enteroviral central nervous system infection by polymerase chain reaction during a large community outbreak. Pediatr Infect Dis J 1994; 13: 177–182.
30. Lina B, Pozzetto B, Andreoletti L, et al. Multicenter evaluating of a commercially available PCR assay for diagnosing enterovirus infection in a panel of cerebrospinal fluid specimens. J Clin Microbiol 1996; 34:3002–3006.
31. Rotbart HA, Ahmed A, Hickey S, et al. Diagnosis of enterovirus infection by polymerase chain reaction of multiple specimen types. Pediatr Infect Dis J 1997; 16:409–411.
32. Glimaker M, Abebe A, Johansson B, et al. Detection of enteroviral RNA by polymerase chain reaction in faecal samples from patients with aseptic meningitis. J Med Virol 1992; 38:54–61.
33. Eid AJ, Brown RA, Patel R, et al. Parvovirus B19 infection after transplantation: a review of 98 cases. Clin Infect Dis 2006; 43:40–48.
34. Gay NJ, Hesketh LM, Cohen BJ, et al. Age specific antibody prevalence to parvovirus B19: how many women are infected in pregnancy? Commun Dis Rep CDR Rev 1994; 4:R104–R107.
35. Erdman DD, Usher MJ, Tsou C, et al. Human parvovirus B19 specific IgG, IgA, and IgM antibodies and DNA in serum specimens from persons with erythema infectiosum. J Med Virol 1991; 35: 110–115.
36. Enders M, Weidner A, Rosenthal T, et al. Improved diagnosis of gestational parvovirus B19 infection at the time of nonimmune fetal hydrops. J Infect Dis 2008; 197:58–62.
37. Xu F, Schillinger JA, Sternberg MR, et al. Seroprevalence and coinfection with herpes simplex virus type 1 and type 2 in the United States, 1988–1994. J Infect Dis 2002; 185:1019–1024.
38. Whitley R, Arvin A, Prober C, et al. A controlled trial comparing vidarabine with acyclovir in neonatal herpes simplex virus infection. Infectious Diseases Collaborative Antiviral Study Group. N Engl J Med 1991; 324:444–449.
39. Kimberlin DW, Lin CY, Jacobs RF, et al. Safety and efficacy of high-dose intravenous acyclovir in the management of neonatal herpes simplex virus infections. Pediatrics 2001; 108:230–238.
40. Whitley RJ, Corey L, Arvin A, et al. Changing presentation of herpes simplex virus infection in neonates. J Infect Dis 1988; 158:109–116.
41. Szinnai G, Schaad UB, Heininger U. Multiple herpetic whitlow lesions in a 4-year-old girl: case report and review of the literature. Eur J Pediatr 2001; 160:528–533.
42. Diamond C, Mohan K, Hobson A, et al. Viremia in neonatal herpes simplex virus infections. Pediatr Infect Dis J 1999; 18:487–489.

43. Kimberlin DW, Lakeman FD, Arvin AM, et al. National Institute of Allergy and Infectious Diseases Collaborative Antiviral Study Group. Application of the polymerase chain reaction to the diagnosis and management of neonatal herpes simplex virus disease. J Infect Dis 1996; 174:1162–1167.
44. Malm G, Forsgren M. Neonatal herpes simplex virus infections: HSV DNA in cerebrospinal fluid and serum. Arch Dis Child Fetal Neonatal Ed 1999; 81:F24–F29.
45. De Tiege X, Heron B, Lebon P, et al. Limits of early diagnosis of herpes simplex encephalitis in children: a retrospective study of 38 cases. Clin Infect Dis 2003; 36:1335–1339.
46. Leong HN, Tuke PW, Tedder RS, et al. The prevalence of chromosomally integrated human herpesvirus 6 genomes in the blood of UK blood donors. J Med Virol 2007; 79:45–51.
47. Tanaka-Taya K, Sashihara J, Kurahashi H, et al. Human herpesvirus 6 (HHV-6) is transmitted from parent to child in an integrated form and characterization of cases with chromosomally integrated HHV-6 DNA. J Med Virol 2004; 73:465–473.
48. Clark DA, Nacheva EP, Leong HN, et al. Transmission of integrated human herpesvirus 6 through stem cell transplantation: implications for laboratory diagnosis. J Infect Dis 2006; 193:912–916.
49. Ward KN, Leong HN, Nacheva EP,et al. Human herpesvirus 6 chromosomal integration in immunocompetent patients results in high levels of viral DNA in blood, sera, and hair follicles. J Clin Microbiol 2006; 44:1571–1574.
50. Ward KN, Leong HN, Thiruchelvam AD, et al. Human herpesvirus 6 DNA levels in cerebrospinal fluid due to primary infection differ from those due to chromosomal viral integration and have implications for diagnosis of encephalitis. J Clin Microbiol 2007; 45:1298–1304.
51. Okuno T, Takahashi K, Balachandra K, et al. Seroepidemiology of human herpesvirus 6 infection in normal children and adults. J Clin Microbiol 1989; 27:651–653.
52. Zerr DM, Meier AS, Selke SS, et al. A population-based study of primary human herpesvirus 6 infection. N Engl J Med 2005; 352:768–776.
53. Hall CB, Long CE, Schnabel KC, et al. Human herpesvirus-6 infection in children. A prospective study of complications and reactivation. N Engl J Med 1994; 331:432–438.
54. Zerr DM, Frenkel LM, Huang ML, et al. Polymerase chain reaction diagnosis of primary human herpesvirus-6 infection in the acute care setting. J Pediatr 2006; 149:480–485.
55. Ward KN, Andrews NJ, Verity CM, et al. Human herpesviruses-6 and -7 each cause significant neurological morbidity in Britain and Ireland. Arch Dis Child 2005; 90:619–623.
56. Chiu SS, Cheung CY, Tse CY, et al. Early diagnosis of primary human herpesvirus 6 infection in childhood: serology, polymerase chain reaction, and virus load. J Infect Dis 1998; 178:1250–1256.
57. Huang LM, Kuo PF, Lee CY, et al. Detection of human herpesvirus-6 DNA by polymerase chain reaction in serum or plasma. J Med Virol 1992; 38:7–10.
58. Suga S, Yazaki T, Kajita Y, et al. Detection of human herpesvirus 6 DNAs in samples from several body sites of patients with exanthem subitum and their mothers by polymerase chain reaction assay. J Med Virol 1995; 46:52–55.
59. Suga S, Yoshikawa T, Kajita Y, et al. Prospective study of persistence and excretion of human herpesvirus-6 in patients with exanthem subitum and their parents. Pediatrics 1998; 102:900–904.
60. Yoshikawa T, Ihira M, Suzuki K, et al. Human herpesvirus 6 infection after living related liver transplantation. J Med Virol 2000; 62:52–59.
61. Norton RA, Caserta MT, Hall CB, et al. Detection of human herpesvirus 6 by reverse transcription-PCR. J Clin Microbiol 1999; 37:3672–3675.
62. Savolainen H, Lautenschlager I, Piiparinen H, et al. Human herpesvirus-6 and -7 in pediatric stem cell transplantation. Pediatr Blood Cancer 2005; 45:820–825.
63. Paterson DL, Singh N, Gayowski T, et al. Encephalopathy associated with human herpesvirus 6 in a liver transplant recipient. Liver Transpl Surg 1999; 5:454–455.
64. Boutolleau D, Duros C, Bonnafous P, et al. Identification of human herpesvirus 6 variants A and B by primer-specific real-time PCR may help to revisit their respective role in pathology. J Clin Virol 2006; 35:257–263.
65. Razonable RR, Fanning C, Brown RA, et al. Selective reactivation of human herpesvirus 6 variant a occurs in critically ill immunocompetent hosts. J Infect Dis 2002; 185:110–113.
66. Ward KN, Gray JJ, Joslin ME, et al. Avidity of IgG antibodies to human herpesvirus-6 distinguishes primary from recurrent infection in organ transplant recipients and excludes cross-reactivity with other herpesviruses. J Med Virol 1993; 39:44–49.
67. Manichanh C, Olivier-Aubron C, Lagarde JP, et al. Selection of the same mutation in the U69 protein kinase gene of human herpesvirus-6 after prolonged exposure to ganciclovir in vitro and in vivo. J Gen Virol 2001; 82:2767–2776.
68. Berneman Z, Ablashi D, Le G, et al. Human herpesvirus-7 is a T-lymphotopic virus and is related to but significantly different from, human herpesvirus-6 and human cytomegalovirus. Proc Natl Acad Sci U S A 1992; 89:10552–10556.

69. Caserta MT, Hall CB, Schnabel K, et al. Primary human herpesvirus 7 infection: a comparison of human herpesvirus 7 and human herpesvirus 6 infections in children. J Pediatr 1998; 133:386–389.
70. Hall CB, Caserta MT, Schnabel KC, et al. Characteristics and acquisition of human herpesvirus (HHV) 7 infections in relation to infection with HHV-6. J Infect Dis 2006; 193:1063–1069.
71. Torigoe S, Kumamoto T, Koide W, et al. Clinical manifestations associated with human herpesvirus 7 infection. Arch Dis Child 1995; 72:518–519.
72. Wyatt LS, Rodriguez WJ, Balachandran N, et al. Human herpesvirus 7: antigenic properties and prevalence in children and adults. J Virol 1991; 65:6260–6265.
73. Clark DA, Freeland JML, Mackie PLK,et al. Prevalence of antibody to human herpesvirus 7 by age. J Infect Dis 1993; 168:251–252.
74. Frenkel N, Schirmer EC, Wyatt LS, et al. Isolation of a new herpesvirus from human D CD4+ T cells. Proc Natl Acad Sci U S A 1990; 87:748–752.
75. Tanaka K, Kondo T, Torigoe S, et al. Human herpesvirus 7: another causal agent for roseola (exanthem subitum). J Pediatr 1994; 125:1–5.
76. Ward KN, Turner DJ, Parada XC, et al. Use of immunoglobulin G antibody avidity for differentiation of primary human herpesvirus 6 and 7 infections. J Clin Microbiol 2001; 39:959–963.
77. Atmar RL, Englund JA, Hammill H. Complications of measles during pregnancy. Clin Infect Dis 1992; 14:217–226.
78. Orenstein WA, Papania MJ, Wharton ME. Measles elimination in the United States. J Infect Dis 2004; 189:S1–S3.
79. CDC. Measels – United States, January 1–April 25, 2008. MMWR Morb Mortal Wkly Rep 2008; 57:494–498.
80. Helfand RF, Heath JL, Anderson LJ, et al. Diagnosis of measles with an IgM capture EIA: the optimal timing of specimen collection after rash onset. J Infect Dis 1997; 175:195–199.
81. Helfand RF, Kebede S, Gary HEJ, et al. Timing of development of measles-specific immunoglobulin M and G after primary measles vaccination. Clin Diagn Lab Immunol 1999; 6:178–180.
82. Garg RK. Subacute sclerosing panencephalitis. Postgrad Med J 2002; 78:63–70.
83. CDC. Recommendations of the Immunization Practices Advisory Committee Mumps Prevention MMWR Morb Mortal Wkly Rep 1989; 38:388–400.
84. CDC. Brief Report: Update: Mumps Activity — United States, January 1–October 7, 2006. MMWR Morb Mortal Wkly Rep 2006; 55:1152–1153.
85. Dayan GH, Quinlisk MP, Parker AA, et al. Recent resurgence of mumps in the United States. N Engl J Med 2008; 358:1580–1589.
86. Glikmann G, Mordhorst CH. Serological diagnosis of mumps and parainfluenza type-1 virus infections by enzyme immunoassay, with a comparison of two different approaches for detection of mumps IgG antibodies. Acta Pathol Microbiol Immunol Scand 1986; 94:157–166.

15 Respiratory Virus Infections

Robert L. Atmar and Stephen B. Greenberg

Department of Medicine and Department of Molecular Virology and Microbiology, Baylor College of Medicine, Houston, Texas, U.S.A.

INTRODUCTION

Respiratory viral infections are common throughout the year in all age groups and in all countries around the world. Respiratory viruses can cause mild illnesses, such as the common cold, or severe illness such as pneumonia. Both immunocompetent and immunocompromised hosts are affected. Although each respiratory virus can infect both the upper and lower respiratory tract, certain viruses are associated with specific clinical syndromes such as croup or laryngotracheobronchitis.

Both DNA and RNA viruses are known to cause respiratory illness (Table 1). In the past 10 years, several new viruses have been reported in association with respiratory illnesses (Table 2). This is a result of newer, more sensitive diagnostic tests, employing nucleic acid technologies, which have significantly increased our ability to detect these viruses (1–10). Although some of the newly identified viruses [e.g., bocavirus (a parvovirus) and WU and KI polyomaviruses] have been identified in respiratory secretions of persons with acute respiratory illnesses, their pathogenic role as causes of such illnesses remains to be determined, and their diagnosis will not be considered further in this chapter. With the use of these newer more sensitive assays, the epidemiology of other respiratory viral infections has been expanded and their clinical importance extended (11–16).

CLINICAL SYNDROMES

Respiratory viruses cause a number of distinct clinical syndromes (Table 3). Individual patients may display symptoms from multiple syndromes.

The Common Cold

The syndrome of acute upper respiratory tract illness has been referred to as the "common cold". Although the respiratory viruses most frequently associated with the common cold are rhinoviruses and coronaviruses (17,18), other viruses such as influenza viruses, parainfluenzaviruses, and adenovirus can manifest similar symptoms. Yearly epidemics occur during the fall and winter in the temperate areas and during the rainy season in the tropics (19,20). Increased crowding indoors during the fall and winter and returning to school in the fall may contribute to the seasonal increases in upper respiratory viral infections. Children experience six to eight colds per year; and adults average two to four colds each year (21,22).

Transmission of respiratory viruses occurs by direct contact, large particles in the air, droplet nuclei suspended in air, or by a combination of more than one mode of spread (19). Rhinovirus has been recovered from hands, and this is thought to be another mode of transmission (20). The spread of common cold viruses has been reported in homes, schools, and daycare centers. Children and mothers appear to have higher secondary attack rates because of more prolonged exposure to school-age children.

The incubation period varies between 12 and 72 hours. Symptoms include nasal discharge and obstruction, sneezing, sore throat, and cough. Fever is uncommon. The median duration of symptoms is approximately seven days, but many cases report symptoms that last for two or more weeks.

Physical examination is usually normal except for nasal discharge and a red nose. The nasal mucosa and pharynx may be erythematous. The chest examination is usually normal. No clinical differences are noted between adults and children.

Table 1 Classification of Respiratory Viruses

Virus family	Genera	Genetic characteristics	Viral particle size	Number of species/ serotypes/genotypes
RNA viruses				
Coronaviridae	*Coronavirus*	ssRNA, positive sense	120–160 nm	3 groups, at least 3 species
Orthomyxoviridae	*Influenzavirus A*	ssRNA, negative sense, segmented	80–120 nm	1 species—multiple HA and NA subtypes
	Influenzavirus B			1 species
	Influenzavirus C			1 species
Paramyxoviridae (subfamily *Paramyxovirinae*)	*Respirovirus*	ssRNA, negative sense	150–300 nm	PIV-1 and PIV-3
	Rubulavirus			PIV-2 and PIV-4
	Pneumovirus			1 human RSV species—two types
(subfamily *Pneumovirinae)*	*Metapneumovirus*			1 HMPV species—two serotypes
Picornaviridae	*Enterovirus*	ssRNA, positive sense	22–30 nm	4 species, >60 serotypes
	Rhinovirus			2 (possibly 3) species, >100 serotypes
	Parechovirus			2 serotypes
DNA viruses				
Adenoviridae	*Mastadenovirus*	dsDNA	70–90 nm	6 species (A–F), 51 serotypes
Herpesviridae (subfamily *Alphaherpesvirinae*)	*Simplexvirus*	dsDNA	~200 nm	2 species (HSV-1, HSV-2)
	Varicellovirus			1 serotype
(subfamily *Betaherpesvirinae*)	*Cytomegalovirus*			1 serotype
(subfamily *Gammaherpesvirinae*)	*Lymphocryptovirus*			1 serotype (EBV)

Table 2 New Viruses Detected in the Respiratory Tract since 2000

Virus	Year reported
Human metapneumovirus	2001
SARS-coronavirus	2003
Coronavirus NL63	2004
Coronavirus HKU1	2005
Human bocavirus	2005
Human rhinovirus C	2007
Human polyomavirus WU and KI	2007

Pharyngitis

Pharyngitis is caused by both virus and bacteria (23). Rhinoviruses and coronaviruses account for many cases of pharyngitis, but adenovirus and herpes simplex virus are causes as well (24). Other common respiratory viruses [influenza viruses and parainfluenza viruses (PIVs)] also can cause acute pharyngitis. Human immunodeficiency virus-1 (HIV-1) has been reported to give acute pharyngitis as a part of the acute retroviral syndrome (25). Epstein–Barr virus, and less commonly cytomegalovirus, cause pharyngitis as part of the infectious mononucleosis syndrome.

It is important to distinguish pharyngitis due to respiratory viruses from group A streptococci. Approximately 25% of pharyngitis cases in children and 10% in adults are reported to be due to streptococci. Streptococcal pharyngitis occurs during the winter and early spring, which is the time that respiratory viruses peak in incidence (26).

Pharyngeal complaints are common with acute respiratory tract infections. However, tonsillar exudates and clinical lymphadenopathy are not common. Herpes simplex virus pharyngitis may be associated with tonsillar exudates and/or palatal vesicles and painful cervical lymphadenopathy (24). Herpangina, an uncommon form of pharyngitis, is caused by coxsackievirus, found in children, and associated with small palatal vesicles. Severe pharyngitis with fever and conjunctivitis is caused by adenoviruses. Sore throat may be a predominant symptom in some patients with acute influenza illness. Infectious mononucleosis secondary to Epstein–Barr virus is frequently diagnosed in young adults with exudative pharyngitis and cervical lymphadenopathy, and it must be differentiated from group A streptococcal pharyngitis.

Acute Otitis Media

Acute otitis media (AOM) is a common illness in young children and results from inflammation and fluid collection in the middle ear (27,28). Most cases occur in children less than three years of age. Most children have no apparent anatomic defect that is responsible for repeated infections. More cases occur in boys than girls. Native Americans, Eskimos, and Australian aborigines have an increased incidence and more severe cases of AOM. Upper respiratory tract viral infections and AOM are closely linked. Up to 60% of episodes of symptomatic upper respiratory illnesses (URIs) among young children have been complicated by AOM and/or otitis media with effusion (OME) (28). In a recently published study, rhinovirus and adenovirus were most frequently detected. AOM occurred in ~30% of children with upper respiratory tract infection due to influenza virus, parainfluenzavirus, enterovirus, or rhinovirus. In contradistinction, Pitkaranta et al. found rhinovirus to be most common in children with AOM (29).

Croup (Acute Laryngotracheobronchitis)

Croup (acute laryngotracheobronchitis) usually begins with a distinctive cough that starts suddenly at night. Nonspecific respiratory symptoms often precede stridor, hoarseness, and respiratory distress. Resolution of the characteristic cough occurs within 48 hours in approximately 60% of children. Croup is usually caused by respiratory viruses and is reported predominantly in children between six months and three years of age (30–32). Adults rarely develop croup. Croup admissions are 50% higher in even-numbered years, which correlate with the prevalence of PIV infections (33). Cases occur throughout the year, but peak in late autumn (34).

Table 3 Association of Common Respiratory Viruses with Different Clinical Syndromes

Virus	Common cold	Pharyngitis	Acute otitis media	Croup	Tracheobronchitis	Bronchiolitis	Pneumonia\ (Pediatric)
RNA viruses							
Rhinovirus	++++	+++	+++	+	++	+++	++
Enterovirus	++	++	+	+	+	++	+
Coronavirus	+++	+++	+	++	++	++	+
Influenza virus	++	++	++	++	++++	++	++
Parainfluenza virus							
Type 1	+	+	+	++++	+	++	+
Type 2	+	+	+	++	+	++	+
Type 3	++	+	+	+++	++	+++	+++
RSV	++	+	+++	++	++	++++	++++
HMPV	++	+	++	++	+	++	++
DNA viruses							
Adenovirus	++	++	+	++	+	++	++

++++, >25%; +++, 10–25%; ++, 1–10%; +, <1%.

Tracheobronchitis

Tracheobronchitis is characterized by a dry cough and is caused by inflammation of the upper tracheobronchial tree. Tracheitis can often be demonstrated by palpation of the trachea during the physical examination. Deep inhalation of air is associated with retrosternal chest discomfort and elicits an episode of coughing. Acute bronchitis is associated with persistent cough for at least five days (35). Physical examination and radiographic findings of pneumonia are absent. Sputum production is usually scant or absent; when present the sputum is mucoid and has a clear or white appearance. Influenza viruses are the most common viruses causing tracheobronchitis, but all of the respiratory viruses can do so. Tracheobronchitis is often accompanied by symptoms in the upper respiratory tract, especially with rhinovirus and coronavirus infection (36).

Bronchiolitis

Bronchiolitis is primarily a disease of infancy and is characterized by fever, cough, tachypnea, rales, wheezing, and hyperinflation of the lungs. Other physical findings can include accessory muscle use, nasal flaring, and nasal discharge. It is the most common reason for hospitalization in young children and a common reason for admission to the pediatric intensive care unit (37). There is a marked seasonality to the occurrence of bronchiolitis, peaking in the winter, and this is a reflection of the primary cause of the disease, RSV. More severe disease is associated with male sex, exposure to cigarette smoke, chronic lung or heart disease, prematurity, and young age (<3 months).

Pneumonia

Pneumonia is an infection of the lung parenchyma and is characterized by fever, cough, dyspnea, rales and rhonchi, and pulmonary infiltrates on chest radiograph. More than any other respiratory clinical syndrome, its etiology is influenced by the patient's age and immune status. In the first several years of life, viruses are the most common cause of pneumonia, with RSV causing the majority of these illnesses (38). PIVs, especially PIV-3, are the next most common, and influenza viruses, rhinoviruses, human metapneumovirus, coronaviruses, and adenoviruses can also cause pneumonia in this age group. Pneumonia in school-age children is less common, and it is less likely to be caused by viruses. In this age group, influenza viruses are the most common cause. In adults, pneumonia is more likely to be due to bacterial causes, although a preceding viral respiratory illness is commonly reported. Influenza A viruses are the most common cause of viral pneumonia in this age group (39). The other respiratory viruses can also cause pneumonia, but are less common. Pneumonia can complicate primary varicella infection in 5% to 50% of adults (40).

Pneumonia is a common complication in immunocompromised patients, including patients who have hematologic malignancies, have received cytotoxic therapy, or have undergone stem cell or solid organ transplantation. All of the common respiratory viruses described above can cause pneumonia in this patient population, and RSV and influenza viruses are the most frequently observed of these viruses (41). In addition, cytomegalovirus is an important pathogen in this group. The common respiratory viruses are not only recognized as causes of pneumonia in immunocompromised patients during the time of year they are circulating in the community, but they can also be found out of season.

VIRAL AGENTS

RNA Viruses

Rhinoviruses and Other Picornaviruses

Rhinoviruses are members of the *Picornaviridae* family. These small, nonenveloped (no lipid coat), positive-sense, single-stranded ribonucleic acid (RNA) viruses have icosahedral symmetry. The capsid is composed of four proteins. Proteins VP1, VP2, and VP3 are on the surface of the viral capsid. Variations in these surface proteins are responsible for antigenic diversity and the host immune response following infection (42,43). VP4 is on the inside of the virus and anchors the RNA core to the viral capsid.

There are more than 100 serotypes of rhinoviruses. Over 90% of these serotypes attach to cells by the intercellular adhesion molecule 1 (ICAM-1) (44). A small number of serotypes use the low-density lipoprotein receptor for attachment and entry into cells (45). One serotype, human rhinovirus (HRV)-87, requires the presence of sialic acid on cellular receptors, unlike the major and minor group serotypes (46).

Many previously uncharacterized rhinovirus strains were identified in 2007 (47). Untypeable or unassigned picornaviruses were identified at the molecular level and given the designation of "HRV-Cs" (48). HRV-C strains are genetically distinct and under consideration by the picornavirus study group as a third species of HRVs (in addition to HRV-A and HRV-B). As yet there has been no cultivation of HRV-C in traditional cell lines. Nevertheless, HRV-C strains appear to be newly identified viruses that cause symptomatic respiratory infections (10,49).

Enterovirus and *Parechovirus* are two additional species in the family *Picornaviridae* that cause a variety of respiratory and nonrespiratory (e.g., aseptic meningitis, conjunctivitis, myocarditis, pleurodynia) syndromes. There are more than 60 serotypes of *Enterovirus* and two serotypes of *Parechovirus*. In the past, parechoviruses were classified as enteroviruses, but sequence analysis of the genome showed that they are a phylogenetically distinct virus group.

Coronaviruses

Coronaviruses are enveloped, positive-sense, single-stranded RNA viruses that replicate in the cytoplasm. Coronaviruses are divided into three groups: group I includes both human (HCoV-229E, HCoV-NL63) and animal pathogens; group II includes both human (HCoV-OC43, HCoV-HKU1, SARS) and animal pathogens; and group III includes only avian pathogens. There is a large surface glycoprotein called the spike (S) protein that functions as the viral cell attachment protein and is a target for the neutralizing antibodies. Most group II coronaviruses (but not the SARS virus) also contain another surface glycoprotein called the hemagglutinin-esterase (HE) protein. CD13 (Human aminopeptidase N) is the cellular receptor for most group I coronaviruses, including HCoV-229E (50). However, HCoV-NL63 does not use CD13 as the receptor cell entry (51). Instead, HCoV-NL63, along with SARS-CoV, uses angiotensin-converting enzyme 2 (ACE2) as the entry receptor. ACE2 is found on ciliated nasal and tracheobronchial epithelial cells (52–54). It has not been possible to cultivate the HCoV-HKU1 viruses, so the receptor is not known (15). The receptor for HCoV-OC43 also has not been identified.

Orthomyxoviruses (Influenza Viruses)

Influenza viruses are enveloped, negative sense, single-stranded RNA viruses with a segmented genome and belong to the family *Orthomyxoviridae*. There are three genera of human viruses, corresponding to the three types (A, B, and C). Influenza A viruses are further divided into subtypes based upon antigenic differences in the two primary surface glycoproteins, the hemagglutinin (HA) and neuraminidase (NA). There are 16 recognized hemagglutinins (H1–H16) and nine neuraminidases (N1–N9), and subtypes are referred to by both their HA and NA designations (e.g., A/H1N1, A/H3N2). The majority of HAs and NAs are not found in viruses that infect humans, but all occur in viruses that infect aquatic birds. However, an increasing number of avian influenza strains are being recognized as sporadic causes of infection in humans (e.g., H5, H7, H9). Influenza B and C viruses are not further divided into subtypes.

The influenza viruses can be divided into their respective genera based upon antigenic differences in two internal viral proteins (the matrix protein and nucleoprotein). These viral proteins serve as targets for many diagnostic assays. A minor surface protein, the M2 protein, is found in influenza A viruses, functions as an ion channel, and is the target of one group of antiviral agents (the adamantanes).

Paramyxoviruses

Paramyxoviruses are enveloped, negative-sense, single-stranded RNA viruses that belong to the family *Paramyxoviridae*. These pleomorphic viruses range from 150 to 300 nm in size, and there are a number of paramyxoviruses that cause respiratory illnesses. Some, like rubeola, cause a distinctive rash illness and will not be considered further in this chapter. PIVs comprise two genera in the subfamily *Paramyxovirinae*. PIV types 1 and 3 belong to the genus *Respirovirus*. PIV types 2, 4A, and 4B belong to the genus *Rubulavirus* (55). Respiratory syncytial virus and human metapneumovirus belong to two genera, *Pneumovirus* and *Metapneumovirus*, respectively, in the subfamily *Pneumovirinae*. The hemagglutinin-neuraminidase, glycoprotein (HN) and the fusion glycoprotein (F) are the major antigens projecting through the lipid envelope of the PIVs. Viruses in the subfamily *Pneumovirinae*, including RSV and HMPV, do not have a hemagglutinin; instead, the major surface proteins are a glycoprotein (G) and fusion glycoprotein (F).

The HN glycoproteins of PIVs attach to sialic acid residues on the surface of the host cell, while the G protein is responsible for attachment of RSV and HMPV to the cell surface. Attachment leads to virus–cell membrane fusion mediated by the F protein. Viral replication takes place in the cytoplasm.

The four major serotypes of human PIV are distinguished based on reaction to complement fixation and hemagglutinating antigens. Because these viruses share common antigens, heterotypic antibody responses are observed frequently with infection. RSV and HMPV are further subdivided into two types (A and B) and many additional subtypes.

DNA Viruses

Adenoviruses

Adenoviruses are members of genus *Mastadenovirus* in the family *Adenoviridae*. Adenoviruses are nonenveloped, icosahedral viruses that are approximately 70 to 90 nm in size. The genome consists of double-stranded DNA and is approximately 32 kb in length. There are six species (formerly called subgroups or subgenera), A through F, that cause human disease, and each species contains one or more serotypes. The species can be separated, in part, by their ability to agglutinate rat or monkey red blood cells. Other characteristics also used to separate adenoviruses into species include phylogenetic analyses of viral genes, GC content of the genome, restriction fragment polymorphisms, antigenic relationships, and oncogenicity in rodents. Species A through E viruses cause respiratory illness, with serotypes within species B and C being the most common. Species F viruses cause gastroenteritis.

Herpes Viruses

Several viruses in the family *Herpesviridae* can cause respiratory clinical syndromes. The herpesviruses are complex, enveloped, double-stranded DNA viruses that are ~200 nm in size. The family has three subfamilies that contain human strains: the *Alphaherpesvirinae*, the *Betaherpesvirinae*, and the *Gammaherpesvirinae*. There are two genera in the alpha herpesviruses that contain human strains. The genus *Simplexvirus* has two species of human virus, *Human herpesvirus 1* (HHV-1) and *Human herpesvirus 2* (HHV-2), while the genus *Varicellovirus* has a single human species, *Human herpesvirus 3* [HHV-3, or varicella-zoster virus (VZV)]. Cytomegalovirus (*Human herpesvirus 5*, CMV) and Epstein–Barr virus (*Human herpesvirus 4*, EBV) are beta and gamma herpesviruses, respectively. While each of these viruses causes distinct clinical syndromes other than those described in the Clinical Syndromes section, their clinical presentation can overlap some of those described above, especially in immunocompromised individuals. These viruses primarily cause disease in the immunosuppressed patient and are discussed further in chapter 24.

EPIDEMIOLOGY

Most of the respiratory viruses have a distinct seasonality (Table 4), although many can be identified as causes of infection throughout the year.

Rhinovirus and Other Picornaviruses

Rhinoviruses cause respiratory illnesses worldwide in all age groups and throughout the year (20). Rhinoviruses are most prevalent during the fall and spring in temperate climates (17). In a prospective study using polymerase chain reaction (PCR) techniques, rhinoviruses accounted for 50% of common colds (17). These viruses are responsible for millions of lost work days, increased school absenteeism, and the majority of physician visits. Most rhinovirus infections are symptomatic. Upper respiratory tract infections caused by respiratory viruses such as rhinoviruses are a frequent reason for overuse of antibiotics (17). Rhinovirus infection also is associated with lower respiratory tract illnesses, including exacerbations of asthma and chronic

Table 4 Epidemiologic Features of Common Respiratory Viruses

Virus	Seasonality	Mode of spread	Incubation period
Rhinovirus	Fall and spring	Droplet, hands	1–5 days
Coronavirus	Winter	Droplet	2–5 days
Adenovirus	Year-round	Direct, aerosol	4–7 days
Influenzavirus A and B	Winter	Aerosol, droplet	1–4 days
Parainfluenzaviruses 1, 2, 3	Fall—types 1, 2; spring—type 3	Droplet	3–6 days
Respiratory syncytial virus	Fall to spring	Hands, $\pm$ droplet	2–8 days
Human metapneumovirus	Fall to spring	Droplet, hands	2–7 days

obstructive lung disease, bronchiolitis, and pneumonia in elderly and immunocompromised persons.

The home is the principal location for transmission of rhinovirus. School-aged children are frequently the introducer of this viral infection. Secondary attack rates in families range from 25% to 70%. Daycare centers and schools are also important locations for spread of rhinovirus (42). Transmission of rhinoviruses can occur by close contact, autoinoculation, fomites, and/or by aerosols (56–60).

Coronavirus

In epidemiologic studies in adults, coronaviruses were estimated to cause ~15% of adult common colds (61,62). Coronaviruses were found to cause epidemics every two to three years with reinfections being common. All ages are susceptible. From epidemiologic studies, coronaviruses were found to be associated with respiratory illnesses, usually in the upper respiratory tract but occasionally causing pneumonia (63). In temperate climates, HCoV-OC43 and HCoV-229E are transmitted primarily during the winter. The elderly are also prone to coronavirus infections in long-term care facilities and these may lead to hospitalization.

Besides HCoV-OC43 and HCoV-229E, other coronavirus strains have been identified recently. A new coronavirus causing severe acute respiratory syndrome (SARS) was reported in 2003 (6,64).

In 2004 and 2005, three closely related coronavirus species were reported (9,52,65–67). NL63 was isolated from an infant with coryza, conjunctivitis, fever, and bronchiolitis (68,69). Patients with HCoV-NL63 have ranged in age from 1 month to 100 years, with the highest infection rate occurring before age five years (70–73).

A newly identified group II human coronavirus, HCoV-HKU1, was found in respiratory specimens of patients with fever and cough (74,75). In a prospective study in Hong Kong, coronaviruses were detected in 2.1% of patients admitted to the hospital with signs and symptoms of acute respiratory illness (76). In a report of 87 infected patients, 13 were positive for HCoV-HKU1, 17 were positive for HCoV-NL63, 53 were positive for HCoV-OC43, and 4 were positive for HCoV-229E (77,78).

In a recent prospective study of respiratory viral infections among hospitalized patients, 5.7% had coronaviruses identified (79). Lower respiratory tract infections were far more common than upper respiratory tract, 75% versus 25%, respectively. Over half of the infections were due to OC43-like strains. Approximately 20% were due to 229E-like strains, and approximately 20% were due to NL63-strains. Coronavirus infections in the first year of life were associated predominantly with OC43-like strains.

Parainfluenza Viruses

PIVs are important causes of a variety of upper and lower respiratory illnesses (80). After respiratory syncytial virus, PIVs are the second leading cause of hospitalization for acute respiratory infections in young children.

Although PIV infections occur throughout the year, seasonal patterns occur with different serotypes. In the United States, PIV-1 infections occur primarily in the fall of odd-numbered years, while PIV-2 infections are most commonly recognized in the fall of even-numbered years

(81). PIV-3 infections occur year-round, but peaks of illness occur each spring. Because PIV-4 infections are infrequently recognized, a clear seasonality has not been established. Seasonal variation of PIVs has not been reported in developing and tropical countries. PIV-3 is the most common cause of PIV infection, with PIV-1, PIV-2, and PIV-4 infections occurring in decreasing frequency.

Over 90% of children have PIV-3 antibodies detected by age five. However, acquisition of antibodies to PIV-1 and PIV-2 is slower. By adulthood, close to 100% of persons have antibodies to PIV-4.

PIV-1 and PIV-2 infections are associated with croup and laryngitis. PIV-3 infects the more distal airways and is associated with bronchiolitis and pneumonia. Croup is caused primarily by PIV-1 and PIV-2 viruses, but it can be seen following PIV-3 and PIV-4 infection (32,34). All of the PIVs cause URIs that include colds, otitis media, and pharyngitis. URIs are the most common illnesses seen in all age groups. Bronchiolitis and pneumonia are more frequently seen in children and in association with PIV-3 infection. Bronchiolitis, pneumonia, and croup are frequently the reasons for hospitalization.

PIV infections can be severe in immunosuppressed children and adults. Infection is most frequently acquired in a community setting, but nosocomial acquisition may occur. The mortality rate with lower respiratory tract infection in immunocompromised patients has been reported to be as high as 35% (82). Asymptomatic infection with PIV-1 and PIV-3 in stem cell transplant recipients has been reported; however, many patients may present only with URIs that do not progress to more severe disease (83). In one study of hematopoietic cell transplant recipients infected with PIV-3, progression from a URI syndrome to an LRI syndrome developed in 13% (84). Progression was associated with the use of corticosteroids in a dose-dependent fashion (85).

PIVs are the third most prevalent virus infections detected in patients with exacerbation of COPD (11). They are also associated with exacerbations of asthma (86). Outbreaks of PIV infection have been noted in hospitals, skilled nursing facilities, and nursing homes. In one outbreak of PIV-3 infection, 50% of residents had a respiratory illness over a four-week period, and 40% of the illnesses were associated with a fever (87). Other clinical syndromes that have been noted to occur with PIV infection include aseptic meningitis, myocarditis, and Guillain–Barre Syndrome.

Respiratory Syncytial Virus

RSV causes annual epidemics from late fall to spring, with the epidemics lasting 20 weeks or more. Infection occurs among all age groups, but primary infection occurs in the first two years of life and is associated with the most serious disease (bronchiolitis and pneumonia). In fact, as many as 40% of first infections are associated with a febrile lower respiratory tract infection (bronchiolitis or pneumonia), and RSV infection is the most common reason for hospitalization in the first year of life (88). In older children and young adults, infection often results in a mild common cold syndrome. In older adults, RSV infection is again associated with lower respiratory tract illness, and it has been estimated to cause approximately 11,000 deaths in the United States each year (89). RSV is also a major cause of pneumonia in severely immunocompromised patients. Hematopoietic stem cell transplant patients are at significantly greater risk of developing symptomatic infection and dying from RSV infection than are solid organ transplant recipients (41). Transmission of infection is by fomites or large droplet aerosol (90). Many nosocomial infections are transmitted from staff or visitors, and strict visitation and health care worker illness-screening policies may be needed to prevent nosocomial infection in very high-risk patients (91).

Human Metapneumovirus

Human metapneumovirus (HMPV) causes illnesses that are similar to those caused by RSV. These viruses mainly circulate from the late fall until the spring, but they can be found throughout the year in some communities (92). Primary infection occurs at an early age, but reinfection can occur throughout life. In some studies, HMPV is the second most common cause of bronchiolitis after RSV, and it can also cause colds and be associated with AOM. Exacerbations of asthma and chronic obstructive pulmonary disease have been described in older children and adults, and pneumonia, sometimes fatal, can occur in severely immunocompromised patients (92).

Influenza Virus

Influenza viruses cause annual epidemics of febrile respiratory illness in temperate climates. The epidemic typically occurs between December and March in the northern hemisphere, and in the southern hemisphere between May and September. Influenza epidemics should be suspected when there is a rapid increase in the number of febrile respiratory illnesses in the community with a concomitant increase in absenteeism from school and work. A community wide epidemic generally lasts three to eight weeks and is associated with a predominant viral subtype (influenza A) or type (influenza B) (93). In addition to annual epidemics, worldwide epidemics (or pandemics) of influenza can also occur, but only with influenza A viruses. Pandemics have occurred in association with the emergence in the human population of a virus strain that has a novel hemagglutinin to which the entire population is susceptible and that is easily transmitted from person-to-person. The potential for pandemic infection has led to interest in avian influenza strains with novel hemagglutinins that can infect humans, such as A/H5N1, A/H7N7, and A/H9N2. Infection with these viruses generally has followed exposure to poultry, but not all cases have had such an exposure.

Infection with epidemic strains is most common in school-age children, hospitalizations are most common in young children and the elderly, and mortality primarily occurs in the elderly. Transmission occurs principally via aerosols or droplets generated during coughing or sneezing. In the United States, there are an average of ~36,000 deaths and ~300,000 hospitalizations annually associated with the epidemics (89,93,94). Influenza A/H3N2 viruses have generally had the greatest impact in recent years, but in 2009 a novel influenza A/H1N1 of swine origin emerged to cause a pandemic (95). By the fall of 2009, the novel 2009 H1N1 strain accounted for the vast majority of influenza infections. Epidemics of influenza B infection also occur and are associated with excess mortality and hospitalizations. Influenza C viruses cause milder infection and are infrequently sought as causes of respiratory illness.

Adenovirus

Adenoviruses cause a variety of illnesses both in the respiratory tract and outside it. The respiratory illnesses include both upper respiratory (pharyngitis and colds) and lower respiratory (bronchiolitis and pneumonia) syndromes. Keratoconjunctivitis, gastroenteritis, meningoencephalitis, hemorrhagic cystitis, and disseminated infection are other diseases associated with adenovirus infection. Adenoviruses cause infections year-round. Peaks of respiratory illness occur during the winter respiratory virus season and peaks of pharyngoconjunctivitis also occur during the summer (e.g., exposure in swimming pools).

DIFFERENTIAL DIAGNOSIS

The differential diagnosis of the clinical syndromes caused by the respiratory viruses includes each of the respiratory viruses (Table 3). Other agents, including bacteria and fungi, are also diagnostic possibilities, especially for clinical syndromes like pharyngitis and pneumonia. The relative probability that an individual respiratory virus is the cause of an illness is influenced by many factors, including the clinical presentation, the age and underlying diseases of the patient, and the season. However, none of these factors allows the identification of the infecting agent, requiring the performance of viral diagnostic assays to achieve a specific diagnosis. The successful identification of infection caused by a specific virus will influence further medical decision making, including the selection of antiviral treatment, the restriction of antibiotic therapy, and the initiation of appropriate infection control precautions.

LABORATORY TESTING

Specimen Types/Handling

Almost any respiratory sample can be used for viral diagnostic testing, but the diagnostic yield varies based upon the type of sample and the diagnostic assay used. Nasopharyngeal (NP) swabs and nasopharyngeal aspirates (NPAs) are frequently used for the rapid detection of most respiratory viruses (96,97). NP swabs are more commonly collected from older children and adults while NPAs are more commonly obtained from infants and younger children. The collection of NP samples is not tolerated well by many patients, and other sample types have also been collected. Nasal wash samples (instillation of saline into the nose and collection of return) and nasal swab samples are used by many clinicians. Nasal swab specimens have been found to be less productive than NPA in some studies but better in others (98). A recently developed flocked-NP swab has been compared with NPA and found to have increased sensitivity for detecting respiratory viruses (99–101). These flocked swabs (Copan Diagnostic, Inc.) are designed for the collection of respiratory specimens. These swabs are made with a spray-on flocked fiber, which attaches to the tip of the swab in a perpendicular manner. This technology increases the number of epithelial cells two- to three-fold, including increasing the number of virus-infected cells (102).

Oropharyngeal samples are another potential source for identification of respiratory viruses. Pharyngeal washes and salivary samples provided a higher diagnostic yield for the early detection of the SARS coronavirus than other clinical samples (103). However, pharyngeal samples, usually collected as a throat swab, are less sensitive for detection of most other respiratory viruses compared to nasal or NP samples (104,105). The collection of a combined nasal and pharyngeal sample allows sampling of both sites.

Respiratory viruses can also be identified in lower respiratory tract samples from patients with lower respiratory tract infection. Sputum or bronchoalveolar lavage (BAL) specimens are the most common samples evaluated from the lower respiratory tract, with BALs being particularly useful in the evaluation of immunocompromised patients.

Once samples have been collected, they should be transported to the clinical laboratory as quickly as possible, especially if virus infectivity is to be measured. The addition of a sucrose- or broth-based transport media can increase virus recovery, especially when swab samples are collected (106). The transport media usually contains antibiotics and antifungals to decrease microbial growth. However, transport media can interfere with the performance of some rapid antigen diagnostic assays, so specimen collection and transport protocols should be put into place based upon the diagnostic strategies (culture, antigen detection, nucleic acid detection, etc.) to be used. A variety of commercial transport media are available.

Clinical samples are transported to the laboratory on wet ice to maintain virus viability. Virus recovery decreases after freezing or if higher transport temperatures are used. If the sample cannot be transported to the laboratory immediately, many respiratory viruses will remain viable for several days while stored at 4°C (107). If longer delays are anticipated, then the sample should be frozen, preferably at −70°C. However, respiratory syncytial virus is particularly labile and virus recovery decreases over time after collection and with freezing the sample.

Direct Examination

Several strategies can be used to directly detect virus or viral antigens in clinical samples. The most common are microscopy, immunofluorescence, enzyme immunoassays (EIA), and nucleic acid detection. Some of these assays identify virus infection in as little as 15 minutes and can be performed as point-of-care tests. A major disadvantage of direct detection approaches compared to cultivation methods is the inability to further characterize the infecting virus (e.g., antigenically, antiviral susceptibility).

Microscopy

Electron microscopy occasionally is used to identify respiratory viruses from clinical samples (108). Viruses can be tentatively identified by morphology and reactivity with immune sera. However, given the need for specialized equipment and personnel experienced in interpreting images, the relatively high costs of the assay, and the lower sensitivity of EM compared to other diagnostic assays, electron microscopy is infrequently used for respiratory virus diagnosis. Evaluation of cytologic or tissue samples with light microscopy may identify viral inclusions or other histopathologic changes suggestive of a specific viral infection (e.g., cytomegalovirus, RSV), and immunohistochemical studies may also be used to identify viral antigens in infected tissues. Appropriate controls (e.g., preimmune sera) should be included in immunohistochemical studies to demonstrate the specificity of positive results.

Immunofluorescence

Fluorescent antibody (FA) assays were the first methods used for the identification of respiratory viruses in the 1950s, and they are still frequently used in many clinical laboratories (109). Two formats are used for virus detection: direct and indirect assays. Virus-specific antibodies are directly labeled with a fluorescent label (e.g., fluorescein) in direct fluorescent antibody (DFA) assays. In contrast, in an indirect fluorescent antibody (IFA) assay the fluorescently labeled antibody is a species-specific anti-immunoglobulin that recognizes the virus-specific antibody that reacts with virus-infected cell. IFA assays are more sensitive than DFA assays because of amplification of the signal with the second antibody, but the specificity is usually lower due to increased background fluorescence. IFA assays take longer to perform than DFA assays.

Immunofluorescent assays require the presence of cellular material to allow interpretation of the test, and thus this diagnostic approach allows an interpretation of the quality of the collected sample. Relative disadvantages include the requirement for a fluorescence microscope, the need for technical expertise in the interpretation of stained slides, and the relatively slower throughput compared to other rapid antigen detection methods. In addition to being used directly on clinical samples, IF assays are used to identify viral antigens in following cell culture, either when cytopathic effect (CPE) is evident or as part of a rapid shell vial assay (see below).

A number of IF assays are commercially available (Table 5). Many of these assays are initially performed with pooled reagents in a multiplex format to identify whether a sample contains any of the target respiratory viruses (influenza A and B, RSV, PIV types 1–3, adenovirus), and if positive, the specific virus is then identified using reagents targeting individual viruses. IF assays for the detection of coronavirus 229E and OC43 have also been described (110). The sensitivity of IF assays in respiratory samples can approach 90–95% compared to cell culture, but many laboratories report the sensitivity to be 70% to 80% (111,112). The specificity of IF assays is generally >95% when interpreted by experienced personnel.

Enzyme Immunoassays and Other Enzyme Assays

Enzyme immunoassays (EIAs) can be performed using a number of different formats, but sandwich EIAs are among the most common for the detection of viral antigens. Virus-specific antisera are used to capture the viral antigen and a second virus-specific antiserum is used for detection. The second antiserum is either directly labeled with an enzyme that then interacts with a substrate to generate a colorimetric or fluorometric signal, or the second antibody is recognized by a third enzyme-labeled antiserum that interacts with a specific substrate to give a positive signal. Alternatively, a nonenzymatic reporter (e.g., isotope, fluorophore) can be used in place of the enzyme. EIAs have been developed for most respiratory viruses, and these assays generally take at least two hours to perform. Sensitivity of the assays is dependent, in part, on the antisera used in the assay and ranges from 50% to 80% for influenza (113).

EIAs have been adapted to rapid formats to provide results in less than 30 minutes without loss in assay performance. The two main methods used are flow-through immunochromatography and lateral flow immunochromatography. In flow-through immunochromatography, viral antigens are concentrated on a membrane and are then identified with enzyme-labeled, virus-specific antibodies that give a colorimetric signal when exposed to the enzyme's substrate. In lateral flow immunochromatography, the clinical sample is applied to a membrane and viral antigen flows across the membrane and is captured by a virus-specific antibody. Captured antigen is then identified by a second labeled antibody that is visualized as a line on the test strip. These rapid immunoassays are commercially available for influenza A and B viruses and for RSV (Table 6). The viral antigens targeted in the rapid immunoassays are conserved antigens (nucleoprotein for influenza viruses, fusion protein for RSV). Assay performance is often different from that described by the manufacturers and varies among different populations, with sensitivity ranging from 60% to 90% (114,115). However, the sensitivity of one rapid antigen assay for the novel 2009 H1N1 pandemic virus strain was only 10% (116). Some are CLIA-waived and are approved as point-of-care tests.

Another rapid diagnosis strategy has been developed for the influenza viruses. The ZStatflu (ZymeTx) assay uses a labeled derivative of sialic acid, the substrate for the viral neuraminidase, to detect the presence of influenza virus. If influenza virus neuraminidase is

Table 5 Commercially Available Immunofluorescence Assay Kits for Respiratory Viruses

Test format	Test (manufacturer)	Viruses detected (sensitivity; specificity)						
		Adenovirus	***Influenza A***	***Influenza B***	**PIV-1**	**PIV-2**	**PIV-3**	**RSV**
DFA	Bartels RSV DFA (Trinity Biotech)	No	No	No	No	No	No	Yes (88–100%; 100%)
	D^3 DFA Respiratory virus Screening and ID Kit (Diagnostic Hybrids, Inc.)	Yes (100%; 100%)	Yes (97–100%; 100%)	Yes (100%; 100%)	Yes (100%; 100%)	Yes (100%; 100%)	Yes (100%; 100%)	Yes (100%; 100%)
	Imagen (Remel Inc.)[a]	Adenovirus DFA (86–100%; 100%)	Influenza A&B DFA (96%; 100%)	Influenza A&B DFA (87%; 99.5%)	Parainfluenza DFA (100%; 100%)	Parainfluenza DFA (100%; 100%)	Parainfluenza DFA (92%; 99%)	RSV DFA (93%; 98%)
	Light Diagnostics Simulfluor Diagnostic Assays (Millipore)	Yes Confirmation only	RSV/Flu A (96%; 99.6%) Flu A/Flu B (80%; 99%)	Flu A/Flu B (50%; 100%)	Yes Confirmation only	Yes Confirmation only	Yes Confirmation only	Yes (98%; 98.9%)
	PathoDx Respiratory Virus Panel (Oxoid)	Yes (92–100%; 100%)	Yes (75–100%; 100%)	Yes (100%; 100%)	Yes (100%; 100%)	Yes (93–100%; 100%)	Yes (100%; 100%)	Yes (90–100%; 100%)
IFA	Bartels Respiratory Viral Detection Kit (Trinity Biotech)	Yes (100%; 99.9%)	Yes (86–100%; 99–99.9%)	Yes (65–100%; 98–100%)	Yes (52–100%; 100%)	Yes (100%; 100%)	Yes (85–100%; 99–99.9%)	Yes (86–100%; 97–100%)
	Imagen Respiratory Virus Screen Kit (Remel Inc.)	Yes	Yes	Yes	Yes	Yes	Yes	Yes
	Light Diagnostics Respiratory Virus Screen (Millipore)	Yes Confirmation only	Yes Confirmation only	Yes Confirmation only	Yes Confirmation only	Yes Confirmation only	Yes Confirmation only	Yes Confirmation only

[a]Also makes an individual DFA kit for human metapneumovirus.

Table 6 Commercially Available Enzyme Immunoassay and Other Enzyme Assay Kits for Respiratory Viruses

Test format	Manufacturer	Kit name	Acceptable clinical samples	Virus(es) detected: Sensitivity (%)	Virus(es) detected: Specificity (%)
Chromatographic EIA (lateral flow)	Fisher Scientific	Sure-Vue RSV	NPS, NA, NW	RSV: >99	99>
	Genzyme	OSOM Flu A+B	NS	Flu A: 74	96
				Flu B: 60	96
	Inverness Medical Professional Diagnostics	BinaxNow Influenza A+B	NA, NPS, NW	Flu A: 81–83	93–97
				Flu B: 53–65	96–100
		BinaxNow RSV	NPS, NW	RSV: 89–93	93–100
		Clearview RSV	NA, NPS	RSV: 93–100	88–97
	Meridian	ImmunoCard STAT RSV Plus	NA, NPS, NS, NW	RSV: 78–94	78–100
	Quidel Corp.	Quickvue Influenza[a]	NA, NPS, NW	Flu combined: 73–81	96–99
		Quickvue Influenza A+B	NA, NPS, NW	Flu A: 72–94	89–99
				Flu B: 62–82	96–99
		Quickvue RSV	NA, NPS, NW	RSV: 83–99	90–92
	Remel Inc.	Xpect A&B	NS, NW, TS	Flu A: 89–100	100
				Flu B: 83–100	100
		Xpect RSV	NPS	RSV: 96	94
	Response Biomedical Corp.	RAMP Influenza A/B Assay	NA, NPS, NW	Flu A: 70–92	94–99
				Flu B: 56–92	95–99
	SA Scientific	SA FluAlert Influenza A Test	NA, NW	Flu A: 76	98
		SA FluAlert Influenza B Test	NA, NW	Flu B: 91	100
Flow-through EIA	Becton-Dickinson	Directigen Flu A	NA, NPS, NW, TS	Flu A: 67–96	88–97
		Directigen Flu A+B	BAL, NA, NPS, NS, NW, TS	Flu A: 76–96	90–100
				Flu B: 71–88	98–100
		Directigen Flu A+B/EZ	NA, NPS, NW, TS	Flu A: 77–91	86–99
				Flu B: 69–80	99–100
		Directigen RSV	NA, NPS, NW, TA	RSV: 93–97	90–97
Optical immunoassay	Inverness Medical Professional Diagnostics	Biostar Flu OIA[a]	NA, NPS, sputum, TS	Flu combined: 62–88	52–80
		Biostar Flu A/B OIA	NA, NPS, sputum, TS	Flu A and Flu B: Same as Biostar Flu OIA	Flu A and Flu B: Same as Biostar Flu OIA
		Biostar RSV OIA	NW, NPS	RSV: 67–87	83–96
Neuraminidase functional assay	ZymeTx	ZstatFlu[a]	NA, NW	Flu combined: 62	99

[a]Does not differentiate influenza A and B virus infections.

Abbreviations: BAL, bronchoalveolar lavage; NA, nasal aspirate; NPS, nasopharyngeal swab; NS, nasal swab; NW, nasal wash; TA, tracheal aspirate; TS, throat swab.

present, the substrate is cleaved by the viral enzyme to yield a product that can be detected visually. The assay does not distinguish between influenza A and B viruses, and it does not detect influenza C virus (which does not have a neuraminidase) or the neuraminidases of the PIVs or bacterial sialidases because these enzymes do not recognize the sialic acid derivative reporter as a substrate and fail to catalyze its cleavage. The ZStatflu assay has performed less well for detection of influenza B infection than for influenza A, although new formats of the test are in development (117,118).

Nucleic Acid Detection

Molecular methods are increasingly being used for the diagnosis and characterization of respiratory viruses. Reverse-transcription polymerase chain reaction (RT-PCR) assays are the most commonly used and target conserved genes within the targeted virus. Monoplex RT-PCR or PCR assays have been developed for every respiratory virus (12). Other molecular assay formats, including nucleic acid sequence-based amplification (NASBA) and loop-mediated isothermal amplification (LAMP), have also been described for some of the respiratory viruses. In general, the molecular assays have performed at least as well as cell culture, and they have offered the advantage of being able to identify poorly cultivatable and nonculturable viruses. However, a drawback of the monoplex assays has been that the large number of different respiratory viruses that cause respiratory illness makes it necessary to use a large number of different assays to detect even the common respiratory viruses. Nevertheless, the monoplex assays have increased the identification of respiratory virus infection in ill patients (11).

The problem of identifying multiple potential causes of infection has been addressed by the development of sensitive multiplex RT-PCR assays (Table 7). Initial attempts at multiplex assay development were complicated by loss of assay sensitivity due to interactions between the different virus-specific primers and probes. Nevertheless, some assays were developed that allowed detection of two to six different viruses. More recently, multiplex assays that utilize microarray technology for distinguishing products have been developed and allow detection of greater than 10 different viral targets. The xTAG™ Respiratory Virus Panel is one such commercially available assay (Table 7). Extracted viral genome is subjected to multiplex RT-PCR amplification by incubation with virus-specific primer pairs for each of the targeted viruses. After treatment of the generated amplicons with exonuclease and alkaline phosphatase to remove the remaining primers, the mixture is exposed to a multiplex pool of virus-specific primers (with each primer having a unique sequence, or Tag), biotinylated deoxynucleotides, and DNA polymerase. Target-specific primer extension occurs if the virus-specific oligonucleotide primers hybridize to the PCR-generated amplicons, and extension of the oligonucleotide primers leads to their labeling with biotin. The biotinylated oligonucleotides are fluorescently labeled with streptavidin–phycoerythrin conjugate. The tagged sequence on the virus-specific oligonucleotide primers are also recognized following hybridization to complementary sequences on the respective virus-specific microbeads, each of which has a different virus-specific fluorescent dye profile. The samples are analyzed using a Luminex 100 flow instrument, which exposes the sample to a green laser and red laser. The green laser identifies the presence and amount of phycoerythrin and the red laser identifies the virus-specific fluorescent dye. Computational software allows determination of the presence or absence of the target virus (119,120). The FDA-approved xTag™ RVP assay can detect 10 different viruses (Table 7) and allows subtyping of influenza A viruses into H1 and H3 subtypes, but the assay can be modified to detect up to 20 different viruses (119).

A number of other multiplex assays for molecular detection of respiratory viruses are under development, are commercially available but not approved by the FDA as a diagnostic reagent, or are commercially available outside the United States (12,119,121–123). One such strategy is the use of microarray technology to detect multiple viruses (124,125). Complementary DNA is made from extracted viral RNA and is amplified in a sequence-independent fashion by random PCR. The amplified DNA is fluorescently labeled and hybridized to a microarray containing virus-specific oligonucleotides. The microarray is more sensitive than DFA and has similar sensitivity to that of individual RT-PCR assays (125). Another strategy combines PCR amplification with electrospray ionization-mass spectrometry to specifically identify viruses present in a sample. In this latter approach, the mass of amplicons is determined with enough

Table 7 Commercially Available Real-time RT-PCR Assays for Respiratory Virus Detection

Name	Test format	Manufacturer	Viruses detected	FDA approved (510 (K) number)
CDC Human Influenza Virus Real-time RT-PCR Detection and Characterization Panel	Multiplex Real-time RT-PCR	CDC	Influenza A and B viruses Subtype identification of A/H1, A/H3, A/H5 (Asian lineage)	Yes (K080570)
Influenza A/H5 (Asian lineage) virus Real-time RT-PCR Primer and Probe Set	Real-time RT-PCR	CDC	Influenza A/H5 (Asian lineage)	Yes (K060159)
MultiCode-PLx RVP	Multiplex Real-time RT-PCR, Target-specific primer extension, Fluidic microbead microarray	EraGen Biosciences	Influenza A and B viruses; RSV subtypes A and B; Parainfluenza viruses types 1–3, 4a, 4b; hMPV; rhinovirus; adenovirus species B, C and E; coronaviruses 229E, OC43, NL63	No
NGEN RVA ASR kit	Multiplex Real-time RT-PCR	Nanogen Inc.	Influenza A and B viruses; RSV; Parainfluenza viruses types 1–3	No
ProFlu+ Assay	Multiplex Real-time RT-PCR	Prodesse	Influenza A and B viruses, RSV	Yes (K073029)
Pro hMPV+ Assay	Real-time RT-PCR	Prodesse	hMPV	Yes (K082688)
Pro Paraflu+ Assay	Multiplex Real-time RT-PCR	Prodesse	Parainfluenza viruses types 1–3	Yes (K091053)
ResPlex II assay	Multiplex Real-time RT-PCR, Target-specific primer extension, Fluidic microbead microarray	Qiagen	Influenza A and B viruses, RSV subtypes A and B, Parainfluenza viruses types 1–4, hMPV, rhinovirus, enterovirus	No
Verigene Respiratory Virus Nucleic Acid Test	Multiplex RT-PCR	Nanosphere, Inc.	Influenza A and B viruses, RSV	Yes (K083088)
xTAG™ Respiratory Virus Panel	Multiplex Real-time RT-PCR, Target-specific primer extension, Fluidic microbead microarray	Luminex Molecular Diagnostics	Influenza A and B viruses, Influenza subtypes A/H1 and A/H3, RSV subtypes A and B, Parainfluenza viruses types 1–3, hMPV, rhinovirus, adenovirus	Yes (K081483)

Table 8 Preferred Cell Lines for Virus Isolation

Virus	Cell line
Rhinovirus	Human embryonic lung: WI-38, MRC-5
Coronavirus	
229E	MRC-5
NL63	LLC-MK2, Vero-B4
SARS	Vero E6
Adenovirus	Human embryonic kidney, A549, HeLa, HEp-2, KB
Influenzavirus A and B	Primary monkey kidney cells, Madin-Darby canine kidney (MDCK), LLC-MK2
Parainfluenzavirus	Primary monkey kidney cells, LLC-MK2
Respiratory syncytial virus	Human heteroploid cells: HEp-2, HeLa, A549
Human metapneumovirus	LLC-MK2

accuracy to predict their sequence, and virus-specific sequences are determined by comparison to a sequence database. This strategy has been successfully applied to the identification and serotyping of adenoviruses and to the identification of respiratory viruses in clinical samples (126,127).

Virus Isolation

A number of different strategies have been used for virus isolation, including inoculation of live animals (e.g., mice), embryonated eggs (e.g., influenza), cell culture, and organ culture (e.g., tracheal rings). Although embryonated eggs are still used in some influenza laboratories, almost all viral diagnostic labs that attempt to cultivate respiratory viruses utilize cell culture. The choice of cell line(s) to be used depends upon the virus(es) being sought. Unfortunately, there is no single cell line that will support the growth of all respiratory viruses (Table 8), and some viruses grow poorly (hMPV, coronaviruses 229E and NL63) or not at all in cell culture (coronaviruses OC43 and HKU-1). However, cells from different sources have been co-cultivated to increase the spectrum of virus strains detected. One such commercially available cell mixture is the R-mix FreshCells line (Diagnostic Hybrids, Athens, OH), a combination of mink lung cells (Mv1Lu) and human adenocarcinoma cells (A549). The R-mix cells have similar sensitivities to those observed with individual cell lines for detection of influenza A and B viruses, RSV, PIVs, and adenovirus (128–130).

Culture conditions need to be modified to allow the growth of certain viruses. For example, continuous cell lines do not produce proteases needed to cleave the hemagglutinin of interpandemic strains of influenza viruses, so exogenous protease must be added (TPCK-trypsin) to support viral growth. The presence of exogenous protein, such as fetal bovine serum, can inactivate the protease and prevent growth, so FBS-containing media needs to be removed and cells rinsed with FBS-free media prior to addition of trypsin-containing media. Some rhinoviruses grow better at lower temperatures, so incubation of cultures at 33–34°C instead of 36–37°C can improve the recovery of rhinoviruses.

A variety of methods are used to monitor cell culture once they are inoculated. Visual inspection for CPE can identify changes characteristic of virus replication, but confirmatory tests (such as immunofluorescence or EIAs) need to be performed for final virus identification. Evaluation for the presence of hemadsorption of red blood cells (RBCs) or hemagglutination activity of the culture supernatant can be performed every two to three days to identify early replication of the hemadsorbing influenza or PIVs. A disadvantage of these approaches is that it can take up to two weeks for a virus to grow to a detectable level in cell culture. To overcome this problem, shell vial assays have been developed to allow earlier detection of respiratory viruses in cell culture. The clinical sample is inoculated onto the cell monolayer by low-speed centrifugation, and after 24 to 48 hours of incubation, the cell line is fixed and stained with labeled, virus-specific antibodies (such as those in listed in Table 5). This approach can achieve comparable sensitivity to standard cell culture using longer incubation times (130), but it has the disadvantage of not providing virus for further characterization (such as antigenic analysis or antiviral testing).

Table 9 Methods for Virus Identification and Typing

Immunologic approaches	Molecular approaches
Immunofluorescence	RT-PCR
Enzyme immunoassay	Hybridization
Neutralization	Sequencing
Hemagglutination inhibition	Microarray

Identification and Typing Systems

A number of different methods can be used to identify and further characterize viral isolates and noncultured viruses in clinical samples (Table 9). The most commonly used are immunologic methods. Isolated viruses are most often identified using either immunofluorescence (applied to infected cells) or EIA. The same commercially available IF reagents and kits used to identify viruses in clinical samples can also be used to identify virus isolates (Table 5). These reagents have sensitivities and specificities of 98% to 100% when used in this manner. On the other hand, although the rapid EIA kits described in Table 6 will identify influenza and RSV from cell cultures, there are limited data available on the use of the assays for this purpose. They may give false-negative results when the virus is present in low concentrations, and the kits are not approved for this application. These assays target shared antigens, such as the influenza virus nucleoprotein or the RSV fusion protein, to initially identify the viruses. Similarly, more specific antisera can be used to further characterize viruses to subtype or even to a specific antigenic type. Such tests are more specialized and are not routinely performed by clinical diagnostic laboratories. Individual virus-specific antisera are also commercially available to identify some of the less common strains, such as parainfluenza types 4a and 4b, which are not included in many respiratory virus identification kits.

Influenza viruses and PIVs each have a viral hemagglutinin, and hemagglutination-inhibition (HAI) is another method for virus identification and typing. HAI assays can be type, subtype, or strain-specific, depending on the antisera used. Neutralization assays can also be used for virus identification, but these assays are more laborious because the isolate must be recultured. Nevertheless, this is the principal immunologic approach for typing rhinoviruses and adenoviruses. HAI can also be used to identify the adenoviruses that are able to hemagglutinate rat or monkey RBCs.

Molecular methods are increasingly being used for virus identification and characterization. The results of RT-PCR testing of clinical samples or cell culture harvests usually identifies a virus to the genus level, but further characterization using RT-PCR can be performed (12). For example, influenza virus and RSV subtypes can be determined by RT-PCR amplification of subtype-specific sequences (131,132). RT-PCR amplicons can be sequenced to determine genotype and to infer serotype, as has been done for the human rhinoviruses (133). Similarly, restriction fragment length polymorphism (RFLP) analysis of amplicons can be used to distinguish virus variants within a subtype when the sequences of both variants are known (134).

Hybridization is another method for virus identification. It is often used in conjunction with RT-PCR assays. In real-time RT-PCR assays the hybridization step is built into the amplification assay (see chap. 5). However, hybridization can also be used to directly identify viruses (135). Dot blot hybridization methods are not more sensitive than immunologic methods (such as ELISA) for virus detection and they are more cumbersome to perform, so direct identification of respiratory viruses by hybridization techniques is not frequently used. As noted earlier, microarray methods are being developed for identification of respiratory viruses in clinical samples. These methods can also be used to further characterize strains by subtype and variant-type by the design and inclusion of subtype- and variant-specific oligonucleotide probes on the microarray (136).

Serology

Serologic methods can be used to identify respiratory virus infections, but are rarely used for clinical diagnosis because they require the evaluation of paired sera (acute and convalescent) collected at least two weeks apart. Diagnosis cannot reliably be made with a single serum

Table 10 Serologic Methods for Different Respiratory Viruses

Virus	Complement fixation	ELISA	HAI	Neutralization
Rhinovirus	NA	NA	Rarely used	Serotype-specific
Coronavirus				
229E	Serotype-specific	Serotype-specific	NA	Serotype-specific
OC43	Serotype-specific	Serotype-specific	Serotype-specific	Serotype-specific
NL63	ND	Serotype-specific	NA	Serotype-specific
SARS	ND	Serotype-specific	NA	Serotype-specific
Adenovirus	Genus-specific	Genus-specific	Serotype-specific	Serotype-specific
Influenzavirus A and B	Type-specific	Type- or subtype-specific	Subtype-specific	Subtype-specific
Parainfluenzavirus	Serotype-specific; cross-reactivity between serotypes	Type-specific	Serotype-specific; cross-reactivity between serotypes	Serotype-specific
Respiratory syncytial virus	Genus-specific	Genus- or type-specific	NA	Genus- or type-specific
Human metapneumovirus	ND	Genus- or type-specific	NA	Serotype-specific

Abbreviations: NA, not available; ND, not described.

sample. Thus, these methods are usually used for epidemiologic and other clinical research studies. A four-fold or greater increase in serum antibody level is considered indicative of infection. A less than four-fold increase in antibody level should not be interpreted as evidence of infection. There are a variety of methods that can be used, and the most common are shown in Table 10.

Complement fixation (CF) assays are relatively easy to perform and assays have been developed for influenza viruses, RSV, parainfluenzaviruses, adenoviruses, and coronaviruses 229E and OC43. These assays target genus- or type-specific epitopes. CF assays are not used for rhinoviruses and enteroviruses because of heterotypic responses and lack of correlation of seroresponse with those measured by neutralization methods. CF assays could be developed but have not been described for the more recently discovered viruses, such as HMPV and the new coronaviruses.

Enzyme-linked immunosorbent assays (ELISAs) are among the most flexible of the serologic assays. The antigen(s) used in these assays can range from whole virus to viral proteins. ELISAs are relatively easy to perform, and they have been applied to measuring antibody responses not only in serum samples but also in respiratory secretions. They can also be modified to measure specific immunoglobulin (IgG, IgA, IgM) responses after infection or vaccination (137,138).

HAI assays can be performed for viruses that have hemagglutination activity. These include influenza viruses, PIV, coronavirus OC43, some adenoviruses and rhinoviruses, and enteroviruses. Not all viruses will hemagglutinate RBCs from a single species, so the source of the RBCs is selected based upon the virus HAI assay being performed. HAI assays are most commonly performed for influenza and PIVs. Chicken or turkey RBCs are commonly used for influenza virus assays, although horse RBCs are preferable when measuring influenza A/H5 antibodies (139). Guinea pig or human RBCs are used for PIV HAI assays. Human serum can contain nonspecific inhibitors of hemagglutinins that will interfere with the performance of HAI assays. The inhibitors can usually be removed by pretreatment of serum with receptor-destroying enzyme (140).

Neutralization assays are based upon the ability of serum (or respiratory secretion) to prevent virus replication, so these assays can only be performed with the cultivable viruses. A variety of different formats have been used, including macroneutralization (in a culture tube), microneutralization (in a 96-well plate), plaque reduction, and fluorescent focus reduction. Virus is mixed and incubated with dilutions of serum, and then inoculated onto the cell culture. The quantity of virus used in the assay depends on the assay format and the virus being used, but varies from <10 TCID50 for the rhinovirus macroneutralization assays to 100 TCID50 for influenza and PIVs. Evidence of neutralization in a serum dilution also varies by assay and includes absence of CPE on the cell culture (141,142), absence of hemagglutination activity in the culture supernatant (143), and reduction in the number of plaques in the cell culture (144).

Antiviral Susceptibility Testing

There are only a limited number of antivirals available for treatment of respiratory virus infections. Ribavirin has been used to treat RSV infections, and resistance has not been recognized to this drug. The adamantanes and neuraminidase inhibitors are the only other antivirals available and they are active against the influenza viruses. Resistance to both classes of drugs has been described and can be measured in vitro.

The adamantanes (amantadine and rimantadine) are only active against influenza A viruses. A single point mutation in the transmembrane portion of one of several amino acids in the viral M2 protein can lead to resistance to this class of drugs. The gold standard for detecting resistance is the plaque inhibition assay (145). RT-PCR amplification of the M2 gene with post-amplification analysis by RFLP or by direct sequencing of the amplicons can also identify genetic mutations that are associated with resistance (146,147). The importance of resistance to the adamantanes is highlighted by the widespread resistance that has emerged among influenza A/H3N2 viruses (148). The pandemic 2009 H1N1 influenza viruses also show a high frequency (>99%) of resistance to the adamantanes (149).

The neuraminidase inhibitors (oseltamivir and zanamivir) are active against both influenza A and influenza B viruses. Resistance to the neuraminidase inhibitors has been described in both groups of viruses (150). Resistance cannot reliably be measured using whole virus in cell culture growth-inhibition assays. Instead, the ability of the drugs to inhibit the enzymatic activity of viral neuraminidase on chemiluminescent or fluorescent substrates is used to identify resistance (151). The neuraminidase inhibition assay has been developed into a commercially available kit (NA-Star, Applied Biosystems) that has been used successfully in surveillance programs to screen influenza virus isolates for resistance (152). Another approach is to detect resistance through the sequencing of the neuraminidase gene following amplification by RT-PCR. This approach can identify specific mutations that have been associated with resistance to oseltamivir, zanamivir, or both drugs (153), but it cannot determine whether a novel mutation in the neuraminidase gene causes resistance. Recently, widespread resistance to oseltamivir has emerged in the United States and Europe in >90% of influenza A/H1N1 viruses, influencing recommendations for the empiric selection of antivirals to treat acute influenza and highlighting the importance of surveillance for antiviral resistance among influenza viruses (154). However, among the novel 2009 H1N1 influenza viruses, resistance to oseltamivir is rare (<1%) through late 2009 (149).

Evaluation and Reporting of Laboratory Results

Each laboratory must determine the goals of its respiratory virus diagnostic program. The use of rapid diagnostic kits (antigen or molecular detection) can quickly identify respiratory virus infection and allow the clinician to initiate targeted antiviral therapy (if appropriate) and to institute appropriate infection control procedures. Cell culture can be performed on samples that are negative by rapid assay while positive samples undergo no additional testing. This approach is cost-effective, but it has the disadvantage that it does not yield virus isolates for further characterization.

A number of factors influence the selection of assays offered by a diagnostic laboratory, including local expertise and the availability of staffing and equipment. There are initial capital expenses associated with the procurement of fluorescent microscopes for IF studies or real-time thermal cyclers used in molecular assays. Expertise is needed to interpret IF assays and to perform and troubleshoot some of the more complex molecular assays. The limited availability of more experienced staff or the need to provide results at all times during the day or night may lead to the decision to initially offer testing with the less complex immunochromatographic assays. Test selection will also influence the manner of collection and the type of clinical sample needed for analysis. Once decisions have been made on the assays that are to be offered, standard operating procedures should be put in place for the collection, transportation, and processing of clinical samples. Such procedures should decrease the likelihood of inaccurate test results due to mishandling of samples. In addition, the performance of each assay should be evaluated periodically with positive and negative controls as part of a quality control program.

REFERENCES

1. Abed Y, Wang D, Boivin G. WU polyomavirus in children, Canada. Emerg Infect Dis 2007; 13(12):1939–1941.
2. Allander T. Human bocavirus. J Clin Virol 2008; 41(1):29–33.
3. Bosis S, Esposito S, Niesters HG, et al. Coronavirus HKU1 in an Italian pre-term infant with bronchiolitis. J Clin Virol 2007; 38(3):251–253.
4. Falsey AR. Human metapneumovirus infection in adults. Pediatr Infect Dis J 2008; 27(suppl 10):S80–S83.
5. Fouchier RA, Hartwig NG, Bestebroer TM, et al. A previously undescribed coronavirus associated with respiratory disease in humans. Proc Natl Acad Sci U S A 2004; 101(16):6212–6216.
6. Ksiazek TG, Erdman D, Goldsmith CS, et al. A novel coronavirus associated with severe acute respiratory syndrome. N Engl J Med 2003; 348(20):1953–1966.
7. Raoult D, La Scola B, Birtles R. The discovery and characterization of Mimivirus, the largest known virus and putative pneumonia agent. Clin Infect Dis 2007; 45(1):95–102.
8. Sloots TP, Whiley DM, Lambert SB, et al. Emerging respiratory agents: new viruses for old diseases? J Clin Virol 2008; 42(3):233–243.
9. van Elden LJ, van Loon AM, van Alphen F, et al. Frequent detection of human coronaviruses in clinical specimens from patients with respiratory tract infection by use of a novel real-time reverse-transcriptase polymerase chain reaction. J Infect Dis 2004; 189(4):652–657.
10. Xiang Z, Gonzalez R, Xie Z, et al. Human rhinovirus group C infection in children with lower respiratory tract infection. Emerg Infect Dis 2008; 14(10):1665–1667.
11. Beckham JD, Cadena A, Lin J, Piedra PA, et al. Respiratory viral infections in patients with chronic, obstructive pulmonary disease. J Infect 2005; 50(4):322–330.
12. Mahony JB. Detection of respiratory viruses by molecular methods. Clin Microbiol Rev 2008; 21(4):716–747.
13. Mazzoncini JP Jr., Crowell CB, Kang CS. Human metapneumovirus: An emerging respiratory pathogen. J Emerg Med 2008.
14. Nicholls JM, Peiris JS. Avian influenza: Update on pathogenesis and laboratory diagnosis. Respirology 2008; 13(suppl 1):S14–S18.
15. Pyrc K, Berkhout B, van der HL. The novel human coronaviruses NL63 and HKU1. J Virol 2007; 81(7):3051–3057.
16. Stempel HE, Martin ET, Kuypers J, et al. Multiple viral respiratory pathogens in children with bronchiolitis. Acta Paediatr 2009; 98(1):123–126.
17. Mackay IM. Human rhinoviruses: the cold wars resume. J Clin Virol 2008; 42(4):297–320.
18. Monto AS, Bryan ER, Ohmit S. Rhinovirus infections in Tecumseh, Michigan: frequency of illness and number of serotypes. J Infect Dis 1987; 156(1):43–49.
19. Gwaltney JM Jr, Heinz BA. Rhinovirus. In: Richman DD, Whitley RJ, Hayden FG, eds. Clinical Virology. Washington, DC: ASM Press, 2002:995–1018.
20. Gwaltney JM Jr. Rhinoviruses. In: Evans AS, Kaslow RA, eds. Viral Infections of Humans, Epidemiology and Control. New York, NY: Plenum Medical Book Company, 1997:815.
21. Dingle KE, Crook D, Jeffery K. Stable and noncompetitive RNA internal control for routine clinical diagnostic reverse transcription-PCR. J Clin Microbiol 2004; 42(3):1003–1011.
22. Gwaltney JM Jr, Hendley JO, Simon G, et al. Rhinovirus infections in an industrial population. I. The occurrence of illness. N Engl J Med 1966; 275(23):1261–1268.
23. McMillan JA, Sandstrom C, Weiner LB, et al. Viral and bacterial organisms associated with acute pharyngitis in a school-aged population. J Pediatr 1986; 109(5):747–752.
24. Glezen WP, Fernald GW, Lohr JA. Acute respiratory disease of university students with special reference to the etiologic role of Herpesvirus hominis. Am J Epidemiol 1975; 101(2):111–121.
25. Hare CB, Kahn JO. Primary HIV infection. Curr Infect Dis Rep 2004; 6(1):65–71.
26. Bisno AL. Acute pharyngitis: Etiology and diagnosis. Pediatrics 1996; 97(6 Pt 2):949–954.
27. Arola M, Ziegler T, Ruuskanen O. Respiratory virus infection as a cause of prolonged symptoms in acute otitis media. J Pediatr 1990; 116(5):697–701.
28. Chonmaitree T, Revai K, Grady JJ, et al. Viral upper respiratory tract infection and otitis media complication in young children. Clin Infect Dis 2008; 46(6):815–823.
29. Pitkaranta A, Virolainen A, Jero J, et al. Detection of rhinovirus, respiratory syncytial virus, and coronavirus infections in acute otitis media by reverse transcriptase polymerase chain reaction. Pediatrics 1998; 102(2 Pt 1):291–295.

30. Bjornson CL, Johnson DW. Croup. Lancet 2008; 371(9609):329–339.
31. Cherry JD. State of the evidence for standard-of-care treatments for croup: are we where we need to be? Pediatr Infect Dis J 2005; 24(suppl 11):S198–S202, discussion.
32. Cherry JD. Clinical practice. Croup. N Engl J Med 2008; 358(4):384–391.
33. Marx A, Torok TJ, Holman RC, et al. Pediatric hospitalizations for croup (laryngotracheobronchitis): Biennial increases associated with human parainfluenza virus 1 epidemics. J Infect Dis 1997; 176(6):1423–1427.
34. Rihkanen H, Ronkko E, Nieminen T, et al. Respiratory viruses in laryngeal croup of young children. J Pediatr 2008; 152(5):661–665.
35. Wenzel RP, Fowler AA III. Clinical practice. Acute bronchitis. N Engl J Med 2006; 355(20):2125–2130.
36. Gonzales R, Bartlett JG, Besser RE, et al. Principles of appropriate antibiotic use for treatment of uncomplicated acute bronchitis: Background. Ann Intern Med 2001; 134(6):521–529.
37. Yanney M, Vyas H. The treatment of bronchiolitis. Arch Dis Child 2008; 93(9):793–798.
38. Durbin WJ, Stille C. Pneumonia. Pediatr Rev 2008; 29(5):147–158.
39. Greenberg SB. Viral pneumonia. Infect Dis Clin North Am 1991; 5(3):603–621.
40. Mohsen AH, McKendrick M. Varicella pneumonia in adults. Eur Respir J 2003; 21(5):886–891.
41. Kim YJ, Boeckh M, Englund JA. Community respiratory virus infections in immunocompromised patients: Hematopoietic stem cell and solid organ transplant recipients, and individuals with human immunodeficiency virus infection. Semin Respir Crit Care Med 2007; 28(2):222–242.
42. Peltola V, Waris M, Osterback R, et al. Rhinovirus transmission within families with children: Incidence of symptomatic and asymptomatic infections. J Infect Dis 2008; 197(3):382–389.
43. Peltola V, Waris M, Osterback R, et al. Clinical effects of rhinovirus infections. J Clin Virol 2008; 43(4):411–414.
44. Greve JM, Davis G, Meyer AM, et al. The major human rhinovirus receptor is ICAM-1. Cell 1989; 56(5):839–847.
45. Hofer F, Gruenberger M, Kowalski H, et al. Members of the low density lipoprotein receptor family mediate cell entry of a minor-group common cold virus. Proc Natl Acad Sci U S A 1994; 91(5):1839–1842.
46. Uncapher CR, DeWitt CM, Colonno RJ. The major and minor group receptor families contain all but one human rhinovirus serotype. Virology 1991; 180(2):814–817.
47. McErlean P, Shackelton LA, Lambert SB, et al. Characterisation of a newly identified human rhinovirus, HRV-QPM, discovered in infants with bronchiolitis. J Clin Virol 2007; 39(2):67–75.
48. Kistler A, Avila PC, Rouskin S, et al. Pan-viral screening of respiratory tract infections in adults with and without asthma reveals unexpected human coronavirus and human rhinovirus diversity. J Infect Dis 2007; 196(6):817–825.
49. Briese T, Renwick N, Venter M, et al. Global distribution of novel rhinovirus genotype. Emerg Infect Dis 2008; 14(6):944–947.
50. Yeager CL, Ashmun RA, Williams RK, et al. Human aminopeptidase N is a receptor for human coronavirus 229E. Nature 1992; 357(6377):420–422.
51. Hofmann H, Pyrc K, van der Hoek L, et al. Human coronavirus NL63 employs the severe acute respiratory syndrome coronavirus receptor for cellular entry. Proc Natl Acad Sci USA 2005; 102(22):7988–7993.
52. van der HL, Pyrc K, Jebbink MF, et al. Identification of a new human coronavirus. Nat Med 2004; 10(4):368–373.
53. van der HL, Sure K, Ihorst G, et al. Croup is associated with the novel coronavirus NL63. PLoS Med 2005; 2(8):e240.
54. van der HL, Pyrc K, Berkhout B. Human coronavirus NL63, a new respiratory virus. FEMS Microbiol Rev 2006; 30(5):760–773.
55. Piedra PA Jr. Englund JA, Glezen WP. Respiratory syncytial virus and parainfluenza viruses. In: Richman DD, Whitley RJ, Hayden FG, eds. Clinical Virology. Washington, DC: ASM Press, 2002:763–790.
56. Cheuk DK, Tang IW, Chan KH, et al. Rhinovirus infection in hospitalized children in Hong Kong: A prospective study. Pediatr Infect Dis J 2007; 26(11):995–1000.
57. Corne JM, Marshall C, Smith S, et al. Frequency, severity, and duration of rhinovirus infections in asthmatic and non-asthmatic individuals: a longitudinal cohort study. Lancet 2002; 359(9309):831–834.
58. Greenberg SB, Allen M, Wilson J, et al. Respiratory viral infections in adults with and without chronic obstructive pulmonary disease. Am J Respir Crit Care Med 2000; 162(1):167–173.
59. Greenberg SB. Respiratory consequences of rhinovirus infection. Arch Intern Med 2003; 163(3):278–284.
60. Wos M, Sanak M, Soja J, et al. The presence of rhinovirus in lower airways of patients with bronchial asthma. Am J Respir Crit Care Med 2008; 177(10):1082–1089.

61. Larson HE, Reed SE, Tyrrell DA. Isolation of rhinoviruses and coronaviruses from 38 colds in adults. J Med Virol 1980; 5(3):221–229.
62. Wenzel RP, Hendley JO, Davies JA, et al. Coronavirus infections in military recruits. Three-year study with coronavirus strains OC43 and 229E. Am Rev Respir Dis 1974; 109(6):621–624.
63. Heugel J, Martin ET, Kuypers J, et al. Coronavirus-associated pneumonia in previously healthy children. Pediatr Infect Dis J 2007; 26(8):753–755.
64. Keightley MC, Sillekens P, Schippers W, et al. Real-time NASBA detection of SARS-associated coronavirus and comparison with real-time reverse transcription-PCR. J Med Virol 2005; 77(4):602–608.
65. Dijkman R, Jebbink MF, El Idrissi NB, et al. Human coronavirus NL63 and 229E seroconversion in children. J Clin Microbiol 2008; 46(7):2368–2373.
66. Drosten C, Gunther S, Preiser W, et al. Identification of a novel coronavirus in patients with severe acute respiratory syndrome. N Engl J Med 2003; 348(20):1967–1976.
67. Vabret A, Mourez T, Dina J, et al. Human coronavirus NL63, France. Emerg Infect Dis 2005; 11(8):1225–1229.
68. Arden KE, Nissen MD, Sloots TP, et al. New human coronavirus, HCoV-NL63, associated with severe lower respiratory tract disease in Australia. J Med Virol 2005; 75(3):455–462.
69. Ebihara T, Endo R, Ma X, et al. Detection of human coronavirus NL63 in young children with bronchiolitis. J Med Virol 2005; 75(3):463–465.
70. Bastien N, Anderson K, Hart L, et al. Human coronavirus NL63 infection in Canada. J Infect Dis 2005; 191(4):503–506.
71. Chiu SS, Chan KH, Chu KW, et al. Human coronavirus NL63 infection and other coronavirus infections in children hospitalized with acute respiratory disease in Hong Kong, China. Clin Infect Dis 2005; 40(12):1721–1729.
72. Vabret A, Dina J, Gouarin S, et al. Human (non-severe acute respiratory syndrome) coronavirus infections in hospitalised children in France. J Paediatr Child Health 2008; 44(4):176–181.
73. Wu PS, Chang LY, Berkhout B, et al. Clinical manifestations of human coronavirus NL63 infection in children in Taiwan. Eur J Pediatr 2008; 167(1):75–80.
74. Esper F, Weibel C, Ferguson D, et al. Coronavirus HKU1 infection in the United States. Emerg Infect Dis 2006; 12(5):775–779.
75. Kupfer B, Simon A, Jonassen CM, et al. Two cases of severe obstructive pneumonia associated with an HKU1-like coronavirus. Eur J Med Res 2007; 12(3):134–138.
76. Lau SK, Woo PC, Yip CC, et al. Coronavirus HKU1 and other coronavirus infections in Hong Kong. J Clin Microbiol 2006; 44(6):2063–2071.
77. Gerna G, Campanini G, Rovida F, et al. Genetic variability of human coronavirus OC43-, 229E-, and NL63-like strains and their association with lower respiratory tract infections of hospitalized infants and immunocompromised patients. J Med Virol 2006; 78(7):938–949.
78. Gerna G, Percivalle E, Sarasini A, et al. Human respiratory coronavirus HKU1 versus other coronavirus infections in Italian hospitalised patients. J Clin Virol 2007; 38(3):244–250.
79. Esper F, Weibel C, Ferguson D, et al. Evidence of a novel human coronavirus that is associated with respiratory tract disease in infants and young children. J Infect Dis 2005; 191(4):492–498.
80. Henrickson KJ. Parainfluenza viruses. Clin Microbiol Rev 2003; 16(2):242–264.
81. Fry AM, Curns AT, Harbour K, et al. Seasonal trends of human parainfluenza viral infections: United States, 1990–2004. Clin Infect Dis 2006; 43(8):1016–1022.
82. Chemaly RF, Ghosh S, Bodey GP, et al. Respiratory viral infections in adults with hematologic malignancies and human stem cell transplantation recipients: a retrospective study at a major cancer center. Medicine (Baltimore) 2006; 85(5):278–287.
83. Chakrabarti S, Collingham KE, Holder K, et al. Parainfluenza virus type 3 infections in hematopoetic stem cell transplant recipients: Response to ribavirin therapy. Clin Infect Dis 2000; 31(6):1516–1518.
84. Elizaga J, Olavarria E, Apperley J, et al. Parainfluenza virus 3 infection after stem cell transplant: Relevance to outcome of rapid diagnosis and ribavirin treatment. Clin Infect Dis 2001; 32(3):413–418.
85. Nichols WG, Corey L, Gooley T, et al. Parainfluenza virus infections after hematopoietic stem cell transplantation: risk factors, response to antiviral therapy, and effect on transplant outcome. Blood 2001; 98(3):573–578.
86. Atmar RL, Guy E, Guntupalli KK, et al. Respiratory tract viral infections in inner-city asthmatic adults. Arch Intern Med 1998; 158(22):2453–2459.
87. Todd Faulks J, Drinka PJ, Shult P. A serious outbreak of parainfluenza type 3 on a nursing unit. J Am Geriatr Soc 2000; 48(10):1216–1218.
88. Yorita KL, Holman RC, Sejvar JJ, et al. Infectious disease hospitalizations among infants in the United States. Pediatrics 2008; 121(2):244–252.
89. Thompson WW, Shay DK, Weintraub E, et al. Mortality associated with influenza and respiratory syncytial virus in the United States. JAMA 2003; 289(2):179–186.

90. Hall CB. Nosocomial respiratory syncytial virus infections: The "Cold War" has not ended. Clin Infect Dis 2000; 31(2):590–596.
91. Raad I, Abbas J, Whimbey E. Infection control of nosocomial respiratory viral disease in the immunocompromised host. Am J Med 1997; 102(3A):48–52.
92. Kahn JS. Epidemiology of human metapneumovirus. Clin Microbiol Rev 2006; 19(3):546–557.
93. Couch RB, Kasel JA, Glezen WP, et al. Influenza: Its control in persons and populations. J Infect Dis 1986; 153(3):431–440.
94. Thompson WW, Shay DK, Weintraub E, et al. Influenza-associated hospitalizations in the United States. JAMA 2004; 292(11):1333–1340.
95. World Health Organization. New influenza A (H1N1) virus: global epidemiological situation, June 2009. Wkly Epidemiol Rec 2009; 84(24):244–248.
96. Ahluwalia G, Embree J, McNicol P, et al. Comparison of nasopharyngeal aspirate and nasopharyngeal swab specimens for respiratory syncytial virus diagnosis by cell culture, indirect immunofluorescence assay, and enzyme-linked immunosorbent assay. J Clin Microbiol 1987; 25(5):763–767.
97. Macfarlane P, Denham J, Assous J, et al. RSV testing in bronchiolitis: Which nasal sampling method is best? Arch Dis Child 2005; 90(6):634–635.
98. Heikkinen T, Marttila J, Salmi AA, et al. Nasal swab versus nasopharyngeal aspirate for isolation of respiratory viruses. J Clin Microbiol 2002; 40(11):4337–4339.
99. Abu-Diab A, Azzeh M, Ghneim R, et al. Comparison between pernasal flocked swabs and nasopharyngeal aspirates for detection of common respiratory viruses in samples from children. J Clin Microbiol 2008; 46(7):2414–2417.
100. Chan KH, Peiris JS, Lim W, et al. Comparison of nasopharyngeal flocked swabs and aspirates for rapid diagnosis of respiratory viruses in children. J Clin Virol 2008; 42(1):65–69.
101. Moore C, Corden S, Sinha J, et al. Dry cotton or flocked respiratory swabs as a simple collection technique for the molecular detection of respiratory viruses using real-time NASBA. J Virol Methods 2008; 153(2):84–89.
102. Daley P, Castriciano S, Chernesky M, et al. Comparison of flocked and rayon swabs for collection of respiratory epithelial cells from uninfected volunteers and symptomatic patients. J Clin Microbiol 2006; 44(6):2265–2267.
103. Wang WK, Chen SY, Liu IJ, et al. Detection of SARS-associated coronavirus in throat wash and saliva in early diagnosis. Emerg Infect Dis 2004; 10(7):1213–1219.
104. Covalciuc KA, Webb KH, Carlson CA. Comparison of four clinical specimen types for detection of influenza A and B viruses by optical immunoassay (FLU OIA test) and cell culture methods. J Clin Microbiol 1999; 37(12):3971–3974.
105. Robinson JL, Lee BE, Kothapalli S, et al. Use of throat swab or saliva specimens for detection of respiratory viruses in children. Clin Infect Dis 2008; 46(7):e61–e64.
106. Johnson FB. Transport of viral specimens. Clin Microbiol Rev 1990; 3(2):120–131.
107. Baxter BD, Couch RB, Greenberg SB, et al. Maintenance of viability and comparison of identification methods for influenza and other respiratory viruses of humans. J Clin Microbiol 1977; 6(1):19–22.
108. Ptakova M, Tumova B. Detection of type A and B influenza viruses in clinical materials by immunoelectronmicroscopy. Acta Virol 1985; 29(1):19–24.
109. Liu C. Rapid diagnosis of human influenza infection from nasal smears by means of fluorescein-labeled antibody. Proc Soc Exp Viol Med 1956; 92:883–887.
110. Sizun J, Arbour N, Talbot PJ. Comparison of immunofluorescence with monoclonal antibodies and RT-PCR for the detection of human coronaviruses 229E and OC43 in cell culture. J Virol Methods 1998; 72(2):145–152.
111. Dominguez EA, Taber LH, Couch RB. Comparison of rapid diagnostic techniques for respiratory syncytial and influenza A virus respiratory infections in young children. J Clin Microbiol 1993; 31(9):2286–2290.
112. Swierkosz EM, Flanders R, Melvin L, et al. Evaluation of the Abbott TESTPACK RSV enzyme immunoassay for detection of respiratory syncytial virus in nasopharyngeal swab specimens. J Clin Microbiol 1989; 27(6):1151–1154.
113. Zambon M. Laboratory diagnosis of influenza. In: Nicholson KG, Webster RG, Hay AJ, eds. Textbook of Influenza. London, UK: Blackwell Science Ltd., 1988:291–313.
114. Selvarangan R, Abel D, Hamilton M. Comparison of BD Directigen EZ RSV and Binax NOW RSV tests for rapid detection of respiratory syncytial virus from nasopharyngeal aspirates in a pediatric population. Diagn Microbiol Infect Dis 2008; 62(2):157–161.
115. Uyeki TM. Influenza diagnosis and treatment in children: A review of studies on clinically useful tests and antiviral treatment for influenza. Pediatr Infect Dis J 2003; 22(2):164–177.
116. Ginocchio CC, Zhang F, Manji R, et al. Evaluation of multiple test methods for the detection of the novel 2009 influenza A (H1N1) during the New York City outbreak. J Clin Virol 2009; 45(3):191–195.

117. Achyuthan KE, Pence LM, Appleman JR, et al. ZstatFlu-II test: A chemiluminescent neuraminidase assay for influenza viral diagnostics. Luminescence 2003; 18(3):131–139.
118. Noyola DE, Clark B, O'Donnell FT, et al. Comparison of a new neuraminidase detection assay with an enzyme immunoassay, immunofluorescence, and culture for rapid detection of influenza A and B viruses in nasal wash specimens. J Clin Microbiol 2000; 38(3):1161–1165.
119. Mahony J, Chong S, Merante F, et al. Development of a respiratory virus panel test for detection of twenty human respiratory viruses by use of multiplex PCR and a fluid microbead-based assay. J Clin Microbiol 2007; 45(9):2965–2970.
120. Merante F, Yaghoubian S, Janeczko R. Principles of the xTAG respiratory viral panel assay (RVP Assay). J Clin Virol 2007; 40(suppl 1):S31–S35.
121. Tillmann RL, Simon A, Muller A, et al. Sensitive commercial NASBA assay for the detection of respiratory syncytial virus in clinical specimen. PLoS One 2007; 2(12):e1357.
122. Li H, McCormac MA, Estes RW, Sefers SE, et al. Simultaneous detection and high-throughput identification of a panel of RNA viruses causing respiratory tract infections. J Clin Microbiol 2007; 45(7):2105–2109.
123. Nolte FS, Marshall DJ, Rasberry C, et al. MultiCode-PLx system for multiplexed detection of seventeen respiratory viruses. J Clin Microbiol 2007; 45(9):2779–2786.
124. Wang D, Coscoy L, Zylberberg M, et al. Microarray-based detection and genotyping of viral pathogens. Proc Natl Acad Sci U S A 2002; 99(24):15687–15692.
125. Chiu CY, Urisman A, Greenhow TL, et al. Utility of DNA microarrays for detection of viruses in acute respiratory tract infections in children. J Pediatr 2008; 153(1):76–83.
126. Blyn LB, Hall TA, Libby B, et al. Rapid detection and molecular serotyping of adenovirus by use of PCR followed by electrospray ionization mass spectrometry. J Clin Microbiol 2008; 46(2):644–651.
127. Lodes MJ, Suciu D, Wilmoth JL, et al. Identification of upper respiratory tract pathogens using electrochemical detection on an oligonucleotide microarray. PLoS One 2007; 2(9):e924.
128. Weinberg A, Brewster L, Clark J, et al. Evaluation of R-Mix shell vials for the diagnosis of viral respiratory tract infections. J Clin Virol 2004; 30(1):100–105.
129. St George K, Patel NM, Hartwig RA, et al. Rapid and sensitive detection of respiratory virus infections for directed antiviral treatment using R-Mix cultures. J Clin Virol 2002; 24(1–2):107–115.
130. Barenfanger J, Drake C, Mueller T, et al. R-Mix cells are faster, at least as sensitive and marginally more costly than conventional cell lines for the detection of respiratory viruses. J Clin Virol 2001; 22(1):101–110.
131. Atmar RL, Baxter BD. Typing and subtyping clinical isolates of influenza virus using reverse transcription-polymerase chain reaction. Clin Diagn Virol 1996; 7(2):77–84.
132. Stockton J, Ellis JS, Saville M, et al. Multiplex PCR for typing and subtyping influenza and respiratory syncytial viruses. J Clin Microbiol 1998; 36(10):2990–2995.
133. Lee WM, Kiesner C, Pappas T, et al. A diverse group of previously unrecognized human rhinoviruses are common causes of respiratory illnesses in infants. PLoS One 2007; 2(10):e966.
134. O'Donnell FT, Munoz FM, Atmar RL, et al. Epidemiology and molecular characterization of co-circulating influenza A/H3N2 virus variants in children: Houston, Texas, 1997–8. Epidemiol Infect 2003; 130(3):521–531.
135. Lehtomaki K, Julkunen I, Sandelin K, et al. Rapid diagnosis of respiratory adenovirus infections in young adult men. J Clin Microbiol 1986; 24(1):108–111.
136. Wang Z, Malanoski AP, Lin B, et al. Resequencing microarray probe design for typing genetically diverse viruses: human rhinoviruses and enteroviruses. BMC Genomics 2008; 9:577.
137. Belshe RB, Gruber WC, Mendelman PM, et al. Correlates of immune protection induced by live, attenuated, cold-adapted, trivalent, intranasal influenza virus vaccine. J Infect Dis 2000; 181(3):1133–1137.
138. Burlington DB, Clements ML, Meiklejohn G, et al. Hemagglutinin-specific antibody responses in immunoglobulin G, A, and M isotypes as measured by enzyme-linked immunosorbent assay after primary or secondary infection of humans with influenza A virus. Infect Immun 1983; 41(2):540–545.
139. Stephenson I, Wood JM, Nicholson KG, et al. Sialic acid receptor specificity on erythrocytes affects detection of antibody to avian influenza haemagglutinin. J Med Virol 2003; 70(3):391–398.
140. Dowdle WR, Kendal AP, Noble GR. Influenza viruses. In: Lennette EH, Schmidt NJ, eds. Procedures for Viral, Rickettsial and Chlamydial Infections. Washington, DC: American Public Health Association, 1979:585–609.
141. Douglas RG Jr., Fleet WF, Cate TR,et al. Antibody to rhinovirus in human sera. I. Standardization of a neutralization test. Proc Soc Exp Biol Med 1968; 127(2):497–502.
142. Piedra PA, Wyde PR, Castleman WL, et al. Enhanced pulmonary pathology associated with the use of formalin-inactivated respiratory syncytial virus vaccine in cotton rats is not a unique viral phenomenon. Vaccine 1993; 11(14):1415–1423.

143. Frank AL, Puck J, Hughes BJ, et al. Microneutralization test for influenza A and B and parainfluenza 1 and 2 viruses that uses continuous cell lines and fresh serum enhancement. J Clin Microbiol 1980; 12(3):426–432.
144. Siber GR, Leombruno D, Leszczynski J, et al. Comparison of antibody concentrations and protective activity of respiratory syncytial virus immune globulin and conventional immune globulin. J Infect Dis 1994; 169(6):1368–1373.
145. Hayden FG, Cote KM, Douglas RG Jr. Plaque inhibition assay for drug susceptibility testing of influenza viruses. Antimicrob Agents Chemother 1980; 17(5):865–870.
146. Englund JA, Champlin RE, Wyde PR, et al. Common emergence of amantadine- and rimantadine-resistant influenza A viruses in symptomatic immunocompromised adults. Clin Infect Dis 1998; 26(6):1418–1424.
147. Klimov AI, Rocha E, Hayden FG, et al. Prolonged shedding of amantadine-resistant influenzae A viruses by immunodeficient patients: detection by polymerase chain reaction-restriction analysis. J Infect Dis 1995; 172(5):1352–1355.
148. Bright RA, Shay DK, Shu B, et al. Adamantane resistance among influenza A viruses isolated early during the 2005–2006 influenza season in the United States. JAMA 2006; 295(8):891–894.
149. http://www.cdc.gov/flu/weekly/weeklyarchives2009-2010/weekly47.htm. Accessed December 11, 2009.
150. Ison MG, Gubareva LV, Atmar RL, et al. Recovery of drug-resistant influenza virus from immunocompromised patients: a case series. J Infect Dis 2006; 193(6):760–764.
151. McKimm-Breschkin J, Trivedi T, Hampson A, et al. Neuraminidase sequence analysis and susceptibilities of influenza virus clinical isolates to zanamivir and oseltamivir. Antimicrob Agents Chemother 2003; 47(7):2264–2272.
152. Sheu TG, Deyde VM, Okomo-Adhiambo M, et al. Surveillance for neuraminidase inhibitor resistance among human influenza A and B viruses circulating worldwide from 2004 to 2008. Antimicrob Agents Chemother 2008; 52(9):3284–3292.
153. Lackenby A, Thompson CI, Democratis J. The potential impact of neuraminidase inhibitor resistant influenza. Curr Opin Infect Dis 2008; 21(6):626–638.
154. CDC. Update: Influenza activity—United States, September 28, 2008–January 31, 2009. MMWR Morb Mortal Wkly Rep 2009; 58(5):115–119.

16 Gastrointestinal Tract Infections

Robert L. Atmar and Mary K. Estes

Department of Medicine and Department of Molecular Virology and Microbiology, Baylor College of Medicine, Houston, Texas, U.S.A.

INTRODUCTION

Gastroenteritis is one of the most common diseases of mankind. Enteric viruses are important causes of gastroenteritis, but the overall contribution of these agents to disease burden is unknown. Gastrointestinal symptoms (diarrhea, abdominal pain, vomiting) can be observed following infection with other viruses (e.g., influenza, SARS coronavirus, dengue, hepatitis), although such infections can usually be suspected based upon the presence of other clinical signs or symptoms. This chapter will focus on the identification and diagnosis of human enteric viruses that primarily cause a gastroenteritis illness.

CLINICAL PRESENTATION

The clinical presentation of infection caused by the gastroenteritis viruses includes both abdominal and systemic symptoms, although subclinical and asymptomatic infection may also occur. Watery diarrhea is a hallmark symptom, and bloody diarrhea is a rare finding. Vomiting is another common symptom, and although it may be associated with diarrhea, vomiting also can occur without diarrhea. Other symptoms include anorexia, nausea, abdominal pain and cramps, malaise, myalgias, headache, and fever.

Although there is considerable overlap in the clinical presentations of infection caused by the gastroenteritis viruses and a diagnosis cannot be made with certainty based upon symptoms alone, certain clinical presentations can suggest a specific etiologic agent as the cause. Severe diarrhea with fever and volume depletion in a child less than two years of age is characteristic of rotavirus infection (1). Outbreaks of acute gastroenteritis associated with food or water consumption, in healthcare facilities, cruise ships, or nursing homes, occurring in adults with vomiting as a predominant complaint suggests norovirus as an etiology (2–4). Both rotavirus and noroviruses have been associated with traveler's diarrhea. Each of the major groups of gastroenteritis viruses have been associated with protracted diarrhea in immunocompromised patients (1,5–8).

VIRAL AGENTS

Viruses from four virus families are causes of gastroenteritis (Table 1). Other viruses that have been detected in fecal samples of persons with diarrhea have been proposed to be causes of gastroenteritis, but there are currently insufficient data available to establish the role of these agents as causes of gastroenteritis in humans. These viruses include agents that cause diarrhea in animals (e.g., coronaviruses, toroviruses, picobirnaviruses, and pestiviruses) and agents for which there are limited data available to assess causality (e.g., Aichi virus). This chapter will consider only those viruses that are clearly established causes of gastroenteritis.

Rotavirus

Rotavirus is a genus within the family *Reoviridae*. The name is derived from the Latin, *rota*, which means "wheel" and refers to the morphological appearance of the virus by electron microscopy (EM). Rotaviruses are non-enveloped, icosahedral viruses that are approximately 75 nm in diameter (100 nm when the VP4 spikes are included). The genome consists of 11 segments of double-stranded RNA, and it is encased in a triple-layered protein capsid. The inner layer is made of VP1, VP2, and VP3, and these proteins are involved in viral replication and mRNA synthesis. The VP6 protein, which constitutes 50% of the mass of the viral particle and is a group-specific antigen, forms the middle layer. The outer layer is made up of VP7 with VP4 spikes emanating through this layer, and both of these proteins are targets for serotype analyses.

Table 1 Characteristics of Viruses that Cause Gastroenteritis

Virus family Virus genus	*Reoviridae* *Rotavirus*	*Caliciviridae* *Norovirus*	 *Sapovirus*	*Adenoviridae* *Adenovirus*	*Astroviridae* *Astrovirus*
Viral genetic characteristics	dsRNA, segmented	ssRNA, positive sense	ssRNA, positive sense	dsDNA	ssRNA, positive sense
Viral particle size	75 nm (100 nm when spikes included)	28–40 nm	28–40 nm	70–90 nm	28–30 nm
# Genotypes or Serotypes	Group A: 19 G serotypes, 32 [P] genotypes;	Genogroup I, 8 genotypes; Genogroup II, 19 genotypes; Genogroups III and IV, 2 genotypes each; Genogroup V, 1 genotype	Genogroup I, 5 genotypes; Genogroup II, 6 genotypes; Genogroups III, IV, and V, 1 genotype each	Group F: 2 serotypes (types 40 and 41)	Eight serotypes (1–8); correspond 1:1 to eight genotypes
Seasonality	Winter	Year round with winter peak	Year round	Year round	Winter months (temperate); rainy season (tropics)
Age groups affected	Predominantly children; also elderly, other adults	All age groups	Predominantly children; occasionally adults	Young children; uncommonly other age groups	Predominantly children; occasionally adults

Reassortment of viral genes can occur when a cell is infected with more than one strain due to the segmented nature of the viral genome.

Rotaviruses are divided into groups based on the presence of cross-reactive epitopes on the VP6 protein. Group A rotaviruses are the most common cause of human infection. Rotaviruses belonging to other groups (B–F) infect animals, although occasionally group B and C rotaviruses can also cause human infection. Group A rotaviruses are classified serotypically based upon properties of the two surface proteins, the glycoprotein VP7 (G type) and the protease sensitive VP4 protein (P type). Alternatively, the sequence of the genes encoding VP7 and VP4 can be used for genotypic classification. Analyses of VP7 serotype and genotype give concordant results, so results of these assays are referred to as G serotype. On the other hand, analyses of VP4 serotype and genotype do not always agree, so VP4 is either classified by the serotype number or genotype number, with the genotype number being written in brackets. There are currently 19 G types and 32 [P] types.

Calicivirus

The family *Caliciviridae* contains two genera, *Norovirus* and *Sapovirus*, that cause infection in humans. The family name is derived from the Latin word, *calix*, which means "cup" or "goblet," and refers to the cup-like depressions seen by EM on the surface of the virus particle. Caliciviruses are nonenveloped, icosahedral viruses that are approximately 28 to 40 nm in size. The genome is a single-stranded, positive sense RNA and encodes nonstructural proteins and two structural proteins, VP1 and VP2. Sapoviruses differ from noroviruses in that the major structural protein, VP1, is in the same open reading frame as that of the nonstructural proteins while the VP1 of noroviruses is encoded in a separate open reading frame. Sapoviruses are also more likely to have the classical "calicivirus" morphology by EM than are noroviruses.

Norovirus

Noroviruses are divided into five genogroups based upon sequence analysis of the viral genome. Genogroups I, II, and IV contain human strains while genogroups III and V contain bovine and murine strains, respectively. Each genogroup is further divided into genotypes (or genetic clusters) based upon the VP1 amino acid sequence. There are currently 8, 19, 2, 2, and 1 genotypes in norovirus genogroups I, II, III, IV, and V, respectively, but there are also additional genotypes within genogroups I and II that have not yet been assigned numbers. The genogroup number is designated with a Roman numeral, while the genotype number is given as an Arabic number. Thus, Norwalk virus, which is the prototype norovirus strain and belongs to genogroup I, genotype 1, is designated as a GI.1 strain. Since human noroviruses cannot be grown in cell culture, it is not clear whether the genotype classification scheme will accurately reflect serotypic differences between groups.

Sapovirus

Sapoviruses are also divided into five genogroups. Genogroup III strains infect pigs, but viruses from all of the other genogroups can infect humans. Similar to the norovirus classification scheme, sapoviruses are further divided into genotypes. Genogroups I and II contain five and six genotypes, respectively, while the other three genogroups each contain only a single genotype.

Adenovirus

Enteric adenoviruses are members of genus *Mastadenovirus* in the family *Adenoviridae*. The family name is derived from the Greek, *aden*, which means "acorn" or "gland" and refers to the adenoid glands from which the prototype strain was isolated. Adenoviruses are nonenveloped, icosahedral viruses that are approximately 70 to 90 nm in size. The genome consists of double-stranded DNA approximately 32 kb in length. There are six species (formerly called subgroups or subgenera), A through F, that cause human disease, and each species contains one or more serotypes. Species can be separated by their ability to agglutinate rat or monkey red blood cells. The enteric adenoviruses belong to species F, which partially hemagglutinate rat red blood cells, and there are two serotypes (types 40 and 41).

Astrovirus

Astroviruses that infect humans are members of the genus *Mamastrovirus* in the family *Astroviridae*. The family and genus names are derived from the Greek word, *astron*, which means "star" and refers to the five- or six-pointed, star-like appearance of the virion by EM. Astroviruses are non-enveloped, icosahedral viruses that are approximately 28 to 30 nm in size. The genome is a single-stranded, positive sense RNA. The viral capsid consists of three proteins that are generated from the proteolytic cleavage of a single larger protein expressed from the open reading frame 2 of the viral genome. There are eight serotypes of human strains as well as additional serotypes that infect animals (calves, chickens, turkeys). Phylogenetic analysis of the capsid gene also groups human astroviruses into eight genotypes. The serological and phylogenetic analyses yield concordant results, so either classification scheme can be used.

EPIDEMIOLOGY

Viral gastroenteritis occurs year round, but there are seasonal differences in prevalence among the virus groups. Rotavirus, the major cause of dehydrating illness worldwide, has a winter peak during the cooler months in temperate climates, but in the tropics no seasonal trend has been observed. Similarly, there is a winter peak in the occurrence of norovirus outbreaks, although endemic disease occurs throughout the year. Astrovirus infections are most common during the winter in temperate climates and during the rainy season in the tropics. No seasonal increase in disease has been reported for sapoviruses or enteric adenoviruses.

Group A rotaviruses cause 600,000 to 800,000 deaths in children younger than five years annually, with most of the deaths occurring in developing countries. Mortality due to rotavirus is uncommon in the United States, but primary infection leads to 55,000 to 70,000 hospitalizations per year. Infection in older children and adults is less common and may occur sporadically, following contact with a pediatric case, or as part of an outbreak [food-borne, waterborne, or in a closed population (e.g., institutionalized)]. Groups B and C rotaviruses are much less common causes of human disease, but may cause sporadic disease or outbreaks of gastroenteritis.

Human caliciviruses infect persons of all ages, and the majority of these infections are caused by noroviruses. Approximately 23 million cases of norovirus infection are estimated to occur annually in the United States, and it is the second most common cause of hospitalization of young children for gastroenteritis (after rotavirus) (2,9). In developed countries, noroviruses have been estimated to cause approximately 64,200 hospitalizations annually, and in developing countries these infections lead to more than one million hospitalizations and approximately 218,000 deaths each year (10). Noroviruses cause both endemic and epidemic infection, with the latter being associated with a seasonal increase in outbreaks of gastroenteritis. Seventy to greater than 95% of the outbreaks of nonbacterial gastroenteritis worldwide are caused by noroviruses. On the other hand, sapoviruses are associated with outbreaks only occasionally and have been most commonly noted to cause gastroenteritis in young children.

Astroviruses predominantly cause infection in young children, but they can also cause disease in immunocompromised patients, the institutionalized elderly, and in otherwise healthy individuals exposed to contaminated food or water. Community-based studies have identified astroviruses to cause approximately 2% to 6% of cases of acute gastroenteritis. The clinical course of disease may be protracted in severely immunocompromised patients, such as bone marrow transplant patients. Astroviruses are only rarely identified as the cause of foodborne or waterborne outbreaks.

Enteric adenoviruses cause 1% to 2% of cases of gastroenteritis in most studies, and children younger than two years are most likely to be symptomatically infected. Chronic diarrhea has also been in association with HIV infection.

The predominant mechanism of transmission for the enteric viruses is the fecal-oral route. Transmission can occur person-to-person, by contact with fomites, or following consumption of contaminated food or water. The relative importance of these routes of transmission differs among the enteric viruses. Transmission of noroviruses has also been associated with exposure to vomitus either directly or through droplets (airborne).

LABORATORY TESTING

Specimen Types/Handling

Enteric viruses are primarily shed in the stool, so fecal samples are the primary specimens used for viral diagnostic testing. Vomitus and serum specimens can also be used in some instances. Greater quantities of virus are present in the first several days after illness onset and decline thereafter, although low levels of virus shedding can persist for several weeks. Thus, diagnostic yields are highest from samples collected in the first several days after illness onset. In general, enteric viruses are less likely to be identified from rectal swabs compared with whole fecal samples, making the latter the preferred specimens for testing (11). Sodium dodecyl sulfate-EDTA-treated chromatography paper strips can also been used to collect, transport, and store samples prior to analysis using RT-PCR methods (12,13). Using this technique, rotavirus and calicivirus nucleic acids remain detectable for up to a month when stored at room temperature and for longer periods of time when stored at −20°C, while the infectivity of rotavirus and feline calicivirus (a surrogate for human caliciviruses) is lost after contact with the paper. This collection method may be of use for the performance of epidemiologic studies in which the subject/patient collects the sample.

A few grams of a fecal sample should be placed into a clean container, which is then sealed tightly for transport to the laboratory. Collection of diarrheal stools from diapered children can be accomplished in a number of ways. The diaper can be lined with a plastic wrap or a disposable diaper can be inverted to have the plastic lining next to the child's skin to prevent the diarrheal stool from soaking into the diaper. Alternatively, a portion of the diarrheal stool can be recovered from the diaper with a wooden tongue depressor or by cutting out a piece of the diaper and soaking the cut fragment in sterile water or phosphate buffered saline.

Media, preservatives, animal serum, or detergents are often added to other clinical samples submitted for viral diagnosis to improve viral stability, but these substances can interfere with the performance of some of the commercially available diagnostic assays (such as enzyme immunoassays (EIAs) and latex agglutination assays) for enteric viruses and should not be added to fecal specimens. Once in the laboratory, the samples should be stored at 4°C and they can be held at that temperature for several days to weeks. Samples that are to be kept for prolonged periods of time should be stored frozen, preferably at −70°C, although some antigen detection kits indicate that freezing at −20°C is acceptable.

Direct Virus Detection

A number of different test formats have been used to directly detect gastroenteritis viruses in clinical specimens, including EM, antigen detection assays, and nucleic acid detection assays. The principles of each of these methods are described in detail in previous chapters. The use of these different approaches varies between diagnostic laboratories, based upon assay availability, local expertise, and laboratory resources.

Electron Microscopy

Electron microscopy was the first method used to detect gastroenteritis viruses. The principal advantages of EM over other test formats are its ability to screen for a large number of different enteric viruses using a single assay and the ability to detect strains not identified using assays that target the most commonly circulating enteric viruses. For example, commercially available antigen detection assays only detect group A rotaviruses and do not detect rotaviruses belonging to other groups that can be detected by EM (e.g., groups B and C rotaviruses). Disadvantages of EM include the capital expense for the equipment, the requirement for expertise in identification of the viruses, and the relatively lower sensitivity of the EM compared with other currently available assays. The sensitivity can be improved somewhat using hyperimmune serum to aggregate viruses (immune EM), but even with this modification the limit of detection is approximately 10^5–10^6 virus particles per milliliter. EM has been reported to be positive in no more than 50% of patients with acute norovirus infection, and it is less sensitive than EIAs for the detection of rotavirus (14,15).

Antigen Detection

Antigen detection assays are commercially available for each of the gastroenteritis viruses, although not all are marketed in the United States (Table 2). A variety of test formats are used, including solid-phase EIAs, immunochromatography, and latex agglutination. Solid-phase EIAs are performed within a clinical laboratory and generally provide results within approximately two hours. The immunochromatography and latex agglutination formats are simple to perform and yield results in less than 30 minutes, although in some reports the sensitivity of latex agglutination is lower (70–90%) than that achieved with the other assay formats (>90%) (16,17).

Antigen detection assays are the principal means currently used to establish the diagnosis of rotavirus infection. The commercially available assays utilize monoclonal or polyclonal antisera that recognize the VP6 protein of group A strains, so these assays will not identify rotavirus strains belonging to other groups. Assay sensitivity is equal to or greater than that achieved by EM. Rotavirus antigen can also be detected in the serum of children with acute infection (18). Although higher sensitivity will be obtained through the evaluation of fecal samples for rotavirus antigen, serum antigen assays will detect infection in some children with antigen-negative stools (19).

Adenoviruses can also be detected in fecal specimens using commercially available antigen detection assays. Most assays do not distinguish between enteric (group F) and adenoviruses belonging to other groups, while the Adenoclone® Type 40/41 assay may fail to detect some group F strains, possibly due to the emergence of new antigenic variants (20).

Commercial antigen detection assays have also been developed for astroviruses and noroviruses, although none are available for routine use in the United States. The norovirus assays have had problems with sensitivity and specificity, and some genotypes are not detected with available assays (21). Nevertheless, these assays have been useful in identifying norovirus-associated outbreaks when multiple (six or more) fecal specimens from the outbreak can be tested (22). Further improvements will be needed before these assays are useful to diagnose norovirus infection in individual patients. No commercial antigen detection assays have been developed for sapoviruses.

Table 2 Commercially Available Antigen Detection Assays for Gastroenteritis Viruses

Test format	Virus(es) detected	Kit name	Manufacturer (website)
Enzyme immunoassay	Rotavirus	Vidas® Rotavirus	bioMérieux, Inc. (www.biomerieux-diagnostics.com)
	Rotavirus	Pathfinder® Rotavirus	Bio-Rad Laboratories (www.bio-rad.com)
	Rotavirus, adenovirus	Premier™ Rotaclone®, Premier™ Adenoclone®, Premier™[a] Adenoclone® Type 40/41	Meridian Bioscience, Inc. (www.meridianbioscience.com)
	Rotavirus, adenovirus	Rotascreen®[b], Adenoscreen®[a,b]	Microgen Bioproducts (http://www.microgenbioproducts.com)
	Rotavirus, adenovirus, astrovirus, norovirus	IDEIA Rotavirus, IDEIA Adenovirus[a,b], Amplified IDEIA Astrovirus[b], IDEIA Norovirus[b]	Oxoid (www.oxoid.com)
	Rotavirus, adenovirus, astrovirus, norovirus	RIDASCREEN® Rotavirus[b], RIDASCREEN® Adenovirus[a,b], RIDASCREEN® Astrovirus[b], RIDASCREEN® Norovirus[b]	R-Biopharm AG (www.r-biopharm.com)
	Norovirus	NV-AD[b]	Denka-Seikin Co., Ltd. (www.denka-seikin.co.jp)
Immunochromatography	Rotavirus, adenovirus	BIOCARD™ Rota stick[b], BIOCARD™ Adeno stick[b]	ANI Biotech Oy (www.anibiotech.fi)
	Rotavirus, adenovirus	Vikia® Rota-Adeno[b]	bioMérieux, Inc. (www.biomerieux-diagnostics.com)
	Rotavirus	ImmunoCard STAT!® Rotavirus	Meridian Bioscience, Inc. (www.meridianbioscience.com)
	Rotavirus, adenovirus	Rotascreen®[b], Adenoscreen®[a,b]	Microgen Bioproducts (http://www.microgenbioproducts.com)
	Rotavirus, adenovirus	RIDA®QUICK Rotavirus, RIDA®QUICK Rotavirus/Adenovirus Combi[a]	R-Biopharm AG (www.r-biopharm.com)
	Rotavirus, adenovirus	Xpect® Rotavirus, SAS™ Adeno Test Kit[a]	Remel (www.remel.com)
Latex agglutination	Rotavirus	Pastorex® Rotavirus	Bio-Rad Laboratories (www.bio-rad.com)
	Rotavirus	Virogen® Rotatest®	Inverness Medical Professional Diagnostics (www.invernessmedicalpd.com)
	Rotavirus, adenovirus	Rotascreen®[b], Adenoscreen®[a,b]	Microgen Bioproducts (http://www.microgenbioproducts.com)
	Rotavirus, adenovirus	Diarlex Rota, Diarlex Adeno, Diarlex Rota-Adeno, Rotalex[b], Adenolex[a,b]	Orion Diagnostica Oy (www.oriondiagnostica.fi)
	Rotavirus	Rotavirus Test Kit	Remel (www.remel.com)

[a]Detects all adenovirus serotypes (not just types 40 and 41).
[b]Not available in the United States.

Nucleic Acid Detection

Several different nucleic acid detection methods have been developed for the detection of gastroenteritis viruses. PCR-based assays are the most commonly employed, although nucleic acid sequence-based amplification (NASBA) assays for the RNA viruses have also been described. Other methods, such as direct detection of viral genomes following chromatography and detection using dot-blot hybridization methods, are not commonly used because of their lower sensitivity compared to amplification-based molecular methods.

PCR-based assays have greater sensitivity than EM and antigen-based diagnostic approaches for each of the gastroenteritis viruses. Complementary DNA must be synthesized from genomic RNA for each of the RNA viruses prior to PCR amplification. Virus-specific primers are then used to amplify conserved portions of the genome. Specificity of the assay is usually established by hybridization with a virus-specific probe or by direct sequencing of the amplicons. The development of real-time RT-PCR assays (Table 3) has considerably shortened the analytic time needed such that results can be obtained within a few hours of assay setup. Real time assays can either detect viral amplicons using a virus-specific probe or by melting curve analysis using a non-specific fluorescent dye, such as SYBR green. The assay format also allows the generation of quantitative data. Another advantage of real time PCR assays is the elimination of the need for post-PCR analyses to confirm assay specificity, decreasing the potential for laboratory contamination with amplified products that would then lead to carry-over contamination and false-positive results.

Substances that inhibit the function of the enzymes used in nucleic acid amplification are frequently present in fecal specimens and can prevent virus detection. A variety of strategies have been pursued to identify and circumvent the problem of sample inhibition. Methods that effectively remove inhibitors from the majority of fecal samples have been developed and vary in their complexity. The simplest approach is dilution of a 10% fecal suspension followed by heating the sample to 95°C release the viral nucleic acid from the capsid (23); with this approach the sample must be assayed immediately after heating to prevent degradation of the nucleic acids. Other methods that utilize detergents and denaturants, such as guanidinium thiocyanate, require additional steps but provide nucleic acid extracts that are less prone to rapid degradation prior to analysis (24). A number of commercial kits are now available to effectively extract viral nucleic acids from fecal samples (24–26). An internal control nucleic acid can be added to the extracted sample and is a common approach for identifying the presence of inhibitors (23).

The relative sensitivity of nucleic acid-based methods compared to antigen detection methods varies by virus. RT-PCR assays have similar or only modestly increased sensitivity compared to antigen detection assays for the identification of group A rotaviruses in clinical samples (11,27). PCR and RT-PCR assays are more sensitive than antigen detection for the identification of enteric adenoviruses and astroviruses, respectively (27,28). RT-PCR assays are the preferred methods for detection of noroviruses and sapoviruses.

Virus Isolation

Virus isolation is not a routine method utilized for identification of gastroenteritis viruses in clinical samples, although it may be performed successfully in some instances. However, none of the human caliciviruses can be grown in cell culture, and the other enteric viruses are relatively fastidious. Group A rotaviruses can be isolated from fecal specimens in several different cell lines (e.g., MA104, CaCo-2, primary monkey kidney), but successful isolation often requires serial passage (29,30). Rectal swabs are less likely than whole fecal samples to yield a positive culture. Virus cultivation methods for group A rotaviruses take longer, are more expensive to perform, and have no better sensitivity than the antigen detection methods (29,31). The successful isolation of group B and C rotaviruses has been described (32).

Enteric adenoviruses were first isolated in Chang conjunctival cells, but Graham 293 cells are now used by many virologists to isolate these viruses (33–35). These viruses often cause little or no cytopathic effect in cell culture and are more difficult to isolate than other human adenoviruses (33).

Astroviruses can be cultivated in several different cell lines, but cell lines derived from human intestinal tissues (CaCo-2, T84) have been the most sensitive (36,37). Rapid detection of positive cultures can be accomplished using a shell vial assay, where virus antigen is detected in

Table 3 Selected Real-Time RT-PCR/PCR Assays for Detection of Gastroenteritis Viruses

Virus	Gene target(s)	Primers	Amplicon detection method	Reference
Rotavirus	NSP3 (group A)	NVP3F, NVP3R	Dual-labeled, virus-specific FRET oligoprobe	(67)
	NSP3 (group A)	NVP3Deg, NVP3R1	Dual-labeled, virus-specific FRET oligoprobe	(68)
	NSP3 (group A)	NSP3F, NSP3R	Dual-labeled, virus-specific FRET oligoprobe	(69)
	VP6 (group A)	VP6-F, VP6-R	SYBR green	(70)
	VP6 (group A)	RotaA-fwd1, RotaA-fwd2, RotaA.rev1, RotaA.rev2	Dual-labeled, virus-specific FRET oligoprobes	(27)
	VP7 (group C)	RotaC.fwd, RotaC.rev	Dual-labeled, virus-specific FRET oligoprobe	(27)
Norovirus	GI and GII—polymerase/capsid	GI—COG1F, COG1R; GII—COG2F, COG2R	Dual-labeled, virus-specific FRET oligoprobe	(71)
	GI—capsid; GII—polymerase	GI—CapA, CapB GII—p110; SR46	SYBR green	(67)
	GIV—polymerase/capsid	Mon 4F, Mon 4R	Dual-labeled, virus-specific FRET oligoprobe	(72)
	GI, GII, and GIV—polymerase	GI—GI.fwd, GI.rev; GII and GIV—GII.fwd, GIV.fwd, GII&IV.rev	Dual-labeled, virus-specific FRET oligoprobes	(73)
Sapovirus	GI, GII, GIV, and GV—polymerase	SaV124F, SaV1F, SaV5F, SaV1245R	Dual-labeled, virus-specific FRET oligoprobes	(74)
	GI, GII, and GIV—polyprotein	sapo.fwdA, sapo.fwdB, sapo.rev	Dual-labeled, virus-specific FRET oligoprobes	(73)
Astrovirus	3′ untranslated region (UTR)	MM2, MM67	SYBR green	(75)
	Capsid protein precursor	Hast.fwd, Hast.rev	Dual-labeled, virus-specific FRET oligoprobes	(73)
Adenovirus	Hexon gene (all adenovirus species)	AQ1, AQ2	Dual-labeled, virus-specific FRET oligoprobes	(76)
	Hexon gene (all adenovirus species)	AdnU-S', AdnU-A2	SYBR green	(77)
	Hexon gene (species F)	Adeno.fwd, Adeno.rev	Dual-labeled, virus-specific FRET oligoprobes	(27)

culture cells by immunofluorescence 18 hours after inoculation (36). Amplification of virus in cell culture followed by detection with RT-PCR is another method used for astrovirus identification (38).

Identification and Typing Systems

A variety of methods are available for the identification and typing of gastroenteritis viruses. These methods can be applied directly to clinical specimens; when a virus is isolated in cell culture, the same methods can be applied to the clinical isolate. Both antigenic and genotypic methods are available. For those viruses that can be grown in cell culture, serotypes can be identified by reactivity in cross-neutralization assays using type-specific antisera.

Rotaviruses

Group A rotaviruses can be classified into both G (for glycoprotein, VP7) and P (for protease sensitive, VP4) serotypes. Monoclonal antibodies have been developed to identify both G and P serotypes, but the monoclonal antibodies are not widely available for use in clinical laboratories (39,40). This has led to development of PCR-based assays where genotypes can be identified using type-specific primers (41–44). These methods have sometimes either failed to genotype strains or led to misclassification due to the emergence of new strains and point mutations in regions targeted by the primers (45,46). A more complete genotyping system has recently been recommended for classifying strains based upon the analysis of the sequences of each of the 11 gene segments (47). This approach allows the identification of reassortants and newly emerging strains.

Caliciviruses

The inability to cultivate human caliciviruses in vitro has prevented the establishment of a serotype classification system for these viruses. Solid-phase immune electron microscopy (SPIEM) employing human convalescent sera has been used to characterize human noroviruses into antigenic types, but the lack of standardized reagents for these assays makes this approach impractical for most laboratories (48). In addition, the biologic significance of different identified antigenic types remains unclear. Instead, both noroviruses and sapoviruses are classified into genogroups and genotypes using molecular methods. The genogroup and genotype are determined based upon the complete sequence of the VP1 gene, but shorter genomic sequences can be used to infer genotype (49–51). Although the polymerase gene has also been used to classify noroviruses into genotypes, the ability of these viruses to undergo recombination can lead to incorrect assignment of genotype when only the polymerase gene sequence information is used (52,53). Obtaining sequence data from both the polymerase and capsid genes allows genotype determination as well as the identification of recombinant strains. Genogroup-specific RT-PCR assays can also identify viral genogroup, and hybridization methods can be used to infer genotype when amplicons sequencing is not feasible (49,54,55). Although assays for sapovirus classification are less well developed, genogroup-specific RT-PCR assays have been described (56).

Enteric adenoviruses

The enteric adenoviruses can be classified either antigenically or genotypically. Human adenoviruses can be separated based upon their ability to agglutinate rat and monkey erythrocytes, and the enteric adenoviruses (species F) partially agglutinate rat erythrocytes, a property shared with species C and E. Species and type-specific epitopes are also found on the hexon and fiber structural proteins, and monoclonal antibodies that recognize these epitopes can be used for classification (57,58). Genotypic classification of the enteric adenoviruses can also be performed by dot-blot hybridization, restriction analysis of genomic DNA, or type-specific PCR assays (27,58).

Astroviruses

Astroviruses are classified into serotypes using type-specific antisera, and these assays can be performed in an ELISA format (59). Astroviruses also can be classified using molecular methods. Genotypes, which correspond directly to serotypes, can be determined by sequence analysis of amplicons generated in RT-PCR assays that utilize primers that amplify all astrovirus strains, or by measurements of amplicon size in RT-PCR assays that utilize type-specific primers that generate amplicons of different sizes based upon strain genotype (59–61).

Serologic Testing

Serologic testing is not routinely used for the diagnosis of individual viral gastroenteritis infections, although a variety of methods are available for such testing. Instead, serologic studies are most commonly used for the performance of epidemiologic and vaccine studies. ELISA methods are the most commonly used methods, with the target antigen being either whole virus, virus-like particles, or individual viral proteins (49,62,63). Four-fold or greater rises in antibody levels between acute and convalescent sera collected two to four weeks apart are indicative of acute infection. In addition to detecting total antibody, the ELISA assay can be modified to detect class (IgM, IgG, and IgA) and subclass specific immune responses. Seroresponses are greatest to homologous antigen, but heterologous responses can also be observed in many assays.

Neutralization antibody assays have been developed for both rotaviruses and astroviruses (64,65). These assays are less likely to detect heterologous responses from strains belonging to other serotypes, but they are more cumbersome to perform because of the requirement to work with live virus.

Antiviral Resistance Testing

There are no licensed antivirals currently available for the treatment or prevention of infections caused by the gastroenteritis viruses described in this chapter.

Evaluation and Reporting of Laboratory Results

As described above, a variety of different assays are available for the diagnosis of infection caused by gastroenteritis viruses. The selection of the assays to be used in a diagnostic laboratory will be affected by the population being evaluated (pediatric, adult, immunocompromised), the likelihood of encountering the different viral pathogens, the availability of necessary equipment, availability of the technical expertise to perform and interpret assay results, and the costs of the assays. Each laboratory must identify the goals of its viral gastroenteritis program and consider the need for more specific diagnosis (e.g., identification of serotypes or genotypes), the need for identification of viruses for which no commercially available rapid antigen tests are available (e.g., noroviruses and sapoviruses), and the need for identification of less common causes of viral gastroenteritis (e.g., group B and C rotaviruses). These assessments will guide the selection of assays to be offered by the laboratory.

The results of an individual diagnostic test must be interpreted in the context of the overall clinical picture. In the appropriate clinical circumstances (e.g., chronic diarrhea in an immunocompromised host), failure to identify a viral pathogen with one assay should not preclude continued evaluation for that pathogen using other assays since none of the available assays has 100% sensitivity. Positive results will have implications for patient management and can influence infection control isolation procedures. For example, because noroviruses have been a common cause of nosocomial outbreaks of gastroenteritis, identification of norovirus infection could lead to increased surveillance for illness among patients and healthcare workers, increased attention to cleaning and disinfection of the patient care environment, and restriction of visitation if an outbreak is identified (66). With the increasing recognition of the impact of enteric viruses as causes of gastroenteritis, it is likely that additional commercialized assays (antigen detection, molecular) will become available in the near future.

REFERENCES

1. Ruuska T, Vesikari T. Rotavirus disease in Finnish children: Use of numerical scores for clinical severity of diarrhoeal episodes. Scand J Infect Dis 1990; 22(3):259–267.
2. Atmar RL, Estes MK. The epidemiologic and clinical importance of norovirus infection. Gastroenterol Clin North Am 2006; 35(2):275–290.
3. Kaplan JE, Feldman R, Campbell DS, et al. The frequency of a Norwalk-like pattern of illness in outbreaks of acute gastroenteritis. Am J Public Health 1982; 72(12):1329–1332.

4. Turcios RM, Widdowson MA, Sulka AC, et al. Reevaluation of epidemiological criteria for identifying outbreaks of acute gastroenteritis due to norovirus: United States, 1998–2000. Clin Infect Dis 2006; 42(7):964–969.
5. Coppo P, Scieux C, Ferchal F, et al. Astrovirus enteritis in a chronic lymphocytic leukemia patient treated with fludarabine monophosphate. Ann Hematol 2000; 79(1):43–45.
6. Gilger MA, Matson DO, Conner ME, et al. Extraintestinal rotavirus infections in children with immunodeficiency. J Pediatr 1992; 120(6):912–917.
7. Lee BE, Pang XL, Robinson JL, et al. Chronic norovirus and adenovirus infection in a solid organ transplant recipient. Pediatr Infect Dis J 2008; 27(4):360–362.
8. Nilsson M, Hedlund KO, Thorhagen M, et al. Evolution of human calicivirus RNA in vivo: Accumulation of mutations in the protruding P2 domain of the capsid leads to structural changes and possibly a new phenotype. J Virol 2003; 77(24):13117–13124.
9. Mead PS, Slutsker L, Dietz V, et al. Food-related illness and death in the United States. Emerg Infect Dis 1999; 5(5):607–625.
10. Patel MM, Widdowson MA, Glass RI, et al. Systematic literature review of role of noroviruses in sporadic gastroenteritis. Emerg Infect Dis 2008; 14(8):1224–1231.
11. Stockman LJ, Staat MA, Holloway M, et al. Optimum diagnostic assay and clinical specimen for routine rotavirus surveillance. J Clin Microbiol 2008; 46(5):1842–1843.
12. Rahman M, Goegebuer T, De Leener K, et al. Chromatography paper strip method for collection, transportation, and storage of rotavirus RNA in stool samples. J Clin Microbiol 2004; 42(4):1605–1608.
13. Wollants E, Maes P, Thoelen I, et al. Evaluation of a norovirus sampling method using sodium dodecyl sulfate/EDTA-pretreated chromatography paper strips. J Virol Methods 2004; 122(1):45–48.
14. Grauballe PC, Vestergaard BF, Meyling A, et al. Optimized enzyme-linked immunosorbent assay for detection of human and bovine rotavirus in stools: Comparison with electron-microscopy, immunoelectro-osmophoresis, and fluorescent antibody techniques. J Med Virol 1981; 7(1):29–40.
15. Thornhill TS, Kalica AR, Wyatt RG, et al. Pattern of shedding of the Norwalk particle in stools during experimentally induced gastroenteritis in volunteers as determined by immune electron microscopy. J Infect Dis 1975; 132(1):28–34.
16. Dennehy PH, Gauntlett DR, Tente WE. Comparison of nine commercial immunoassays for the detection of rotavirus in fecal specimens. J Clin Microbiol 1988; 26(9):1630–1634.
17. Wilhelmi I, Colomina J, Martin-Rodrigo D, et al. New immunochromatographic method for rapid detection of rotaviruses in stool samples compared with standard enzyme immunoassay and latex agglutination techniques. Eur J Clin Microbiol Infect Dis 2001; 20(10):741–743.
18. Blutt SE, Kirkwood CD, Parreno V, et al. Rotavirus antigenaemia and viraemia: A common event? Lancet 2003; 362(9394):1445–1449.
19. Blutt SE, Matson DO, Crawford SE, et al. Rotavirus antigenemia in children is associated with viremia. PLoS Med 2007; 4(4):e121.
20. Moore PL, Steele AD, Alexander JJ. Relevance of commercial diagnostic tests to detection of enteric adenovirus infections in South Africa. J Clin Microbiol 2000; 38(4):1661–1663.
21. Burton-MacLeod JA, Kane EM, Beard RS, et al. Evaluation and comparison of two commercial enzyme-linked immunosorbent assay kits for detection of antigenically diverse human noroviruses in stool samples. J Clin Microbiol 2004; 42(6):2587–2595.
22. Gray JJ, Kohli E, Ruggeri FM, et al. European multicenter evaluation of commercial enzyme immunoassays for detecting norovirus antigen in fecal samples. Clin Vaccine Immunol 2007; 14(10):1349–1355.
23. Schwab KJ, Estes MK, Neill FH, et al. Use of heat release and an internal RNA standard control in reverse transcription-PCR detection of Norwalk virus from stool samples. J Clin Microbiol 1997; 35(2):511–514.
24. Rasool NB, Monroe SS, Glass RI. Determination of a universal nucleic acid extraction procedure for PCR detection of gastroenteritis viruses in faecal specimens. J Virol Methods 2002; 100(1–2):1–16.
25. Ishida S, Yoshizumi S, Ikeda T, et al. Sensitive and rapid detection of norovirus using duplex TaqMan reverse transcription-polymerase chain reaction. J Med Virol 2008; 80(5):913–920.
26. Iturriza-Gomara M, Xerry J, Gallimore CI, et al. Evaluation of the Loopamp (loop-mediated isothermal amplification) kit for detecting Norovirus RNA in faecal samples. J Clin Virol 2008; 42(4):389–393.
27. Logan C, O'Leary JJ, O'Sullivan N. Real-time reverse transcription-PCR for detection of rotavirus and adenovirus as causative agents of acute viral gastroenteritis in children. J Clin Microbiol 2006; 44(9):3189–3195.
28. Rohayem J, Berger S, Juretzek T, et al. A simple and rapid single-step multiplex RT-PCR to detect Norovirus, Astrovirus and Adenovirus in clinical stool samples. J Virol Methods 2004; 118(1):49–59.
29. Cumino AC, Giordano MO, Martinez LC, et al. Culture amplification in human colon adenocarcinoma cell line (CaCo-2) combined with an ELISA as a supplementary assay for accurate diagnosis of rotavirus. J Virol Methods 1998; 76(1–2):81–85.

30. Ward RL, Knowlton DR, Pierce MJ. Efficiency of human rotavirus propagation in cell culture. J Clin Microbiol 1984; 19(6):748–753.
31. Cicek C, Karatas T, Altuglu I, et al. Comparison of ELISA with shell vial cell culture method for the detection of human rotavirus in fecal specimens. New Microbiol 2007; 30(2):113–118.
32. Sanekata T, Kuwamoto Y, Akamatsu S, et al. Isolation of group B porcine rotavirus in cell culture. J Clin Microbiol 1996; 34(3):759–761.
33. Cromeans TL, Lu X, Erdman DD, et al. Development of plaque assays for adenoviruses 40 and 41. J Virol Methods 2008; 151(1):140–145.
34. Kidd AH, Madeley CR. In vitro growth of some fastidious adenoviruses from stool specimens. J Clin Pathol 1981; 34(2):213–216.
35. Wigand R, Baumeister HG, Maass G, et al. Isolation and identification of enteric adenoviruses. J Med Virol 1983; 11(3):233–240.
36. Brinker JP, Blacklow NR, Herrmann JE. Human astrovirus isolation and propagation in multiple cell lines. Arch Virol 2000; 145(9):1847–1856.
37. Willcocks MM, Carter MJ, Laidler FR, et al. Growth and characterisation of human faecal astrovirus in a continuous cell line. Arch Virol 1990; 113(1–2):73–81.
38. Mustafa H, Palombo EA, Bishop RF. Improved sensitivity of astrovirus-specific RT-PCR following culture of stool samples in CaCo-2 cells. J Clin Virol 1998; 11(2):103–107.
39. Ramachandran M, Gentsch JR, Parashar UD, et al. Detection and characterization of novel rotavirus strains in the United States. J Clin Microbiol 1998; 36(11):3223–3229.
40. Taniguchi K, Urasawa T, Morita Y, et al. Direct serotyping of human rotavirus in stools by an enzyme-linked immunosorbent assay using serotype 1-, 2-, 3-, and 4-specific monoclonal antibodies to VP7. J Infect Dis 1987; 155(6):1159–1166.
41. Das BK, Gentsch JR, Cicirello HG, et al. Characterization of rotavirus strains from newborns in New Delhi, India. J Clin Microbiol 1994; 32(7):1820–1822.
42. Gentsch JR, Glass RI, Woods P, et al. Identification of group A rotavirus gene 4 types by polymerase chain reaction. J Clin Microbiol 1992; 30(6):1365–1373.
43. Gouvea V, Glass RI, Woods P, et al. Polymerase chain reaction amplification and typing of rotavirus nucleic acid from stool specimens. J Clin Microbiol 1990; 28(2):276–282.
44. Simmonds MK, Armah G, Asmah R, et al. New oligonucleotide primers for P-typing of rotavirus strains: Strategies for typing previously untypeable strains. J Clin Virol 2008; 42(4):368–373.
45. Iturriza-Gomara M, Kang G, Gray J. Rotavirus genotyping: Keeping up with an evolving population of human rotaviruses. J Clin Virol 2004; 31(4):259–265.
46. Rahman M, Sultana R, Podder G, et al. Typing of human rotaviruses: Nucleotide mismatches between the VP7 gene and primer are associated with genotyping failure. Virol J 2005; 2:24.
47. Matthijnssens J, Ciarlet M, Rahman M, et al. Recommendations for the classification of group A rotaviruses using all 11 genomic RNA segments. Arch Virol 2008; 153(8):1621–1629.
48. Lewis D, Ando T, Humphrey CD, et al. Use of solid-phase immune electron microscopy for classification of Norwalk-like viruses into six antigenic groups from 10 outbreaks of gastroenteritis in the United States. J Clin Microbiol 1995; 33(2):501–504.
49. Atmar RL, Estes MK. Diagnosis of noncultivatable gastroenteritis viruses, the human caliciviruses. Clin Microbiol Rev 2001; 14(1):15–37.
50. Hansman GS, Oka T, Katayama K, Takeda N. Human sapoviruses: Genetic diversity, recombination, and classification. Rev Med Virol 2007; 17(2):133–141.
51. Zheng DP, Ando T, Fankhauser RL, et al. Norovirus classification and proposed strain nomenclature. Virology 2006; 346(2):312–323.
52. Bull RA, Tanaka MM, White PA. Norovirus recombination. J Gen Virol 2007; 88(Pt 12):3347–3359.
53. Fankhauser RL, Monroe SS, Noel JS, et al. Epidemiologic and molecular trends of "Norwalk-like viruses" associated with outbreaks of gastroenteritis in the United States. J Infect Dis 2002; 186(1):1–7.
54. Vinje J, Vennema H, Maunula L, et al. International collaborative study to compare reverse transcriptase PCR assays for detection and genotyping of noroviruses. J Clin Microbiol 2003; 41(4):1423–1433.
55. Vinje J, Koopmans MP. Simultaneous detection and genotyping of "Norwalk-like viruses" by oligonucleotide array in a reverse line blot hybridization format. J Clin Microbiol 2000; 38(7):2595–2601.
56. Okada M, Yamashita Y, Oseto M, et al. The detection of human sapoviruses with universal and genogroup-specific primers. Arch Virol 2006; 151(12):2503–2509.
57. Herrmann JE, Perron-Henry DM, Blacklow NR. Antigen detection with monoclonal antibodies for the diagnosis of adenovirus gastroenteritis. J Infect Dis 1987; 155(6):1167–1171.
58. Wold WSM, Horwitz MS. Adenoviruses. In: Knipe DM, Howley PM, eds. Fields Virology. Philadelphia, PA: Lippincott, Willliams & Wilkins, 2007: 2395–2436.
59. Noel JS, Lee TW, Kurtz JB, et al. Typing of human astroviruses from clinical isolates by enzyme immunoassay and nucleotide sequencing. J Clin Microbiol 1995; 33(4):797–801.

60. Guix S, Bosch A, Pinto RM. Human astrovirus diagnosis and typing: Current and future prospects. Lett Appl Microbiol 2005; 41(2):103–105.
61. Matsui M, Ushijima H, Hachiya M, et al. Determination of serotypes of astroviruses by reverse transcription-polymerase chain reaction and homologies of the types by the sequencing of Japanese isolates. Microbiol Immunol 1998; 42(8):539–547.
62. Estes MK. Rotaviruses. In: Knipe DM, Howley PM, eds. Fields Virology. Philadelphia, PA: Lippincott Williams & Wilkins, 2007:1917–1974.
63. Noel J, Cubitt D. Identification of astrovirus serotypes from children treated at the Hospitals for Sick Children, London 1981–1993. Epidemiol Infect 1994; 113(1):153–159.
64. Gerna G, Battaglia M, Milenesi G, et al. Serotyping of cell culture-adapted subgroup 2 human rotavirus strains by neutralization. Infect Immun 1984; 43(2):722–729.
65. Hudson RW, Herrmann JE, Blacklow NR. Plaque quantitation and virus neutralization assays for human astroviruses. Arch Virol 1989; 108(1–2):33–38.
66. Greig JD, Lee MB. Enteric outbreaks in long-term care facilities and recommendations for prevention: A review. Epidemiol Infect 2009; 137(2):145–155.
67. Pang X, Lee B, Chui L, et al. Evaluation and validation of real-time reverse transcription-PCR assay using the LightCycler system for detection and quantitation of norovirus. J Clin Microbiol 2004; 42(10):4679–4685.
68. Freeman MM, Kerin T, Hull J, et al. Enhancement of detection and quantification of rotavirus in stool using a modified real-time RT-PCR assay. J Med Virol 2008; 80(8):1489–1496.
69. Zeng SQ, Halkosalo A, Salminen M, et al. One-step quantitative RT-PCR for the detection of rotavirus in acute gastroenteritis. J Virol Methods 2008; 153(2):12–23.
70. Kang G, Iturriza-Gomara M, Wheeler JG, et al. Quantitation of group A rotavirus by real-time reverse-transcription-polymerase chain reaction: Correlation with clinical severity in children in South India. J Med Virol 2004; 73(1):118–122.
71. Kageyama T, Kojima S, Shinohara M, et al. Broadly reactive and highly sensitive assay for Norwalk-like viruses based on real-time quantitative reverse transcription-PCR. J Clin Microbiol 2003; 41(4):1548–1557.
72. Trujillo AA, McCaustland KA, Zheng DP, et al. Use of TaqMan real-time reverse transcription-PCR for rapid detection, quantification, and typing of norovirus. J Clin Microbiol 2006; 44(4):1405–1412.
73. Logan C, O'Leary JJ, O'Sullivan N. Real-time reverse transcription PCR detection of norovirus, sapovirus and astrovirus as causative agents of acute viral gastroenteritis. J Virol Methods 2007; 146(1–2):36–44.
74. Oka T, Katayama K, Hansman GS, et al. Detection of human sapovirus by real-time reverse transcription-polymerase chain reaction. J Med Virol 2006; 78(10):1347–1353.
75. Zhang Z, Mitchell DK, Afflerbach C, et al. Quantitation of human astrovirus by real-time reverse-transcription-polymerase chain reaction to examine correlation with clinical illness. J Virol Methods 2006; 134(1–2):190–196.
76. Heim A, Ebnet C, Harste G, et al. Rapid and quantitative detection of human adenovirus DNA by real-time PCR. J Med Virol 2003; 70(2):228–239.
77. Miura-Ochiai R, Shimada Y, Konno T, et al. Quantitative detection and rapid identification of human adenoviruses. J Clin Microbiol 2007; 45(3):958–967.

17 Infections of the Skin and Mucosa

Parisa Ravanfar
Department of Dermatology, Dartmouth-Hitchcock Medical Center, Lebanon, New Hampshire, U.S.A.

Natalia Mendoza, Anita Satyaprakash, Rosella Creed, and Stephen Tyring
Center for Clinical Studies, Houston, Texas, U.S.A.

INTRODUCTION

Most viruses capable of causing human infection have been associated with some form of mucocutaneous manifestation. Although some viral infections are clinically recognized by their characteristic cutaneous appearance, other viruses are less commonly associated with cutaneous lesions and may not display unique clinical characteristics. Also, depending on the host, the clinical manifestations may vary greatly. Laboratory diagnosis is therefore imperative in many clinical cases of viral infections involving the skin and mucosa.

DNA VIRUSES

Human Papillomavirus

Human papillomaviruses (HPV) are members of the Papillomavirus family that cause disease in humans. They consist of double-stranded, circular DNA that encodes for early and late proteins. The early proteins (E1–E7) allow for DNA replication and RNA transcription, while the late proteins (L1–L2) are viral capsids that form into virions. The three major cutaneous lesions exhibited are plantar warts caused by HPV-1, verruca vulgaris caused by HPV-2, and verruca plana caused by HPV-3. Cutaneous HPV lesions typically present as flesh-colored exophytic papules; they are a common finding and mostly occur on the hands and fingers of patients, although they can occur on any area of the skin. Cutaneous warts occur most frequently in children, affecting up to 20% of school-aged children. The morphology of the lesions varies depending on the type of HPV and the body location of the virus. The majority of warts are verruca vulgaris, or common warts, followed by verruca plana, or plantar warts. HPV may also present as anogenital lesions; the HPV types responsible for these lesions will be further covered in Chapter 22.

HPV lesions are generally diagnosed clinically; however, biopsies of verrucae due to HPV will often reveal tissue patterns such as acanthosis, papillomatosis, hyperkeratosis, parakeratosis, and thrombosed dermal capillaries. Another typical finding on tissue biopsy is the presence of koilocytes, which are large keratinocytes with an eccentric, pyknotic nucleus with a perinuclear halo. Biopsies are often conducted to determine if lesions are neoplastic or dysplastic. Cytopathologic analysis of cervical specimens is often performed through analysis of a Papanicolaou smear.

A highly specific laboratory method for detection of HPV is immunohistochemical staining for HPV capsid antigens. However, this method has low sensitivity as dysplastic or neoplastic lesions contain few, if any, capsid antigens, leading to false negatives.

Other techniques for the diagnosis of HPV that are both highly specific and sensitive are DNA or RNA detection methods. The hybrid capture assay is a highly sensitive method of HPV detection. The hybrid capture assay is used to detect high-risk HPV types on thin-preparation, liquid-based cervical specimens (1). The hybrid capture assay starts with combining the clinical specimen with a base solution and thus releasing the nucleic acids. Next, the released target DNA merges with virus-specific RNA probes, creating DNA:RNA hybrids. These nucleic acid hybrids are then combined onto a solid phase coat and "captured" by antibodies specific for that viral RNA:DNA hybrid. The captured hybrids are detected with antibodies conjugated to alkaline phosphatase resulting in chemiluminescence that can be amplified and detected. The sensitivity of the hybrid capture assay is over 90% (1).

PCR is both very sensitive and specific for viral detection, but is currently used primarily for research purposes.

POXVIRUSES

Variola Virus (Smallpox)

Variola virus belongs to the family Poxviridae, subfamily Chordopoxvirinae, genus Orthopoxvirus. The virus consists of a single, linear, double-stranded DNA, and replicates in the host cell's cytoplasm.

The last naturally occurring smallpox outbreak was in 1977 in Somalia and the last case of smallpox infection in the United States was in 1949. The World Health Organization declared global eradication of smallpox in 1980 and currently only two laboratories contain variola virus isolates. There is significant concern that if this virus were to fall into the wrong hands it could be used as a dangerous weapon for biologic warfare.

Due to the eminent public health threat of smallpox infection, quick and reliable laboratory diagnosis is crucial. If there is a high suspicion of smallpox, public health officials must be notified immediately. Scrapings of skin lesions, blood samples, and tonsillar swabs must be sent to the CDC. If there is low to moderate suspicion of smallpox infection, other infections should be ruled out, such as VZV, HSV, and enterovirus (2). If specimens are high risk for variola infection, then laboratory tests must be performed in an enhanced BSL-3 laboratory to ensure safety (2).

There are many laboratory methods to confirm the diagnosis of smallpox. Some of these methods are specific for the variola virus, while others are for the general diagnosis of orthopoxviruses. According to the CDC, the laboratory tests required are PCR testing for variola, orthopoxvirus including variola, and nonvariola orthopoxviruses. PCR testing for variola virus allows for rapid detection of variola viral infection and is highly sensitive. Pan-orthopoxvirus PCR will also identify variola virus.

One specific method is to examine specimens directly for the presence of virions under electron microscopy (EM), which is recommended by the CDC if it is available. Negative stain EM can detect orthopoxvirus particles in approximately 95% of specimens from patients with variola or monkeypox infections and 65% of patients with vaccinia infections (3). Under EM, the variola virions measure approximately 225×300 nm and appear rectangular or brick-shaped when viewed lengthwise and ovoid when viewed on end, a distinguishing characteristic (3). However, EM visualization of such virions does not necessarily confirm smallpox infection, as vaccinia, monkeypox, and molluscum are poxviruses that share the same morphology as variola. Although EM can distinguish orthopoxviruses from other viruses, it is a not as sensitive as PCR.

Multiple PCR-based laboratory tests have been developed including generic orthopoxvirus assays, orthopoxvirus-specific real-time PCR assays, and TaqMan® assays. Such methods are rapid and highly specific; however, attention must be given to the fact that the prevalence of smallpox worldwide is zero, which may lead to high false-positive results. Therefore, PCR-based methods are best used when there is a high clinical suspicion for smallpox infection.

Confirmation of smallpox infection may be achieved by isolation of the virus in live-cell cultures with a subsequent nucleic acid–based method such as PCR to detect organism-specific DNA or RNA sequences extracted from the microorganism. Smallpox infection may also be confirmed by culturing the virus on egg chorioallantoic membrane (CA) and identification of characteristic pock lesions. CA was extensively used as the method for identification of the virus prior to the eradication of smallpox. CA is specific, but as it does not allow for rapid diagnosis and has low sensitivity, it is best used as a confirmatory test.

Due to the relatively large size of the variola virus, direct examination of vesicular or pustular material for viral particles is another method that has been commonly employed. Aggregations of the virus known as Guarnieri bodies may be seen in the cytoplasm of Hematoxylin-Eosin or silver-stained material under light microcopy (4).

Immunohistochemical studies can also be performed to detect the presence of the viral antigen. Serologic studies may be performed, but these results do not differentiate between the various orthopoxviruses (5).

Vaccinia

Vaccinia virus is a large, enveloped virus consisting of linear, double-stranded DNA, and a member of the Poxviridae family. It is best known for its use in smallpox vaccination; allowing for the worldwide eradication of smallpox. As a live-virus vaccine, there are potential clinical complications, especially in immunocompromised patients. Such complications include vaccinia necrosum, bacterial superinfections, generalized vaccinia, eczema vaccinatum, erythema multiforme, and encephalitis (6). Although routine vaccination for smallpox is no longer practiced, there have been multiple reports of human infection with vaccinia over the past decade (7). This zoonotic disease has been associated with human contact with cows (7). Vaccinia infection results in skin lesions on areas of contact in the form of vesicules and ulcers.

Identification of the *vaccinia virus* can be achieved by methods such as detection of pock morphology on the chorioallantoic membrane of chick embryos. Other methods include electron microscopy and atomic force microscopy, PCR and sequencing. Real-time PCR assays have also been proven as effective and sensitive not only for detection, but also for identification of variants of the *Vaccinia virus* (7).

Monkeypox

Monkeypox was first isolated in 1958 from captive primate rash specimens. Human monkeypox is a zoonotic disease that is acquired by contact with diseased animals, but can also be transmitted amongst humans. Clinical manifestations of human monkeypox infection are very similar to that of smallpox. Most humans have a prodrome of headache, fever, and diaphoresis; followed by a vesiculopustular rash that often begins on the trunk and spreads peripherally. Prominent lymphadenopathy is a defining aspect of this infection that separates it from smallpox infection.

Between 1970 and 1986 over 400 cases of human monkeypox infections were reported in Africa; 95% of which were within Zaire. The first cases of human monkeypox in the western hemisphere were reported as an outbreak in June 2003 in the Midwestern United States. The outbreak was associated with human contact with infected pet prairie dogs that had been housed together with Gambian rats (8).

Laboratory testing for monkeypox detection is often required for diagnosis. Orthopoxviral antigen may be detected in specimens by immunohistochemical staining with rabbit antivaccinia polyclonal antibody. Cytopathologic changes may also be identified within one to four days by viral culture, demonstrating plaques of elongated and rounded cells with prominent cytoplasmic bridging and formation of syncytium (8). Negative-stain EM may reveal the brick-shaped virions indicative of poxvirus infection. As previously mentioned this does not distinguish between vaccinia, molluscum, variola, or monkeypox infection and therefore is not particularly useful when the above viral infections are also in question.

Multiple nucleic acid tests have been developed for identification of monkeypox virus. Real-time PCR detection of monkeypox is a fast and sensitive method. The real-time PCR assays target two different orthopox virus genes; one is DNA polymerase (E9L) and the other is the envelope protein (B6R). Li et al. calculated the viral detection limit using real-time PCR assay E9L, with 95% confidence, at 2.54 fg viral DNA ($\sim$12.5 genomes). Overall, the E9L-Non-variola (E9L-NVAR) assay can reliably detect as few as 12.5 genomes of purified vaccine or Monkeypox virus DNA without giving false-positive results (9).

Molluscum Contagiosum

Molluscum contagiosum virus is a DNA poxvirus that causes a benign infection of the epidermis and mucosa. There are four subtypes of molluscum contagiosum, MCV1, MCV2, MCV3, and MCV4, and multiple variants within each subtype. MCV1 is the most common subtype found in infections.

Molluscum contagiosum is principally a clinical diagnosis, made by visualization of multiple 3–6 mm skin-colored papules with central umbilication. Each papule contains a core of cellular material, which can be expressed through curettage; typically, diagnosis is confirmed by obtaining a crushed prep of this material and staining it with either Wright, Giemsa, Gram, or Papanicolaou stains so as to visualize Henderson–Paterson intracytoplasmic viral inclusion bodies. In addition, Hematoxylin–Eosin stained formalin-fixed tissue from a biopsy specimen would also reveal these diagnostic findings. There are also reports of the use of 10% potassium hydroxide solution applied after curettage of a molluscum and crushing the core material between two glass slides (10). This technique allows for visualization of clusters of Henderson–Paterson bodies within the specimen. Laboratory confirmation of molluscum contagiosum can also be achieved by molecular methods such as DNA hybridization and restriction mapping techniques. Another laboratory method that is rapid, highly specific and sensitive, and minimally invasive is real-time PCR and subsequent DNA sequencing.

Orf

The Orf virus is a subspecies of the genus Parapoxvirus and the causal agent of contiguous ecthyma in livestock. In humans, infection occurs through skin abrasions, and lesions progress from macular, then papular, to large nodular and sometimes papillomatous lesions. Lesions often occur on the hands, arms, or face and are frequently solitary rather than multiple (11).

Orf lesions are often diagnosed clinically, but laboratory diagnosis may be required. Electron microscopy is the most common laboratory method of diagnosis. Orf virus is identified though EM by the characteristic ovoid shape that parapoxviruses exhibit, as well as the spiral arrangement of the surface filaments (3). Virus isolation is another method, but it requires primary ovine or bovine cells. Serologic methods are also available. PCR diagnosis of Orf infection has been found to be highly sensitive and specific (12).

HUMAN HERPESVIRUSES (HHV)

Herpes Simplex Viruses 1 and 2 (HHV1 & HHV2)

Herpes simplex virus (HSV) is a double-stranded DNA virus that is classified into two types, HSV-1 and HSV-2. HSV-1 is associated with oral lesions (herpes labialis), and HSV-2 typically causes genital lesions (herpes genitalis). HSV-2 will be further discussed in Chapter 22.

HSV-1 is an enveloped virus that replicates in the nucleus. It weighs 96×10^6 kDa and is composed of three main structural components: a nucleocapsid containing the genome, the envelope, and the tegument, which is a proteinaceous layer between the capsid and the envelope.

Infection with the herpes simplex virus has two phases; first is the primary infection in which the virus positions itself in a nerve ganglion, and next is the secondary phase in which the virus causes recurrent disease in that nerve distribution.

Herpes labialis is ubiquitous, with up to 90% of adults exhibiting serologic evidence of HSV-1 (13). The majority of primary infections are asymptomatic and can be detected only by elevated IgG antibody titer levels (14).

Primary infection with HSV-1 occurs through contact with infected body fluids, vesicular fluid from lesions, or direct contact with lesions. Primary infection is often asymptomatic, but 10% to 30% of people may present with signs and symptoms (15,16). The incubation period can range from 2 to 20 days, and generalized symptoms such as myalgias, headache, fever, and malaise may occur, followed one to three days later by mucocutaneous, grouped, vesicular lesions. HSV-1 lesions are characterized as 1- to 2-mm grouped, painful, vesicles with surrounding erythema in the oral mucosa that rupture and coalesce to form irregular shallow ulcers that eventually crust and resolve in two to four weeks, occasionally with postinflammatory hypo- or hyperpigmentation. In the majority of individuals, the virus is initially asymptomatic and lies dormant in the neural ganglia; it periodically reactivates, often during periods of stress. Recurrent symptomatic episodes occur in approximately one-third of infected individuals. Other than herpes labialis, HSV-1 can also cause herpetic whitlow (herpetic paronychia), ocular infections (blepharitis, conjunctivitis, epithelial keratitis, stomal keratitis, iridocyclitis, retinitis), genital infection similar to HSV-2, and subsequent neonatal herpetic infections by vertical transmission. Other diseases that have been associated with HSV-1 infection are erythema multiforme,

Behcet's disease, Bell's palsy, and Meniere's disease. Furthermore, HSV is the most common cause of viral encephalitis in adults outside of tropic regions (15).

Clinical history and physical appearance are the basis of diagnosis of typical HSV-1 herpes labialis lesions. However, confirmatory laboratory diagnosis is often required for atypical lesions.

Viral cultures are often performed as laboratory confirmation. Such cultures are obtained by swabbing the base of the vesicle with specific HSV viral culture kits. HSV cultures can be obtained within 24 to 48 hours, allowing for identification of the virus and its type (17). Tzanck smears are another common laboratory confirmation method in which the fluid from a vesicle is collected with a cotton swab and smeared onto a glass slide, then stained. A positive smear reveals characteristic multinucleated giant cells. However, Tzanck smears do not differentiate between HSV-1, HSV-2, or VZV infections.

Lesions can also be tested for HSV by rapid direct fluorescent antibody testing. Serological testing can be performed to discern between HSV-1 and HSV-2, as well as to distinguish between primary and secondary infection.

Virologic Tests

Cytology

The Tzanck prep is used to detect cytologic changes in specimens from suspected HSV lesions. The characteristic findings include syncytial giant cells, Cowdry Type A intranuclear inclusions, and ballooning cytoplasm. Although this test is inexpensive, it is running out of favor in the clinical setting due to its low sensitivity (less than 60%) and lack of specificity, as it cannot identify the type of virus involved (15).

Culture

Detection of HSV-1 by isolation of the virus in tissue culture is a commonly used method that has high specificity and allows for typing of the virus. The virus can be detected through culture within 24 to 48 hours and is relatively inexpensive. However, the sensitivity of viral culture is low, especially for recurrent lesions, and even lower for healing lesions (18,19). Culture of CSF for HSV has very low sensitivity.

Direct Immunofluorescence

Direct immunofluorescence (DFA) or immunoperoxidase assay allows for rapid detection of HSV, but has low sensitivity and a high false-negative rate. Some tests to not discriminate between the types of virus (15). Furthermore, negative or indeterminate results often require confirmation by culture or PCR (15).

PCR

PCR for HSV DNA is the most sensitive method for diagnosis of HSV (20) (Fig. 1). Real-time PCR allows for rapid diagnosis and for identification of the type of HSV. The ability of real-time PCR to distinguish between HSV types through melting curve analysis minimizes the number of reactions needed (20). Nucleic acid amplification is the method of choice for detection of HSV CNS infections (21).

Serologic Tests

After infection with HSV, both type-specific and nontype-specific antibodies are formed and persist throughout patients' lives. Western blot is the gold standard for antibody detection and has both a sensitivity and specificity of over 99% for symptomatic infections established over six months (23). Accurate type-specific HSV-1 serologic assays that test for HSV-1 glycoprotein G1 became commercially available in 1999 (23) and are particularly accurate (21,23). HSV-1 IgG-specific assays approved by the FDA include HerpeSelectTM-1® enzyme-linked immunosorbent assay (ELISA) IgG and HerpesSelectTM-1 Immunoblot IgG. The sensitivities of these assays range from 80% to 98% and specificities range from 96% to 100% (23). False positives are more likely to occur in patients with a low clinical suspicion of HSV infection.

Testing for HSV IgM antibodies does not increase the specificity of the serological diagnosis in patients with clinical signs of HSV, and recurrent HSV infections are not always associated

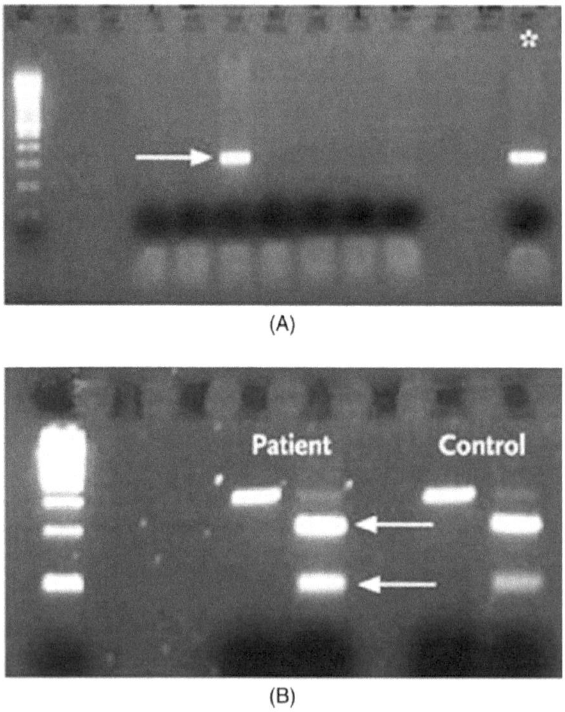

Figure 1 Specimen of cerebrospinal fluid subjected to polymerase-chain reaction assay for herpes simplex virus DNA. PCR products are fractionated over a 2% agarose gel (Sigma-Aldrich), along with a DNA molecular weight standard (Panel **A**). The expected 290-bp product amplified from the patient's cerebrospinal fluid is indicated by an arrow and the positive control by an asterisk. Negative controls flank the patient's specimen. Digestion of the amplified product from the patient's specimen with the restriction endonuclease ApaI (New England Biolabs) results in the formation of two cleavage products, 190 bp and 100 bp (Panel **B** arrows). Digestion of the positive control results in an identical pattern. *Source*: From Ref. 22.

with a significant rise in antibody titer since most individuals are already serologically HSV positive (15).

Varicella Zoster Virus (HHV-3)

Varicella (Chickenpox)

Primary infection with VZV presents as varicella, which manifests as low-grade fever, malaise, and disseminated pruritic lesions in all stages (vesicles, pustules, and crusts). Clinical diagnosis of varicella may be confirmed with viral culture or Tzanck smear. A rapid and sensitive procedure is direct immunofluorescence of fluid from the base of a vesicle (24). Varicella is further covered in Chapter 14.

Herpes Zoster (Shingles)

In 1888, von Bokay discovered the clinical association between varicella and herpes zoster when he observed that children without known immunity to varicella experienced chickenpox after contact with herpes zoster (25). Subsequently, in 1954, Thomas Weller isolated VZV from the vesicular fluid of both varicella and herpes zoster lesions (25).

VZV is a DNA virus in the herpesvirus family and as is the case with other alpha herpesviruses, VZV persists in the body after primary infection and remains latent in sensory nerve ganglia.

Up to 20% of immunocompetent individuals and up to 50% of immunocompromised individuals experience reactivation of the VZV as herpes zoster, commonly known as shingles, involving the skin of single or adjacent dermatomes. Individuals who live to the age of 85 years have a 50% chance of developing zoster in their lifetime. Reactivation of VZV has been associated with certain risk factors such as older age, immunosuppression, intrauterine exposure, history of primary varicella infection occurring at less than 18 months of age, and a positive family history of herpes zoster (26). Immunocompromised patients are also at risk for disseminated herpes zoster, which can cause generalized skin lesions and have CNS, pulmonary, and hepatic involvement. Postherpetic neuralgia is a common cause of morbidity due to zoster and is characterized as pain in the area of zoster lesions that persists greater than three months after the lesions have resolved.

Virologic tests

Laboratory diagnosis of zoster is not commonly required, but is useful in atypical cases.

VZV may be isolated in tissue culture, and laboratory techniques allow for differentiation between wild-type strains and vaccine strains of VZV. Although viral culture is specific, it is not very sensitive, as the virus is difficult to isolate.

A Tzanck smear can also be performed on herpes zoster vesicles to look for multinucleated giant cells, but as previously mentioned, Tzanck preps are not specific and have been mostly replaced by more useful laboratory methods.

PCR is the test of choice for rapid diagnosis of VZV infection. Real-time PCR for VZV, which produces results in a matter of a few hours, is also available and is the most sensitive and specific laboratory method available for the diagnosis of VZV (25).

Direct immunofluorescence (DFA) can also be utilized for laboratory confirmation when PCR is not available (Fig. 2); however, DFA has lower sensitivity than PCR and results are dependent on specimen collection and handling. Ideally, a specimen is collected by unroofing a fluid-filled vesicle and rubbing the base of the opened lesion with a swab (25). Crusted lesions are also a good source of specimen collection for PCR analysis. PCR and DFA allow for detection of viral components even in the setting of negative cultures because DNA and viral proteins are still present even after the disappearance of culturable virus (25).

Serologic Tests

Complement fixation (CF) assays may be used for retrospective diagnosis of herpes zoster. Other serologic tests include indirect fluorescent antibody (IFA), fluorescent antibody to membrane

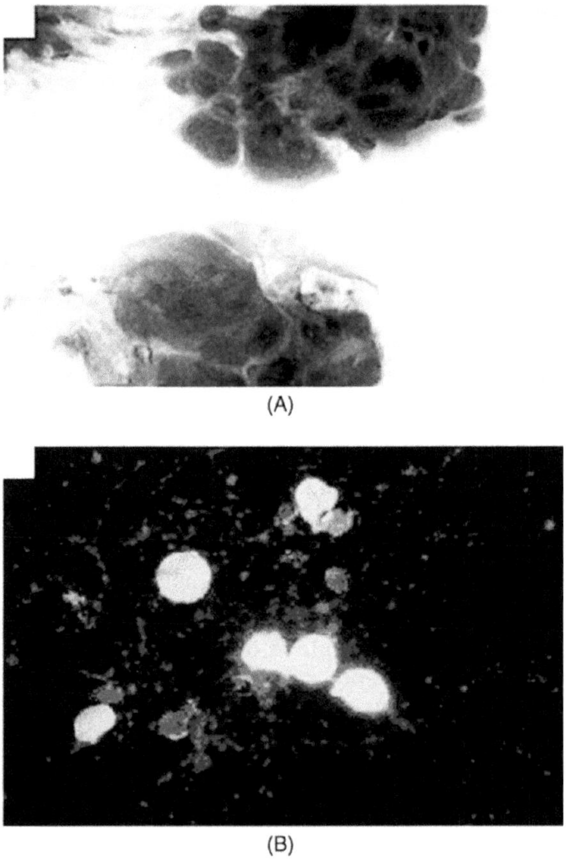

(A)

(B)

Figure 2 (*See color insert*). Diagnosis of Herpes Zoster. Panel (**A**) shows a positive Tzanck smear (×400). Wright's stain demonstrates multinucleated giant cells. Panel (**B**) shows a positive direct immunofluorescence assay (×400). Cells are stained with fluorescein-conjugated monoclonal antibodies against varicella zoster virus; green fluorescence indicates the presence of varicella zoster virus antigens. *Source*: From Ref. 27.

antigen (FAMA), neutralization, indirect hemagglutination (IHA), immune adherence hemagglutination (IAHA), radioimmunoassay (RIA), latex agglutination (LA), and ELISA. ELISA is a sensitive and specific test, and is commercially available. LA is more sensitive than ELISA, but also has a higher rate of false-positives (25). The false-positives can be decreased by performing LA as a dilution series (25). Both ELISA and LA may be useful in screening patients for varicella immunity. VZV T-cell mediated immunity is significantly more accurate in identifying the immune response to VZV and is directly correlated with clinical outcome, as opposed to humoral immunity which is not as strongly correlated with clinical immunity (28).

Epstein–Barr Virus (HHV-4)

Epstein–Barr virus (EBV) is the causal agent of infectious mononucleosis and is also associated with numerous different malignancies, such as Hodgkin's lymphoma, non-Hodgkin's lymphoma, nasopharyngeal carcinoma, oral hairy leukoplakia, gastric carcinoma, inflammatory pseudotumor, and posttransplant lymphoproliferative disorder. Peripheral smears from patients with infectious mononucleosis often exhibit over 10% atypical lymphocytes.

Infectious mononucleosis typically presents with signs and symptoms of malaise, headache, and fatigue, followed by fever, sore throat, and lymphadenopathy. It is not uncommon to find hepatosplenomegaly. A viral exanthem due to EBV most often presents as red papules in the background of pink patches, and less often as morbilliform, scarlantiniform, erythema, vesicles, and purpura. The viral exanthem occurs during the first week of illness. Small oral petechiae at the border of the hard and soft palate can be present in one-third of patients. If the infection is treated inappropriately with ampicillin or other penicillins, a benign erythematous morbilliform eruption often develops on the trunk and extremities that begins to desquamate after a week.

Laboratory confirmation of infectious mononucleosis should start with the Monospot test. The Monospot test is a commonly used serological assay; it is a heterophile antibody test that was pioneered in 1932 before EBV was even identified as the causative agent of infectious mononucleosis (29). The original test was based on the observation that serum or plasma obtained from patients with infectious mononucleosis would agglutinate horse or sheep erythrocytes. Today, the modern variation of the test utilizes latex-beads coated with bovine heterophile antigens instead of horse or sheep erythrocytes. The Monospot test allows for rapid serologic confirmation of clinical cases with signs and symptoms of infectious mononucleosis.

If the Monospot test is suspected to be falsely negative, as may be the case in young children or elderly patients, further laboratory tests can be performed. For example, indirect fluorescent antibody tests can be prepared and interpreted by patterns of antibody localization to complexes of viral proteins jointly identified as early antigen (EA), viral capsid antigen (VCA), or EBNA (30); however, this test is subjective and labor intensive. Other tests that are more rapid and objective include ELISA, PCR, and real-time PCR assays. EBV DNA testing is often reserved for atypical presentations or for immunosuppressed patients (29).

EBER In Situ Hybridization

The gold standard for detecting and localizing latent EBV in tissue samples is EBV-encoded RNA (EBER) in situ hybridization (31). EBER are small, nonencoding, viral RNA genes. EBER hybridization is used to confirm the diagnosis of posttransplant lymphoproliferative disorders due to EBV. In EBV-associated Hodgkin's disease, EBER is generally found solely in Reed–Sternberg cells and mononuclear variants; however, in mononucleosis one would typically find small and large EBER-positive cells, including immunoblasts surrounding necrotic zones (29).

Other laboratory tests for EBV include EBV clonality assay by Southern blot analysis, EBV DNA amplification, EBV viral load, immunohistochemistry, culture of EBV or EBV-infected B lymphocytes, electron microscopy, serology, and real-time PCR assays.

Cytomegalovirus (HHV-5)

Cytomegalovirus (CMV) is found in all geographic regions and amongst all cultural groups. It is estimated that between 50% and 80% of adults in the United States are infected with CMV by the age of 40 years (32). Most infections with CMV are asymptomatic and often go undiagnosed; however, CMV infection can cause severe morbidity and mortality in immunocompromised

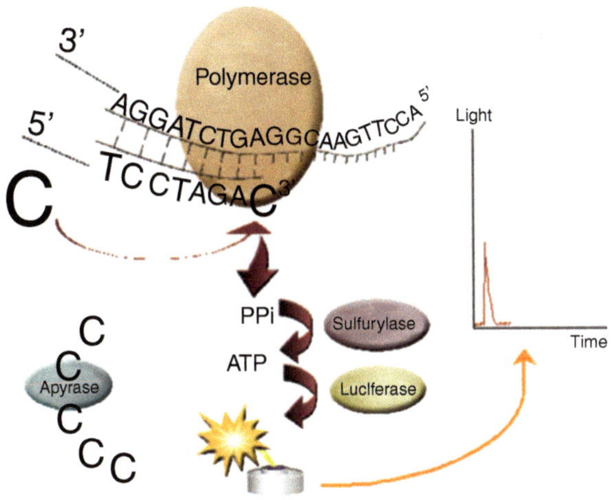

Figure 4.3 The pyrosequencing reaction, demonstrating release of pyrophosphate (PPi) accompanying the addition of each nucleotide. This in turn releases ATP which drives luciferase expression.

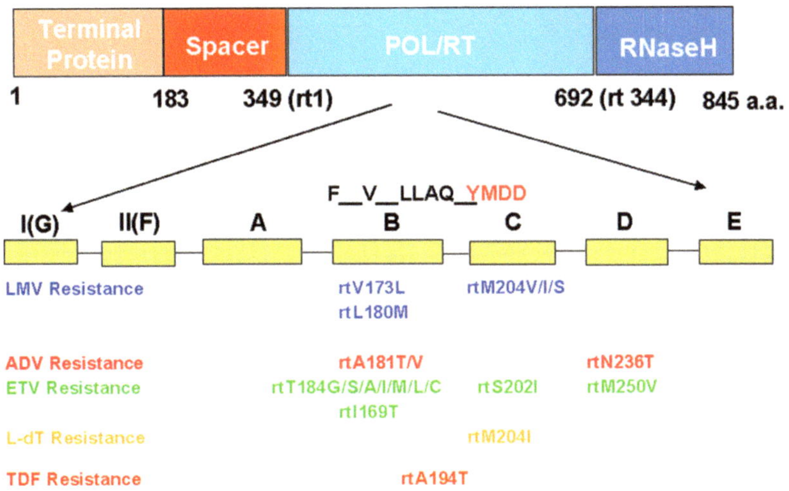

Figure 4.4 Primary HBV polymerase mutations (domains A to G) associated with antiviral drug resistance in chronic HBV infection. *Abbreviations*: LMV, lamivudine; ADV, adefovir; ETV, entecavir; L-dT, telbivudine; TDF, tenofovir. The YMDD motif associated with lamivudine and telbivudine resistance is located at residues 203 to 206.

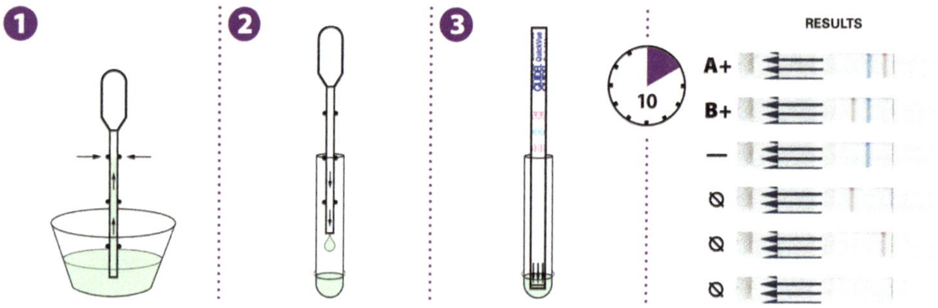

(A)

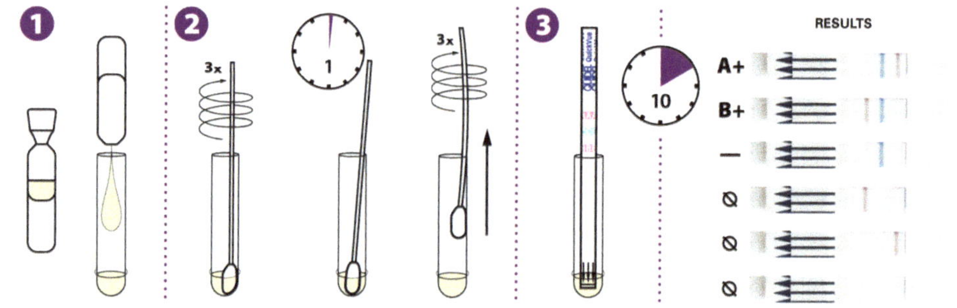

(B)

Figure 8.6 Quidel's QuickView test for influenza virus (**A**) and (**B**), using either nasal wash (*top*) or/and nasal swabs (*bottom*).

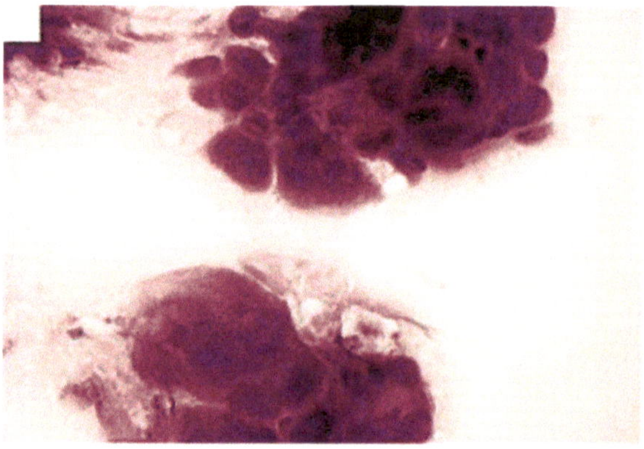

(A)

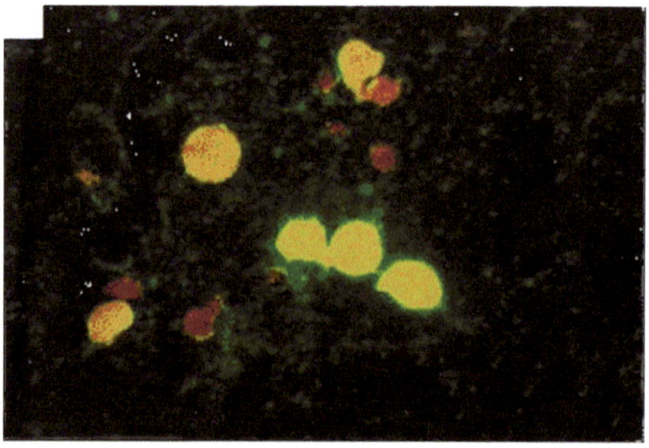

(B)

Figure 17.2 Diagnosis of Herpes Zoster. Panel (**A**) shows a positive Tzanck smear (x400). Wright's stain demonstrates multinucleated giant cells. Panel (**B**) shows a positive direct immunofluorescence assay (×400). Cells are stained with fluorescein-conjugated monoclonal antibodies against varicella zoster virus; green fluorescence indicates the presence of varicella zoster virus antigens. *Source*: From Ref. 27.

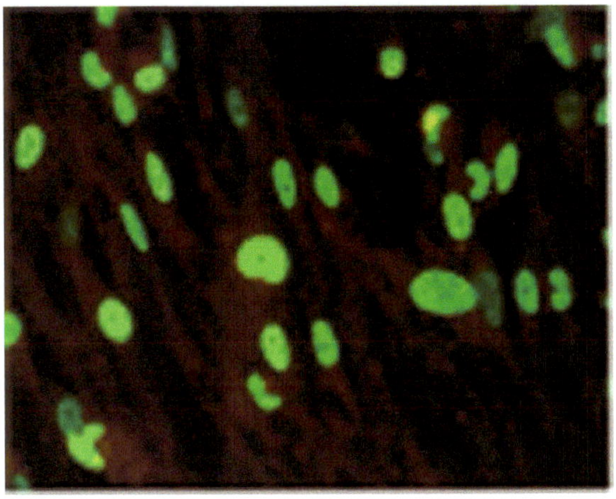

Figure 17.3 Shell vial assay for cytomegalovirus (CMV) (immunofluorescence stain, ×400). *Source*: From Ref. 36.

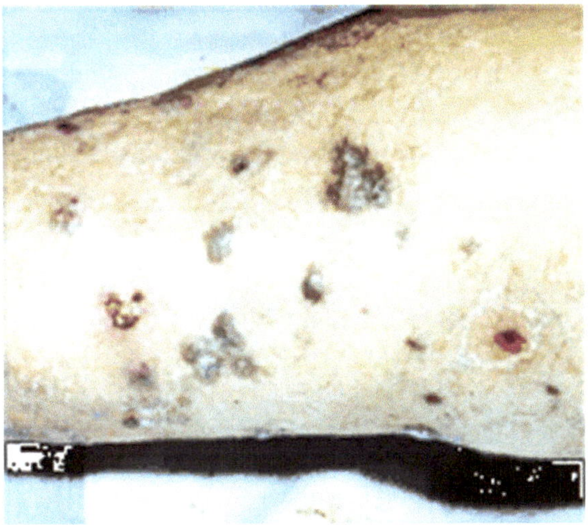

(A)

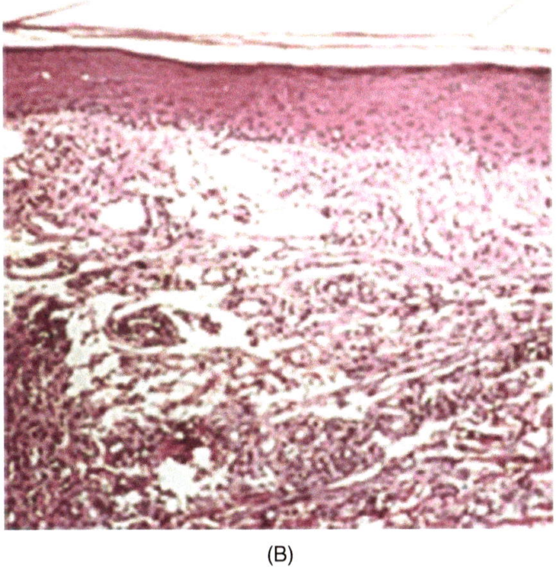

(B)

Figure 17.4 Kaposi's sarcoma. Panel (**A**) shows the lesions of classic Kaposi's sarcoma. Panel (**B**) shows the characteristic histologic features (hematoxylin and eosin, × 20). The proliferation of spindle-shaped tumor cells has led to the formation of abnormal vascular slits, some of which contain red cells. Mitotic activity is absent in this lesion, and the degree of pleomorphism of the tumor cells is mild. *Source*: From Ref. 50.

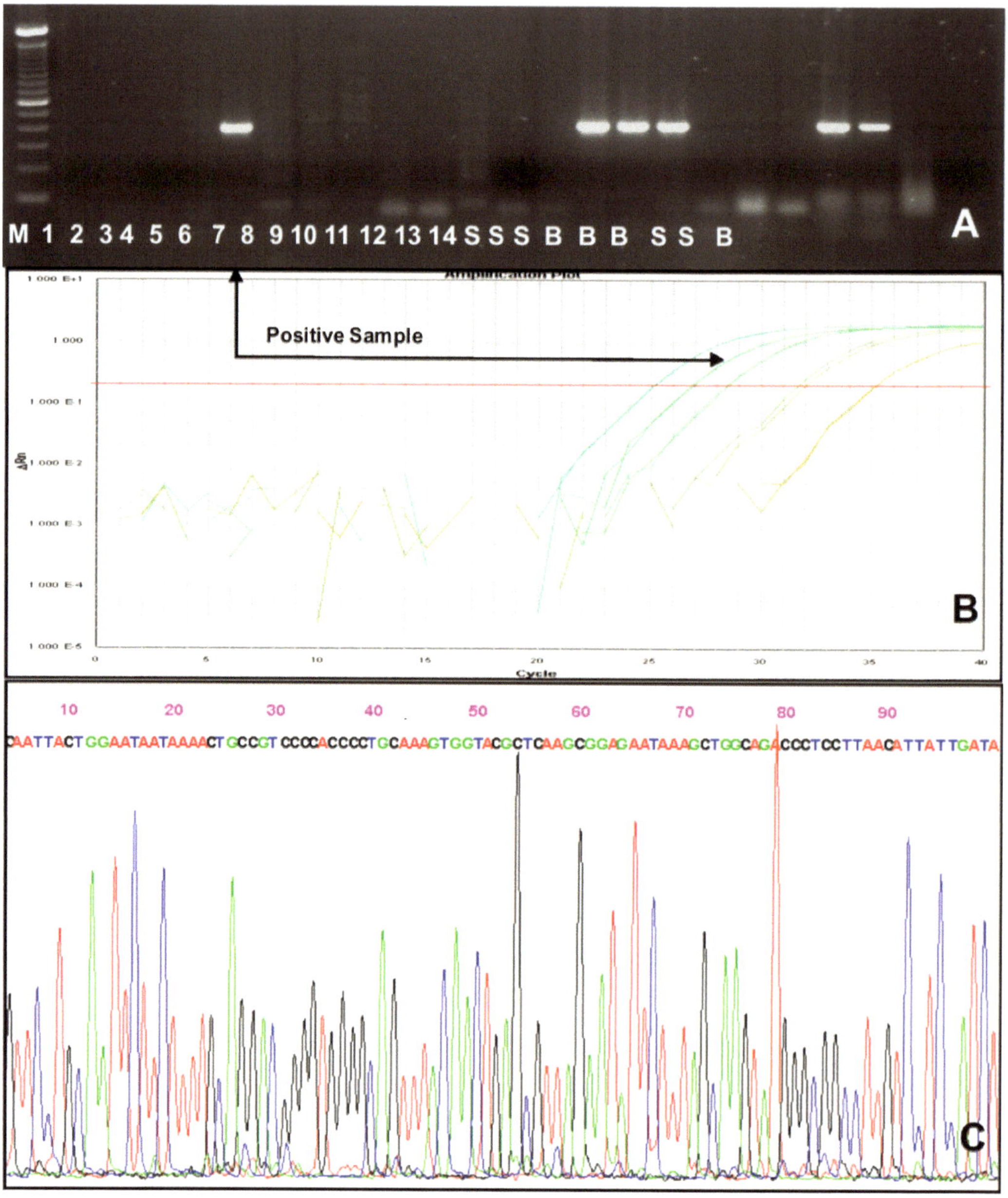

Figure 18.3 Molecular diagnostics in heart tissue.

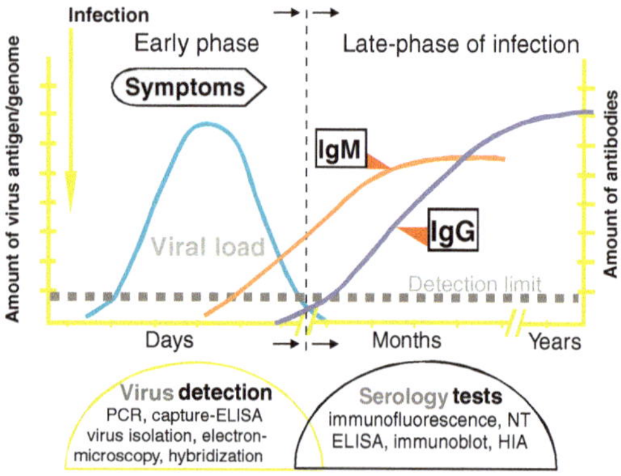

Figure 26.1 Schema for diagnosis of viral infections.

patients and in pregnant women with primary CMV infection. Immunocompetent patients may experience CMV mononucleosis, a similar clinical presentation as EBV mononucleosis, with fever, fatigue, sore throat, lymphadenopathy, and possibly organomegaly. Of those patients who experience CMV mononucleosis, up to one-third of them can present with a generalized morbilliform rash, or less commonly, perifollicular papulopustules, vesiculobullous eruptions, and nodular or ulcerative lesions.

Maternal CMV infection is the most common congenital viral infection. Primary CMV infection in pregnant women has been associated with neonatal CNS injury, hearing loss, congenital malformations, thrombocytopenia, hepatosplenomegaly, and intrauterine growth retardation. Cutaneous lesions in the neonate are typically referred to as "blueberry muffin" due to the presence of purpuric macules and papules secondary to persistent dermal hematopoiesis. The majority of CMV infections are asymptomatic and therefore not diagnosed; however, if symptomatic CMV infection is suspected there are a variety of laboratory diagnostic tests available.

Serologic Tests

The diagnosis of primary CMV infection is simple if seroconversion can be demonstrated. However, seroconversion documentation is not always possible given that there is not a screening program for CMV infection, which would be impractical. CMV IgM is used to detect active or recent infection and has sensitivity ranging from 30% to 88% depending on the kit (33). Although anti-CMV IgM antibodies may indicate acute or recent infection, it does not necessarily indicate primary infection and false-positives are common. Anti-CMV IgG avidity assays are the most reliable method of identification of primary infection in pregnant women (33). Sensitivity of IgG avidity assays ranges from 92.8% to 100% and specificity ranges from 82.5% to 100% (33). Antibody avidity indicates the strength with which a multivalent antibody binds to a multivalent antigen. Therefore, the antibodies produced during a primary immune response have a lower avidity than the antibodies of a matured immune response from a nonprimary infection. Avidity assays are useful for confirmation of primary infection. Immunoblot is the gold standard when confirming the presence of IgM antibodies in the serum with very high sensitivity and specificity.

Virologic Tests

During primary CMV infection the virus can be found in body fluids such as saliva, urine, and vaginal secretions; however, viral shedding may also occur during reinfection or reactivation of the virus, therefore making recovery of CMV in secretions a unreliable method of diagnosing primary infection (34). Various laboratory methods have been developed to detect and quantify CMV in the blood.

Methods to detect and quantify viremia have been classically time consuming due to dependence on the appearance of cytopathic effect, determination of 50% tissue culture infectious doses, and plaque assays (34). However, this method has largely been replaced by a "shell vial" assay that provides results within 24 hours. This assay is based on the postulation that each immediate-early antigen (p72)-positive cell in a human fibroblast monolayer has been infected by a single leukocyte carrying CMV (35). The shell vial monolayer is stained using either immunofluorescence or immunoperoxidase methods and a monoclonal antibody reactive to the CMV major immediate-early protein (35) (Fig. 3). The number of positive nuclei is then tallied.

Detection of antigenemia identifies and quantifies peripheral blood leukocytes that are positive for the CMV matrix phosphoprotein pp65. It is a very sensitive and rapid method (results can be available within a matter of a few hours); however, it is limited by the subjectivity of reading slides.

Detection and quantification of CMV DNA in blood may be achieved through PCR assays or hybridization techniques. Identification of viral genetic material is highly specific and sensitive with the use of PCR. Detection of CMV RNA in the blood indicates CMV replication in vivo, and late viral transcripts may better represent viral replication dissemination (37,38).

Recognition of circulating cytomegalic endothelial cells (CEC) in the blood of neonates and fetuses with symptomatic congenital CMV infection allows for diagnosis of congenital

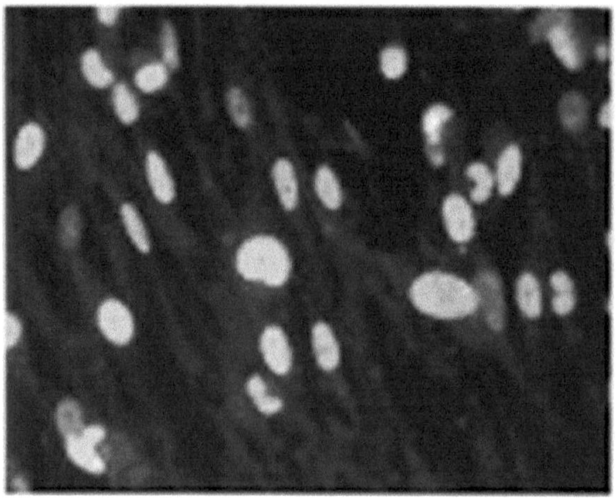

Figure 3 (*See color insert*). Shell vial assay for cytomegalovirus (CMV) (immunofluorescence stain, ×400). *Source*: From Ref. 36.

CMV, but identification of endotheliemia is not particularly sensitive in the diagnosis of CMV infection.

Congenital CMV in the neonate is diagnosed if the virus is detected in the urine, saliva, or blood of the neonate within the first three weeks after birth. Diagnosis is dependent on viral detection rather than antibody detection.

HUMAN HERPESVIRUS-6 (HHV-6)

HHV-6 was first isolated in 1986 and was subsequently found to be linked to roseola infantum (exanthum subitum) in 1988 by Yamanishi et al. Over 90% of humans are infected by HHV-6 by early childhood and can experience reactivation of the virus regardless of immune status. Primary infection with HHV-6 often occurs by two years of age and is associated with an indistinctive febrile illness. A fraction of these patients may experience the classic roseola rash. The characteristic rash is revealed either during the illness or following defervescence in approximately 20% of patients experiencing primary HHV-6 infection (39). HHV-6 may cause severe disease in immunocompromised patients.

Skin diseases associated with HHV-6 include exanthum subitum, "glove-and-socks" syndrome, hypersensitivity drug reactions, Gianotti–Crosti syndrome, pityriasis rosea, and lymphoproliferative malignancies.

Laboratory diagnosis of HHV-6 is not commonly performed, as the virus is ubiquitous and often asymptomatic. There is no currently accepted test for diagnosis of HHV-6; however, the virus can be cultured in lymphocytes that have been stimulated by phytohemagglutin and interleukin-2 (40). Nevertheless, a positive viral culture is not particularly valuable because most individuals carry the virus asymptomatically.

ELISA may also be utilized for confirmatory diagnosis; however, it does not differentiate between variants A and B and may give false-positive results due to cross-reactivity with other herpesviruses.

An HHV-6 antibody panel is another available method. An IgM titer of 20 or more is indicative of recent infection or reactivation of the virus. An IgG titer of 10 or more is indicative of prior infection with HHV-6. Due to the ubiquitous nature of HHV-6, the correlation between specific disease and a single titer is of little use. A fourfold rise or seroconversion of IgG or IgM titers represents acute infection or virus reactivation.

HUMAN HERPESVIRUS-7 (HHV-7)

Herpesvirus-7 was first isolated in 1990 and is very similar to HHV-6. Like HHV-6, it is highly prevalent worldwide and greater than 90% of humans experience primary infection by the age of 10 years. The mode of transmission is likely through salivary fluid. Even less is known of the pathogenicity of HHV-7 than of HHV-6. HHV-7 has been postulated as the causative agent of pityriasis rosea and has also been proposed as another causative agent for exanthema subitum.

In addition, HHV-7 infection has been implicated in chronic fatigue syndrome, posttransplant skin eruptions, reactivation of HHV-6 infection, and febrile illness of infancy (41).

Diagnosis of HHV-7 is similar to that of HHV-6, with laboratory clinical diagnosis rarely performed. There are currently no standardized assays available for HHV-7. HHV-7-specific antibodies are available and can be utilized to differentiate between HHV-7 and HHV-6. As with HHV-6, the virus may be cultured and rises in IgG titer may be demonstrated.

HUMAN HERPESVIRUS-8 (HHV-8)

HHV-8 is a member of the *Gammaherpesvirinae* subfamily, which is further divided into gamma-1/lymphocryptoviruses and gamma-2/rhadinoviruses. HHV-8 is the only virus in the rhadinovirus genus (42).

HHV-8 was first identified in Kaposi's sarcoma lesions and was initially thought to be a sexually transmitted disease due to the relatively high prevalence of the disease in AIDS patients. Unlike many other herpesviruses, HHV-8 is not ubiquitous and seroprevalence varies by region and HIV status. For instance, the seroprevalence in northern countries is estimated to be 1% to 5% in healthy individuals, but the seroprevalence is as high as 50% in some areas in Africa (43,44).

HHV-8 is shed primarily through saliva but can also be transmitted through other body fluids. HHV-8 is associated with Kaposi's sarcoma (KS), a vascular neoplasm that occurs most frequently in homosexual and bisexual men with AIDS (45). The less common, but classic variant of KS is seen an elderly men of Mediterranean and Middle European descent as well as in men in Sub-Saharan Africa, and the endemic or African variant arises in some parts of Africa in children and young adults. KS lesions typically present as deep red–purple macules that evolve into papules, plaques, and tumors that can occur in any organ. KS is the most common AIDS-related malignancy in developed nations. Castleman's disease is another lymphoproliferative disease associated with HHV-8 due to hyperproliferation of B-cells and is characterized by tumors in lymph nodes throughout the body. Primary effusion lymphoma, which is also associated with HHV-8 infection, is a rare B-cell lymphoma found almost solely in HIV-infected persons.

PCR is highly sensitive in detection of HHV-8 DNA in infected lesions, especially in KS. However, owing to the high sensitivity of PCR, there is also a risk of cross-contamination and false-positive results (46). PCR is a very reliable and rapid laboratory method for detection of HHV-8 infection, and real-time PCR allows for rapid detection and quantification of the virus.

Skin biopsy may be required for confirmation of KS (Fig. 4). A valuable method for detecting HHV-8 in tissues is the use of immunohistochemistry with antilatent nuclear antigen 1 (LANA1) antibodies (47,48). LANA1 is a latent protein expressed in the nucleus of cells infected with HHV-8 (47,48). Antibodies recognize a repeated protein sequence that allows for straightforward detection of the protein (49). Use of immunochemistry for laboratory diagnosis of HHV-8 is ideal for confirmation of KS, multicentric Castleman disease, and HHV-8 related lymphomas (45).

First generation serologic assays are available for HHV-8 antibody detection (51). Second-generation immunoblotting assays have also been developed and are reported to be both sensitive and specific in detection of HHV-8 infection (52).

PARVOVIRUS B-19

Human parvovirus B-19 was discovered in 1975 and is a member of the family Parvoviridae, genus *Parvovirus*. It lacks an envelope and consists of a linear single-stranded DNA that encodes for three proteins: two structural or capsid proteins (VP1 and VP2) and one nonstructural protein (NS1).

Although Parvovirus B19 infections occur in all ages and all seasons, they are most common in school-aged children during the late winter and early spring months (53). Prevalence of immunity to B19 increases with age, with up to 75% of adults over 40 years of age exhibiting B19 IgG antibody (54).

Clinical manifestations of B19 infection are broad with possible dermatologic, rheumatologic, or hematologic effects. Erythema infectiosum, or fifth disease, is associated with B19 infection and characterized by a rash occurring in three stages after the contagious period. The first stage is recognized by a "slapped-cheek" appearance, with an erythematous, nontender,

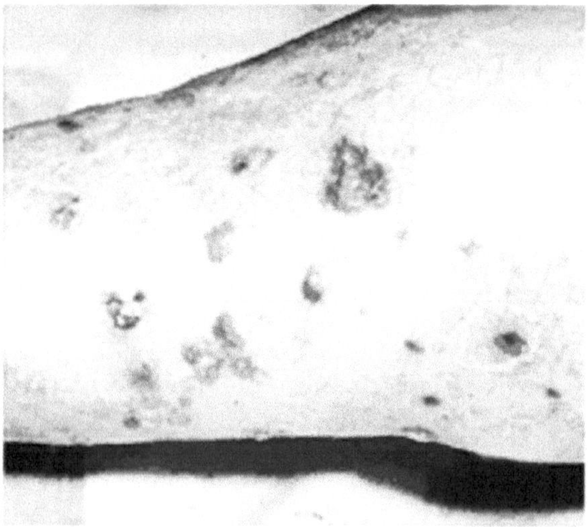

(A)

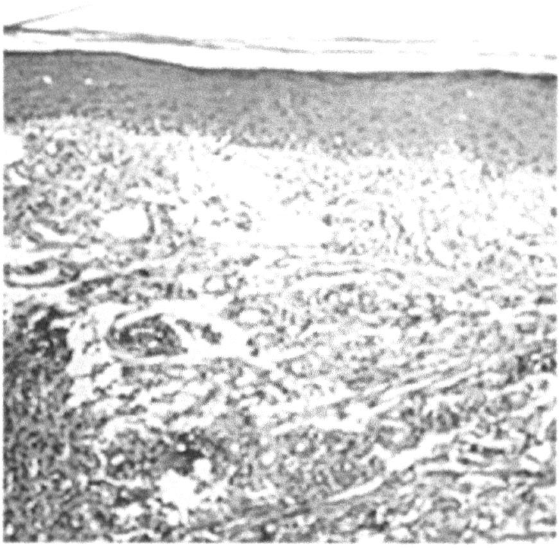

(B)

Figure 4 (*See color insert*). Kaposi's sarcoma. Panel (**A**) shows the lesions of classic Kaposi's sarcoma. Panel (**B**) shows the characteristic histologic features (hematoxylin and eosin, ×20). The proliferation of spindle-shaped tumor cells has led to the formation of abnormal vascular slits, some of which contain red cells. Mitotic activity is absent in this lesion, and the degree of pleomorphism of the tumor cells is mild. *Source*: From Ref. 50.

warm exanthem on both cheeks. The second stage is characterized by an erythematous papular rash with central clearing, resulting in a reticular pattern on the proximal extremities, trunk, and buttocks. The last stage involves recurrence of the rash related to exposure to sun, heat, or stress. Another dermatologic manifestation of B19 infection is papular, purpuric "gloves-and-socks" syndrome. B19 is also linked to arthropathy. Transient aplastic crisis is a well-known manifestation of B19 infection in patients with hematologic disorders.

Diagnosis of B19 is often based on clinical features; however, laboratory diagnosis may be required for atypical cases or for confirmatory reasons. Enzyme immunoassays to test for B19 IgG and IgM are available and have been shown to have high sensitivity (97–100%) and high specificity (79–99%) (55).

Because of the inability to routinely culture parvovirus B19, laboratory diagnosis of B19 infection is achieved through either direct detection of B19 antigens in specimens, or through detection of a specific antibody response in serum.

B19 antigen detection can be performed using counterimmunoelectrophoresis (CIE), immunoelectromicroscopy (IEM), radioimmunoassay (RIA), enzyme immunoassays (EIA), blot immunoassays, and receptor-mediated hemagglutination (RHA).

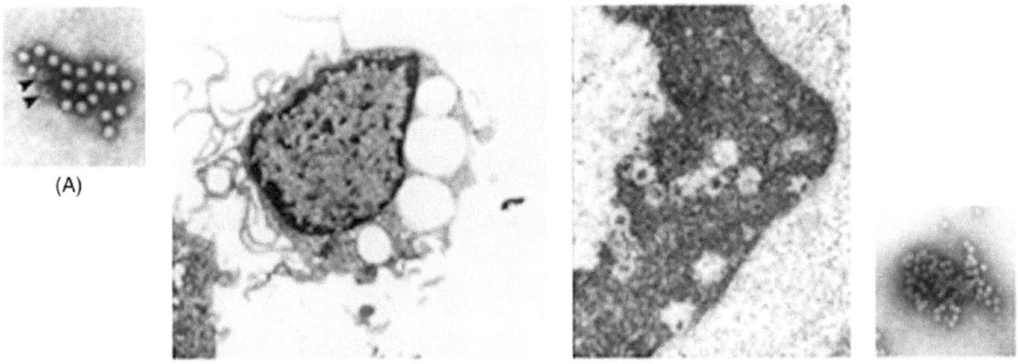

(A) (B) (C) (D)

Figure 5 Transmission electron micrographs showing native parvovirus B19 in serum and cells and recombinant capsids. Panel (**A**) shows symmetric, icosahedral particles, about 25 nm in diameter, and empty capsids (arrowheads) in serum from an infected person. Panel (**B**) shows human erythroid progenitor cells infected in vitro with the virus; vacuoles and cytoplasmic pseudopods are present (×10,000). In Panel (**C**), which shows Panel (**B**) at higher magnification, marginated chromatin contains assembled capsids (×100,000). Panel (**D**) shows empty recombinant parvovirus capsids produced in baculovirus system (×154,000). *Source*: From Ref. 60.

EIA that detects specific IgM antibodies (Bitrim Parvovirus B19) has a reported sensitivity of 89.1% and specificity of 99.4% (56). IEM is primarily used to confirm positive results from other B19 assays (57).

Serologic diagnosis of parvovirus B19 infection may be achieved with detection of IgG antibody with ELISA, RIA, or IFA and may be confirmed with western blot. IgM antibody can also be detected by ELISA, RIA, or IFA as an indicator of recent infection and may be confirmed by western blot.

Virologic laboratory diagnosis of B19 may be accomplished through detection of B19 viral antigen by ELISA, CIE, or IFA. B19 DNA may be detected by hybridization. Furthermore, PCR can be used to amplify as well as detect B19 DNA.

B19 infection can also be detected by EM visualization and identification of B19 viral particles. EM of B19 infected cells illustrates cytoplasmic vacuolization, pseudopod extensions, swollen mitochondria, and crystalline assortments of B19 viral particles within marginated nuclear chromatin clumps (Fig. 5) (58,59). Light microscopy may be utilized to visualize intranuclear inclusions or giant pronormoblasts due to B19 infection. In situ immunohistochemistry to detect B19 antigens and in situ hybridization to detect B19 DNA may also be utilized to further enhance microscopic diagnosis of B19 and increase sensitivity (53).

RNA VIRUSES

Enteroviruses

Enteroviruses belong to the family *Picornaviridae,* genus Enterovirus, and consist of many of the most common viral infections in humans. Enteroviruses exhibit a vast array of clinical manifestations. Among the most common mucocutaneous manifestations are Hand-Foot-and-Mouth disease (HFMD), herpangina, macular and papular rashes, roseola-like lesions, Boston exanthema disease, urticarial rashes, and eruptive pseudoangiomatosis.

HFMD findings include oral aphthea-like erosions and red macules on the skin that transform into white oval vesicles with surrounding erythema. The vesicles have a characteristic rhomboid shape and occur mainly on the palms, soles, dorsal aspects of the digits and seldom on the face, buttocks, and legs (61). The vesicles break down into shallow, painful ulcers that heal without sequelae within approximately a week.

Coxsackievirus A-16 is the enterovirus most commonly associated with HFMD; however, outbreaks due to Enterovirus-71 (EV-71) have also been described. Coxsackievirus Group A (CVA) serotypes CVA-4, -5, -6, -7, -9, -10 and Coxsackievirus Group B (CVB) types CVB-5 and -13 may also be associated with HFMD.

Herpangina is an acute self-limiting illness that often affects children aged one to seven years and is associated with high fever, sore throat, dysphagia, and malaise (62) in addition to mucosal manifestations. Herpangina lesions are characterized by small, grey-white vesicles with a red halo on the posterior palate, uvula, and tonsils. The vesicles ulcerate and heal spontaneously in approximately seven days. Herpangina has been commonly associated with the Coxsackie A viruses (A2, A4, A5, A6, A8, A10), but may also be due to Group B Coxsackie, echoviruses, and nonspecific enteroviruses.

Diagnosis is often clinical, but multiple laboratory methods for the diagnosis of enterovirus infections are available and have evolved over time. Initially, the gold standard for enterovirus detection was by viral culture and identification by neutralization reaction through the use of intersecting pools of type-specific antisera. Immunofluorescent assays with monoclonal antibodies for enteroviruses are also available. Panenterovirus-PCR assay is a method that has largely replaced viral culture, but it does not allow for distinction between the viruses. The current laboratory diagnostic method of choice for enterovirus typing is sequencing of the VP-1 gene. The VP1 gene encodes type-specific epitopes and thus correlates with serotype.

Other Enteroviruses are further discussed in Chapter 16.

SUMMARY

Many mucocutaneous viral infections are diagnosed clinically; however, some may require laboratory confirmation. The most commonly utilized lab tests include viral cultures, direct immunofluorescence, and serology. PCR is also becoming increasingly popular with its high sensitivity and specificity; however, it is not yet commercially available for many viruses. The type of laboratory test utilized depends on the clinical presentation, the degree of suspicion for that specific viral infection, and the sensitivity and specificity of the test desired for the particular virus involved.

REFERENCES

1. Vernick JP, Steigman CK. The HPV DNA virus hybrid capture assay: what is it–and where do we go from here? MLO Med Lab Obs 2003; 35(3):8–10, 3; quiz 4–5.
2. Acute, Generalized Vesicular or Pustular Rash Illness Testing Protocol in the United States. http://emergency.cdc.gov/agent/smallpox/diagnosis/pdf/poxalgorithm11-14-07.pdf. CDC, 2007. Accessed 2007.
3. Negative Staining Electron Microscopy Protocol for Rash Illness. http://emergency.cdc.gov/agent/smallpox/lab-testing/pdf/em-rash-protocol.pdf CDC, 2006. Accessed August 2008.
4. IDSA. Smallpox: Current, comprehensive information on pathogenesis, microbiology, epidemiology, diagnosis, treatment, and prophylaxis. www.cidrap.umn.edu/idsa/bt/smallpox/biofacts/smllpx-summary.html. IDSA, 2008. Accessed August 26, 2008.
5. Breman JG, Henderson DA. Diagnosis and management of smallpox. N Engl J Med 2002; 346(17): 1300–1308.
6. Guharoy R, Panzik R, Noviasky JA, et al. Smallpox: clinical features, prevention, and management. Ann Pharmacother 2004; 38(3):440–447.
7. Trindade GD, Li Y, Olson VA, et al. Real-time PCR assay to identify variants of Vaccinia virus: Implications for the diagnosis of bovine vaccinia in Brazil. J Virol Methods 2008; 152(1–2):63–71.
8. Reed KD, Melski JW, Graham MB, et al. The detection of monkeypox in humans in the Western Hemisphere. N Engl J Med 2004; 350(4):342–350.
9. Li Y, Olson VA, Laue T, et al. Detection of monkeypox virus with real-time PCR assays. J Clin Virol 2006; 36(3):194–203.
10. Bauer JH, Miller OF, Peckham SJ. Medical Pearl: confirming the diagnosis of molluscum contagiosum using 10% potassium hydroxide. J Am Acad Dermatol 2007; 56(suppl 5):S104–S105.
11. Murphy FA Gibbs EPJ, Horzinek MC, et al. Poxviridae. In: Murphy FA Gibbs EPJ, Horzinek MC, Studdert MJ, eds. Veterinary virology. San Diego, CA: Academy Press, 1999:277–291.
12. Torfason EG, Gunadottir S. Polymerase chain reaction for laboratory diagnosis of orf virus infections. J Clin Virol 2002; 24(1–2):79–84.
13. Bader C, Crumpacker CS, Schnipper LE, et al. The natural history of recurrent facial-oral infection with herpes simplex virus. J Infect Dis 1978; 138(6):897–905.
14. Habif T CJ, Chapman MS, Dinulos J, et al. Genital Herpes Simplex. In: Habit TP, ed. Skin Disease Diagnosis and Treatment, 2nd ed. Philadelphia, PA: Elsevier Mosby, 2005:176–179.

15. Arduino PG, Porter SR. Herpes Simplex Virus Type 1 infection: overview on relevant clinicopathological features. J Oral Pathol Med 2008; 37(2):107–121.
16. Chauvin PJ, Ajar AH. Acute herpetic gingivostomatitis in adults: a review of 13 cases, including diagnosis and management. J Can Dent Assoc 2002; 68(4):247–251.
17. Habif TP, Campbell JL, Chapman MS, et al. Herpes Zoster (Shingles). In: Habif TP, ed. Skin Disease Diagnosis and Treatment. Philadelphia, PA: Elsevier Mosby, 2005:210–215.
18. Scoular A, Gillespie G, Carman WF. Polymerase chain reaction for diagnosis of genital herpes in a genitourinary medicine clinic. Sex Transm Infect 2002; 78(1):21–25.
19. Wald A, Huang ML, Carrell D, et al. Polymerase chain reaction for detection of herpes simplex virus (HSV) DNA on mucosal surfaces: comparison with HSV isolation in cell culture. J Infect Dis 2003; 188(9):1345–1351.
20. Whiley DM, Syrmis MW, Mackay IM, et al. Preliminary comparison of three LightCycler PCR assays for the detection of herpes simplex virus in swab specimens. Eur J Clin Microbiol Infect Dis 2003; 22(12):764–767.
21. Schloss L, van Loon AM, Cinque P, et al. An international external quality assessment of nucleic acid amplification of herpes simplex virus. J Clin Virol 2003; 28(2):175–185.
22. Harris LN, et al. Massachusetts Medical Society. "Case 26–2003—A 50 Year-Old Colombian Man with Fever and Seizures". N Eng J Med 2003; 349:789–796.
23. Song B, Dwyer DE, Mindel A. HSV type specific serology in sexual health clinics: use, benefits, and who gets tested. Sex Transm Infect 2004; 80(2):113–117.
24. Habif TP, Campbell JL, Chapman MS, et al. In: Habif TP, ed. Skin Disease Diagnosis and Treatment. Philadelphia, PA: Elsevier Mosby, 2005:204–209.
25. Atkinson WHJ, McIntyre L, Wolfe S, eds. Center of Disease Control. Epidemiology and Prevention of Vaccine-Preventable Diseases. The Pink Book. In: 10th ed. Varicella. Atlanta: Public Health Foundation, 2008.
26. Hicks LD, Cook-Norris RH, Mendoza N, et al. Family history as a risk factor for herpes zoster: a case-control study. Arch Dermatol 2008; 144(5):603–608.
27. Gnann JW, Whitley RJ; Massachusetts Medical Society. "Herpes Zoster." N Eng J Med 2002; 347: 340–346.
28. Levin MJ, Oxman MN, Zhang JH, et al. Varicella-zoster virus-specific immune responses in elderly recipients of a herpes zoster vaccine. J Infect Dis 2008; 197(6):825–835.
29. Gulley ML, Tang W. Laboratory assays for Epstein-Barr virus-related disease. J Mol Diagn 2008; 10(4):279–292.
30. Zeytinoglu A, Altuglu I, Karatas E, et al. Comparison of immunofluorescence assay and multiplexed microparticle-based immunoassay for detecting Epstein-Barr virus viral capsid antigen antibodies. J Virol Methods 2008; 148(1–2):300–302.
31. Gulley ML, Pulitzer DR, Eagan PA, et al. Epstein-Barr virus infection is an early event in gastric carcinogenesis and is independent of bcl-2 expression and p53 accumulation. Hum Pathol 1996; 27(1):20–27.
32. http://www.cdc.gov/cmv/clinicians.htm. Accessed August 3, 2008.
33. Lazzarotto T, Guerra B, Lanari M, et al. New advances in the diagnosis of congenital cytomegalovirus infection. J Clin Virol 2008; 41(3):192–197.
34. Revello MG, Gerna G. Diagnosis and management of human cytomegalovirus infection in the mother, fetus, and newborn infant. Clin Microbiol Rev 2002; 15(4):680–715.
35. Gerna G, Revello MG, Percivalle E, et al. Quantification of human cytomegalovirus viremia by using monoclonal antibodies to different viral proteins. J Clin Microbiol 1990; 28(12):2681–2688.
36. Modlin JF, et al. Massachusetts Medical Society. Case 25–2003—A newborn boy with petechiae and thrombocytopenia. N Eng J Med 2003; 349:691–700.
37. Gozlan J, Laporte JP, Lesage S, et al. Monitoring of cytomegalovirus infection and disease in bone marrow recipients by reverse transcription-PCR and comparison with PCR and blood and urine cultures. J Clin Microbiol 1996; 34(9):2085–2088.
38. Lam KM, Oldenburg N, Khan MA, et al. Significance of reverse transcription polymerase chain reaction in the detection of human cytomegalovirus gene transcripts in thoracic organ transplant recipients. J Heart Lung Transplant 1998; 17(6):555–565.
39. Pruksananonda P, Hall CB, Insel RA, et al. Primary human herpesvirus 6 infection in young children. N Engl J Med 1992; 326(22):1445–1450.
40. Clark DA. Human herpesvirus 6. Rev Med Virol 2000; 10(3):155–173.
41. De Araujo T, Berman B, Weinstein A. Human herpesviruses 6 and 7. Dermatol Clin 2002; 20(2): 301–306.
42. Moore PS, Chang Y. Molecular virology of Kaposi's sarcoma-associated herpesvirus. Philos Trans R Soc Lond B Biol Sci 2001; 356(1408):499–516.

43. Gao SJ, Kingsley L, Li M, et al. KSHV antibodies among Americans, Italians and Ugandans with and without Kaposi's sarcoma. Nat Med 1996; 2(8):925–928.
44. Schulz TF. The pleiotropic effects of Kaposi's sarcoma herpesvirus. J Pathol 2006; 208(2):187–198.
45. Beral V, Peterman TA, Berkelman RL, et al. Kaposi's sarcoma among persons with AIDS: a sexually transmitted infection? Lancet 1990; 335(8682):123–128.
46. Laurent C, Meggetto F, Brousset P. Human herpesvirus 8 infections in patients with immunodeficiencies. Hum Pathol 2008; 39(7):983–993.
47. Kellam P, Boshoff C, Whitby D, et al. Identification of a major latent nuclear antigen, LNA-1, in the human herpesvirus 8 genome. J Hum Virol 1997; 1(1):19–29.
48. Dupin N, Fisher C, Kellam P, et al. Distribution of human herpesvirus-8 latently infected cells in Kaposi's sarcoma, multicentric Castleman's disease, and primary effusion lymphoma. Proc Natl Acad Sci U S A 1999; 96(8):4546–4551.
49. Kellam P, Bourboulia D, Dupin N, et al. Characterization of monoclonal antibodies raised against the latent nuclear antigen of human herpesvirus 8. J Virol 1999; 73(6):5149–5155.
50. Antman K, Chang Y; Massachusetts Medical Society. Medical progress: Kaposi's sarcoma. N Eng J Med 2000; 342:1027–1038.
51. Moore PS, Gao SJ, Dominguez G, et al. Primary characterization of a herpesvirus agent associated with Kaposi's sarcomae. J Virol 1996; 70(1):549–558.
52. Gao SJ, Moore PS. Molecular approaches to the identification of unculturable infectious agents. Emerg Infect Dis 1996; 2(3):159–167.
53. Centers for Disease Control and Prevention. Current Trends Risks Associated with Human Parvovirus B19 Infection. MMWR Morb Mortal Wkly Rep 1989: 38(6):81–88, 93–97. http://www.cdc.gov/mmwr/preview/mmwrhtml/00001348.htm.
54. Kelly HA, Siebert D, Hammond R, et al. The age-specific prevalence of human parvovirus immunity in Victoria, Australia compared with other parts of the world. Epidemiol Infect 2000; 124(3):449–457.
55. Sloots T, Devine PL. Evaluation of four commercial enzyme immunoassays for detection of immunoglobulin M antibodies to human parvovirus B19. Eur J Clin Microbiol Infect Dis 1996; 15(9):758–761.
56. Ramirez MM, Mastrobattista JM. Diagnosis and management of human parvovirus B19 infection. Clin Perinatol 2005; 32(3):697–704.
57. Zerbini M, Gallinella G, Cricca M, et al. Diagnostic procedures in B19 infection. Pathol Biol (Paris) 2002; 50(5):332–338.
58. Field AM, Cohen BJ, Brown KE, et al. Detection of B19 parvovirus in human fetal tissues by electron microscopy. J Med Virol 1991; 35(2):85–95.
59. Young N, Harrison M, Moore J, et al. Direct demonstration of the human parvovirus in erythroid progenitor cells infected in vitro. J Clin Invest 1984; 74(6):2024–2032.
60. Young NS, Brown KE; Massachusetts Medical Society. Parvovirus B19. N Eng J Med 2004; 350:586–597.
61. Habif TP, Campbell JI, Chapman S, et al. Food and mouth disease. In: Habit TP, ed. Skin Diseases Diagnosis and Treatment. Philadelphia, PA: Elsevier Mosby, 2005:216–219.
62. Nakayama T, Urano T, Osano M, et al. Outbreak of herpangina associated with Coxsackievirus B3 infection. Pediatr Infect Dis J 1989; 8(8):495–498.

18 Cardiovascular Viral Infections

Heinz-Peter Schultheiss and Uwe Kühl

Medizinische Klinik II, Cardiology und Pneumonology, Charité—Universitätsmedizin Berlin, Berlin, Germany

INTRODUCTION

Inflammatory processes induced by viral or bacterial infections are believed to be one of the major pathogenetic mechanisms in myocarditis and inflammatory cardiomyopathy (dilated cardiomyopathy, inflammatory; DCMI). Although virtually any microbial agent can cause myocardial inflammation and dysfunction, bacterial infections are rare in these conditions in western countries and viral forms are considered the most common cause of acquired cardiomyopathies nowadays. While Coxsackievirus involvement is well established in pediatric myocarditis due to direct isolation of infectious virus, its involvement in adult heart disease rests primarily upon serological evidence and direct detection of enterovirus RNA in heart muscle by nucleic acid hybridization and RT-PCR. Subsequent molecular biological studies have identified distinct genotypes of different viruses and virus subtypes in myocardial tissues of patients with acute, chronic, and end-stage heart diseases. Spontaneous improvement of ventricular function following virus clearance, complete or partial reversibility of ventricular dysfunction after antiviral treatment, progression of myocardial dysfunction in patients with persisting viral infections, and adverse prognosis in virus-positive patients on the other hand have led to the assumption that dilated cardiomyopathy (DCM) may be a late sequela of viral myocarditis.

Although the cases documented so far indicate that 12% of patients with clinically suspected myocarditis and 40% of biopsy-proven myocarditis cases develop DCM, and five- to six-year mortality rates of 20% to 56% have been reported in acute myocarditis, comprehensive studies on the long-term prognosis of patients with chronic viral heart disease and unambiguous proof of virus persistence as the single cause for the progression of myocarditis to DCM are still lacking. In the past, most studies have neither consequently analyzed myocardial tissue for possible infectious causes nor unambiguously proven virus persistence as a cause of the reported clinical outcome (1–5).

The often incoherent results obtained so far can be attributed to temporal changes of virus epidemics, regional differences in the etiological profiles of viruses, differences in the diagnostic tests with respect to timing of biopsy and involvement of virus subtypes, and clinical differences between studied cohorts. Since diagnostic accuracy is mandatory for epidemiological and therapeutical considerations, the available data on viral heart disease is reviewed in this chapter in the view of current pathogenetic concepts and the possibilities and limits of currently available diagnostic tools are discussed, which are used for the diagnosis of cardiovascular infections.

EPIDEMIOLOGY OF VIRAL HEART DISEASE

The overall incidence of myocarditis in viral infections is estimated at 3% to 6% (6). The actual incidence of virus-induced myocarditis or cardiomyopathy is less well established because viral heart disease can be inapparent, is difficult to diagnose, and can vary with different viruses as a function of circulating virus populations (7–15). Since the introduction of the more sensitive and rapid molecular biological techniques used to analyze endomyocardial biopsy (EMB) specimens, the incidence of detected viral genomes has increased constantly, but geographical and temporal changes have been recognized. Apart from enteroviruses, analysis of endomyocardial biopsies using molecular biology techniques has also identified (with geographical differences and varying degrees of frequency) distinct genotypes of erythroviruses (parvovirus B19), human herpesvirus 6 (HHV6), adenoviruses (ADVs), human immune deficiency virus (HIV), cytomegalovirus (CMV), herpes simplex type 2 virus, and hepatitis C virus (HCV) (16–39). If the substantial number of cases involving a clinically suspected diagnosis (recent onset of arrhythmia, contraction disorders, cardiac enlargement) is included; however, the figures are likely to be even higher (40,41).

In a comprehensive study by Bowles et al., nested PCR amplified a viral product in 40% of samples of 773 mostly younger American patients below 18 years of age with myocarditis (n = 624) or DCM (n = 149), with ADVs and enteroviruses predominating in the PCR analysis. Only 1% tested positive for parvovirus (20). In different German studies, viral genomes have been documented in 30% to 73% of EMB of patients with left ventricular dysfunction, and parvovirus genomes were detected as frequently as enteroviral genome (25,36,42). Some more recent studies from Western Europe and the United States have demonstrated a decrease in the prevalence of enteroviruses and ADVs as pathogens, while erythroviral genomes have become detectable in higher frequencies (24,25,37). HCV-associated cardiomyopathy, on the other hand, has primarily been detected in Asian countries, especially in Japan (39,43). Chagas cardiomyopathy, a frequent protozoon cause of DCM, is limited to Central and South America.

Although viral infections can cause serious human diseases, the majority of viral infections is asymptomatic or oligosymptomatic, and therefore such infections are frequently not recognized as possible causes of delayed onset of heart disease (44). In the past, viral myocarditis and chronic viral heart disease have therefore more often been a clinically derived diagnosis of exclusion, rather than a specifically proven diagnosis. Several factors have hampered an early identification of afflicted patients, including the temporal changes of virus epidemics and the geographical differences in the etiological profiles of viruses; the presence of myocarditic and nonmyocarditic virus variants; the enormous variability of clinical symptoms of viral heart disease, which may range from asymptomatic presentation to manifest heart failure; and the lack of consequent diagnostic efforts for complete virus analysis. In many cases these factors have prevented the generation of valid epidemiological data.

PATHOGENESIS

In many viral infections, viruses are processed in lymphoid organs and may proliferate within immune cells such as lymphocytes or macrophages. Subsequently, they achieve target organ infection through hematogenous or lymphangitic spread. The early phase of myocardial disease is initiated by infection of cardiac myocytes, fibroblasts, or endothelial cells (ECs) through receptor-mediated endocytosis (23,45–47).

Enteroviruses, which traditionally have been considered the most common agent in myocarditis and acute or endstage DCM, directly infect cardiomyocytes in animal models and human disease (48). The host cells are entered after viral binding to the coxsackie-adenoviral receptor (CAR) and the decay accelerating factor (DAF; CD55) that serves as a coreceptor for enterovirus internalization. CAR is a tight junction protein that is localized in the cardiovascular, immune, and neurological systems, and which is critical for internalization of the virus (49). Its de novo induction on the myocyte surface of 60% of the DCM hearts and its colocalization with the coreceptors for adenovirus internalization avβ3 and avβ5 indicate that CAR is an important molecular determinant for the cardiotropism of both coxsackievirus and adenovirus (49,50).

By contrast, erythroviruses do not infect myocytes or other interstitial cells of the cardiac tissue. Their genomes have been localized in ECs of venuoles, small arteries, or arterioles in fulminant myocarditis or sudden onset heart failure (35,48,51). In chronic inflammatory cardiomyopathy, PVB19 infection is predominantly detected in ECs of small capillaries (48,51). Preliminary data furthermore indicate that herpesvirus 6 may infect both cardiac myocytes and ECs (52), while infection sites of other cardiotropic viruses have not been identified in detail in human disease.

Viral myocarditis develops with three pathologically distinct phases (Fig. 1) (43,53). Most information on this issue is known from enteroviral infections for which excellent small animal models exist that parallel human disease. After enterovirus internalization, the negative strand RNA is reverse transcribed into a positive strand for subsequent virus replication (22,54). A direct virus-related cytolysis of cardiomyocytes is already detected before any inflammatory infiltrate develops and appears to be decisive in fulminant cases of myocarditis (*phase 1*). Resulting myocyte necrosis may cause a significant loss of contractile tissue, which is accompanied by rapidly developing cardiac failure and early death of the host. Early antiviral defense mechanisms of the innate immune system are triggered by foreign molecular antigens through the

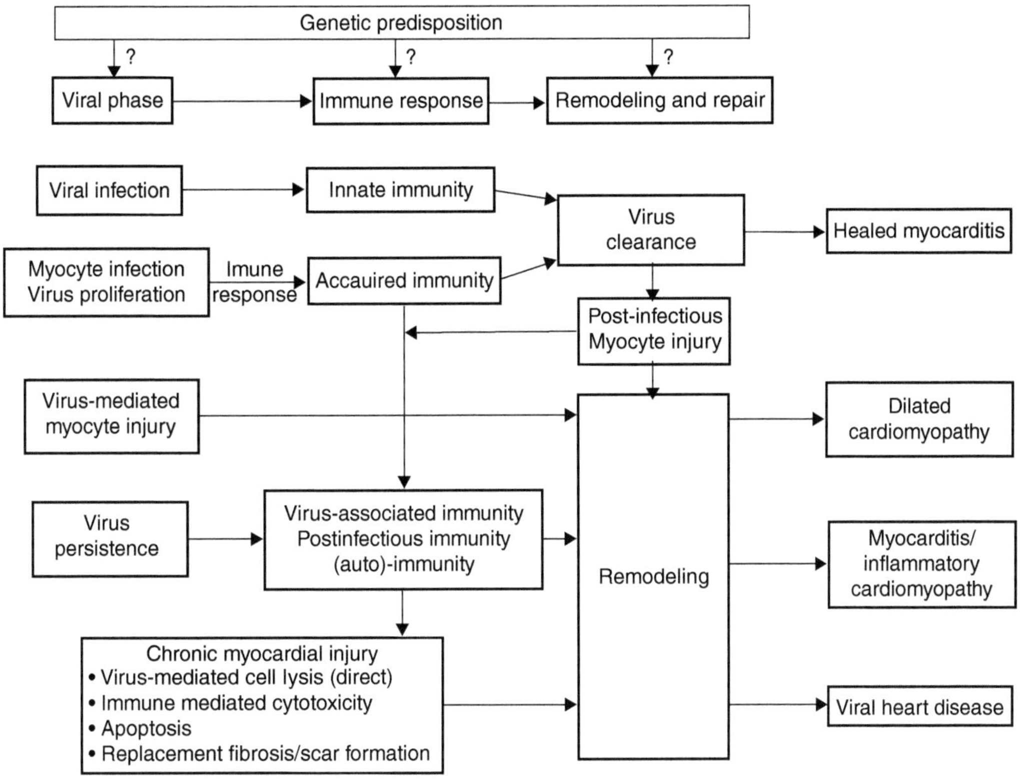

Figure 1 Three-phase model of cardiovascular infection.

ubiquitous toll-like receptors (55–57). Cytokines released by macrophages and activation of natural killer cells that directly kill virus-infected heart cells through perforin and granzyme mediated lysis contribute to early myocardial lesions and impaired myocardial function (58–62).

The activation of antigen-specific cell mediated immunity initiates the *second phase* of virus clearance (63–68). Because virus-infected cells are destroyed by immune effector cells of the emerging acquired in flammatory response, virus clearance will occur at the expense of further loss of infected myocytes. The ensuing myocardial damage depends on the scale of the cellular virus infection and increases with growing virus dispersion, which in addition to the early virus- and immune-mediated injury (phase 1), contributes to tissue remodeling and possible progression of the disease. The healing process therefore results primarily from a partial destruction of myocardial tissue that is not capable of regeneration.

Negative immune modulation, which is an important property of an intact immune system to prevent excessive tissue damage by an overwhelming immune response, normally occurs rapidly after successful elimination of the infectious pathogens. Under certain circumstances, chronic immune stimulation and autoimmunity may result from incompletely cleared virus infection or in response to the virus- and immune-mediated chronic tissue damage, respectively. Both the ongoing antigenic trigger from continuously synthesized viral proteins and continuously released intracellular proteins from necrotic or apoptotic myocardial cells may contribute to these phenomena.

Postulated pathogenetic mechanisms sufficient to initiate and propagate such processes include latent virus infection without detectable virus replication, low virus loads, molecular mimicry, release of virus encoded or intracellular myocyte proteins, auto-antibodies, activation of matrix-degrading proteases, and subsequent matrix remodeling with reparative fibroses (69–74). Such mechanisms may initially damage some individual cells but ultimately affect the whole myocardium. This finally may account for a clinical picture that is consistent with an often irreversible DCM (1).

In the *third remodeling phase*, virus infection may have been cleared completely and antiviral immune responses may have been resolved regularly. The extent of the myocardial damage determines the further clinical course of these patients whose biopsy results at that time will be that of idiopathic DCM. A postinfectious disease can only be diagnosed if history or previous diagnostics had proven a preceding infectious or inflammatory state. On the other hand, latent virus infection, virus-associated low-grade immune reactivity, and/or autoimmune processes may continuously assert negative effects on myocardial performance. Biopsy would then be consistent with inflammatory cardiomyopathy or persisting viral heart disease. Without sophisticated diagnostic analyses, the clinical picture is indistinguishable from that of DCM.

CLINICAL COURSE AND PROGNOSIS OF CHRONIC VIRAL HEART DISEASE

If the antiviral immunity has elaborated fast and efficiently with subsequent rapid resolution of cellular processes, residual damage of the myocardium may be minor and the remaining myocardium can compensate sufficiently for the partial loss of contractile tissue. Consequently, these patients may recover completely with no or only minor residual clinical signs of heart injury. Follow-up biopsy would indicate healed myocarditis.

Depending on the severity of initial cardiac damage, other patients may retain residual myocardial damage. Moderate loss of contractile tissue with more pronounced remodeling of the myocardial matrix accounts for the course of those patients who only partially recover. Within weeks or months after acute onset of disease, these patients may experience significant improvement of myocardial function and symptoms, although complete recovery is rare. It is difficult to estimate, however, whether such an improvement represents true myocardial recovery at the cellular level or whether the clinical course is better attributed to newly administered heart failure medication. In the longer run, many of these patients experience progression of heart failure despite regular heart failure medication. At this time point, idiopathic DCM is diagnosed, histologically.

The resulting clinical presentation is, however, not only influenced by the severity of irreversible matrix alterations and the potential of the myocardium to compensate for these processes. It may also depend on the effects that are exerted on the cardiac tissue by a persisting virus infection, low-grade inflammatory processes, and autoimmune mechanisms. Under these circumstances, biopsy-derived findings will be compatible with inflammatory cardiomyopathy or chronic viral heart disease.

One of the main difficulties that result from biopsy analysis in acute disease is caused by the sampling error, because early disease processes are often focal and therefore can easily be missed in cases in which an insufficient number of biopsy specimens is analyzed. In chronic disease, in which anamnestic data often do not point to a specific disease entity, e.g., a postinfectious state, only a positive biopsy result is diagnostic. Although replacement fibrosis may be the result of a preceding viral infection or inflammation, it is impossible to identify the true cause of the myocardial alterations in biopsies of patients with "idiopathic" DCM. One has to keep in mind, however, that a positive proof of virus infection only provides a diagnostic hint for a possible pathogenetic cause of the disease. It does not prove that the virus infection is the only pathogenetic cause of heart failure or progression of the disease.

The transition of myocarditis into DCM following direct virus- or immune-mediated myocardial damage is generally accepted and supported by the literature. Continuous myocardial damage caused by persisting virus infection and/or ongoing immune processes, however, has not been proven unambiguously in human disease. A great deal of skepticism stems from the inconsistency of currently available data. This inconsistency is derived from insufficiently diagnosed and inconsequently followed cohorts of patients. There are, however, a few sound clinical reports that suggest that such mechanisms in fact may directly contribute to the progression of heart failure and adverse prognosis in human disease. The mechanisms by which latent virus infections and low-grade inflammatory or autoimmune processes might deliver harmful effects to the myocardium, however, cannot be proven by any of these data and are currently not completely understood.

The clinical importance of persistent enteroviral genomes in the myocardium was investigated by Why and colleagues who demonstrated a higher mortality at 25 months (25% vs. 4%) in 41 patients with persistent enteroviral infection (75). The data reported by Frustaci et al.

from a retrospective analysis of immunosuppressed patients with inflammatory cardiomyopathy point to a similar direction (37). Patients with persistence of different viruses (except for HCV) did not improve, or even deteriorated upon immunosuppression while virus-negative patients improved significantly. Seven of the nonresponders died or were transplanted within 9 months. In a recent paper, Caforio and coworkers reported a two-year follow-up of patients with active ($n = 85$) and borderline myocarditis ($n = 89$) in which virus persistence was a univariate predictor of adverse prognosis, in addition to antiheart autoantibodies and clinical signs/symptoms of left- and right-heart failure (74).

In rapidly resolving myocarditis, virus clearance seems to occur very early within the first two weeks after onset of symptoms. In studies analyzing chronic heart disease, virus persistence has mostly been postulated because viral genomes were detected at the late stage of the disease. An attempt to definitely prove virus persistence for the whole study period by follow-up PCR analysis has never been carried out. It is therefore still unknown whether adverse prognosis has to be attributed to the early and more pronounced tissue damage in initially virus-positive patients or whether it is directly caused by a latent or smoldering viral infection.

In an attempt to gain more information on this important issue, we followed 172 consecutive patients with left-ventricular dysfunction and biopsy-proven viral infection by reanalysis of biopsies and hemodynamic measurements after a median period of 6.8 months (range, 5.4 to 11.9) (5). Viral genomes persisted in 64% of patients with single virus infections. Spontaneous clearance was observed in 50% of infections with enterovirus, versus 36% with adenovirus, 22% with parvovirus B19, and 44% with HHV6. These data demonstrate that clearance of the virus infection may occur late in the course of the disease and therefore, a single biopsy analysis can never prove virus persistence unambiguously. Clearance of viral genomes was associated with a significant decrease in left-ventricular dimensions and improvement in left-ventricular ejection fraction (LVEF) by 8% ($p < 0.001$). In contrast, LV function decreased by 3% ($p < 0.01$) in patients with persisting viral genomes. Upon interferon-β treatment of patients with chronic enterovirus and adenovirus persistence (median history: 44 months), clearance of viral genomes was observed in all 22 patients of a nonrandomized study (76). Virus clearance again was paralleled by a significant decrease of ventricular dimensions and clinical complaints. LVEF improved by 8.5% ($p < 0.001$). Comparison of histological, immunohistological, and molecular biological data furthermore highlighted that myocardial inflammation was often associated with the virus infection and resolved when the viruses were cleared spontaneously or during antiviral treatment of patients with chronic EV and ADV infection (5,76). This indicates that the course of the virus infection predetermines the clinical course rather than the virus-associated inflammation.

These are the first data that more directly point to the fact that the spontaneous or treatment associated course of a chronic viral infection does reflect the clinical course of viral heart disease, and furthermore indicate that ventricular dysfunction in chronic viral heart disease is not solely caused by a virus-induced irreversibly damaged myocardium. Further detailed analyses are necessary in order to elucidate mechanisms responsible for the reversibly depressed myocardium. With respect to the clinical management of patients, the outlined data argue for the importance of recognizing patients at an early and still reversible stage of virus-associated heart disease. Only patients with still minor or moderate alterations of the heart tissue may benefit from early and specific treatment, and progression of heart failure might be prevented by timely therapy (Fig. 2).

DIAGNOSTIC VIROLOGICAL TOOLS IN CARDIOLOGY

Virus Serology and Direct Virus Isolation

In 1974, Grist and Bell presented the first comprehensive serological evidence that enterovirus infection might be associated with myocarditis (77). The role of these viruses in DCM, however, was less well established, and based mainly on the detection of high titers of neutralizing antibody in cases of sudden onset of disease. This led to the first proposal that myocarditis is an infectious disease. Although serological and clinical evaluations have often confirmed a positive association between increased or changing enterovirus virus titers and a clinical presentation suggestive of myocarditis, later simultaneous analyses of serological and molecular

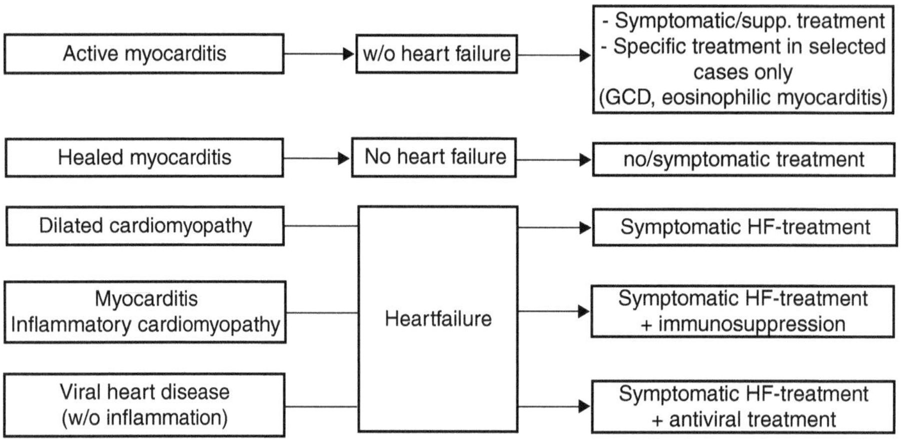

Figure 2 Stages of myocarditis and appropriate management.

biological data showed that serologic testing does not prove heart tissue infection in patients with myocarditis or DCM and therefore is without diagnostic accuracy (78,79).

Virus Isolation

In spite of clinical and serological evidence of a preceding infection, isolation of replicating infectious virus is rarely successful in adult patients with either clinically suspected myocarditis or DCM. In accordance with animal experimental data, viruses have only occasionally been isolated from heart tissues within the first one or two weeks after onset of acute disease in children (80,81). Direct virus isolation does not work well in chronic disease, and given the often subacute clinical presentation and timing of first diagnosis of most patients with suspected virus-associated disease, this approach is without any diagnostic relevance.

MOLECULAR BIOLOGICAL DETECTION OF VIRAL GENOMES IN THE MYOCARDIUM

In Situ Hybridization

The low diagnostic accuracy of the standard virological methods has promoted the development of molecular biological detection assays for viral genomes in cardiac tissues. In situ hybridization of enteroviral RNA in EMB specimens was first demonstrated in 1986 by Bowles (82,83) and Kandolf (84). Nowadays, the major advantage of in situ hybridization is the exact localization of viral genome at the cellular level in different compartments of tissues. Apart from the information on direct localization of viral infections, in situ hybridization is a time-consuming procedure not suitable for routine clinical diagnosis, which demands rapid etiologic information.

QUALITATIVE VIRUS DETECTION BY NESTED POLYMERASE CHAIN REACTION

The introduction of the more rapid nested polymerase chain reaction (nPCR) in the routine molecular biological analysis of viral infections has substantially increased our knowledge of possible cardiotropic viruses in patients with acquired heart disease. Detected viral genomes include that of enteroviruses, ADVs), erythroviruses including parvovirus B19 (B19V), CMV, influenza A virus, respiratory syncitial virus (RSV), herpes simplex virus (HSV), Epstein–Barr virus (EBV), HHV6, HIV, and HCV (16–39).

Virus Replication

Analysis of virus replication in the myocardium (using plus–minus strand detection) and preliminary data from follow-up analyses suggest that the clinical course of enterovirus-, adenovirus-, erythrovirus-, or human herpes virus-6-positive patients may be linked with the course of the viral infection, and that persistence of these viruses may provide an adverse impact on the prognosis of afflicted patients (5,22,24,29,54,74,75,85,86). Of note, we are currently not aware of the replicative pattern or metabolic activity of cardiotropic viruses others than Coxsackievirus B (CVB).

QUANTITATIVE VIRUS DETECTION BY REAL-TIME PCR

Meanwhile, a more quantitative approach to the majority of cardiotropic viruses has been introduced in specialized laboratories. Virus loads, if analyzed by real-time PCR, have been reported to be low in PVB19-positive patients with chronic disease (Fig. 3) (87,88). Detected viral copy numbers range between 50 and 500,000 copies/μg of myocardial DNA, and even lower numbers of viral copies are usually detected for EV, ADV, EBV, and HHV6. The quantitative approach and its clinical interpretation are hampered by the fact that actual virus numbers are often missed due to sampling error, especially in focal disease. If the timing of the biopsy is

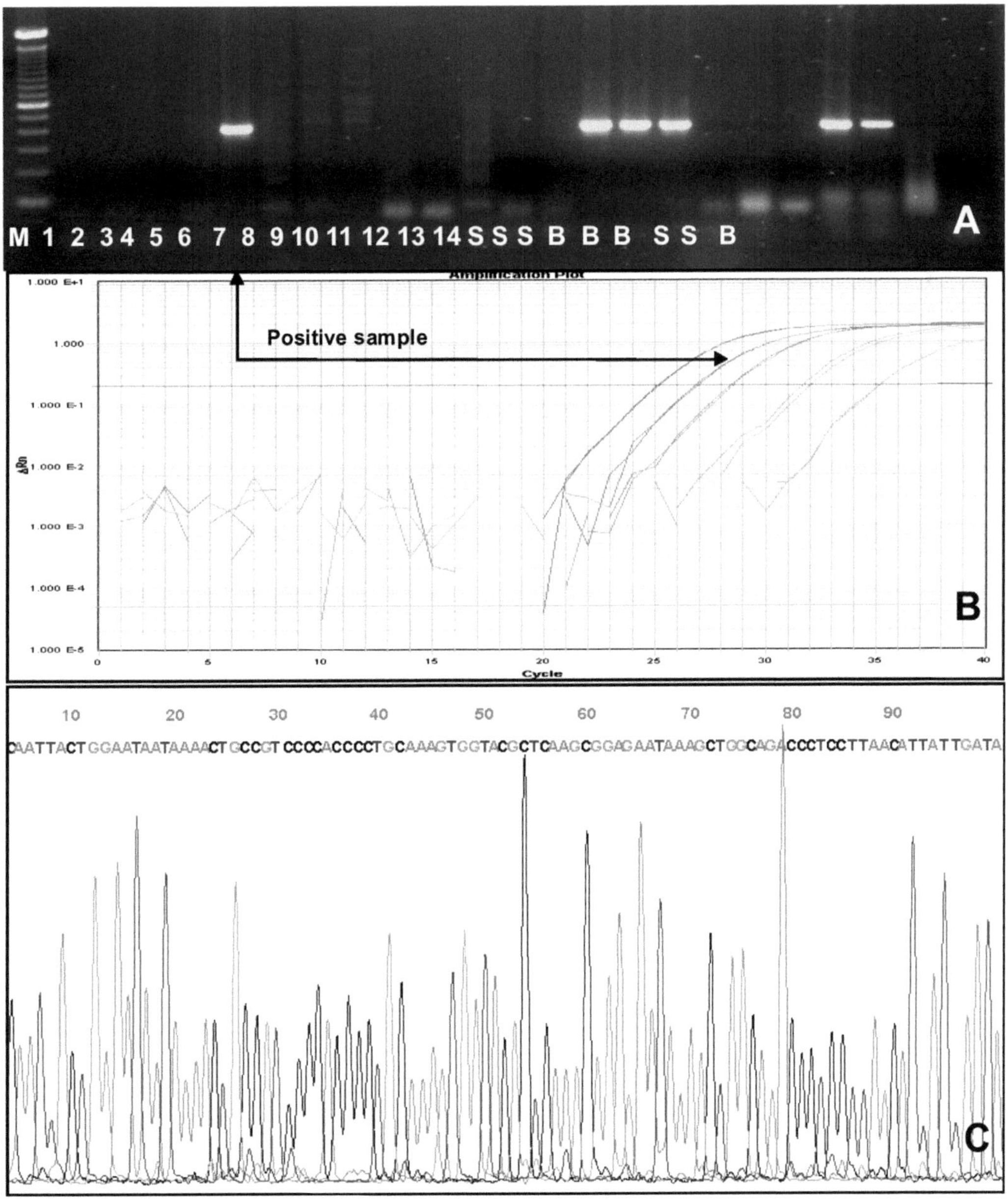

Figure 3 (*See color insert*). Molecular diagnostics in heart tissue.

late with respect to the onset of virus-induced chronic disease, it may only detect residues of incompletely cleared viruses. Erythroviruses can persist in cardiac tissue for many years, and low levels of myocardial PVB19 of approximately 10^2 to 10^4 genome equivalents/μg isolated nucleic acid are consistent with a persistent type of infection (34,87–89). Low virus copies at diagnosis neither exclude virus-associated myocardial injury nor necessarily do they correlate with the magnitude of myocardial damage at delayed diagnosis, because such copy numbers may only constitute remnants of earlier cardiopathogenetic disease stages. Tissue injury is furthermore caused by the initial antiviral immune response, which may have outlived viral clearance. In this setting, impact on disease may be completely independent from the detection of virus or the actual number of viral copies detected at the time of biopsy. At present, information on sampling error and follow-up information comparing quantitative virus data and virus-associated myocardial inflammation with the clinical course are not available.

IDENTIFICATION OF VIRUS SUBTYPES BY GENOMIC SEQUENCING

PCR analysis followed by sequencing can allow the identification of specific virus subtypes, and the exclusion of contaminating PCR products that may interfere with diagnostic accuracy. For reasons of methodological suitability and rapid clinical information, primer pairs are generally designed to simultaneously amplify the genomic sequence of ADVs encoding the hexon protein, the 5′ nontranslated region of the enteroviruses, or constant regions of different erythrovirus and HHV6 variants, respectively (20,22,23,90). Using this approach most subtypes of enteroviruses including CVB and echoviruses, most adenovirus subtypes, and the different PVB19 and HHV6 genotypes can be recognized in a single PCR reaction. Further information on different virus subtypes can be obtained from sequence analysis of amplification products (Fig. 3).

Myocarditis can be caused by all six coxsackie B viruses, although CVB 1–5, genetically related strains such as CV A9, and some echoviruses, are more frequently detected than the other 63 human enteroviruses (Table 1) (13). Recently, we have shown that erythrovirus genotype 2, previously not described in human heart tissues, is highly prevalent in the hearts of DCM patients while the less prevalent genotype 1 (parvovirus B19) is associated with more severely disturbed cardiac function (88). Cardiac dysfunction seems to be less frequently present in tissues with latent PVB19 infection lacking myocardial inflammation, which is observed in about 40% of PVB19-positive hearts. Therefore, clinical consequences of any molecular biological virus analyses have to be interpreted in connection with histologic and immunohistologic analysis of the inflammatory state of the heart (26,48,91). In another biopsy-based analysis, distinct human herpesvirus 6 variants (A and B) were detected in tissue specimens of patients with clinically suspected cardiomyopathy and persistent heart failure symptoms (Kühl et al., submitted). Electron microscopy furthermore demonstrated replicating HHV-6 virus particles in cardiac vascular ECs and cardiac myocytes indicating direct cardiac cell involvement. These findings

Table 1 Infectious Pathogens in Myocarditis and Inflammatory Cardiomyopathy

		Cardiotopic subtypes/ variants
RNA viruses	Picornaviruses (Coxsackie, Echo, Polio) Orthomyxovirus (Influenza A, B, C) Paramyxoviruses (Mumps) Togaviruses (Rubella) Hepatitis C virus Flaviviruses (Dengue fever, Yellow Fever)	CVB 1–6, Echo 30 Influenza A
DNA viruses	Adenoviruses Herpesviruses (cytomegalovirus, Epstein–Barr, HIV, human herpesvirus 6) erythroviruses	Ad1,2,3 and 5 HHV 6 A/B B19V, genotype 2
Nonviral pathogens	Bacteria, spirochetes, rickettsiae, protozoae, parasites, fungi, mycobacteriae, legionellae, borreliosis	*B. burgendorfferii*

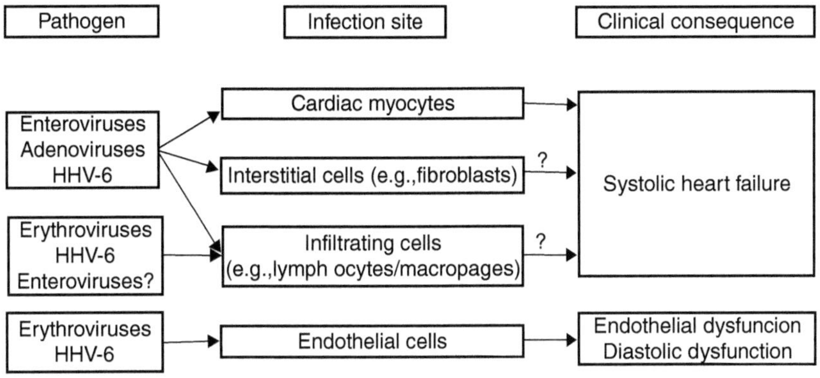

Figure 4 Cardiac dysfunctions caused by different cell tropism of cardiotropic viruses.

may explain endothelial and diastolic dysfunction associated with the presence of erythrovirus and herpesvirus infections, and furthermore indicate that different virus infection sites may be associated with distinct clinical entities such as endothelial and diastolic dysfunction or overt systolic heart failure (Fig. 4) (26,92).

It is currently not known how different virus subtypes or variants cause disease-specific cardiac pathogenicity, nor what determines a cardiovirulant phenotype. Differences in virus replication rates, differences in antigenic epitopes causing a modified impact on the antiviral immune response, distinct induction of myocyte apoptosis, and single nucleotide exchanges within distinct coding regions may render a cardiovirulant strain into an attenuated phenotype (93–96). The existence of such different virus variants with distinct pathogenetic potentials may explain the often observed discrepancies between PCR results and unpredictable clinical courses.

In addition to these virus-related factors, host-specific differences may influence the course of infectious heart diseases. At least 25% of DCM patients in Western populations have evidence for familial disease (97). In this context, the genetic background may be responsible for immune alterations and thereby influence antiviral immunity. Furthermore, the clinical effects of mutations of myocardial proteins such as actin or dystrophin could be aggravated by certain infections if virus encoded proteins effect cardiac components. A possible example of direct virus mediated effect is the cleavage of dystrophin by the enterovirus protease A2, thereby interfering with an essential part of the cardiomyocyte cytoskeleton responsible for force transmission (70,98).

INTERPRETATION OF MOLECULAR BIOLOGICAL FINDINGS

Recent advances in quantitative (qPCR) and qualitative (nested PCR) molecular techniques provide the basis for sensitive detection of less than 10 gene copies of various viral pathogens in the myocardium. This high sensitivity of diagnostic technologies provides both challenges and opportunities. Its clinical impact on prognosis and decisions for treatment largely depends on a standardized set of diagnostic methods and a careful interpretation of obtained data. It is of utmost interest to both the patient and the physician that an invasive procedure such as biopsy yields the most relevant diagnostic and prognostic information. With respect to data interpretation, it is therefore mandatory that a complete work-up of biopsy samples is carried out in a highly standardized fashion, including histological, immunohistological, and molecular biological methods. A biopsy-based diagnosis that only relies on the histological or immunohistological analysis of myocardial inflammation cannot yield a clinically relevant diagnosis. An incomplete diagnostic workup will only result in an incomplete diagnosis with insufficient clinical information and insight into the pathogenesis of the disease, and thus cannot provide the basis for any biopsy-based therapeutic decision.

Owing to the sensitivity of PCR analysis, specific precautions are warranted to ensure sample preservation and to exclude sample contamination. These precautions include rapid and proper handling of samples from the cathlab to the diagnostic laboratory, using pathogen-free biopsy devices and storage vials and conditions that avoid sample degradation and contamination. Although PCR analysis may be conducted from formalin-fixed tissues or paraffin-embedded tissue sections, sensitivity can be increased using tissue samples shock frozen with liquid nitrogen. Large-scale studies comparing these different procedural methods, however, do not exist and will have to be established. Spatial separation of the different steps of RNA and DNA preparation and diagnostic procedures furthermore reduces possible contamination and improves diagnostic accuracy. Finally, contaminating products can only be excluded by sequencing of sample amplicons.

The major limitation for the interpretation of EMB data remains the reduced sensitivity due to sampling error, particularly in the setting of early, focal disease (99). Therefore, it is mandatory that a sufficient number of biopsy specimens, each of 2–3 mm^2 in size, from different regions of the interventricular septum or the left ventricular myocardium, is obtained for the PCR analysis. This is especially true for rarely detected viruses such as EV, ADV, or EBV, but also may influence the diagnostic accuracy of the more commonly detected infectious agents such as PVB19 or HHV6. Unfortunately, data elucidating this important diagnostic issue do not exist. Preliminary information suggests that at least four to six tissue specimens should be processed for PCR analysis of RNA and DNA viral genomes in order to reduce the sampling error associated with low diagnostic yields (Kühl, unpublished data).

In spite of this diagnostic limitation, the proof of viral genome persistence in >65% of DCM patient in a biopsy-based, six months follow-up analysis suggests that viral genomes can be detected in a clinically reasonable percentage, at least in the chronic state of the disease. Although the diffuse distribution of viruses may allow a clinical diagnosis with close correlation to the clinical course (5,37,75,76), it has to be remembered that—similar as for histological analysis of inflammation--only a positive biopsy result is diagnostic while a negative PCR does not exclude any infectious origin of the disease. A reasonable interpretation of biopsy results therefore needs the knowledge of the clinical course of the disease.

REFERENCES

1. D'Ambrosio A, Patti G, Manzoli A, et al. The fate of acute myocarditis between spontaneous improvement and evolution to dilated cardiomyopathy: A review. Heart 2001; 85(5):499–504.
2. Grogan M, Redfield MM, Bailey KR, et al. Long-term outcome of patients with biopsy-proved myocarditis: Comparison with idiopathic dilated cardiomyopathy. J Am Coll Cardiol 1995; 26(1):80–84.
3. Kodama M, Oda H, Okabe M, et al. Early and long-term mortality of the clinical subtypes of myocarditis. Jpn Circ J 2001; 65(11):961–964.
4. Mangnani JW, Suk-Danik HJ, Dec GW, et al. Survival in biopsy-proven myocarditis: A long-term retrospective analysis of the histopathological, clinical, and hemodynamic predictors. Am Heart J 2006; 151:463–470.
5. Kühl U, Pauschinger M, Seeberg B, et al. Viral persistence in the myocardium is associated with progressive cardiac dysfunction. Circulation 2005; 112(13):1965–1970.
6. Kereiakes DJ, Parmley WW. Myocarditis and cardiomyopathy. Am Heart J 1984; 108(5):1318–1326.
7. Guillen-Ortega F, Soto ME, Reyes PA. Primary acute myocarditis. A 10- years institutional experience. Arch Cardiol Mex 2005; 75(suppl 3):81–88.
8. Karjalainen J, Heikkila J. Incidence of three presentations of acute myocarditis in young men in military service. A 20-year experience. Eur Heart J 1999; 20(15):1120–1125.
9. Dechkum N, Pangsawan Y, Jayavasu C, et al. Coxsackie B virus infection and myopericarditis in Thailand, 1987–1989. Southeast Asian J Trop Med Public Health 1998; 29(2):273–276.
10. Friman G, Wesslen L, Fohlman J, et al. The epidemiology of infectious myocarditis, lymphocytic myocarditis and dilated cardiomyopathy. Eur Heart J 1995; 16(suppl O):36–41.
11. Martin AB, Webber S, Fricker FJ, et al. Acute myocarditis. Rapid diagnosis by PCR in children. Circulation 1994; 90(1):330–339.
12. Gravanis MB, Sternby NH. Incidence of myocarditis. A 10-year autopsy study from Malmo, Sweden. Arch Pathol Lab Med 1991; 115(4):390–392.
13. Riecansky I, Schreinerova Z, Egnerova A, et al. Incidence of Coxsackie virus infection in patients with dilated cardiomyopathy. Cor Vasa 1989; 31(3):225–230.
14. Wakafuji S, Okada R. Twenty year autopsy statistics of myocarditis incidence in Japan. Jpn Circ J 1986; 50(12):1288–1293.
15. Karjalainen J, Nieminen MS, Heikkila J. Influenza A1 myocarditis in conscripts. Acta Med Scand 1980; 207(1–2):27–30.

16. Jin O, Sole MJ, Butany JW, et al. Detection of enterovirus RNA in myocardial biopsies from patients with myocarditis and cardiomyopathy using gene amplification by polymerase chain reaction [see comments]. Circulation 1990; 82(1):8–16.
17. Grasso M, Arbustini E, Silini E, et al. Search for Coxsackievirus B3 RNA in idiopathic dilated cardiomyopathy using gene amplification by polymerase chain reaction. Am J Cardiol 1992; 69(6):658–664.
18. Weiss LM, Liu XF, Chang KL, et al. Detection of enteroviral RNA in idiopathic dilated cardiomyopathy and other human cardiac tissue. J Clin Invest 1992; 90:156–159.
19. Muir P, Nicholson F, Jhetam M, et al. Rapid diagnosis of enterovirus infection by magnetic bead extraction and polymerase chain reaction detection of enterovirus RNA in clinical specimens. J Clin Microbiol 1993; 31(1):31–38.
20. Bowles NE, Ni J, Kearney DL, et al. Detection of viruses in myocardial tissues by polymerase chain reaction. evidence of adenovirus as a common cause of myocarditis in children and adults. J Am Coll Cardiol 2003; 42(3):466–472.
21. Bowles NE, Bayston TA, Zhang HY, et al. Persistence of enterovirus RNA in muscle biopsy samples suggests that some cases of chronic fatigue syndrome result from a previous, inflammatory viral myopathy. J Med 1993; 24(2–3):145–160.
22. Pauschinger M, Doerner A, Kuehl U, et al. Enteroviral RNA replication in the myocardium of patients with left ventricular dysfunction and clinically suspected myocarditis. Circulation 1999; 99(7):889–895.
23. Pauschinger M, Bowles NE, Fuentes-Garcia FJ, et al. Detection of adenoviral genome in the myocardium of adult patients with idiopathic left ventricular dysfunction. Circulation 1999; 99(10):1348–1354.
24. Kühl U, Pauschinger M, Bock T, et al. Parvovirus B19 infection mimicking acute myocardial infarction. Circulation 2003; 108(8):945–950.
25. Kühl U, Pauschinger M, Noutsias M, et al. High prevalence of viral genomes and multiple viral infections in the myocardium of adults with "idiopathic" left ventricular dysfunction. Circulation 2005; 111(7):887–893.
26. Tschope C, Bock CT, Kasner M, et al. High Prevalence of Cardiac Parvovirus B19 Infection in Patients with Isolated Left Ventricular Diastolic Dysfunction. Circulation 2005; 111:879–886.
27. Papadogiannakis N, Tolfvenstam T, Fischler B, et al. Active, fulminant, lethal myocarditis associated with parvovirus B19 infection in an infant. Clin Infect Dis 2002; 35(9):1027–1031.
28. Rohayem J, Dinger J, Fischer R, et al. Fatal myocarditis associated with acute parvovirus B19 and human herpesvirus 6 coinfection. J Clin Microbiol 2001; 39(12):4585–4587.
29. Dec GW Jr., Palacios IF, Fallon JT, et al. Active myocarditis in the spectrum of acute dilated cardiomyopathies. Clinical features, histologic correlates, and clinical outcome. N Engl J Med 1985; 312(14):885–890.
30. Dec GW Jr., Waldman H, Southern J, et al. Viral myocarditis mimicking acute myocardial infarction. J Am Coll Cardiol 1992; 20(1):85–89.
31. Ambinder RF, Lambe BC, Mann RB, et al. Oligonucleotides for polymerase chain reaction amplification and hybridization detection of Epstein-Barr virus DNA in clinical specimens. Mol Cell Probes 1990; (4):397–407.
32. Aurelius E, Johansson B, Skoldenberg B, et al. Rapid diagnosis of herpes simplex encephalitis by nested polymerase chain reaction assay of cerebrospenal fluid. Lancet 1991; 337:189–192.
33. Muir SW, Murray J, Farquharson MA, et al. Detection of cytomegalovirus in upper gastrointestinal biopsies from heart transplant recipients: Comparison of light microscopy, immunocytochemistry, in situ hybridisation, and nested PCR. J Clin Pathol 1998; 51(11):807–811.
34. Nigro G, Bastianon V, Colloridi V, et al. Human parvovirus B19 infection in infancy associated with acute and chronic lymphocytic myocarditis and high cytokine levels: Report of 3 cases and review. Clin Infect Dis 2000; 31(1):65–69.
35. Bultmann BD, Klingel K, Sotlar K, et al. Fatal parvovirus B19-associated myocarditis clinically mimicking ischemic heart disease: An endothelial cell-mediated disease. Hum Pathol 2003; 34(1): 92–95.
36. Pankuweit S, Lamparter S, Schoppet M, et al. Parvovirus B19 genome in endomyocardial biopsy specimen. Circulation 2004; 109(14):e179.
37. Frustaci A, Chimenti C, Calabrese F, et al. Immunosuppressive therapy for active lymphocytic myocarditis: Virological and immunologic profile of responders versus nonresponders. Circulation 2003; 107(6):857–863.
38. Matsumori A, Ohashi N, Hasegawa K, et al. Hepatitis C virus infection and heart diseases: A multicenter study in Japan. Jpn Circ J 1998; 62(5):389–391.
39. Matsumori A. Hepatitis C virus infection and cardiomyopathies. Circ Res 2005; 96(2):144–147.
40. Drory Y, Turetz Y, Hiss Y, et al. Sudden unexpected death in persons less than 40 years of age. Am J Cardiol 1991; 68(13):1388–1392.

41. Fabre A, Sheppard MN. Sudden adult death syndrome and other non-ischaemic causes of sudden cardiac death. Heart 2006; 92(3):316–320.
42. Lotze U, Egerer R, Tresselt C, et al. Frequent detection of parvovirus B19 genome in the myocardium of adult patients with idiopathic dilated cardiomyopathy. Med Microbiol Immunol (Berl) 2004; 193(2–3):75–82.
43. Liu PP, Schultheiss HP. Myocarditis. In: Baunwald E, ed. Heart Disease, 8 edn. Philadelphia, PA: W B Saunders Co, 2008:1775–1792.
44. Modlin FJ. Picornaviridae. In: Mandell GL, Douglas RG Jr., Bennett JE, eds. Principles and Practice of Infectious Diseases. New York, NY: Churchill Livingstone, 1990:721–732.
45. Martino TA, Petric M, Brown M, et al. Cardiovirulent coxsackieviruses and the decay-accelerating factor (CD55) receptor. Virology 1998; 244(2):302–314.
46. Liu PP, Opavsky MA. Viral myocarditis: Receptors that bridge the cardiovascular with the immune system? Circ Res 2000; 86(3):253–254.
47. Bergelson JM, Cunningham JA, Droguett G, et al. Isolation of a common receptor for Coxsackie B viruses and adenoviruses 2 and 5. Science 1997; 275(5304):1320–1323.
48. Klingel K, Sauter M, Bock CT, et al. Molecular pathology of inflammatory cardiomyopathy. Med Microbiol Immunol (Berl) 2004; 193(2–3):101–107.
49. Bergelson JM. Receptors mediating adenovirus attachment and internalization. Biochem Pharmacol 1999; 57(9):975–979.
50. Noutsias M, Fechner H, de Jonge H, et al. Human coxsackie-adenovirus receptor is colocalized with integrins alpha(v)beta(3) and alpha(v)beta(5) on the cardiomyocyte sarcolemma and upregulated in dilated cardiomyopathy: Implications for cardiotropic viral infections. Circulation 2001; 104(3): 275–280.
51. Bock CT, Klingel K, Aberle S, et al. Human Parvovirus B19: A new emerging pathogen of inflammatory cardiomyopathy. J Vet Med B Infect Dis Vet Public Health 2005; 52:340–343.
52. Krueger GRF, Rojo J-, Buja LM, et al. Human herpesvirus-6 (HHV-6) is a possible cardiac pathogen: An immunihistological and ultrastructural study. Hospital General 2008; 71:187–191.
53. Mason JW. Myocarditis and dilated cardiomyopathy: An inflammatory link. Cardiovasc Res 2003; 60(1):5–10.
54. Fujioka S, Kitaura Y, Ukimura A, et al. Evaluation of viral infection in the myocardium of patients with idiopathic dilated cardiomyopathy. J Am Coll Cardiol 2000; 36(6):1920–1926.
55. Triantafilou K, Orthopoulos G, Vakakis E, et al. Human cardiac inflammatory responses triggered by Coxsackie B viruses are mainly Toll-like receptor (TLR) 8-dependent. Cell Microbiol 2005; 7(8): 1117–1126.
56. Stewart MJ, Smoak K, Blum MA, et al. Basal and reovirus-induced beta interferon (IFN-beta) and IFN-beta-stimulated gene expression are cell type specific in the cardiac protective response. J Virol 2005; 79(5):2979–2987.
57. Fuse K, Chan G, Liu Y, et al. Myeloid differentiation factor-88 plays a crucial role in the pathogenesis of Coxsackievirus B3-induced myocarditis and influences type I interferon production. Circulation 2005; 112(15):2276–2285.
58. Liu CC, Young HY, Young JDE. Lymphocyte-mediated cytolysis and disease. N Engl J Med 1996; 335:1651–1659.
59. Liu CC, Persechini PM, Young JD. Perforin and lymphocyte-mediated cytolysis. Immunol Rev 1995; 146:145–175.
60. Gebhard JR, Perry CM, Harkins S, et al. Coxsackievirus B3-induced myocarditis: Perforin exacerbates disease, but plays no detectable role in virus clearance. Am J Pathol 1998; 153(2):417–428.
61. Deonarain R, Cerullo D, Fuse K, et al. Protective Role for Interferon-{beta} in Coxsackievirus B3 Infection. Circulation 2004; 110:3540–3543.
62. El-Menyar AA. Cytokines and myocardial dysfunction: State of the art. J Card Fail 2008; 14(1):61–74.
63. Opavsky MA, Penninger J, Aitken K, et al. Susceptibility to myocarditis is dependent on the response of alphabeta T lymphocytes to coxsackieviral infection. Circ Res 1999; 85(6):551–558.
64. Liu F, Whitton JL, Slifka MK. The rapidity with which virus-specific CD8+ T cells initiate IFN-gamma synthesis increases markedly over the course of infection and correlates with immunodominance. J Immunol 2004; 173(1):456–462.
65. Afanasyeva M, Georgakopoulos D, Rose NR. Autoimmune myocarditis: Cellular mediators of cardiac dysfunction. Autoimmun Rev 2004; 3(7–8):476–486.
66. Tam PE. Coxsackievirus Myocarditis: Interplay between Virus and Host in the Pathogenesis of Heart Disease. Viral Immunol 2006; 19(2):133–146.
67. Fairweather D, Frisancho-Kiss S, Yusung SA, et al. IL-12 protects against coxsackievirus B3-induced myocarditis by increasing IFN-gamma and macrophage and neutrophil populations in the heart. J Immunol 2005; 174(1):261–269.

68. Streitz M, Noutsias M, Volkmer R, et al. NS1 specific CD8+ T-cells with effector function and TRBV11 dominance in a patient with parvovirus B19 associated inflammatory cardiomyopathy. PLoS one 2008; 3(6):e2361.
69. Klingel K, Hohenadl C, Canu A, et al. Ongoing enterovirus-induced myocarditis is associated with persistent heart muscle infection: Quantitative analysis of virus replication, tissue damage, and inflammation. Proc Natl Acad Sci U S A 1992; 89(1):314–318.
70. Badorff C, Lee GH, Lamphear BJ, et al. Enteroviral protease 2 A cleaves dystrophin: Evidence of cytoskeletal disruption in an acquired cardiomyopathy. Nat Med 1999; 5(3):320–326.
71. Fairweather D, Kaya Z, Shellam GR, et al. From infection to autoimmunity. J Autoimmun 2001; 16(3):175–186.
72. Pauschinger M, Chandrasekharan K, Schultheiss HP. Myocardial remodeling in viral heart disease: Possible interactions between inflammatory mediators and MMP-TIMP system. Heart Fail Rev 2004; 9(1):21–31.
73. Heymans S, Pauschinger M, De Palma A, et al. Inhibition of urokinase-type plasminogen activator or matrix metalloproteinases prevents cardiac injury and dysfunction during viral myocarditis. Circulation 2006; 114(6):565–573.
74. Caforio AL, Calabrese F, Angelini A, et al. A prospective study of biopsy-proven myocarditis: Prognostic relevance of clinical and aetiopathogenetic features at diagnosis. Eur Heart J 2007; 28(11):1326–1333.
75. Why HJ, Meany BT, Richardson PJ, et al. Clinical and prognostic significance of detection of enteroviral RNA in the myocardium of patients with myocarditis or dilated cardiomyopathy. Circulation 1994; 89(6):2582–2589.
76. Kühl U, Pauschinger M, Schwimmbeck PL, et al. Interferon-beta treatment eliminates cardiotropic viruses and improves left ventricular function in patients with myocardial persistence of viral genomes and left ventricular dysfunction. Circulation 2003; 107(22):2793–2798.
77. Grist NR, Bell EJ. A six-year study of coxsackievirus B infections in heart disease. J Hyg (Lond) 1974; 73(2):165–172.
78. Tracy S, Chapman NM, McManus BM, et al. A molecular and serologic evaluation of enteroviral involvement in human myocarditis. J Mol Cell Cardiol 1990; 22(4):403–414.
79. Keeling PJ, Tracy S. Link between enteroviruses and dilated cardiomyopathy: Serological and molecular data. Br Heart J 1994; 72(suppl 6):S25–S29.
80. El-Hagrassy MM, Banatvala JE, Coltart DJ. Coxsackie-B-virus-specific IgM responses in patients with cardiac and other diseases. Lancet 1980; 2(8205):1160–1162.
81. Muir P, Nicholson F, Tilzey AJ, et al. Chronic relapsing pericarditis and dilated cardiomyopathy: Serological evidence of persistent enterovirus infection. Lancet 1989; 1(8642):804–807.
82. Bowles NE, Richardson PJ, Olsen EG, et al. Detection of Coxsackie-B-virus-specific RNA sequences in myocardial biopsy samples from patients with myocarditis and dilated cardiomyopathy. Lancet 1986; 1(8490):1120–1123.
83. Bowles NE, Rose ML, Taylor P, et al. End-stage dilated cardiomyopathy. Persistence of enterovirus RNA in myocardium at cardiac transplantation and lack of immune response. Circulation 1989; 80(5):1128–1136.
84. Kandolf R, Ameis D, Kirschner P, et al. In situ detection of enteroviral genomes in myocardial cells by nucleic acid hybridization: An approach to the diagnosis of viral heart disease. Proc Natl Acad Sci U S A 1987; 84(17):6272–6276.
85. Baboonian C, Treasure T. Meta-analysis of the association of enteroviruses with human heart disease. Heart 1997; 78(6):539–543.
86. Frustaci A, Chimenti C, Pieroni M, et al. Cell death, proliferation and repair in human myocarditis responding to immunosuppressive therapy. Mod Pathol 2006; 19(6):755–765.
87. Klein RM, Jiang H, Niederacher D, et al. Frequency and quantity of the parvovirus B19 genome in endomyocardial biopsies from patients with suspected myocarditis or idiopathic left ventricular dysfunction. Z Kardiol 2004; 93(4):300–309.
88. Kühl U, Lassner D, Pauschinger M, et al. Prevalence of erythrovirus genotypes in the myocardium of patients with dilated cardiomyopathy. J Med Virol 2008; 80(7):1243–1251.
89. Schowengerdt KO, Ni J, Denfield SW, et al. Association of parvovirus B19 genome in children with myocarditis and cardiac allograft rejection: Diagnosis using the polymerase chain reaction. Circulation 1997; 96(10):3549–3554.
90. Kandolf R, Hofschneider PH. Molecular cloning of the genome of a cardiotropic Coxsackie B3 virus: Full-length reverse-transcribed recombinant cDNA generates infectious virus in mammalian cells. Proc Natl Acad Sci U S A 1985; 82(14):4818–4822.
91. Pankuweit S, Ruppert V, Eckhardt H, et al. Pathophysiology and aetiological diagnosis of inflammatory myocardial diseases with a special focus on parvovirus b19. J Vet Med B Infect Dis Vet Public Health 2005; 52(7–8):344–347.

92. Vallbracht KB, Schwimmbeck PL, Kuhl U, et al. Differential aspects of endothelial function of the coronary microcirculation considering myocardial virus persistence, endothelial activation, and myocardial leukocyte infiltrates. Circulation 2005; 111(14):1784–1791.
93. Gauntt CJ, Trousdale MD, LaBadie DR, et al. Properties of coxsackievirus B3 variants which are amyocarditic or myocarditic for mice. J Med Virol 1979; 3(3):207–220.
94. Mikami S, Kawashima S, Kanazawa K, et al. Expression of nitric oxide synthase in a murine model of viral myocarditis induced by coxsackievirus B3. Biochem Biophys Res Commun 1996; 220(3):983–989.
95. Reetoo KN, Osman SA, Illavia SJ, et al. Quantitative analysis of viral RNA kinetics in coxsackievirus B3-induced murine myocarditis: Biphasic pattern of clearance following acute infection, with persistence of residual viral RNA throughout and beyond the inflammatory phase of disease. J Gen Virol 2000; 81(Pt 11):2755–2762.
96. Huber SA. T cells expressing the gamma delta T cell receptor induce apoptosis in cardiac myocytes. Cardiovasc Res 2000; 45(3):579–587.
97. Elliott P, Andersson B, Arbustini E, et al. Classification of the cardiomyopathies: A position statement from the European Society Of Cardiology Working Group on Myocardial and Pericardial Diseases. Eur Heart J 2008; 29(2):270–276.
98. Badorff C, Knowlton KU. Dystrophin disruption in enterovirus-induced myocarditis and dilated cardiomyopathy: From bench to bedside. Med Microbiol Immunol (Berl) 2004; 193(2–3):121–126.
99. Mahrholdt H, Goedecke C, Wagner A, et al. Cardiovascular magnetic resonance assessment of human myocarditis: A comparison to histology and molecular pathology. Circulation 2004; 109(10):1250–1258.

19 Viral Hepatitis

John Scott
Department of Medicine, University of Washington, Seattle, Washington, U.S.A.

Atif Zaman
Department of Medicine, Oregon Health and Sciences University, Portland, Oregon, U.S.A.

Michael Chang
Department of Medicine, Portland VA Medical Center, Portland, Oregon, U.S.A.

INTRODUCTION

Hepatitis refers to inflammation of the liver, which can result from a variety of agents, either infectious (such as viruses) or noninfectious (such as alcohol, medications, or fatty liver). This chapter reviews hepatitis caused by viruses that primarily affect the liver, including hepatitis A virus (HAV), hepatitis B virus (HBV), hepatitis C virus (HCV), hepatitis D or delta virus (HDV), and hepatitis E virus (HEV). Historically, "infectious" hepatitis was differentiated from "serum" hepatitis by the different routes of transmission: primarily fecal-oral versus blood exposures and sexual intercourse. Subsequent work identified the etiologic viral agents as HAV and HBV, respectively.

Owing to a combination of a safer blood supply, better patient education, and vaccination against HAV and HBV, the incidence of new HAV, HBV, and HCV infections has declined precipitously in the last 10 years. In 2007, there were an estimated 13,000 cases of acute hepatitis A, 13,000 cases of acute hepatitis B, and 2800 cases of acute hepatitis C in the United States (1). An estimated four million Americans are believed to be chronically infected with hepatitis C (2) and 1.25 million Americans (3) chronically infected with hepatitis B.

CLINICAL PRESENTATION

Acute infectious hepatitis is commonly acquired in the absence of any symptoms. The development of symptoms is dependent on age and host factors. Common symptoms of infection, when they are present, include fever, right upper quadrant pain, malaise, nausea, vomiting, and anorexia. This prodrome can then progress over days to jaundice, dark urine, and clay-colored stools. In severe cases, fulminant liver failure may result, as evidenced by confusion and coagulopathy. The most common symptom of chronic hepatitis is fatigue. Advanced cases of chronic hepatitis can result in cirrhosis and decompensated liver disease. Complications of cirrhosis include jaundice, ascites, lower extremity edema, confusion, and gastrointestinal bleeding.

VIRAL AGENTS

Hepatitis A is a small, nonenveloped, positive sense, single-stranded RNA of the genus Hepatovirus within the Picornaviridae family (4). Only one serotype exists, although there are several genetically distinct forms of HAV (genotypes). Infection from any one of the different genotypes confers immunity to all genotypes (5). Hepatitis B is a partially double-stranded DNA, enveloped virus. It utilizes a low-fidelity reverse transcriptase, resulting in numerous mutations with each replication cycle. Hepatitis C is a positive sense, single-stranded RNA virus of the family Flaviviridae (6). It is composed of a highly conserved 5′ noncoding region which is essential for viral replication, and is the main region for PCR analysis. The lack of proofreading during RNA replication, combined with rapid viral replication (10^{10} to 10^{12} virions/day) leads to significant genomic variation (7). Hepatitis D is a small, single-stranded RNA virus, classified in the *Deltavirus* genus. It is unique in that it is only seen in the context of HBV infection, relying on HBV for nucleocapsid assembly and the surface antigen for encapsidation (8). Hepatitis E is another small, positive sense, single-stranded RNA virus, similar to Hepatitis A.

Table 1 Epidemiology, Prevention, and Therapy of Viral Hepatitis

Viral agent	Acute or chronic	Route of transmission	Vaccine available	Treatment available
HAV	Acute only	Fecal-oral	Yes	No
HBV	Both	Percutaneous, blood transfusion, perinatal, sexual	Yes	Yes
HCV	Both	Percutaneous, blood transfusion, infrequently sexual, rarely perinatal	No	Yes
HDV	Both	Percutaneous, blood transfusion, sexual, household	No, but HBV vaccination can prevent HDV	Yes
HEV	Acute only	Fecal-oral	No	No

Hepatitis E is currently unclassified, as it is unique in sharing similarities to several different classes of viruses. Like HAV, there is one serotype with several known genotypes. Infection with any one genotype confers immunity to all genotypes (9).

EPIDEMIOLOGY

Table 1 summarizes the similarities and differences in the epidemiology of viral hepatitis. HAV has long been associated with overcrowding and poverty and is now most prevalent in regions with suboptimal sanitation systems. HAV is spread via fecal-oral contamination. Prior to the widespread use of the HAV vaccine, HAV struck at-risk populations (such as Native American and Hispanic children) in periodic epidemics. Since the institution of pediatric HAV immunization in 1996 for high-risk groups and the expansion in 2006 to include all children, the incidence of new HAV cases in the United States has declined precipitously with a disappearance of the previously seen health disparities (10,11). HAV usually follows a benign course with the majority of people having a full recovery by two months. At-risk populations for developing a more fulminant course of illness include those over the age of 40 and individuals coinfected with other viral hepatitides (12).

Approximately 350 million persons worldwide are chronically infected with HBV. Nearly one in three persons has been exposed to HBV, making it one of the most common chronic viral infections in the world (13). In the United States, 1.25 million people are chronically infected, with the majority being foreign born (3). Eastern and southern Asia, as well as sub-Saharan Africa, are endemic regions where seroprevalence may exceed 8% (14). In developed nations, HBV is primarily transmitted through injection drug use, sexual intercourse, or occupational percutaneous exposures. In Asia, most HBV is spread perinatally, from an infected mother to her child. In Africa and circumpolar regions, transmission is primarily horizontal, in the first five years of life. The likelihood of chronic HBV infection depends on the age of exposure. For example, children exposed perinatally have a 95% chance of becoming chronically infected, whereas exposed adults have a 3% to 5% probability of developing chronic infection (14). HBV is a major cause of hepatocellular carcinoma (HCC) worldwide; overall, approximately one million deaths annually are directly attributable to HBV infection (14).

HCV is one of the most common chronic viral infections in the United States. It is primarily spread through exposure to contaminated blood. Risk factors for transmission include needle sharing through intravenous drug abuse, use of nasal cocaine, blood product transfusion prior to 1992, receipt of clotting factors prior to 1987, tattoos, and less commonly through high-risk sexual activity and perinatal transmission. There are an estimated 170 million cases worldwide, with nearly four million cases in the United States (2). Approximately 75% to 80% of people exposed to HCV will develop chronic liver infection. The time course from date of infection to the development of cirrhosis is variable, ranging between 20 and 40 years (15). Although the incidence of new HCV infections is falling, given the long prodrome of this illness before the development of cirrhosis, the peak incidence in morbidity and mortality from this disease is predicted to occur around 2030 (16). HCV is currently the most common etiology for liver transplantation in the United States and Europe. HCV-related mortality in the United States is expected to climb from 13,000 cases in 2000 to up to 39,000 cases by 2030 (16,17).

HDV is common in areas where HBV is endemic, particularly in the Middle East, Mediterranean, Africa, parts of South America and Central Asia. Five percent of those with chronic HBV also have HDV coinfection (8). In the United States, HDV is mainly seen in injection drug users or those with frequent blood transfusions, although sexual and household contact may be the route of transmission less commonly. HDV is associated with much higher rates, and more rapid development, of cirrhosis (60–70% after 5–10 years) (18). The overall mortality rate is 10 times greater than that with HBV infection alone (19). HDV may be acquired at the time of HBV infection (coinfection) or after chronic HBV infection is established (super-infection). The latter situation more commonly leads to chronic HDV infection.

Hepatitis E is the most common cause of epidemic, enterically transmitted hepatitis worldwide. HEV is passed via the fecal–oral route, similar to HAV infection. HEV is endemic in most equatorial countries, leading to sporadic outbreaks (20). The largest known outbreak occurred in Xinjiang, China, between 1986 and 1988, affecting almost 120,000 individuals. Population studies from this outbreak revealed that individuals aged 15 to 40 developed the most severe symptoms, and that there was a high mortality rate associated with late-term pregnancy and HEV infection (up to 20%) (21).

DIFFERENTIAL DIAGNOSIS

Hepatitis can result from other viruses (cytomegalovirus, herpes simplex virus, adenovirus, Epstein–Barr virus), bacteria (pyogenic abscesses), fungi (candida), and parasites (echinococcus, schistosomiasis). In addition, noninfectious causes include alcohol, medications, fatty liver disease, toxins, autoimmune injury, hereditary abnormalities (Wilson's disease, hemochromatosis, alpha1-antitrypsin deficiency), and right-sided heart failure.

LABORATORY TESTING

Frequently, chronic viral hepatitis is diagnosed because routine "liver function tests" (total bilirubin, alanine aminotransferase, aspartine aminotransferase, alkaline phosphatase, and gamma-glutamyl transferase) are abnormal. Both alanine aminotransferase (ALT) and aspartine aminotransferase (AST) are intracellular enzymes that are released when hepatocytes die. Because the ALT is made primarily by hepatocytes, an elevated level is most specific for inflammation of the liver. Despite the misnomer "liver function test," high levels of AST and ALT represent hepatocyte destruction. Importantly, clinical laboratories often list a higher range of "normal"; the cutoff for a normal ALT should be <19 U/L in women and <30 U/L in men (22).

SPECIMEN TYPES/HANDLING

In general, serologies should be performed on serum or plasma separated from whole blood within 24 hours. These samples may be stored for up to five days at 2–8°C or frozen indefinitely at −70°C. Collection tubes should not contain heparin, as this compound can interfere with the performance of the assays. Repeated freeze/thaw cycles should be avoided. For nucleic acid assays (viral levels, genotype, resistance testing), blood samples should be collected in tubes containing EDTA (lavender top) or citrate dextrose (yellow top). Processing should occur within six hours; testing should be performed within 24 hours or the sample should be frozen at −70°C.

DIRECT EXAMINATION (MICROSCOPY, ANTIGEN DETECTION, AND NUCLEIC ACID DETECTION)

As discussed later in this chapter, HAV and HEV are both detected using commercially available test kits detecting antibody to viral specific antigen. As both viruses have a single serotype, diagnostic tests looking for virus-specific IgM and IgG are the gold standard for diagnosing these illnesses. Identification of IgM is correlated to recent infection, while the presence of IgG has traditionally represented prior infection. PCR for HEV is available in some areas; however, it is not widely available. In the United States, testing for HEV is only available in research laboratories, which can be located through the CDC. Tools such as cell culture, enzyme immunoassay (EIA), immune electron microscopy, and reverse transcriptase polymerase chain reaction (RT-PCR) are all available, but clinically impractical and more appropriate for research settings.

HBV can be detected by electron microscopy, although this is expensive and laborious, and is therefore only performed in research settings. Three distinct morphologic entities can be visualized: the spherical surface antigens, the filamentous form of surface antigens, and the double-shelled spherical particle containing the HBV virion, known as the Dane particle. The surface antigen (HBsAg) is produced in both acute and chronic infections. The e antigen (HBeAg) is indicative of active replication and has been associated with a higher risk of HCC (23). More detailed information on antigen detection and interpretation can be found in the section on serologic testing.

As with other chronic viral infections, the level of HBV DNA has become increasingly important in the natural history and outcome of treatment. For example, in one large Taiwanese study, researchers found that HBV DNA levels were highly predictive of developing HCC: patients with baseline HBV DNA levels >1 million copies/mL were 10 times more likely to develop HCC over the next decade compared to those with <300 copies/mL (24). A similar study by the same group showed that patients with higher viral loads were more likely to develop cirrhosis (25). The viral level is also used for determining whether treatment should be initiated and for monitoring the response to therapy (26).

There are three commonly used methods for detecting HBV DNA: polymerase chain reaction (PCR), branched chain DNA amplification (bDNA), and hybrid capture. Among PCR tests, the older Monitor assay has a more limited range and poorer sensitivity compared to the real-time assay (27). Originally developed by National Genetics Institute, but now available through LabCorp, the SuperQuant assay has a range from 100 to 10^9 genome copies/mL (19). Real-time PCR quantification of HBV DNA is the most popular assay because of its improved sensitivity and simplicity compared to bDNA (Versant v 3.0, Bayer) (28,29). Using the TaqMan platform, real-time PCR quantification is sensitive and linear from a range of 1.7–8.0 $\log_{10}$ IU/mL and is equally efficacious for genotypes A–H (30). Using specifically designed RNA probes that bind target DNA and then amplify the signal, the hybrid capture technology can detect DNA levels to 10^3 copies/mL, but is limited to a 4 log range (Ultrasensitive HBV Hybrid Capture II) (31). The only FDA-approved assay for HBV DNA is the qualitative PCR test for the screening of HBV from blood donors (HBV AmpliScreen, Roche Molecular Systems, Pleasanton, CA). As with the quantification of HCV, HBV DNA quantification by real-time PCR should incorporate the WHO International Standard as an internal standard; reporting of results in IU/mL ensures better reproducibility and comparability between labs.

Hepatitis C virus detection is normally conducted in two phases: a screening enzyme immunoassay (EIA) detecting antibodies to HCV proteins, followed by a confirmatory assay which employs either a more specific recombinant immunoblot assay (RIBA) antibody test or direct detection of HCV RNA (the latter is recommended for most cases).

The second-generation screening test (EIA-2) detected antibodies to three HCV-encoded proteins: C22 (core protein), C33 nonstructural region 3 protein (NS3), and C100 nonstructural region 4 protein (NS4). This test has a sensitivity of 92% to 95%. The third generation EIA test (EIA-3) is able to detect a fourth HCV protein (NS5), resulting in excellent sensitivity (97–99%) and specificity (>99%) (32,33). False negatives can occur in immunosuppressed populations, such as HIV-positive patients, and in hemodialysis patients. In these populations, a sensitive nucleic acid test for HCV should be used if there is no other explanation for liver function test abnormalities. The false-positive rate can be as high as 35% in low-risk populations, underscoring the need for focused screening and confirmatory testing (34).

Once an initial screening test returns as positive, a confirmatory assay can be performed to rule out a false-positive screen. Two methods are currently available for use as a confirmatory test: RIBA-2 and molecular testing for HCV RNA. RIBA-2 is more expensive and more technically demanding to perform. The RIBA-2 test also detects the C22, C33, and C100 proteins and is interpreted as positive if two or more antibodies are present, indeterminate if 1 is detected, and negative if no antibodies are present.

The most sensitive and specific method for detecting HCV is the direct testing method that relies on polymerase chain reaction to detect HCV RNA in the serum. Molecular testing, also known as nucleic acid testing (NAT), detects actual viral presence and not immune surrogates of infection. NAT also allows for earlier identification of infection in acute HCV, when antibody levels are undetectable but active virus is present in the blood. Two types of NAT are available today: a qualitative NAT and a quantitative NAT (35).

The qualitative NATs include the AMPLICOR 2.0 and Ampliscreen 2.0 by Roche Diagnostics, Indianapolis, IN, and the UltraQual, which is run at a reference lab at the National Genetics Institute in Los Angeles, CA. These three tests RT-PCR to amplify HCV RNA to allow for detection of active infection, with a lower limit of detection of 50 IU/mL (36). Another qualitative technique, known as transcription-mediated amplification (TMA) is employed by the Versant HCV RNA Qualitative Assay (Bayer Diagnostics, Emeryville, CA) and the Procleix HIV-1/HCV assay (GenProbe, SanDiego, CA), and both have a lower limit of detection down to 5 IU/mL, with even better sensitivities (>98%) (37).

Quantitative measurement of HCV RNA is done through RT-PCR, real-time PCR, or the branched chain DNA method (bDNA). SuperQuant (National Genetics Institute) and MONITOR 2.0 (Roche Diagnostics) employ RT-PCR techniques, while VERSANT bDNA3.0 (Bayer Corp, Tarrytown, NY) utilizes a branched chain technique (38,39). Real-time PCR using TaqMan (Roche) technology is the newest advancement to date with detection limits similar to the qualitative assays (detection as low as 10 IU/mL and as high as 100 million IU/mL) in a completely automated system (31). Because of the rapid turnaround, automation, and test characteristics, real-time PCR has become the most popular NAT for HCV.

The HDV antigen can be visualized microscopically from liver biopsy specimens. However, this process requires specialized staining, which is neither rapid nor simple to perform. Similarly, detection of HDV Ag using EIA and Western blotting is technically difficult, although more sensitive than the above technique. Specialized laboratories (such as Focus Diagnostics) are able to perform serologic tests for HDV. Detection of HDV RNA by PCR is a more reliable diagnostic test than serologies because it overcomes the pitfall of antigen sequestration in immune complexes (40). Although only performed in research laboratories, quantification of HDV RNA by real-time PCR may be important because higher levels may correlate with more severe disease (41) and HDV viral kinetics are predictive of treatment success (42). The sensitivity of this assay is 10 to 100 copies/mL (40). Interestingly, HDV RNA and HBV DNA levels can vary in an inverse relationship. A qualitative HDV RNA assay is available commercially at specialty labs. This assay uses primers that target the most highly conserved C terminus of the HDV Ag-encoding region; however, it has been difficult to design primers that accommodate the extensive diversity, accounting for the qualitative nature of this test.

VIRAL ISOLATION

In general, hepatitis viruses have been very difficult to grow in culture. Although HBV can grow in either healthy adult or fetal human hepatocytes, this culture system does not support propagation of infectious virions (43). Thus, it is not an option as a diagnostic test. Hepatitis C viral culture has only recently been established within the last several years and is exclusively a research tool at this time (44). Both HAV and HEV have cell culture models; however, all are plagued with poor adaptation and slow growth making them useful only as research tools (45,46).

IDENTIFICATION

HAV and HEV are both identified using standard EIA for IgM and IgG. Identification of HBV and HDV is based on the results of serological and nucleic amplification assays. HCV, as described earlier, can be diagnosed using both EIA and direct molecular identification with NAT.

TYPING SYSTEMS

Clinically, HAV has one serotype despite having at least four genetically identifiable genotypes (47). Immunity to one genotype confers immunity to all genotypes and no clinically significant difference exists among the different genotypes. HEV appears to also have four major genotypes and one serotype. The genotypes are distributed geographically with up to 20% variation in homology at the nucleic acid level between genotypes (9). There again appears to be one dominant epitope resulting in a single serotype across genotypes, conferring immunity to all genotypes after infection.

HBV has been classified into eight major genotypes (A–H), which differ by >8% in the whole viral sequence (48). Genotypes have unique geographic niches. For example, genotypes B and C are most common in Asia. In the United States, genotypes A2, B, C, and D are most common. Knowledge of HBV genotype is increasingly important because genotype can predict the response to interferon (A responds better than D) (49) and certain genotypes (namely, C and F) are much more likely to lead to hepatocellular carcinoma (50,51). Detection of genotype is performed by identifying nucleotide sequences in the most highly conserved region of the genome, usually the pre-S or S region. Genotyping can be performed by line probe analysis (Inno-Lipa HBV DRv2, Innogenetics, Gent, Belgium, or Quest Diagnostics) or in-house DNA sequencing, although neither assay is FDA approved.

Because HCV has both a high replication rate (10^{12} virions/day) and lacks proofreading activity during RNA replication, there is significant divergence in HCV RNA sequences. There are six genetically distinct groups of HCV called genotypes, with greater than 30% variability in genomic sequencing between types. Genotype 1 infection is the most common in the United States, comprising 71.5% of HCV infections in the United States, followed by genotypes 2 and 3 (~19%) (52,53). Quasispecies are divergent HCV sequences within an individual that are attributed to viral mutation. Although quasispecies have no clear prognostic role in the treatment or natural history of the disease, they may explain viral resistance and breakthrough during treatment with interferon and ribavirin (52).

Genotype testing is important for infected individuals as it has a great impact on treatment duration and success. Treatment response for genotype 1 infection is estimated to be between 40% and 50% with one year of treatment with pegylated interferon and ribavirin, whereas genotype 2 and 3 patients have an 80% to 90% response rate with treatment durations as short as 24 weeks (54,55).

There are three major genotypes for HDV. Genotype 1 is more common in Western countries and can be associated with fulminant hepatitis (8). Genotype 2 is more often seen in Asia and follows a less aggressive course. Genotype 3 is found in the Amazon River basin and is also associated with fulminant hepatitis. Genotyping is performed for study purposes only, either by restriction fragment length polymorphisms, sequencing, or immunohistochemical staining using genotype-specific anti-HD antibodies (8).

SEROLOGICAL TESTING

Serologic testing for acute infectious hepatitis should always include a battery of tests. A broad differential is necessary to acknowledge unlikely sources of infection prior to ruling them out based on history and physical exam. Given the significant overlap in symptoms and duration of prodrome of the infectious hepatitides, a standard panel should include Hepatitis A IgG and IgM, Hepatitis B SAg, SAb, core IgG, core IgM, and Hepatitis C antibody and RNA evaluation. If there is a history of travel to endemic areas where HEV is present, then HEV IgG and IgM should also be evaluated. Likewise, if there is a suspicion for prior HBV infection or a person resides in an endemic region for HBV, PCR should be considered, as well as testing for HDV.

Typically, patients are exposed to HAV via fecal–oral passage. The virus is absorbed in the intestine where it is passed to the liver. The incubation period is 15 to 49 days. Once symptoms occur, jaundice usually develops within one to two weeks. Asymptomatic infection is common in up to 90% of children younger than 5 years, and even in 20% to 30% of adults. Symptoms usually abate at the onset of the jaundice period, which usually lasts about two weeks. Complete recovery occurs in 60% of individuals by two months, and in nearly 100% by six months (56). Serologic testing for HAV consists of serologic evaluation for HAV IgM and IgG using either EIA or RIA test kits. Although stool testing and HAV PCR are available, most patients are no longer shedding virus in their stool on presentation and HAV PCR is unnecessary in the proper clinical setting with positive HAV IgM. HAV IgM is usually detectable two weeks into the prodrome period and stays elevated for up to six months. HAV IgG will begin to rise at the same time as the IgM, but will stay elevated for years after infection (57).

The HBV surface antigen (HBsAg) is a protein on the surface of HBV, which is produced in both acute and chronic HBV. The total core antibody (HBcAb) is indicative of past or current

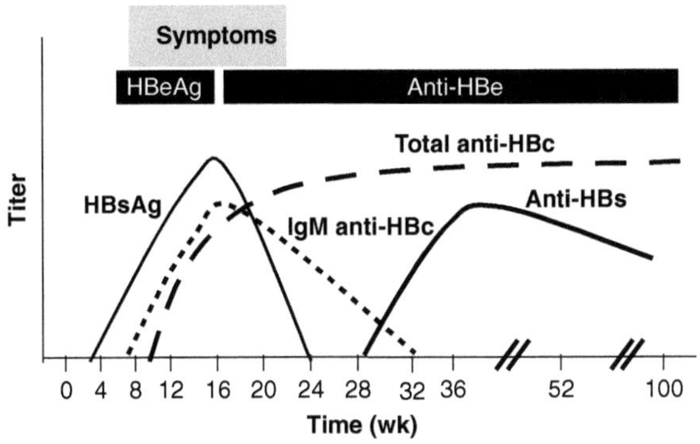

Figure 1 Typical sequence of serologic markers in patients with acute HBV and resolution. *Source*: Adapted from Ref. 71.

infection. Core IgM is produced in acute infections and thus is a marker of acute HBV. Core IgG indicates past exposure to HBV. The typical sequence of serologic markers for a patient who has acute HBV and resolution is shown in Figure 1. The HBsAg is the first serologic marker of infection, appearing as early as one-week postexposure, but usually between 6 and 10 weeks. Shortly thereafter, the HBeAg is found in the blood. Several weeks after the HBeAg and HBsAg appear, the aminotransferases will peak (more often the ALT is higher than the AST). Approximately 10 weeks after exposure, the IgM core antibody typically becomes positive.

After acute exposure, there are three possible outcomes (26):

1. Resolved hepatitis B infection. Evidence of recovery is manifested by normalization of the ALT, clearance of the HBeAg and development of anti-HBe antibodies, very low or undetectable levels of HBV DNA, HBsAg clearance and development of anti-HBsAg (or surface antibody).
2. Inactive carrier state. The surface antigen remains positive, but the ALT normalizes, HBeAg is negative, and HBV DNA levels are very low (<2000 IU/mL). The total core antibody is positive but IgM is negative. Importantly, between 4% and 20% of patients in this stage will have periodic flares of hepatitis with reversion back to HBeAg positivity (26).
3. Chronic hepatitis B. This is characterized by high viral levels, >2000 IU/mL. The HBeAg can remain positive and the ALT can fluctuate between normal and abnormally high levels. When the HBV DNA levels are very high (>20,000 IU/mL), patients are at higher risk for development of hepatic complications and transmission to other persons. A subset of patients will have a negative eAg due to core and pre-core promoter mutations. The HBV DNA level is usually 1–2 log lower than in HBVeAg+ patients. The key feature in distinguishing chronic inactive carriers from HBVeAg− patients is the HBV DNA level. Patients with an HBV DNA >2000 IU/mL are considered to be HBVeAg negative, chronically infected. Patients with elevated ALT and HBeAg+ status may seroconvert at a rate of 8% to 12% per year (26).

Occasionally, an isolated total core antibody is observed. There are four possible explanations for this situation: (*i*) the patient is in the "window period" in acute infection when the surface antigen has been cleared but surface antibody has not developed yet; (*ii*) there is chronic infection but the HBsAg is not detected; (*iii*) the infection is resolved but HBsAb titers are too low; (*iv*) the core antibody test is a false positive. In most cases, the reason is a false-positive test or remote infection; however, 20% of patients will have detectable HBV DNA (58). There are more than a dozen commercially available assays for the detection of these serologic markers. The sensitivity, specificity, and reproducibility are excellent. Most tests employ enzyme immunoassay or chemiluminescence technologies.

Acute HCV infection is associated with an early rise in HCV RNA levels as early as two weeks after initial infection. Jaundice occurs in up to 25% of individuals with acute HCV and may be associated with severe hepatitis, with ALT levels exceeding 1000 U/L in the first month of infection. Unfortunately, the majority of HCV infections are only mildly symptomatic and

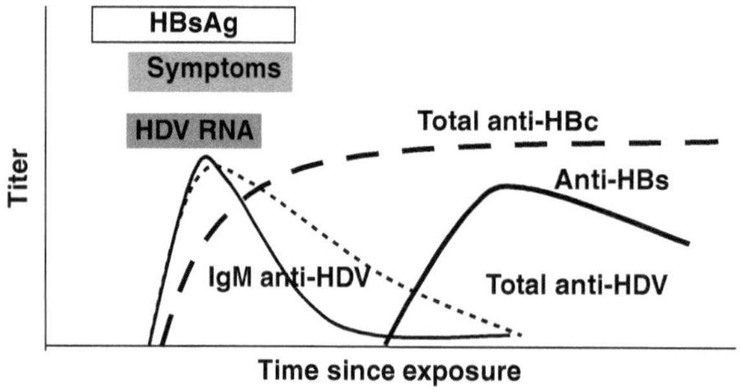

Figure 2 Serologic course of acute HDV infection, when acquired as a coinfection with HBV. *Source*: Adapted from Ref. 71.

patients rarely present to their physician during the acute illness. Up to 85% of individuals exposed to HCV will develop chronic infection (15). Testing for HCV in the acute setting should be by PCR and by HCV Ab to detect active viral replication and to determine if HCV Ab is already present at the time of presentation. HCV antibody will traditionally become detectable around week 12 and checking an HCV Ab would be important early on to determine the time frame at which a patient is presenting. The HCV PCR should be repeated at week 4 and week 12 to look for spontaneous clearance of the virus. If HCV still persists at 12 weeks after initial infection, strong consideration should be given to treating the infection at this early stage (59,60).

Reference laboratories are able to detect antibodies to HDAg, both in the IgG and IgM classes. Interpretation of acute and chronic infection is not as straightforward as in HBV infection, in part because HDV can result from coinfection or superinfection of HBV. In coinfection, the HBsAg, IgG, and IgM HBVcAb, total and IgM anti-HDV will all be positive. If HDV is cleared, then total anti-HDV will wane over time (Fig. 2). In acute HDV coinfection, IgM is produced before IgG, so an isolated IgM antibody is indicative of acute HDV. In HDV superinfection, the key difference is that IgM component of HBcAb is negative—HBsAg, total HB core antibody, and total anti-HDV are all positive. In chronic HDV infection, both IgM and IgG are produced. However, in chronic HDV, the IgM is found in a monomeric form, whereas in acute HDV, the IgM is primarily pentameric (61).

HEV infection has a similar incubation period to HAV of 15 to 60 days, as well as a similar prodrome phase followed by an icteric phase. Like HAV, symptoms tend to abate by six weeks (62). HEV IgM levels rise near the peak of the ALT level and coincide with the disappearance of virus in the blood. HEV IgM may remain elevated for up to six months, whereas HEV IgG can remain elevated for years after infection. HEV is associated with fulminant hepatitis in at-risk populations, particularly pregnant women, with higher case fatalities and fetal mortality with late trimester infection (20).

ANTIVIRAL RESISTANCE

Antiviral resistance is a major consideration in the treatment of chronic HBV infection. There are currently five FDA-approved medications available in oral form (lamivudine, adefovir, telbivudine, tenofovir, and entecavir) and two injectables (standard and pegylated interferon). Lamivudine, an oral nucleoside analog, was a life-saving drug when made available on a compassionate basis in the mid-1990s and after approval for treatment of HBV in 1998. Lamivudine is safe, relatively inexpensive, and well tolerated even with long-term therapy. However, resistance develops frequently with prolonged use of the drug: 17–32% after one year and 67% after four years of therapy (63,64). Moreover, long-term studies demonstrate that lamivudine resistance accelerates the progression of liver disease (65). Lower rates of resistance are observed for the other medications, but this is still a problem given the long-term nature of HBV therapy. Certain mutations in the polymerase region have been well characterized as leading to resistance. Direct DNA sequencing, line probe assay, and restriction fragment length polymorphism (RFLP) analysis have been utilized, but all have pitfalls. DNA sequencing has the advantage of being able to detect novel mutations, but suffers from a relative inability to detect minor variants, compared to line probe (66). The line probe is commercially available, but can sometimes produce

Table 2 Interpretation of HBV Serological Tests

Tests	Results	Interpretation
HBsAg anti-HBc anti-HBs	Negative Negative Negative	Susceptible
HBsAg anti-HBc anti-HBs	Negative Positive Positive	Immune due to natural infection
HBsAg anti-HBc anti-HBs	Negative Negative Positive	Immune due to hepatitis B vaccination
HBsAg anti-HBc IgM anti-HBc anti-HBs	Positive Positive Positive Negative	Acutely infected
HBsAg anti-HBc IgM anti-HBc anti-HBs	Positive Positive Negative Negative	Chronically infected
HBsAg anti-HBc anti-HBs	Negative Positive Negative	Interpretation unclear; four possibilities: 1. Resolved infection but low titers anti-HBs (most common) 2. False-positive anti-HBc, thus susceptible 3. "Low level" chronic infection, HBsAg is undetectable 4. Resolving acute infection, in window period when HBsAg is cleared

Source: From Ref. 72.

indeterminate results, has a lower specificity, and can only detect known mutations, compared to sequencing. RFLP is labor intensive and requires knowledge of proper endonucleases.

Hepatitis C virus resistance to interferon and ribavirin therapy is multifactorial, based on host and viral factors. Host factors that affect viral response to treatment include race, age, gender, weight, coinfection with HIV or other viral hepatitis, immune suppression, alcohol consumption, the presence of diabetes, and the degree of liver disease. Viral factors include mainly genotype and viral load (67). Although one study suggested that a large number of mutations in the NS5A region (the interferon sensitivity determining region) correlated with the poorer treatment response in genotype 1-infected patients (68), this finding has not been confirmed in either clinical or molecular studies in Western countries (69). Both HAV and HEV are self-limited infections without a chronic infectious state.

EVALUATION AND REPORTING OF LABORATORY RESULTS

A positive HAV IgM test is indicative of recent infection (<6 months), whereas an isolated positive HAV IgG test should be interpreted as past exposure or vaccination but current immunity to HAV. With widespread use of the HAV vaccine, outbreaks of acute HAV are becoming much less common. Therefore, a positive IgM in the context of no known exposures, clinical symptoms, or ALT elevation should be interpreted with caution, as there is no confirmatory test for HAV and false positives have been reported in such situations (70). Acute HAV is a reportable disease in the United States and many developed countries.

Interpretation of HBV serological results is shown in Table 2. Levels of HBsAb $\geq$10 IU/mL are considered protective. Acute HBV, a positive HBsAg in pregnant women, and the initial diagnosis of chronic HBV in all patients are reportable conditions in most US jurisdictions.

An algorithm for the testing and result reporting of HCV is shown in Figure 3. The EIA is the initial screening test for HCV. EIA test results are interpreted by the strength of the optical density reading, as compared to a standard. On the basis of this ratio, called the signal to cutoff

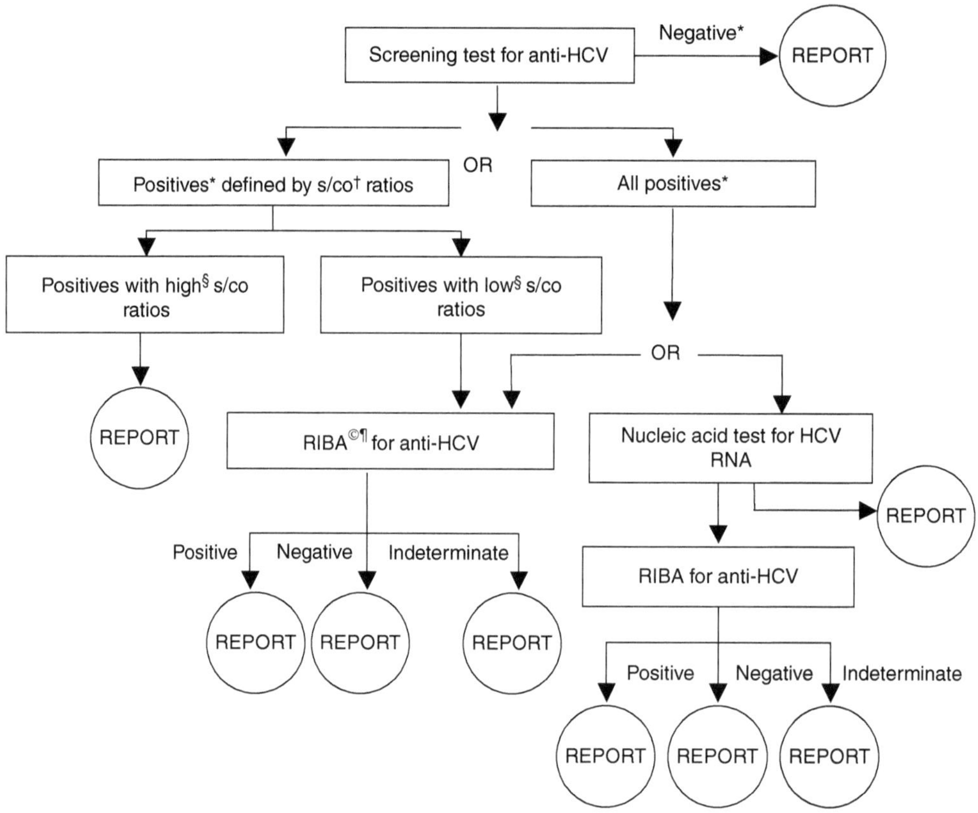

Figure 3 Laboratory algorithm for antibody to hepatitis C virus (anti-HCV) testing and result reporting. *Interpretation of screening immunoassay test results based on criteria provided by the manufacturer. †Signal to cut off. §Screening test-positive results are classified as having high s/co ratios if their ratios are at or above a predetermined value that predicts a supplemental test-positive result ≥95% of the time among all populations tested; screening test-positive results are classified as having low s/co ratios if their ratios are below this value. ¶Recombinant immunoblot assay. *Source*: From Ref. 34.

ratio (S/Co), a specimen is considered positive if the S/Co ≥1 and negative if <1. However, specimens just over 1 are likely to be false positives. A study by the CDC found that specimens with a S/Co greater than 3.8 have a 95% chance or greater of being positive on confirmatory testing (34). A positive test for HCV antibody in at-risk individuals indicates past exposure to HCV. Because 15% to 25% of individuals may clear HCV, confirmatory testing with a sensitive HCV RNA test (Amplicor or real-time PCR) is recommended. If this is positive, then the patient is determined to have chronic HCV, which is a reportable condition. If negative, the patient most likely is not chronically infected. However, low level or transient viremia can occur and therefore a follow-up NAT in 6 to 12 months is recommended. The RIBA is infrequently used but may have a role in diagnosing high-risk patients with a positive EIA but negative NAT. In such cases, a RIBA-positive result may indicate intermittent viremia and subsequent NAT should be performed in 6 to 12 months.

Detection of total anti-HDV is indicative of HDV infection. Patients who were exposed to HDV and have since resolved the infection may have low or undetectable levels of total anti-HDV. HDV IgM can be detected in both acute and chronic HDV, although in different forms (see above). The disappearance of HDV IgM is correlated with resolution of chronic HDV infection. PCR tests are used mainly for following the response to antiviral therapy. No direct treatment for HDV exists; treatment is aimed at controlling HBV replication and inhibiting HBV sAg production, which is necessary for HDC replication.

A positive test result for IgM to HEV is evidence of acute HEV infection. This highly sensitive and specific assay is the method of choice for diagnosis in low-prevalence countries, such as the United States, and in endemic countries. A rapid point of care test is now available for resource-limited countries.

REFERENCES

1. Centers for Disease Control. Disease burden from hepatitis A, B and C in the United States. http://www.cdc.gov/hepatitis/statistics.htm. Accessed December 7, 2009.
2. Armstrong GL, Wasley A, Simard EP, et al. The prevalance of hepatitis C virus infection in the United States, 1999 through 2002. Ann Intern Med 2006; 144:705–714.
3. Margolis HS, Coleman PJ, Brown RE, et al. Prevention of hepatitis B virus transmission by immunization: An economic analysis of current recommendations. JAMA 1995; 274:1201–1208.
4. Lemon SM, Jansen RW, Brown EA. Genetic, antigenic and biological differences between strains of hepatitis A virus. Vaccine 1992; 10:S40–S44.
5. Ping LH, Jansen RW, Stapleton JT, et al. Identification of an immunodominant antigenic site involving the capsid protein VP3 of hepatitis A virus. Proc Natl Acad Sci U S A 1988; 85:8281–8285.
6. He LF, Alling D, Popkin T, et al. Determining the size of non-A, non-B hepatitis virus by filtration. J Infect Dis 1987; 156:636–640.
7. Zeuzem S, Schmidt JM, Lee JH, et al. Effect of interferon alfa on the dynamics of hepatitis C virus turnover in vivo. Hepatology 1996; 23:366–371.
8. Farci P. Delta hepatitis: An update. J Hepatol 2003; 39:S212–S219.
9. Bradley DW, Purdy MA, Reyes GR. Hepatitis E genome. Molecular features, expression of immunoreactive proteins and sequence divergence. J Hepatol 1991; 13:S152–S154.
10. Bialek SR, Thoroughman DA, Hu D, et al. Hepatitis A incidence and Hepatitis A vaccination among American Indians and Alaska Natives, 1990–2001. Am J Public Health 2004; 94:996–1001.
11. Vogt TM, Wise ME, Bell BP, et al. Declining hepatitis A mortality in the United States during the era of hepatitis A vaccination. J Infect Dis 2008; 197:1282–1288.
12. Vento S, Garofano T, Renzini C, et al. Fulminant hepatitis associated with hepatitis A virus superinfection in patients with chronic hepatitis C. N Engl J Med 1998; 338:286–290.
13. Purcell RH. The discovery of the hepatitis viruses. Gastroenterol 1993; 104:955–963.
14. Lee WM. Hepatitis B virus infection. N Engl J Med 1997; 337:1733–1744.
15. Lauer GM, Walker BD. Hepatitis C virus infection. N Engl J Med 2001; 345:41–52.
16. Davis GL, Albright JE, Cook SF, et al. Projecting future complications of chronic hepatitis C in the United States. Liver Transpl 2003; 9:331–338.
17. Wise M, Bialek S, Finelli L, et al. Changing trends in hepatitis C-related mortality in the United States, 1995–2004. Hepatology 2008; 47:1128–1135.
18. Saracco G, Rosina F, Brunetto MR, et al. Rapidly progressive HBVsAg-positive hepatitis in Italy: The role of hepatitis delta virus infection. J Hepatol 1987; 5:274–281.
19. Horvat RT, Tegtmeier GE. Hepatitis B and D viruses. In: Murray PR, ed. Manual of Clinical Microbiology, Vol. 2. 9th ed. Washington, DC: ASM Press, 2007:1641–1659.
20. Khuroo MS, Rustgi VK, Dawson GJ, et al. Spectrum of hepatitis E virus infection in India. J Med Virol 1994; 43:281–286.
21. Zhuang H, Cao XY, Liu CB, et al. Epidemiology of hepatitis E in China. Gastroenterol Jpn 1991; 26(suppl 3):135–138.
22. Prati D, Taioli E, Zanella A, et al. Updated definitions of healthy ranges for serum alanine aminotransferase levels. Ann Intern Med 2002; 137:1–10.
23. Yang HI, Lu SN, Liaw YF, et al. Hepatitis B e antigen and the risk of hepatocellular carcinoma. N Engl J Med 2002; 347:168–174.
24. Chen CJ, Yang HI, Su J, et al. Risk of hepatocellular carcinoma across a biological gradient of serum hepatitis B virus DNA level. JAMA 2006; 295:65–73.
25. Iloeje UH, Yang HI, Su J, et al. Predicting cirrhosis risk based on the level of circulating hepatitis B viral load. JAMA 2006; 130:678–686.
26. Lok AS, McMahon BJ. Chronic hepatitis B. Hepatology 2007; 45:507–537.
27. Sum SS, Wong DK, Yuen JC, et al. Comparison of the COBAS TaqMan HBV test with the COBAS Amplicor monitor test for measurement of hepatitis B virus DNA in serum. J Med Virol 2005; 77: 486–490.
28. Garbuglia AR, Angeletti C, Lauria FN, et al. Comparison of Versant HBV DNA 3.0 and COBAS AmpliPrep-COBAS TaqMan assays for hepatitis B DNA quantification: Possible clinical implications. J Virol Methods 2007; 146:274–280.
29. Ronsin C, Pillet A, Bali C, et al. Evaluation of the COBAS AmpliPrep-total nucleic acid isolation-COBAS TaqMan hepatitis B virus (HBV) quantitative test and comparison to the VERSANT HBV DNA 3.0 assay. J Clin Microbiol 2006; 44:1390–1399.

30. Chevaliez S, Bouvier-Alias M, Laperche S, et al. Performance of the COBAS AmpliPrep/COBAS TaqMan real-time PCR assay for hepatitis B virus DNA quantification. J Clin Microbiol 2008; 46:1716–1723.
31. Konnick EQ, Erali M, Ashwood ER, et al. Evaluation of the COBAS Amplicor HBV Monitor assay and comparison with the ultrasensitive HBV Hybrid Capture 2 assay for quantification of hepatitis B virus DNA. J Clin Microbiol 2005; 43:596–603.
32. Colin C, Lanoir D, Touzet S, et al. Sensitivity and specificity of third-generation hepatitis C virus antibody detection assays: An analysis of the literature. J Viral Hepat 2001; 8:87–95.
33. Gretch DR. Diagnostic tests for hepatitis C. Hepatology 1997; 26:43S–47S.
34. Alter MJ, Kuhnert WL, Finelli L; Centers for Disease Control and Prevention. Guidelines for laboratory testing and result reporting of antibody to hepatitis C virus. MMWR Morb Mortal Wkly Rep 2003; 52:1–15.
35. Scott JD, Gretch DR. Molecular diagnostics of hepatitis C virus infection: A systematic review. JAMA 2007; 297:724–732.
36. Nolte FS, Fried MW, Shiffman ML, et al. Prospective multicenter clinical evaluation of AMPLICOR and COBAS AMPLICOR hepatitis C virus tests. J Clin Microbiol 2001; 39:4005–4012.
37. Ross RS, Viazov S, Hoffmann S, et al. Performance characteristics of a transcription-mediated nucleic acid amplification assay for qualitative detection of hepatitis C virus RNA. J Clin Lab Anal 2001; 15:308–313.
38. Elbeik T, Surtihadi J, Destree M, et al. Multicenter evaluation of the performance characteristics of the Bayer VERSANT HCV RNA 3.0 assay (bDNA). J Clin Microbiol 2004; 42:563–569.
39. Morishima C, Chung M, Ng KW, et al. Strengths and limitations of commercial tests for hepatitis C virus RNA quantification. J Clin Microbiol 2004; 42:421–425.
40. Smedile A, Ciancio A, Rizzetto M. Hepatitis D virus. In: Richman DD, Whitley RJ, Hayden FG, eds. Clinical virology. Washington, DC: ASM Press, 2002:1227–1240.
41. Yamashiro T, Nagayama K, Enomoto N, et al. Quantification of the level of hepatitis delta virus RNA in serum, by real-time polymerase chain reaction and its possible correlation with the clinical stage of liver disease. J Infect Dis 2004; 189:1151–1157.
42. Castelnau C, Le Gal F, Ripault MP, et al. Efficacy of peginterferon alpha-2b in chronic hepatitis delta: Relevance of quantitative RT-PCR for follow-up. Hepatology 2006; 44:728–735.
43. Gripon P, Rumin S, Urban S, et al. Infection of a human hepatoma cell line by hepatitis B virus. Proc Natl Acad Sci USA 2002; 99:15655–15660.
44. Lindenbach BD, Evans MJ, Syder AJ, et al. Complete replication of hepatitis C virus in cell culture. Science 2005; 309:623–626.
45. Binn LN, Lemon SM, Marchwicki RH, et al. Primary isolation and serial passage of hepatitis A virus strains in primate cell cultures. J Clin Microbiol 1984; 20:28–33.
46. Divizia M, Gabrieli R, Degener AM, et al. Evidence of hepatitis E virus replication on cell cultures. New Microbiol 1999; 22:77–83.
47. Robertson BH, Averhoff F, Cromeans TL, et al. Genetic relatedness of hepatitis A virus isolates during a community-wide outbreak. J Med Virol 2000; 62:144–150.
48. Norder H, Courouce AM, Coursaget P, et al. Genetic diversity of hepatitis B virus strains derived worldwide: Genotypes, subgenotypes, and HBsAg subtypes. Intervirology 2004; 47:289–309.
49. Janssen HLA, van Zonneveld M, Senturk H, et al. Pegylated interferon alfa-2b alone or in combination with lamivudien for HBeAg-positive chronic hepatitis B: A randomised trial. Lancet 2005; 365: 123–129.
50. Chan HL, Hui AY, Wong ML, et al. Genotype C hepatitis B virus infection is associated with an increased risk of hepatocellular carcinoma. Gut 2004; 53:1494–1498.
51. Livingston SE, Simonetti JP, McMahon BJ, et al. Hepatitis B virus genotypes in Alaska Native people with hepatocellular carcinoma: Preponderance of genotype F. J Infect Dis 2007; 195:5–11.
52. Farci P, Purcell RH. Clinical significance of hepatitis C virus genotypes and quasispecies. Semin Liver Dis 2000; 20:103–126.
53. Lau JY, Davis GL, Prescott LE, et al; Hepatitis Interventional Therapy Group. Distribution of hepatitis C virus genotypes determined by line probe assay in patients with chronic hepatitis C seen at tertiary referral centers in the United States. Ann Intern Med 1996; 124:868–876.
54. Fried MW, Shiffman ML, Reddy R, et al. Peginterferon alfa-2a plus ribavirin for chronic hepatitis C virus infection. N Engl J Med 2002; 347:975–982.
55. Hadziyannis SJ, Sette HJ, Morgan TR, et al. Peginterferon-a2a and ribavirin combination therapy in chronic hepatitis C. Ann Intern Med 2004; 140:346–355.
56. Lednar WM, Lemon SM, Kirkpatrick JW, et al. Frequency of illness associated with epidemic hepatitis A virus infections in adults. Am J Epidemiol 1985; 122:226–233.
57. Zuckerman AJ. The clinical and laboratory features of acute hepatitis in the Royal Air Force. Mon Bull Minist Health Public Health Lab Serv 1965; 24:340–346.

58. Servoss JC, Friedman LS. Serologic and molecular diagnosis of hepatitis B virus. Clin Liver Dis 2004; 8:267–281.
59. Licata A, Di Bona D, Schepis F, et al. When and how to treat acute hepatitis C? J Hepatol 2003; 39:1056–1062.
60. National Institutes of Health. National Institutes of Health consensus development conference statement: Management of hepatitis C: 2002. Hepatology 2002; 5:S1–S20.
61. Macagno S, Smedile A, Caredda F, et al. Momomeric (7S) immunoglobulin M antibodies to hepatitis delta virus in hepatitis type D. Gastroenterol 1990; 98:1582–1586.
62. Mast EE, Krawczynski K, Purdy MA. Hepatitis E: An overview. Annu Rev Med 1996; 47:257–266.
63. Chang TT, Lai CL, Chien RN, et al. Four years of lamivudine treatment in Chinese patients with chronic hepatitis B. J Gastroenterol Hepatol 2004; 19:1276–1282.
64. Dienstag JL, Schiff ER, Wright TL, et al. Lamivudine as initial treatment for chronic hepatitis B in the United States. N Engl J Med 1999; 341:1256–1263.
65. Liaw Y-F, Sung JJY, Chow WC, et al. Lamivudine for patients with chronic hepatitis B and advanced liver disease. N Engl J Med 2004; 351:1521–1531.
66. Olivero A, Ciancio A, Abate ML, et al. Performance of sequence analysis, INNO-LiPA line probe assays and AFFIGENE assays in the detection of hepatitis B virus polymerase and precore/core promoter mutations. J Viral Hepat 2006; 13:355–362.
67. Poynard T, Bedossa P, Opolon P. Natural history of liver fibrosis progression in patients with chronic hepatitis C. Lancet 1997; 349:825–832.
68. Enomoto N, Sakuma I, Asahina Y, et al. Mutation in the nonstructural protein 5A gene and response to interferon in patients with chronic hepatitis C virus 1b infection. N Engl J Med 1996; 334:77–81.
69. He Y, Katze MG. To interfere or to anti-interfere: The interplay between hepatitis C and interferon. Viral Immunol 2002; 15:95–119.
70. Centers for Disease Control. Positive test results for acute hepatitis A virus infection among persons with no recent history of acute hepatitis – United States. MMWR Morb Mortal Wkly Rep 2005; 54: 453–456.
71. CDC. http://www.cdc.gov/hepatitis/Resources/Professionals/Training/Serology/training.htm#one. Accessed December 7, 2009.
72. http://www.cdc.gov/hepatitis/HBV/HBVfaq.htm#general. Accessed December 7, 2009.

20 Viral Hemorrhagic Fevers[1]

Jens H. Kuhn
NIAID Integrated Research Facility at Fort Detrick (IRF-Frederick), Fort Detrick, Frederick, Maryland, and Tunnell Consulting, Inc., King of Prussia, Pennsylvania, U.S.A.

Peter B. Jahrling
NIAID Integrated Research Facility at Fort Detrick (IRF-Frederick), Fort Detrick, Frederick, Maryland, U.S.A.

Sheli R. Radoshitzky
United States Army Medical Research Institute of Infectious Diseases (USAMRIID), Fort Detrick, Frederick, Maryland, U.S.A.

INTRODUCTION

Viral hemorrhagic fevers (VHFs) are acute clinical syndromes caused by diverse viruses that have a distinct vasotropism. Capillaropathy, coagulation abnormalities, fever, hemorrhages, and high lethality are hallmarks of the clinical presentation of VHFs (1,2). Hemorrhages typically occur because of increased endothelial permeability or potentially endothelial destruction due to viral replication. Contrary to widespread belief, however, hemorrhages are almost never severe enough to cause life-threatening hypovolemia (3). Instead, lethal disease outcomes are usually a direct consequence of aberrant innate immune responses (4). Dendritic cells, which are probably an initial target of most VHF-causing viruses, become rapidly impaired. Lymphocytes undergo mass apoptosis, and infected macrophages and other cells release an abundance of diverse cytokines (5). Together, these events lead to decreased directed immune responses to infected cells, altered vascular function, coagulopathy frequently resulting in focal organ necroses due to clogged microvasculature, and, eventually, death due to multiorgan failure (6).

Human VHFs are caused by enveloped and single-stranded RNA viruses classified in the four families *Arenaviridae*, *Bunyaviridae*, *Flaviviridae*, and *Filoviridae* (7). With the exception of dengue viruses, VHF-causing viruses are distinctly endemic to particular geographic locations allowing for conditions to maintain the populations of their arthropod, rodent, insectivore or chiropteran hosts (2). Human infections typically occur through direct contact with infected hosts (bites, scratches, consumption of contaminated meat) or through contact with their excreta and secreta (inhalation of or contact of abrasions with dried or fresh urine and feces) (6). Transmission of the viruses among humans usually occurs by direct contact or shared clinical utensils, whereas aerosol spread of VHF agents is rare during natural outbreaks. VHFs are therefore bona fide nosocomial diseases especially in the mostly underdeveloped countries where they typically occur, as proper disinfection regimens, patient treatment, patient isolation, and the distribution of single-use medical supplies is often impossible (6).

The etiological agents of human VHFs are priority research pathogens in many nations (see http://www3.niaid.nih.gov/topics/BiodefenseRelated/Biodefense/research/CatA.htm) because increased air travel and higher population densities heighten the risk of their inadvertent import into and spread among nonendemic areas. More importantly, most of the VHF-causing agents have the potential to be used for the construction of biological weapons because they can easily be grown in tissue culture to relatively high titers, are quite stable and highly infectious as respirable aerosols, and cause high morbidity and lethality in target populations (8). Overall research progress on VHF-causing pathogens is impeded as many are classified as WHO Risk Group III and IV agents, thereby requiring high containment

[1] Opinions, interpretations, conclusions, and recommendations are those of the authors and are not necessarily endorsed by the U.S. Department of the Army, U.S. Department of Defense or the Department of Health and Human Services.

laboratories (US: BSL-3 and BSL-4) (9). For most VHF agents, FDA-licensed vaccines are unavailable and treatment is limited to supportive care. Although specific and sensitive diagnostic tests are increasingly available, most physicians cannot diagnose VHF agents routinely, especially because their initial clinical presentation is easily confused with much more common infectious diseases, such as shigellosis or malaria (6).

This chapter will review the general characteristics of and specific diagnostic tests for human VHF agents requiring BSL-4 procedures in the United States (9).

EPIDEMIOLOGY

Arenaviral Hemorrhagic Fevers

The family *Arenaviridae* contains one genus, *Arenavirus*, which comprise at least 30 species (10–12). Serologically and genomically, arenaviruses cluster into two phylogenetic lineages, which are commonly referred to as the Old and New World arenaviruses (10,11,13,14). The latter group is further subdivided into the four clades A, B, A/Rec, and C. Arenaviral hemorrhagic fevers are caused by two Old World arenaviruses (Lassa and 'Lujo'), and by five New World clade B pathogens (Chapare, Guanarito, Junín, Machupo, and Sabiá) (10–14). The viruses chronically infect specific rodents of the Muroidea superfamily (Old World arenaviruses: family Muridae; New World arenaviruses: family Cricetidae) without development of overt clinical symptoms, and the geographic distribution of arenaviruses is determined by the range of these rodents. Outbreaks of arenaviral hemorrhagic fevers are usually related to perturbations in ecosystems that bring humans in contact with rodents or rodent excreta or secreta. This transmission mode explains why farm workers are at particular risk of arenavirus infection.

Old World Arenaviral Hemorrhagic Fevers

Lassa fever, caused by Lassa virus (LASV), was initially described in 1969 when an American missionary nurse fell ill in Lassa, Nigeria, and started a chain of nosocomial infections that extended from health care workers in Africa to laboratory workers in the United States (15). Subsequent major Lassa fever outbreaks occurred all over West Africa (Liberia, Republic of Guinea, and Sierra Leone) and could be traced back to contact with the host of the virus, the African soft-furred rat (*Praomys natalensis*). Approximately 100,000 to 500,000 people are infected with LASV per year. Of those, 5000 to 10,000 die, and approximately 30,000 suffer from long-term sequelae such as deafness (16,17). Lassa fever has occasionally been imported to Canada, Europe, Japan, and the United States by travelers from West Africa (18–22), but in most cases these were single, albeit often fatal, infections and no person-to-person transmission occurred.

'Lujo virus (LUJV)' is a recently described arenavirus that branches off the ancestral node of the Old World arenaviruses. It was discovered in 2008 in a man from Zambia, who infected four health care workers during his hospital stay in South Africa. Four out of the five infections were fatal (12,23).

New World Arenaviral Hemorrhagic Fevers

Argentinian (Junín) hemorrhagic fever (AHF) cases were first reported from Argentina in 1955 (24). Junín virus (JUNV), the etiological agent of the disease (25) is maintained by the drylands vesper mouse (*Calomys musculinus*) (26). AHF is a seasonal disease, with the peak of infections recorded during the corn-harvesting season in Argentina. Infections typically occur by inhalation of blood mist produced from rodents caught in mechanical harvesting machines, and direct person-to-person transmission is rare (27). Approximately 30,000 cases of AHF have been recorded, but case numbers have dropped to <100/year after the distribution of a live-attenuated (non-FDA-approved) vaccine. In the absence of treatment, AHF lethality reaches 20% (28).

Bolivian (Machupo) hemorrhagic fever (BHF) was recognized among people in Bolivia in 1959. The etiological agent, Machupo virus (MACV), was isolated from patients and from healthy large vesper mice (*Calomys callosus*) (29). Between 1962 and 1964, more than 1000 patients became infected and 180 died. Control of *C. callosus* terminated the outbreaks (30). Transmission of MACV occurs mainly through food and water contaminated with infected rodent excreta during harvest season. Human-to-human transmission is atypical.

Guanarito virus (GTOV) emerged in 1989 as the cause of "Venezuelan hemorrhagic fever" rampant among settlers that had moved into a cleared forest area in Venezuela. By 1991, 104 infections had been reported, ~25% of which were fatal (31,32). The cotton rat (*Sigmodon alstoni*) is the principle host of GTOV. Reminiscent of JUNV, GTOV predominantly infects agricultural workers during harvesting season. The latest recognized outbreak occurred in 2002, bringing the total case number of all GTOV infections since 1989 to ~200 (26,33).

Sabiá virus (SABV) caused one fatal case of "Brazilian hemorrhagic fever" in Brazil in 1990 (34), followed by two nonfatal laboratory infections in Brazil in 1992 and in the United States in 1994 (35). The host of SABV is unknown.

'Chapare virus' caused a small VHF outbreak in Bolivia in 2003. Initial studies suggest that this virus, whose host remains unknown, is most closely related to SABV (14).

Bunyaviral Hemorrhagic Fevers

The family *Bunyaviridae* contains five genera, *Hantavirus, Nairovirus, Orthobunyavirus, Phlebovirus,* and *Tospovirus* and more than 100 species (36). VHF outbreaks could be traced back to viruses of all genera, with the exception of *Tospovirus*. However, only one virus, the nairovirus Crimean-Congo hemorrhagic fever virus, is a WHO Risk Group IV virus.

Crimean-Congo Hemorrhagic Fever

The first medical description of Crimean-Congo hemorrhagic fever dates back to 1945, when Soviet researchers described an unusual number of deaths among troops and peasants of the Steppe Region of western Crimea (now in Ukraine) (37). A similar illness was noted in a child in the Belgian Congo (now Democratic Republic of the Congo) in 1956, and it was later established that the same virus, Crimean-Congo hemorrhagic fever virus (CCHFV), was the underlying cause of all these cases (38). CCHFV is endemic to Asia, Eastern Europe, the Middle East, and central and southern Africa (39), where it is predominantly transmitted through the bite of ticks (*Hyalomma* sp.), by contact with infected animals or animal products, or through blood or bodily secretions of infected people (39). CCHFV is frequently associated with explosive nosocomial outbreaks in hospitals.

Flaviviral Hemorrhagic Fevers

At least eight flaviviruses (family *Flaviviridae*, genus *Flavivirus*) can cause VHFs in humans, among them dengue viruses 1–4 and yellow fever virus. Three flaviviruses, 'Alkhurma hemorrhagic fever virus (AHFV),' Kyasanur Forest disease virus (KFDV), and Omsk hemorrhagic fever virus (OHFV) are WHO Risk Group IV viruses (40).

KFDV is endemic only in India's Karnataka State, where it was discovered in 1957 after a VHF epizootic among Hanuman langurs (*Semnopithecus entellus*) and Bonnet macaques (*Macaca radiata*) (41), and possibly in China (42). KFDV is predominantly transmitted by ticks of the species *Haemaphysalis spinigera* to rodents, insectivores, birds, and humans (43). KFDV infected approximately 2500 humans between 1999 and 2004, and its lethality ranges from 2% to 10% (43).

'AHFV' was isolated in 1995 from Saudi Arabian VHF patients (44) and recognized as a close relative of KFDV (45). It is probably maintained in sand tampans (*Ornithodoros savignyi*) (46). Human infections occurred after contamination of skin wounds with the blood of infected sheep and camels and by ingestion of unpasteurized contaminated milk. Approximately 20 human infections have been described to date with a lethality of >30% (46).

OHFV is endemic only to a small area around Omsk in Siberian Russia. It is transmitted by *Dermacentor reticulates* ticks among water voles (*Arvicola terrestris*) and muskrats (*Ondatra zibethica*) (47,48). Humans become infected after tick bites, contact with bodily fluids of infected animals, and by consumption of milk from infected caprids and ovids (47,48). Omsk hemorrhagic fever has a lethality of <3%.

Filoviral Hemorrhagic Fevers

The family *Filoviridae* contains two genera, *Marburgvirus* and *Ebolavirus*; five of the six filoviruses [Lake Victoria marburgvirus (MARV), 'Bundibugyo ebolavirus (BEBOV),' Côte d'Ivoire ebolavirus (CIEBOV), Sudan ebolavirus (SEBOV), and Zaire ebolavirus (ZEBOV)] cause

VHFs in humans that are fatal in ~72% of the cases (49,50). MARV was discovered in 1967 among laboratory workers and veterinarians with VHF in Germany and Yugoslavia (51–53). The infections traced back to a supplier of infected African green monkeys (*Chlorocebus aethiops*) in Uganda. SEBOV and ZEBOV were discovered in 1976 during almost simultaneous human VHF outbreaks in Sudan and Zaire (now Democratic Republic of the Congo), respectively (54). CIEBOV was responsible for two deadly epizootics among Western chimpanzees (*Pan troglodytes verus*) and one nonfatal human infection in Côte d'Ivoire in 1994 (55). Finally, 'BEBOV' was discovered in 2007 in Uganda (50), but detailed epidemiological descriptions have not yet been published. Altogether, filoviruses caused ~3000 human cases (49), and killed potentially tens of thousands of great apes (56). The natural hosts of filoviruses remain unknown, although recent findings suggest that frugivorous bats are prime host candidates (57).

CLINICAL PRESENTATION

Lassa Fever

LASV causes a wide spectrum of disease, ranging from subclinical to fatal infection. The average incubation time of Lassa fever in humans is 2 to 16 days. Initial symptoms are unspecific and reminiscent of influenza. Patients experience often sudden onsets of (sometimes biphasic) fever with chills, arthralgia, headaches, myalgia, malaise, and general weakness. With ongoing progression of the disease, patients develop conjunctivitis, sore throats, cough, chest pain, pneumonitis, epigastric pain, nausea, vomiting, and diarrhea. Palatal, pharyngeal, and tonsillar erythemata develop in ~30% of the cases. Facial and truncal flushing, petechiae, purpura, ecchymoses, epistaxis, gastrointestinal and genitourinary bleeding are among the hemorrhagic symptoms that develop occasionally during the late stages of severe infections. Central nervous system symptoms are infrequent findings. They include blurred vision, disorientation, dizziness, convulsions, seizures and coma, and are associated with a poor prognosis. Alopecia, fatigue, tinnitus and partial or total deafness are typical Lassa fever sequelae in reconvalescents, and deafness often is permanent. Death occurs from shock, succeeding bradycardia and hypotension, respiratory insufficiency, or cardiac arrest. Lassa fever is particularly severe in pregnant women, often resulting in the death of the mother and the unborn. The clinical chemistry is characterized by elevated levels of CPK, LDH, and SGOT (15,58,59).

New World Arenaviral Hemorrhagic Fevers

The clinical presentation of the New World arenaviral hemorrhagic fevers is severe and very similar (60,61). Subclinical infections are rare. The incubation periods range from 1 to 2 weeks, followed by gradually increasing fever, malaise, headache, myalgia, epigastric pain, and anorexia over several days. Patients further develop back pains, retroorbital pains accompanied by photophobia, dizziness, coughing, constipation, or mild diarrhea. During the second week of illness, ~15% to 30% of the patients develop hemorrhagic and/or neurologic symptoms, including bleeding from the mucous membranes, petechiae, ecchymoses, bleeding from puncture sites, melena, hematemesis, irritability, lethargy, tremors of the hands and tongue, convulsions, delirium, and coma. Death occurs from shock 7 to 12 days after onset of disease. Convalescence can last weeks and be complicated by fatigue, weakness, dizziness, deafness, and alopecia. The hematological abnormalities of acutely ill patients are characterized by leukocytopenia, thrombocytopenia, and, sometimes, signs of disseminated intravascular coagulation (33,34,62,63).

Crimean-Congo Hemorrhagic Fever

Crimean-Congo hemorrhagic fever develops in four distinct phases (incubation period, prehemorrhagic phase, hemorrhagic phase, and convalescence). The incubation period is as short as 1 to 9 days after tick bites and 5 to 9 days after exposure to infected blood. The onset of the prehemorrhagic phase is sudden, including such nonspecific symptoms as fever, headaches, dizziness, myalgia, photophobia, nausea, sore throat, conjunctivitis, and diarrhea with severe abdominal pain. A probably large number of patients then convalesces and therefore remains undiagnosed. Some patients progress and enter the hemorrhagic phase, which starts on day 5 to 7 post disease onset. Crimean-Congo hemorrhagic fever is the VHF with the most severe bleeding manifestations. Next to petechiae, large hematomas are typical, as well as gastrointestinal and cerebral hemorrhages, hematemesis, hemoptysis, menometrorrhagia, melena, and hematuria. Hepatomegaly in the absence of jaundice and splenomegaly are among the typical findings upon patient examination. Death occurs on days 5 to 14 post onset of disease. Convalescence is sometimes characterized by labile pulse, tachycardia, and deafness (64,65).

Kyasanur Forest Disease, "Alkhurma Hemorrhagic Fever," and Omsk Hemorrhagic Fever

Kyasanur Forest disease begins with an abrupt onset of fever, myalgia, headache, lymphadenopathy, retroorbital pain, conjunctivitis, and vesicular lesions on the upper palate after an incubation period of 3 to 8 days. Hemorrhagic manifestations occur as early as day 3 post disease onset, and include gastrointestinal bleeding, menorrhagia, petechiae, purpura, ecchymoses, and epistaxis. CNS involvement, including tremors, abnormal reflexes, confusion, hemi- or paraparesis, and coma is common. "Alkhurma hemorrhagic fever" and Omsk hemorrhagic fever generally resemble Kyasanur Forest disease, although arthralgia and a more pronounced thrombocytopenia have been observed in patients infected with 'Alkhurma virus.' Omsk hemorrhagic fever patients often present with bronchial pneumonia and fever to absent CNS symptoms (43,44,48,66,67).

Filoviral Hemorrhagic Fever

The very similar clinical presentation of marburgvirus and ebolavirus infections occurs in two phases (49). After an incubation time of 3 to 7 days, patients first present with influenza-like symptoms, such as abdominal pain, anorexia, arthralgia, asthenia, back pain, diarrhea, fever, headaches, enlarged lymph nodes, myalgia, nausea, pyrexia or vomiting. A maculopapular rash usually develops after approximately seven days on the face, buttocks, trunk, or arms, and later generalizes over almost the entire body. Patients then either recover (often with sequelae, such as alopecia, prolonged weight loss, arthralgia, conjunctivitis, loss of vision or hearing, parotitis, psychosis, orchitis, dysesthesias or pericarditis), or progress to the second phase characterized by anuria, hiccups, terminal tachypnea, and hemorrhagic manifestations, such as bleeding from the gums, hematemesis, hemoptysis, melena or hematuria. Neurological involvement is infrequent and involves confusion, convulsions, meningitis, tinnitus, hearing loss, sudden bilateral blindness or dysesthesias. Secondary bacterial and fungal infections are common. Death occurs 8 to 16 days after infection from shock after multiorgan failure (68–70). Clinical chemistry is characterized by elevated levels of sGOT, sGPT, glutamate dehydrogenase, sorbitol dehydrogenase, and γ-GT, indicating liver damage. Creatinine and urea levels increase prior to renal failure, and hypokalemia is typical because of diarrhea and vomiting. Leukopenia with a left shift of the granulocytes is characteristic during the first days of disease, accompanied by severe thrombocytopenia. The second clinical stage is characterized by leukocytosis and a decrease of clotting factors and prolonged thrombin and cephalin times, indicating disseminated intravascular coagulation (49,68–71).

MOLECULAR CHARACTERISTICS

Arenaviruses

Arenavirions are round to pleomorphic enveloped particles that range in size from 50 to 300 nm in diameter. Their noninfectious, ambisense single-stranded RNA genome is bisegmented. Each segment encodes two proteins. The small or S segment (~3.5 kb) encodes the nucleoprotein NP and the spike-protein precursor GPC, and the large or L segment (~7.2 kb) encodes the RNA-dependent RNA polymerase L and the matrix protein Z (72). GPC, which is processed into the three virion surface-associated proteins SSP, GP1, and GP2 (73,74), contains the most important antibody epitopes. Arenaviral replication occurs exclusively in the cytoplasm (72), where viral RNAs can be detected easily by PCR (see below).

Crimean-Congo Hemorrhagic Fever Virus

CCHFV particles are spherical or pleomorphic particles with a diameter of ~80 to 120 nm. They contain a negative-stranded trisegmented genome (~18 kb) that is not infectious. The small (S) RNA encodes the nucleoprotein NP, the medium(M) RNA encodes the two virion spike

Figure 1 Zaire ebolavirus particles harvested from supernatant of infected MA-104 cells. The virions averaged 1280 to 1300 nm in length and were characteristically longer than Lake Victoria marburgvirus particles similarly treated. Preparation was negatively contrasted with 1% phosphotungstic acid (pH = 6.6) (×49,400).

proteins Gn and Gc and a nonstructural protein (NSm), and the large (L) RNA encodes the RNA-dependent RNA polymerase L (75). NP and the spike proteins are exploited primarily in diagnostic assays (see below).

Flaviviruses

Flavivirions are spherical particles that are 50 nm in diameter. They contain an infectious monopartite and positive-stranded RNA (~11 kb), which encodes a single polyprotein that is co- and post-translationally cleaved into the three structural proteins C, prM, and E, and seven nonstructural proteins (40,45,76). E represents the flaviviral surface spike protein and is the predominant surface-exposed structure. Flavivirus replication occurs in the cytoplasm. Virions assemble in the endoplasmic reticulum and are then released from the cell by exocytosis (40,77).

Filoviruses

Filovirions are the only human filamentous viruses, which facilitates their identification by electron microscopy (see Figure 1). The particles are often torus-, horseshoe- or 6-shaped and are between 795 and 1086 nm long and ~80 nm in diameter (78). Filoviral genomes (~19 kb) are monopartite, single-stranded, negative-sense RNAs (79) that are noninfectious and contain seven genes (49). The filoviral genomes encode seven structural proteins: nucleoprotein (NP), polymerase cofactor (VP35), matrix protein VP40, spike protein precursor (preGP), transcription factor (VP30), secondary matrix protein (VP24), and RNA-dependent RNA polymerase (L) (80). In contrast to marburgviruses, ebolaviruses synthesize two additional, nonstructural, proteins from the GP gene, sGP and ssGP (81). preGP is post-translationally cleaved into the subunits GP_1 and GP_2, which stay associated as a heterodimer that then trimerizes and becomes exposed on the virion surface (82).

CLINICAL DIAGNOSIS OF VIRAL HEMORRHAGIC FEVERS

VHF should be taken into consideration for patients presenting with a severe febrile illness and vascular abnormalities (subnormal blood pressure, postural hypotension, petechiae, hemorrhagic diathesis, flushing of the face and chest, nondependent edema), who have recently traveled to rural areas especially in Africa, South America, or Asia, or when intelligence suggests a biological attack (8,83). Obtaining a detailed travel history is crucial for establishing a preliminary probable diagnosis since most VHF pathogens are endemic to particular areas only (83,84). Likewise, it is critical to establish the patient's exposure to animals since a history of tick or mosquito bites or the exposure to rodents or bats may steer suspicions toward particular VHFs (84). A purely clinical diagnosis of particular VHF is next to impossible as pathognomonic markers are absent and the individual VHFs present with rather unspecific symptoms. Numerous viral (influenza, measles, fulminant hepatitides), bacterial (gram-negative septicemias, plague, rickettsioses, typhoid fever), parasitic (malaria), and even fungal infections (histoplasmosis) mimic those symptoms, as do certain intoxications (snake envenomation) and noninfectious diseases (leukemias, hemolytic-uremic syndrome, Kawasaki disease) (84,85). Therefore, laboratory diagnosis is necessary to confirm VHF (86).

LABORATORY DIAGNOSIS OF VIRAL HEMORRHAGIC FEVERS

Safety Concerns

All viruses discussed in this chapter are WHO Risk Group IV pathogens. Consequently, they must be handled in maximum-containment facilities (biosafety level 4 in the United States) (9). However, these viruses are most often encountered in nature where such facilities are not available. Therefore, at a minimum, barrier-nursing procedures (caps, gowns, disposable gloves, face masks with respirators) must be enforced among first responders to an outbreak or clinical health care personnel in such areas (84). Manipulation of clinical samples (sera, tissues, etc.) should be minimized in the field and postponed until they have been shipped to a maximum-containment facility. Samples should be packaged according to the recommendations of the International Air Transport Association (IATA) and be shipped to a laboratory of the International High Security Laboratory Network (IHSLN). Use of sharps (especially needles) should be avoided if possible both in the laboratory as well as in the field since the infectious dose of most VHF agents is minimal and accidental needle pricks often prove fatal. Likewise, all procedures potentially prone to aerosol production (centrifugation, autopsies, necropsies) should be minimized (9,87,88). Whenever possible, virus should be inactivated. Addition of Triton X-100, chloroform, diethyl ether, sodium deoxycholate, SDS, acetic acid, or β-propionolactone is efficient in reducing or abolishing viral activity since all discussed hemorrhagic fever-causing viruses are enveloped. Such inactivation is recommended prior to serological assays, whereas heat inactivation (60°C for 1 hour) permits the safe determination of serum electrolytes, blood urea nitrogen, and creatinine in clinical samples in the field. γ-irradiation (^{60}Co) on dry ice is the ideal inactivation method but is rarely available in VHF outbreak areas (89–91).

Collection of Clinical Samples

For virus isolation, serum, plasma, or, less ideally, whole blood should be collected from VHF patients during the acute febrile stage, and frozen on dry ice or in liquid nitrogen vapor as lower temperatures lead to rapid viral inactivation. The choice of anticoagulant is important as citrate interferes with IFA, EDTA may interfere with ELISA, and both citrate and oxalate cause nonspecific cytopathic effects in cells routinely used for virus isolation. Heparin is the anticoagulant of choice for antibody and antigen detection assays, and EDTA for PCR. LASV, but not MACV, is also relatively easily isolated from throat washings for several weeks after disease onset and less frequently from urine for approximately 30 days post onset. Filoviruses have been isolated from throat washings, saliva, urine, semen, and anterior eye fluid even during the convalescence phase (49). Throat washings and urine samples should be mixed with buffered diluents containing 10% FBS or rabbit serum prior to freezing.

Tissues may be fixed in formaldehyde or embedded in paraffin for virus detection by immunohistochemical (IHC) methods (92,93). Impression smears of tissues infected with VHF agents may be fixed by immersion in cold acetone and stored frozen at −20°C until examination for antigen with IFA or ELISA. RNA extraction media containing guanidinium thiocyanate generally inactivate VHF viruses, but the extraction should be conducted in a laminar flow hood as precaution. It is also important to remember that flaviviral genomic RNA is infectious by itself, requiring caution when isolating or handling it.

Laboratory Diagnosis

Virus Isolation

Cell culture

Most VHF patients are viremic at presentation. Virus isolation should always be the ultimate goal of all diagnostic efforts and can be achieved by inoculating African green monkey kidney cells (CV-1, Vero clones, MA104) with serially diluted fresh or frozen serum/plasma/whole blood or clarified tissue homogenates (biopsies/necropsies) in the case of all discussed VHF agents. The cells should be observed for the development of cytopathic effects and for the presence of viral antigen using IFA or ELISA. Vero cells inoculated with high-titered samples express detectable antigen within 1 to 2 days after inoculation, whereas cells inoculated with low-titered samples may accumulate antigen within one week. Blind passaging of tissue culture supernatant is called for in the absence of antigen after one week to confirm the absence of virus. Cocultivation of hypaque-ficol-separated PBMCs with Vero cells may result in enhanced antigen expression (94).

Animal inoculation

VHF agent isolation may be facilitated by inoculation of animals when cell culture is impossible or difficult, and sometimes allows to establish a preliminary diagnosis of an agent based on animal susceptibility (95). For instance, MACV and JUNV can be isolated by inoculation of newborn hamsters and mice. Peripheral injection of Old and New World hemorrhagic fever arenaviruses is successful in young adult guinea pigs (death occurs after 7–18 days) (96–98). Guinea pigs are also the animals of choice for the isolation of filoviruses, although they mostly produce an exclusively febrile (nonlethal) disease. Serial passage leads to increasingly severe and finally fatal disease. ZEBOV, but not the other filoviruses, is pathogenic for newborn mice after intracranial inoculation (99–101). Suckling mice are the traditional animals for isolation of CCHFV, and KFDV and OHFV can be isolated in adult BALB/c mice (47,102,103).

Electron Microscopy (EM)

Electron microscopy, performed on heparinized blood, urine, or tissue culture supernatant, is most useful for the diagnosis of filoviruses, because these agents are the only human viral pathogens with a filamentous morphology (104) (see Figure 1). Filoviruses and arenaviruses also induce morphologically unique intracytoplasmic viral inclusions that can be differentiated by EM. Immunoelectron microscopy (IEM), employing specific murine monoclonal antibodies or guinea pig polyclonal sera, can be used for the diagnosis of individual filoviruses (104), which more or less resemble each other by eye. IEM has also been employed for the diagnosis of arenavirus infections, but the spherical morphology of arenaviruses, bunyaviruses, and flaviviruses is not sufficiently characteristic to permit a definite identification based on EM alone.

Detection of Antigen

Antigen capture enzyme-linked immunosorbent assays (ELISAs)

Sensitive and specific antigen capture ELISA systems have been successfully applied for the detection of arenaviral (105–107), CCHFV (108), and filoviral antigens (109,110) in γ-irradiated or β-propionolactone-inactivated viremic sera, tissue culture supernatants, urine, and throat washes. Plate wells are coated with a mixture of antibodies and incubated with sample in fourfold dilutions in SerDil. Virus-specific polyclonal rabbit antisera are added to the wells, as such sera are more sensitive and demonstrate higher avidity than monoclonal antibodies. After incubation, the wells are exposed to antirabbit IgG and antigen detected using, for instance, the horseradish peroxidase system. Samples are considered positive if the OD_{410} exceeds the mean plus 3 standard deviations for the normal controls. The threshold sensitivity is $\sim$1.3–3.2 $\times$ 10^2 pfu/mL—sufficient for the detection of viral antigen in most clinical samples (105,111–113).

Immunohistochemistry (IHC)

Immunohistochemical methods for the detection of arenaviral and filoviral antigens have become increasingly important (92,93). Infected tissue embedded in paraffin is used as starter material. The paraffin block is sectioned and the sections mounted on silane-coated slides. The sections are then deparaffinized, hydrated, digested with protease, and stained with immune sera or monoclonal antibody cocktails and subsequently with biotinylated antiserum. The biotin probe is then detected using the streptavidin-alkaline phosphatase system (92,114). IHC is exceptionally useful for the detection of filoviral antigen in skin biopsies (92) and allows for the retrospective diagnosis of filovirus infections in archived tissue samples (115). A sensitive IHC assay is also available for the detection of CCHFV (116), but has not been widely applied yet.

Detection of Virus-Specific Antibodies

Indirect fluorescence assay (IFA)

Although increasingly being replaced by ELISA and other diagnostic tests for VHF agents, IFA remains popular because of its simplicity. For IFA, "spot slides" can be established with uninfected control and virus-infected cells, or cells expressing recombinant antigen. Patient (or animal) sera have to be diluted serially (1:4–1:8 and higher) and incubated with the spots, followed by incubation with appropriate reporter-conjugated antibodies. Most experts consider the endpoint to be the highest dilution producing "typical" fluorescence that is clearly positive relative to uninfected cells, but this interpretation is subjective. Specific and sensitive IFA systems have been developed for the detection of IgG and IgM antibodies against arenaviruses (107,117,118), CCHFV (119–121), OHFV (122), and filoviruses (123). In the case of filoviruses, backgrounds may be high, so that dilutions of 1:64 or 1:80 are adopted as cut-off titers.

Antibody capture enzyme-linked immunosorbent assays (ELISAs)

ELISAs have mostly replaced the more subjective IFA tests. For IgG detection, these systems employ γ-irradiation-inactivated lysates of virus-infected cells or cells expressing recombinant antigen coated into microwell plates. The wells are immersed in test sera or plasma in serial fourfold dilution usually starting at 1:100, followed by incubation with species-specific secondary antibodies coupled to, for instance, horseradish peroxidase. Samples are considered positive if the OD_{410} exceeds the mean plus 3 standard deviations for the normal serum controls. For IgM detection, plates are coated with antihuman IgM and incubated with the clinical sample. Then, inactivated viral cell-slurry antigen is added to the well and the antigen detected using polyclonal antiserum. ELISAs are the diagnostic test of choice for LASV (105) and are available for human and animal samples potentially infected with JUNV and MACV (124–126). Virtually all Lassa fever patients can be diagnosed within hours of hospital admission using a combination of Lassa antigen capture ELISA and Lassa IgM and IgG ELISA (105,127). In the case of CCHFV and LASV, the most promising ELISAs are based on recombinant NP (107,120,128,129). An ELISA system is also available for the detection of antibodies to KFDV (130).

Plaque-reduction neutralization tests (PRNT)

Plaque-reduction neutralization tests are the most specific of the serological tests for arenavirus infections. The principle of PRNT is to test for the presence of virus-neutralizing antibodies in a patient's serum by exposing it to a virus-containing culture followed by a standard plaque assay to quantify the virus. The higher the titer of neutralizing antibodies, the fewer plaques the virus will cause in a cell lawn. The generation of neutralizing antibodies in human sera often requires many weeks, but they do persist for years, therefore allowing for testing of convalescent sera. In the case of LASV, test sera are diluted 1:10 in medium containing a complement source (10% guinea pig serum) and mixed with serial dilutions of challenge virus because LASV-neutralizing antibody activity is rapidly lost upon dilution. Neutralizing antibody titers are expressed as a $\log_{10}$ neutralization index (LNI), defined as [($\log_{10}$ pfu in control) minus ($\log_{10}$ pfu in test serum)]. In the case of New World hemorrhagic fever arenaviruses, sera are serially diluted, whereas the virus titer is kept constant. The serum dilution calculated (by probit analysis) to reduce the control number of plaques by 50% or 80% ($PRNT_{50}$ or $PRNT_{80}$) is usually taken as the endpoint. $PRNT_{80}$ is used to distinguish JUNV from MAVC. Vero cells are used for the plaque assay for all arenaviruses. PRNT for filoviruses have not yet proven useful.

Detection of Nucleic Acids

Reverse transcription polymerase chain reaction (RT-PCR)

Detection of viral genomic RNA by reverse transcription polymerase chain reaction (RT-PCR) is evolving to become the gold standard of VHF agent diagnosis. RT-PCR is relatively easy to perform, highly sensitive and specific, and can be performed on virus-inactivated specimens. It is particularly useful in cases where isolation of the infectious virus is difficult. Classical RT-PCR begins with the extraction of viral RNA from tissues using commercially available kits. The RT and amplification steps are ideally performed in one-tube reactions, again using commercially available systems. Primers for both reactions are designed in silico against either particular virus strains (targeting known regions in highly diverse genes, such as those coding for the viral envelope glycoproteins) or entire virus groups (targeting regions conserved in all viruses of one group, such as VP40 of filoviruses). A confirmatory test, such as virus isolation, serology or a second RT-PCR targeting a different region, should be performed after a positive RT-PCR test to rule out the only notorious drawback of RT-PCR, cross contamination. The standard threshold for detection of RT-PCR is 10^5 genomic-sense RNA copies/mL (131–133). Real-time RT-PCR assays, which are more sensitive than standard RT-PCR protocols, have been established for the detection and quantification of LASV, CCHFV, and filoviruses in clinical samples (112,134,135). The assays are based on a Superscript RT/Platinum Taq polymerase mixture, and rely on the incorporation of a viral sequence-specific fluorogenic probe into the amplified product and its detection through 5′ nuclease action or fluorescence resonance energy transfer (FRET) using a LightCycler instrument. Alternatively, fluorescent dyes, such as SYBR Green I, intercalate nonspecifically into the amplified product (112,134,135). The 95% detection limit of some of these assays is in the range of 1545 to 2835 viral genome equivalents/mL of serum (134). Multiplex PCR assays can be used to differentiate up to 10 different VHF pathogens (136). RT-PCR has been successfully used for the detection of CCHFV in clinical samples, such as saliva and urine (137) or acute-phase sera (138–140), and TaqMan minor groove-binding protein assays are under evaluation (141). Simple RT-PCR assays can detect arenaviruses in whole blood samples (142), and real-time RT-PCRs are available for the detection and differentiation of New World hemorrhagic fever arenaviruses (143).

In situ hybridization (ISH)

ISH has been successfully used to identify ZEBOV and CCHFV infection in tissue samples (116,144), but no such assays have been reported for hemorrhagic fever arenaviruses or flaviviruses.

CONCLUSIONS

The rapid diagnosis of VHF and the particular etiological agent is critical in the field to allow for adequate outbreak intervention, including quarantine measures, education of health care providers and affected population, and initiation of treatment regimens. IFA on slides spotted with inactivated viral antigens, as well as ELISA, have been used in field hospitals to detect seroconversion in humans, but the development of antibodies is often to slow or in many cases even absent because of virus-induced immunosuppression. Therefore, once specific IgG or IgM antibodies are detected, one can relatively safely assume an ongoing infection, whereas the absence of antibodies does not rule it out. However, one should not forget that cross-reactive antibodies can complicate matters, especially regarding filoviruses, against which antibodies were found around the world even in areas where filoviruses are not known to be endemic (49). Consequently, field diagnosis today relies on the detection of viral antigen, with antigen capture ELISA being the preferred method. PCR assays are increasingly used to detect viral nucleic acids even in the field, and provide a powerful tool to confirm ELISA results. The final diagnosis can then be confirmed in maximum-containment facilities by PRNT, ISH, IHC, and EM, and, finally, virus isolation. At this point in time, diagnostics for hemorrhagic fever arenaviruses and filoviruses are most advanced, whereas improvements are needed for CCHFV and hemorrhagic fever-causing flaviviruses.

REFERENCES

1. Gajdusek DC. Virus hemorrhagic fevers. Special reference to hemorrhagic fever with renal syndrome (epidemic hemorrhagic fever). J Pediatr 1962; 60(6):841–857.
2. Smorodintsev AA, Kazbintsev LI, Chudakov VG. Virus Hemorrhagic Fevers [Russian]. Leningrad, Leningrad Region, USSR: Gosudarstvennoe Izdatelstvo Meditsinskoi Literatury, 1963.
3. Aleksandrowicz P, Wolf K, Falzarano D, et al. Viral haemorrhagic fever and vascular alterations. Hämostaseologie 2008; 28(1–2):77–84.
4. Zampieri CA, Sullivan NJ, Nabel GJ. Immunopathology of highly virulent pathogens: insights from Ebola virus. Nat Immunol 2007; 8(11):1159–1164.
5. Mohamadzadeh M, Chen L, Olinger GG, et al. Filoviruses and the balance of innate, adaptive, and inflammatory responses. Viral Immunol 2006; 19(4):602–612.
6. Marty AM, Jahrling PB, Geisbert TW. Viral hemorrhagic fevers. Clin Lab Med 2006; 26(2):345–386, viii.

7. Howard CR. Viral Haemorrhagic Fevers, Vol 11. Amsterdam, The Netherlands: Elsevier, 2005.
8. Borio L, Ingelsby T, Peters CJ, et al. Hemorrhagic fever viruses as biological weapons—medical and public health management. J Am Med Assoc 2002; 287(18):2391–2405.
9. US Department of Health and Human Services, Centers for Disease Control and Prevention, National Institutes of Health. Biosafety in Microbiological and Biomedical Laboratories (BMBL), 5th ed. Washington, DC: US Government Printing Office, 2007. HHS Publication No. (CDC) 93–8395.
10. Charrel RN, de Lamballerie X, Emonet S. Phylogeny of the genus Arenavirus. Curr Opin Microbiol 2008; 11(4):362–368.
11. Salvato MS, Clegg JCS, Buchmeier MJ, et al. Family Arenaviridae. In: Fauquet CM, Mayo MA, Maniloff J, et al., eds. Virus Taxonomy—Eighth Report of the International Committee on Taxonomy of Viruses. San Diego, CA: Elsevier/Academic Press, 2005:725–733.
12. Briese T, Paweska JT, McMullan LK, et al. Genetic detection and characterization of Lujo virus, a new hemorrhagic fever-associated arenavirus from southern Africa. PLoS Pathog 2009; 5(5): e1000455.
13. Bowen MD, Peters CJ, Nichol ST. The phylogeny of New World (Tacaribe complex) arenaviruses. Virology 1996; 219(1):285–290.
14. Delgado S, Erickson BR, Agudo R, et al. Chapare virus, a newly discovered arenavirus isolated from a fatal hemorrhagic fever case in Bolivia. PLoS Pathog 2008; 4(4):e1000047.
15. Frame JD, Baldwin JM Jr., Gocke DJ,et al. Lassa fever, a new virus disease of man from West Africa. I. Clinical description and pathological findings. Am J Trop Med Hyg 1970; 19(4):670–676.
16. Ogbu O, Ajuluchukwu E, Uneke CJ. Lassa fever in West African sub-region: an overview. J Vector Borne Dis 2007; 44(1):1–11.
17. McCormick JB, Webb PA, Krebs JW, et al. A prospective study of the epidemiology and ecology of Lassa fever. J Infect Dis 1987; 155(3):437–444.
18. Holmes GP, McCormick JB, Trock SC, et al. Lassa fever in the United States. Investigation of a case and new guidelines for management. N Engl J Med 1990; 323(16):1120–1123.
19. Mahdy MS, Chiang W, McLaughlin B, et al. Lassa fever: the first confirmed case imported into Canada. Can Dis Wkly Rep 1989; 15(39):193–198.
20. Haas WH, Breuer T, Pfaff G, et al. Imported Lassa fever in Germany: surveillance and management of contact persons. Clin Infect Dis 2003; 36(10):1254–1258.
21. Hirabayashi Y, Oka S, Goto H, et al. An imported case of Lassa fever with late appearance of polyserositis. J Infect Dis 1988; 158(4):872–875.
22. Schmitz H, Kohler B, Laue T, et al. Monitoring of clinical and laboratory data in two cases of imported Lassa fever. Microbes Infect 2002; 4(1):43–50.
23. Paweska JT, Sewlall NH, Ksiazek TG, et al. Nosocomial outbreak of novel arenavirus infection, southern Africa. Emerg Infect Dis 2009; 15(10):1598–1602.
24. Arribalzaga RA. New epidemic disease due to unidentified germ: nephrotoxic, leukopenic and enanthematous hyperthermia. Dia Med 1955; 27(40):1204–1210 [Spanish].
25. Parodi AS, Greenway DJ, Rugiero HR, et al. Concerning the epidemic outbreak in Junin. Dia Med 1958; 30(62):2300–2301 [Spanish].
26. Charrel RN, de Lamballerie X. Arenaviruses other than Lassa virus. Antiviral Res 2003; 57(1–2):89–100.
27. Maiztegui JI. Clinical and epidemiological patterns of Argentine haemorrhagic fever. Bull World Health Organ 1975; 52(4–6):567–575.
28. Enria DA, Feuillade MR. Argentine haemorrhagic fever (Junin virus, Arenaviridae): a review on clinical, epidemiological, ecological, treatment and preventive aspects of the disease. In: Travassos de Rosa APA, Vasconcelos PFC, Travassos de Rosa JFS, eds. An Overview of Arbovirology in Brazil and Neighboring Countries. Belem, Brazil: Instituto Evandro Chaggas, 1998:219–232.
29. Johnson KM, Wiebenga NH, Mackenzie RB, et al. Virus isolations from human cases of hemorrhagic fever in Bolivia. Proc Soc Exp Biol Med 1965; 118:113–118.
30. Peters CJ, Kuehne RW, Mercado RR, et al. Hemorrhagic fever in Cochabamba, Bolivia, 1971. Am J Epidemiol 1974; 99(6):425–433.
31. Salas R, de Manzione N, Tesh RB, et al. Venezuelan haemorrhagic fever. Lancet 1991; 338(8774):1033–1036.
32. Tesh RB, Jahrling PB, Salas R, et al. Description of Guanarito virus (Arenaviridae: Arenavirus), the etiologic agent of Venezuelan hemorrhagic fever. Am J Trop Med Hyg 1994; 50(4):452–459.
33. de Manzione N, Salas RA, Paredes H, et al. Venezuelan hemorrhagic fever: clinical and epidemiological studies of 165 cases. Clin Infect Dis 1998; 26(2):308–313.
34. Lisieux T, Coimbra M, Nassar ES, et al. New arenavirus isolated in Brazil. Lancet 1994; 343(8894): 391–392.

35. Barry M, Russi M, Armstrong L, et al. Brief report: treatment of a laboratory-acquired Sabia virus infection. N Engl J Med 1995; 333(5):294–296.
36. Nichol ST, Beaty BJ, Elliott RM, et al. Family Bunyaviridae. In: Fauquet CM, Mayo MA, Maniloff J, et al., eds. Virus Taxonomy–Eighth Report of the International Committee on Taxonomy of Viruses. San Diego, CA: Elsevier/Academic Press, 2005:695–716.
37. Chumakov MP. A new tick-borne disease-Crimean hemorrhagic fever. In: Sokolov AE, Chumakov MP, Kolachev AA, eds. Crimean Hemorrhagic Fever (Acute Infectious Capillary Toxicosis) [Russian]. Simferopol, U.S.S.R.: Izdanie Otdel'noj Primorskoj Armii, 1945:13–43.
38. Casals J. Antigenic similarity between the virus causing Crimean hemorrhagic fever and Congo virus. Proc Soc Exp Biol Med 1969; 131(1):233–236.
39. Ergonul O, Whitehouse CA. Crimean-Congo Hemorrhagic Fever—A Global Perspective. Dordrecht, The Netherlands: Springer, 2007.
40. Thiel H-J, Collett MS, Gould EA, et al. Family Flaviviridae. In: Fauquet CM, Mayo MA, Maniloff J, et al., eds. Virus Taxonomy—Eighth Report of the International Committee on Taxonomy of Viruses. San Diego, CA: Elsevier/Academic Press, 2005:981–998.
41. Work TH, Trapido H. Kyasanur forest disease—a new virus disease in India. Summary of preliminary report of investigations of the Virus Research Centre on an epidemic disease affecting forest villagers and wild monkeys of Shimoga District, Mysore. Indian J Med Sci 1957; 11:341–342.
42. Wang J, Zhang H, Fu S, et al. Isolation of Kyasanur Forest disease virus from febrile patient, Yunnan, China. Emerg Infect Dis 2009; 15(2):326–328.
43. Pattnaik P. Kyasanur forest disease: an epidemiological view in India. Rev Med Virol 2006; 16(3):151–165.
44. Qattan I, Akbar N, Afif H, et al. A novel flavivirus: Makkah Region 1994–1996. Saudi Epidemiol Bull 1996; 3(1):1–3.
45. Charrel RN, Zaki AM, Attoui H, et al. Complete coding sequence of the Alkhurma virus, a tick-borne flavivirus causing severe hemorrhagic fever in humans in Saudi Arabia. Biochem Biophys Res Commun 2001; 287(2):455–461.
46. Charrel RN, Fagbo S, Moureau G, et al. Alkhurma hemorrhagic fever virus in Ornithodoros savignyi Ticks. Emerg Infect Dis 2007; 13(1):153–155.
47. Kharitonova NN, Leonov YA. Omsk Hemorrhagic Fever [Russian]. Novosibirsk, USSR: Nauka, 1978.
48. Busygin FF. Omsk hemorrhagic fever—current status of the problem [Russian]. Vopr Virusol 2000; 45:4–9.
49. Kuhn JH. Filoviruses—A Compendium of 40 Years of Epidemiological, Clinical, and Laboratory Studies. Archives of Virology Supplement. Vol 20. Vienna, Austria: Springer, 2008.
50. Towner JS, Sealy TK, Khristova ML, et al. Newly discovered Ebola virus associated with hemorrhagic fever outbreak in Uganda. PLoS Pathog 2008; 4(11):e1000212.
51. Siegert R, Shu H-L, Slenczka W, et al. The etiology of an unknown human infection transmitted by monkeys. Dtsch Med Wochenschr 1967; 92(51):2341–2343.
52. Stille W, Böhle E, Helm E, et al. An infectious disease transmitted from green monkeys. Dtsch Med Wochenschr 1968; 93(12a):572–582.
53. Todorović K, Mocić M, Klašnja R, et al. An unknown virus disease transmitted from infected green-monkeys to men [Serbo-Croatian]. Glas Srp Akad Nauka Umetn Odelj Med Nauka 1969; CCLXXV(22):91–101.
54. Bowen ETW, Lloyd G, Harris WJ, et al. Viral haemorrhagic fever in southern Sudan and northern Zaire. Preliminary studies on the aetiological agent. Lancet 1977; i(8011 Pt 1):571–573.
55. Formenty P, Boesch C, Wyers M, et al. Ebola virus outbreak among wild chimpanzees living in a rain forest of Côte d'Ivoire. J Infect Dis 1999; 179(suppl 1):S120–S126.
56. Walsh PD, Abernethy KA, Bermejo M, et al. Catastrophic ape decline in western equatorial Africa. Nature 2003; 422(6932):611–614.
57. Towner JS, Amman BR, Sealy TK, et al. Isolation of genetically diverse Marburg viruses from Egyptian fruit bats. PLoS Pathog 2009; 5(7):e1000536.
58. Monath TP, Maher M, Casals J, et al. Lassa fever in the Eastern Province of Sierra Leone, 1970–1972. II. Clinical observations and virological studies on selected hospital cases. Am J Trop Med Hyg 1974; 23(6):1140–1149.
59. Knobloch J, McCormick JB, Webb PA, et al. Clinical observations in 42 patients with Lassa fever. Tropenmed Parasitol 1980; 31(4):389–398.
60. Peters CJ. Human infection with arenaviruses in the Americas. Curr Top Microbiol Immunol 2002; 262:65–74.
61. Vainrub B, Salas R. Latin American hemorrhagic fever. Infect Dis Clin North Am 1994; 8(1):47–59.
62. Harrison LH, Halsey NA, McKee KT Jr., et al. Clinical case definitions for Argentine hemorrhagic fever. Clin Infect Dis 1999; 28(5):1091–1094.

63. Stinebaugh BJ, Schloeder FX, Johnson KM, et al. Bolivian hemorrhagic fever. A report of four cases. Am J Med 1966; 40(2):217–230.
64. Ergonul O. Clinical and pathologic features of Crimean-Congo hemorrhagic fever. In: Ergonul O, Whitehouse CA, eds. Crimean-Congo Hemorrhagic Fever—A Global Perspective. Dordrecht, Netherlands: Springer, 2007:207–220.
65. Swanepoel R, Gill DE, Shepherd AJ, et al. The clinical pathology of Crimean-Congo hemorrhagic fever. Rev Infect Dis 1989; 11(suppl 4):S794–S800.
66. Pavri K. Clinical, clinicopathologic, and hematologic features of Kyasanur Forest disease. Rev Infect Dis 1989; 11(suppl 4):S854–S859.
67. Belov GF, Tofaniuk YV, Kurzhukov GP, et al. Clinical-epidemiological characterization of Omsk hemorrhagic fever during 1988 through 1992 [Russian]. Zh Mikrobiol Epidemiol Immunobiol 1995; (4):88–91.
68. Bausch DG, Nichol ST, Muyembe-Tamfum JJ, et al. Marburg hemorrhagic fever associated with multiple genetic lineages of virus. N Engl J Med 2006; 355(9):909–919.
69. Bwaka MA, Bonnet M-J, Calain P, et al. Ebola hemorrhagic fever in Kikwit, Democratic Republic of the Congo: clinical observations in 103 patients. J Infect Dis 1999; 179(suppl 1):S1–S7.
70. Formenty P, Hatz C, le Guenno B, et al. Human infection due to Ebola virus, subtype Côte d'Ivoire: clinical and biologic presentation. J Infect Dis 1999; 179(suppl 1):S48–S53.
71. Sureau PH. Firsthand clinical observations of hemorrhagic manifestations in Ebola hemorrhagic fever in Zaire. Rev Infect Dis 1989; 11(suppl 4):S790–S793.
72. Buchmeier MJ, de la Torre J-C, Peters CJ. Arenaviridae: the viruses and their replication. In: Knipe DM, Howley PM, eds. Fields Virology, Vol 2. 5th ed. Philadelphia: Lippincott Williams & Wilkins, 2007:1791–1827.
73. Gallaher WR, DiSimone C, Buchmeier MJ. The viral transmembrane superfamily: possible divergence of Arenavirus and Filovirus glycoproteins from a common RNA virus ancestor. BMC Microbiol 2001; 1(1):article 1.
74. York J, Romanowski V, Lu M, et al. The signal peptide of the Junin arenavirus envelope glycoprotein is myristoylated and forms an essential subunit of the mature G1-G2 complex. J Virol 2004; 78(19):10783–10792.
75. Schmaljohn CS, Nichol ST. Bunyaviridae. In: Knipe DM, Howley PM, eds. Fields Virology, Vol 2. 5th ed. Philadelphia: Lippincott Williams & Wilkins, 2007:1741–1789.
76. Lin D, Li L, Dick D,et al. Analysis of the complete genome of the tick-borne flavivirus Omsk hemorrhagic fever virus. Virology 2003; 313(1):81–90.
77. Gubler D, Kuno G, Markoff L. Flaviviruses. In: Knipe DM, Howley PM, eds. Fields Virology, Vol 1. 5th ed. Philadelphia: Lippincott Williams & Wilkins, 2007:1153–1252.
78. Geisbert TW, Jahrling PB. Differentiation of filoviruses by electron microscopy. Virus Res 1995; 39(2–3): 129–150.
79. Regnery RL, Johnson KM, Kiley MP. Virion nucleic acid of Ebola virus. J Virol 1980; 36(2):465–469.
80. Sanchez A, Kiley MP, Holloway BP, et al. Sequence analysis of the Ebola virus genome: organization, genetic elements, and comparison with the genome of Marburg virus. Virus Res 1993; 29(3): 215–240.
81. Sanchez A, Trappier SG, Mahy BWJ,et al. The virion glycoproteins of Ebola viruses are encoded in two reading frames and are expressed through transcriptional editing. Proc Natl Acad Sci U S A 1996; 93(8):3602–3607.
82. Volchkov VE, Feldmann H, Volchkova VE, et al. Processing of the Ebola virus glycoprotein by the proprotein convertase furin. Proc Natl Acad Sci U S A 1998; 95(10):5762–5767.
83. Jahrling PB, Marty AM, Geisbert TW. Viral hemorrhagic fevers. In: Dembek ZF, ed. Medical Aspects of Biological Warfare. Washington, D.C.: Borden Institute, Walter Reed Army Medical Center, 2007: 271–310.
84. Gear JHS. The diagnosis of hemorrhagic fever. In: Gear JHS, ed. CRC Handbook of Viral and Rickettsial Hemorrhagic Fevers. Boca Raton, FL: CRC Press, 1988:231–239.
85. Gear JHS. Clinical aspects of African viral hemorrhagic fevers. Rev Infect Dis 1989; 11(suppl 4):S777–S782.
86. Shepherd AJ. Viral hemorrhagic fevers: laboratory diagnosis. In: Gear JHS, ed. CRC Handbook of Viral and Rickettsial Hemorrhagic Fevers. Boca Raton, FL: C.R.C. Press, 1988:241–250.
87. Centers for Disease Control. Management of patients with suspected viral hemorrhagic fever. MMWR Morb Mortal Wkly Rep 1988; 37(suppl 3):1S–16S.
88. Centers for Disease Control and Prevention. Update: management of patients with suspected viral hemorrhagic fever—United States. MMWR Morb Mortal Wkly Rep 1995; 44(25):475–479.
89. Mitchell SW, McCormick JB. Physicochemical Inactivation of Lassa, Ebola and Marburg Viruses and Effect on Clinical Laboratory Analysis. J Clin Microbiol 1984; 20(3):486–489.

90. van der Groen G, Elliott LH. Use of betapropionolactone inactivated Ebola, Marburg and Lassa intracellular antigens in immunofluorescent antibody assay. Ann Soc Belg Med Trop 1982; 62(1): 49–54.
91. Bhagat CI, Lewer M, Prins A, et al. Effects of heating plasma at 56°C for 30 min and at 60°C for 60 min on routine biochemistry analytes. Ann Clin Biochem 2000; 37(6):802–804.
92. Zaki SR, Shieh W-J, Greer PW, et al. A novel immunohistochemical assay for the detection of Ebola virus in skin: implications for diagnosis, spread, and surveillance of Ebola hemorrhagic fever. J Infect Dis 1999; 179(suppl 1):S36–S47.
93. Centers for Disease Control and Prevention. Imported Lassa fever—New Jersey, 2004. MMWR Morb Mortal Wkly Rep 2004; 53(38):894–897.
94. Ambrosio AM, Enria DA, Maiztegui JI. Junin virus isolation from lympho-mononuclear cells of patients with Argentine hemorrhagic fever. Intervirology 1986; 25:97–102.
95. Gowen BB, Holbrook MR. Animal models of highly pathogenic RNA viral infections: hemorrhagic fever viruses. Antiviral Res 2008; 78(1):79–90.
96. Jahrling PB, Smith S, Hesse RA, et al. Pathogenesis of Lassa virus infection in guinea pigs. Infect Immun 1982; 37(2):771–778.
97. Oubina JR, Carballal G, Videla CM, et al. The guinea pig model for Argentine hemorrhagic fever. Am J Trop Med Hyg 1984; 33(6):1251–1257.
98. Hall WC, Geisbert TW, Huggins JW, et al. Experimental infection of guinea pigs with Venezuelan hemorrhagic fever virus (Guanarito): a model of human disease. Am J Trop Med Hyg 1996; 55(1): 81–88.
99. Lupton HW, Lambert RD, Bumgardner DL, et al. Inactivated vaccine for Ebola virus efficacious in guineapig model. Lancet 1980; ii(8207):1294–1295.
100. Connolly BM, Steele KE, Davis KJ, et al. Pathogenesis of experimental Ebola virus infection in guinea pigs. J Infect Dis 1999; 179(suppl 1):S203–S217.
101. Hevey M, Negley D, Pushko P, et al. Marburg virus vaccines based upon alphavirus replicons protect guinea pigs and nonhuman primates. Virology 1998; 251(1):28–37.
102. Tignor GH, Hanham CA. Ribavirin efficacy in an in vivo model of Crimean-Congo hemorrhagic fever virus (CCHF) infection. Antiviral Res 1993; 22(4):309–325.
103. Holbrook MR, Aronson JF, Campbell GA, et al. An animal model for the tickborne flavivirus—Omsk hemorrhagic fever virus. J Infect Dis 2005; 191(1):100–108.
104. Geisbert TW, Jahrling PB. Use of immunoelectron microscopy to show Ebola virus during the 1989 United States epizootic. J Clin Pathol 1990; 43(10):813–816.
105. Bausch DG, Rollin PE, Demby AH, et al. Diagnosis and clinical virology of Lassa fever as evaluated by enzyme-linked immunosorbent assay, indirect fluorescent-antibody test, and virus isolation. J Clin Microbiol 2000; 38(7):2670–2677.
106. Nakauchi M, Fukushi S, Saijo M, et al. Characterization of monoclonal antibodies to Junin virus nucleocapsid protein and application to the diagnosis of hemorrhagic fever caused by South American arenaviruses. Clin Vaccine Immunol 2009; 16(8):1132–1138.
107. Saijo M, Georges-Courbot MC, Marianneau P, et al. Development of recombinant nucleoprotein-based diagnostic systems for Lassa fever. Clin Vaccine Immunol 2007; 14(9):1182–1189.
108. Larichev VF, Manzeniuk IN, Naidenova EV, et al. ELISA and PCR test systems used to detect Crimean-Congo hemorrhagic fever virus. Vopr Virusol 2007; 52(4):43–46.
109. Saijo M, Georges-Courbot MC, Fukushi S, et al. Marburgvirus nucleoprotein-capture enzyme-linked immunosorbent assay using monoclonal antibodies to recombinant nucleoprotein: detection of authentic Marburgvirus. Jpn J Infect Dis 2006; 59(5):323–325.
110. Sherwood LJ, Osborn LE, Carrion R Jr., et al. Rapid assembly of sensitive antigen-capture assays for Marburg virus, using in vitro selection of llama single-domain antibodies, at biosafety level 4. J. Infect Dis 2007; 196(suppl 2):S213–S219.
111. Ksiazek TG, Rollin PE, Williams AJ, et al. Clinical virology of Ebola hemorrhagic fever (EHF): virus, virus antigen, and IgG and IgM antibody findings among EHF patients in Kikwit, Democratic Republic of the Congo, 1995. J Infect Dis 1999; 179(suppl 1):S177–S187.
112. Grolla A, Lucht A, Dick D, et al. Laboratory diagnosis of Ebola and Marburg hemorrhagic fever. Bull Soc Pathol Exot 2005; 98(3):205–209.
113. Shepherd AJ, Swanepoel R, Gill DE. Evaluation of enzyme-linked immunosorbent assay and reversed passive hemagglutination for detection of Crimean-Congo hemorrhagic fever virus antigen. J Clin Microbiol 1988; 26(2):347–353.
114. Jaax NK, Davis KJ, Geisbert TW, et al. Lethal experimental infection of rhesus monkeys with Ebola-Zaire (Mayinga) virus by the oral and conjunctival route of exposure. Arch Pathol Lab Med 1996; 120(2):140–155.

115. Geisbert TW, Jaax NK. Marburg hemorrhagic fever: report of a case studied by immunohistochemistry and electron microscopy. Ultrastruct Pathol 1998; 22(1):3–17.
116. Burt FJ, Swanepoel R, Shieh WJ, et al. Immunohistochemical and in situ localization of Crimean-Congo hemorrhagic fever (CCHF) virus in human tissues and implications for CCHF pathogenesis. Arch Pathol Lab Med 1997; 121(8):839–846.
117. Wulff H, Lange JV. Indirect immunofluorescence for the diagnosis of Lassa fever infection. Bull World Health Organ 1975; 52(4–6):429–436.
118. Peters CJ, Webb PA, Johnson KM. Measurement of antibodies to Machupo virus by the indirect fluorescent technique. Proc Soc Exp Biol Med 1973; 142(2):526–531.
119. Papa A, Bozovi B, Pavlidou V, et al. Genetic detection and isolation of Crimean-Congo hemorrhagic fever virus, Kosovo, Yugoslavia. Emerg Infect Dis 2002; 8(8):852–854.
120. Qing T, Saijo M, Lei H, et al. Detection of immunoglobulin G to Crimean-Congo hemorrhagic fever virus in sheep sera by recombinant nucleoprotein-based enzyme-linked immunosorbent and immunofluorescence assays. J Virol Methods 2003; 108(1):111–116.
121. Saijo M, Qing T, Niikura M, et al. Immunofluorescence technique using HeLa cells expressing recombinant nucleoprotein for detection of immunoglobulin G antibodies to Crimean-Congo hemorrhagic fever virus. J Clin Microbiol 2002; 40(2):372–375.
122. Nikolaev VP. Comparative assessment of methods for detecting the Omsk hemorrhagic fever virus. Vopr Virusol 1989; 34(3):366–368.
123. Elliott LH, Bauer SP, Perez-Oronoz G, et al. Improved specificity of testing methods for filovirus antibodies. J Virol Methods 1993; 43(1):85–100.
124. Ivanov AP, Rezapkin GV, Dzagurova TK, et al. Indirect solid-phase immunosorbent assay for detection of arenavirus antigens and antibodies. Acta Virol 1984; 28(3):240–245.
125. Riera LM, Feuillade MR, Saavedra MC, et al. Evaluation of an enzyme immunosorbent assay for the diagnosis of Argentine haemorrhagic fever. Acta Virol 1997; 41(6):305–310.
126. Morales MA, Calderon GE, Riera LM, et al. Evaluation of an enzyme-linked immunosorbent assay for detection of antibodies to Junin virus in rodents. J Virol Methods 2002; 103(1):57–66.
127. Niklasson BS, Jahrling PB, Peters CJ. Detection of Lassa virus antigens and Lassa-specific IgG and IgM by enzyme-linked immunosorbent assays. J Clin Microbiol 1984; 20:239–244.
128. Saijo M, Tang Q, Shimayi B, et al. Recombinant nucleoprotein-based serological diagnosis of Crimean-Congo hemorrhagic fever virus infections. J Med Virol 2005; 75(2):295–299.
129. Ure AE, Ghiringhelli PD, Possee RD, et al. Argentine hemorrhagic fever diagnostic test based on recombinant Junin virus N protein. J Med Virol 2008; 80(12):2127–2133.
130. Geetha PB, Ghosh SN, Gupta NP, et al. Enzyme linked immunosorbent assay (ELISA) using β-lactamase for the detection of antibodies to KFD virus. Indian J Med Res 1980; 71:329–332.
131. Trappier SG, Conaty AL, Farrar BB, et al. Evaluation of the polymerase chain reaction for diagnosis of Lassa virus infection. Am J Trop Med Hyg 1993; 49(2):214–221.
132. Lozano ME, Enria D, Maiztegui JI, et al. Rapid diagnosis of Argentine hemorrhagic fever by reverse transcriptase PCR-based assay. J Clin Microbiol 1995; 33(5):1327–1332.
133. Zhai J, Palacios G, Towner JS, et al. Rapid molecular strategy for filovirus detection and characterization. J Clin Microbiol 2007; 45(1):224–226.
134. Drosten C, Göttig S, Schilling S, et al. Rapid detection and quantification of RNA of Ebola and Marburg viruses, Lassa virus, Crimean-Congo hemorrhagic fever virus, Rift Valley fever virus, dengue virus, and yellow fever virus by real-time reverse transcription-PCR. J Clin Microbiol 2001; 40(7):2323–2330.
135. Duh D, Saksida A, Petrovec M, et al. Novel one-step real-time RT-PCR assay for rapid and specific diagnosis of Crimean-Congo hemorrhagic fever encountered in the Balkans. J Virol Methods 2006; 133:175–179.
136. Palacios G, Briese T, Kapoor V, et al. MassTag polymerase chain reaction for differential diagnosis of viral hemorrhagic fevers. Emerg Infect Dis 2006; 12(4):692–695.
137. Bodur H, Akinci E, Onguru P, et al. Detection of Crimean-Congo hemorrhagic fever virus genome in saliva and urine. Int J Infect Dis 2009.
138. Wolfel R, Paweska JT, Petersen N, et al. Low-density macroarray for rapid detection and identification of Crimean-Congo hemorrhagic fever virus. J Clin Microbiol 2009; 47(4):1025–1030.
139. Yapar M, Aydogan H, Pahsa A, et al. Rapid and quantitative detection of Crimean-Congo hemorrhagic fever virus by one-step real-time reverse transcriptase-PCR. Jpn J Infect Dis 2005; 58(6):358–362.
140. Burt FJ, Leman PA, Smith JF, et al. The use of a reverse transcription-polymerase chain reaction for the detection of viral nucleic acid in the diagnosis of Crimean-Congo haemorrhagic fever. J Virol Methods 1998; 70(2):129–137.
141. Garrison AR, Alakbarova S, Kulesh DA, et al. Development of a TaqMan minor groove binding protein assay for the detection and quantification of Crimean-Congo hemorrhagic fever virus. Am J Trop Med Hyg 2007; 77(3):514–520.

142. Lozano ME, Ghiringhelli PD, Romanowski V, et al. A simple nucleic acid amplification assay for the rapid detection of Junin virus in whole blood samples. Virus Res 1993; 27(1):37–53.
143. Vieth S, Drosten C, Charrel R, et al. Establishment of conventional and fluorescence resonance energy transfer-based real-time PCR assays for detection of pathogenic New World arenaviruses. J Clin Virol 2005; 32(3):229–235.
144. Steele K, Crise B, Kuehne A, et al. Ebola virus glycoprotein demonstrates differential cellular localization in infected cell types of nonhuman primates and guinea pigs. Arch Pathol Lab Med 2001; 125(5):625–630.

21 Infections of the Central Nervous System

Julie E. Reznicek
Department of Medicine, Division of Infectious Diseases, Vanderbilt University School of Medicine, Nashville, Tennessee, U.S.A.

Karen C. Bloch
Departments of Medicine and Preventive Medicine, Vanderbilt University School of Medicine, Nashville, Tennessee, U.S.A.

Yi-Wei Tang
Departments of Pathology and Medicine, Vanderbilt University School of Medicine, Nashville, Tennessee, U.S.A.

INTRODUCTION

Viral infections of the central nervous system (CNS) often present a diagnostic dilemma. The number of causative agents is vast, much greater than a century ago, when the established etiologies of CNS infections were mainly limited to rabies virus and polio virus. The physical findings of CNS infection are often nonspecific, and initial laboratory results may provide few additional clues as to the etiology. Molecular diagnostic testing has dramatically improved the ability to detect viral CNS infections, but requires expertise in the attributes and limitations of these techniques. Even discriminating infectious from noninfectious causes may be challenging, as metabolic, autoimmune, neoplastic, toxic, and endocrinologic entities may mimic meningitis or encephalitis. CNS infections represent syndromes where close collaboration between clinicians and laboratorians is crucial in providing a rapid diagnostic evaluation and an appropriate interpretation of results.

This chapter concentrates on viral infections of the CNS in immunocompetent adults, with a focus on assessment and laboratory evaluation. The close anatomical proximity of meninges and brain parenchyma often blurs the distinction between meningitis and encephalitis, causing an overlap syndrome termed meningoencephalitis. For the purposes of this chapter, meningoencephalitis will be considered a subset of encephalitis, while meningitis will refer to isolated inflammation of the meninges. There is significant geographic variability with respect to the microbiology of CNS infections, and the following discussion will highlight the most significant viral pathogens in the United States. A brief overview of laboratory diagnostic techniques will initially be provided followed by a more detailed discussion specifically relating to the various viral pathogens.

CLINICAL PRESENTATION

Patients with CNS infections may display a wide range of nonspecific symptoms, and in many cases viral meningitis and encephalitis are indistinguishable at presentation. Both encephalitis and meningitis typically present with fever and headache. Altered mental status is almost a universal finding in encephalitis and is usually present at presentation. In contrast, cognition frequently remains intact in viral meningitis, and this group is more likely to have objective findings of meningeal inflammation such as nuchal rigidity (1).

Acute Viral Meningitis

Meningitis is typically classified by acuity of onset, cerebrospinal fluid (CSF) parameters, and underlying etiology. Viral meningitis is almost without exception an acute illness, with variable clinical signs of meningeal irritation evolving over a period of hours to days, but without the presence of neurologic dysfunction (1). Often, patients will recount a recent viral syndrome, followed by the abrupt onset of a high fever and headache (2). Various exanthems may also be present, suggestive of a specific pathogen such as varicella zoster virus or enteroviruses. The laboratory hallmark of viral meningitis is an increase in the number of white blood cells in the CSF. In contrast to bacterial meningitis, patients with viral meningitis typically manifest a mononuclear rather than neutrophilic pleocytosis and the absolute white cell count is typically less than 300 cells/mm^3 (range <100–1000 cells/mm^3) (3).

Viral Encephalitis

Encephalitis is the presence of an inflammatory process of the brain parenchyma, with an altered level of consciousness, representing clinical evidence of brain dysfunction (4). Severe impairment, such as coma, can occur secondary to diffuse cerebral cortex involvement. Some organisms show neurotropism for particular anatomic sites; for example, HSV-1 infection almost universally involves the temporal lobe, with EEG evidence of periodic lateralized epileptiform discharges arising from this area (5). Rarely, patients with inflammation localized to extra-cerebral portions of the CNS may have intact cognition. For instance, primary varicella infection is associated with cerebellar inflammation, with findings of ataxia and nystagmus but no cognitive deficits (6,7). Other neurologic manifestations may include seizures, behavioral changes (such as psychosis), focal paresis or paralysis, cranial nerve palsies, or movement disorders such as chorea (8).

EPIDEMIOLOGY

Limited data exist regarding the incidence of CNS infections. While a number of specific pathogens are nationally notifiable infections in the United States (e.g., West Nile virus, rabies), the syndromes of encephalitis, meningitis, and myelitis are not reportable. In a large population-based study performed over 32 years, the incidence of aseptic meningitis and encephalitis was found to be 10.9/100,000 person years and 7.4/100,000 person years, respectively (9). More recently, a study using national discharge data for encephalitis identified an incidence rate of 7.3/100,000 population, accounting for 230,000 hospital days, and 1400 deaths annually (10). This estimate is remarkably similar to studies in Finnish children, where a rate of 8.8/100,000 was reported (11).

Viral Meningitis

Enteroviruses are the most common cause of viral meningitis. Data from the National Enterovirus Surveillance System (NESS) suggests that enteroviruses cause an estimated 10 to 15 million symptomatic infections in the United States annually, including 30,000 to 75,000 cases of meningitis (12). Enteroviral infections account for 80% to 92% of aseptic meningitis cases in which an agent is identified (13). The remainder of cases are attributed to arboviruses and various human herpes viruses, with HSV-2 predominating. Other viruses implicated in the aseptic meningitis syndrome are specific to unique host populations, such as mumps virus in nonimmunized individuals, or lymphocytic choriomeningitis virus in those with rodent contact. Even with the advent of more sophisticated and extensive diagnostic testing, such as nucleic acid amplification methods, the etiology remains unknown in many cases of aseptic meningitis (32–75%) (14).

Viral Encephalitis

The microbiology of viral encephalitis has evolved over the past decade. While vaccine-preventable conditions such as measles, mumps, rubella, and varicella have declined in incidence (15,16), newer pathogens have emerged to replace these viruses. West Nile virus (WNV), which was first identified as a cause of encephalitis in the United States in 1999, now accounts for >1000 cases/year (17). Increasing numbers of immunocompromised patients translates into a higher number of susceptible hosts at risk for opportunistic viruses such as CMV, EBV, JC virus, and HHV-6.

While HSV-1, WNV, and the enteroviruses are the most commonly identified etiologies of encephalitis in the United States, a large number of additional agents have been reported to cause encephalitis (14,18). Conceptually, these can be categorized based on the strength of association between the specific agent and CNS disease. Neurotropic viruses that are relatively frequent causes of encephalitis, and have known neurotropic potential, are listed in Table 1. Viruses that are strongly neurotropic but relatively uncommon causes of encephalitis are enumerated in

Table 1 Relatively Common Viral Causes of Meningoencephalitis in the United States

Etiology	Epidemiology	Clinical features	Diagnosis
Enteroviruses	Peak incidence in late summer and early fall; more common in children	Aseptic meningitis (most common) to encephalitis	CSF PCR or viral culture. Stool or throat swab PCR or culture suggestive, but not diagnostic of CNS involvement
Epstein–Barr virus	Either during acute infection or reactivation	Cerebella ataxia, sensory distortion, primary CNS lymphoma	CSF PCR (indicative of infection or reactivation), frequently seen as a dual infection, serology (IgM positive in acute infection)
Herpes simplex virus (HSV) 1 and 2	HSV-1 accounts for 5–10% of all cases of encephalitis, HSV-2 primarily causes aseptic meningitis, although may cause encephalitis in neonates	Temporal lobe seizures (apraxia, lip smacking), behavioral abnormalities. Unilateral or bilateral temporal lobe enhancement on MRI	CSF PCR, CSF serologies if >1 week of symptoms
LaCrosse virus	Mosquito-borne, endemic in Midwestern and eastern United States, peak incidence in school aged-children	Varies from subclinical seizures to coma, generally full recovery	Serology
St. Louis encephalitis virus (SLE)	Mosquito-borne, endemic to Western United States, with periodic outbreaks in central/eastern United States, peak incidence in adults >50 years	Tremors, seizures, paresis, urinary symptoms, SIADH variably present	Serology (cross reacts with other flaviviruses)
Varicella zoster virus (VZV)	Acute infection (chickenpox) or reactivation (shingles)	Vesicular rash (disseminated or dermatomal), cerebellar ataxia, large vessel vasculitis	DFA or PCR of skin lesions, CSF PCR, serum IgM (acute infection)
West Nile virus (WNV)	Mosquito-borne cause of epidemic encephalitis throughout United States, Europe. Peak incidence adults >50 years	Weakness and acute flaccid paralysis, tremors, myoclonus, Parkinsonian features, MRI with basal ganglia and thalamic lesions	CSF IgM, serology (cross-reactivity with SLE)

Abbreviations: CSF, cerebrospinal fluid; PCR, polymerase chain reaction.

Table 2. Finally, Table 3 lists the most problematic group; viruses that have been anecdotally associated with encephalitis, but which are poorly neurotropic and direct proof of causation is limited.

Despite extensive evaluation, a specific pathogen is identified in less than 50% of cases (14,19,20). For instance, during the first two years of the California Encephalitis Project, a prospective study using a combination of serologic and molecular diagnostic techniques to improve pathogen recovery, in 208/334 (62%) patients an underlying cause of encephalitis was not identified despite extensive testing (20). This group obtained similar results in a much larger follow up report of 1570 cases, where a confirmed, probable, or even possible etiology was identified in only 29% of cases of encephalitis (8).

DIFFERENTIAL DIAGNOSES

Noninfectious conditions may present in an almost identical manner to viral meningitis or encephalitis. Important noninfectious causes of aseptic meningitis include malignancy (e.g., lymphomatous meningitis), rheumatologic diseases (e.g., vasculitis), or granulomatous disease (e.g., sarcoidosis). Drug-induced aseptic meningitis poses a particular dilemma, as this is a diagnosis of exclusion. Although a number of drugs have been anecdotally linked with aseptic meningitis, the most common associations are with nonsteroidal antiinflammatory drugs and sulfa antibiotics (21). The diagnosis of drug-induced aseptic meningitis should be suspected if the patient's lumbar puncture results normalize after the discontinuation of the putative offending agent.

Encephalitis must be further differentiated from two clinically similar noninfectious entities, namely, encephalopathy and postinfectious or parainfectious encephalomyelitis. Encephalopathy, a syndrome of depressed consciousness without direct CNS parenchymal involvement, can be seen in a number of conditions, including metabolic abnormalities, hypoxia, ischemia, endocrinologic disorders, extra-CNS infections, or toxic ingestions. Encephalopathy can often be distinguished from viral encephalitis by the absence of fever, more gradual onset, lack of a CSF pleocytosis and normal neuroimaging (22). Treatment of encephalopathy is aimed at ameliorating the underlying cause or condition.

Postinflammatory encephalomyelitis is a demyelinating disease of the CNS that typically follows a mild infectious illness or immunization. Fever, headache, and focal neurologic signs are variably present, making this syndrome clinically indistinguishable from viral encephalitis (23). The most frequent subtype of postinflammatory encephalomyelitis is acute disseminated encephalomyelitis (ADEM), seen almost exclusively in children and adolescents, and characterized by poorly defined white matter lesions on MRI that enhance following gadolinium administration (24). Postinflammatory encephalomyelitis is presumed to be mediated by an immunologic response to an antecedent antigenic stimulus, and accounts for 5% to 15% of cases of encephalitis (10,25). Treatment typically involves immunotherapy (e.g., high-dose corticosteroids or immunoglobulin infusion).

DIAGNOSTIC WORKUP

Viral CNS infections represent a diagnostic challenge given the large number of potential infectious causes and the multitude of molecular and serologic tests available. Guidelines for the diagnosis and management of encephalitis have recently been published, which address many of these issues (14). Although most viral causes of encephalitis are not treatable (with the notable exception of herpes simplex encephalitis), a thorough diagnostic evaluation is important. Identification of a causative agent allows discontinuation of potentially toxic empiric antimicrobials, may affect prognosis, and can be of use for guiding public health interventions such as mosquito control strategies. The sections below describe the general diagnostic considerations in CNS viral infections. More detailed discussion of the specific techniques is provided in the technical chapters devoted to viral culture, serology, and molecular diagnosis.

History and Physical

The history and physical examination play a crucial role in guiding diagnostic testing. Epidemiologic or clinical considerations that may increase suspicion for a specific pathogen are provided in Tables 1–3. By actively eliciting information on the time course of events, animal or arthropod

Table 2 Less Common Viral Causes of Meningoencephalitis in the United States

Etiology	Epidemiology	Clinical features	Diagnosis
Eastern equine encephalitis (EEE) virus	Coastal states (Atlantic and Gulf); children and elderly disproportionately affected	Ranges from subclinical to fulminant encephalitis. Mortality 50–70%	Serology
Herpes B virus	Transmitted by bite of old-world macaque, infections in lab techs and veterinarians	Vesicular eruption at site of bite followed by neurologic symptoms, including transverse myelitis	Culture and PCR of vesicles and CSF
Lymphocytic choriomeningitis (LCM) virus	Peak incidence in fall and winter, rodents (including pets such as hamsters, guinea pigs) serve as reservoir	Orchitis, parotitis	Serology
Measles virus	Vaccine preventable; measles inclusion body encephalitis onset 1–6 months after infection; SSPE a late manifestation (>5 years after infection)	Measles encephalitis nonspecific; SSPE has a subacute onset with progressive dementia, myoclonus, seizures, and ultimately, death	CSF antibodies, brain tissue PCR; EEG changes often diagnostic
Mumps virus	Vaccine preventable	Parotiditis, orchitis, hearing loss frequent	Serology, throat swab PCR, CSF culture or PCR
Powassan virus	Tick-borne; endemic in New England, Great Lakes Region, and Canada	Nonspecific	Serology
Rabies virus	Vaccine preventable. Most common vector is the bat, and bites are often unrecognized. Dogs important in developing countries	Numbness or neuropathic pain at bite site, progressing to hydrophobia (with drooling), agitation, delirium, autonomic instability, coma. Paralytic form with ascending paralysis in <30%	Antibodies (serum, CSF), PCR of saliva or CSF, IFA of nuchal biopsy or CNS tissue. Coordinate testing with local health department.
Rubella	Vaccine preventable	Neurologic findings typically occur at same time as rash and fever	Serology, CSF antibodies
Western equine encephalitis (WEE) virus	Summer and early fall onset; Western US and Canada, Central and South America	Nonspecific	Serology

Abbreviations: CSF, cerebrospinal fluid; PCR, polymerase chain reaction; SSPE, subacute sclerosing panencephalitis.

Table 3 Selected Viral Pathogens of Unknown Neurotropic Potential that are Anecdotally Associated with Meningoencephalitis

Etiology	Epidemiologic and clinical features	Diagnosis
Adenovirus	Sporadic cases; children and immunocompromised at greatest risk; variably associated respiratory symptoms	Viral culture or PCR from respiratory site, CSF, or brain tissue
Human herpes virus-6	Usually immunocompromised, particularly bone marrow transplant recipients; latent infection of neural tissues making significance of detection in brain tissue difficult to determine	CSF PCR
Hepatitis C virus	Hepatitis C seropositive patient	CSF PCR
Human metapneumovirus	Newly described pathogen almost exclusively in children	Respiratory tract PCR
Human parechovirus 1 and 2	Previously echovirus 22 and 23, causes neonatal sepsis and meningitis	CSF PCR
Influenza A and B viruses	Sporadic disease in children, with most reports from Japan and southeast Asia; upper respiratory symptoms; acellular CSF, 10% with bilateral thalamic necrosis; high mortality	Respiratory tract culture, PCR, or rapid antigen. CSF and brain PCR infrequently positive
Parvovirus B-19	Sporadic causes, variably associated with a skin rash	IgM antibody, CSF PCR
Rotavirus	Typically children, winter months, usually with diarrhea	Stool antigen, CSF PCR (CDC)

Abbreviations: CSF, cerebrospinal fluid; PCR, polymerase chain reaction; CDC, Centers for Disease Control and Prevention.

exposures, travel, and extra-CNS manifestations, clinicians may be able to narrow down the vast differential diagnosis. This becomes even more important in the occasional patient with recent international travel, or unusual occupational exposure (e.g., veterinarian), leading the physician to focus on uncommon agents.

The physical exam may also provide important diagnostic clues. For example, the patient's skin should be thoroughly examined for exanthems, animal bites, or arthropod exposures. The presence of localized or generalized lymphadenopathy may suggest specific pathogens such as EBV, CMV, or HIV. Inferences regarding the patient's overall immune system can be made if an opportunistic infection, such as oral candidiasis, is found.

General Diagnostic Evaluation

Patients with suspected CNS infection should receive a standard admission laboratory evaluation. A peripheral blood cell count with differential, renal, and hepatic function tests, coagulation studies, and blood cultures should all be obtained (14). An initial chest x-ray should be obtained to evaluate for pulmonary involvement suggestive of respiratory pathogens (e.g., influenza, adenovirus), or hilar adenopathy. Other laboratory testing (e.g., HIV testing) can be considered, however is not mandatory if there is a low clinical suspicion for coinfection.

Neuroimaging

Imaging should be performed in any patient presenting with a possible CNS infection. The possible exception is in the young, previously healthy patient presenting with fever and headache, but no focal neurologic signs and normal cognition (26). Typically, a noncontrasted head CT is the initial study. Although less sensitive than magnetic resonance imaging (MRI) (27,28), CT provides rapid evaluation for CNS mass lesions contraindicating lumbar puncture. Newer MRI techniques, beyond conventional T1 and T2 spin-echo sequences, are now available, and may be useful in certain settings. For example, diffusion-weighted MRI (DWI) allows differentiation of cytotoxic from vasogenic edema and distinguishes recent from an old insult (4). DWI may be more sensitive than traditional MRI in detecting early signal abnormalities in viral encephalitis caused by HSV-1, enterovirus 71, and West Nile virus (29).

Abnormalities noted on neuroimaging are often relatively nonspecific; however, occasionally radiologic changes may be suggestive of a specific agent. For instance, more than 90%

Table 4 Radiologic Abnormalities in Encephalitis and Associated Viral Etiologies

Neuroimaging finding	Possible viral agent
Arteritis and infarctions	Varicella zoster virus; Nipah virus
Cerebellar lesions	Varicella zoster virus; Epstein–Barr virus
Focal lesions in basal ganglia, thalamus, and/or brainstem	Epstein–Barr virus; Eastern equine encephalitis virus; St. Louis encephalitis virus; West Nile virus; Enterovirus 71; Influenza virus (acute necrotizing encephalopathy), human herpes virus-6
Temporal lobe	Herpes simplex virus
White matter abnormalities	Varicella zoster virus; Epstein–Barr virus; human herpesvirus 6, JC virus; acute disseminated encephalomyelitis secondary to infection or immunization

of patients with confirmed HSV-1 encephalitis will have abnormalities of one or both temporal lobes visualized on MRI (30). Table 4 summarizes specific radiologic patterns associated with viral etiologic agents. Other neurologic diagnostic aids such as electroencephalography (EEG) and fluorine-18 fluorodeoxyglucose positron imaging (FDG-PET) have been studied in encephalitis, but the findings are usually nonspecific, and therefore these tests are not routinely recommended (14).

Cerebrospinal Fluid Analysis

Patients with suspected CNS infection should undergo lumbar puncture for CSF analysis. Contraindications to lumbar puncture include increased intracranial pressure due to edema or mass lesion or severe coagulopathy. Ideally, lumbar puncture should be performed while the patient is in the lateral decubitus position to obtain an accurate opening pressure. At least 10 to 15 mL of spinal fluid should be obtained for testing. This relatively small volume is replaced by production of additional CSF in about 30 minutes (31).

CSF analysis should routinely include cell count, glucose, and protein measurement. Protein levels >200 mg/dL are rarely found in viral CNS infections, and are suggestive of a bacterial process (32). Similarly, a CSF glucose of <40 mg/dL or CSF to serum glucose ratio of <0.4 is strongly indicative of a bacterial rather than a viral meningitis (33).

A CSF WBC count ≥5 cells/mm^3, while not specific for infection, confirms meningeal inflammation. Accurate measurement of pleocytosis requires prompt testing of CSF, as the neutrophil concentration declines by as much as 50% by two hours post-LP (34). If a significant delay in processing or analyzing the sample is anticipated, the CSF should be frozen to preserve specimen integrity.

Viral CNS infections typically have a mild (50–1000 cells/mm^3) pleocytosis, with a mononuclear predominance. In contrast, bacterial meningitis presents with a neutrophilic pleocytosis, with the total CSF WBC count frequently >1000 cells/mm^3. Of note, early in viral meningitis, there may be a transient neutrophilic predominance; in 87% of cases, this reverts to a lymphocytic pleocytosis if lumbar puncture is repeated within eight hours (35). Even when the presentation is strongly suggestive of a viral process, bacterial cultures of CSF, and in some cases fungal and mycobacterial cultures, are indicated to exclude an alternative, potentially treatable infection (14).

Direct Microscopic Examination (CSF)

Direct microscopic examination of CSF has a limited role in the diagnosis of viral CNS infections; however, it is an essential diagnostic step in order to exclude bacteria, mycobacteria, yeasts, molds, and occasionally parasites as possible etiologies. Rarely, "Mollaret cells," or large, friable cells with faintly staining vacuolated cytoplasm, may be visualized in patients with recurrent episodes of aseptic meningitis ("Mollaret's meningitis") due to HSV-2 infection. These cells, originally categorized as endothelial cells, are now thought to be activated macrophages (36).

Viral Cultures (CSF)

Viral cultures of CSF are typically performed in four different cell lines; African green monkey cells, Vero cells, human amniotic epithelial cells, and human embryonic skin fibroblasts, with cells evaluated daily for cytopathic effects (4). With the availability of nucleic acid amplification testing (NAAT), viral cultures provide limited additional information and are no longer recommended as a routine component in the evaluation of viral CNS infections (14). In a retrospective study of >22,000 CSF viral cultures, a positive result was found in only 5.7% of samples and resulted in an estimated healthcare expenditure of over one million dollars (37).

Serology (CSF)

Detection of intrathecal antibody production provides important diagnostic information for nonherpes group viruses, and has been best studied for flavivirus infections. For instance, the presence of virus-specific IgM antibody in CSF is considered diagnostic of West Nile virus, although not necessarily of an acute infection as detectable intrathecal antibody can persist for >500 days (38). The presence of intrathecal IgG antibody is less specific, as this can passively diffuse across the blood–brain barrier. Measuring the index of CSF to serum organism-specific antibody titers may be useful in confirming the presence of a systemic or nervous system infection in selected circumstances (39,40).

Molecular Assays (CSF)

Nucleic acid amplification-based molecular diagnostic clinical procedures, such as PCR, provide a rapid method for diagnosis of CNS viral infections with high sensitivity and accuracy. Table 5 describes the application of molecular methods to specific CNS pathogens. PCR and other PCR-derived techniques, including reverse transcriptase (RT)-PCR, multiplex PCR, nested PCR, broad-range PCR, and real-time PCR have collectively revolutionized the diagnosis and monitoring of CNS infections (41–51). In addition to PCR, transcription-mediated amplification (TMA) or nucleic acid sequence-based amplification (NASBA) begins with the synthesis of a DNA molecule complementary to the target nucleic acid (usually RNA). The NASBA-based method for enterovirus detection possesses good sensitivity and specificity (52,53). Other non-PCR nucleic acid amplification techniques include strand displacement amplification, ligase chain reaction, cycling probe technology, branched DNA technology, hybrid capture system, and Invader technology. Several studies have indicated their applications in the diagnosis of CNS infections (54–56).

Virus concentrations in the CSF are usually low; therefore, techniques have been developed to further enhance test sensitivity at the amplification product detection and identification stage. A colorimetric microtiter plate PCR system as well as an Invader Plus system (Third Wave Technologies, Madison, WI) incorporate an additional signal amplification into the detection step, reaching an analytical sensitivity of 10 HSV copies per reaction in CSF (51,54,57). A duplex real-time TaqMan PCR device has been developed to simultaneously detect and differentiate HSV and enteroviruses in one tube (42). The system provides a rapid and sensitive procedure by enhancing the 5′-exonuclease activity of the *Taq* polymerase. Differentiating these two agents may be useful during summer and early fall months when both HSV and enteroviruses are circulating in the community. The GeneXpert Dx system (Cepheid, Sunnyvale, CA, USA) is a fully integrated and automated nucleic acid sample preparation, amplification, and real-time detection system. This system-based enterovirus assay, which is used to provide rapid and on-demand diagnosis of aseptic meningitis, has recently received approval by the Food and Drug Administration (FDA). The complete automation and rapid-result capability of the enterovirus assay make it uniquely suited for urgent ("stat") testing (45).

EVALUATION AND REPORTING OF LABORATORY RESULTS

The interpretation of a positive PCR result requires clinical correlation. A positive qualitative CSF result indicates that pathogen-specific nucleic acid is detected in CSF, but is not necessarily indicative of acute infection. PCR can detect latent or low-grade persistent infection as well as active disease (58). For instance, Epstein–Barr virus (EBV) survives latently in human macrophages, and the virus may reactivate in the setting of an alternative CNS infection (59). Similarly, latent HHV-6 virus can be amplified from brain tissue in healthy hosts (60).

In addition, molecular diagnostic tests are not standardized, resulting in significant heterogeneity in testing methodology. Blinded proficiency testing for HSV by PCR performed by nine European reference laboratories produced concordant results for only 28% to 30% of samples

Table 5 Application of Molecular Methods in Detecting Common Viral Pathogens Causing CNS Infections

Organisms	Clinical diseases/ disorders	Molecular test applicability[a]	Comments
Adenoviruses	Meningoencephalitis	B	Rare cause of CNS infection. Molecular testing of CSF or brain tissue if clinical suspicion based on symptoms (e.g., keratoconjunctivitis) or non-CNS viral culture
Arboviruses	Meningoencephalitis	B	Includes EEE, LAC, SLE, WEE, WNV, VEE, JE, POW, and RVF. Serology is the test of choice, although CSF RT-PCR has a role in diagnosis of WNV encephalitis of the immunocompromised host (see comments in text)
CMV	Encephalitis, myelitis	A	CSF PCR is the test of choice. Quantitative PCR is useful to diagnose the primary CMV encephalopathy, and follow response to treatment. Direct detection of resistance-related mutations has been reported
Enteroviruses	Meningitis, rarely encephalitis	A	Molecular method is the test of choice. A stat test providing results in hours can significantly shorten hospital stay (see text); One step PCR including real-time formats may be less sensitive
Epstein–Barr virus	Encephalitis, meningitis, cerebellar ataxia, primary CNS lymphoma	A	Molecular method is the test of choice. EBV reactivation or detection of latent virus in lymphocytes may result in positive findings by PCR of limited clinical relevance. Presence of EBV in setting of dual infection also of unclear significance. Presence of EBV in CSF may be suggestive of primary CNS lymphoma (see text).
HHV-6, HHV-7	Encephalitis	A	Molecular method is the test of choice. The clinical relevance of a positive results is uncertain, as may reflect detection of latent virus. Encephalitis most clearly demonstrated with HHV-6 in stem cell transplant recipients.
HIV-1	Meningoencephalitits, myelitis	A	HIV-1 viral load may be discordant between peripheral blood and CSF

HSV-1, HSV-2	Encephalitis, meningitis (Mollaret's meningitis)	A	Molecular method is the test of choice. One step PCR including real-time formats may have inadequate sensitivity.
HTLV-1	Myelitis	B	Serology is the test of choice. Molecular method may enhance sensitivity
Influenza and parainfluenza viruses	Encephalitis	B	Serology or antigen detection is the test of choice. Most cases with influenza-associated encephalopathy are PCR negative, suggesting pathophysiology not due to direct invasion
JC virus	Progressive multifocal leukoencephalopathy	A	Molecular method is the test of choice. False-positive result may happen due to high level of BKV
LCMV	Meningoencephalitis	B[b]	Serology is the test of choice. Molecular method may enhance sensitivity
Measles virus	Subacute sclerosing panencephalitis	B	Serology is the test of choice. Molecular method may enhance sensitivity. Measles viral load may monitor SSPE disease progression
Mumps virus	Encephalitis, meningitis	B	Serology is the test of choice. Molecular method may enhance sensitivity
Parvovirus B-19	Meningoencephalitis	B	CSF PCR useful for confirming infection in patients with serologic evidence of acute infection
Rabies virus	Encephalitis	B[b]	DFA or serology remains the test of choice. Molecular method used for non-CNS specimens (see text)
VZV	Meningitis, myelitis, encephalitis	A	Molecular testing diagnostic method of choice, although may be negative late in infection or with vasculitis. Skin manifestations may be absent (see text)
West Nile virus	Encephalitis, myelitis	B[b]	IgM serology is the test of choice. Molecular method may enhance sensitivity, particularly in immunocompromised hosts.
Nipah and Hendra virus	Meningitis, encephalitis	B	IgM serology is the test of choice. Molecular method confirms CNS infection

[a]A, test is generally useful for the indicated diagnosis; B, test is useful under certain circumstances or for the diagnosis of specific forms of infection, as delineated in the right-hand column; C, test is seldom useful for general diagnostic purposes but may be available in reference laboratories for epidemiological studies or for the diagnosis of unusual conditions.
[b]Molecular methods were used in these cases associated with transplant of solid organ form infected donor.

tested (61). In this study, 8% to 18% of samples tested falsely positive, while the remainder tested falsely negative, with a linear relationship between quantitative viral load and the probability of detecting virus. Similar discordant results have been demonstrated on enteroviral proficiency panels (62).

Molecular Diagnostic Studies Outside of the CNS

Clinical signs and symptoms aid in determining when diagnostic testing of specimens from extraneural sites (e.g., oropharynx, stool, and cutaneous vesicles) is appropriate. For example, PCR or antigen detection of a nasopharyngeal aspirate for influenza may assist in diagnosis of a patient presenting during the appropriate season with neurologic findings, fevers, and respiratory complaints. Fluid obtained from vesicular skin lesions can be tested by direct fluorescent antibody (DFA) or PCR techniques to diagnose cutaneous HSV or VZV infections.

Ancillary PCR testing may provide diagnostic leads; however, positive results are not always indicative of CNS disease. For instance, during acute infection, enterovirus can frequently be isolated from the respiratory or gastrointestinal tracts, with low-level viral shedding from these extra-CNS sites persisting for several weeks (63). Because of this prolonged window for detection, identification of enterovirus (or other viruses) outside of the CNS may be an incidental finding, and caution is needed in interpreting these results.

Brain Biopsy

Prior to the widespread availability of HSV PCR, brain biopsy was the gold standard diagnostic test for HSE (64). Nucleic acid amplification tests have essentially replaced routine brain biopsy for the diagnosis of this syndrome (5), and limited data exists regarding the yield of this invasive procedure for alternative diagnoses (65). Brain biopsy may be indicated in patients who continue to deteriorate rapidly despite acyclovir, with a negative noninvasive workup (14). In the rare case where brain biopsy is pursued, the highest yield is obtained by sampling areas of focal abnormality on neuroimaging or gross examination by the neurosurgeon. Fresh brain tissue should be sent for viral culture, specific PCR studies, and immunofluorescence and a second portion placed in formalin for routine histopathology with additional staining for infectious agents. One of the challenges in pursing brain biopsy is determining the optimal timing for this procedure, as detection of virus is maximized early in the course of the infection.

SPECIFIC VIRAL PATHOGENS

Herpes Viruses

The members of the human herpes virus (HHV) family are all DNA viruses that can be further subclassified as alpha, beta, and gamma viruses (66). Although many herpes viruses are neurotropic, some clinically impact only specific populations (e.g., the immunocompromised), and will be discussed in other chapters. This chapter specifically discusses HSV-1 and 2, VZV, and EBV, which can all cause CNS complications as a result of primary infection or reactivation.

Herpes Simplex Virus

Clinical Presentation

The term "herpes" dates back to 400 BC, and is attributed to Hippocrates, derived from the Greek word *herpein*, meaning "to creep." Herpes simplex virus may be further subtyped into HSV-1 and HSV-2 (5). HSV-1 typically causes orolabial lesions ("fever blisters"). Following acute infection, the virus resides latently in the trigeminal ganglion, where it can intermittently reactivate to cause recurrent skin lesions in a perioral distribution. HSV-2 predominately causes recurrent anogenital lesions.

While both HSV subtypes can cause neurologic disease, the clinical spectrum differs significantly (67). HSV encephalitis (HSE) accounts for 10% to 20% of all viral encephalitides in the United States (68). Beyond infancy, more than 95% of HSE is due to HSV-1 rather than HSV-2 (5). HSV-1 causes a focal necrotizing encephalitis of the temporal lobes, and patients typically present with headache, fever, personality changes, aphasia, and seizures (69). MRI is superior to CT scan, and demonstrates either unilateral or bilateral temporal lobe involvement (30).

In contrast, HSV-2 is a frequent cause of lymphocytic meningitis, but an uncommon cause of HSE (70). Approximately 20% of patients with an initial episode of HSV-2 meningitis will develop a recurrent aseptic meningitis, previously termed Mollaret's meningitis (71). Patients typically have an abrupt onset of fever and meningismus, with resolution of symptoms within three to four days without treatment (36). Attacks can reoccur sporadically over many years at varying intervals.

Laboratory Evaluation

PCR Detection

CSF HSV PCR has replaced brain biopsy as the primary method for diagnosis of HSE, and is now a recommended component of the diagnostic evaluation for all patients with encephalitis (14). Molecular testing offers the advantage of a minimally invasive approach coupled with a turnaround time of less than one day if colorimetric enzyme immunoassay or real-time PCR is used (72). In experienced laboratories, the sensitivity of CSF HSV PCR among adults approaches 100%, with a specificity of 94% (64). The sensitivity of CSF HSV PCR for neonates and infants is lower, ranging from 75% to 100% (73).

Consensus guidelines recommend that all patients with encephalitis receive empiric acyclovir at presentation (14). Even with antiviral therapy, there is no appreciable loss of sensitivity of HSV PCR for up to one week, although sensitivity falls close to 50% by the second week of treatment (64). Typically, empiric acyclovir is discontinued when results of HSV PCR return negative; this approach has been called into question by reports of false-negative PCR results from CSF specimens obtained>72 hours after the onset of symptoms (74). For patients with a clinical presentation suggestive of HSE (e.g., temporal lobe enhancement on MRI), repeat testing on a second specimen obtained three to seven days after onset of symptoms is recommended prior to discontinuation of acyclovir (70). An additional cause of a falsely negative HSV PCR result is the presence of large numbers of red blood cells in the CSF, as porphyrin can inhibit the PCR reaction even with as many as 20,000 copies of HSV DNA (64).

Many uncertainties remain regarding optimal use of PCR for prognosis and monitoring therapy. Quantification of HSV in CSF has been reported to be a predictor of prognosis in HSE; however, this test is not widely available and its utility remains unknown (30). Some authorities recommend repeating HSV PCR on a second CSF specimen obtained 10 to 14 days after initiation of acyclovir, with prolongation of therapy for persistently positive results (75). In neonates, persistent detection of HSV by PCR at the end of treatment has been shown to be a poor prognostic sign (76).

While HSV PCR is frequently performed on patients with presumed neurologic infection, in practice a small fraction of CSF samples test positive (77). For this reason, there has been interest in developing a screening system to improve laboratory efficiency. Retrospective (78) and prospective (79) studies have demonstrated low yield to testing CSF specimens with normal WBC counts and protein levels. While it is tempting to apply these criteria to limit laboratory testing, atypical cases of HSV neurologic infections in the setting of normal CSF cell counts led to concern regarding standardized screening algorithms (80,81).

Serologic Evaluation

CSF and serum antibody testing has minimal utility in the diagnosis of HSE; however, for different reasons. The seroprevalence of HSV-1 IgG is so high among adults that a positive test has limited use (70). Since only 1/3 of cases of HSE are associated with primary HSV-1 infection, the absence of HSV IgM does not exclude neurologic disease (82). Detection of intrathecal HSV antibody is typically delayed at least two weeks into the course of the illness, and therefore while CSF antibody testing is not recommended for diagnosis at the time of presentation, antibody testing may have a role in making a retrospective diagnosis if convalescent CSF is available (83). In the rare setting where CSF antibody testing is performed, serum testing should also be done, as a ratio of serum to CSF HSV IgG titers of $\leq$20 is suggestive of intrathecal antibody production (4).

Varicella Zoster Virus

Clinical Presentation

Primary infection with varicella zoster virus (VZV) results in varicella, commonly known as "chickenpox." After infection, the virus remains latent in cranial-nerve and dorsal-root ganglia, with reactivation causing dermatomal, or in the immunocompromised host, disseminated zoster ("shingles") (84). Before 1995, when the varicella vaccine became widely available, approximately 90% of primary VZV infections occurred before adolescence (85).

Meningoencephalitis following primary varicella infection or zoster has been described in older literature; however, prior to the availability of PCR, VZV was implicated solely on the temporal relationship between a typical rash and neurologic symptoms. With the widespread availability of molecular testing, a wider clinical spectrum of VZV-related neurologic disease has been identified, often in patients without cutaneous findings (1,85). In a multicenter Finnish study of 174 patients with confirmed or probable VZV CNS infection, 27% and 65% of patients with encephalitis or meningitis, respectively, did not have skin lesions, a disease termed *herpes sine herpete* (86). In one study of immunocompetent adults, a delayed rash was described, appearing a median of six days after the onset of signs and symptoms of meningitis (87).

Several discreet neurologic syndromes have been attributed to VZV infection. Acute cerebellar ataxia is a complication of primary varicella infection, and does not occur with viral reactivation (1). Patients with acute cerebellar ataxia develop acute gait ataxia, nystagmus, vomiting, tremor, and headache, although usually have intact cognition. Symptoms typically begin in the 10 days following cutaneous eruption; however, rarely there may be up to a three week latency between rash and onset of cerebellar symptoms (7). Full recovery generally occurs within weeks to months.

Another neurologic syndrome suggestive of VZV is CNS vasculitis of either large or small blood vessels (84). PCR testing has identified virus particles in blood vessels in patients with encephalitis symptoms as well as in those with cerebrovascular accident. For example, large vessel unifocal granulomatous arteritis is mainly a disease of the elderly and is characterized by acute focal deficit (motor or sensory) that develops weeks or months after zoster (or varicella) in a contralateral distribution (84). Recurrent ischemic episodes have been documented with this syndrome, and the mortality rate is approximately 25% (88). Small vessel multifocal vasculopathy is a subacute condition, seen more frequently in immunocompromised patients (89,90).

Laboratory Evaluation

PCR Detection

The test characteristics of VZV PCR vary based on the neurologic syndrome, duration of symptoms, and the host immune status. In one study of patients with a clinical suspicion for CNS VZV infection, the sensitivity of CSF VZV PCR was 44% for patients with a dermatomal rash and meningoradiculitis compared to 66% for patients with a generalized rash and encephalitis (91). Other investigators have demonstrated that 25% of patients with neurologic manifestations associated with either primary VZV infection or reactivation had a positive CSF PCR (86). Among the subset with primary infection, PCR was positive only among children <10 years of age. In a small study of children with acute cerebellar ataxia, three of five cases (60%) had a positive CSF VZV PCR result (92). Quantitative VZV PCR is not widely available; however, in one study of patients with meningitis or encephalitis due to VZV reactivation, higher levels of virus in the CSF were found with encephalitis compared to meningitis, and were predictive of a more severe illness (93).

The time from the onset of rash to CSF sampling appears to be an important determinant in the sensitivity of VZV PCR. Among patients with zoster and meningitis, CSF PCR was positive for 62% of CSF samples collected <7 days after the onset of rash, compared to 25% of those collected >7 days after skin eruption (91). Similar results have been reported with primary VZV infection, with CSF PCR positive only among cases sampled <9 days after dermatologic onset (86). The diagnostic yield of CSF PCR among patients with VZV vasculopathy has been reported to be 28%, with the decreased sensitivity attributed to the significant delay (average 4.2 months) between the onset of neurologic symptoms and molecular testing in these cases (90).

VZV DNA can be amplified from the CSF in 2.5% to 7% of HIV-infected patients with new neurologic signs or symptoms (94,95). The significance of this finding is not always clear, as VZV may accompany other CNS pathogens, suggesting subclinical reactivation in some cases (95). Cutaneous findings are variably present in HIV-infected patients with VZV CNS infection (96,97). Among patients with VZV vasculopathy, PCR was positive in more than half of immunocompromised patients, as compared to only 16% of immunocompetent patients (90).

Serologic Evaluation

Serologic evaluation of patients with VZV CNS disease is useful for discriminating primary infection from reactivation. While dermatomal disease is highly suggestive of reactivation, disseminated zoster may be seen in immunocompromised individuals, and may be indistinguishable from the skin eruption of primary varicella. In these cases, a positive serum VZV IgM, particularly in the absence of IgG antibody, is diagnostic of new infection (86).

Measurement of intrathecal antibodies may play an important role in diagnosis, particularly when there is a significant delay between the onset of rash and acquisition of CSF. While intrathecal antibodies are identified in only a third of all of the patients with dermatomal zoster and meningoradiculitis, among the subset with CSF obtained more than seven days into the illness detectable antibody was found in 83% (91). This finding was even more pronounced among patients with VZV vasculopathy, in whom 93% had detectable CSF IgG antibody, with only one quarter of patients with detectable intrathecal antibody also having a positive CSF VZV PCR (90). In children, a combination of PCR and IgM tests is the best approach. In adults, PCR together with the measurement of intrathecal antibody production is the most sensitive combination of diagnostic tests (86).

Epstein–Barr Virus

Clinical Presentation

By adulthood, over 90% of the general population has serologic evidence of prior Epstein–Barr virus (EBV) infection (83). The most common clinical presentation of acute EBV infection is infectious mononucleosis, characterized by pharyngitis, cervical lymphadenopathy, and fever; however, asymptomatic seroconversion occurs in the majority of acute infections (98). Following acute infection, EBV establishes latency in B-lymphocytes, with symptomatic reactivation occurring primarily in individuals with significant impairment in T-cell immunity (99).

The frequency of neurologic complications with EBV infection remains low (<0.5% of infected patients), even when more sensitive molecular testing is used for diagnosis (13). Commonly associated symptoms of acute mononucleosis, such as headache and neck stiffness, do not necessarily imply invasive neurologic infection, as these are frequent accompaniments to viremia. Neurologic complications can result from acute infection or reactivation, and include symptoms spanning meningitis, encephalitis, myelitis, and acute cerebellar ataxia (100). A unique neurologic symptom of EBV infection is a sensory distortion, known as the "Alice in Wonderland" syndrome, with patients experiencing a vivid sense of enlargement or shrinkage of the body (101,102). Primary CNS lymphoma (PCNSL), a malignancy seen among HIV infected individuals, is due to localized reactivation of EBV in the CNS, with clonal proliferation of lymphocytes (103).

Laboratory Evaluation

PCR Detection

Amplification of EBV DNA in CSF represents a diagnostic challenge, as the presence of virus may reflect either CNS infection or detection of activated virus as part of the inflammatory response induced by another agent or condition (59,83). Quantitative viral loads may prove to be useful in discriminating infection from reactivation due to an alternative pathogen; however, a precise cutoff value has not been determined (104). It is also unknown whether dual infection with EBV in addition to an alternative neurotrophic agent changes the clinical course or outcome of the primary infection.

The significance of detection of EBV in the CSF of patients with HIV deserves special mention. PCNSL was a relatively common complication of HIV infection in the era predating highly active antiretroviral therapy (105). In HIV-infected patients, EBV amplification from the

CSF has been proposed as a marker for PCNSL among patients with CNS mass lesions (106), and as a predictor of the later development of PCNSL among patients without radiologic lesions (107,108). This approach has been called into question by more recent studies documenting a positive predictive value of CSF EBV PCR in HIV-infected patients with neurologic disease of between 10% and 29% (109,110). While the test characteristics may be improved slightly by using a quantitative cut-point of 10,000 copies/mL (109), EBV viral load does not reliably discriminate between PCNSL and EBV encephalitis (104), and pathologic review of brain tissue is still required for definitive diagnosis of PCNSL.

Serologic Evaluation

Serologic testing for EBV is an important ancillary approach to molecular testing. Prior to the availability of PCR, serology was the sole means of confirming acute EBV infection among patients with new-onset neurologic symptoms, many of whom lacked the classic signs and symptoms of infectious mononucleosis (111,112). It is unknown how frequently patients with neurologic complications of serologically confirmed acute EBV infection have detectable virus in the CSF (113), but it appears that, at least in adults, the majority of patients with positive CSF PCR for EBV have serologies consistent with viral reactivation (59). Measurement of intrathecal antibody production as a marker of EBV infection has been largely replaced by PCR (114).

Arboviruses

Arboviruses are a heterogeneous group of pathogens that share a common route of transmission through the bite of an infected arthropod (mosquito or tick) vector (115). More than 60 arboviruses indigenous to North America have been indentified (116), although only six are considered significant causes of human disease in the United States, and represent nationally notifiable human infections. West Nile virus has emerged in the last decade as the most common cause of epidemic encephalitis in the Western hemisphere. Given the public health importance of this pathogen, there has been significant interest in developing molecular methods for rapid diagnosis and screening. For this reason, WNV will be discussed separately from the other arboviruses.

West Nile Virus

West Nile Virus (WNV) is a single-stranded RNA virus of the family *Flaviviridae*, belonging to the Japanese encephalitis virus serogroup that includes the St. Louis encephalitis (SLE) virus, Japanese encephalitis (JE) virus, and Murray Valley encephalitis virus. WNV is transmitted by mosquito vectors, with avian species serving as zoonotic amplifiers and reservoirs. Humans play a limited role in the transmission cycle given the relatively low level and short duration of viremia.

Historically, WNV has been a relatively minor human and animal pathogen, causing infrequent epidemics associated with a mild febrile illness and variable neurologic manifestations (17). Beginning in the 1990s, a more virulent subtype emerged, which caused a high case fatality rate among birds, and was responsible for large human outbreaks of encephalitis in Romania and Israel (117). In 1999, a cluster of five elderly patients hospitalized with encephalitis all living in a single borough of New York City was ultimately attributed to WNV (118). Since this time, more than 16,000 human cases have been reported to the Centers for Disease Control, with >600 deaths, primarily among patients with neuroinvasive disease (17).

Clinical Manifestations

The majority of human WNV infections are subclinical; however, approximately 20% of infected persons will develop a flu-like illness with high fever, chills, malaise, headache, arthralgias, and myalgias, termed West Nile Fever (WNF) (119). About half of these patients will develop a morbillaform or maculopapular rash, with a higher incidence in children (120). Persistent fatigue and malaise, lasting a median of 36 days has been reported following WNF (121).

Approximately 1 in 150 infected individuals will develop West Nile neurologic disease (WNND), characterized as meningitis, myelitis, or encephalitis, either separately or as an overlap syndrome (122). Severe neurologic disease is more common in the elderly, and among immunocompromised patients (119,120,123). The case fatality rate for patients with encephalitis ranges from 4% to 15%, with mortality increasing linearly with age (17). Long-term complications of WNND are reported in 60% of survivors, with fatigue, weakness, depression, personality changes, gait problems, and memory deficits persisting one year after infection (124).

Laboratory Evaluation

PCR Detection

The appropriate use of a given diagnostic test for WNV is best understood in the context of viral kinetics and the host immune response to infection. WNV viremia begins within one to three days of inoculation, but is typically short-lived (125). Constitutional symptoms typically precede neurologic involvement, which develops seven to nine days after infection, at which time the virus has been cleared from the bloodstream (126,127). For this reason, the primary niche for nucleic acid amplification testing (NAAT) is in the screening of asymptomatic blood donors, where RT-PCR or nucleic acid sequence-based amplification (NASBA) can detect as few as 50 viral copies/mL (128).

A recent study performed during a large community outbreak of WNV in Canada found 45% of plasma samples from patients with WNV infection had detectable virus (129). Predictors of a positive plasma WNV PCR included testing within eight days of symptom onset and WNF (36% positive) rather than WNND (9.5% positive). Plasma WNV viral load did not correlate with either the duration of symptoms or the presence of CNS involvement.

Amplification of WNV from CSF is of limited utility, with only 57% of patients with serologically confirmed WNND having positive CSF PCR by real-time PCR, and 0% of specimens positive by conventional PCR (130). The exception to this is the immunocompromised host, a population at high risk for severe neurologic disease and adverse outcomes. Case reports have documented positive CSF WNV NAAT in transplant patients and those with hematologic malignancies (131,132). In these cases, virus in the blood and CSF may persist for a prolonged period of time, congruent with the delay in detectable neutralizing antibody (133).

Serologic Evaluation

Serum testing for WNV-specific IgM antibody via commercially available antibody-capture enzyme-linked immunoassays remains the mainstay of diagnosis (134). Evidence of intrathecal IgM WNV antibody in the setting of a clinically compatible illness is highly suggestive of WNND. Elevated WNV IgM antibody in serum is also suggestive of acute infection, with demonstration of a fourfold increase in antibody titer between acute and convalescent samples a definitive laboratory confirmation. The finding of an elevated WNV titer on a single serum specimen should be interpreted with caution, as even IgM antibody may persist for >500 days following clinical illness (38).

A second consideration in interpreting antibody levels is that WNV cross-reacts serologically with other members of the Japanese encephalitis virus subgroup and related flaviviruses, including Yellow Fever and Dengue viruses. The likelihood of flavivirus cross-reactivity can be minimized by testing for concomitant IgM titers against other Japanese serocomplex members based on travel or vaccine history, or by measuring plaque reduction neutralization titers against WNV and other flaviviruses (126).

Other Arboviruses

Other endemic arboviruses may be suspected based on geographic distribution, host characteristics, and, to a lesser extent, clinical presentation (Table 5). For instance, La Crosse virus (LAC) is responsible for nearly all infections caused by the California serogroup in the United States (135). Most cases have been reported in the Midwestern and mid-Atlantic states (1), during the classic arboviral encephalitis season of late summer to early fall. More than 90% of symptomatic infections occur in individuals <15 years of age, with a male predominance (116).

An age disparity is also seen in infections secondary to St. Louis encephalitis virus (SLE); however, with this agent 87% of encephalitis cases occur in patients over the age of 60 (1), and the case fatality rate approaches 25% in those >70 years of age (136). SLE is marked by periodic epidemics against a background of low-level endemic disease. The last outbreak occurred in 1993, with 1095 reported cases and 201 deaths (136). More recently, SLE infections have been

Table 5 Differentiating Features of Arboviruses Endemic to North America Causing Human Disease

Virus	Family	Vector	Host	Geographic distribution (North America)	Other
Western equine encephalitis (WEE) virus	*Togaviridae*	*Culex tarsalis* mosquito	Children > adults	West of the Mississippi river	During epidemics a high % of adult population displays asymptomatic seroconversion
Eastern equine encephalitis (EEE) virus	*Togaviridae*	*Culiseta melanura* mosquito	Elderly, children	Eastern seaboard and Gulf coast	Highest case fatality rate of all North American arboviruses
West Nile virus (WNV)	*Flaviviridae*	*Culex sp.* mosquito	Adults (rare in children), with elderly most at risk for neurologic diseases	Throughout the continental United States and Canada	Reports of transmission through laboratory accidents, transfusions, and organ transplantation
St. Louis encephalitis (SLE) virus	*Flaviviridae*	*Culex sp.* mosquito	80% of encephalitis cases occurring in individuals > 60 years of age	Endemic in western United States with periodic outbreaks east of the Mississippi River	Patients may experience urinary symptoms (e.g., incontinence) early in disease
Powassan encephalitis (POW) virus	*Flaviviridae*	*Ixodes cookei* tick	Children and adolescents, 2/3 are male	Maine, Michigan, New York, and Wisconsin; Canada	Case fatality rate 10%
LaCrosse (LAC) virus (California serogroup)	*Bunyaviridae*	*Aedes triseriatus* mosquito	Most symptomatic infections occur in individuals <15 years of age	Forested regions along the basins of Mississippi and the Ohio rivers	Responsible for nearly all infections caused by California serogroup

reported sporadically during the summer and fall season in almost every region of the United States.

Eastern equine encephalitis (EEE) virus causes the most severe CNS disease and has the highest fatality rate among the North American arboviruses (1,137). Since 1955, fewer than 36 human cases per year of EEE have been reported (116), with most cases occurring in the coastal regions of the Gulf and mid-Atlantic states (138). The geographic variability between western equine encephalitis (WEE) and EEE is directly related to the range of their respective primary mosquito vectors (139). WEE virus is a rare cause of encephalitis, with only seven cases reported between 1987 and 2002, with most occurring in children (1). Although the mortality rate of WEE encephalitis remains low (3–4%), morbidity secondary to seizure disorders and developmental delays remains significant (140).

In comparison to the aforementioned arboviruses which are transmitted by mosquito vectors, the vector for transmission of Powassan (POW) virus is an *Ixotid* tick. According to the Center for Disease Control and Prevention, Division of Vector Borne Infectious Diseases, only 12 cases of confirmed or probable Powassan human infections have been reported, with all of them occurring in Maine, Michigan, New York, and Wisconsin. This may represent either the rarity of this infection, or more likely, infrequent diagnostic testing for this pathogen.

Laboratory Evaluation

PCR Detection

Highly sensitive nucleic acid amplification-based techniques do exist for many of the medically important arboviruses. However, their role in routine diagnostics remains limited to the research setting at this time (138). The main factor limiting the utility of PCR is the relatively short period of viremia, with undetectable viral levels at the onset of CNS or systemic disease. The incidence of many of these CNS infections among humans is so small that large-scale, prospective studies are limited to mosquito pools, making clinical interpretation of PCR results difficult (141). RT-PCR in CSF has been most extensively studied among members of the Bunyaviridae, specifically La Crosse virus (142,143) but this test is not widely available, and has infrequently been validated on clinical samples.

Serologic Evaluation

The mainstay of diagnostic testing for the non-WNV arboviruses is through demonstration of serum antibodies. Identification of a single IgG or IgM elevated titer in the setting of a clinically compatible disease is strongly suggestive of disease, with a fourfold increase between acute and convalescent titers diagnostic of infection (144). As discussed under the WNV section, SLE virus is cross-reactive with WNV, and further testing is required to definitively differentiate between these two infections.

Enteroviruses

Prior to the 1960s, the taxonomy for enteroviruses was based on humans and animal model systems (echoviruses, coxsackieviruses group A and B, and polioviruses). The overlap between groups led to a great deal of confusion, leading to the joining of the above-mentioned viruses into the *Enterovirus* genus, comprising five groups containing >80 subtypes differentiated by a numeric designation (e.g., enterovirus 71) (145,146). Two former members of this genus, echoviruses 22 and 23, have been reassigned as human parechovirus 1 and 2, which have been demonstrated to cause meningitis in young children (147). Humans are the only natural hosts for the enteroviruses, and the principal mode of human-to-human transmission is fecal-oral (1).

Clinical Presentation

The clinical syndromes attributable to enteroviruses range from possibly a mild febrile illness to potentially fatal conditions. Enteroviral infection remains the most common cause of aseptic meningitis in the United States, accounting for 80% to 92% of cases in which an etiologic agent is identified (13), with a seasonal predilection extending from June to October (148). Adolescents and adults can experience a myriad of symptoms, such as myalgias, headache, vomiting, anorexia, and various exanthems. The fever may be biphasic; presenting initially with constitutional symptoms, then briefly resolving, only to return with the onset of neurologic symptoms (1).

While aseptic meningitis due to the nonpolio enteroviruses tends to be mild and self-limiting, rarely these viruses can be associated with more severe, life-threatening manifestations, such as encephalitis, paralysis, myopericarditis, and neonatal sepsis (148). Recent outbreaks of enterovirus 71 in Southeast Asia have been associated with a brain stem encephalitis in pediatric patients, causing a high mortality and considerable cognitive morbidity among survivors (149,150).

Laboratory Diagnosis

PCR Detection

EV-specific reverse transcriptase polymerase chain reaction (EV-PCR) has a sensitivity and specificity that approach 100%, and with new colorimetric techniques can provide results in approximately five hours (48). A positive PCR has been associated with cost savings due to the shorter length of hospital stay and decreased diagnostic testing in children with confirmed enteroviral meningitis (13,151). Most EV PCR assays target the highly conserved 5′-nontranslated region (NTR), allowing amplification of all human enteroviruses (152). The disadvantage of this approach is that further typing through sequencing is rarely performed clinically, which may impede early recognition of a localized outbreak.

Although the definitive diagnosis of enteroviral meningoencephalitis is through detection of virus in CSF specimens, detection of the virus in other samples (e.g., blood, stool, urine, and throat) is suggestive of CNS infection if the clinical syndrome is compatible and CSF is unavailable. In one study of 34 patients who received a diagnosis of enteroviral meningitis on the basis of viral isolation and/or antibody detection, the specificity and sensitivity of fecal PCR was reported to be 96% (63). This group also reported that the CSF diagnostic yield by PCR was significantly lower for samples obtained >2 days after symptom onset, a previously unpublished finding. The advantage of a rectal swab includes a high enteroviral load in stool as well as prolonged viral shedding via the gastrointestinal tract that may persist for greater than two weeks (153). Unfortunately, the latter can also rarely result in a false-positive stool PCR result in a patient with prior enteroviral gastroenteritis and an alternative pathogen causing neurologic dysfunction.

Rabies

Rabies virus is an RNA rhabdovirus that originally was thought to be transmitted solely through the documented bite of an infected animal. Since the early 1980s, less direct mechanisms of transmission have been documented (e.g., aerosolized virus, unrecognized bat puncture wound), prompting the Centers for Disease Control to change their recommendations for post-exposure prophylaxis (154). Transmission through organ donation has also been documented (155). According to the World Health Organization, 55,000 cases of rabies are estimated to occur worldwide, mainly from wild animal bites (particularly dogs). On average, there have been one to three human cases per year in the United States during the past 20 years (156).

Clinical Presentation

Human rabies can present in two general forms, encephalitic or furious rabies (80–85% of cases), or the less common paralytic form. The former will initially have a clinical syndrome of headache, fever, malaise, nausea, and vomiting, which may be indistinguishable from any other encephalitis, or simply resemble a self-limiting gastrointestinal illness. Patients frequently complain of paresthesias at the site of inoculation. This prodrome may last only a couple of days, before an acute neurologic syndrome consisting of excessive salivation, agitation, hydrophobia, and nuchal rigidity begins, often accompanied by autonomic nervous system involvement (157). Antirabies vaccine or rabies immune globulin (RIG) is not effective once symptoms are present, and death typically occurs within one to two weeks of symptom onset.

Laboratory Diagnosis

PCR Detection

Ideally, the diagnosis of rabies should be made antemortem to assist in prognosis and to facilitate postexposure treatment of contacts. For this reason, coupled with the fact that diagnostic testing is usually limited to specialized laboratories, all testing for suspected cases of rabies should be coordinated with the local or state health department. A small study evaluating rabies virus RT-PCR on both CSF and saliva among patients with confirmed rabies identified a sensitivity of 9% and 30% respectively (158). A more recent study comparing conventional PCR compared to real-time PCR of saliva found a sensitivity of 37% for the former, compared to 75% for the latter (159).

Serologic Evaluation

A combination of testing approaches on different tissue specimens may be required for diagnosis, as no single test is uniformly sensitive for antemortem diagnosis. A retrospective review of various testing modalities prior to death from rabies identified detection of antigen through immunofluorescence of skin from the nape of the neck as the most sensitive test (69%), followed by detection of serum antibodies (39%), CSF antibodies (21%), and immunofluorescence of corneal imprints (14%) (158). Seroconversion or antigen identification may occur later in the course of the illness, so that testing of serial samples may be indicated (160).

CONCLUSIONS

Viral infections of the CNS are a significant cause of morbidity and mortality. While in many cases no specific pathogen is identified, prompt and thorough diagnostic testing is essential to facilitate rapid diagnosis, assist with prognosis and clinician decision making, and allow rapid mobilization of public health resources to combat diseases such as arboviruses or rabies. As the epidemiology of CNS viruses continues to shift with a decline in vaccine-preventable illnesses matched by a rise in new or emerging pathogens, a coordinated approach between clinicians and laboratorians is required for the most expeditious and cost-effective approach to diagnosis.

REFERENCES

1. Romero JR, Newland JG. Viral meningitis and encephalitis: Traditional and emerging viral agents. Semin Pediatr Infect Dis 2003; 14(2):72–82.
2. Coyle PK. Overview of acute and chronic meningitis. Neurol Clin 1999; 17(4):691–710.
3. Fitch MT, van de Beek D. Emergency diagnosis and treatment of adult meningitis. Lancet Infect Dis 2007; 7(3):191–200.
4. Steiner I, Budka H, Chaudhuri A, et al. Viral encephalitis: A review of diagnostic methods and guidelines for management. Eur J Neurol 2005; 12(5):331–343.
5. Whitley RJ. Herpes simplex encephalitis: Adolescents and adults. Antiviral Res 2006; 71(2–3):141–148.
6. Kleinschmidt-DeMasters BK, Gilden DH. Varicella-Zoster virus infections of the nervous system: Clinical and pathologic correlates. Arch Pathol Lab Med 2001; 125(6):770–780.
7. Connolly AM, Dodson WE, Prensky AL, et al. Course and outcome of acute cerebellar ataxia. Ann Neurol 1994; 35(6):673–679.
8. Glaser CA, Honarmand S, Anderson LJ, et al. Beyond viruses: Clinical profiles and etiologies associated with encephalitis. Clin Infect Dis 2006; 43(12):1565–1577.
9. Beghi E, Nicolosi A, Kurland LT, et al. Encephalitis and aseptic meningitis, Olmsted County, Minnesota, 1950–1981: I. Epidemiology. Ann Neurol 1984; 16(3):283–294.
10. Khetsuriani N, Holman RC, Anderson LJ. Burden of encephalitis-associated hospitalizations in the United States, 1988–1997. Clin Infect Dis 2002; 35(2):175–182.
11. Rantala H, Uhari M. Occurrence of childhood encephalitis: A population-based study. Pediatr Infect Dis J 1989; 8(7):426–430.
12. Strikas RA, Anderson LJ, Parker RA. Temporal and geographic patterns of isolates of nonpolio enterovirus in the United States, 1970–1983. J Infect Dis 1986; 153(2):346–351.
13. Ramers C, Billman G, Hartin M, et al. Impact of a diagnostic cerebrospinal fluid enterovirus polymerase chain reaction test on patient management. JAMA 2000; 283(20):2680–2685.
14. Tunkel AR, Glaser CA, Bloch KC, et al. The management of encephalitis: Clinical practice guidelines by the Infectious Diseases Society of America. Clin Infect Dis 2008; 47(3):303–327.

15. Koskiniemi M, Vaheri A. Effect of measles, mumps, rubella vaccination on pattern of encephalitis in children. Lancet 1989; 1(8628):31–34.
16. Barnes DW, Whitley RJ. CNS diseases associated with varicella zoster virus and herpes simplex virus infection. Pathogenesis and current therapy. Neurol Clin 1986; 4(1):265–283.
17. Gubler DJ. The Continuing Spread of West Nile Virus in the Western Hemisphere. Clin Infect Dis 2007; 45:1039–1046.
18. Koskiniemi M, Rantalaiho T, Piiparinen H, et al. Infections of the central nervous system of suspected viral origin: A collaborative study from Finland. J Neurovirol 2001; 7(5):400–408.
19. Khetsuriani N, Holman RC, Lamonte-Fowlkes AC,et al. Trends in encephalitis-associated deaths in the United States. Epidemiol Infect 2007; 135(4):583–591.
20. Glaser CA, Gilliam S, Schnurr D, et al. In search of encephalitis etiologies: Diagnostic challenges in the California Encephalitis Project, 1998–2000. Clin Infect Dis 2003; 36(6):731–742.
21. Jolles S, Sewell WA, Leighton C. Drug-induced aseptic meningitis: Diagnosis and management. Drug Saf 2000; 22(3):215–226.
22. Solomon T, Hart IJ, Beeching NJ. Viral encephalitis: A clinician's guide. Pract Neurol 2007; 7(5): 288–305.
23. Garg RK. Acute disseminated encephalomyelitis. Postgrad Med J 2003; 79(927):11–17.
24. Young NP, Weinshenker BG, Lucchinetti CF. Acute disseminated encephalomyelitis: Current understanding and controversies. Semin Neurol 2008; 28(1):84–94.
25. Johnson RT. Acute encephalitis. Clin Infect Dis 1996; 23(2):219–224; quiz 25–26.
26. Tunkel AR, Hartman BJ, Kaplan SL, et al. Practice guidelines for the management of bacterial meningitis. Clin Infect Dis 2004; 39(9):1267–1284.
27. Marchbank ND, Howlett DC, Sallomi DF, et al. Magnetic resonance imaging is preferred in diagnosing suspected cerebral infections. BMJ 2000; 320(7228):187–188.
28. Dale RC, de Sousa C, Chong WK, et al. Acute disseminated encephalomyelitis, multiphasic disseminated encephalomyelitis and multiple sclerosis in children. Brain 2000; 123(Pt 12):2407–2422.
29. Maschke M, Kastrup O, Forsting M, et al. Update on neuroimaging in infectious central nervous system disease. Curr Opin Neurol 2004; 17(4):475–480.
30. Domingues RB, Fink MC, Tsanaclis AM, et al. Diagnosis of herpes simplex encephalitis by magnetic resonance imaging and polymerase chain reaction assay of cerebrospinal fluid. J Neurol Sci 1998; 157(2):148–153.
31. de Almeida SM, Nogueira MB, Raboni SM, et al. Laboratorial diagnosis of lymphocytic meningitis. Braz J Infect Dis 2007; 11(5):489–495.
32. Spanos A, Harrell FE Jr., Durack DT. Differential diagnosis of acute meningitis. An analysis of the predictive value of initial observations. JAMA 1989; 262(19):2700–2707.
33. Bonadio WA. The cerebrospinal fluid: Physiologic aspects and alterations associated with bacterial meningitis. Pediatr Infect Dis J 1992; 11(6):423–431.
34. Steele RW, Marmer DJ, O'Brien MD, et al. Leukocyte survival in cerebrospinal fluid. J Clin Microbiol 1986; 23(5):965–966.
35. Feigin RD, Shackelford PG. Value of repeat lumbar puncture in the differential diagnosis of meningitis. N Engl J Med 1973; 289(11):571–574.
36. Ginsberg L. Difficult and recurrent meningitis. J Neurol Neurosurg Psychiatry 2004; 75(suppl 1): i16–i21.
37. Polage CR, Petti CA. Assessment of the utility of viral culture of cerebrospinal fluid. Clin Infect Dis 2006; 43(12):1578–1579.
38. Roehrig JT, Nash D, Maldin B, et al. Persistence of virus-reactive serum immunoglobulin m antibody in confirmed west nile virus encephalitis cases. Emerg Infect Dis 2003; 9(3):376–379.
39. Denne C, Kleines M, Dieckhofer A, et al. Intrathecal synthesis of anti-viral antibodies in pediatric patients. Eur J Paediatr Neurol 2007; 11(1):29–34.
40. Monteyne P, Albert F, Weissbrich B, et al. The detection of intrathecal synthesis of anti-herpes simplex IgG antibodies: Comparison between an antigen-mediated immunoblotting technique and antibody index calculations. European Union Concerted Action on Virus Meningitis and Encephalitis. J Med Virol 1997; 53(4):324–331.
41. Ahmed A, Brito F, Goto C, et al. Clinical utility of the polymerase chain reaction for diagnosis of enteroviral meningitis in infancy. J Pediatr 1997; 131(3):393–397.
42. Doong H, Xu J, McCormac M, et al. Duplex eQ-PCR for simultaneous detection and differentiation of herpes simplex viruses and enteroviruses in cerebrospinal fluid. In: 108th General Meeting of the American Society for Microbiology Boston, Massachusetts, 2008.
43. Espy MJ, Uhl JR, Sloan LM, et al. Real-time PCR in clinical microbiology: Applications for routine laboratory testing. Clin Microbiol Rev 2006; 19(1):165–256.

44. Gautheret-Dejean A, Manichanh C, Thien-Ah-Koon F, et al. Development of a real-time polymerase chain reaction assay for the diagnosis of human herpesvirus-6 infection and application to bone marrow transplant patients. J Virol Methods 2002; 100(1–2):27–35.
45. Kost CB, Rogers B, Oberste MS, et al. Multicenter beta trial of the GeneXpert Enterovirus assay. J Clin Microbiol 2007; 45(4):1081–1086.
46. Powell KF, Anderson NE, Frith RW, et al. Non-invasive diagnosis of herpes simplex encephalitis. Lancet 1990; 335(8685):357–358.
47. Rabenau HF, Clarici A, Mühlbauer G, et al. Rapid detection of enterovirus infection by automated RNA extraction and real-time fluorescence PCR. J Clin Virol 2002; 25(2):155–164.
48. Rotbart HA, Sawyer MH, Fast S, et al. Diagnosis of enteroviral meningitis by using PCR with a colorimetric microwell detection assay. J Clin Microbiol 1994; 32(10):2590–2592.
49. Rouah E, Rogers BB, Wilson DR, et al. Demonstration of Epstein-Barr virus in primary central nervous system lymphomas by the polymerase chain reaction and in situ hybridization. Hum Pathol 1990; 21(5):545–550.
50. Rowley AH, Whitley RJ, Lakeman FD, et al. Rapid detection of herpes-simplex-virus DNA in cerebrospinal fluid of patients with herpes simplex encephalitis. Lancet 1990; 335(8687):440–441.
51. Tang YW, Rys PN, Rutledge BJ, et al. Comparative evaluation of colorimetric microtiter plate systems for detection of herpes simplex virus in cerebrospinal fluid. J Clin Microbiol 1998; 36(9):2714–2717.
52. Fox JD, Han S, Samuelson A, et al. Development and evaluation of nucleic acid sequence based amplification (NASBA) for diagnosis of enterovirus infections using the NucliSens Basic Kit. J Clin Virol 2002; 24(1–2):117–130.
53. Landry ML, Garner R, Ferguson D. Real-time nucleic acid sequence-based amplification using molecular beacons for detection of enterovirus RNA in clinical specimens. J Clin Microbiol 2005; 43(7):3136–3139.
54. Allawi HT, Li H, Sander T, et al. Invader plus method detects herpes simplex virus in cerebrospinal fluid and simultaneously differentiates types 1 and 2. J Clin Microbiol 2006; 44(9):3443–3447.
55. Debiasi RL, Tyler KL. Molecular methods for diagnosis of viral encephalitis. Clin Microbiol Rev 2004; 17(4):903–925, table of contents.
56. van der Meer JT, Drew WL, Bowden RA, et al. Summary of the International Consensus Symposium on advances in the diagnosis, treatment and prophylaxis and cytomegalovirus infection. Antiviral Res 1996; 32(3):119–140.
57. Mantero G, Zonaro A, Albertini A, et al. DNA enzyme immunoassay: General method for detecting products of polymerase chain reaction. Clin Chem 1991; 37(3):422–429.
58. Kleinschmidt-DeMasters BK, DeBiasi RL, Tyler KL. Polymerase chain reaction as a diagnostic adjunct in herpesvirus infections of the nervous system. Brain Pathol 2001; 11(4):452–464.
59. Weinberg A, Bloch KC, Li S, et al. Dual infections of the central nervous system with Epstein-Barr virus. J Infect Dis 2005; 191(2):234–237.
60. Boutolleau D, Duros C, Bonnafous P, et al. Identification of human herpesvirus 6 variants A and B by primer-specific real-time PCR may help to revisit their respective role in pathology. J Clin Virol 2006; 35(3):257–263.
61. Schloss L, van Loon AM, Cinque P, et al. An international external quality assessment of nucleic acid amplification of herpes simplex virus. J Clin Virol 2003; 28(2):175–185.
62. Van Vliet KE, Muir P, Echevarria JM, et al. Multicenter proficiency testing of nucleic acid amplification methods for the detection of enteroviruses. J Clin Microbiol 2001; 39(9):3390–3392.
63. Kupila L, Vuorinen T, Vainionpaa R, et al. Diagnosis of enteroviral meningitis by use of polymerase chain reaction of cerebrospinal fluid, stool, and serum specimens. Clin Infect Dis 2005; 40(7):982–987.
64. Lakeman FD, Whitley RJ; National Institute of Allergy and Infectious Diseases Collaborative Antiviral Study Group. Diagnosis of herpes simplex encephalitis: Application of polymerase chain reaction to cerebrospinal fluid from brain-biopsied patients and correlation with disease. J Infect Dis 1995; 171(4):857–863.
65. Whitley RJ, Cobbs CG, Alford CA Jr., et al; NIAD Collaborative Antiviral Study Group. Diseases that mimic herpes simplex encephalitis. Diagnosis, presentation, and outcome. JAMA 1989; 262(2): 234–239.
66. Mori I, Nishiyama Y. Herpes simplex virus and varicella-zoster virus: Why do these human alphaherpesviruses behave so differently from one another? Rev Med Virol 2005; 15(6):393–406.
67. O'Sullivan CE, Aksamit AJ, Harrington JR, et al. Clinical spectrum and laboratory characteristics associated with detection of herpes simplex virus DNA in cerebrospinal fluid. Mayo Clin Proc 2003; 78(11):1347–1352.
68. Corey L, Spear PG. Infections with herpes simplex viruses (2). N Engl J Med 1986; 314(12):749–757.
69. Whitley RJ, Soong SJ, Linneman C Jr., et al. Herpes simplex encephalitis. Clinical Assessment. JAMA 1982; 247(3):317–320.

70. Tyler KL. Herpes simplex virus infections of the central nervous system: Encephalitis and meningitis, including Mollaret's. Herpes 2004; 11(suppl 2):57A–64A.
71. Bergstrom T, Vahlne A, Alestig K, et al. Primary and recurrent herpes simplex virus type 2-induced meningitis. J Infect Dis 1990; 162(2):322–330.
72. Tang YW. Laboratory Diagnosis of CNS Infections by Molecular Amplification Techniques. Expert Opin Med Diagn 2007; 1(4):489–509.
73. Kimberlin DW, Lakeman FD, Arvin AM, et al; National Institute of Allergy and Infectious Diseases Collaborative Antiviral Study Group. Application of the polymerase chain reaction to the diagnosis and management of neonatal herpes simplex virus disease. J Infect Dis 1996; 174(6):1162–1167.
74. Weil AA, Glaser CA, Amad Z, et al. Patients with suspected herpes simplex encephalitis: Rethinking an initial negative polymerase chain reaction result. Clin Infect Dis 2002; 34(8):1154–1157.
75. Cinque P, Cleator GM, Weber T, et al. The role of laboratory investigation in the diagnosis and management of patients with suspected herpes simplex encephalitis: A consensus report. The EU Concerted Action on Virus Meningitis and Encephalitis. J Neurol Neurosurg Psychiatry 1996; 61(4):339–345.
76. Malm G, Forsgren M. Neonatal herpes simplex virus infections: HSV DNA in cerebrospinal fluid and serum. Arch Dis Child Fetal Neonatal Ed 1999; 81(1):F24–F29.
77. Simko JP, Caliendo AM, Hogle K, et al. Differences in laboratory findings for cerebrospinal fluid specimens obtained from patients with meningitis or encephalitis due to herpes simplex virus (HSV) documented by detection of HSV DNA. Clin Infect Dis 2002; 35(4):414–419.
78. Tang YW, Hibbs JR, Tau KR, et al. Effective use of polymerase chain reaction for diagnosis of central nervous system infections. Clin Infect Dis 1999; 29(4):803–806.
79. Hanson KE, Alexander BD, Woods C, et al. Validation of laboratory screening criteria for herpes simplex virus testing of cerebrospinal fluid. J Clin Microbiol 2007; 45(3):721–724.
80. Fodor PA, Levin MJ, Weinberg A, et al. Atypical herpes simplex virus encephalitis diagnosed by PCR amplification of viral DNA from CSF. Neurology 1998; 51(2):554–559.
81. Auyeung P, Dunn A. Atypical case of Herpes simplex encephalitis. Intern Med J 2008; 38(4):294–295.
82. Nahmias AJ, Whitley RJ, Visintine AN, et al. Herpes simplex virus encephalitis: Laboratory evaluations and their diagnostic significance. J Infect Dis 1982; 145(6):829–836.
83. Gilden DH, Mahalingam R, Cohrs RJ, et al. Herpesvirus infections of the nervous system. Nat Clin Pract Neurol 2007; 3(2):82–94.
84. Gilden DH, Kleinschmidt-DeMasters BK, LaGuardia JJ, et al. Neurologic complications of the reactivation of varicella-zoster virus. N Engl J Med 2000; 342(9):635–645.
85. Steiner I, Kennedy PG, Pachner AR. The neurotropic herpes viruses: Herpes simplex and varicella-zoster. Lancet Neurol 2007; 6(11):1015–1028.
86. Koskiniemi M, Piiparinen H, Rantalaiho T, et al. Acute central nervous system complications in varicella zoster virus infections. J Clin Virol 2002; 25(3):293–301.
87. Ihekwaba UK, Kudesia G, McKendrick MW. Clinical features of viral meningitis in adults: Significant differences in cerebrospinal fluid findings among herpes simplex virus, varicella zoster virus, and enterovirus infections. Clin Infect Dis 2008; 47(6):783–789.
88. Hilt DC, Buchholz D, Krumholz A, et al. Herpes zoster ophthalmicus and delayed contralateral hemiparesis caused by cerebral angiitis: Diagnosis and management approaches. Ann Neurol 1983; 14(5):543–553.
89. Gilden D. Varicella zoster virus and central nervous system syndromes. Herpes 2004; 11(suppl 2):89A–94A.
90. Nagel MA, Cohrs RJ, Mahalingam R, et al. The varicella zoster virus vasculopathies: Clinical, CSF, imaging, and virologic features. Neurology 2008; 70(11):853–860.
91. Gregoire SM, van Pesch V, Goffette S, et al. Polymerase chain reaction analysis and oligoclonal antibody in the cerebrospinal fluid from 34 patients with varicella-zoster virus infection of the nervous system. J Neurol Neurosurg Psychiatry 2006; 77(8):938–942.
92. Puchhammer-Stockl E, Popow-Kraupp T, Heinz FX, et al. Detection of varicella-zoster virus DNA by polymerase chain reaction in the cerebrospinal fluid of patients suffering from neurological complications associated with chicken pox or herpes zoster. J Clin Microbiol 1991; 29(7):1513–1516.
93. Aberle SW, Aberle JH, Steininger C, et al. Quantitative real time PCR detection of Varicella-zoster virus DNA in cerebrospinal fluid in patients with neurological disease. Med Microbiol Immunol 2005; 194(1–2):7–12.
94. Burke DG, Kalayjian RC, Vann VR, et al. Polymerase chain reaction detection and clinical significance of varicella-zoster virus in cerebrospinal fluid from human immunodeficiency virus-infected patients. J Infect Dis 1997; 176(4):1080–1084.
95. Cinque P, Bossolasco S, Vago L, et al. Varicella-zoster virus (VZV) DNA in cerebrospinal fluid of patients infected with human immunodeficiency virus: VZV disease of the central nervous system or subclinical reactivation of VZV infection? Clin Infect Dis 1997; 25(3):634–639.

96. Iten A, Chatelard P, Vuadens P, et al. Impact of cerebrospinal fluid PCR on the management of HIV-infected patients with varicella-zoster virus infection of the central nervous system. J Neurovirol 1999; 5(2):172–180.
97. De La Blanchardiere A, Rozenberg F, Caumes E, et al. Neurological complications of varicella-zoster virus infection in adults with human immunodeficiency virus infection. Scand J Infect Dis 2000; 32(3):263–269.
98. Crawford DH, Macsween KF, Higgins CD, et al. A cohort study among university students: Identification of risk factors for Epstein-Barr virus seroconversion and infectious mononucleosis. Clin Infect Dis 2006; 43(3):276–282.
99. Straus SE, Cohen JI, Tosato G, et al. NIH conference. Epstein-Barr virus infections: Biology, pathogenesis, and management. Ann Intern Med 1993; 118(1):45–58.
100. Kleinschmidt-DeMasters BK, Gilden DH. The expanding spectrum of herpesvirus infections of the nervous system. Brain Pathol 2001; 11(4):440–451.
101. Evans RW. Case studies of uncommon headaches. Neurol Clin 2006; 24(2):347–362.
102. Liaw SB, Shen EY. Alice in Wonderland syndrome as a presenting symptom of EBV infection. Pediatr Neurol 1991; 7(6):464–466.
103. Hochberg FH, Miller G, Schooley RT, et al. Central-nervous-system lymphoma related to Epstein-Barr virus. N Engl J Med 1983; 309(13):745–748.
104. Weinberg A, Li S, Palmer M, et al. Quantitative CSF PCR in Epstein-Barr virus infections of the central nervous system. Ann Neurol 2002; 52(5):543–548.
105. Cinque P, Scarpellini P, Vago L, et al. Diagnosis of central nervous system complications in HIV-infected patients: Cerebrospinal fluid analysis by the polymerase chain reaction. AIDS 1997; 11(1): 1–17.
106. Cinque P, Vago L, Dahl H, et al. Polymerase chain reaction on cerebrospinal fluid for diagnosis of virus-associated opportunistic diseases of the central nervous system in HIV-infected patients. AIDS 1996; 10(9):951–958.
107. Cinque P, Brytting M, Vago L, et al. Epstein-Barr virus DNA in cerebrospinal fluid from patients with AIDS-related primary lymphoma of the central nervous system. Lancet 1993; 342(8868):398–401.
108. al-Shahi R, Bower M, Nelson MR, et al. Cerebrospinal fluid Epstein-Barr virus detection preceding HIV-associated primary central nervous system lymphoma by 17 months. J Neurol 2000; 247(6): 471–472.
109. Corcoran C, Rebe K, van der Plas H, et al. The predictive value of cerebrospinal fluid Epstein-Barr viral load as a marker of primary central nervous system lymphoma in HIV-infected persons. J Clin Virol 2008; 42(4):433–436.
110. Ivers LC, Kim AY, Sax PE. Predictive value of polymerase chain reaction of cerebrospinal fluid for detection of Epstein-Barr virus to establish the diagnosis of HIV-related primary central nervous system lymphoma. Clin Infect Dis 2004; 38(11):1629–1632.
111. Grose C, Henle W, Henle G, et al. Primary Epstein-Barr-virus infections in acute neurologic diseases. N Engl J Med 1975; 292(8):392–395.
112. Domachowske JB, Cunningham CK, Cummings DL, et al. Acute manifestations and neurologic sequelae of Epstein-Barr virus encephalitis in children. Pediatr Infect Dis J 1996; 15(10):871–875.
113. Imai S, Usui N, Sugiura M, et al. Epstein-Barr virus genomic sequences and specific antibodies in cerebrospinal fluid in children with neurologic complications of acute and reactivated EBV infections. J Med Virol 1993; 40(4):278–284.
114. Joncas JH, Chicoine L, Thivierge F, et al. Epstein-Barr virus antibodies in the cerebrospinal fluid. A case of infectious mononucleosis with encephalitis. Am J Dis Child 1974; 127(2):282–285.
115. Anonymous. Arboviral infections of the central nervous system–United States, 1996–1997. MMWR Morb Mortal Wkly Rep 1998; 47(25):517–522.
116. Calisher CH. Medically important arboviruses of the United States and Canada. Clin Microbiol Rev 1994; 7(1):89–116.
117. Lanciotti RS, Roehrig JT, Deubel V, et al. Origin of the West Nile virus responsible for an outbreak of encephalitis in the northeastern United States. Science 1999; 286(5448):2333–2337.
118. Nash D, Mostashari F, Fine A, et al. The outbreak of West Nile virus infection in the New York City area in 1999. N Engl J Med 2001; 344(24):1807–1814.
119. Gould EA, Solomon T. Pathogenic flaviviruses. Lancet 2008; 371(9611):500–509.
120. Johnson RT. Emerging viral infections of the nervous system. J Neurovirol 2003; 9(2):140–147.
121. Watson JT, Pertel PE, Jones RC, et al. Clinical characteristics and functional outcomes of West Nile Fever. Ann Intern Med 2004; 141(5):360–365.
122. Voelker R. Effects of West nile virus may persist. JAMA 2008; 299(18):2135–2136.
123. Chowers MY, Lang R, Nassar F, et al. Clinical characteristics of the West Nile fever outbreak, Israel, 2000. Emerg Infect Dis 2001; 7(4):675–678.

124. Murray KO, Sierra M, Resnick M, et al. Long-term Clinical Sequelae and Rates of Recovery Following Infection with West Nile VIrus. In: International Conference on Emerging Infectious Diseases, March 17, 2008, Atlanta, Georgia, U.S.A.
125. Dauphin G, Zientara S. West Nile virus: Recent trends in diagnosis and vaccine development. Vaccine 2007; 25(30):5563–5576.
126. Tyler KL. West Nile virus infection in the United States. Arch Neurol 2004; 61(8):1190–1195.
127. Hindiyeh M, Shulman LM, Mendelson E, et al. Isolation and characterization of West Nile virus from the blood of viremic patients during the 2000 outbreak in Israel. Emerg Infect Dis 2001; 7(4): 748–750.
128. Busch MP, Caglioti S, Robertson EF, et al. Screening the blood supply for West Nile virus RNA by nucleic acid amplification testing. N Engl J Med 2005; 353(5):460–467.
129. Tilley PA, Fox JD, Jayaraman GC, et al. Nucleic acid testing for west nile virus RNA in plasma enhances rapid diagnosis of acute infection in symptomatic patients. J Infect Dis 2006; 193(10):1361–1364.
130. Lanciotti RS, Kerst AJ, Nasci RS, et al. Rapid detection of west nile virus from human clinical specimens, field-collected mosquitoes, and avian samples by a TaqMan reverse transcriptase-PCR assay. J Clin Microbiol 2000; 38(11):4066–4071.
131. Hiatt B, DesJardin L, Carter T, et al. A fatal case of West Nile virus infection in a bone marrow transplant recipient. Clin Infect Dis 2003; 37(9):e129–e131.
132. Huang C, Slater B, Rudd R, et al. First isolation of West Nile virus from a patient with encephalitis in the United States. Emerg Infect Dis 2002; 8(12):1367–1371.
133. Iwamoto M, Jernigan DB, Guasch A, et al. Transmission of West Nile virus from an organ donor to four transplant recipients. N Engl J Med 2003; 348(22):2196–2203.
134. Petersen LR, Marfin AA, Gubler DJ. West Nile virus. JAMA 2003; 290(4):524–528.
135. Roos KL. Encephalitis. Neurol Clin 1999; 17(4):813–833.
136. Day JF. Predicting St. Louis encephalitis virus epidemics: Lessons from recent, and not so recent, outbreaks. Annu Rev Entomol 2001; 46:111–138.
137. Whitley RJ, Gnann JW. Viral encephalitis: Familiar infections and emerging pathogens. Lancet 2002; 359(9305):507–513.
138. Bloch KC, Glaser C. Diagnostic approaches for patients with suspected encephalitis. Curr Infect Dis Rep 2007; 9(4):315–322.
139. Lambert AJ, Martin DA, Lanciotti RS. Detection of North American eastern and western equine encephalitis viruses by nucleic acid amplification assays. J Clin Microbiol 2003; 41(1):379–385.
140. Bale JF Jr. Viral encephalitis. Med Clin North Am 1993; 77(1):25–42.
141. Read SJ, Jeffery KJ, Bangham CR. Aseptic meningitis and encephalitis: The role of PCR in the diagnostic laboratory. J Clin Microbiol 1997; 35(3):691–696.
142. Cinque P, Bossolasco S, Lundkvist A. Molecular analysis of cerebrospinal fluid in viral diseases of the central nervous system. J Clin Virol 2003; 26(1):1–28.
143. Lambert AJ, Nasci RS, Cropp BC, et al. Nucleic acid amplification assays for detection of La Crosse virus RNA. J Clin Microbiol 2005; 43(4):1885–1889.
144. Centers for Disease Control and Prevention. Case definitions for infectious conditions under public health surveillance. MMWR Recomm Rep 1997; 46(RR-10):1–55.
145. Khetsuriani N, Lamonte-Fowlkes A, Oberst S, et al. Enterovirus surveillance–United States, 1970–2005. MMWR Surveill Summ 2006; 55(8):1–20.
146. Pallansch MA, Roos RP. Enteroviruses: Polioviruses, Coxsackieviruses, Echoviruses, and New Enteroviruses, 4th ed. Philadelphia, PA: Lippincott Williams and Wilkins, 2001.
147. Wolthers KC, Benschop KS, Schinkel J, et al. Human parechoviruses as an important viral cause of sepsis like illness and meningitis in young children. Clin Infect Dis 2008; 47(3):358–363.
148. Anonymous. Outbreaks of aseptic meningitis associated with echoviruses 9 and 30 and preliminary surveillance reports on enterovirus activity–United States, 2003. MMWR Morb Mortal Wkly Rep 2003; 52(32):761–764.
149. Huang CC, Liu CC, Chang YC, et al. Neurologic complications in children with enterovirus 71 infection. N Engl J Med 1999; 341(13):936–942.
150. Chang LY, Huang LM, Gau SS, et al. Neurodevelopment and cognition in children after enterovirus 71 infection. N Engl J Med 2007; 356(12):1226–1234.
151. King RL, Lorch SA, Cohen DM, et al. Routine cerebrospinal fluid enterovirus polymerase chain reaction testing reduces hospitalization and antibiotic use for infants 90 days of age or younger. Pediatrics 2007; 120(3):489–496.
152. Tavakoli NP, Wang H, Nattanmai S, et al. Detection and typing of enteroviruses from CSF specimens from patients diagnosed with meningitis/encephalitis. J Clin Virol 2008; 43(2):207–211.
153. Chung PW, Huang YC, Chang LY, et al. Duration of enterovirus shedding in stool. J Microbiol Immunol Infect 2001; 34(3):167–170.

154. Anonymous. Human rabies prevention–United States, 1999. Recommendations of the Advisory Committee on Immunization Practices (ACIP). MMWR Recomm Rep 1999; 48:1–21.
155. Lapierre V, Tiberghien P. Transmission of rabies from an organ donor. N Engl J Med 2005; 352(24):2552; author reply.
156. Fishbein DB, Robinson LE. Rabies. N Engl J Med 1993; 329(22):1632–1638.
157. Hankins DG, Rosekrans JA. Overview, prevention, and treatment of rabies. Mayo Clin Proc 2004; 79(5):671–676.
158. Crepin P, Audry L, Rotivel Y, et al. Intravitam diagnosis of human rabies by PCR using saliva and cerebrospinal fluid. J Clin Microbiol 1998; 36(4):1117–1121.
159. Nagaraj T, Vasanth JP, Desai A, et al. Ante mortem diagnosis of human rabies using saliva samples: Comparison of real time and conventional RT-PCR techniques. J Clin Virol 2006; 36(1):17–23.
160. Anonymous. Human rabies–Tennessee, 2002. MMWR Morb Mortal Wkly Rep 2002; 51(37):828–829.

22 Sexually Transmitted Infections

Anna Wald
Departments of Medicine, Laboratory Medicine, Epidemiology, University of Washington, and Vaccine and Infectious Disease Institute, Fred Hutchinson Cancer Research Center, Seattle, Washington, U.S.A.

Patti E. Gravitt
Department of Epidemiology, Johns Hopkins Bloomberg School of Public Health, Baltimore, Maryland, U.S.A.

Rhoda A. Morrow
Department of Laboratory Medicine, University of Washington, and Seattle Children's, Seattle, Washington, U.S.A.

INTRODUCTION

Infections with herpes simplex virus and human papilloma virus cause the most common viral sexually transmitted infections. Despite sexual intercourse as the most common route of transmission for both viruses, HSV and HPV have distinct clinical manifestations, and different approaches are used for laboratory diagnosis. Therefore, these two infections will be discussed separately. Molluscum contagiosum, a viral skin infection without serious health sequelae that often affects the genital area is also briefly discussed.

HSV and HPV are not the only sexually transmitted viral pathogens. Worldwide, sexual intercourse is also the predominant mode of spread for HIV and hepatitis B; these are covered in chapters 23 and 19. Hepatitis C and hepatitis A can also be transmitted during sexual activity, although for neither virus this is a common or efficient mode of transmission (chap. 19).

HERPES SIMPLEX VIRUS

Clinical Presentation

Human herpes simplex virus type 1 and type 2 are among the most ubiquitous human infections. Both are alphaherpesviruses, and cause syndromes that are indistinguishable without laboratory testing despite a propensity for somewhat different clinical manifestations caused by each virus, or different natural history. Table 1 shows the clinical syndromes caused by HSV-1 and HSV-2 and the relative frequency of each type.

Genital herpes is caused by HSV-1 and HSV-2. Clinical episodes of genital herpes can reflect either new infection in an immunologically naïve host or recurrent infection that results from viral reactivation. While the severity of infection is in general greater during newly acquired infection, the exact scenario in a given person cannot be determined without laboratory testing. New genital HSV-1 or HSV-2 infection in a person that is seronegative for both viruses is termed primary infection and is clinically the most severe. New genital HSV-2 infection in a person previously infected with HSV-1, termed nonprimary initial infection, tends to be less severe, and occurs with comparable frequency to primary infection, as HSV-1 appears to give only marginal protection against HSV-2. The converse, nonprimary HSV-1 in a person previously infected with HSV-2 is rare, as HSV-2 appears to protect against HSV-1 acquisition. Recurrent infection is most mild and results from reactivation of the previously acquired infection and reinfection of the genital epithelium.

Newly acquired genital HSV-1 and HSV-2 infections result from contact with infected oral or genital secretions during sexual activity. Fever, malaise, headache are common and last one to two weeks (1). Most patients who seek medical care have genital pain or notice lesions. The initial lesions begin as erythematous papules, and progress rapidly to small vesicles. These ulcerate, causing shallow erosions with an erythematous margin. The number of lesions varies widely from a few to more than a hundred. In women, the ulcer stage can be prolonged as

Table 1 Predilection of HSV Types to Cause Clinical Syndromes

Syndrome	HSV-1	HSV-2
Genital herpes, initial episode	+++	+++
Genital herpes, recurrent	+	+++
Orolabial herpes	+++	+
HSV encephalitis	+++	
HSV meningitis	+	+++
Neonatal herpes	++	++
Herpes gladiatorum	++	
HSV keratitis/iritis	+++	
Acute retinal necrosis	+	++
Whitlow	+	++

vesicles are often not noted, and the ulcers can coalesce. The vulva, perineum, and perianal area are frequently involved; bilateral involvement is characteristic of primary infection. Women often have cervicitis and occasionally present with isolated vaginal discharge without other genital symptoms. Urethritis is also common and may result from irritation of vulvar lesions by the urinary stream or from internal urethral lesions. In men, the lesions tend to involve the penile shaft; perianal involvement is common among men who have sex with men. Urethritis and proctitis can also occur. In men, the ulcers may dry rapidly and crust before healing occurs. The genital lesions during primary infection can last three weeks, with new lesions continuing to form for first two weeks. Bilateral inguinal lymphadenopathy is common and often markedly tender. Neurological symptoms, such as aseptic meningitis or urinary retention often begin during second and third week of illness, and resolve slowly but almost always completely in the following weeks. Rare patients report persistent neuropathic pain in the genital area or other areas enervated by sacral nerves.

Recurrent genital herpes has a milder course with less numerous lesions and no or few systemic symptoms or adenopathy. Usually the lesions are unilateral and heal within 5 to 10 days without therapy. In many patients, some episodes are preceded by neuropathic pain in the area of the recurrence, or distally in the distribution of the affected nerve. This "prodrome" is often the most bothersome symptom of recurrence, and one least responsive to antiviral therapy.

The natural history of genital HSV-1 and HSV-2 differ. Almost all patients with genital HSV-2 infection will have recurrences, especially during the initial years of infection (2). In a well-characterized cohort of 326 patients with documented acquisition of HSV-2, the median rate of recurrences was four among women and five among men in the first year; 20% had more than 10 recurrences during that time. Over time, there is a slow decrease in the frequency of recurrences, although about a quarter of patients report more recurrences in year 5 than in year 1 of infection (3). In contrast, many persons with genital HSV-1 infection will have a single recurrence during the initial year of infection, and only a few recurrences in subsequent years (4).

In immunocompromised patients, the lesions can be extensive, extend deeply into the submucosa, and persist for many months without treatment. They may also be located in places infrequently affected by HSV in immunocompetent persons or extend from the mucosa into keratinized skin. The atypical appearance and location of the lesions can delay diagnosis; occasionally a biopsy is performed that demonstrates typical histopathology. However, such lesions usually have a high load of virions, and swabs of ulcers will easily yield virus in laboratory tests.

Viral Shedding

Intermittent shedding from mucosa is a hallmark feature of herpes simplex virus infections. Following initial replication in the epithelium, HSV enters the nerve cell and travels to the ganglia. In animal models, ganglionitis ensues, with resultant spread of the virus to adjacent and contralateral ganglia. The virus then persists in the ganglia for the life of the host (5). At this time, viral replication is limited and the repertoire of expressed genes differs from productive infection. Animal and limited human data indicate that this latent state reflects active control of viral replication by CD8 cells within the ganglia. However, the immune control at the ganglion level is leaky, as virus ends up traveling centrifugally to the mucosa at frequent intervals. Therefore, HSV-specific CD8 cells also cluster around nerve endings in the mucosa and limit epithelial viral replication (6).

The frequency of viral reactivation in the human host has been studied using daily swabs of the genital mucosa. Overall, the shedding rate in immunocompetent persons varies widely, with a mean about 12% to 20%, depending on population under study (7). However, some people can have HSV present on genital mucosa up to 70% to 90% of days sampled, especially during the first year following acquisition of HSV-2 (8). Others have very infrequent shedding. Studies that use frequent sampling (four times a day) have found that many of the episodes of reactivation are short, with approximately 50% lasting 12 hours or less (9). Almost all of the short lasting episodes are asymptomatic, and the risk of a clinically recognized recurrence increases with the length of the shedding. Similarly, the copy number found in samples is higher when a lesion is present and the duration of shedding is longer. Yet even during short episodes of reactivation, the viral copy number often exceeds 10^3 per 1 mL of specimen, which has been associated with viral isolation and transmission of HSV to the neonate at delivery (10). In persons who have clinically evident genital HSV-2 disease, about one-third of the viral shedding occurs in between recognized genital lesions. The asymptomatic viral shedding reflects the biology of the virus rather than lack of appreciation of difficult to find lesions, as days of viral detection without accompanying lesions have been noted in research studies in which the participants are carefully examined by experienced clinicians.

The pattern of viral shedding differs in persons who are HSV-2 seropositive but without a history of genital herpes. Overall, the shedding rate in such persons is lower compared with those who have a history of genital herpes, because such persons predominantly have asymptomatic viral shedding (11). Once diagnosed with a serologic test, or after transmitting to a partner, most asymptomatic persons will begin to recognize recurrent herpes, albeit their recurrences are short and infrequent, probably explaining why they have not been diagnosed with herpes.

Clinically, viral shedding is important as a predictor of the risk of transmission. In about 85% of transmission events, the source partner does not have a history of genital herpes; in these cases it is difficult to say whether transmission occurred in the absence of lesions, or the lesions were mild and unrecognized. However, even among persons who have a clear history of genital herpes, ~75% of transmissions will occur during asymptomatic shedding (12). The relative risk of transmission in the presence of viral shedding has been quantified best during maternal to child HSV transmission. The risk of neonatal herpes is 345-fold higher when HSV is isolated from genital secretions at the time of birth compared with lack of HSV isolation (13). Because sexual transmission involves multiple exposures, the risk associated with viral shedding is more difficult to quantify. In addition, sexual behavior of persons with frequent genital lesions is likely different than those who are asymptomatic. However, even among persons with newly diagnosed herpes, 20% engage in sexual activity prior to the healing of lesions (14).

Epidemiology

Herpes simplex virus infections are ubiquitous in human populations. Because the clinical manifestations are absent or variable, epidemiologic studies have relied on serological surveys to define the distribution of HSV in various populations. Historically, oral HSV-1 was acquired in childhood through close contact within the family, and almost all persons were HSV-1 seropositive. Universal HSV-1 infection is still observed in serosurveys conducted in developing countries of Asia or Africa. Improvement in the standard of living and hygiene has resulted in later age of acquisition of HSV-1 in North America and Europe, so that overall, only 60% of persons in the United States are HSV-1 seropositive (15). In the last decade, an increasing proportion of genital herpes has been attributed to HSV-1 infection, perhaps due to lack of HSV-1 infection in childhood, and first encounter with HSV-1 during initiation of sexual activity.

In contrast to HSV-1, HSV-2 infection affects a smaller proportion of the population, especially in developed countries. Most recent estimates show that in the United States 17% of persons aged 14 to 49 are HSV-2 seropositive (15). Among women of childbearing age, 22% are

HSV-2 seropositive; the rates are about 15% in men of comparable age (16). In Europe, most countries report an HSV-2 seroprevalence of 10% to 20%. In contrast, studies among general population in Africa report a higher prevalence. For example, HSV-2 prevalence was over 50% among women and over 25% in men in a survey of four urban areas in sub-Saharan Africa (17). Fewer studies are available from Asia: a study among pregnant women in China showed HSV-2 prevalence of 11% (18).

The risk of HSV-2 in a person reflects both the background HSV-2 seroprevalence in the community of sexual partners and the individual sexual behavior. Thus, people who practice high-risk sexual behavior, including men who have sex with men (MSM) and commercial sex workers, have higher HSV-2 prevalence, ranging from 70% to 100% in some surveys. Teenage girls in South Africa acquire HSV-2 very rapidly after initiation of sexual activity, with prevalence exceeding 60% by the age of 21 (19).

Serosurveys for HSV-2 not only allow for monitoring of sexual mores of the population but also herald the spread of HIV within the community. The risk of HIV acquisition is increased two- to fourfold in people who also have HSV-2 infection; the elevated risk has been noted among HSV-2 seropositive women (RR = 3), heterosexual men (RR = 3), and men who have sex with men (RR = 1.7) (20,21). The attenuation of the relative risk among MSM probably reflects the independent risk for HIV infection conferred by receptive anal intercourse.

Neonatal HSV is a severe infection acquired by infants through exposure to infected genital secretions during birth. Women who are HSV-2 infected, regardless of history of genital herpes, are at risk for transmitting HSV-2 to their newborns. However, despite fairly frequent exposure to genital shedding of HSV during birth, less than 1% of infants born to women with established HSV-2 infection develop neonatal HSV infection. In contrast, women who acquire genital HSV toward the end of pregnancy are at very high risk (30–50%) of transmitting HSV to the neonate (22). The reasons probably include lack of maternal antibodies that are transferred to the fetus in the last month of gestation, and perhaps the high viral load that accompanies HSV infection in an immunologically naïve host. Recently, HSV-1 has become a frequent cause of neonatal HSV, paralleling the rise in genital HSV-1 infection (23).

Laboratory Testing and Differential Diagnosis

The diagnosis of genital herpes is difficult to establish without laboratory testing. While the classical presentation of multiple vesicular or ulcerative lesions is sometimes diagnostic, this presentation accounts for minority of patients who have genital herpes. Most persons with genital HSV-2 and HSV-1 infection have mild, nonspecific symptoms that may not be bothersome and often do not result in a visit to a medical provider. The difficulty in clinical diagnosis of genital herpes was illustrated in a study of a candidate vaccine, in which people at known risk for HSV-2 were prospectively followed for acquisition (24). Among 155 persons who acquired HSV-2, only 39% were diagnosed at the time of the acquisition. Of note, 20% of persons who were told that they have genital herpes at the time of the presentation did not have the infection after careful laboratory evaluation. Thus, the clinical diagnosis is neither sensitive nor specific, and laboratory confirmation should be sought in every case, including in persons who have had the clinical diagnosis without prior laboratory tests.

In a patient presenting with genital lesions, viral culture has been the gold standard for diagnosis. However, this technique is highly dependent on (*i*) stage of the lesions, with vesicular or early ulcerative lesions having the highest yield, (*ii*) type of infection, with primary infection yielding virus in ~80% compared with ~40% for recurrent lesions, (*iii*) technique of swabbing and transport time to the laboratory. When viral culture is used as the only laboratory test, and the result is negative, the patient (and the physician) may believe that he or she does not have genital HSV infection. As such, the utility of viral culture is limited and amplification of the viral genome is likely to replace culture in many settings. In comparison to viral culture, HSV DNA is up to four times more sensitive, is resistant to degradation in the environment, and can be performed rapidly (25). The appreciation of HSV as the most common cause of genital ulcers worldwide did not occur until the development of molecular diagnostic techniques, since viral culture has especially low yield in tropical countries. Direct antigen detection is available in some settings and it is similar in sensitivity to viral culture.

Many university and commercial laboratories offer HSV DNA PCR for diagnostic purposes. Initially, the test was developed for use in central nervous system infections with HSV in adults and neonates, as viral culture is almost never positive from the spinal fluid, and availability of HSV PCR has reduced the need for brain biopsies in patients with encephalitis (26). In the last decade, the use of HSV PCR has expanded to genital ulcer disease, and is likely to continue to replace viral culture. The test has not been commercialized, and therefore, the sensitivity of the assay is likely to vary between labs.

The differential diagnosis of herpes simplex includes both infectious and noninfectious etiologies of genital lesions. Among the STDs, chancroid and syphilis also present with genital ulcers. While certain morphologic characteristics are more typical of syphilitic or chancroidal ulcerations, clinical acumen is not sufficient for accurate diagnosis, and laboratory confirmation is desirable. The difficulty in establishing a clinical diagnosis is even greater among patients who have concomitant HIV infection. Knowledge of local epidemiology, and patients' travel and sexual history are needed to establish whether syphilis or chancroid should be in the differential. Mixed infections are also possible. Other genital infections that may mimic HSV include candidal infections and scabies. Trauma is probably the most common noninfectious reason for genital lesions. Systemic illness such as Crohn's disease, erythema multiforme, Behcet's disease, or fixed drug eruption can also mimic genital herpes.

Type-Specific Serologic Tests

In many circumstances, a diagnosis of HSV infection must be made in the absence of genital lesions. Examples include patients who have recurrent lesions, but whose cultures are negative, patients with vague or atypical genital symptoms, patients whose partners have genital herpes, or those that request comprehensive evaluation for STIs. In such settings, serologic assays for HSV-1 and HSV-2 antibodies are the recommended approach. In addition, in a patient who presents with their initial genital lesions, serologic assays will help distinguish whether the infection is newly acquired or recurrent. In newly acquired infection, antibody to the type of HSV found in the genital tract will be negative. Conversely, if a patient presents with lesions that contain HSV-2 and is seropositive for HSV-2, the infection represents reactivation and not new acquisition. Alternatively, HSV-1 or HSV-2 acquisition can be established by seroconversion in a previously seronegative patient. Once acquired, antibodies persist for life and can be used as marker for the infection. In an asymptomatic patient with HSV-2 antibody, genital infection can be presumed, and most of such patients will recognize recurrences and have viral shedding, once counseled about the disease. In an asymptomatic patient with HSV-1 antibody, the site of infection is either oral or genital, or both. Among adults who seroconvert to HSV-1, a third of infections present with genital lesions, a third with oral lesions, and a third remain asymptomatic.

Antibodies to HSV develop during the first several weeks after infection and persist indefinitely (27). Although antibodies to most viral proteins are highly cross-reactive between HSV-1 and HSV-2, antibodies to glycoprotein G (gG-1 or gG-2) are predominantly type specific. Because prior antibody to HSV-1 can effectively mask detection of responses to HSV-2 in tests that use viral antigen mixtures, type-specific glycoprotein G (gG)-based assays should be specifically requested when antibody status to HSV-2 is sought.

Type-specific serologic tests based on glycoprotein G have been commercially available for over 10 years and several test formats are now available (Table 2). The most widely available and most cost-effective for laboratories with moderate volumes are enzyme immunoassays. A strip immunoblot ("HerpeSelect 1&2 Differentiation Immunoblot") is available from Focus. This test detects both HSV-1 and HSV-2 and can be quickly and easily run as a single test in low volume laboratories without need for special equipment. Point-of-care tests from Biokit and Focus can detect HSV-2 antibodies from capillary blood or serum; results from either test are obtained by visual inspection of a color change within minutes of sampling. These tests depend on subjective reading of color change and their accuracy against Western blot can vary by location and experience of the reader (28). Very high throughput tests for centralized high volume laboratories are now available from Focus and Diasorin. Performance of these tests correlates well with that of commercial ELISAs (29,30). The gold standard test for HSV

Table 2 Commercially Available Tests in the United States for HSV-2 Type-Specific Antibodies

Name	Manufacturer/year of FDA approval	Comment	HSV-1 test available
HerpeSelect® HSV-2 ELISA	Focus Diagnostics (2000/2002)	Semi-automated platforms available	Yes; separate ELISA plate with gG-1
CAPTIA™ HSV-2 IgG Type Specific ELISA	Trinity Biotech USA (2004)	Semi-automated platforms available	Yes; separate ELISA plate with gG-1
Euroimmun Anti-HSV-2 ELISA (IgG) and Anti-HSV-1 ELISA (IgG)	Euroimmun US LLC (2007)		Yes; separate ELISA plate
biokitHSV-2 Rapid Test	Biokit USA (1999)	Point-of care Also sold as SureVue* HSV-2 Rapid Test by Fisher HealthCare	No
HerpeSelect® Express™ HSV-2	Focus Diagnostics (2007)	Point-of-care	No
HerpeSelect® 1 & 2 Differentiation Immunoblot	Focus Diagnostics (2000)	Strip immunoblot; does not require optical density reader in the lab.	Yes; gG-1 band on each strip
Plexus™HerpeSelect® 1 & 2 IgG	Focus Diagnostics	High-throughput Luminex xMAP based test	Yes, multiplex format
Liaison® HSV-2 Type Specific IgG	DiaSorin Inc (2008)	High-throughput chemiluminescence	Yes
UW Western blot	University of Washington Virology Laboratory (N/A)	Reference tests, useful for confirming positive test results	Yes
HSV-2 Inhibition ELISA	Focus Diagnostics (N/A)		No

type-specific serology remains the University of Washington Western blot, which detects antibodies to multiple viral proteins, including gG (31).

Reported performance of serologic tests varies with the gold standard used and with the time elapsed between infection and serum sampling. False-negative test results can occur especially within six weeks of infection (27). Unfortunately, IgM testing is not necessarily helpful in identifying early HSV-2 infection. Type-specific IgM tests are not on the market and nontyping HSV IgM tests are not useful in those with HSV-1 antibodies due to frequently reappearing IgM long after primary infection. Prototype type-specific HSV-2 IgM tests based on gG-2 have not been able to discriminate between new and recurrent infections (32). Repeat testing on a later specimen may be helpful if recent acquisition of genital herpes is suspected.

False-positive HSV antibody test results, while not frequent with gG-2–based assays, should be considered in patients with a low likelihood of HSV infection or in patients with prior HSV-1 infection (33). Established reference tests are available in cases where confirmatory testing of positive specimens is indicated (Table 2). Some studies have shown that using a biokit HSV-2 Rapid assay as a confirmatory test for sera with low positive ELISA results improves positive predictive values substantially (34,35).

Serology for HSV-2 is useful in the following scenarios: (*i*) recurrent genital symptoms or atypical symptoms with negative HSV cultures; (*ii*) a clinical diagnosis of genital herpes without laboratory confirmation; and (*iii*) a partner with genital herpes. Knowledge of HSV-2 infection status in each partner reduces the risk of HSV-2 transmission by 50% in those couples (36).

Because nearly all HSV-2 infections are sexually acquired, the presence of type-specific HSV-2 antibody indicates anogenital infection. The presence of HSV-1 antibody, alone, is more difficult to interpret since serology cannot discriminate between HSV-1 antibodies elicited by oral infections and anogenital infections. For serologic detection of primary HSV-1, it should be noted that HSV-1 ELISAs have been shown to be notably less sensitive than Western blot for HSV-1 antibody in pediatric sera (37). Use of more than one test provides optimal sensitivity for HSV-1 (38).

Type-specific HSV-2 tests are not recommended for pediatric sera and false-positive tests have been reported (34,37). Because of the high prevalence of asymptomatic HSV-2 and the varied time course of seroconversion among patients, use of serology for medical–legal cases is seldom useful. Testing of victims of sexual assault is controversial and should commence on a case-by-case basis after careful counseling and consideration of sexual history and clinical presentation.

HUMAN PAPILLOMAVIRUSES

Human papillomaviruses (HPVs) belong to the virus family papillomaviridae and represent a diverse array of viruses. HPVs are classified according to genetic sequence homology, and phylogenetically group into two broad genera, alpha and beta, which have a unique epithelial predisposition to the mucosogenital and cutaneous epithelium, respectively. This chapter covers the sexually transmitted HPVs that belong to the alpha genus. These types are further subdivided into species groups based again on nucleic acid sequence relatedness. The species groupings correlate with some phenotypic characteristics of the virus, including cancer risk and epithelial tropism (39).

Clinical Presentation

The overwhelming majority of oro-ano-genital HPV infections are asymptomatic and detectable only by nucleic acid–based technologies. Productive infection with both high- and low-risk HPV infections can cause cellular changes, which are microscopically diagnosed as low-grade squamous intraepithelial lesions (LSILs). Like HPV infection in general, these LSILs are usually transient and resolve without intervention. High-risk HPVs can cause more severe precancerous lesions, or high-grade squamous intraepithelial lesions (HSILs). These lesions, classified histologically as moderate or severe cervical intraepithelial neoplasia (CIN2/3), are the targets of cervical cancer screening. Treatment of CIN2/3 lesions via excisional therapy prevents the progression to life-threatening invasive cervical cancer and is the basis for the success of cervical cancer screening programs in the United States and elsewhere.

Certain types of low-risk HPVs (e.g., HPV6 and HPV11) are the cause of 90% of external genital warts, which can be found in both men and women in many areas of the lower genital tract (e.g., cervix, vulva, vagina, perineum, anus, penis). Treatment of anogenital warts may include surgical intervention or topical application of medications, including podofilox or imiquimod. In very rare cases, perinatal HPV6/11 infection of the larynx can result in juvenile onset recurrent respiratory papillomatosis. Treatment involves repeated surgical debulking, with or without adjuvant chemotherapy, and relapses are common.

Epidemiology

HPV is not a reportable STI and are therefore not monitored in the United States. The National Health and Nutrition Examination Survey (NHANES), a representative sample of the US population, reported a prevalence of high risk (HR) HPV of 15.2% (40). The prevalence of HR-HPV was higher (23%) in a sentinel surveillance survey from women attending sexually transmitted disease, family planning, or primary care clinics (41). The age and city adjusted prevalence of HR-HPV in STD clinics was higher than primary care clinics (26% vs. 17%). Both studies found a decrease in HR-HPV prevalence with increasing age.

HPV is transmitted predominately via sexual intercourse. While this is the major mode of transmission, epithelial abrasion during nonpenetrative intercourse is likely sufficient for transmission and might explain detection of HPV in self-reported virgins (42). HPV is easily transmitted; over one-half of women have been shown to acquire HPV within four years of sexual debut (42) and cumulative lifetime risk likely approaches 80% to 90% in sexually active populations (43). Approximately 90% of infections are subclinical and become undetectable without intervention within one to two years (44), facilitated by a robust cell-mediated immune response. Approximately 50% to 70% of women with incident HPV infection will develop type-specific serum antibodies to the infection within 18 months (45). While high titer serum antibodies against HPV have been shown to prevent future type-specific infections and associated disease among vaccinated individuals (46), the protection associated with natural infection is not clear (47,48). The 10% of women who do not clear HPV within the first two years of infection are at a significantly increased risk of developing cervical cancer precursor lesions, or cervical intraepithelial neoplasia (CIN) (49).

Men also suffer from HPV-associated disease, including anogenital warts and cancers of the anus, oropharynx, and penis. However, since cervical cancer rates outside of well-screened populations are significantly higher than HPV-associated cancers in men, less attention has been focused on HPV in men, despite their being an important reservoir of infection. Recently, studies have reported a similarly high prevalence of penile HR-HPV in men in the United States (23.3%) (50). Interestingly, the prevalence of penile HR-HPV did not decrease with age as seen with cervical HPV infection. Because there is no direct treatment for HPV infection in men (except for treatment of external genital warts), screening for HPV in males is currently not recommended.

Laboratory Diagnosis

Specimen Types

In routine clinical practice, the most common and efficient sample for HPV diagnosis is a cervical brush sample placed in liquid-based cytology (LBC) medium. This sampling allows the same sample to be used for routine cytologic diagnosis, and residual cells to be processed for HPV testing. Sampling methods may vary, and include use of the Cervex Brush, spatula, and endocervical brush. The digene HC2 DNA collection device is another alternative for cervical specimen collection compatible with commercial testing methods. Both methods direct sampling to the cervical os, and the devices are designed to obtain both endo- and ecto-cervical samples.

Self-collection of cervicovaginal samples using Dacron swabs has been extensively evaluated and agreement in HR-HPV detection between self-collected and physician-collected swabs is good to excellent in most studies (51,52). While self-sampling is not currently standard-of-care, it should be considered as a potential means for cervical cancer screening in situations where a speculum exam is not feasible (53).

Nucleic Acid Detection

Nucleic acid detection is the gold standard for diagnosis of HPV infection. At present, three assays are approved by the US Food and Drug Administration (FDA): the digene HPV test (Qiagen, Gaithersburg, MD), and the Cervista^{TM} HPV HR and HPV 16/18 tests (Hologic, Bedford, MA). Indications for HPV testing are to triage patients with Pap smear results indicating atypical squamous cells of undetermined significance (ASCUS) and to use as a co-test with Pap smears in all women over age 30 years (54). It is important to note that HPV and Pap co-testing in women younger than 30 years is NOT recommended due to the high prevalence of transient HPV infections in this age group.

Qiagen's digene HPV test, or the Hybrid Capture 2 (hc2) test, detects the presence of one or more of 13 HR-HPV types, including HPV16, 18, 31, 33, 35, 39, 45, 51, 52, 56, 58, 59, and 68. The hc2 test is based on target amplification technology. Essentially, extracted and NaOH-denatured DNA from the cervical sample is added to a hybridization buffer containing full-length RNA probes specific to the 13 targeted HR-HPV genotypes. Specimen DNA and probe RNA hybrids are captured in anti-DNA:RNA antibody-coated microwells. After unhybridized nucleic acid is washed away, the captured DNA:RNA hybrids are detected by a second anti-DNA:RNA antibody conjugated to alkaline phosphatase and detected with a chemiluminescent substrate. The signal is amplified by virtue of the ability of multiple conjugated detection antibodies to bind to each DNA:RNA hybrid. The alkaline phosphatase cleaves the substrate, emitting light that is read by a luminometer and measured as relative light units (RLUs). The assay is scored as HPV positive if the RLU from the sample is equal to or greater than the test positive control (CO) reaction mean (e.g., $RLU/CO \geq 1.0$).

It should be noted that some low-risk (LR) HPV genotypes at high copy number will result in a false-positive HR-HPV test result, including types 40, 42, 53, 66, 67, 70, 73, 81, 82, and 82v (55). In addition, hc2 is not able to discriminate the specific HPV type(s) present in the clinical sample.

The Cervista™ HR-HPV test detects the presence of one or more of 14 high-risk HPV genotypes: HPV16, 18, 31, 33, 35, 39, 45, 51, 52, 56, 58, 59, 66, and 68. This test is based on the Invader® chemistry of isothermal signal amplification (56), utilizing probe, invader, and fluorescence energy transfer (FRET) oligonucleotides. In a primary reaction, the target-specific probe and invader oligo bind the target sequence creating a single base overlap, which is recognized by a proprietary Cleavase® enzyme that cleaves the 5′ end of the target probe that overlaps the invader probe-binding region. These cleaved products then interact with the FRET probe and a second cleavase reaction to release fluorophores that are measured using standard fluorescence plate readers. The Cervista™ HPV16/18 test utilizes similar technology but targets detection of HPV16 and 18 alone. Because of the unique risk of CIN2/3 following a single HPV16/18 positive test (57), current guidelines suggest immediate colposcopic referral of HPV16/18 positive, cytologically normal patients versus retesting by HR-HPV tests and Pap smear in 12 months if cytologically normal and positive for non-HPV16/18 genotypes (54,58).

Virus Isolation

HPV cannot be cultured via traditional diagnostic culture methods; therefore, isolation of intact virus is not required. Rather, specimens are processed for purification of viral nucleic acid. HPV DNA isolation is performed according to standard methodologies, depending on the downstream assay to be used.

For hc2 testing, specimens collected using the digene sampler kit are simply denatured in NaOH and added directly to the hybridization reaction. If residual LBC medium is used, the sample must be mixed with the manufacturer supplied Sample Conversion Buffer and cells pelleted by centrifugation. Following removal of the supernatant, the cells are resuspended and digested in standard transport medium (STM) and denaturation solution by vortex and 65°C incubation prior to hybridization.

Both LBC residual specimens and the digene STM samples are compatible with the PCR-based genotyping methods described below, though a more thorough DNA purification including protease digestion and EtOH precipitation may be required. DNA can be isolated directly from the specimens using the Qiagen MinElute Media DNA extraction kit (Qiagen, Valencia, CA).

It should be noted that one of the greatest sources of interlaboratory variability in HPV testing comes from variability in specimen processing. The protocols described above have been extensively validated. Deviations in specimen collection medium and the volume of specimen processed or assayed can result in significant variation in the sensitivity of HPV detection. Modifications to specimen preparation protocols are discouraged, but if needed, should be tested in parallel to validated methods to ensure comparable performance.

Typing

HPV genotyping can be performed using either consensus primer or type-specific PCR-based methods. Two consensus PCR systems have been validated and are commercially available as research use only products; the Roche HPV Linear Array (LA) Test (Roche Diagnostics, Indianapolis, IN) and the Innogenetics INNO-LiPA HPV Genotyping Extra test (Innogenetics NV, Belgium). These assays utilize HPV type-specific primer pools; the 18-oligo PGMY primer pool in the LA test (59), and the 10-oligo SPF10 primer pool in the INNO-LiPA test. These primers afford single reaction amplification across a broad range of genital HPV types. HPV type discrimination is achieved by reverse line blot hybridization, wherein the labeled PCR products are hybridized to a single probe strip containing separate probes for 37 and 25 genotypes in the LA and InnoLiPA test, respectively. The shorter PCR product length using the SPF10 primers (62bp vs. 450bp for LA) makes this a superior assay for HPV diagnosis in highly degraded samples, such as paraffin-embedded, formalin-fixed tissues. These systems have comparable performance for detection of type-specific HPV, with the LA detecting more multiple HPV infections, on average (60,61). Other consensus HPV PCR systems have been widely used (e.g., MY09/11 and GP5+/6+) in large epidemiologic studies, but are not currently available as commercial assays. Development of genotyping assays using bead array probe technologies may increase the throughput for consensus PCR genotyping methods (62).

Several type-specific real-time PCR assays have been described for genotyping HPV (reviewed in Ref. 60). These assays may offer slightly greater sensitivity at a type-specific level compared to the consensus PCR methods, but require a separate amplification for each type, decreasing the efficiency of genotyping. These assays allow for HPV viral load quantification; however, the large degree of overlap in HPV viral load by lesion severity precludes any significant clinical utility to a HPV viral load determination (60,63).

Serologic Detection

Approximately 50% to 60% of HPV DNA positive women will develop type-specific antibodies to the L1 capsid protein. Titers are generally very low, particularly when compared to titers induced by the L1 virus-like particle (VLP) vaccine (64). HPV antibodies recognize conformational epitopes of the capsid protein. Standard serologic assay formats, including ELISA and competitive RIA, utilize type-specific L1 VLPs as the capture antigen. Sero-reactivity to HPV L1 capsid is type-specific; therefore, comprehensive serology requires multiple assays. The presence of L1 HPV antibody is therefore a specific, but relatively insensitive marker of cumulative HPV type-specific exposure, and as such has no clinical utility.

Immunohistochemistry

Cellular expression of the E7 oncoprotein from HR-HPV infection results in loss of pRB tumor suppressor function, with upregulation of p16 as a downstream consequence (65). Immunohistochemical detection of p16 from cytology or histology slides is therefore a good marker of active viral infection with HR-HPV (66).

MOLLUSCUM CONTAGIOSUM

Molluscum contagiosum is a benign viral infection of the skin. While in children molluscum appears usually on the face, trunk, and extremities, in young adults genital skin involvement is common, suggesting sexual transmission. Lack of serious outcomes associated with this infection, and frequent spontaneous resolution, have led to a paucity of natural history and treatment studies.

Molluscum contagiosum is a human virus that belongs to the pox family of viruses, although its relationship to variola is distant (67). In tropical climates, it frequently causes an eruption in childhood, while in developed countries it appears more frequently as a sexually transmitted infection (68). Transmission occurs by skin-to-skin contact, and incubation period appears to take several months. Infection also can appear following skin trauma, such as tattooing and shaving, or through fomites, such as sharing contaminated gymnastic equipment or towels. A several-fold increase in the frequency of molluscum has been noted in STD clinics in developed countries in the last 30 years (68). Since eczema is a risk factor for molluscum, the rise may parallel the increase in atopic disorders rather than indicate spread in novel populations.

The lesions present as opaque, skin-colored papules, usually 3 to 5 mm in diameter, and have a central umbilication. They number 5 to 20 in immunocompetent persons. In contrast to HPV, which is in the differential diagnosis of such lesions, molluscum shows preference for skin rather mucosal surfaces. Because clearance of molluscum depends on intact cellular immunity, patients with advanced HIV infection, or iatrogenic immunocompromise can present with crops of several hundred lesions, often on the face.

The diagnosis of molluscum is made on clinical grounds. If the presentation is atypical, a biopsy can confirm with diagnosis. The infection is limited to the epithelium, with characteristic changes demonstrating hyperplastic keratinocytes with molluscum bodies. The viral particles can also be demonstrated on electron microscopy where they appear as typical pox virus, or by amplification of the viral genome. While serologic testing has been developed for research purposes, clinical utility has not been demonstrated, and the test is not available for clinical use. In immunocompetent host, genital warts are the most common similarly appearing genital lesions. In patients with advanced HIV disease, disseminated fungal infection, especially Cryptococcus, can present with lesions that morphologically mimic molluscum contagiosum (69).

In disseminated cryptococcal disease, the lesions are numerous and tend to occur on face and trunk; signs of systemic illness are usually present.

REFERENCES

1. Corey L, Adams HG, Brown ZA, et al. Clinical course of genital herpes simplex virus infections in men and women. Ann Intern Med 1983; 48:973.
2. Benedetti JK, Corey L, Ashley R. Recurrence rates in genital herpes after symptomatic first-episode infection. Ann Intern Med 1994; 121:847–854.
3. Benedetti JK, Zeh J, Corey L. Clinical reactivation of genital herpes simplex virus infection decreases in frequency over time. Ann Intern Med 1999; 131(1):14–20.
4. Engelberg R, Carrell D, Krantz E, et al. Natural History of Genital Herpes Simplex Virus Type 1 Infection. Sex Transm Dis 2003; 30(2):174–177.
5. Baringer JR. Recovery of herpes simplex virus from human sacral ganglia. N Engl J Med 1974; 291:828–830.
6. Zhu J, Koelle DM, Cao J, et al. Virus-specific CD8+ T cells accumulate near sensory nerve endings in genital skin during subclinical HSV-2 reactivation. J Exp Med 2007; 204(3):595–603.
7. Corey L, Wald A, Patel R, et al. Once-daily valacyclovir to reduce the risk of transmission of genital herpes. N Engl J Med 2004; 350(1):11–20.
8. Gupta R, Wald A, Krantz E, et al. Valacyclovir and Acyclovir for Suppression of Shedding of Herpes Simplex Virus in the Genital Tract. J Infect Dis 2004; 190(8):1374–1381.
9. Mark KE, Wald A, Magaret AS, et al. Rapidly cleared episodes of herpes simplex virus reactivation in immunocompetent adults. J Infect Dis 2008; 198(8):1141–1149.
10. Cone R, Hobson A, Brown Z, et al. Frequent detection of genital herpes simplex virus DNA by polymerase chain reaction among pregnant women. JAMA 1994; 272:792–796.
11. Wald A, Zeh J, Selke S, et al. Reactivation of genital herpes simplex virus type 2 infection in asymptomatic seropositive persons. N Engl J Med 2000; 342(12):844–850.
12. Mertz GJ, Benedetti J, Ashley R, et al. Risk factors for the sexual transmission of genital herpes. Ann Intern Med 1992; 116:197–202.
13. Brown ZA, Wald A, Morrow RA, et al. Effect of serologic status and cesarean delivery on transmission rates of herpes simplex virus from mother to infant. JAMA 2003; 289(2):203–209.
14. Richards J, Krantz E, Selke S, et al. Healthcare seeking and sexual behavior among patients with symptomatic newly acquired genital herpes. Sex Transm Dis 2008; 35(12):1015–1021.
15. Xu F, Sternberg MR, Kottiri BJ, et al. Trends in herpes simplex virus type 1 and type 2 seroprevalence in the United States. JAMA 2006; 296(8):964–973.
16. Xu F, Markowitz LE, Gottlieb SL, et al. Seroprevalence of herpes simplex virus types 1 and 2 in pregnant women in the United States. Am J Obstet Gynecol 2007; 196(1):43, e1–e6.
17. Weiss H, Buve A, Robinson N, et al. The epidemiology of HSV-2 infection and its association with HIV infection in four urban African populations. AIDS 2001; 15(suppl 4):S97–S108.
18. Chen XS, Yin YP, Chen LP, et al. Herpes simplex virus 2 infection in women attending an antenatal clinic in Fuzhou, China. Sex Transm Infect 2007; 83(5):369–370.
19. Auvert B, Ballard R, Campbell C, et al. HIV infection among youth in a South African mining town is associated with herpes simplex virus-2 seropositivity and sexual behaviour. AIDS 2001; 15(7):885–898.
20. Sobngwi-Tambekou J, Taljaard D, Lissouba P, et al. Effect of HSV-2 serostatus on acquisition of HIV by young men: results of a longitudinal study in Orange Farm, South Africa. J Infect Dis 2009; 199(7):958–964.
21. Freeman EE, Weiss HA, Glynn JR, et al. Herpes simplex virus 2 infection increases HIV acquisition in men and women: systematic review and meta-analysis of longitudinal studies. AIDS 2006; 20(1):73–83.
22. Brown ZA, Selke S, Zeh J, et al. The acquisition of herpes simplex virus during pregnancy. N Engl J Med 1997; 337(8):509–515.
23. Kropp RY, Wong T, Cormier L, et al. Neonatal herpes simplex virus infections in Canada: results of a 3-year national prospective study. Pediatrics 2006; 117(6):1955–1962.
24. Langenberg A, Corey L, Ashley R, et al. A Prospective Study of New Infections with Herpes Simplex Virus Type 1 and Type 2. N Engl J Med 1999; 341:1432–1438.
25. Wald A, Huang ML, Carrell D, et al. Polymerase Chain Reaction for Detection of Herpes Simplex Virus (HSV) DNA on Mucosal Surfaces: Comparison with HSV Isolation in Cell Culture. J Infect Dis 2003; 188(9):1345–1351.
26. Kimberlin DW. Herpes simplex virus infections of the central nervous system. Semin Pediatr Infect Dis 2003; 14(2):83–89.
27. Ashley Morrow R, Krantz E, Friedrich D, et al. Clinical correlates of index values in the focus HerpeSelect ELISA for antibodies to herpes simplex virus type 2 (HSV-2). J Clin Virol 2006; 36(2):141–145.

28. Saville M, Brown D, Burgess C, et al. An evaluation of near patient tests for detecting herpes simplex virus type-2 antibody. Sex Transm infect 2000; 76(5):381–382.
29. Chen F CJ, Prince H, Wagner M, et al. Evaluation of a Multiplex Bead-based Assay for Herpes Simplex Virus-1 and-2 (HSV-1 and HSV-2) Type-specific Serology. 2006 National STD Prevention Conference; 2006; Jacksonville, FL.
30. Bobba G PA, Vaught D, Tuttle A, et al. Comparison of LIAISON HSV-1 and HSV-2 type specific IgG fully automated chemiluminescent immunoassays to HerpeSelect 1 and 2 and Captia HSV-1 and HSV-2 ELISA assays. 48th Ann Intersci Conf Antimicrob Agents and Chem, 2008 and 46th Infect Dis Soc of America Ann Meeting, 2008.
31. Ashley RL, Militoni J, Lee F, et al. Comparison of Western blot (immunoblot) and glycoprotein G-specific immunodot enzyme assay for detecting antibodies to herpes simplex virus types 1 and 2 in human sera. J Clin Microbiol 1988; 26(4):662–667.
32. Morrow R, Friedrich D. Performance of a novel test for IgM and IgG antibodies in subjects with culture-documented genital herpes simplex virus-1 or -2 infection. Clin Microbiol Infect 2006; 12(5):463–469.
33. Golden MR, Ashley-Morrow R, Swenson P, et al. Herpes simplex virus type 2 (HSV-2) Western blot confirmatory testing among men testing positive for HSV-2 using the focus enzyme-linked immunosorbent assay in a sexually transmitted disease clinic. Sex Transm Dis 2005; 32(12):771–777.
34. Ramos S, Lukefahr JL, Morrow RA, et al. Prevalence of herpes simplex virus types 1 and 2 among children and adolescents attending a sexual abuse clinic. Pediatr Infect Dis J 2006; 25(10):902–905.
35. Morrow RA, Friedrich D, Meier A, et al. Use of "biokit HSV-2 Rapid Assay" to improve the positive predictive value of Focus HerpeSelect HSV-2 ELISA. BMC Infect Dis 2005;5:84.
36. Wald A, Krantz E, Selke S, et al. Knowledge of partners' genital herpes protects against herpes simplex virus type 2 acquisition. J Infect Dis 2006; 194(1):42–52.
37. Leach CT, Ashley RL, Baillargeon J, et al. Performance of two commercial glycoprotein G-based enzyme immunoassays for detecting antibodies to herpes simplex viruses 1 and 2 in children and young adolescents. Clin Diagn Lab Immunol 2002; 9(5):1124–1125.
38. Ashley-Morrow R, Krantz E, Wald A. Time course of seroconversion by HerpeSelect ELISA after acquisition of genital herpes simplex virus type 1 (HSV-1) or HSV-2. Sex Transm Dis 2003; 30(4):310–314.
39. Schiffman M, Herrero R, Desalle R, et al. The carcinogenicity of human papillomavirus types reflects viral evolution. Virology 2005; 337(1):76–84.
40. Dunne EF, Unger ER, Sternberg M, et al. Prevalence of HPV infection among females in the United States. JAMA 2007; 297(8):813–819.
41. Datta SD, Koutsky LA, Ratelle S, et al. Human papillomavirus infection and cervical cytology in women screened for cervical cancer in the United States, 2003–2005. Ann Intern Med 2008; 148(7):493–500.
42. Winer RL, Lee SK, Hughes JP, et al. Genital human papillomavirus infection: incidence and risk factors in a cohort of female university students. Am J Epidemiol 2003; 157(3):218–226.
43. Brown DR, Shew ML, Qadadri B, et al. A longitudinal study of genital human papillomavirus infection in a cohort of closely followed adolescent women. J Infect Dis 2005; 191(2):182–192.
44. Franco EL, Villa LL, Sobrinho JP, et al. Epidemiology of acquisition and clearance of cervical human papillomavirus infection in women from a high-risk area for cervical cancer. J Infect Dis 1999; 180(5):1415–1423.
45. Carter JJ, Koutsky LA, Hughes JP, et al. Comparison of human papillomavirus types 16, 18, and 6 capsid antibody responses following incident infection. J Infect Dis 2000; 181(6):1911–1919.
46. Joura EA, Kjaer SK, Wheeler CM, et al. HPV antibody levels and clinical efficacy following administration of a prophylactic quadrivalent HPV vaccine. Vaccine 2008; 26(52):6844–6851.
47. Viscidi RP, Schiffman M, Hildesheim A, et al. Seroreactivity to human papillomavirus (HPV) types 16, 18, or 31 and risk of subsequent HPV infection: results from a population-based study in Costa Rica. Cancer Epidemiol Biomarkers Prev 2004; 13(2):324–327.
48. Viscidi RP, Snyder B, Cu-Uvin S, et al. Human papillomavirus capsid antibody response to natural infection and risk of subsequent HPV infection in HIV-positive and HIV-negative women. Cancer Epidemiol Biomarkers Prev 2005; 14(1):283–288.
49. Schlecht NF, Kulaga S, Robitaille J, et al. Persistent human papillomavirus infection as a predictor of cervical intraepithelial neoplasia. JAMA 2001; 286(24):3106–3114.
50. Giuliano AR, Lu B, Nielson CM, et al. Age-specific prevalence, incidence, and duration of human papillomavirus infections in a cohort of 290 US men. J Infect Dis 2008; 198(6):827–835.
51. Winer RL, Feng Q, Hughes JP, et al. Concordance of self-collected and clinician-collected swab samples for detecting human papillomavirus DNA in women 18 to 32 years of age. Sex Transm Dis 2007; 34(6):371–377.

52. Petignat P, Faltin DL, Bruchim I, et al. Are self-collected samples comparable to physician-collected cervical specimens for human papillomavirus DNA testing? A systematic review and meta-analysis. Gynecol Oncol 2007; 105(2):530–535.
53. Sowjanya AP, Paul P, Vedantham H, et al. Suitability of self-collected vaginal samples for cervical cancer screening in periurban villages in Andhra Pradesh, India. Cancer Epidemiol Biomarkers Prev 2009; 18(5):1373–1378.
54. Solomon D, Papillo JL, Davey DD. Statement on HPV DNA test utilization. Am J Clin Pathol 2009; 131(6):768–769; discussion 70–73.
55. Wentzensen N, Gravitt PE, Solomon D, et al. A study of Amplicor human papillomavirus DNA detection in the atypical squamous cells of undetermined significance-low-grade squamous intraepithelial lesion triage study. Cancer Epidemiol Biomarkers Prev 2009; 18(5):1341–1349.
56. de Arruda M, Lyamichev VI, Eis PS, et al. Invader technology for DNA and RNA analysis: principles and applications. Expert Rev Mol Diagn 2002; 2(5):487–496.
57. Khan MJ, Castle PE, Lorincz AT, et al. The elevated 10-year risk of cervical precancer and cancer in women with human papillomavirus (HPV) type 16 or 18 and the possible utility of type-specific HPV testing in clinical practice. J Natl Cancer Inst 2005; 97(14):1072–1079.
58. Wright TC Jr., Massad LS, Dunton CJ, et al. 2006 consensus guidelines for the management of women with cervical intraepithelial neoplasia or adenocarcinoma in situ. Am J Obstet Gynecol 2007; 197(4):340–345.
59. Gravitt PE, Peyton CL, Alessi TQ, et al. Improved amplification of genital human papillomaviruses. J Clin Microbiol 2000; 38(1):357–361.
60. Gravitt PE, Coutlee F, Iftner T, et al. New technologies in cervical cancer screening. Vaccine. 2008; 26(suppl 10):K42–K52.
61. Castle PE, Porras C, Quint WG, et al. Comparison of two PCR-based human papillomavirus genotyping methods. J Clin Microbiol 2008; 46(10):3437–3445.
62. Schmitt M, Bravo IG, Snijders PJ, et al. Bead-based multiplex genotyping of human papillomaviruses. J Clin Microbiol 2006; 44(2):504–512.
63. Fontaine J, Hankins C, Mayrand MH, et al. High levels of HPV-16 DNA are associated with high-grade cervical lesions in women at risk or infected with HIV. AIDS 2005; 19(8):785–794.
64. Harro CD, Pang YY, Roden RB, et al. Safety and immunogenicity trial in adult volunteers of a human papillomavirus 16 L1 virus-like particle vaccine. J Natl Cancer Inst 2001; 93(4):284–292.
65. Nakao Y, Yang X, Yokoyama M, et al. Induction of p16 during immortalization by HPV 16 and 18 and not during malignant transformation. Br J Cancer 1997; 75(10):1410–1416.
66. Klaes R, Friedrich T, Spitkovsky D, et al. Overexpression of p16(INK4A) as a specific marker for dysplastic and neoplastic epithelial cells of the cervix uteri. Int J Cancer 2001; 92(2):276–284.
67. Gubser C, Hue S, Kellam P, et al. Poxvirus genomes: a phylogenetic analysis. J Gen Virol 2004; 85(Pt 1):105–117.
68. Tyring SK. Molluscum contagiosum: the importance of early diagnosis and treatment. Am J Obstet Gynecol 2003; 189(suppl 3):S12–S16.
69. Ramdial PK, Calonje E, Sing Y, et al. Molluscum-like cutaneous cryptococcosis: a histopathological and pathogenetic appraisal. J Cutan Pathol 2008; 35(11):1007–1013.

23 Human Immunodeficiency Viruses: HIV-1 and HIV-2

Robert W. Coombs
Department of Laboratory Medicine, University of Washington, Seattle, Washington, U.S.A.

INTRODUCTION

It cannot be understated that infection with human immunodeficiency virus type 1 (HIV-1; HIV infection) has become the defining viral infection of the past three decades, if not of the past century. Although endogenous retrovirus infection per se has contributed to the silent evolution of our human primate ancestors, it was the emergence of HIV-1 as a fatal viral infection with its attendant immunodeficiency that pushed the envelop of medical research to embrace a totally new dimension of laboratory diagnostics and therapeutics. No other contemporary viral infection has furthered our understanding of the important biological relationship between viruses and the immune system to the same degree, as has HIV-1. Importantly, from the perspective of this chapter, HIV-1 has provided an important impetus for introducing the complexities of serological and molecular diagnostics to both the general public and the health care provider, in addition to the laboratory specialist. The introduction of quantitative nucleic acid testing (HIV-1 RNA and to a lesser extent DNA) and drug susceptibility genotyping as routine molecular diagnostics for the clinical management of HIV-1 infection, along with the acceptance of point-of-care HIV-1 diagnostics, has enriched the interactions among the Public Health Service (United States Centers for Disease Control, CDC), the clinical laboratory, and the clinical practice community. Contemporary reviews of the pathogenesis, epidemiology, clinical features, and treatment of HIV-1 infection are available elsewhere (1). This chapter will focus on recent developments in the laboratory diagnosis and management of HIV infection to enhance the optimum use of laboratory resources for these purposes and updates a recent review of the subject (2).

CLINICAL PRESENTATION AND LABORATORY DIAGNOSIS

General Considerations

Shortly after exposure to HIV-1, virus replication can be detected in the blood plasma or cells and is followed by a predictable cellular and humoral immune response. The development of virological and serological markers following HIV-1 infection has been codified as Fiebig Stages I to VI, which is summarized in Table 1 (3). For practical purposes, most persons infected with HIV-1 are identified by the immune response, through a series of sensitive and specific tests that demonstrate HIV-1-specific antibodies. However, HIV-1 infection can also be detected by nucleic acid target or nucleic acid target signal amplification assays of viral RNA or proviral DNA in nearly all HIV-1 seropositive persons, and such tests may be particularly important in the person with very recent infection (before confirmatory antibodies develop, Fiebig Stages I–IV). Current United States CDC laboratory criteria now allow for the use of nucleic acid, antigen, and viral culture to confirm HIV-1 infection in conjunction with a reactive HIV-1 enzyme immunoassay (EIA) (4). However, as a nuanced caveat for HIV-1/-2 screening, the CDC recommends that HIV-1/2 virological (nonantibody) tests should not be used in lieu of approved HIV-1/2 antibody screening tests because a negative result (i.e., undetectable or nonreactive) from an HIV-1/2 virological test does not rule out the diagnosis of HIV infection. As such, a negative HIV-1 test in the context of suspected exposure to HIV-1 should be followed by repeat HIV-1/-2 testing at scheduled intervals up to 6 to 12 months following any high-risk exposure.

For the clinician, suspicion of new (acute) HIV infection is raised by appropriate signs and symptoms, such as a self-limited mononucleosis-like illness, prolonged fever, fatigue or weight loss, onset of opportunistic infections, and/or high-risk behaviors: male-to-male sex, injection

Table 1 Laboratory Stages of Primary HIV-1 Infection

	Laboratory assay test for HIV-1					
			Antibody (enzyme immunoassay)			
Stage	RNA	P24 antigen	1st generation	2nd or 3rd generation	Western blot	Cumulative time stage in days to (95% CI)
I	+	−	−	−	−	5 (3, 8)
II	+	+	−	−	−	10 (7, 14)
III	+	+	−	+	−	14 (10, 17)
IV	+	±	−	+	Indeterminate	19 (15, 23)
V	+	±	±	+	+[a]	89 (47, 130)
VI	+	±	+	+	+[b]	Indefinite

+, present; −, absent; ±, variable.
[a]Confirmed Western blot without p31 band.
[b]Confirmed Western blot with p31 band.
Source: From Ref. 3.

drug use, receipt of multiple blood product transfusions (in resource-constrained settings), heterosexual or homosexual contact with a high-risk or HIV-1 seropositive partner, residents of certain areas of Africa, Asia, or the Caribbean where HIV prevalence is exceptionally high, male or female prostitution, or a child born to parents who are members of these risk groups. However, physicians face a major challenge in diagnosing HIV infection in a person who may be acutely infected, who is asymptomatic, who has atypical signs and symptoms, or who denies being in a high-risk group.

Current CDC recommendations are to increase HIV screening of patients, including pregnant women, in health care settings; and once the diagnosis is made, to initiate antiretroviral therapy, to decrease vertical transmission, reduce HIV-related morbidity and mortality, improve quality of life, restore and preserve immunological function, and maximally and durably suppress viral load (1,5). These recommendations raise additional challenges to the clinician and the clinical laboratory not only to diagnose infection as early as possible but also to assess quantitatively viral RNA, which provides prognosis for disease progression and the eventual success or failure of therapy.

The current approach to the clinical management of HIV-1 infection is driven by clinical trial data that plasma viral RNA level and CD4 cell count are the most suitable biological markers of clinical progression and therapy efficacy. A single HIV-1 RNA measurement is still the strongest baseline predictor of time to AIDS and death and generally explains about half of the variability in these clinical outcomes (6).

For descriptive and technical purposes, laboratory detection of HIV-1 or HIV-2 infection can be stratified into assays that identify HIV-specific antibodies and those that identify infectious HIV virus, viral antigen, or viral nucleic acids.

Acute Infection

There are approximately two million new HIV infections per year worldwide, of which 38,000 to 68,000 occurs in the United States. The precise role of acute infection in the sexual transmission of HIV is not well described but epidemiologic studies suggest that half or more of transmissions occur during acute infection because of high viral levels in blood and genital secretions (7). Whether epidemiological, diagnostic, and therapeutic efforts to identify and treat these persons at this early stage of infection will influence transmission or the clinical outcome remains unknown. Recent improvements in the sensitivity and specificity of anti-HIV assays have resulted in a significant shortening of the preseroconversion window period from 42 to 45 days (the mean infectious window period documented in the post-1987 look-back studies) to less than 20 days with EIA assays that measure both anti-HIV IgM and IgG (Table 1) (3). Introduction of direct virus detection assays into the testing algorithm reduces the window period by 9 days for p24 antigen or DNA PCR and 11 days for RNA PCR (Table 2) (8). Plasma pooling

Table 2 HIV-1 Testing Assays Ranked According to Their Approximate Negative Test Window Periods and Average Reduction in Time to a Reactive Test from Longest to Shortest

HIV laboratory test	Summary of assay procedure	Average time to a reactive test: negative test widow (weeks)	Average reduction in time to a reactive test (days)[a]
First-generation EIA[b]	Viral particles used to bind patient HIV antibody, detected by marker conjugated to antihuman antibody	~6	...
Second-generation EIA	Same as first-generation EIA except uses purified HIV-1 antigen or recombinant virus	~4–6	10
Third-generation EIA	"Antigen sandwich": synthetic peptide used to bind patient HIV-1 antibody followed by marker conjugated to additional HIV antigen: detects both IgM and IgG anti-HIV-1	~3–4	6
Fourth-generation EIA	Uses third-generation EIA methodology plus monoclonal antibody to detect patient-associated HIV-1 p24 antigen	~2	5
Pooled HIV NAT[c]	First combines multiple individual samples into one common pool, then uses HIV-1 RNA PCR or other amplification techniques to detect patient viral nucleic acids (RNA)	<1–2	3
Individual HIV-1 NAT	As above, except that samples are tested individually rather than diluted by pooling; also used to confirm and identify a patient from a positive NAT pool	<1–2	3

[a]Compared with the preceding less sensitive assay.
[b]Enzyme immunoassay.
[c]Nucleic acid test for RNA.
Source: From Ref. 8.

methods have been used to screen for acute infection in several population-based studies (9,10). Although the majority of seroconversions occur within two months of exposure, delayed seroconversion is well established, with approximately 5% of occupational infections taking more than six months after the exposure to seroconvert (11).

Viremia, as detected by p24 antigen or HIV RNA, precedes anti-HIV seroconversion consistently by one to three weeks. This delay between viremia and seroconversion (the viral "eclipse" phase) likely represents a period of localized viral replication at the mucosal site of inoculation and possibly in lymphoid tissues that drain inoculation sites, prior to systemic viremia and subsequent seroconversion (12).

HIV ANTIBODY ASSAYS

General Considerations

The HIV-1 testing algorithm recommended by the United States Public Health Service comprises initial screening by a Federal Drug Administration (FDA)-licensed EIA followed by confirmatory antibody testing of repeatedly reactive specimens with an FDA-licensed supplemental test, for example, immunoblot (Western blot) or immunofluorescence assay (IFA) or one of several virological (i.e., nonantibody) tests: either HIV nucleic acid, HIV 24 antigen with neutralization, or HIV isolation by viral culture (4). Although the EIA is highly sensitive and specific, the positive predictive value of the EIA is highly dependent on the seroprevalence of HIV-1 antibody in the population to which the individual being tested belongs. Therefore, use of both the EIA and a further supplementary test increases the accuracy of detecting HIV infection. Clinical laboratory practice dictates that the results of a repeatedly reactive HIV EI assay should never be released to the clinician or patient by the clinical laboratory without the results of the more specific confirmatory testing. However, with the advent of simple/rapid point-of-care (s/r POC) HIV testing, there is now a compelling need for the clinical laboratory to report timely "presumptive-positive" EIA results in conjunction with a reactive rapid HIV antibody test; confirmatory WB testing can occur later and a final report issued. Importantly, for the maximum diagnostic accuracy, the laboratory HIV test results should be interpreted by the clinician in conjunction with the clinical and epidemiological history of the person being tested.

Enzyme Immunoassays

There are four basic formats for commercially available EIA. First generation assays use antigens derived from whole disrupted virus and an enzyme-conjugated antihuman IgG sandwich technique for capture and detection of anti-HIV antibodies. Second generation assays use recombinant (rDNA) viral protein (antigen) and conjugated antihuman IgG, or they use rDNA antigen for both capture of anti-HIV antibodies and detection of these antibodies, using enzyme-conjugated rDNA proteins as probe. Third generation assays use synthetic peptides; if IgG or IgM antibodies to either HIV-1 or HIV-2 are present, they bind to these peptides and are recognized by an enzyme-conjugated antihuman IgG sandwich technique. Fourth generation assays combine the attributes of the third generation assay along with detection of HIV-1 p24 antigen, which further enhances the detection of early HIV-1 infection.

The more sensitive third and fourth generation assays have shortened the estimated antibody-negative "window period" of primary infection to less than 20 days compared to a median of 3 months for first and second generation assays (Table 2) (8). The specificities of the current commercial EIAs are above 99.5%. For the first generation assays, which are still in use, false-positive reactions result from nonspecific cross-reacting antibodies in persons with underlying immunological disease, gravidity, multiple transfusions, or recent immunization. For the other generation assays, cross-reacting antibodies to the yeast and bacteria that produce the commercial peptides are responsible. Several commercially available EIAs screen for both HIV-1 and HIV-2 antibody. For epidemiological purposes, specialized EIAs are available to evaluate the level of anti-HIV antibody and estimate HIV-1 seroincidence (13).

Immunoblot

As already mentioned, the positive predictive value of the EIA is highly dependent on the seroprevalence of HIV antibody in the population from which the person being tested comes. Therefore, to prevent a false-positive diagnosis of HIV infection, confirmation of a reactive EIA is required using an independent testing method with high specificity. The immunoblot (or Western blot) is the most commonly used confirmatory test in the United States, although other confirmatory tests are available, such as the IFA or supplemental HIV-1 RNA qualitative assays (e.g., Gen-Probe APTIMA RNA Qualitative Assay).

The Western blot detects the serum antibodies directed against specific HIV proteins of varying molecular weights following their separation by gel electrophoresis and blotting onto nitrocellulose paper. The Western blot detects antibodies to the following specific HIV-1 proteins: core (p17, p24, and the gag precursors p40 and p55); polymerase (p31, p51, p66); and envelope (gp41, gp120/160). The reported analytic specificity of the immunoblot assay is 97.8%. The Western blot is interpreted as negative when no antibody-antigen band is present, and positive when antibodies are present to core (p24) and envelope (gp41 or gp120/160) and, in some cases, integrase (p31). Although several organizations have proposed criteria for interpreting Western blot reactivity, the Centers for Disease Control and Prevention (CDC) endorses interpretative criteria that require the presence of antibodies to at least two of three HIV-1 antigens: p24, gp41, or gp120/160 (14).

Regardless of the HIV-1 antibody seroprevalence, a reactive EIA and confirmatory Western blot together have a positive predictive value of greater than 99.99% (15). In the blood donor population, approximately 10 of 10,000 persons (0.1%) without risk for HIV-1 infection will be repeatedly reactive by the HIV-1 EIA. However, 8/10 low-risk persons with repeatedly reactive HIV-1 EI assays will be negative by the HIV-1 Western blot and 2/10 will be Western blot indeterminate. False-positive results for HIV-1 antibody, when both the EI assay and Western blot are reactive in a person who is not infected with HIV-1, are extremely rare (less than 1 in 100,000 persons screened). Therefore, indeterminate Western blots are more common than false-positive Western blots in screening persons from populations with low HIV-1 antibody seroprevalence (16). Fortunately, other HIV-1 detection tests (see below) allow for resolution of confusing serological results.

Other Confirmatory Methods

As already mentioned, the indirect IFA is also approved for confirmatory testing, but it is used less commonly than the immunoblot confirmation method. Other confirmation strategies have been reported; for example, the use of a first or second generation EI assay as a screen followed by a native HIV-1 gp160 EI assay, or a recombinant DNA-derived antigen-based peptide EI assay. These are attractive alternatives to the WB and have reportedly comparable results, but are not FDA approved for this purpose; however, their full potential has not been exploited for improving on the current diagnostic algorithm (17–19). The FDA-approved Gen-Probe APTIMA HIV-1 RNA Qualitative Assay could also be used for confirmation but is of limited value if HIV-1 is suppressed because of antiretroviral therapy.

Nevertheless, the combination of rapid tests could be a cheaper and faster alternative to the conventional testing algorithm in developing countries (20). To meet the demand for a same-day reporting of HIV test results, in my own clinical laboratory, we immediately confirm the reactivity of an HIV-1/2 EI assay with the Multispot HIV-1/2 Assay to report a "preliminary HIV infection" result, and follow this with the traditional confirmatory WB test result later (author's personal written communication, 9/1/09).

Indeterminate Immunoblots

With the increased use of HIV-1 antibody screening in low-risk populations, it is essential for the primary care provider to be able to interpret HIV-1 test results accurately. Between 4% and 20% of serum samples that are repeatedly reactive by HIV-1 EI assay are interpreted as indeterminate by Western blot (21). Indeterminate Western blots (IWBs) in HIV-1-infected persons may result from early antibody formation against viral core antigens during primary infection, from early detection of HIV-1 antibody by the more sensitive IgM-detecting third and p24 antigen-detecting fourth generation EIA before there is confirmation by immunoblot, and rarely, from the loss of core-specific antibody late in infection due to severe immunosuppression. In HIV-1-negative persons, cross-reacting antibody to HIV-2 has been implicated (22). False-positive immunoblots may occur following immunization with experimental HIV-1 vaccines but otherwise are extremely uncommon and occur with a frequency of <1 per 135,000 tests (15,23).

In summary, the following recommendations are made for the clinical management of patients with an IWB (24). Low-risk individuals with a nonreactive EI assay upon repeat testing do not need further follow-up. High-risk individuals should be followed serologically for at least six months, especially those with a p24 band on Western blot. The early, selective use of supplemental tests such as HIV-1 p24 antigen, HIV-1 culture (now rarely used but available in some research laboratories), HIV-1 proviral DNA, or plasma RNA may help determine the infectious status of high-risk individuals before full seroconversion occurs. Negative supplemental tests may also help alleviate the anxiety associated with an indeterminate HIV-1 serology.

Simple/Rapid Serologic Testing

Simple, rapid, reliable, and less expensive alternatives to the EIA with confirmatory immunoblot have been sought for use in acute care settings, emergency rooms, sexually transmitted disease clinics, medical field settings, and developing countries. Test designs are based on three formats: immunoconcentration (flow-through devices), immunochromatography (lateral-flow devices), and particle agglutination (25). Currently, a total of six rapid HIV tests have been approved by the Food and Drug Administration (FDA) and are available in the United States; however, only three, the OraQuick Advance Rapid HIV-1/2 Antibody Test device (OraSure Technologies), the Uni-Gold Recombigen HIV Test (Trinity Biotechnologies), and the Clearview HIV-1/2 Stat-Pak (Chembio Diagnostics Systems) are CLIA-waived and suitable for POC testing; the others are of moderate complexity and thus more suitable for the laboratory setting. These test procedures generally take less than 30 minutes, and negative results are available immediately. Positive tests must be confirmed with a Western blot or IFA. Generally the positive predictive value of these assays is comparable to the standard EIA (>80%) and the negative predictive values approach 100% for some but not others. The relative sensitivity (99.9%) and specificity (99.6%) of these tests and their ability to detect both IgM and IgG antibody may make them particularly useful for detecting early infection. However, as some of these simple/rapid assays rely on detecting antibody to gp41 (e.g., OraQuick), any delay in producing anti-gp41 could reduce the sensitivity of these assays in early HIV infection (26).

The CDC now recommends that diagnostic HIV testing and opt-out HIV screening be part of routine clinical care in all health care settings as this will greatly enhance testing programs by preventing the need for delayed counseling of seronegative patients and by providing preliminary results to seropositive patients thus ensuring optimal clinical and preventive care (5). These preliminary results may encourage patients to return for confirmatory test results and to adopt risk-reducing behaviors sooner compared to currently accepted test-reporting algorithms (27). In addition, the rapid HIV-1 screening of source contacts following occupational exposures to blood will minimize the duration of antiretroviral prophylaxis therapy for the exposed health care worker, thus minimizing cost and alleviating anxiety following the exposure sooner if the test is negative.

Detection of HIV-1 Antibody in Saliva

HIV-1 antibodies can be detected reliably in the oral fluids of HIV-1-infected persons (28). There are a number of obvious advantages to collecting specimens for HIV-1 testing using a noninvasive specimen collection procedure; for example, greater safety, increased patient compliance, and an alternative to phlebotomy. Earlier problems with low sensitivity have been corrected by using special collection devices that concentrate and stabilize the salivary-associated immunoglobulins. Modification of the EIA and Western blot has increased the sensitivity to 97% to 100% and the specificity to 98% to 100% depending on the study (29). In June 2004, the FDA approved the OraQuick® Advance Rapid HIV-1/2 Antibody Test (OraSure Technologies) for the detection of HIV-1/2 antibodies in oral fluid as a POC test.

HIV Antibody Testing in Resource-Limited Settings

Several challenges exist for HIV antibody testing in the resource-constrained setting; either domestically or internationally. For example, 21% of HIV-infected persons in the United States are unaware of their HIV status and may be responsible for 50% of transmissions (5,30); moreover, approximately 80% of people living with HIV in low- and middle-income countries do not know that they are HIV-positive, and recent surveys in sub-Saharan Africa showed that just 12% of men and 10% of women have been tested for HIV and received their test results (http://www.who.int/mediacentre/news/releases/2007/pr24/en/print.html; last accessed 10/06/09).

In this setting, HIV testing should follow recommended CDC-UNAIDS-WHO HIV testing strategies and relevant national HIV testing algorithms, for which a brief summary follows (http://www.who.int/diagnostics_laboratory/publications/HIVRapidsGuide.pdf; last accessed 10/06/09).

Testing algorithms involve either sequential (serial) testing or parallel testing (http://www.who.int/diagnostics_laboratory/en/; last accessed 10/06/09). EIA-based algorithms are generally serial and require confirmation by a second positive test result that uses either different antigens or testing platform (or both) from the first test. A second reactive test result is considered confirmatory of the first reactive test result if the seroprevalence is 5% or more. In low seroprevalence settings, where false-positives are more likely, a third confirmatory test may be required (e.g., HIV-1 RNA or DNA). Serial testing is less expensive and a second test is only required when the initial test is reactive.

Parallel (simultaneous) testing is only recommended when using whole blood finger stick samples rather than venous blood and is suitable for simple/rapid testing, although serial testing also may be used. A parallel testing algorithm uses two tests based on either different antigens or platforms (or both) and the assays are conducted simultaneously. Concordant negative or positive results are considered as true negatives or positives, respectively.

When two test results are discordant, specialist laboratory advice may be required. The WHO and UNAIDS also recommend that specific tests have sensitivities of at least 99% and specificities of at least 98%, and that combination testing algorithms need to be evaluated in the context of use before wide-scale implementation. To illustrate this point, a recent study from Uganda suggested that weak bands detected by simple/rapid antibody assays and which otherwise should be reported as positive according to the manufacturer, decreased the positive predictive value of the assay algorithm used in this study (31). As such, the presence of weak positive bands should be confirmed by EI assay and Western blot before releasing the results, and quality control of the simple/rapid assays should use standard serological assays.

Detection of HIV-1 Subtypes

The envelope protein of HIV-1 isolates from different geographic locations worldwide can differ in more than 35% of amino acid positions (32). As a consequence of this diversity, HIV-1 strains are divided into three groups, M (major, which is responsible for most of the infection in the Americas and Europe), O (outlier, a rare form found in Cameroon and Gabon), N (non-M-non-O), and most recently a new lineage P, which was isolated from a Cameroonian female and is related to gorilla simian immunodeficiency virus. Within group M, 11 major subtypes (or clades) designated A-H, J, K, and U, and 29 major circulating recombinant forms have been defined (http://www.hiv.lanl.gov/; last accessed 10/06/09). Subtype B is the most common subtype in the United States and Europe, while subtypes A, C, and AE are prominent in Africa and Asia. Non-B subtypes are of increasing importance in the United States and may comprise approximately 5% of HIV infections nationally (33). Recombinant forms are prevalent in specific geographical areas of the world; for example, AE is prevalent in Southeast Asia, AG from west and central Africa, AB from Russia, FD from Democratic Republic of Congo, BC from China, BF from South America, and several additional unique recombinants that combine three or more subtypes.

The group O virus stain has been isolated from persons of west-central African origins, with scattered reports of group O virus from Europe and the United States. Diagnostic kit reagents have been modified to ensure optimal sensitivity and specificity for group O virus antibody.

Since most of the primer pairs for HIV-1 RNA polymerase chain reaction (PCR) amplification have been optimized for group B viruses (see below), it is not surprising that HIV-1 RNA PCR may also fail to detect HIV-1 group O and some group M subtypes (34,35). To accommodate this deficiency, primer pair modifications for the HIV-1 *gag* target region have been incorporated into the Roche Amplicor™ HIV-1 DNA and Roche Monitor™ HIV-1 RNA version 1.5 assays and Roche COBAS Ampliprep/COBAS TaqMan HIV-1 (version 1.0) real-time PCR assay. However, difficulty detecting some clade B and non-clade B HIV-1 has lead to a new version of the Roche COBAS Ampliprep/COBAS TaqMan assay (version 2.0) with targeting of both *gag* and *LTR* regions to expand assay sensitivity. Because of the large number of *pol*-specific synthetic oligonucleotide target probes used by the bDNA assay (Bayer Versant HIV-1 RNA 3.0 assay), detection of group O and different group B subtypes has not been a quantitative problem for the bDNA assay.

Detection of HIV-2 Antibodies

In the United States, only a relatively few cases of HIV-2 have been reported, but surveillance varies from state to state, and thus HIV-2 infection is likely underappreciated (36). HIV-2 among U.S. blood donors is extremely rare, with only three cases detected from screening 74 million donations up to June, 1995 (37). Of the 62 persons reported with HIV-2 infection in the United States, 44 (77%) were born in, had traveled to, or had a sex partner from western Africa. Nevertheless, diagnosis of HIV-2 infection continues to be an emerging problem in the United States and is probably underreported; thus, antibody screening for both viruses is warranted.

HIV-1 and HIV-2 genomes share about 60% homology in conserved genes such as *gag* and *pol* and 35% to 45% homology in the *env* genes. The core proteins of HIV-1 and HIV-2 display frequent cross-reactivity, whereas the envelope proteins are more type specific. Despite this cross-reactivity, anti-HIV-1 EIAs used for screening blood donors in the United States are estimated to detect 55% to 91% of HIV-2 infections (38). Western blots for HIV-1 antibodies may be positive, negative, or indeterminate with HIV-2-positive sera. For the confirmation of HIV-2 EIA reactivity, p26 and gp36 correspond to their HIV-1 counterparts p24 and gp41, respectively. Busch et al. tested 913 anti-HIV-1-reactive blood donor sera using an anti-HIV-2 screening EIA, with confirmation by an anti-HIV-2 env-peptide EIA and an anti-HIV-2 Western blot. These 913 sera were derived from anti-HIV-1 screening of approximately 242,000 donations over a three-year period. No HIV-2 infections were identified.

When HIV testing is indicated, tests for antibodies to both HIV-1 and HIV-2 should be obtained if epidemiological risk factors for HIV-2 infection are present, if clinical evidence exists for HIV disease in the absence of a positive test for antibodies to HIV-1, or if HIV-1 immunoblot results exhibit the unusual indeterminate pattern of *gag* plus *pol* bands in the absence of *env* bands.

The following procedures are recommended if testing for both HIV-1 and HIV-2 is performed by means of a combination HIV-1/HIV-2 EIA (22): A repeatedly reactive specimen by HIV-1/HIV-2 EI assay should be tested by HIV-1 immunoblot (or another licensed HIV-1 supplemental test). A positive result by HIV1 immunoblot confirms the presence of antibodies to HIV-1, and testing for HIV-2 is recommended only if HIV-2 risk factors are present. If the HIV-1 Western blot result is negative or indeterminate, an HIV-2 EI assay should be performed. If the HIV-2 EI assay is reactive, an HIV-2 supplemental test such as an HIV-1-specific Western blot should be performed. In addition, HIV-2 DNA PCR has been used to determine infection with HIV-1, HIV-2 or both viruses (see below). As mentioned above, the University of Washington Clinical Retrovirology Laboratory evaluates all reactive HIV-1/2 EI assays immediately with a Multispot HIV-1/2 rapid test. Thus far, we have noted complete concordance with HIV-1 or HIV-2-specific WB results, which generally identify one or two HIV-2 infections per year using this approach—that is, approximately 1% of all HIV-1 WB confirmed tests at the University of Washington are HIV-1 RNA negative, Multispot HIV-2 reactive, and HIV-2 WB confirmed (author's personal written communication, 10/1/09).

DETECTION OF HUMAN IMMUNODEFICIENCY VIRUS

Culture

The detection of HIV-1 by mixed-lymphocyte coculture is a specialized procedure that has extremely high specificity but lower sensitivity in patients with high CD4+ cell counts and low viral levels below 10^{4-5} $\log_{10}$ RNA copies/mL compared to viral nucleic acid detection methods (see below) (39,40). The lower sensitivity of HIV-1 coculture (other than for pediatric diagnosis) compared to currently available nucleic acid detection methods, as well as its greater cost, time requirements, and highly specialized technical nature, leaves HIV-1 culture restricted primarily to research laboratories. However, there may be a rekindled interest in using HIV-1 coculture for assessing viral containment following potent antiretroviral therapy (41–43).

HIV-1 P24 Antigen

The primary use for p24 antigen detection is for identifying subjects in the antibody-negative window period of acute HIV-1 infection, but this has more or less been supplanted by HIV nucleic acid testing (NAT). Although antigen detection is a less expensive alternative to viral RNA detection in this setting, both viral RNA and peripheral blood mononuclear cell culture are significantly more sensitive than detection of p24 antigenemia, even with the added sensitivity of p24 antigen acid dissociation (44). However, a tyramide signal amplification-boosted EI assay for quantification of heat-dissociated p24 antigen reportedly has equivalent sensitivity to viral RNA reverse transcriptase polymerase chain reaction (RT-PCR) amplification at 200 to 400 RNA copies/mL. The reactivity of the p24 antigen EI assay requires confirmation by a neutralization assay (45).

Viral Nucleic Acid

The detection of viral nucleic acid (proviral DNA or viral RNA) by commercially available amplification technologies provides a specific and sensitive direct detection method to identify persons who are infected but who have not seroconverted, to identify infected infants, and to resolve indeterminate HIV-1 antibody serologies (46–48). In addition, the quantification of plasma viral RNA has assumed a critically important role in assessing disease prognosis and response to antiretroviral therapy.

Viral DNA in Peripheral Blood Mononuclear Cells

Qualitative HIV-1 DNA PCR amplification is a commonly used assay method for the diagnosis of HIV-1 infection in neonates and infants (49). The Roche Amplicor^{TM} HIV-1 test kit (Roche Diagnostic Systems, Inc., Branchburg, NJ) is FDA-licensed for this clinical use and a research-use-only real-time COBAS AP/TM HIV-1 DNA assay is available.

The major advantages of HIV-1 DNA PCR over culture are its increased sensitivity and more rapid reporting time; that is, one day compared to two to four weeks. However, the diagnostic performance of HIV RNA detection may match or exceed that of culture and HIV DNA detection (50). There is always a possible risk of false-positive reactivity due to contamination of the specimen with amplicons (so-called carry-over product contamination), although this is decreased somewhat by the use of the uracil *N*-glycolsylase enzyme in the commercial assay (51,52). False negatives can also occur because of inhibition of the PCR reaction by hemoglobin or heparin or when there are fewer target cells in the assay than expected. To control for the latter, and to improve the precision of the assay, testing for HIV-1 DNA should also include concurrent amplification of a cell-associated host gene such as HLA-DQa or globin locus. Participation in a quality assurance program will also ensure that problems with sensitivity and specificity are quickly identified. The use of HIV-1 DNA PCR for the diagnosis of infection in adults should be limited to situations in which antibody tests are known to be insufficient or as a confirmation test when low levels of HIV RNA (<5000 RNA copies/mL) are detected with suspected HIV primary infection. There is no established role for monitoring HIV-1 DNA levels during therapy.

Viral RNA in Plasma

The detection of plasma HIV-1 RNA by reverse transcription-polymerase chain (RT-PCR) amplification, nucleic acid sequence-based amplification (NASBA), bDNA signal amplification or transcription-mediated amplification (TMA) is more sensitive than p24 antigen EI assay or culture for detecting virus (Table 3). There is, therefore, much interest in using plasma HIV-1 RNA as a diagnostic test and there is a strong rationale for incorporating NAT into diagnostic

Table 3 FDA-Approved Quantitative Commercial HIV-1 RNA Assays Available in United States (2009)[a]

HIV-1 RNA test	Manufacturer	Amplification method
RT-PCR amplification	Amplicor HIV-1 Monitor (Roche) v1.5; most widely used in clinical practice but being replaced with a real-time PCR assay (see below)	RNA gag target
Nucleic acid sequence-based amplification (NASBA)	NucliSens HIV-1 RNA QT (bioMerieux)	RNA target
Nucleic acid hybridization and branched DNA signal amplification (bDNA)	VERSANT HIV-1 RNA 3.0 (Siemens)	Signal; multiple *pol* targets enhance non-clade B detection
Real-time RT-PCR amplification[b]	COBAS Ampliprep/COBAS TaqMan HIV-1 Test[c] (Roche); RealTime HIV-1 viral load test (Abbott)	RNA Target (*gag* or *pol integrase, respectively*); assays have enhanced group M, N, and O detection and quantification

[a]Other nucleic acid target-amplification assays include: (*i*) DNA hybridization and colorimetric detection (Digene assay); (*ii*) multiplex transcription-mediated amplification (Gen-Probe APTIMA HIV-1 or Procleix HIV-1/HCV assay, both of which are FDA-approved for screening of blood products) among others.

[b]Most common independently validated (in-house) assay platform.

[c]A second version of this assay that targets *gag* and *LTR* regions is approved for use in Europe but is pending FDA-approval in the United States.

algorithms, particularly when acute infection is suspected. To avoid false-positive diagnosis, the HIV-1 RNA assay should be used diagnostically only as a supplemental test for detecting antibody-negative acute infection. Thus, in this particular diagnostic setting, a reactive HIV-1 RNA assay (particularly one with a low viral RNA copy number, <5000 copies/mL) should be confirmed by another nucleic acid technology preferably HIV-1 DNA PCR or HIV-1 p24 antigen if available. Alternatively, one can retest for the development of HIV-1-specific antibody, which should generally occur within two or three weeks after viral RNA is detected (Table 1). The presence of HIV-1 RNA alone requires a correlation with the medical and epidemiological history and, importantly, a repeat blood draw for confirmatory HIV-1 testing.

HIV-1 RNA Quantitative Assays

Five different FDA-approved commercial assays are available to detect and quantify viral RNA in plasma. These assays quantify HIV-1 RNA by either amplifying the target RNA or the signal (Table 3). The limits of quantification provide an acceptable sensitivity and range for most clinical purposes. The limit of quantification represents the level at which the intra-assay variation is less than 0.15 $\log_{10}$ RNA copies/mL such that the 95% confidence limits for the difference between two estimates are equivalent to $\pm$ 0.5 $\log_{10}$ RNA copies/mL or approximately a threefold difference in viral RNA that can be reliably measured (53). This interpretation differs from the kit manufacturers who claim a lower level of quantification base on less strict criteria.

Two automated specimen preparation and real-time PCR amplification assay instruments, the Roche COBAS AmpliPrep/COBAS TaqMan HIV-1 Test (based on amplification of *gag* targets) and the Abbott m2000 System RealTime HIV-1 assay [based on amplification of a *pol* (integrase) target], have been FDA-approved for monitoring HIV-1 infection and should facilitate the processing of large numbers of clinical specimens for HIV-1 RNA quantification. Although both automated assays detect and quantify HIV-1 subtypes across a broad dynamic HIV-1 RNA range (50–10 million RNA copies/mL of plasma), mutations in *gag* have lead to design changes in target primers and probes to accommodate these mutations and the expanding global repertoire of recombinant HIV-1 (35).

Specimen Considerations for HIV-1 RNA Quantification

In order to minimize the variability of quantitative HIV-1 RNA test results, samples collected for a particular assay should be processed at the same time post blood draw, using the same anticoagulant blood draw tube type (54). In general, EDTA is the preferred anticoagulant. Based on the work of Holodniy et al., the general recommendation has been to separate and store plasma at −70°C within six hours of collection. This rapid specimen processing may place a considerable burden on the laboratory. Moreover, a six-hour processing requirement may be too stringent, and a recent study is reassuring in this regard (55). For the Amplicor HIV-1 Monitor™ assay, viral RNA copy numbers were maintained within 0.5 $\log_{10}$ (threefold) in both blood and plasma samples held at ambient temperature or 4°C for up to three days and remained stable despite limited freezing and thawing (56).

The use of filter paper to collect and store whole blood for later analysis of viral nucleic acid is an attractive alternative to phlebotomy and appears to be suitable for both quantification and sequencing of HIV specimens obtained under field conditions (57–61).

Detection of HIV-1 Subtypes

As HIV-1 subtypes establish within different geographical areas, primer pairs and probes used for HIV nucleic acid detection and quantification have been modified to detect non-B subtypes. For example, the APTIMA HIV-1 RNA Qualitative Assay (Gen-Probe, San Diego, CA) is reported to show sensitive detection of all major HIV-1 group M subtypes, in addition to variants from groups N and O (62). Primer pair modifications have been incorporated into the Roche Amplicor HIV-1 DNA and Monitor HIV-1 RNA assays (version 1.5), COBAS AP/TM and Abbott RealTime m2000 System assays. Because of the large number of *pol*-specific synthetic oligonucleotide target probes used by the bDNA assay (VERSANT HIV-1 RNA 3.0 Assay, Bayer Corporation, Norwood, MA), detection of group O and different non-B subtypes has been less of a problem for the bDNA assay (63).

Detection of HIV-2

Specific primers and probes are necessary to detect HIV-2 nucleic acid (64). Significant genetic diversity of HIV-2 divides this virus into five genetic subtypes, A to E, with subtype A the most common. Similarly to HIV-1, recombination of phylogenetically distinct HIV-2 viruses occurs but genetic recombination between HIV-1 and HIV-2 has not been reported (65). In contrast to HIV-1, HIV-2 RNA levels in plasma and semen are generally lower, corresponding to a slower immunological deterioration and lower transmission rates (66). The unappreciated increase in the number of HIV-2 infections in the United States argues for the development of quantitative HIV-2 RNA assays to assist with the clinical management of these patients.

ANTIRETROVIRAL DRUG SUSCEPTIBILITY GENOTYPE AND PHENOTYPE

Incomplete inhibition of HIV-1 replication in vivo may arise because of poor drug absorption, patient noncompliance with therapy, variations in host antiretroviral drug pharmacokinetics and compartmentalization or infection with drug-resistant virus variants (termed primary drug resistance). This incomplete inhibition may result in the emergence of drug-resistant HIV-1 variants (secondary drug resistance) and thus is an important cause of therapy failure. An assessment of drug resistance may be helpful in selecting antiretroviral therapy, but this has not been rigorously proven (67). Nevertheless, commercial assays for antiretroviral drug resistance are available and clinical studies suggest that viral drug resistance is often associated with poor virological response to therapy. Expert interpretation is recommended given the complexity of results and assay limitations (68). A particularly useful algorithm is available for this purpose at http://hivdb.stanford.edu, last accessed 10/01/09.

Antiretroviral drug susceptibility is determined either genotypically by assessing for mutations that confer resistance or phenotypically by assessing for the susceptibility of the virus isolate (or *pol*-recombinant) ex vivo. Genotypic methods to detect HIV resistance include DNA sequencing of the entire viral population or clones, selective PCR assay, determination of point mutations, differential probe hybridization; enzyme immunoassay modification of the oligoligase detection reaction assay; and the commercially available HIV-1 reverse transcription line probe assay (69). Several kits are commercially available for genotypic resistance testing. HIV-1 Trugene (Visible Genetics/Bayer Healthcare, Suwanee, GA) and Viroseq (Celera/Abbott Laboratories, Rockville, GA and San Fransisco, CA) are FDA-approved. However, genotypic changes may not always correlate with changes in drug susceptibility of the clinical isolate (70). Advisory panels have recommended routine use of susceptibility testing in clinical practice but many clinicians continue to base decisions to start or change therapy on the viral RNA level, CD4+ cell count, and previous antiretroviral drug history with careful attention to patient education about adherence to the prescribed therapy regimen (1,68).

To determine the drug susceptibility phenotype, PCR amplified *pol* gene amplicons containing reverse transcriptase and protease are obtained from the plasma or serum-associated virus (vRNA) or from the cell-associated provirus (vDNA). These amplicons are inserted into a laboratory HIV DNA clone that has the RT and protease genes deleted. The infectious HIV DNA clone is then propagated in a permissive cell line to create a pool of infectious recombinant virus. This recombinant virus is used to determine the susceptibility to single antiretroviral drugs. A modification of this approach uses a recombinant test vector (RTV) HIV DNA that contains the patient's viral *pol* gene and an indicator gene (luciferase) that is inserted into the *env* gene thus preventing the RTV from expressing HIV-1 envelope protein. Cotransfection of permissive cells with RTV DNA and a plasmid that expresses the envelope proteins of amphotropic murine leukemia virus (MuLV) results in a pseudo-typed virus. The ability of these pseudo-typed virus particles to complete a single round of replication is assessed by measuring luciferase production in susceptible target cells. The antiviral activity of a protease inhibitor (PI), for example, is measured by adding a PI to the cotransfected cells, which results in the production of noninfectious pseudo-typed virions incapable of infecting new target cells. The antiviral activity of a reverse transcriptase inhibitor (RTI), for example, is measured by adding a RTI to target cells, which prevents the infection of these cells by the pseudo-typed virus that arises from the original cotransfected cells (70).

It should be noted that primary HIV-1 resistance can also be observed in viruses recovered from a substantial number of people with newly diagnosed infection; this is called "transmitted drug resistance." Transmitted drug resistance is a clear demonstration of the failure of HIV-1 prevention efforts. Although the prevalence of transmitted drug resistance is declining because of improved viral suppression from better clinical care and more effective antiretroviral therapy, the risk of such resistance preventing sustained viral suppression clearly behooves the clinician to determine the drug susceptibility of the patient's HIV-1 before starting therapy (71).

USE OF HIV-1 RNA TO MONITOR INFECTION

HIV-1 RNA Level in Infected Adults

The assessment of immune dysfunction as evaluated by the CD4 T-cell count and the level of virological containment as assessed by the plasma HIV-1 RNA level provide prognostic information for patients with HIV-1 infection. One of the most important concepts to emerge from our understanding of HIV-1 disease pathogenesis is that the magnitude of HIV-1 replication in infected persons is associated with the rate of disease progression (72–74). The level of plasma HIV RNA reflects the infected person's ability to contain viral replication such that the replication and clearance of virus reaches a quasi-steady state and thus defines, in part, the subsequent rate of disease progression (75,76). This quasi-steady state has been referred to as the viral "set-point" and appears to be established in the first three to six months following primary infection, during which time an HIV-specific humoral and cytotoxic lymphocyte response is established. The viral steady state represents the nadir of viral containment, after which time the plasma viral RNA level may increase slowly, conferring additional risk for the development of AIDS. However, some patients may continue to have a decline in plasma viral RNA from the steady state, conferring additional clinical benefit.

In general, patients who are more likely to progress rapidly have a higher plasma viral RNA steady state than do those who progress more slowly (77). However, the predictive value of high plasma viral RNA levels decreases over time, while the predictive value of low CD4+ cell count and CD4+ cell function increases over time. Thus, in the later stages of infection, immune deficiency (i.e., CD4 cell count) is most predictive of disease progression (78).

Importantly, because there is a continuous gradient of risk of disease progression associated with the viral RNA steady-state level, one of the objectives of antiretroviral therapy is to "reset" this plasma viral RNA steady-state level to one with a lower risk of disease progression. Results from HIV-1 therapy trials show that inhibition of HIV-1 replication (as assessed by plasma HIV-1 RNA level) is associated with a delay in clinical disease progression. In summarizing the data from several early large randomized clinical trials involving subjects who received primarily nucleoside therapies, Marschner et al. showed that a 10-fold decrease in plasma viral RNA level from baseline to week 24 yielded a 72% reduction in the risk of progression (95% CI, 61–81%, $p < 0.001$) and that large reductions in plasma viral RNA level were the most desirable (79). Importantly, any reduction in excess of the natural variability of plasma HIV-1 RNA measurement (approximately threefold, or 0.5 $\log_{10}$) is associated with a delay in disease progression. However, in this study and others the prognostic interpretation of any given plasma viral RNA reduction also depended on the treatment response of the CD4+ cell count. Even though the change in plasma viral RNA is a better predictor of clinical progression than is the CD4+ cell response, together the viral RNA and CD4+ cell count responses more fully characterize the risk of disease progression than does either one alone (80). These earlier clinical trial data indicate that a more complete assessment of a patient's prognosis is achieved by monitoring both the plasma viral RNA level and CD4+ cell count, which defines the basic goal for the contemporary laboratory monitoring of HIV-1 infection.

Monitoring HIV-1 RNA in Pregnant Women

Monitoring of viral RNA levels in pregnant women is no different than for nonpregnant women or men (1,81). Although both the plasma viral RNA level and the CD4 cell count are independently predictive of vertical transmission risk, the change in plasma viral RNA level only explains, at most, 50% of the benefit of zidovudine therapy (82). These data strongly suggest that there is a prophylactic benefit from antiretroviral therapy on vertical transmission. The strong association between vertical transmission and maternal plasma viral RNA level indicates that plasma viral RNA levels should be suppressed to <1000 copies/mL and preferably to undetectable to reduce the risk of vertical transmission during pregnancy to <1% (81). Furthermore, because transmission may occur when plasma HIV-1 RNA is not detectable, plasma HIV-1 RNA levels should not be the determining factor when deciding when to use antiretroviral prophylaxis. As such, antiretroviral therapy is recommended in all pregnant women, regardless of virologic, immunologic, or clinical parameters, for the purpose of preventing mother-to-child transmission (81).

Use of Plasma Viral RNA to Define Virological Failure

A precise definition of therapeutic failure based on viral RNA level alone has not been developed and varies depending on the clinical trial that relies on this definition for a primary endpoint. Such a definition should embrace the clinical status of the patient, the CD4+ cell count, and the plasma viral RNA level. The failure of plasma viral RNA to decline by at least 30-fold (1.5 $\log_{10}$) or more from baseline following four to eight weeks of therapy is generally considered to represent a suboptimal virological response (79). In addition, many clinicians would also consider the inability to achieve undetectable plasma viral RNA by 12 to 24 weeks of therapy as evidence for therapeutic failure (1). The consideration of undetectable viral RNA as a benchmark of success is based on the recognition that viral replication in the presence of selective antiretroviral drug pressure could potentially result in the development of drug resistance. However, many patients fail to achieve undetectable viral RNA levels or they experience a rebound in viral RNA after starting antiretroviral therapy (83).

Viral Resistance

It has been somewhat arbitrarily defined that any sustainable 0.5 $\log_{10}$ (threefold) rise in plasma viral RNA above the therapy-induced plasma viral RNA nadir that is not attributable to intercurrent infection, vaccination, incomplete adherence to the antiretroviral therapy regimen, decreased absorption of antiretroviral drugs, altered drug metabolism, drug–drug interactions, or testing methodology, likely represents viral failure due to the emergence of drug-resistant HIV variants or potential superinfection with a new drug-resistant strain of HIV (84). Although genotypic and phenotypic changes associated with drug resistance in vitro are not always synonymous with clinical drug failure, retrospective and prospective clinical trials on the predictive value of these tests have supported their adjunctive use for selecting the next antiretroviral regimen during virological failure (1).

Uncertainty in Measuring HIV-1 RNA Level

There is uncertainty in assigning a value to a single plasma viral RNA measurement and it is helpful for both the clinician and laboratorian to appreciate this measurement uncertainty (85). This uncertainty arises from specimen handling, the performance characteristics of the assay, the technical variability of the assay, whether the different specimens are tested by batch or real-time, and the infected person's natural variation in virus level (86,87). In total, these factors define, with 95% confidence, a variability in the estimated plasma viral RNA copy number of at least fivefold (0.7 $\log_{10}$) for single RNA measurements. Consequently, a single measurement of plasma viral RNA is associated with a defined range of values above or below the measured value at least 95% of the time (2.5% of the time values may be greater than and 2.5% of the time less than this fivefold range). For example, a person with a plasma viral RNA value of 5000 copies/mL obtained from a single plasma specimen taken today may have a measured viral RNA value anywhere from 1000 to 25,000 copies/mL on repeat testing of another blood draw taken within the next few days to weeks, falling within this range 95% of the time.

From both laboratory and clinical perspectives, a rigorous virology quality assurance program is critically important and has shown that the intra-assay standard deviation for quantitative HIV-1 RNA assays ranges from less than 0.1 to 0.2 $\log_{10}$ HIV-1 RNA copies/mL of plasma (88). This precision enables the assays to distinguish reliably three- to eightfold changes in plasma viral RNA for batched testing and 4- to 19-fold changes for real-time testing. Obviously, the uncertainty in defining the true plasma viral RNA level contributes important *uncertainty* for changing antiretroviral therapy based on a single plasma viral RNA value.

In addition to the above considerations, variability in interpreting absolute plasma viral RNA levels across different clinical studies arises because of the patient population studied, the use of serum or plasma to assess the viral RNA level, different viral RNA assay methods, and different anticoagulants and storage conditions. For example, viral RNA levels are generally one-half log value less for serum than for plasma depending on the assay method used; bDNA values are generally twofold less than those for RT-PCR; and heparin interferes with the detection of viral RNA by both bDNA and RT-PCR assays but not for NASBA (55,85).

Thus, some patients could have successful therapy regimens inappropriately changed based on estimates of plasma viral RNA levels that are associated with considerable uncertainty both in their measurement and in the clinical meaning of the plasma viral RNA value, particularly when the plasma viral RNA and CD4+ cell responses are discordant. This is of particular concern for aggressive therapy management when decisions to change therapy are based on the quantification of viral RNA near the reliable limit of detection for an assay. Plasma viral RNA assessments by both the bDNA and RT-PCR assays are usually concordant for most patients below the level of quantification. However, these assessments will be discordant in approximately 20% of patients, and the decision to either maintain or switch antiviral therapy based on the assay quantification limit will be affected by the choice of the viral RNA assay used (89).

CONCLUSION

The diagnosis of HIV-1 infection by detecting and confirming the presence of HIV-1/2-specific antibodies is now augmented by more simple/rapid antibody detection assays. These rapid assays are being used in sexually transmitted disease clinics, urgent care settings, employee heath clinics, and at the time of labor and delivery for pregnant women without prior prenatal HIV testing. By providing a more rapid assessment of infection status, clinicians can offer more immediate and thus better HIV-1 counseling for patients.

The application of molecular diagnostics such as PCR and other nucleic acid amplification technology provides earlier supplementary confirmation for the serologic determination of HIV-1 infection, and in particular a more definitive and timely resolution of indeterminate immunoblot results (10). With repeat specimen collection and testing, the nucleic acid amplification technologies have a defined role for diagnosing HIV infection in the absence of antibody (i.e., acute infection) or in the presence of acquired antibody (e.g., neonatal infection) (8).

The direct quantification of HIV-1 RNA in plasma has revolutionized the clinical management of HIV-1-infected patients over the past several years. However, there are other factors, such as the host's immune status, HLA haplotype, and the viral genotype and phenotype that contribute important prognostic information about disease progression, which is reflected in but not necessarily captured by a single viral RNA value. Testing for HIV-1 drug resistance has become common for assessing virological failure and guiding antiretroviral therapy.

ACKNOWLEDGMENTS

Grant support: University of Washington Center for AIDS Research AI-27757, AI-38858.

REFERENCES

1. Panel on Clinical Practices for Treatment of HIV Infection. Guidelines for the Use of Antiretroviral Agents in HIV-Infected Adults and Adolescents. Washington, D.C.: U.S. Public Health Service, 2008:1–139.
2. Coombs R. Clinical laboratory diagnosis of HIV-1 and use of viral RNA to monitor infection. In: Holmes K, Sparling PF, Lemon SM, et al., eds. Sexually Transmitted Diseases, 4th ed. New York: McGraw-Hill, 2008:1329–1347.
3. Fiebig EW, Wright DJ, Rawal BD, et al. Dynamics of HIV viremia and antibody seroconversion in plasma donors: implications for diagnosis and staging of primary HIV infection. AIDS 2003; 17:1871–1879.
4. Schneider E, Whitmore S, Glynn KM, et al. Revised surveillance case definitions for HIV infection among adults, adolescents, and children aged <18 months and for HIV infection and AIDS among children aged 18 months to <13 years—United States, 2008. MMWR Recomm Rep 2008; 57:1–12.
5. Centers for Disease Control and Prevention (CDC). Revised recommendations for testing of adults, adolescents, and pregnant women in health-care settings. MMWR Morb Mortal Wkly Rep 2006; 55(No RR-14):1–17.

6. Mellors JW, Margolick JB, Phair JP, et al. Prognostic value of HIV-1 RNA, CD4 cell count, and CD4 cell count slope for progression to AIDS and death in untreated HIV-1 infection. JAMA 2007; 297:2349–2350.
7. Stekler J, Sycks BJ, Holte S, et al. HIV dynamics in seminal plasma during primary HIV infection. AIDS Res Hum Retroviruses 2008; 24:1269–1274.
8. Stekler J, Maenza J, Stevens CE, et al. Screening for acute HIV infection: lessons learned. Clin Infect Dis 2007; 44:459–461.
9. Pilcher CD, Fiscus SA, Nguyen TQ, et al. Detection of acute infections during HIV testing in North Carolina. N Engl J Med 2005; 352:1873–1883.
10. Stekler J, Swenson PD, Wood RW, et al. Targeted screening for primary HIV infection through pooled HIV-RNA testing in men who have sex with men. AIDS 2005; 19:1323–1325.
11. Busch MP, Satten GA. Time course of viremia and antibody seroconversion following human immunodeficiency virus exposure. Am J Med 1997; 102:117–124; discussion 125–126.
12. Zhang ZQ, Wietgrefe SW, Li Q, et al. Roles of substrate availability and infection of resting and activated CD4+ T cells in transmission and acute simian immunodeficiency virus infection. Proc Natl Acad Sci U S A 2004; 101:5640–5645.
13. Dobbs T, Kennedy S, Pau CP, et al. Performance characteristics of the immunoglobulin G-capture BED-enzyme immunoassay, an assay to detect recent human immunodeficiency virus type 1 seroconversion. J Clin Microbiol 2004; 42:2623–2628.
14. Centers for Disease Control and Prevention (CDC). Model performance evaluation program: testing for human immunodeficiency virus type-1 (HIV-1) infection. Washington, D.C.: Department of Health and Human Services, 1991.
15. MacDonald KL, Jackson JB, Bowman RJ, et al. Performance characteristics of serologic tests of human immunodeficiency virus type-1 (HIV-1) antibody among Minnesota blood donors. Ann Intern Med 1989; 110:617–621.
16. Celum CL, Coombs RW. Indeterminate HIV-1 Western blots: implications and considerations for widespread HIV testing. J Gen Intern Med 1992; 7:640–645.
17. Fonseca K, Anand CM. Predicting human immunodeficiency virus type 1-positive sera by using two enzyme immunoassay kits in a parallel testing format. J Clin Microbiol 1991; 29:2507–2512.
18. Nair BC, Ford G, Kalyanaraman VS, et al. Enzyme immunoassay using native envelope glycoprotein (gp160) for detection of human immunodeficiency virus type 1 antibodies. J Clin Microbiol 1994; 32:1449–1456.
19. Ng VL, Chiang CS, Debouck C, et al. Reliable confirmation of antibodies to human immunodeficiency virus type 1 (HIV-1) with an enzyme-linked immunoassay using recombinant antigens derived from the HIV-1 gag, pol, and env genes. J Clin Microbiol 1989; 27:977–982.
20. Sato PA, Maskill WJ, Tamashiro H, et al. Strategies for laboratory HIV testing: an examination of alternative approaches not requiring Western blot. Bull World Health Organ 1994; 72:129–134.
21. Celum CL, Coombs RW, Lafferty W, et al. Indeterminate human immunodeficiency virus type 1 Western blots: seroconversion risk, specificity of supplemental tests, and an algorithm for evaluation. J Infect Dis 1991; 164:656–664.
22. O'Brien TR, George JR, Epstein JS, et al. Testing for antibodies to human immunodeficiency virus type 2 in the United States. MMWR Morb Mortal Wkly Rep 1992; 41(No RR-12):1–9.
23. Burke DS, Brundage JF, Redfield RR, et al. Measurement of the false positive rate in a screening program for human immunodeficiency virus infections. N Engl J Med 1988; 319:961–964.
24. Celum CL, Coombs RW, Jones M, et al. Risk factors for repeatedly reactive HIV-1 EIA and indeterminate Western blots. A population-based case-control study. Arch Intern Med 1994; 154:1129–1137.
25. Greenwald JL, Burstein GR, Pincus J, et al. A rapid review of rapid HIV antibody tests. Curr Infect Dis Rep 2006; 8:125–131.
26. Stekler JD, Swenson PD, Coombs RW, et al. HIV testing in a high-incidence population: is antibody testing alone good enough? Clin Infect Dis 2009; 49:444–453.
27. Irwin K, Olivo N, Schable CA, et al. Performance characteristics of a rapid HIV antibody assay in a hospital with a high prevalence of HIV infection. CDC-Bronx-Lebanon HIV Serosurvey Team. Ann Intern Med 1996; 125:471–475.
28. Malamud D. Oral diagnostic testing for detecting human immunodeficiency virus-1 antibodies: a technology whose time has come. Am J Med 1997; 102(suppl 4A):9–14.
29. Emmons WW. Accuracy of oral specimen testing for human immunodeficiency virus. Am J Med 1997; 102(suppl 4A):15–20.
30. Marks G, Crepaz N, Janssen RS. Estimating sexual transmission of HIV from persons aware and unaware that they are infected with the virus in the USA. AIDS 2006; 20:1447–1450.
31. Gray RH, Makumbi F, Serwadda D, et al. Limitations of rapid HIV-1 tests during screening for trials in Uganda: diagnostic test accuracy study. BMJ 2007; 335(7612):188–190.

32. Korber B, Hoelsher M. HIV-1 subtypes: implications for epidemiology, pathogenicity, vaccines and diagnostics—Workshop report from the European Commission (DG XII, INCO-DC) and the joint United Nations programme on HIV/AIDS. AIDS 1997; 11:UNAIDS17–UNAIDS36.
33. Cartwright CP. The changing epidemiology of HIV/AIDS at a Minnesota hospital: impact of demographic change and viral diversity. J Med Virol 2006; 78(suppl 1):S19–S21.
34. Elbeik T, Alvord WG, Trichavaroj R, et al. Comparative analysis of HIV-1 viral load assays on subtype quantification: Bayer Versant HIV-1 RNA 3.0 versus Roche Amplicor HIV-1 Monitor version 1.5. J Acquir Immune Defic Syndr 2002; 29:330–339.
35. Korn K, Weissbrich B, Henke-Gendo C, et al. Single-point mutations causing more than 100-fold underestimation of human immunodeficiency virus type 1 (HIV-1) load with the Cobas TaqMan HIV-1 real-time PCR assay. J Clin Microbiol 2009; 47:1238–1240.
36. Onorato IM, O'Brien TR, Schable CA, et al. Sentinel surveillance for HIV-2 infection in high-risk US populations. Am J Public Health 1993; 83:515–519.
37. Centers for Disease Control and Prevention (CDC). Testing for antibodies to human immunodeficiency virus type 2 in the United States. MMWR Morb Mortal Wkly Rep 1992; 41(R-12):1–8.
38. Busch MP, Petersen L, Schable C, et al. Monitoring blood donors for HIV-2 infection by testing anti-HIV-1 reactive sera. Transfusion 1990; 30:184–187.
39. Coombs RW, Collier AC, Allain JP, et al. Plasma viremia in human immunodeficiency virus infection. N Engl J Med 1989; 321:1626–1631.
40. Nicholson JK, Spira TJ, Aloisio CH, et al. Serial determinations of HIV-1 titers in HIV-infected homosexual men: association of rising titers with CD4 T cell depletion and progression to AIDS. AIDS Res Hum Retroviruses 1989; 5:205–215.
41. Demeter LM, Bosch RJ, Coombs RW, et al. Detection of replication-competent human immunodeficiency virus type 1 (HIV-1) in cultures from patients with levels of HIV-1 RNA in plasma suppressed to less than 500 or 50 copies per milliliter. J Clin Microbiol 2002; 40:2089–2094.
42. Finzi D, Hermankova M, Pierson T, et al. Identification of a reservoir for HIV-1 in patients on highly active antiretroviral therapy. Science 1997; 278:1295–1300.
43. Wong JK, Hezareh M, Gunthard HF, et al. Recovery of replication-competent HIV despite prolonged suppression of plasma viremia. Science 1997; 278:1291–1295.
44. Bollinger RC Jr., Kline RL, Francis HL,et al. Acid dissociation increases the sensitivity of p24 antigen detection for the evaluation of antiviral therapy and disease progression in asymptomatic human immunodeficiency virus-infected persons. J Infect Dis 1992; 165:913–916.
45. Alter HJ, Epstein JS, Swenson SG, et al. Prevalence of human immunodeficiency virus type 1 p24 antigen in U.S. blood donors: an assessment of the efficacy of testing in donor screening. N Engl J Med 1990; 323:1312–1317.
46. Mariotti M, Rouger P, Thauvin M, et al. Failure to detect evidence of human immunodeficiency virus type 1 (HIV-1) infection by polymerase chain reaction assay in blood donors with isolated core antibodies (anti-p24 or -p17) to HIV-1. Transfusion 1990; 30:704–706.
47. Phair JP, Margolick JB, Jacobson LP, et al. Detection of infection with human immunodeficiency virus type 1 before seroconversion: correlation with clinical symptoms and outcome. J Infect Dis 1997; 175:959–962.
48. Shearer WT, Quinn TC, LaRussa P, et al. Viral load and disease progression in infants infected with human immunodeficiency virus type 1. Women and Infants Transmission Study Group. N Engl J Med 1997; 336:1337–1342.
49. Owens DK, Holodniy M, McDonald TW, et al. A meta-analytic evaluation of the polymerase chain reaction for the diagnosis of HIV infection in infants. JAMA 1996; 275:1342–1348.
50. Lambert JS, Harris DR, Stiehm ER, et al. Performance characteristics of HIV-1 culture and HIV-1 DNA and RNA amplification assays for early diagnosis of perinatal HIV-1 infection. J Acquir Immune Defic Syndr 2003; 34:512–519.
51. Longo MC, Berninger MS, Hartley JL. Use of uracil DNA glycosylase to control carry-over contamination in polymerase chain reactions. Gene 1990; 93:125–128.
52. Mansky LM, Temin HM. Lower in vivo mutation rate of human immunodeficiency virus type 1 than that predicted from the fidelity of purified reverse transcriptase. J Virol 1995; 69:5087–5094.
53. Yen-Lieberman B, Brambilla D, Jackson B, et al. Evaluation of a quality assurance program for quantitation of human immunodeficiency virus type 1 RNA in plasma by the AIDS Clinical Trials Group virology laboratories. J Clin Microbiol 1996; 34:2695–2701.
54. Holodniy M, Mole L, Yen-Lieberman B, et al. Comparative stabilities of quantitative human immunodeficiency virus RNA in plasma from samples collected in VACUTAINER CPT, VACUTAINER PPT, and standard VACUTAINER tubes. J Clin Microbiol 1995; 33:1562–1566.
55. Lew J, Reichelderfer P, Fowler M, et al. Determinations of levels of human immunodeficiency virus type 1 RNA in plasma: reassessment of parameters affecting assay outcome. TUBE Meeting Workshop

Attendees. Technology Utilization for HIV-1 Blood Evaluation and Standardization in Pediatrics. J Clin Microbiol 1998; 36:1471–1479.
56. Sebire K, McGavin K, Land S, et al. Stability of human immunodeficiency virus RNA in blood specimens as measured by a commercial PCR-based assay. J Clin Microbiol 1998; 36:493–498.
57. Cassol S, Salas T, Gill MJ, et al. Stability of dried blood spot specimens for detection of human immunodeficiency virus DNA by polymerase chain reaction. J Clin Microbiol 1992; 30:3039–3042.
58. Cassol S, Weniger BG, Babu PG, et al. Detection of HIV type 1 env subtypes A, B, C, and E in Asia using dried blood spots: a new surveillance tool for molecular epidemiology. AIDS Res Hum Retroviruses 1996; 12:1435–1441.
59. Comeau AM, Pitt J, Hillyer GV, et al. Early detection of human immunodeficiency virus on dried blood spot specimens: sensitivity across serial specimens. J Pediatr 1996; 129:111–118.
60. Fiscus SA, Brambilla D, Grosso L, et al. Quantitation of human immunodeficiency virus type 1 RNA in plasma by using blood dried on filter paper. J Clin Microbiol 1998; 36:258–260.
61. Mitchell C, Jennings C, Brambilla D, et al. Diminished human immunodeficiency virus type 1 DNA yield from dried blood spots after storage in a humid incubator at 37 degrees C compared to −20 degrees C. J Clin Microbiol 2008; 46:2945–2949.
62. Linnen JM, Gilker JM, Menez A, et al. Sensitive detection of genetic variants of HIV-1 and HCV with an HIV-1/HCV assay based on transcription-mediated amplification. J Virol Methods 2002; 102: 139–155.
63. Coste J, Montes B, Reynes J, et al. Comparative evaluation of three assays for the quantitation of human immunodeficiency virus type 1 RNA in plasma. J Med Virol 1996; 50:293–302.
64. Schutten M, van den Hoogen B, van der Ende ME, et al. Development of a real-time quantitative RT-PCR for the detection of HIV- 2 RNA in plasma. J Virol Methods 2000; 88:81–87.
65. Curlin ME, Gottlieb GS, Hawes SE, et al. No evidence for recombination between HIV type 1 and HIV type 2 within the envelope region in dually seropositive individuals from Senegal. AIDS Res Hum Retroviruses 2004; 20:958–963.
66. Gottlieb GS, Hawes SE, Agne HD, et al. Lower levels of HIV RNA in semen in HIV-2 compared with HIV-1 infection: implications for differences in transmission. AIDS 2006; 20:895–900.
67. Panidou ET, Trikalinos TA, Ioannidis JP. Limited benefit of antiretroviral resistance testing in treatment-experienced patients: a meta-analysis. AIDS 2004; 18:2153–2161.
68. Hirsch MS, Gunthard HF, Schapiro JM, et al. Antiretroviral drug resistance testing in adult HIV-1 infection: 2008 recommendations of an International AIDS Society-USA panel. Clin Infect Dis 2008; 47:266–285.
69. Schuurman R. State of the art of genotypic HIV-1 drug resistance. Curr Opin Infect Dis 1997; 10:480–484.
70. Petropoulos CJ, Parkin NT, Limoli KL, et al. A novel phenotypic drug susceptibility assay for human immunodeficiency virus type 1. Antimicrob Agents Chemother 2000; 44:920–928.
71. Vercauteren J, Wensing AM, van de Vijver DA, et al. Transmission of drug-resistant HIV-1 is stabilizing in Europe. J Infect Dis 2009; 200:1503–1508.
72. Henrard DR, Phillips JF, Muenz LR, et al. Natural history of HIV-1 cell-free viremia. JAMA 1995; 274:554–558.
73. Mellors JW, Kingsley LA, Rinaldo CR, et al. Quantitation of HIV-1 RNA in plasma predicts outcome after seroconversion. Ann Intern Med 1995; 122:573–579.
74. O'Brien TR, Blattner WA, Waters D, et al. Serum HIV-1 RNA levels and time to development of AIDS in the Multicenter Hemophilia Cohort Study. JAMA 1996; 276:105–110.
75. Ho DD, Neumann AU, Perelson AS, et al. Rapid turnover of plasma virions and CD4 lymphocytes in HIV-1 infection. Nature 1995; 373:123–126.
76. Wei X, Ghosh SK, Taylor ME, et al. Viral dynamics in human immunodeficiency virus type 1 infection. Nature 1995; 373:117–122.
77. Ioannidis JP, Cappelleri JC, Lau J, et al. Predictive value of viral load measurements in asymptomatic untreated HIV-1 infection: a mathematical model. AIDS 1996; 10:255–262.
78. de Wolf F, Spijkerman I, Schellekens PT, et al. AIDS prognosis based on HIV-1 RNA, CD4+ T-cell count and function: markers with reciprocal predictive value over time after seroconversion. AIDS 1997; 11:1799–1806.
79. Marschner IC, Collier AC, Coombs RW, et al. Use of changes in plasma levels of human immunodeficiency virus type 1 RNA to assess the clinical benefit of antiretroviral therapy. J Infect Dis 1998; 177:40–47.
80. Hughes MD, Johnson VA, Hirsch MS, et al. Monitoring plasma HIV-1 RNA levels in addition to CD4+ lymphocyte count improves assessment of antiretroviral therapeutic response. ACTG 241 Protocol Virology Substudy Team. Ann Intern Med 1997; 126:929–938.
81. Perinatal HIV Guidelines Working Group. Public Health Service Task Force Recommendations for Use of Antiretroviral Drugs in Pregnant HIV-Infected Women for Maternal Health and Interventions

to Reduce Perinatal HIV Transmission in the United States. Washington, D.C.: U. S. Public Health Service, 2009:1–90.
82. Sperling RS, Shapiro DE, Coombs RW, et al. Maternal viral load, zidovudine treatment, and the risk of transmission of human immunodeficiency virus type 1 from mother to infant. N Engl J Med 1996; 335:1621–1629.
83. Egger M, May M, Chene G, et al. Prognosis of HIV-1-infected patients starting highly active antiretroviral therapy: a collaborative analysis of prospective studies. Lancet 2002; 360:119–129.
84. Saag MS. Use of virologic markers in clinical practice. J Acquir Immune Defic Syndr Hum Retrovirol 1997; 16(suppl 1):S3–S13.
85. Brambilla D, Leung S, Lew J, et al. Absolute copy number and relative change in determinations of human immunodeficiency virus type 1 RNA in plasma: effect of an external standard on kit comparisons. J Clin Microbiol 1998; 36:311–314.
86. Paxton WB, Coombs RW, McElrath MJ, et al. Longitudinal analysis of quantitative virologic measures in human immunodeficiency virus-infected subjects with > or = 400 CD4 lymphocytes: implications for applying measurements to individual patients. National Institute of Allergy and Infectious Diseases AIDS Vaccine Evaluation Group. J Infect Dis 1997; 175:247–254.
87. Raboud JM, Montaner JSG, RAE S, et al. Issues in the design of trials of therapies for subjects with human immunodeficiency virus infection that use plasma RNA level as an outcome. J Infect Dis 1997; 175:576–582.
88. Lin HJ, Myers LE, Yen-Lieberman B, et al. Multicenter evaluation of methods for the quantitation of plasma HIV-1 RNA. J Infect Dis 1994; 170:553–562.
89. Yerkovich M, Sampoleo R, Dragavon J, et al. Low HIV-1 RNA copy number assessed by bDNA and RT-PCR assays: implications for managing patients above or below the viral RNA quantitation limit. In: Fifth National Conference on Human Retroviruses; February 1–5, 1998; Chicago. Abstract 303.

24 Diagnosis of Polyomavirus Infection, Replication, and Disease

Paola Cinque
Clinic of Infectious Diseases, San Raffaele Hospital, Milan, Italy

Alexis Dumoulin
Transplantation Virology and Division of Diagnostics, Institute for Medical Microbiology, Department of Biomedicine, University of Basel, Basel, Switzerland

Hans H. Hirsch
Transplantation Virology and Division of Diagnostics, Institute for Medical Microbiology, Department of Biomedicine, University of Basel, and Division of Infectious Diseases and Hospital Epidemiology, University Hospital of Basel, Basel, Switzerland

INTRODUCTION

Polyomavirus (PyV) discovery was inaugurated in the 1950s with the report of a transmissible contamination of murine leukemia virus preparations that caused multiple tumors (Greek: poly; -oma) in newborn mice (1,2). In the 1960s, the simian virus (SV)-40 was discovered as a contamination of polio- and adenovirus vaccines raised in rhesus monkey kidney cells. The transforming properties in nonpermissive host cells and experimental tumor models could be attributed to the viral early gene product called the large T (tumor) antigen. SV40 became not only a paradigm of DNA tumor viruses, but also an important model of virus–host interactions. Despite occasional reports of SV40 detection in human specimens, a consistent role in human disease has not been demonstrated.

In the 1970s, BK virus (BKV) and JC virus (JCV) were the first PyVs detected in human specimens. BKV was isolated from urine of a kidney transplant patient *B.K.* with ureteric stenosis and urinary shedding of "decoy cells" (3). Its pathogenic potential remained initially less well defined, although an increased vulnerability of kidney transplant recipients had been noted. JCV was isolated from brain tissue of a patient *J.C.* with progressive multifocal leukoencephalopathy (PML) following its first electron microscopic visualization in 1965 (4,5). PML had been recognized in the 1950s as a rare, mostly fatal disease of the central nervous system of patients with hematologic malignancies. However, most cases were encountered during the HIV/AIDS era before combination antiretroviral therapy. Since 2005, PML has resurfaced in patients treated with immunomodulatory and lymphocyte depleting monoclonal antibodies for autoimmune diseases such as inflammatory bowel disease, rheumatoid arthritis, and multiple sclerosis (6–8).

In the 1980s, BKV was linked to PyV-associated hemorrhagic cystitis (PyVHC) occurring in 5% to 15% of allogenic hematopoietic stem cell transplant (HSCT) recipients (9–11). Despite very high urine BKV loads, PyVHC is not fully explained by high-level BKV replication alone, but appears to require urotoxic and immunologic co-factors for the clinical presentation of PyVHC. In the 1990s, BKV was identified as the key etiologic agent of polyomavirus-associated nephropathy (PyVAN), a complication increasingly encountered in kidney transplants receiving potent immunosuppression (12–16). Currently, 1% to 10% of kidney transplant patients are at risk of loosing their graft due to PyVAN. Lately, however, cases of PyV-associated multifocal leukoencephalopathy (PyVML) due to BKV (17) and JCV-mediated PyVAN have been reported (18,19), which abrogate the classic etiologic attribution of one PyV to one disease.

In 2006 and 2007, the Karolinska Institute virus (KIV) (20) and the Washington University virus (WUV) (21) were identified in human respiratory secretions screened by molecular cloning strategies. The clinical significance of either virus in human disease is still under study. In 2008, Merkel cell virus (MCV) was identified by digital transcriptome substraction in Merkel cell carcinoma, a rare aggressive skin cancer of elderly and immunocompromised patients (22,23).

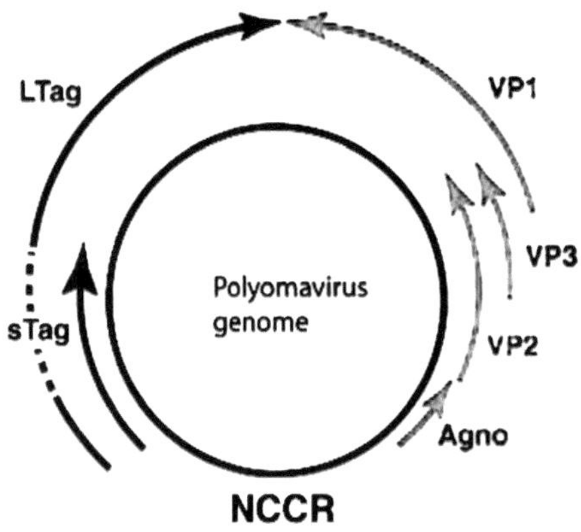

Figure 1 Genetic organisation of the polyomavirus genome. The genome size is approximately 5100 base pairs. Viral early genes (large and small T-antigen) are shown in black arrows, viral genes (VP1, VP2, VP3 and Agno) are shown in gray arrows.

Thus, clinical and laboratory experts are challenged today by a rising number of PyVs and associated diseases in an increasingly heterogeneous population of profoundly immunosuppressed patients.

VIROLOGICAL ASPECTS

PyVs are nonenveloped icosahedral particles of 40 to 45 nm in diameter (2) and are fairly resistant to environmental inactivation, ether, acid, or heat e.g. 50°C for 1 hour (1,2,24). The virions contain a circular double-stranded DNA genome of approximately 5100 base pairs complexed with histones (1). The overall genome structure is conserved and consists of three parts (Fig. 1): (*i*) the noncoding control region (NCCR) bearing the origin of viral DNA replication as well as enhancer/promoter elements coordinating viral early and late gene expression; (*ii*) the viral early genes encoding the large and small T-antigen on one strand; (*iii*) the viral late genes encoding the capsid proteins VP1, VP2, VP3, as well as the regulatory agnoprotein on the other strand. The six PyVs detected in human specimens share a high degree of homology (50–80%) at the nucleic acid and at the amino acid level (22). BKV, JCV, and SV40 cluster together, as well as KIV and WUV, whereas MCV appears more closely related to monkey and rodent PyVs (Fig. 2). There are currently six BKV subtypes (Ia, Ib, Ic, II, III, IV) that correspond to at least four different VP1 serotypes (25,26).

PyVs are characterized by a narrow host cell range, which is determined not only by cell surface receptors, but also by host cell restriction of viral gene expression from the NCCR. BKV

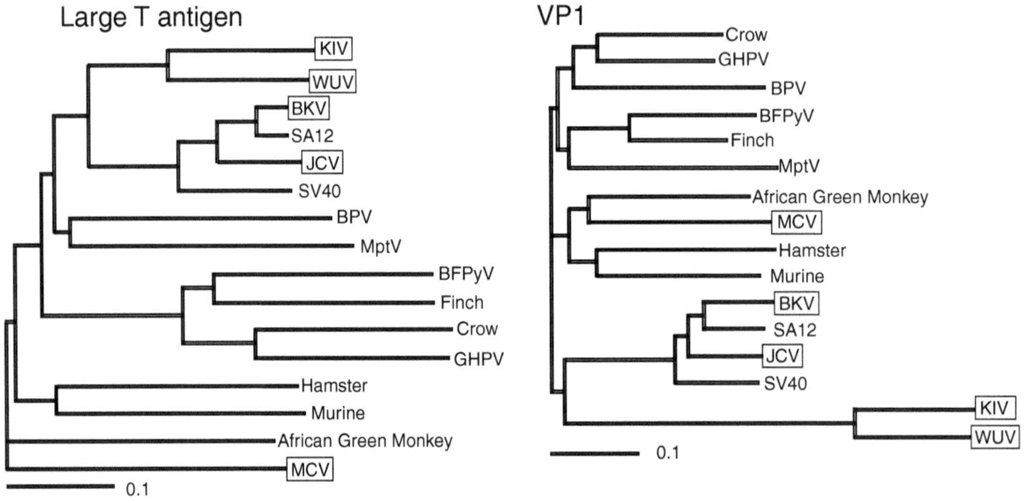

Figure 2 Phylogenetic trees of the polyomaviruses based on large T-antigen and VP1 amino acid sequences. *Source*: Adapted from Ref. 22.

is taken up by caveolae after interacting with α(2,3) sialic acid–bearing and gangliosides (GD1b and GT1b) receptors and activation of signaling events (27,28). JCV is taken up by clathrin-coated pits following the interaction with α(2,6) sialic acid–surface structures and by specific receptors such as the $5HT_{2A}$ serotonin receptor (27). The receptors for KIV and WUV are as yet unknown. The human PyV life cycle is well described for SV40, JCV, and BKV in vitro in specific host cells. For BKV, primary human renal tubular epithelial cells (RPTECs) are thought to correspond best to its natural target host cell. Early gene expression occurs at 12 to 36 hours postinfection (p.i.). Large T-antigen is a multifunctional protein that mediates host cell activation and inhibition of apoptosis through pRB and p53 inactivation, viral genome replication by helicase activity and recruiting cellular DNA polymerase, and also activates viral late gene expression. Late gene expression occurs from 24 to 72 hours p.i. followed by virion assembly in the nucleus and release of infectious progeny (29). Recent studies on PyVAN tissues reported similar kinetics of the BKV life cycle in vivo (30). Host cell tropism at the level of the NCCR is mediated in synergy with host cell activation and signaling events that may involve cytokines, growth factors, and hormones. Propagation of archetype NCCR BKV and JCV is almost impossible without using adapted transformed cell lines. Moreover, following propagation in tissue culture, viral variants with rearranged NCCR are rapidly selected indicating that this part of the PyV genome is not stable (31). NCCR rearrangements have also been detected in vivo in JCV genomes from PyVML (32) and more recently in BKV genomes from PyVAN (33). From this rearranged quasispecies, BKV variants with increased early gene expression, higher viral replication, and accelerated cytopathology are selected (33).

The oncogenic potential of PyVs has been linked to the expression of viral early genes, in particular the large T-antigen. Uncoupling of large T-antigen expression from late viral life cycle with virion assembly and host cell lysis is the hallmark of oncogenic transformation (34,35). Failure to activate viral late gene expression may result from rearranged NCCRs or from chromosomal integration or mutation. Of note, large T-expression may cause genetic instability. The detection of PyV DNA or of large T-antigen expression in tumor tissues may be an important diagnostic indicator, but is by itself not sufficient to resolve the issue of whether or not PyV is a driver of oncogenesis or an innocent passenger that preferentially infects and persists in neoplastic cells. Thus, the potential role of SV40, BKV, and JCV in human malignancies such as mesothelioma, neuroblastoma, and carcinomas of the bladder, colon, and prostate remains under investigation. For MCV, however, convincing evidence for genetic uncoupling of viral early and late gene expression is available that is comparable to HHV-8 and Kaposi sarcoma.

EPIDEMIOLOGY OF POLYOMAVIRUS INFECTION, REPLICATION, AND DISEASE

The seroprevalence rates of PyV infections in healthy adults have been determined for BKV (80–90%), JCV (35–58%), WUV (69%), KIV (55%), MCV (25–42%), and SV40 (2–9%) (36–38). Of note, significant cross-reactivity has been reported between SV40 and BKV. Thus, although six PyVs have been detected in diverse human specimens, SV40 does not seem to circulate efficiently in human populations, despite documented exposure through vaccines, monkey facilities, and animal parks.

The route of natural transmission of BKV and JCV is not resolved and may be oral and/or respiratory. Primary BKV infection is acquired early during childhood, increasing to >90% in young adults. Primary JCV infection follows later, and seroprevalence rates continuously increase throughout adult life to approximately 60% (36–38). Both BKV and JCV establish a state of latent, nonreplicative infection in the renourinary tract and possibly other tissues. Thus, BKV- or JCV-seropositive individuals must be considered infected, even in the absence of detectable viral replication. A population-based study of 2345 sera from England in 1991 using hemagglutination inhibition assays (HIA) indicated an overall seropositivity of 81% for BKV and of 35% for JCV (36). Similar rates were reported in other studies using virus-like particles (VLPs) generated from recombinant expression of the capsid VP1 (38).

In a comprehensive study of 400 healthy immunocompetent blood donors in Switzerland, BKV IgG was detected in 81% and JCV IgG in 58% using enzyme immunoassays (EIA) with BK and JC VP1-based VLPs as antigens (37). Asymptomatic replication with urinary BKV and JCV shedding was detectable in 7% and in 19%, with median urine viral loads of 3.5 and 4.6 log geq/mL, respectively (37). All individuals shedding BKV and/or JCV were IgG seropositive. BKV- or JCV-specific IgM were detected in less than 1% and occurred only in IgG-seropositive individuals. No BKV or JCV DNA was detected in plasma under these conditions. Thus, the qualitative detection of BKV or JCV DNA in urine is not sufficient to diagnose impaired immunity or even disease. Recent studies report the detection of BKV DNA in 38% of stool and rectal swabs of hospitalized children, and in that study, SV40-positive DNA rates were reported in 8% (39).

In immunodeficient patients, the rate of urinary BKV shedding increases to more than 50% and urine viral loads mostly exceed 7 log 10 geq/mL. Levels and rates of urinary JCV shedding seem to be less affected by immune status. In bone marrow transplant recipients, BKV can be detected in 60% to 80% of urine samples and in 40% of stool samples, the latter being more frequent in patients with documented urinary shedding. While these observations underline the importance of persistent T-cell surveillance in BKV- and JCV-seropositive individuals (15,37), it is also evident that reliable quantitative PCR assays are more suited for identifying correlations with PyV disease. The detection of BKV DNA in blood or of JCV DNA in cerebrospinal fluid has been used as surrogate markers of disease for defining patients with presumptive PyVAN and laboratory confirmed PyVML, respectively (40).

DIAGNOSTIC TESTS FOR BKV

BKV Serology

The detection of BKV-specific antibodies identifies individuals with previous exposure to BKV. In general, rising titers indicate recent exposure but do not distinguish between exogenous or endogenous BKV. BKV-specific IgM titers result from antigen priming of naïve B-cells and are typically observed during primary BKV infection. BKV-specific IgM can also be detected in IgG-positive individuals following secondary exposure if antigen levels exceed the corresponding IgG activity. Such serological responses also occur in immunodeficient patients, but may be delayed.

HIA, EIA, indirect immunofluorescence, and neutralization assays have been used for detecting BKV-specific antibodies. HIA has been the traditional test format and is based on the observation that BKV virions, like JCV, agglutinate human blood group 0 erythrocytes. Anti-BKV titers are defined as the highest serial serum dilution able to inhibit a positive reaction between a fixed ratio of virions and washed red blood cells. The antibodies measured by HIA are hence directed against the three-dimensional conformation of virion capsid and correlate well with neutralizing activities. Although HIA is very specific, HIA titers are of limited sensitivity and cannot distinguish between IgG, IgM, and IgA antibody classes. The test principle is straight forward, but the reagents are not standardized, in particular the BKV virions that need to be prepared from tissue culture in expert laboratories.

Indirect immunofluorescence for BKV antibodies has been described by different researchers (41,42) using BKV infected cell lines. Data in kidney transplant patients and in bone marrow transplant patients report increasing antibody titers following BKV reactivation. Indirect immunofluorescence is a standard technique in diagnostic laboratories, but producing BKV-infected cells for diagnostic purposes requires considerable expertise. In addition, VP1 expression and virion assembly are coupled to pronounced cytopathic effects with detachment and lysis of cells. This affects the quality of the slides and may require cytospins of detached cells with considerable morphological heterogeneity (42). Owing to the nuclear localization of VP1 and large T-antigen, distinction from interfering anti-nuclear antibody activities may be difficult. Serum dilutions starting from 1:10 are typically used. A recent study reported that HSCT recipients with titers higher than 1:10 before transplantation were more likely to develop high-level urine BKV loads posttransplant, but this requires validation (42).

EIA using VLPs generated from recombinant BKV VP1 have been found to be more sensitive than HIA, but equally specific (43–45), and can differentiate IgG, IgM, and IgA responses. Technically, the three-dimensional capsid conformation must be reconstituted and be present in excess of the antibody titers for a (semi-)quantitative read-out. Accordingly, we commonly coat 75 to 150 ng of reconstituted VLP (confirmed by electron microscopy) per plate well and use 1:400 serum dilutions for screening. Recent studies suggest that kidney transplant recipients with low or undetectable BKV antibodies pretransplant may be at higher risk of BKV viremia and PyVAN posttransplant than seropositive recipients with higher titers (46–49). A higher risk has also been reported for recipients failing to mount a BKV-specific IgA response at week 1 posttransplant (50). While high anti-BKVLP levels in donors and low anti-BKVLP levels in recipients may be markers of increased risk for BKV replication after kidney transplantation, longitudinal studies posttransplant indicate that rises in BKVLP IgG and IgM frequently coincide with BKV viruria and viremia, without conferring identifiable protection. Similarly, IgA antibodies to BKVLP posttransplant may be a marker of recent pretransplant exposure (44). This suggests that the BKV-specific antibody activity may be a surrogate marker of the BKV-specific cellular immunity.

Recombinant BKV VP1 not assembled into VLP has also been used to detect IgG and IgM antibody titer increases in kidney transplant patients with BKV replication and PyVAN (51). However, nonassembled VP1 antigens are less sensitive than VLP, resulting in an increased proportion of false BKV seronegatives among dialysis patients (45). In vitro denaturation experiments of VLP indicate that anti-VP1 and anti-VLP are indeed different antibody populations with little cross-reactivity (45). These observations have to be kept in mind when considering the reported seroprevalence rates for KIV, WUV, and MCV, because these studies did not use reconstituted VLPs as antigens (38). EIA responses to other BKV proteins have yielded mixed results. The BKV agnoprotein, which is an abundant cytoplasmic protein expressed late in the viral lifecycle in tissue culture as well as in PVAN biopsies, does not elicit significant antibody responses (51,52). Thus, anti-agno responses do not seem to be useful markers of concurrent or recent BKV exposure or immunity. Antibody responses against the BKV large T-antigen have been systematically studied in kidney transplant patients and were detected in only 10% to 40% of individuals with anti-BKVLP responses. Antibody responses against the amino-terminal domain of large T-antigen were specifically increased in kidney transplant patients clearing plasma BKV loads after reducing maintenance immunosuppression (45). Thus, anti-BKV large T-antigen may be an indicator of emerging immune control, similar to Epstein–Barr virus nuclear antigen (EBNA-1) antibodies (45).

BKV-Specific Cellular Immunity

In the last five years, BKV-specific T-cell assays have been investigated for their potential to predict the risk for BKV replication and disease in kidney transplant patients (53–55). Most frequently, interferon-γ has been used as a read-out following stimulation of T-cells with BKV-antigen preparations or with BKV peptides (53,54). For the clinical diagnostic laboratory, it should be noted that the frequency of the BKV-specific interferon-γ responses in the peripheral blood was 1–2 orders of magnitude lower compared to responses elicited with CMV antigens. This makes it difficult to use flow cytometry and intracellular cytokine staining directly from peripheral blood mononuclear cells. To overcome this problem, T-cells from peripheral blood have been expanded following stimulation with BKV-specific antigen, which allowed the identification of specific HLA-I restricted BKV epitopes in tetramer assays. Interestingly, conserved epitopes between BKV and JCV VP1 and large T-antigen have been identified, suggesting some degree of cross-protection. Interferon-γ ELISpot assays have been used, but it has been difficult to identify significant differences between kidney transplant patients progressing to BKV viremia and those protected. By contrast, significant differences could be detected between patients with ongoing and clearing BKV viremia (55–58). Thus, ELISpot assays may be useful for clinical guidance of whether or not immunosuppression has been sufficiently reduced (59). However, calcineurin-inhibitor concentrations inhibiting T-cell activation signal 1 were critical determinants for the magnitude of the interferon-γ responses, whereas sirolimus, mycophenolate, or leflunomide had no direct effect (59). Further work is needed to establish the value of BKV-specific cellular immunoassays for clinical management.

BKV Cell Culture

In vitro, BKV replicates in primary human renal proximal tubular epithelial cells, human umbilical cord venous endothelial cells, human embryonic kidney cell lines, and WI-38, but early studies in the diagnostic laboratory used African green monkey kidney, for example, Vero cells. An important drawback of tissue culture is the selection of viral variants with rearranged NCCR.

Cell culture has been instrumental in the isolation of PyV and in early laboratory studies, but for diagnostic purposes, BKV isolation has been effectively replaced by molecular testing.

BKV Antigen Detection

The detection of BKV antigens by immunohistochemistry has become the pivotal test in the diagnosis of PyVAN. Cross-reacting antibodies raised against the SV40 large T-antigen have been successfully used (60–62), and are recommended as confirmatory adjunct test for diagnosing PyVAN (40). These antibodies cannot distinguish between large T-antigen derived from SV40, BKV, or JCV. For tissue diagnosis, this may actually be an advantage, since rare cases of JCV-mediated PyVAN have been encountered in kidney transplants and in HIV/AIDS (15). In situ hybridization is used in some centers for a specific tissue diagnosis and has the potential to distinguish between BKV, JCV, and SV40 if probes of sufficient sequence difference are used under stringent conditions. Detection of BKV agnoprotein or VP1 may represent an alternative (30,51,63), but since late gene expression is associated with considerable cell lysis, background staining may be increased.

BKV Urine Microscopy

Urine cytology for detecting altered cells with intranuclear inclusions has been instrumental in identifying BKV (3). The presence of "decoy cells" in phase-contrast microscopy or Papanicoulaou staining is a sign of high-level PV replication in urothelial and/or tubular epithelial cells and must be distinguished from neoplasia. In expert cytology laboratories, the absence of decoy cells has a high negative predictive value of >95% for excluding PyVAN in kidney transplant patients. In patients with baseline renal function, the positive predictive value of decoy cells is less than 20%, but may reach 50% to 60% when higher numbers of >10 are present per cytospin (64) or high power field (65), when shedding lasts for more than two months, or when there are signs of inflammation or decoy cell casts. Staining of urine cytospin preparations for large T-antigen expression has been explored for specific differentiation from other altered cells. Electron microscopy of native urine preparations can detect PyV virions of 40 to 50 nm diameter and typically requires >7 log particles per milliliter. The negative predictive value of electron microscopy is lower than that of urine cytology or urine PCR, and the positive predictive value is around 50%. Recently, PyV aggregates have been proposed as a more specific sign of PyVAN (66). As an important caveat, neither urine cytology nor electron microscopy can distinguish between BKV or JCV replication, both of which can be associated with PyVAN. The role of urine microscopy has not been conclusively studied for other patients at risk for BKV disease.

BKV Molecular Genetic Testing

BKV DNA can be detected in various specimens including urine, plasma, renal biopsies, stool, cerebrospinal fluid, aqueous humor, and brain tissue. The diagnostic significance depends on the patient characteristics, the type of specimen, and the viral loads. Testing for BKV DNA load in urine and blood has become a pivotal laboratory assay in the management of kidney transplants. Negative urine BKV PCR allows PVAN to be ruled out with a high negative predictive value of >95%, whereas plasma BKV loads of >4 log persisting for >3 weeks have positive predictive value of 50% to 80% (40,67). Numerous PCR protocols for the detection of BKV DNA have been published and a selection is shown in Table 1. In a multicenter study, tissue cultured BKV, BKV genome-containing plasmids, and paired plasma and urine samples from kidney transplant patients were examined in a blinded fashion using routine tests in multiple laboratories (68). Although good overall sensitivity and specificity were obtained, the results revealed consistent differences of up to 1 log between different laboratories. Moreover, at low BKV loads close to the limit of detection, excess competitor JCV DNA seemed to reduce the sensitivity of some assays (Fig. 3). A systematic single-center study compared four published and three newly designed BKV load assays targeting different sequences of the large T-antigen and the VP1 gene. Marked variability was noted which was associated with polymorphisms particularly among the less-frequent BKV subtype III and IV isolates. A composite assay simultaneously targeting both large T-antigen and VP1-sequences performed better by detecting approximately 10% of previously missed subtypes (69). Together, the data indicate that external quality control and in particular external quantification references are needed to allow conversion of BKV loads obtained in different laboratories for threshold definitions, patient management, and multicenter studies.

Table 1 PCR Assays for the Detection of BK and JC

Reference	Method	Target region	Specificity	Study size
(126)	Nested PCR	VP1/LT	JC	Quality control study including 7 laboratories
(87)	Nested PCR	LT	JC	26 HIV patients with focal lesions
(127)	Quantitative competitive PCR	VP2	JC	11 PML patients
(128)	Taqman	LT-ag	BK and JC	103 bone marrow transplant patients and 11 healthy individuals
(129)	Taqman	LT-ag	BK	20 kidney transplant patients including 4 with PVAN
(71)	Taqman	LT-ag	BK	Time course of a single kidney transplant patient with PVAN
(130)	Taqman	LT-ag	BK and JC	46 pediatric kidney transplant patients
(131)	LightCycler FRET	VP2	BK, JC, SV40	11 brain biopsies from PML patients
(91)	Taqman	LT	JC	61 HIV patients with PML
(132)	Taqman MGB	LT-ag	BK	n.s.
(133)	LightCycler SybrGreen	LT-ag	BK	n.s.
(134)	Taqman	LT-ag	BK	855 adult kidney transplant patients
(135)	Standard PCR, RFLP	LT	BK, JC	56 HIV patient with focal lesions
(136)	Taqman	LT	JC	45 HIV patients with PML
(69)	Taqman	VP1/LT	BK	230 consecutive clinical urine/serum specimens
(94)	Nested PCR	VP2	BK, JC	42 AIDS patients and 55 controls

Abbreviation: n.s., not specified.

Recently, an international external quality assessment study for BKV and JCV was conducted as a first step to address these issues (70).

The protocol used for routine detection of BKV DNA in Basel, Switzerland, has been described previously (14,19,33,71). The overall performance of this assay was robust (69), but recently published sequence polymorphisms point to under-quantification of rare variants (Fig. 4). For the forward primer, 137 sequences were identical; single-mismatch mutations

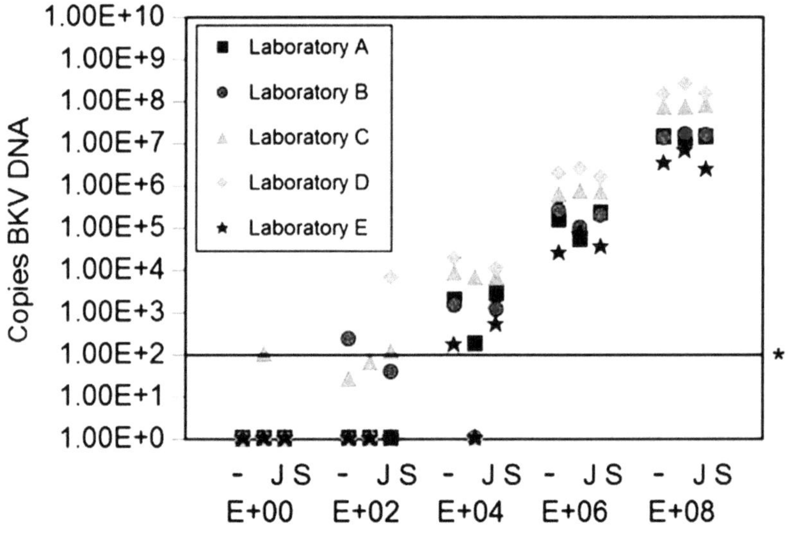

Figure 3 Blinded comparison of BKV load quantitation using 5 different quantitative BKV PCR assays in transplant centers A to E. X-axis denotes input BKV genomes: E+00, none; E+02, 100; E+04, 10'000; E+06, 1'000'000; (-) no competitor; (J) addition of 6 log10 geq JCV; (S) addition of 6 log10 geq SV40. *Source*: Ref. 68.

Accession number	Position of 1st nt	Alignment	Nr of identical sequences	% of total
Forward primer				
		ATACACAGCAAAGCAGGCAAGGGTTCTATTACTAAATACAG	**220 total**	**100**
EU681739	107	....	137	62
EU681749	107	.....................A...................	60	27
EU681746	47	............A........T...................	12	5
AB301092	4198	..................G......................	11	5
JCV		............AG.A....CA..............C....		
SV40		...............A.....A..............C....		
Probe				
		TAGAAGACCCTAAAGACTTTCCCTCTGATCTACACCAGTTT	**221 total**	**100**
EU681746	140	....	33	15
EU681775	183	......................T..................	181	82
AB263924	4291	......................T.G................	4	1.8
EU681743	108	.............................G...........	1	0.4
AB263926	4291	........................G................	1	0.4
AB269864	4294	......................T..............A...	1	0.4
JCV		.......................GTA.....G..TGCA..C		
SV40		..........C..G...........A..AT.G.TA_.....		
Reverse primer				
		CAGTGATGAAGAAGCAACAGCAGATTCTCAACACTCAACAC	**205 total**	**100**
V01109	2630	....	138	67
EU681749	260	.........G..G..............C.............	48	23
EU681746	200	...........................C.............	12	6
AB269839	4343	.........G..G.......G......C.............	6	3
AB269825	4354	........-..G...............C............. (\ A)	1	0.5
JCV		.........T...__...._...GA..C.............		
SV40		.........T..G..T..T..T..C................		

Figure 4 Alignment of the oligonucleotides used for the BKV Taqman PCR in Basel. The accession number of a representative database entry for each sequence variant, the position of the first nucleotide aligning with the query sequence, the number of database entries with identical sequence as well as the percentage of the total number of sequences for each sequence variant are indicated. Dots represent nucleotide identical to the query sequence; capital letters represent variations. The alignment was performed with the BLAST algorithm, searching against the NCBI nucleotide sequence databases using standard parameters in September 2008.

in center positions were observed in another 60 and 11 sequences, which were unlikely to affect the assay. Similarly, in another 12 sequences (5%), two point mutations were observed in the center and 5′-end of the primer. The probe sequence was identical to 33 sequences of the NCBI database. However, single point mutations in the center or the 3′-end of the probe were found in 183 sequences accounting for 83% of available sequences. Considering the length of the probe and the position of the mutations, these are unlikely of the affect the performance of the assay. However, for five sequences (2%) with double mutations, nucleotide degeneration might be considered. For the reverse primer, single point mutations were found in 12 sequences and dual point mutations in 48 sequences (30%). Six sequences bearing a fourth mutation were found. Since the same three mutations are found in roughly one-third of the sequences, degeneration at positions 3 and 6 (both purines), and 21 (pyrimidine) should be considered. Thus, continuous analysis and assay adaptation is recommended for clinical laboratories.

The standard assay in Basel tests 5 μL quadruplicates of twofold concentrated eluates obtained by DNA extraction from of 200 μL plasma or urine using the Magnapure™ reagents and robotics (Roche Diagnostics, Basel, Switzerland) or the Corbett X-tractor Gene and the Corbett VX reagents (Qiagen, Hombrechtikon, Switzerland). CSF samples are extracted manually with Qiaamp Blood kit (Qiagen, Hombrechtikon, Switzerland). One of four replicates routinely spiked with 1000 geq of the reference plasmid to monitor for PCR inhibition. We use 300 nM of both primers and 200 nM of the FAM-labeled probe in 12.5 μL of a twofold concentrated amplification master mix (Eurogentec, Seraing, Belgium) containing the polymerase, 10 mM $MgCl_2$, dNTP (including dUTP), and uracil-*N*-glycosylase (total volume 25 μL). Quantification is performed by using a standard curve generated by three concentrations (10e4, 10e6, and 10e8 cp/mL) of a reference plasmid as well as one additional quantification control of 3000 geq/mL. The temperature profile consists of a preincubation step at 50°C, two minutes to allow for enzymatic decontamination of potential uracyl-containing amplicons, followed by 95°C; 15 minutes for hot-start ampli-Taq activation and 45 cycles of 95°C; 15 seconds; 60°C; 60 seconds. The threshold of the PCR assays is defined by the lowest dilution yielding 50% of positive results in 10 replicas that correspond to 3 geq per assay or 300 Ggeq/mL of extracted specimen fluid. In order to increase sensitivity of the assay for CSF samples, a total volume of 400 μL is extracted and eluted in 100 μL, and 10 μL of the eluate is used in a 50 μL PCR. This allows lowering the limit of detection to 75 GEq/mL. In case of inhibition or inconsistent results, we repeat DNA extraction for the analysis. Each PCR assay is monitored by one contamination control consisting of water, which is taken through the entire process of DNA extraction to identify contamination at the level of extraction, and nontemplate controls to identify contamination at the level of amplification, all done in triplicate.

DIAGNOSIS OF POLYOMAVIRUS-ASSOCIATED NEPHROPATHY

The key renal disease associated with BKV is PyVAN. BKV-mediated PyVAN has been encountered sporadically in native kidneys of patients with inherited, acquired, or pharmacologic immunodeficiency, but consistently up to 10% of kidney transplant patients are at risk (15). PyVAN pathogenesis is driven by persistent high-level replication in renal tubular epithelial cells. According to BKV replication dynamics, BKV replication starts in the allograft and is then amplified in the urothelial compartment with back-feeding into the allograft causing cytopathic loss of 6–7 log tubular epithelial cells each day. The cytopathic effects consist of cell enlargement, rounding, and detachment, with areas of denuded basement membrane. This process elicits an inflammatory response with granulocytic and lymphocytic infiltrates progressively accumulating in the interstitial space and invading tubules. With persisting cell turnover and inflammation, tubular atrophy and fibrosis ensue causing irreversible graft damage (62).

The natural course of BKV-mediated PyVAN in kidney transplant recipients is characterized by a paradigmatic progression of high urine BKV loads of >7 log geq/mL, followed by increasing plasma BKV loads of >4 log per mL and histologically confirmed PyVAN in allograft biopsy (14). The time between each step is variable, but a median of approximately six weeks has been the rule of thumb in a large number of retrospective and prospective studies (19,67,72–74). High-level urinary BKV shedding ("decoy cells," high urine BKV loads, BKV VP1 mRNA) and detection of BKV DNA in plasma were recognized as sensitive and specific surrogate markers of PyVAN, respectively (13,14,19,75).

The following diagnostic definitions have been proposed for kidney transplant recipients (40):

1. Possible PyVAN: urine BKV loads of >7 log per mL and undetermined or undetectable plasma BKV loads.
2. Presumptive PyVAN: plasma BKV loads of >4 log per mL for >3 weeks, and undetermined or negative histopathology of PyVAN.
3. Definitive PyVAN: histopathology and detection of PV by immunohistochemistry (large T-antigen expression) or in situ hybridization. The histological presentation should be graded as PyVAN pattern A (predominant cytopathic), PyVAN pattern B (predominant inflammatory-cytopathic), or PyVAN pattern C (predominant tubular-atrophy, fibrosis) as described (40).

In kidney transplant patients, routine screening for BKV replication should be performed at least every three months during the first two years posttransplant, and then annually until the fifth year posttransplant (40). Testing urine for BKV replication is recommended to rule out PyVAN, and can be performed by cytology or real-time PCR. Patients with high-level urinary BKV replication should be tested for plasma BKV DNA load. In patients with plasma BKV DNA loads of >4 log geq/mL for >3 weeks, a diagnosis of presumptive PyVAN is made and an allograft biopsy should be considered for a diagnosis of definitive PyVAN (40). It is also recommended to test urine and/or plasma BKV DNA loads, when a work-up for allograft dysfunction is indicated or when an allograft biopsy is performed for any indication including for protocol biopsies (40). The incidence of high-level viruria ranges from 5% to 25% in heart, liver, and lung transplant recipients, but PyVAN is a rare complication. Routine screening is currently not recommended in nonrenal solid organ transplants, but testing for BKV DNA in plasma should be considered in the diagnostic work-up case of creeping renal failure. This may also be appropriate for other types of immunodeficient patients. Similar approaches are currently being studied for PyVHC in allogenic hematopoietic stem cell transplant patients (42), but the current data are insufficient to recommend general screening and preemptive intervention (76).

DIAGNOSTIC TESTS FOR JCV

JCV Serology

HIA and EIA detecting JCV capsid antigens have been used in clinical research studies, but are currently not available to diagnostic virology laboratories. Serum serology is generally considered to be of limited value for the diagnosis of PyVML, since this disease primarily affects severely immunodeficient patients who are known to have suboptimal primary or secondary antibody responses. However, a recent study reported that HIV-infected patients surviving PyVML had significantly higher antibody titers than nonsurvivors or HIV-infected control patients (77). Intrathecal antibody responses to JCV may be of independent value to diagnose PyVML in cases where JCV DNA has been below the level of detection. CSF antibodies against VP1 were found in 76% of PyVML cases compared to 11% of HIV-infected patients without PyVML, and the specific intrathecal antibody index was positive (78). Also, the JCV-specific intrathecal antibody index increased in patients with HIV-related PyVML who underwent disease remission following initiation of HAART and in parallel with JCV DNA decline in CSF (79). Nevertheless, the detection of JCV DNA in CSF by PCR represents the primary diagnostic approach.

JCV-Specific Cellular Immunity

JCV-specific T-cell responses have been studied in PyVML patients by assays measuring the JCV-specific CD4+ or/and CD8+ T-cell responses using lymphoproliferation, ELISpot, and intracellular cytokine staining assays. In addition, cytotoxic T cells have been characterized using chromium release and flow cytometry for annexin V. Most of the characterized JCV epitopes were derived from JCV VP-1 capsid protein (80). Because of the low frequency in peripheral blood, JCV-specific T cells have been expanded in vitro over one to three weeks following stimulation with JCV peptides loaded on activated monocytes or dendritic cells. Although T-cell activity following in vitro expansion may not be directly translatable into frequencies in the peripheral blood of PyVML patients, JCV-specific CD8+ T cells have been linked to improved survival (80–82). In another study, JCV-specific cytotoxic T cells could be detected in CSF in approximately half of PyVML patients with favorable outcome (83). In a case-control study of the Swiss HIV Cohort, PyVML survivors tended to have more interferon-γ releasing T cells than nonsurvivors (77). Clearly, the significance of assessing JCV-specific cell-mediated immune responses in the clinical routine requires more structured studies.

JCV Isolation in Cell Cultures

JCV has a very narrow host cell range which likely depends on several factors, including the availability of specific cellular receptors, the presence of cell type-specific proteins that regulate JCV DNA replication and transcription, and the architecture of the regulatory region (84,85). Only a few cell systems are fully permissive to the lytic growth of JCV. Primary human fetal glial cells and primary glia-derived astrocytes are the most permissive system for JCV propagation in cell culture. Some cell lines allow for JCV replication, such as SVG cells derived from fetal human glial cells transformed with the SV40 T-Ag, KG-1 cells, or COS-7 cells, derived from monkey cells transformed with SV40 T-Ag. Because of the long time required to demonstrate viral growth and the low sensitivity, the use of cell cultures to study JCV infection is limited to the research setting.

JCV Molecular Genetic Testing

Detection of JCV DNA in CSF by nucleic acid amplification techniques, primarily by polymerase chain reaction (PCR), is currently recommended as the first line diagnostic test for suspected PyVML diagnosis. The diagnostic sensitivity of qualitative (frequently nested) JCV DNA PCR in CSF ranged from 70% to >90% and the specificity from 80% to almost 100% (86). In recent years, these have been progressively replaced by quantitative assays, which also provide an estimate of JCV DNA level in CSF. However, approximately one-fourth of patients with PyVML will in fact have a negative test at the time of the first CSF examination (87). The probability of detecting JCV DNA in CSF may increase with progression of PyVML reflecting more extensive viral replication in the brain lesions. Thus, repeat lumbar puncture should be considered if the initial CSF analysis is negative for JCV DNA but clinical suspicion of PyVML remains high. In HIV-1/AIDS patients with PML receiving combination antiretroviral therapy, the likelihood of detecting JCV in CSF may decline over time, which most likely reflects the curtailing of JCV replication by the recovering immune system (88,89). The detection rate of JCV DNA may be higher in native noncentrifuged CSF than in CSF supernatant from which cells have been removed (87). Given the pivotal role of CSF analysis for PyVML diagnosis, quality control issues regarding DNA preparation and the characteristics of the JCV PCR assay become essential in the diagnostic virology laboratory. Timely revision of published target sequences of PCR primers and probes is recommended to identify the need for adaptation as more sequences become available. As shown in Figure 5, the primers used for routine detection of JCV DNA in Basel bind to a highly conserved region of the large T-gene. The primers and probes used in Milano have similar characteristics (Fig. 6).

Measuring JCV DNA loads in CSF by quantitative techniques may provide additional information. In untreated HIV-1/AIDS patients, higher JCV copy numbers are predictive of shorter survival (90,91). Higher CSF copy numbers were also observed in patients with lower CD4 cell counts. These associations suggest that the degree of immune deficiency may affect the rate of JCV replication in the CNS and that this, in turn, may correlate with disease progression. Following initiation of combination antiretroviral therapy, approximately half of the HIV-1/AIDS patients with PyVML show disease remission (92,93), which is associated with clearance of the virus from the CSF (91). By contrast, persisting JCV DNA detection in CSF, even at low levels, is associated with clinical progression of PyVML.

JCV DNA sequencing studies have investigated the NCCR and the major viral capsid protein VP-1. An archetype form of the JCV NCCR is consistently present in urine of healthy and immunosuppressed subjects, whereas rearranged sequences, carrying deletions and insertions, most of which are partial sequence duplications, are typically found in the CSF and brain tissues of patients with PyVML. Occasionally, rearranged JCV NCCR may also be detectable in blood of patients with PyVML. NCCR sequencing may help to confirm the uniqueness of the rearrangement indicating a true positive result as opposed to contamination, which was occasionally observed in earlier studies, particularly those using nested PCR. Moreover,

```
Accession  Position of                                                        Nr of identical   % of
number     1st nt                      Alignment                              sequences         total
------------------------------------------------------------------------------------------------------

Forward primer

                      AGTTCTATTACTAAACACAGCTTGACTGAGGAATGCATGCA   470 total   100
AB372038   4193       .........................................   459          98
AF281614   3926       .................................G.......     6           1
AB127027   4193       ............G............................     1           0.2
AF295731   3926       ............................G............     1           0.2
AF015536   3926       ...........................A.............     1           0.2
AB077872   4188       ...........................T.............     1           0.2
AB048546   4191       ..............................A..G.......     1           0.2
BKV                   ...............T...........A..A..CTGG..T.
SV40                  .....................A.....T.AA..ACT..A..

Probe

                      GGGGTAGAGTGTTGGGATCCTGTGTTTTCATCATCACTGGC   476 total   100
AB372038   4283       .........................................   434          91
AB262413   4221       ...........C.............................    25           5
AY378086   4016       ...................................C.....     6           1
AF015537   4016       .....G...................................     5           1
AY386377   4016       ....................C....................     1           0.2
AF396422   4016       ............................G............     1           0.2
U61771     4273       ......................A..A...............     1           0.2
AM712489     26       ...............C.........................     1           0.2
AM712484     26       .........G...............................     1           0.2
AB372037   4281       .........................................     1           0.2
                               \
                             GGA
BKV                   ..T..T........A..AT..........T...........
                            \
                            C
SV40                  .....T..A.....A.....A..AGCC..C..........A
                                \ \
                                G AG
```

In the original figure, the primer and probe sequences are underlined and shaded.

Figure 5 Alignment of the oligonucleotides used for the JCV Taqman PCR in Basel, Switzerland (see Fig. 3).

Accession number	Position of 1st nt	Alignment	Nr of identical sequences	% of total
Reverse primer				
		`TGGTGGAATACATTTAATGAGAAGTGGGATGAAGACCTGTT`	**318 total**	**100**
AB372038	4384	`.........................................`	301	95
AF300967	4117	`....................A....................`	11	3
AY382185	4117	`.................................A.......`	2	0.6
AB126987	4384	`..........G..............................`	1	0.3
AF281615	4155	`.................................T.......`	1	0.3
AB048582	4384	`.........................................`	1	0.3
AY342299	4117	`..............A.....A............A.......`	1	0.3
BKV		`.........T.C........A..A...........TT.A..`		
SV40		`....TCA.........CCAT..........G..GGCT..T.` (insertion `\ CT` after GGCT)		

Figure 5 (*Continued*)

sequence definition of viral variants in different body sites over time might help elucidate steps of PyVML pathogenesis and their relation to JCV replication in the CNS. The JCV capsid protein VP-1 is likely a main target of both humoral and cell-mediated immune responses. Sequencing of the VP1 gene from urine of healthy subjects has led to the identification of at least 12 distinct JCV subtypes, which seem to have evolved in distinct geographic regions. JCV sequence variability may impact on primer targets and should therefore be consulted in the design and testing of PCR assays.

JCV DNA may also be detected in other specimens including blood, mononuclear cells, plasma, urine, brain, and respiratory secretions (94). However, detection rates vary widely among published studies. Among patients with HIV-related PyVML, JCV DNA has been reported in PBMC of 10% to 60% and in plasma of 15% to 40%. However, JCV DNA is also observed in a significant proportion of immunocompromised subjects without the disease. These figures suggest that JCV DNA detection in blood is unlikely to provide a sensitive and specific diagnostic tool for identifying patients with PyVML, although it may prove useful for the genetic characterization of extracerebral virus. In urine, JCV DNA is detected in approximately one-third of healthy individuals at levels that do not allow discrimination between patients with or without PML. However, neither JCV DNA detection in urine nor measurement of virus level is relevant for the diagnosis of PyVML (95–101). The detection of JCV DNA has also been reported in brain tissue of PyVML and non-PyVML cases as well as in a variety of extracerebral tissues—including kidney, liver, lung, lymph nodes, spleen, heart, and gastrointestinal tract—by blot hybridization (102) and nucleic acid amplification methods (103,104). The clinical significance of these findings is unclear, especially in the absence of tissue damage or evidence of virion production. Therefore, the biopsy or autopsy should be primarily evaluated by histology, immunohistochemistry for large T-antigen expression, or by in-situ hybridization.

Accession number	Position of 1st nt	Alignment	Nr of identical sequences	% of total
Forward primer				
		GGGGTAGAGTGTTGGGATCCTGTGTTTTCA	**481 total**	**100**
AB372038	4280	....	449	93
AB262413	4278	...........C..................	25	5
AY386377	4013	....................C.........	1	0.2
AF396422	4013	............................G.	1	0.2
AF295736	4013	........................G.....	1	0.2
U61771	4270	......................A..A....	1	0.2
AM712484	23	.........G....................	1	0.2
AM712489	23	...............C..............	1	0.2
AB372037	4278	.... \\ GGA	1	0.2
BKV		..T..T...........AT..CCTG..G.C		
SV40		..A.....A.....A..GT.A.CAG.AG.C		
Probe				
		ATCATCACTGGCAAACATTTCTTCATGGCA	**470 total**	**100**
AB372038	4313	....	376	78
AY536239	4363	..................C...........	80	17
AY382188	4046	......C.............T.........	5	1
AY386376	4046	....................T.........	3	0.6
AY378086	4046	......C.......................	2	0.4
AF396428	4047	..........C...................	1	0.2
AF281604	4046	............G.................	1	0.2
AB113137	4313	.....................C........	1	0.2
AB048582	4313	..................C..C........	1	0.2
BKV		C.................A...........		
SV40		.........A.ATGG.........TGA...		

Figure 6 Alignment of the oligonucleotides used for the JCV Taqman PCR in Milano, Italy (see Fig. 3).

Accession number	Position of 1st nt	Alignment	Nr of identical sequences	% of total
Reverse primer				
		AAAACAGGTCTTCATCCCACTTCTCATTAA	**470 total**	**100**
AB372038	4338	....	454	97
AF300967	4071	......................T.......	11	2
AY382185	4071	.........T....................	2	0.4
AF281615	4071	.........A....................	1	0.2
AB048582	4340	..........C...................	1	0.2
AY342299	4071	.........T............T.....T.	1	0.2
BKV		.G..T.AA...........T..T.......		
SV40		.........T...C..ATTAAAGG....CC		

Figure 6 (*Continued*)

DIAGNOSIS OF POLYOMAVIRUS-ASSOCIATED MULTIFOCAL LEUKOENCEPHALOPATHY

The key disease caused by JCV is PyVML. The pathogenesis is characterized by uncontrolled high-level replication of JCV in oligodendrocytes of the white matter, for example, subcortical areas of the central nervous system. The common denominator of PyVML patients is profound and prolonged immune dysfunction as encountered in HIV/AIDS, malignancies, chemotherapy, transplantation, and exposure to monoclonal antibodies targeting lymphocyte surface markers. However, the onset of PyVML is somewhat erratic suggesting that additional risk factors must be operative.

In suspected PyVML cases in which PCR fails to detect JCV DNA in CSF, a definitive diagnosis of PyVML requires brain biopsy. PyVML can be recognized by the presence of typical histopathology features, including enlarged oligodendrocytes with intranuclear inclusions, bizarre astrocytes, and lipid-laden macrophages. PyVML lesions are classically described as devoid of inflammatory cells, although "inflammatory" forms of PyVML are increasingly being observed. These forms are characterized by perivascular and parenchymal mononuclear infiltrates, mainly consisting of CD8+ T lymphocytes and monocyte/macrophages and are usually encountered in less severely immunocompromised patients, such as patients with HIV-related PyVML following the initiation of HAART. The presence of JCV needs be confirmed by immunohistochemistry, in situ nucleic acid hybridization, or electron microscopy. Examination of oligodendrocytes by electron microscopy shows viral particles with a diameter of about 40 to 50 nm typically aggregating into filamentous or crystal-like structures. In lesions characterized by high viral loads, virions can also be observed in other cell types, including astrocytes and occasionally neurons, as well as within vacuoles of macrophages, most likely resulting from phagocytosis. Although neurons are usually spared by the infection, JCV may occasionally infect neuronal cells in the granular layer of the cerebellum (105). This finding has been observed both in the context of typical PyVML lesions (106) and in HIV-infected patients without classical PyVML lesions (105,107). There are no noninvasive screening tests that would identify patients at increased risk for PyVML. Limited case series indicate that detecting JCV DNA in blood has a poor sensitivity even in patients with definitive PyVML, and blood JCV loads are low, if at all detectable. Therefore, the diagnosis of PyVML depends primarily on early identification of neurological deficits in an immunodeficient individual followed by magnetic resonance imaging and CSF sampling for JCV DNA by PCR.

According to a recent consensus proposal (108), the diagnosis of PyVML has been categorized as

1. Possible PyVML, in cases of patients with typical neurologic deficits and compatible radiological signs on magnetic resonance imaging.
2. Laboratory-confirmed PyVML in cases of JCV detection in cerebrospinal fluid (CSF) by PCR.
3. Definitive PyVML in cases of histological proof in brain tissues from biopsies or autopsies.

DIAGNOSIS OF POLYOMAVIRUS-ASSOCIATED RESPIRATORY TRACT INFECTION

The respiratory route has been implicated in BKV transmission, but it is presently uncertain if BKV is a relevant cause of respiratory tract infections. However, BKV has been detected in the respiratory tract of profoundly immunosuppressed HIV/AIDS patients with overwhelming systemic BKV replication and multiorgan failure, and was also isolated from a child undergoing HSCT for acute leukemia (15). BKV was not identified in any of 727 respiratory tract samples from 499 patients, whereas JCV was found in 5 (1%) cases, MCV in 1 (0.2%), KIV and WUV in 8 (1.6%) each (94). Using KIV VP1-specific PCR, positive results were obtained in 1% (6 of 637) nasopharyngeal aspirates from patients with respiratory tract disease and 0.5% (1 of 192) stool samples from patients with gastroenteritis (20). KIV could not be detected in urine, serum, whole blood, or isolated leukocytes. KIV-positive patients were mainly children, and in all but one case, other respiratory viruses were also found in KIV-positive samples (109). WUV was identified in 43 samples obtained from 2135 patients with acute respiratory tract infections. Of those, 31 were also positive for other respiratory viruses (21). In a retrospective study, Norja et al. reported that the prevalence of KIV and WUV was not higher in patients with respiratory tract infections compared to a control group (94). Thus, despite several pilot studies, the clinical significance of KIV and WUV is still unclear.

KIV and WUV Molecular Diagnostic Assays

KIV and WUV diagnostics currently rely exclusively on amplification and detection of DNA. Bialasiewicz et al. developed real-time PCR assay allowing detection of KIV and WUV DNA from respiratory samples (110). The assay KIV-A and WUV-B appeared to be most suitable for routine diagnostics. DNA was extracted using the High Pure Nucleic Acid kit (Roche Diagnostics, Rotkreuz, Switzerland) according to the manufacturer's instructions. The primers KI-A-141-F and KI-A-200-R for KIV and WU-B-2729-F and WU-B-2808-R for WUV were used at a final concentration of 400 nM (Table 2). The FAM-labeled probes KI-A-182-TM and WU-B-2997-TM were used at a final concentration of 160 nM. The assay used 2 μL of extracted DNA in a total volume of 25 μL. The twofold concentrated amplification mix Quantitect Probe Master Mix (Qiagen, Hombrechtikon, Switzerland) was used for both the KIV and WUV PCR. The temperature profile consisted of an incubation at 95°C; 15 minutes for hot-start activation and 55 cycles of 95°C; 15 seconds; 60°C; 60 seconds. The limit of detection of both assays, defined as the lowest concentration allowing detection in 100% of the replicates, was determined to be 10 geq per reaction. An alternative to this protocol was described by Sharp et al. (94). The authors used a nested PCR detecting KI and WU, using alternative inner primers allowing specific detection of the viruses (Table 2). Although this protocol allows semi-quantification by limiting dilution of the DNA extracts, the complexity of this method might not be suitable for routine diagnostics.

Diagnosis of MCV-Associated Diseases

MCV-DNA could be amplified in 8 of 10 MCC tumors by using MCV-specific PCR and southern hybridization. In a control group, only 5 of 59 samples were positive, indicating a positive association of MCV-DNA and MCC (22). In a recent study of autopsy samples from 42 AIDS patients, only one sample tested positive for MCV, indicating a possible absence of reactivation

Table 2 Primers and Probes for the Amplification and Detection of KIV and WUV DNA

Description	Sequence	Target	Reference
KI-A-141-F	ACC TGA TAC CGG CGG AAC T	KI NCCR, forward primer	(110)
KI-A-200-R	CGC AGG AAG CTG GCT CAC	KI NCCR, reverse primer	(110)
KI-A-182-TM	FAM-CCA CAC AAT AGC TTT CAC TCT TGG CGT GA-TAMRA	KI NCCR, probe	(110)
WU-B-2729-F	CTA CTG TAA ATT GAT CTA TTG CAA CTC CTA	WU large T-antigen, forward primer	(110)
WU-B-2808-R	GGG CCT ATA AAC AGT GGT AAA ACA ACT	WU large T-antigen, reverse primer	(110)
WU-B-2997-TM	FAM-CCT TTC CTC CAC AAA GGT CAA GTA AA-TAMRA	WU large T-antigen, probe	(110)
WUKI_OS	ATC TRT AGC TGG AGG AGC AGA G	WU and KI VP2, outer PCR, forward primer	(94)
WUKI_OAS	CCY TGG GGA TTG TAT CCT GMG G	WU and KI VP2, outer PCR, reverse primer	(94)
WUKI_IS	RTC AAT TGC TGG WTC TGG AGC TGC	WU and KI VP2, inner PCR, forward primer	(94)
WUKI_IAS	TCC ACT TGS ACT TCC TGT TGG G	WU and KI VP2, inner PCR, reverse primer	(94)
WU_IAS	CTG TTA CAC CTT GTG TTA CAG TT	WU VP2, inner PCR, reverse primer	(94)
KI_IAS	GTT ACA GCT TGG GTA GCT TGA	KI VP2, inner PCR, reverse primer	(94)

of MCV under immune suppression or a low overall prevalence of MCV (94). The association between MCV and Merkel cell carcinoma was confirmed by other independent studies. No association with other malignancies such as nonmelanoma skin cancer or prostate cancer tissue has been reported so far (111,112). BKV DNA has been associated with a variety of malignancies in the past, and most recently with precursor stages of prostate cancer; decisive studies are ongoing.

MCV Molecular Diagnostic Assays

MCV diagnostics currently rely exclusively on PCR techniques. The protocol by Feng et al. has been used in most published studies (22). This protocol consists of three PCR reactions (LT1, LT3, and VP1) performed in parallel (Table 3). DNA from tumor tissues is isolated by phenol-chloroform extraction and 100 ng is used in the PCR reactions. The genomic DNA is amplified using Taq DNA Polymerase (Invitrogen). The initial denaturation step at 94°C for 3 minutes is followed by 31 cycles of 94°C; 45 seconds, 58°C; 30 seconds and 72°C; 45 seconds, and then 15 minutes at 72°C. The sensitivity of detection can be increased by southern blot hybridization using specific internal probes. An assay recently published by Sharp et al., targeting the small T-antigen of MCV, may constitute an interesting alternative (94).

Diagnosis of SV40 Associated Diseases

Although it is now generally accepted that human populations have been exposed to SV40 through contaminated polio- and adenovirus vaccines, its potential role as a human pathogen is still controversial. Several studies have shown that SV40 can cause various tumors in rodents. Also, SV40 can cause a PyVML-like encephalitis following primary intrathecal inoculation of rhesus monkeys (113,114). The detection of SV40 DNA by PCR in human mesothelioma samples at high frequency seemed to support a role of SV40 in the development of this malignant tumor in humans (115). However, this conclusion was challenged by the fact that laboratory contamination by plasmid-borne viral DNA might account for a large proportion of the positive SV40 PCR (116). The controversy regarding the role of SV40 in human tumors has not been resolved so far (117). Besides the controversial role in human malignancies, SV40 DNA was reported to be detectable in 9% of tonsil samples from immunocompetent children (118), in

Table 3 Primers for the Amplification and Detection of MCV DNA

Name	Sequences	Target	Reference
LT1 forward	TAC AAG CAC TCC ACC AAA GC	MCV large T-antigen, forward primer	(22)
LT1 reverse	TCC AAT TAC AGC TGG CCT CT	MCV large T-antigen, reverse primer	(22)
LT3 forward	TTG TCT CGC CAG CAT TGT AG	MCV large T-antigen, forward primer	(22)
LT3 reverse	ATA TAG GGG CCT CGT CAA CC	MCV large T-antigen, reverse primer	(22)
VP1 forward	TTT GCC AGC TTA CAG TGT GG	MCV VP1, forward primer	(22)
VP1 reverse	TGG ATC TAG GCC CTG ATT TTT	MCV VP1, reverse primer	(22)
MCPyV_OS	GGC AAC ATC CCT CTG ATG AAA GC	MCV small T-antigen, outer PCR, forward primer	(94)
MCPyV_OAS	CCA CCA GTC AAA ACT TTC CCA AGT AGG	MCV small T-antigen, outer PCR, reverse primer	(94)
MCPyV_IS	CTT AAA GCA TCA CCC TGA TAA AGG	MCV small T-antigen, inner PCR, forward primer	(94)
MCPyV_IAS	AAA CCA AAG AAT AAA GCA CTG ATA GCA	MCV small T-antigen, inner PCR, reverse primer	(94)

16% of blood samples from healthy donors (119), in 3% of urine samples from healthy children (120), in 8% of stool samples from hospitalized children (39), and in 6% of urine samples from lung-transplant recipients (121). Also, SV40 has been associated with a potential role in kidney disease, including PyVAN and focal segmental glomerulosclerosis (122–124). A number of diagnostic tools have become available from experimental model systems including cross-reactive antibodies raised against the SV40 large T-antigen, which have been instrumental in PyVAN diagnosis. Antibody assays using SV40 VLP have been generated. However, few studies combined these tools in a convincing manner. Clearly, independent evidence is needed, preferably with novel diagnostic approaches to shed light into the clinical role of SV40.

SV40 Molecular Diagnostic Assays

Many assays for detection of SV40 DNA target the large T-antigen. A widely used assay was described by Bergsagel (125) (Table 4). However, the presence of large T-sequence in numerous expression plasmids enhances the risk of laboratory contamination, as described by Lopez-Rios (116). Great care must therefore be taken to strictly separate diagnostic procedures from any work involving plasmid in high concentrations. It is recommended to use primer pairs directed at regions less prone to contamination, such as the primers SVINTfor and SVINTrev that bind to the intron sequence of the large T-antigen (Table 4). DNA from tumor tissues was isolated with Trizol and 1 μg were used in the 50 μL PCR reactions. The primers were used at a concentration of 400 nM. The DNA was amplified using HotStar Taq DNA Polymerase (Qiagen) and the reaction mix included 10 mM of each dNTP, 25 mM $MgCl_2$, and PCR buffer. The temperature profile consisted of an initial activation step at 95°C; 15 minutes and 44 cycles of 94°C; 60 seconds, annealing temperature; 30 seconds and 72°C; 60 seconds followed by 10 minutes at

Table 4 Primers for the Amplification and Detection of SV-40 DNA

Name	Sequences	Target	Reference
SVINTfor	AAG TAA GGT TCC TTC ACA AAG	SV-40 large T-antigen, intron	(116)
SVINTrev	AAG TGA GGT ATT TGC TTC TTC	SV-40 large T-antigen, intron	(116)
SV.for3	TGA GGC TAC TGC TGA CTC TCA ACA	SV-40 large T-antigen, intron	(125)
SV.rev	GCA TGA CTC AAA AAA CTT AGC AAT TCT G	SV-40 large T-antigen, intron	(125)

72°C. The annealing temperatures were 62°C for the cycles 1 and 2, 60°C for the cycles 3 and 4, and 58°C for the cycles 5 to 44.

CONCLUSION

The clinical significance of PyV infections in humans has become increasingly challenging since the discovery of BK virus and JC virus more than 25 years ago. On the one hand, the population at risk for PyV infections seems to have increased in parallel with the more complex and widespread use of potent immunosuppressive and immunomodulating therapies. On the other hand, new PyVs have been discovered, although their clinical significance is not conclusively defined. This burden is passed on to the clinical virology laboratory. The most relevant diagnostic tests are based on PCR, but despite high sensitivity and specificity, PCR assays are challenged by the limited knowledge of natural virus variants and quality control issues.

REFERENCES

1. Cole CN, *Polyomavirinae: The Viruses and Their Replication*, in *Fundamental Virology*, DMK Bernard N. Fields, Peter M. Howley, Editor. 1996, Lippincott-Raven Publishers. p. 917–945.
2. Büchen-Osmond C, International Committee on Taxonomy of Viruses Database (ICTVdB), 2008. **www.ncbi.nlm.nih.gov/ICTVdb/.**
3. Gardner SD, Field AM, Coleman DV, Hulme B, *New human papovavirus (B.K.) isolated from urine after renal transplantation.* Lancet, 1971. **1**(7712): p. 1253–57.
4. Zurhein G, Chou SM, *Particles Resembling Papova Viruses in Human Cerebral Demyelinating Disease.* Science, 1965. **148**: p. 1477–9.
5. Padgett BL, Walker DL, ZuRhein GM, et al., *Cultivation of papova-like virus from human brain with progressive multifocal leucoencephalopathy.* Lancet, 1971. **1**(7712): p. 1257–60.
6. Kleinschmidt-DeMasters BK, Tyler KL, *Progressive multifocal leukoencephalopathy complicating treatment with natalizumab and interferon beta-1a for multiple sclerosis.* N Engl J Med, 2005. **353**(4): p. 369–74.
7. Langer-Gould A, Atlas SW, Green AJ, et al., *Progressive multifocal leukoencephalopathy in a patient treated with natalizumab.* N Engl J Med, 2005. **353**(4): p. 375–81.
8. Van Assche G, Van Ranst M, Sciot R, et al., *Progressive multifocal leukoencephalopathy after natalizumab therapy for Crohn's disease.* N Engl J Med, 2005. **353**(4): p. 362–8.
9. Arthur RR, Shah KV, Baust SJ, et al., *Association of BK viruria with hemorrhagic cystitis in recipients of bone marrow transplants.* N Engl J Med, 1986. **315**(4): p. 230–234.
10. Apperley JF, Rice SJ, Bishop JA, et al., *Late-onset hemorrhagic cystitis associated with urinary excretion of polyomaviruses after bone marrow transplantation.* Transplantation, 1987. **43**(1): p. 108–12.
11. Bedi A, Miller CB, Hanson JL, et al., *Association of BK virus with failure of prophylaxis against hemorrhagic cystitis following bone marrow transplantation.* J Clin Oncol, 1995. **13**(5): p. 1103–1109.
12. Binet I, Nickeleit V, Hirsch HH, et al., *Polyomavirus disease under new immunosuppressive drugs: a cause of renal graft dysfunction and graft loss.* Transplantation, 1999. **67**(6): p. 918–922.
13. Nickeleit V, Klimkait T, Binet IF, et al., *Testing for polyomavirus type BK DNA in plasma to identify renal-allograft recipients with viral nephropathy.* N Engl J Med, 2000. **342**(18): p. 1309–15.
14. Hirsch HH, Knowles W, Dickenmann M, et al., *Prospective study of polyomavirus type BK replication and nephropathy in renal-transplant recipients.* N Engl J Med, 2002. **347**(7): p. 488–496.
15. Hirsch HH, Steiger J, *Polyomavirus BK.* Lancet Infect Dis, 2003. **3**(10): p. 611–623.
16. Randhawa PS, Finkelstein S, Scantlebury V, et al., *Human polyoma virus-associated interstitial nephritis in the allograft kidney.* Transplantation, 1999. **67**(1): p. 103–109.
17. Hix JK, Braun WE, Isada CM, *Delirium in a renal transplant recipient associated with BK virus in the cerebrospinal fluid.* Transplantation, 2004. **78**(9): p. 1407–1408.
18. Kazory A, Ducloux D, Chalopin JM, et al., *The first case of JC virus allograft nephropathy.* Transplantation, 2003. **76**(11): p. 1653–1655.
19. Drachenberg CB, Hirsch HH, Papadimitriou JC, et al., *Polyomavirus BK versus JC replication and nephropathy in renal transplant recipients: a prospective evaluation.* Transplantation, 2007. **84**(3): p. 323–30.
20. Allander T, Andreasson K, Gupta S, et al., *Identification of a third human polyomavirus.* J Virol, 2007. **81**(8): p. 4130–6.
21. Gaynor AM, Nissen MD, Whiley DM, et al., *Identification of a novel polyomavirus from patients with acute respiratory tract infections.* PLoS Pathog, 2007. **3**(5): p. e64.
22. Feng H, Shuda M, Chang Y, Moore PS, *Clonal integration of a polyomavirus in human Merkel cell carcinoma.* Science, 2008. **319**(5866): p. 1096–100.

23. Becker JC, Houben R, Ugurel S, et al., *MC polyomavirus is frequently present in Merkel cell carcinoma of European patients.* J Invest Dermatol, 2009. **129**(1): p. 248–50.
24. Bofill-Mas S, Formiga-Cruz M, Clemente-Casares P, et al., *Potential transmission of human polyomaviruses through the gastrointestinal tract after exposure to virions or viral DNA.* J Virol, 2001. **75**(21): p. 10290–9.
25. Luo C, Bueno M, Kant J, Randhawa P, *Biologic diversity of polyomavirus BK genomic sequences: Implications for molecular diagnostic laboratories.* J Med Virol, 2008. **80**(10): p. 1850–7.
26. Luo C, Bueno M, Kant J, et al., *Genotyping Schemes for Polyomavirus BK using Gene Specific Phylogenetic Trees and SNP Analysis.* J Virol, 2008. **24**: p. 24.
27. Dugan AS, Eash S, Atwood WJ, *Update on BK virus entry and intracellular trafficking.* Transpl Infect Dis, 2006. **8**(2): p. 62–7.
28. Low JA, Magnuson B, Tsai B, Imperiale MJ, *Identification of gangliosides GD1b and GT1b as receptors for BK virus.* J Virol, 2006. **80**(3): p. 1361–6.
29. Bernhoff E, Gutteberg TJ, Sandvik K, et al., *Cidofovir Inhibits Polyomavirus BK Replication in Human Renal Tubular Cells Downstream of Viral Early Gene Expression.* Am J Transplant, 2008. **8**: p. 1413–1422.
30. Seemayer CA, Seemayer NH, Durmuller U, et al., *BK virus large T and VP-1 expression in infected human renal allografts.* Nephrol Dial Transplant, 2008. **23**(12): p. 3752–61.
31. Hanssen Rinaldo C, Hansen H, Traavik T, *Human endothelial cells allow passage of an archetypal BK virus (BKV) strain–a tool for cultivation and functional studies of natural BKV strains.* Arch Virol, 2005. **150**(7): p. 1449–58.
32. Loeber G, Dorries K, *DNA rearrangements in organ-specific variants of polyomavirus JC strain GS.* J Virol, 1988. **62**(5): p. 1730–5.
33. Gosert R, Rinaldo CH, Funk GA, et al., *Polyomavirus BK with rearranged noncoding control region emerge in vivo in renal transplant patients and increase viral replication and cytopathology.* J Exp Med, 2008. **205**(4): p. 841–52.
34. Rinaldo CH, Myhre MR, Alstad H, et al., *Human polyomavirus BK (BKV) transiently transforms and persistently infects cultured osteosarcoma cells.* Virus Res, 2003. **93**(2): p. 181–7.
35. Hirsch HH, *BK virus: opportunity makes a pathogen.* Clin Infect Dis, 2005. **41**(3): p. 354–60.
36. Knowles WA, Pipkin P, Andrews N, et al., *Population-based study of antibody to the human polyomaviruses BKV and JCV and the simian polyomavirus SV40.* J Med Virol, 2003. **71**(1): p. 115–23.
37. Egli A, Infanti L, Dumoulin A, et al., *Prevalence of Polyomavirus BK and JC Infection and Replication in 400 Healthy Blood Donors.* J Infect Dis, 2009. **199**: p. 837–46.
38. Kean JM, Rao S, Wang M, Garcea RL, *Seroepidemiology of human polyomaviruses.* PLoS Pathog, 2009. **5**(3): p. e1000363.
39. Vanchiere JA, White ZS, Butel JS, *Detection of BK virus and simian virus 40 in the urine of healthy children.* J Med Virol, 2005. **75**(3): p. 447–54.
40. Hirsch HH, Brennan DC, Drachenberg CB, et al., *Polyomavirus-associated nephropathy in renal transplantation: interdisciplinary analyses and recommendations.* Transplantation, 2005. **79**(10): p. 1277–1286.
41. Noss G, *Human polyoma virus type BK infection and T antibody response in renal transplant recipients.* Zentralbl Bakteriol Mikrobiol Hyg [A], 1987. **266**(3–4): p. 567–74.
42. Wong AS, Chan KH, Cheng VC, et al., *Relationship of Pretransplantation Polyoma BK Virus Serologic Findings and BK Viral Reactivation after Hematopoietic Stem Cell Transplantation.* Clin Infect Dis, 2007. **44**(6): p. 830–7.
43. Hamilton RS, Gravell M, Major EO, *Comparison of antibody titers determined by hemagglutination inhibition and enzyme immunoassay for JC virus and BK virus.* J Clin Microbiol, 2000. **38**(1): p. 105–9.
44. Randhawa PS, Gupta G, Vats A, et al., *Immunoglobulin G, A, and M responses to BK virus in renal transplantation.* Clin Vaccine Immunol, 2006. **13**(9): p. 1057–63.
45. Bodaghi S, Comoli, P., Boesch, R., Azzi, A., Gosert, R., Leuenberger, D., Ginevri, F., Hirsch, H.H., *Antibody Responses to Recombinant Polyomavirus BK Large T and VP1 Proteins in Pediatric Kidney Transplant Patients.* J Clin Micro, 2009. **(in press)**.
46. Bohl DL, Storch GA, Ryschkewitsch C, et al., *Donor origin of BK virus in renal transplantation and role of HLA C7 in susceptibility to sustained BK viremia.* Am J Transplant, 2005. **5**(9): p. 2213–21.
47. Ginevri F, De Santis R, Comoli P, et al., *Polyomavirus BK infection in pediatric kidney-allograft recipients: a single-center analysis of incidence, risk factors, and novel therapeutic approaches.* Transplantation, 2003. **75**(8): p. 1266–1270.
48. Smith JM, McDonald RA, Finn LS, et al., *Polyomavirus nephropathy in pediatric kidney transplant recipients.* Am J Transplant, 2004. **4**(12): p. 2109–17.
49. Bohl DL, Brennan DC, Ryschkewitsch C, et al., *BK virus antibody titers and intensity of infections after renal transplantation.* J Clin Virol, 2008. **43**(2): p. 184–9.
50. Randhawa P, Bohl D, Brennan D, et al., *Longitudinal analysis of levels of immunoglobulins against BK virus capsid proteins in kidney transplant recipients.* Clin Vaccine Immunol, 2008. **15**(10): p. 1564–71.

51. Leuenberger D, Andresen PA, Gosert R, et al., *Human polyomavirus type 1 (BK virus) agnoprotein is abundantly expressed but immunologically ignored.* Clinical and Vaccine Immunology, 2007. **14**(8): p. 959–68.
52. Rinaldo CH, Traavik T, Hey A, *The agnogene of the human polyomavirus BK is expressed.* J Virol, 1998. **72**(7): p. 6233–6.
53. Comoli P, Binggeli S, Ginevri F, Hirsch HH, *Polyomavirus-associated nephropathy: update on BK virus-specific immunity.* Transpl Infect Dis, 2006. **8**(2): p. 86–94.
54. Comoli P, Hirsch HH, Ginevri F, *Cellular immune responses to BK virus.* Curr Opin Organ Transplant, 2008. **13**(6): p. 569–74.
55. Binggeli S, Egli A, Schaub S, et al., *Polyomavirus BK-Specific Cellular Immune Response to VP1 and Large T-Antigen in Kidney Transplant Recipients.* Am J Transplant, 2007. **7**(5): p. 1131–9.
56. Comoli P, Basso S, Azzi A, et al., *Dendritic cells pulsed with polyomavirus BK antigen induce ex vivo polyoma BK virus-specific cytotoxic T-cell lines in seropositive healthy individuals and renal transplant recipients.* J Am Soc Nephrol, 2003. **14**(12): p. 3197–3204.
57. Chen Y, Trofe J, Gordon J, et al., *BKV and JCV large T antigen-specific CD8(+) T cell response in HLA A*0201(+) kidney transplant recipients with polyomavirus nephropathy and patients with progressive multifocal leukoencephalopathy.* J Clin Virol, 2008. **42**(2): p. 198–202.
58. Ginevri F, Azzi A, Hirsch HH, et al., *Prospective monitoring of polyomavirus BK replication and impact of pre-emptive intervention in pediatric kidney recipients.* Am J Transplant, 2007. **7**(12): p. 2727–35.
59. Egli A, Koehli, S., Dickenmann, M., Hirsch, H.H., *Inhibition of Polyomavirus BK- and Cytomegalovirus-specific T-cell responses by immunosuppressive drugs.* Transplantation, 2009. **in press**.
60. Purighalla R, Shapiro R, McCauley J, Randhawa P, *BK virus infection in a kidney allograft diagnosed by needle biopsy.* Am J Kidney Dis, 1995. **26**(4): p. 671–3.
61. Nickeleit V, Hirsch HH, Binet IF, et al., *Polyomavirus infection of renal allograft recipients: from latent infection to manifest disease.* J Am Soc Nephrol, 1999. **10**(5): p. 1080–1089.
62. Drachenberg CB, Papadimitriou JC, Hirsch HH, et al., *Histological patterns of polyomavirus nephropathy: correlation with graft outcome and viral load.* Am J Transplant, 2004. **4**(12): p. 2082–92.
63. Jorgensen GE, Hammarin AL, Bratt G, et al., *Identification of a unique BK virus variant in the CNS of a patient with AIDS.* J Med Virol, 2003. **70**(1): p. 14–9.
64. Drachenberg CB, Beskow CO, Cangro CB, et al., *Human polyoma virus in renal allograft biopsies: morphological findings and correlation with urine cytology.* Hum Pathol, 1999. **30**(8): p. 970–7.
65. Singh HK, Bubendorf L, Mihatsch MJ, et al., *Urine cytology findings of polyomavirus infections.* Adv Exp Med Biol, 2006. **577**: p. 201–12.
66. Singh HK, Andreoni KA, Madden V, et al., *Presence of urinary haufen accurately predicts polyomavirus nephropathy.* J Am Soc Nephrol, 2009. **20**(2): p. 416–27.
67. Viscount HB, Eid AJ, Espy MJ, et al., *Polyomavirus polymerase chain reaction as a surrogate marker of polyomavirus-associated nephropathy.* Transplantation, 2007. **84**(3): p. 340–5.
68. Gordon J, Brennan D., Limaye, A., Randhawa, P., Storch, G., Trofe, J., Weck, K., and Hirsch, H.H., *Multicenter validation of polyomavirus BK quantification for screening and monitoring of renal transplant recipients.* Am J Transplant, 2005. **5**(Suppl. 11): p. S381–S382.
69. Hoffman NG, Cook L, Atienza EE, et al., *Marked variability of BK virus load measurement using quantitative real-time PCR among commonly used assays.* J Clin Microbiol, 2008. **46**(8): p. 2671–80.
70. MacKay W, Paola C, Hirsch H, et al., *Nucleic acid tests for JCV and BKV in the management of immunocompromised patients: results of an international external quality assessment study Quality Control for Molecular Diagnostics (QCMD) (Abstract P5-03).* ESCV 2008 - Clinical Virology Annual Meeting, Saariskä, Finnland, 2008: p. 39.
71. Hirsch HH, Mohaupt M, Klimkait T, *Prospective monitoring of BK virus load after discontinuing sirolimus treatment in a renal transplant patient with BK virus nephropathy.* J Infect Dis, 2001. **184**(11): p. 1494–5; author reply 1495–6.
72. Wu SW, Chang HR, Hsieh MC, et al., *Early diagnosis of polyomavirus type BK infection in tailoring immunosuppression for kidney transplant patients: screening with urine qualitative polymerase chain reaction assay.* Transplant Proc, 2008. **40**(7): p. 2389–91.
73. Almeras C, Foulongne V, Garrigue V, et al., *Does reduction in immunosuppression in viremic patients prevent BK virus nephropathy in de novo renal transplant recipients? A prospective study.* Transplantation, 2008. **85**(8): p. 1099–104.
74. Giraldi C, Noto A, Tenuta R, et al., *Prospective study of BKV nephropathy in 117 renal transplant recipients.* New Microbiol, 2007. **30**(2): p. 127–30.
75. Ding R, Suthanthiran M, *Noninvasive diagnosis of BK virus nephritis by measurement of messenger RNA for BK VP1 virus in urine.* Transplantation, 2003. **76**(2): p. 446.
76. Dropulic LK, Jones RJ, *Polyomavirus BK infection in blood and marrow transplant recipients.* Bone Marrow Transplant, 2008. **41**(1): p. 11–8.

77. Khanna N, Wolbers, M., Mueller, N.J., Garzoni, C., DuPasquier, R.A., Fux, C.A., Vernazza, P., Bernasconi, E., Viscidi, R., Battegay, M., Hirsch, H.H., *JCV-Specific Immune Responses in HIV-1 Patients with Progressive Multifocal Leukoencephalopathy.* J Virol, 2009. **[Epub ahead of print]**.
78. Weber T, Trebst C, Frye S, et al., *Analysis of the systemic and intrathecal humoral immune response in progressive multifocal leukoencephalopathy.* J Infect Dis, 1997. **176**(1): p. 250–4.
79. Giudici B, Vaz B, Bossolasco S, et al., *Highly active antiretroviral therapy and progressive multifocal leukoencephalopathy: effects on cerebrospinal fluid markers of JC virus replication and immune response.* Clin Infect Dis, 2000. **30**(1): p. 95–9.
80. Du Pasquier RA, Kuroda MJ, Zheng Y, et al., *A prospective study demonstrates an association between JC virus-specific cytotoxic T lymphocytes and the early control of progressive multifocal leukoencephalopathy.* Brain, 2004. **127**(Pt 9): p. 1970–8.
81. Du Pasquier RA, Clark KW, Smith PS, et al., *JCV-specific cellular immune response correlates with a favorable clinical outcome in HIV-infected individuals with progressive multifocal leukoencephalopathy.* J Neurovirol, 2001. **7**(4): p. 318–22.
82. Koralnik IJ, Du Pasquier RA, Kuroda MJ, et al., *Association of prolonged survival in HLA-A2+ progressive multifocal leukoencephalopathy patients with a CTL response specific for a commonly recognized JC virus epitope.* J Immunol, 2002. **168**(1): p. 499–504.
83. Du Pasquier RA, Autissier P, Zheng Y, et al., *Presence of JC virus-specific CTL in the cerebrospinal fluid of PML patients: rationale for immune-based therapeutic strategies.* Aids, 2005. **19**(18): p. 2069–76.
84. Sabath BF, Major EO, *Traffic of JC virus from sites of initial infection to the brain: the path to progressive multifocal leukoencephalopathy.* J Infect Dis, 2002. **186 Suppl 2**: p. S180–6.
85. Sweet TM, Del Valle L, Khalili K, *Molecular biology and immunoregulation of human neurotropic JC virus in CNS.* J Cell Physiol, 2002. **191**(3): p. 249–56.
86. Cinque P, Scarpellini P, Vago L, et al., *Diagnosis of central nervous system complications in HIV-infected patients: cerebrospinal fluid analysis by the polymerase chain reaction.* Aids, 1997. **11**(1): p. 1–17.
87. Hirsch HH, Meylan PR, Zimmerli W, et al., *HIV-1-infected patients with focal neurologic signs: diagnostic role of PCR for Toxoplasma gondii, Epstein-Barr virus, and JC virus.* Clin Microbiol Infect, 1998. **4**(10): p. 577–584.
88. Antinori A, Cingolani A, Lorenzini P, et al., *Clinical epidemiology and survival of progressive multifocal leukoencephalopathy in the era of highly active antiretroviral therapy: data from the Italian Registry Investigative Neuro AIDS (IRINA).* J Neurovirol, 2003. **9 Suppl 1**: p. 47–53.
89. Marzocchetti A, Di Giambenedetto S, Cingolani A, et al., *Reduced rate of diagnostic positive detection of JC virus DNA in cerebrospinal fluid in cases of suspected progressive multifocal leukoencephalopathy in the era of potent antiretroviral therapy.* J Clin Microbiol, 2005. **43**(8): p. 4175–7.
90. Yiannoutsos CT, Major EO, Curfman B, et al., *Relation of JC virus DNA in the cerebrospinal fluid to survival in acquired immunodeficiency syndrome patients with biopsy-proven progressive multifocal leukoencephalopathy.* Ann Neurol, 1999. **45**(6): p. 816–21.
91. Bossolasco S, Calori G, Moretti F, et al., *Prognostic significance of JC virus DNA levels in cerebrospinal fluid of patients with HIV-associated progressive multifocal leukoencephalopathy.* Clin Infect Dis, 2005. **40**(5): p. 738–44.
92. De Luca A, Pezzotti, P., Gasnault, J., Cinque, P., Berenguer, J., Di Giambenedetto, S., Cingolani, A., Taofik, Y., Pedale, R., Miralles, P., Larussa, D., Sinha, S., Cauda, R., Marra, C. M., Ammassari, A., and Antinori, A., *Survival and neurological outcome with or without cidofovir (CDV) in AIDS-related progressive multifocal leukoencephalopathy (PML) on HAART: A multicohort analysis.* J Neurovirol, 2005. **11** ((S2)): p. S16–S124.
93. Khanna N, Elzi, L., Mueller, N.J., Garzoni, C., Cavassini, M., Fux, C.A., Vernazza, P., Bernasconi, E., Battegay, M., Hirsch, H.H., for the Swiss HIV Cohort Study, *Incidence and Outcome of Progressive Multifocal Leukoencephalopathy in 20 years of the Swiss HIV Cohort Study.* Clinical Infectious Diseases, 2009. **48**: p. 1459–66.
94. Sharp CP, Norja P, Anthony I, et al., *Reactivation and Mutation of Newly Discovered WU, KI, and Merkel Cell Carcinoma Polyomaviruses in Immunosuppressed Individuals.* J Infect Dis, 2009. **199**(3): p. 398–404.
95. Knowles WA, Pillay D, Johnson MA, et al., *Prevalence of long-term BK and JC excretion in HIV-infected adults and lack of correlation with serological markers.* J Med Virol, 1999. **59**(4): p. 474–479.
96. Koralnik IJ, Boden D, Mai VX, et al., *JC virus DNA load in patients with and without progressive multifocal leukoencephalopathy.* Neurology, 1999. **52**(2): p. 253–60.
97. Lednicky JA, Vilchez RA, Keitel WA, et al., *Polyomavirus JCV excretion and genotype analysis in HIV-infected patients receiving highly active antiretroviral therapy.* Aids, 2003. **17**(6): p. 801–7.
98. Markowitz RB, Thompson HC, Mueller JF, et al., *Incidence of BK virus and JC virus viruria in human immunodeficiency virus-infected and -uninfected subjects.* J Infect Dis, 1993. **167**(1): p. 13–20.
99. Sundsfjord A, Flaegstad T, Flo R, et al., *BK and JC viruses in human immunodeficiency virus type 1-infected persons: prevalence, excretion, viremia, and viral regulatory regions.* J Infect Dis, 1994. **169**(3): p. 485–490.

100. Behzad-Behbahani A, Klapper PE, Vallely PJ, et al., *Detection of BK virus and JC virus DNA in urine samples from immunocompromised (HIV-infected) and immunocompetent (HIV-non-infected) patients using polymerase chain reaction and microplate hybridisation.* J Clin Virol, 2004. **29**(4): p. 224–229.
101. Shah KV, Daniel RW, Strickler HD, Goedert JJ, *Investigation of human urine for genomic sequences of the primate polyomaviruses simian virus 40, BK virus, and JC virus.* J Infect Dis, 1997. **176**(6): p. 1618–21.
102. Grinnell BW, Martin JD, Padgett BL, Walker DL, *Naturally occurring and passage-induced variation in the genome of JC virus.* Prog Clin Biol Res, 1983. **105**: p. 61–77.
103. Caldarelli-Stefano R, Vago L, Omodeo-Zorini E, et al., *Detection and typing of JC virus in autopsy brains and extraneural organs of AIDS patients and non-immunocompromised individuals.* J Neurovirol, 1999. **5**(2): p. 125–33.
104. Newman JT, Frisque RJ, *Detection of archetype and rearranged variants of JC virus in multiple tissues from a pediatric PML patient.* J Med Virol, 1997. **52**(3): p. 243–52.
105. Koralnik IJ, Wuthrich C, Dang X, et al., *JC virus granule cell neuronopathy: A novel clinical syndrome distinct from progressive multifocal leukoencephalopathy.* Ann Neurol, 2005. **57**(4): p. 576–80.
106. Richardson EP, Jr., *Progressive multifocal leukoencephalopathy.* N Engl J Med, 1961. **265**: p. 815–23.
107. Du Pasquier RA, Corey S, Margolin DH, et al., *Productive infection of cerebellar granule cell neurons by JC virus in an HIV+ individual.* Neurology, 2003. **61**(6): p. 775–82.
108. Cinque P, Koralnik IJ, Clifford DB, *The evolving face of human immunodeficiency virus-related progressive multifocal leukoencephalopathy: defining a consensus terminology.* J Neurovirol, 2003. **9 Suppl 1**: p. 88–92.
109. Bialasiewicz S, Whiley DM, Lambert SB, et al., *A newly reported human polyomavirus, KI virus, is present in the respiratory tract of Australian children.* J Clin Virol, 2007. **40**(1): p. 15–8.
110. Bialasiewicz S, Whiley DM, Lambert SB, et al., *Development and evaluation of real-time PCR assays for the detection of the newly identified KI and WU polyomaviruses.* J Clin Virol, 2007. **40**(1): p. 9–14.
111. Ridd K, Yu S, Bastian BC, *The presence of polyomavirus in non-melanoma skin cancer in organ transplant recipients is rare.* J Invest Dermatol, 2009. **129**(1): p. 250–2.
112. Bluemn EG, Paulson KG, Higgins EE, et al., *Merkel cell polyomavirus is not detected in prostate cancers, surrounding stroma, or benign prostate controls.* J Clin Virol, 2009. **5**: p. 5.
113. Dang X, Axthelm MK, Letvin NL, Koralnik IJ, *Rearrangement of simian virus 40 regulatory region is not required for induction of progressive multifocal leukoencephalopathy in immunosuppressed rhesus monkeys.* J Virol, 2005. **79**(3): p. 1361–6.
114. Axthelm MK, Koralnik IJ, Dang X, et al., *Meningoencephalitis and demyelination are pathologic manifestations of primary polyomavirus infection in immunosuppressed rhesus monkeys.* J Neuropathol Exp Neurol, 2004. **63**(7): p. 750–8.
115. Strickler HD, Goedert JJ, Devesa SS, et al., *Trends in U.S. pleural mesothelioma incidence rates following simian virus 40 contamination of early poliovirus vaccines.* J Natl Cancer Inst, 2003. **95**(1): p. 38–45.
116. Lopez-Rios F, Illei PB, Rusch V, Ladanyi M, *Evidence against a role for SV40 infection in human mesotheliomas and high risk of false-positive PCR results owing to presence of SV40 sequences in common laboratory plasmids.* Lancet, 2004. **364**(9440): p. 1157–66.
117. Rivera Z, Strianese O, Bertino P, et al., *The relationship between simian virus 40 and mesothelioma.* Curr Opin Pulm Med, 2008. **14**(4): p. 316–21.
118. Patel NC, Vilchez RA, Killen DE, et al., *Detection of polyomavirus SV40 in tonsils from immunocompetent children.* J Clin Virol, 2008. **43**(1): p. 66–72.
119. Pancaldi C, Balatti V, Guaschino R, et al., *Simian virus 40 sequences in blood specimens from healthy individuals of Casale Monferrato, an industrial town with a history of asbestos pollution.* J Infect, 2008.
120. Vanchiere JA, Nicome RK, Greer JM, et al., *Frequent detection of polyomaviruses in stool samples from hospitalized children.* J Infect Dis, 2005. **192**(4): p. 658–64.
121. Thomas LD, Vilchez RA, White ZS, et al., *A prospective longitudinal study of polyomavirus shedding in lung-transplant recipients.* J Infect Dis, 2007. **195**(3): p. 442–9.
122. Li RM, Branton MH, Tanawattanacharoen S, et al., *Molecular identification of SV40 infection in human subjects and possible association with kidney disease.* J Am Soc Nephrol, 2002. **13**(9): p. 2320–30.
123. Li RM, Mannon RB, Kleiner D, et al., *BK virus and SV40 co-infection in polyomavirus nephropathy.* Transplantation, 2002. **74**(11): p. 1497–504.
124. Chhajed PN, Bubendorf L, Hirsch H, et al., *Mesothelioma after lung transplantation.* Thorax, 2006. **61**(10): p. 916–7.
125. Bergsagel DJ, Finegold MJ, Butel JS, et al., *DNA sequences similar to those of simian virus 40 in ependymomas and choroid plexus tumors of childhood.* N Engl J Med, 1992. **326**(15): p. 988–93.
126. Weber T, Klapper PE, Cleator GM, et al., *Polymerase chain reaction for detection of JC virus DNA in cerebrospinal fluid: a quality control study. European Union Concerted Action on Viral Meningitis and Encephalitis.* J Virol Methods, 1997. **69**(1-2): p. 231–7.
127. Drews K, Bashir T, Dorries K, *Quantification of human polyomavirus JC in brain tissue and cerebrospinal fluid of patients with progressive multifocal leukoencephalopathy by competitive PCR.* J Virol Methods, 2000. **84**(1): p. 23–36.

128. Biel SS, Held TK, Landt O, et al., *Rapid quantification and differentiation of human polyomavirus DNA in undiluted urine from patients after bone marrow transplantation.* J Clin Microbiol, 2000. **38**(10): p. 3689–95.
129. Limaye AP, Jerome KR, Kuhr CS, et al., *Quantitation of BK virus load in serum for the diagnosis of BK virus-associated nephropathy in renal transplant recipients.* J Infect Dis, 2001. **183**(11): p. 1669–72.
130. Herman J, Van Ranst M, Snoeck R, et al., *Polyomavirus infection in pediatric renal transplant recipients: evaluation using a quantitative real-time PCR technique.* Pediatr Transplant, 2004. **8**(5): p. 485–92.
131. Beck RC, Kohn DJ, Tuohy MJ, et al., *Detection of polyoma virus in brain tissue of patients with progressive multifocal leukoencephalopathy by real-time PCR and pyrosequencing.* Diagn Mol Pathol, 2004. **13**(1): p. 15–21.
132. McNees AL, White ZS, Zanwar P, et al., *Specific and quantitative detection of human polyomaviruses BKV, JCV, and SV40 by real time PCR.* J Clin Virol, 2005. **34**(1): p. 52–62.
133. Sehbani L, Kabamba-Mukadi B, Vandenbroucke AT, et al., *Specific and quantitative detection of human polyomaviruses BKV and JCV by LightCycler real-time PCR.* J Clin Virol, 2006. **36**(2): p. 159–62.
134. Si-Mohamed A, Goff JL, Desire N, et al., *Detection and quantitation of BK virus DNA by real-time polymerase chain reaction in the LT-ag gene in adult renal transplant recipients.* J Virol Methods, 2006. **131**(1): p. 21–27.
135. Fink MC, Penalva de Oliveira AC, Milagres FA, et al., *JC virus DNA in cerebrospinal fluid samples from Brazilian AIDS patients with focal brain lesions without mass effect.* J Infect, 2006. **52**(1): p. 30–6.
136. Marzocchetti A, Sanguinetti M, Giambenedetto SD, et al., *Characterization of JC virus in cerebrospinal fluid from HIV-1 infected patients with progressive multifocal leukoencephalopathy: insights into viral pathogenesis and disease prognosis.* J Neurovirol, 2007. **13**(4): p. 338–46.

25 Ocular Viral Infections

Thomas J. Liesegang
Mayo Clinic, Jacksonville, Florida, U.S.A.

There is a large and increasing number of viruses recognized as causing ocular disease, either as ocular disease alone or in associated with other systemic manifestations of the infection. Some of the infections are minor, but many are sight threatening or associated with significant morbidity or mortality. The major viruses discussed sequentially in this chapter are listed in Table 1.

ADENOVIRUS

The Adenoviridae (ADV) has over 50 serotypes that tend to infect the epithelium causing infections of the upper respiratory tract. ADVs produce infection worldwide in all age groups but more commonly in children, without the definite seasonality seen with other respiratory viruses. The specific symptom complex is affected by the age and immune status of the host as well as the site of the infection because of unique tissue tropisms. The most common serotypes associated with respiratory tract disease are serotypes 1, 2, 3, 5, 7, 8, and 21. Military recruits are most commonly infected with serotypes 4 and 7. ADVs are transmitted by close contact with respiratory or ocular secretions, fomites, or contaminated swimming pools. Transmission may also occur by contaminated instruments or eye drops in physicians' offices. Most ADV diseases are self-limiting, although infections can be fatal.

Adenovirus Ocular Disease

At least 18 serotypes have been associated with human conjunctivitis. Most ADV eye disease presents clinically as distinct syndromes including simple follicular conjunctivitis (multiple serotypes); pharyngoconjunctival fever (PCF; most commonly serotype 3 or 7); and epidemic keratoconjunctivitis (EKC; usually serotype 8, 19, or 37, subgroup D). Simple ADV follicular conjunctivitis is self-limited without systemic disease and with minimal symptoms. Punctate epithelial keratitis may be present and mild. Adenoviral PCF is seen predominantly in children with a constellation of conjunctivitis, fever, pharyngitis, and cervical or preauricular lymphadenopathy. PCF tends to occur in outbreaks, such as at children's summer camps (swimming pool conjunctivitis), daycare centers, or health care settings, and is associated with types 3 and 7. PCF is highly contagious and, following five- to eight-day incubation period, can be spread by contact with the eyes and mouth for one to two weeks after the onset of symptoms. PCF duration is one to two weeks. The acute follicular conjunctivitis develops in one eye with progression to the second, usually less involved, eye (Fig. 1). Other findings include chemosis, conjunctival hemorrhages, watery discharge, photophobia, and mild periorbital pain. Sequelae are rare although a mild epithelial keratitis or subepithelial infiltrates may develop which is much less marked than in EKC. Adenoviral EKC, caused mainly by types 8, 19, and 37, can be a more severe eye disease than PCF and usually occurs in adults. It is usually not accompanied by pharyngitis but occurs in epidemics and is highly contagious with significant economic losses in the workplace. EKC caused by type 8 spread rapidly in 1941 from the Pacific war theatre to the west coast of the United States primarily through shipyards (hence "shipyard eye") and then across the United States. More recently, adenovirus types 19 and 37 have caused epidemics of typical EKC. Outbreaks of conjunctivitis have also been traced to ophthalmologists' offices and were presumably caused by contaminated ophthalmic solutions or diagnostic equipment. Following a 5- to 14-day incubation period, EKC presents in one eye and then spreads to the other eye. The clinical course includes foreign body sensation, photophobia, impaired vision, swelling of conjunctiva and eyelids (chemosis), subconjunctival hemorrhage, follicular or papillary conjunctivitis, and preauricular adenopathy. Some patients may have a prominent subconjunctival hemorrhage resembling enteroviral acute hemorrhagic conjunctivitis (AHC).

Table 1 Selected Viruses Causing Ocular Infection

Virus	Virus family
DNA viruses (all double stranded)	
Adenovirus	Adenoviridae
Herpes Simplex virus 1 (HHV1)	Herpesviridae
Herpes Simplex virus 2 (HHV2)	Herpesviridae
Varicella-Zoster Virus (HHV3)	Herpesviridae
Epstein–Barr virus (HHV4)	Herpesviridae
Cytomegalovirus (HHV5)	Herpesviridae
Herpes Virus 8 (HHV8)	Herpesviridae
Papillomavirus	Papovaviridae
Variola virus (smallpox)	Poxviridae
Vaccinia virus	Poxviridae
Molluscum Contagiosum virus	Poxviridae
RNA viruses (all single stranded)	
Mumps virus	Paramyxoviridae
Measles (rubeola) virus	Paramyxoviridae
Human Immunodeficiency virus	Retroviridae
Enterovirus (includes Polio, Coxsackie, Echo and Entero viruses)	Picornaviridae
Dengue virus	Flaviviridae
West Nile virus	Flaviviridae
Hepatitis C	Flaviviridae
Rubella virus	Togaviridae
Chikungunya virus	Togaviridae
Rabies virus	Rhabdoviridae

Keratitis eventually develops in most patients and is usually noted about one week into the illness. Diffuse corneal epithelial infection can progress to punctate epithelial lesions followed by stromal keratitis, which may persist. Pseudomembranes may be present predominantly on the tarsal conjunctiva. Viral shedding can persist for up to two weeks. The adenovirus replication occurs within the corneal epithelium but the corneal infiltrates are likely caused by an immunopathologic response to viral infection. Occasionally, the light sensitivity and reduced vision from EKC subepithelial infiltrates may persist for months to years but usually resolve. Chronic complications may include conjunctival scarring and dry eye.

Diagnosis and Prevention of Adenovirus Disease

Diagnosis is usually made on the basis of clinical findings in the presence of an epidemic of conjunctivitis. Cell culture of virus is considered conclusive evidence of ADV infection and also permits later serotype identification. Cultures from conjunctival swabs should be taken early although excretion from the conjunctiva may occur for a week in PCF and perhaps two weeks in EKC. Routine culture may take up to two weeks but shell-vial techniques provide more rapid viral isolation within three days. The typical cytopathic effects include rounding and clustering of swollen cells. Histologic examination of tissue may detect adenoviral intranuclear inclusions.

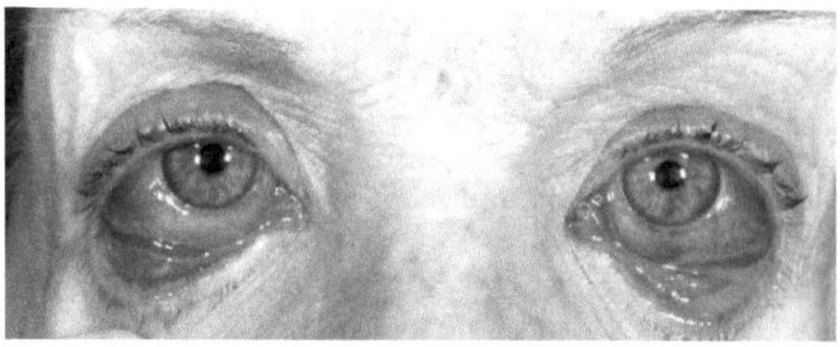

Figure 1 Adenoviral pharynoconjunctival fever (PCF) in an adult with acute follicular conjunctivitis of both eyes. The patient has conjunctival injection, chemosis, watery discharge, and photophobia and is highly contagious.

ADV antibody titer is too delayed for clinical usage but can confirm the diagnosis with a fourfold increase. Rapid assays to detect ADV in clinical ocular specimens are available, including polymerase chain reaction (PCR) assays, hybridization, restriction endonuclease digestion, fluorescent antibody microscopy, enzyme immunoassay, immunoelectron microscopy, time-resolved fluoroimmunoassay, and type-specific enzyme immunoassay, although these methods are not routinely available to the clinician (1–4). Direct fluorescent antibody assay is most commonly used, but PCR is more sensitive and rapid although limited by availability and equipment. A multiplex PCR assay for simultaneous detection of known viral and chlamydial pathogens associated with follicular keratoconjunctivitis has been developed (5).

ADV diseases are very contagious. Many patients with EKC carry the virus on their hands, and ADV can remain viable for several weeks on sinks and hand towels as sources of transmission. Personal hygiene, including isolation of the person from family and others, is mandatory. A temporary leave of absence from school or work is appropriate until the disease subsides; this may be up to two weeks. Ophthalmologists and ophthalmology offices are frequent sources of spread (6); ophthalmology practices should have procedure policies in place for implementation when cases are identified. Disinfection of equipment is inadequate with alcohol, detergents, or chlorhexidine, so instruments should be disinfected by immersion in a 1% solution of sodium hypochlorite (bleach) for 10 minutes or by steam autoclaving. Handwashing does not reliably remove adenoviruses from contaminated fingers. Gloves should be used to examine patients with EKC. Inadequate chlorination of swimming pools can lead to outbreaks of adenovirus infections. The adenoviruses induce effective and long-lasting immunity against reinfection.

HERPES SIMPLEX VIRUS

Herpesviruses are responsible for a wide spectrum of common acute, latent, and chronic human infections. Of the eight known human herpesviruses, those that affect the eye include herpes simplex virus (HSV) types 1 and 2, varicella-zoster virus (VZV), Epstein–Barr virus (EBV), cytomegalovirus (CMV), and Kaposi sarcoma–associated herpesvirus/human herpesvirus 8 (KSHV). HSV infections occur worldwide in both developed and underdeveloped countries and are transmitted to susceptible individuals during close personal contact with mucosal surfaces or abraded skin. HSV has two major antigenic types causing different epidemiologic patterns of infection. HSV-1 more commonly causes infection above the waist (orofacial and ocular infection), and HSV-2 below the waist (genital infection), but either virus can cause disease in either location and HSV-1 and HSV-2 may coinfect the same nerve ganglia. Patterns are shifting in most developed countries with HSV-1 antibodies now more commonly acquired in adolescence than in childhood; at the same time the more sexually active are increasing in the prevalence of HSV-2 antibodies. (7) About one-third of people suffer recurrent HSV infections. Reactivation of HSV from the ganglia (predominantly trigeminal or sacral) may be associated with asymptomatic excretion or with the development of mucosal herpetic lesions. HSV infection is spread by direct contact with infected lesions or their secretions but most commonly occurs as a result of exposure to viruses shed asymptomatically. HSV can be transmitted to neonates as they pass through the birth canal of a mother with genital infection and, in the newborn, can cause disease confined to the skin and mucous membranes or systemic infection including encephalitis.

Herpesvirus Ocular Disease

Primary HSV-1 infection in humans manifests as a nonspecific upper respiratory tract infection and is not usually recognized as HSV. Ocular involvement may be part of primary HSV infection and manifests as vesicles on the skin or eyelid margin and a unilateral follicular blepharoconjunctivitis with a palpable preauricular lymph node. Patients with primary ocular HSV infection can develop epithelial keratitis but stromal keratitis and uveitis are uncommon in this primary ocular infection.

Recurrent HSV infection is caused by reactivation of the latent virus in the sensory ganglion, transport of virus down the nerve axon to sensory nerve endings, and subsequent infection of ocular surface epithelia. The triggers for recurrence of HSV are long debated, although there may be increased recurrence of ocular infection associated with HIV infection (8). Recurrence rates for ocular HSV infections are 25% to 50% within two years. Recurrent HSV can affect almost

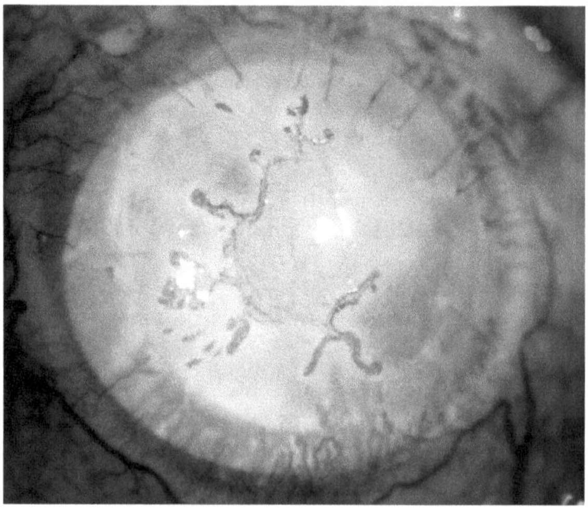

Figure 2 HSV infection of human corneal epithelium with several large arborizing epithelial dendritic lesions in a patient with a prior corneal transplant (see sutures). There is also profound reduction in corneal sensation.

any ocular tissue, including the eyelid, conjunctiva, cornea, iris, trabecular meshwork within the eye, and retina. The most common presentations of clinically recognizable recurrent ocular HSV infection include blepharoconjunctivitis, epithelial corneal infection (epithelial keratitis), stromal corneal infection (stromal keratitis), and iridocyclitis (or uveitis).

HSV blepharoconjunctivitis consists of eyelid vesicles or conjunctival involvement. HSV corneal epithelial keratitis causes symptoms of foreign-body sensation, light sensitivity, blurred vision, and redness. HSV infection of human corneal epithelium manifests as areas of punctate epithelial keratitis that may coalesce into one or more arborizing dendritic epithelial swollen cells with an associated ulcer among the swollen cells (Fig. 2). The dendritic keratitis may coalesce further and enlarge into a more expansive geographic epithelial ulcer. There may be focal or diffuse reduction in corneal sensation following HSV epithelial keratitis. Herpetic stromal keratitis can be necrotizing or nonnecrotizing (interstitial or disciform) and have different clinical manifestations. Some of these features are related to viral replication but major features are related to immunologic manifestations (9). Variations include unifocal or multifocal stromal haze, whitening, necrosis and suppuration, edema, scarring, or thinning as well as corneal neovascularization and anterior chamber inflammation manifest by cells, flare, and keratic precipitates (inflammatory cell precipitates on the back of the cornea). HSV iridocyclitis manifests with an anterior chamber inflammatory response that may accompany any of the forms of stromal or epithelial keratitis, or may occur independently of corneal disease. Cells, flare, elevated intraocular pressure, and iris transillumination defects may be found. Infectious HSV has been cultured from the anterior chamber of some patients.

Acute retinal necrosis (10) is a serious rare retinal infection that can be caused by HSV or VZV and manifests as decreased vision in either immunocompetent or immunocompromised hosts (11). The prominent features include a necrotizing retinitis, vitritis, and retinal vasculitis, often leading to retinal detachment. Approximately one-third of patients develop bilateral involvement.

Diagnosis and Prevention of HSV Infection

HSV ocular disease is usually a clinical diagnosis. Viral culture is useful for the diagnosis, but the fragile viral envelope requires careful handling. The sensitivity of all detection methods depends on the stage of the lesions (with higher sensitivity with the earlier vesicular lesion than the older ulcerative lesions), on whether the patient has a first or a recurrent episode of the disease (with higher sensitivity in first than in recurrent episodes), and on whether the sample is from an immunosuppressed or an immunocompetent patient (with more antigen or DNA in immunosuppressed patients). Vesicular fluid can be aspirated or an alcohol-wiped vesicle can be gently "unroofed" with a sterile needle and a sterile cotton swab rubbed in the lesion and then placed in viral culture medium. The characteristic cytopathic effects appear in the selected cell line within 24 to 48 hours. The efficiency of primary virus isolation and speed of diagnosis can be improved using the "shell vial" culture. The Tzanck smear, which exhibits the cytopathic effect within infected cells, can be prepared from lesion scrapings on slides that are fixed with methanol and stained with Giemsa or Wright stain preparations. The presence of multinucleated giant cells indicates infection with either HSV or VZV. Papanicolaou cervicovaginal stains also can demonstrate intranuclear inclusions. Herpes virus can be detected by electron microscopy or direct immunofluorescent antibody staining for antigen and immune electron microscopy using virus-specific monoclonal antibodies that can distinguish HSV 1, 2, and VZV. Studies have confirmed the value of the Herpchek testing kit for HSV antigen testing and the ELVIS (Enzyme Linked Virus Inducible System) as well as PCR (12). DNA amplification by PCR is gaining greater acceptance for routine diagnostic purposes (11,13). Serology by virus neutralization, complement fixation, passive hemagglutination, ELISA, complement-mediated cytolysis, antibody-dependent cellular lysis, or radioimmunoassay demonstrates a rising antibody titer during primary infection but is of no diagnostic assistance during recurrent episodes since the majority of adults is latently infected with HSV and already has positive serology.

No vaccine tested completely prevents infection or establishment of latency, and thus prevention of HSV infection depends on reducing transmission and perhaps reactivation. Prophylactic acyclovir has been demonstrated to reduce ocular recurrences during the period of time oral antivirals are taken (14). Host-to-host transmission can be lessened for most human herpesviruses by simple hygiene and avoiding contact with a person who has evidence of recurrent infection. Fomites, including toilet seats and towels, are not important modes of transmission. Because infectious virus is often excreted before the appearance of overt symptoms of recurrent infection and because "silent" recurrent episodes of shedding occur, no method can be fully preventative. Specifically, spread of infection through contact with oral secretions may be an occupational hazard for respiratory care and dental care providers and thus gloves are recommended for these workers. Herpesviruses are readily inactivated by a variety of physical and chemical agents including the common disinfectants. Standard methods of sterilization are all adequate for decontaminating medical equipment.

VARICELLA-ZOSTER VIRUS

VZV is a member of the alpha Herpesviridae subfamily with worldwide geographic distribution and the cause of two distinct viral syndromes, varicella and zoster. While HSV latency occurs in ganglion neurons, VZV may establish latency instead in ganglion satellite cells rather than neurons, or in both.

Varicella

Before the licensure of varicella vaccine, varicella used to occur in annual spring epidemics among susceptible children in temperate climates with the number of cases of equivalent to the annual birth cohort, and serological tests confirmed that almost the entire population had evidence of prior VZV infection even if they do not recall the event. The varicella vaccine and zoster vaccine will alter the course of both VZV diseases, though perhaps in a complex fashion (15,16). Varicella usually occurs in childhood as a generalized vesicular rash of the skin and mucous membrane accompanied by mild constitutional symptoms of fever and malaise. Varicella is contracted by direct contact with VZV skin lesions or respiratory secretions through airborne droplets and is highly contagious for naive individuals. VZV infection is usually a self-limited infection of childhood rarely associated with long-term sequelae; infection of neonates, adults, or immunosuppressed individuals, however, can be associated with severe complications.

Varicella Ocular Disease

Ocular involvement is usually mild and self-limited. Eyelid and conjunctival vesicles or conjunctival follicles, and internal ophthalmoplegia can occur. The cornea may be involved with a dendritic keratitis, somewhat similar to HSV including punctate epithelial keratitis, or stromal clouding or scarring from keratitis. Although subepithelial infiltrates, stromal keratitis, disciform keratitis, uveitis, and elevated intraocular pressure are rare, recurrent varicella keratouveitis may cause significant morbidity in some patients.

Congenital Varicella Syndrome

Varicella acquired during pregnancy can have severe consequences for both the mother and the fetus. Because most adults born in the United States are immune by prior varicella or varicella vaccine, the incidence of maternal varicella is now low. The most striking anomalies of the congenital varicella syndrome are unusual cutaneous defects with cicatricial skin scars, atrophy of an extremity, and evidence of damage to the autonomic nervous system. Many affected infants have microcephaly and cortical atrophy. The ocular findings of the congenital varicella syndrome consist of chorioretinitis, microophthalmia, and cataracts.

Diagnosis and Prevention of Varicella

Laboratory evaluation is not usually necessary. The definitive diagnosis of VZV infection can be made using tissue culture methods, but the virus is quite labile and must be stored at −70°C if cultures cannot be inoculated immediately. Ideally, vesicular fluid should be collected in unheparinized capillary tubes and transferred directly into human embryonic lung fibroblasts. The virus produces typical focal intranuclear inclusion bodies. Cytologic methods by the Tzank technique can be used to detect multinucleated giant cells often containing eosinophilic intranuclear (Cowdry type A) inclusions in lesion specimens or tissue sections. Rapid diagnosis of cutaneous VZV infection can be accomplished by antigen detection in epithelial cells from the base of a vesicle with immunohistochemical techniques such as enzyme immunoassay methods or in situ hybridization. PCR techniques are the most sensitive for the detection of VZV. Antibodies can be measured in low concentrations at the time of onset of the varicella exanthem. VZV IgG antibodies persist for life after primary infection and assays for IgG antibodies to VZV are valuable to determine the immune status of individuals whose clinical history of varicella is unknown or equivocal. The immune status of contacts can be determined most easily with the latex agglutination test. Because complement-fixing antibody is lost rapidly after infection, it cannot be used for determining susceptibility.

Varicella is a highly contagious and persons shed the virus in respiratory secretions before the onset of the characteristic rash, so that avoidance of infected individuals is not always possible. Fomites are not an important mode of transmission. A single attack of chickenpox usually confers lifetime immunity for varicella. Two live attenuated VZV vaccines are available in the United States for the prevention of varicella: a single-antigen varicella vaccine (Varivax) and a combination measles, mumps, rubella, and varicella vaccine (MMRV). Because of recent "breakthrough" infections, the new recommendations include a routine two-dose varicella vaccination program for children and a second dose catch-up varicella vaccination for others who received the vaccine, and routine vaccination of all healthy persons 13 years of age or older who are without evidence of immunity. The incidence of varicella diminished 90% after implementation of the varicella vaccine in 1995, and the vaccine has permanently changed the epidemiology of varicella in the United States. The vaccine appears to also reduce the incidence of recurrent VZV infection.

Herpes Zoster Disease

Latent VZV can be detected in most trigeminal, thoracic, and geniculate ganglia. The virus may reactivate later in life, causing herpes zoster (HZ), either in response to many stimuli or to an alteration of cell-mediated immunity (CMI) associated with aging or with immunosuppression. HZ occurs in all ages, including otherwise healthy individuals, and eventually occurs in 10% to 20% of the population. Recurrent HZ is exceedingly rare except in immunocompromised hosts, especially those with AIDS. People younger than 50 years presenting with HZ should be queried about risk factors for HIV or evaluated for HIV. In immunosuppressed individuals, HZ is more likely to be severe, prolonged, and lead to viremia and disseminated disease, which can result in visceral or neurologic infection, with increased morbidity and mortality. Postherpetic neuralgia (PHN) occurs after HZ infection in approximately 50% of patients older than 50 years. The pain of PHN can be severe and debilitating and may persist for months or even years.

Herpes Zoster Ophthalmicus

Although the majority of HZ cases involve the thoracic dermatomes, about 15% involve the ophthalmic division of the trigeminal (fifth cranial nerve); it can also involve the mandibular

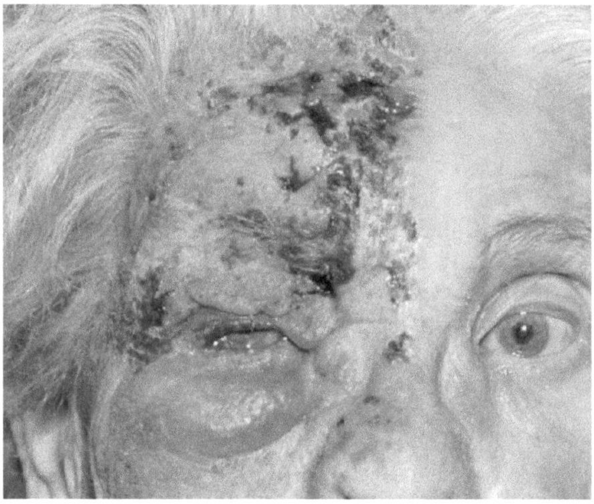

Figure 3 Herpes zoster ophthalmicus (HZO) vesicular and scabbing rash over distribution of the first division of the trigeminal nerve. The patient demonstrates Hutchinson's sign (skin lesions at the tip, side, or root of the nose), a strong predictor of ocular inflammation, and corneal denervation in HZO. Potential eyelid complications include eyelid scarring, marginal notching, loss of cilia, trichiasis, and cicatricial entropion or ectropion.

and maxillary divisions of the trigeminal nerve. Herpes zoster ophthalmicus (HZO) produces a vesicular rash over distribution of the first division of the trigeminal nerve (Fig. 3). With HZ new crops of skin lesions appear in the same area within seven days. A maculopapular rash, followed by vesicles and then pustules, is characteristic; HZ dermatitis may result in large scabs that resolve slowly and leave significant scarring. Although seemingly localized, there is usually a viremia, confirmed by PCR, associated with each recurrence of HZ as demonstrated by distant vesicles and occasionally serious disseminated or visceral disease in otherwise healthy individuals.

Hutchinson's sign is defined as HZ skin lesions at the tip, side, or root of the nose and is a strong predictor of ocular inflammation and corneal denervation in HZO. The pathophysiology of the diffuse and severe ocular complications of HZO includes components of virus infection, inflammatory and immune reactions, vascular and neural inflammation, and tissue scarring with the vascular and neural inflammation well developed even before the rash is evident and certainly before antiviral therapy can be instituted (17).

Eyelid involvement may result in residual eyelid scarring, marginal notching, loss of cilia, trichiasis, and cicatricial entropion or ectropion. Conjunctival injection is very common. Episcleral (episcleritis) or scleral inflammation (scleritis) associated with zoster may be nodular, zonal, or diffuse. Corneal complications occur in 65% of individuals with HZO including dendriform epithelial lesions, neurotrophic keratitis (numb cornea), nummular corneal infiltrates, interstitial stromal keratitis, and disciform keratitis; chronic corneal stromal inflammation can lead to corneal vascularization, lipid keratopathy, and corneal opacification. Anterior uveitis with increased intraocular pressure, orbital edema, papillitis, or retrobulbar optic neuritis are other features of HZO in some patients. Common neurological complications of HZO include acute neuralgia and PHN; rare complications include a delayed contralateral hemiplegia, encephalitis, and myelitis.

Necrotizing herpetic retinopathy is a continuous spectrum of posterior segment inflammation induced by several herpes viruses, most commonly VZV (18). Its two most recognizable clinical patterns are acute retinal necrosis (ARN) (10) and progressive outer retinal necrosis (PORN). Usually, the former occurs in healthy persons and HIV patients with only mild immune dysfunction and higher CD4 counts, whereas the latter usually develops in those who are severely immunosuppressed. Patients with ARN usually present with acute unilateral loss of vision, photophobia, floaters, and pain. Fellow eye involvement may occur in about 1/3 of cases, usually within six weeks of disease onset. The essential clinical findings are a triad of occlusive retinal arteriolitis, vitritis, and a multifocal yellow-white peripheral retinitis (19). PORN (20) is essentially a morphologic variant of acute necrotizing herpetic retinitis, occurring most often in patients with advanced stages of HIV/AIDS or in patients who are otherwise profoundly immunocompromised. The most common cause of PORN is VZV; however, HSV has also been implicated. The retinitis is similar to ARN except that the posterior pole may be involved early in the course of the disease, vitreous inflammatory cells are typically absent, and the retinal vasculature is minimally involved. A previous history of cutaneous HZ is present in most patients.

Diagnosis and Prevention of Herpes Zoster Disease

HZ infection can be confirmed by virus isolation, direct immunofluorescence of antigen in tissue scrapings (such as immunoperoxidase assay), PCR detection of viral DNA, or demonstration of a fourfold rise in antibodies to viral antigens (21). Because it is a strongly cell-associated member of the HSV group, VZV is difficult to detect or isolate in cell-free specimens. The Tzanck smear of the scraping of the base of a lesion may rapidly confirm multinucleated giant cells but the sensitivity of this method is low. The most frequently employed serologic tools for assessing host response are the immunofluorescent detection of antibodies to VZV membrane antigens, the fluorescent antibody to membrane antigen (FAMA) test, immune adherence hemagglutination, and enzyme-linked immunosorbent assay (ELISA).

With the herpetic retinopathies, the diagnosis is usually made clinically. In select circumstances, intraocular fluid analysis of aqueous and/or vitreous samples can be taken (22). Intraocular antibody production as a measure of the host response to VZV can be computed using the Goldmann–Witmer (GW) coefficient: the ratio of specific antibody (aqueous or vitreous)/total IgG (aqueous or vitreous) to specific antibody (serum)/total IgG (serum), as measured by ELISA or radioimmunoassay. A ratio of greater than four is considered diagnostic of local antibody production. PCR may increase the diagnostic yield and quantitative PCR may add additional information with respect to viral load, disease activity, and response to therapy. Endoretinal biopsy may also be diagnostic.

Specific CMI to VZV is the major determinant of the risk and severity of HZ in both elderly patients and patients receiving immunosuppressive therapy. Antibody levels to VZV only confirm a prior VZV infection and cannot be used as a marker for susceptibility to HZ, although the presence of antibodies does continue to protect from a recurrence of varicella. Unfortunately, determining the activity of VZV-specific T-cells is a cumbersome laboratory procedure and cannot be used for clinical testing. In 2006, the FDA approved a higher-dose VZV vaccine (Zostavax) protective against HZ on the basis of results of a large randomized clinical study in more than 38,000 individuals. This study, conducted in individuals 60 years of age and older, showed that the vaccine reduced the incidence of HZ by 51%, compared with placebo. In addition, in those individuals who did develop HZ, PHN was reduced by 39%. In 2008, the CDC also recommended the vaccine.

EPSTEIN–BARR VIRUS

EBV was discovered in 1964 from tissue samples of patients with Burkitt lymphoma and the causal relationship between EBV and infectious mononucleosis was first observed in 1968. In developing countries, most of the population is exposed to EBV at an early age with EBV infection resulting in subclinical infection; if acquired later in life, it causes the clinical condition of infectious mononucleosis with fever, tonsillar pharyngitis, and lymphadenopathy. EBV antibodies are found in almost all adults. As in the case of other herpesviruses, infection with EBV is lifelong with the virus residing in B lymphocytes and perhaps nasopharyngeal mucosal epithelial cells. Many infected B lymphocytes have EBV DNA present within their nucleus in a circular nonintegrated form. EBV confers on infected B lymphocytes the ability to grow continuously in cell culture, a process termed *immortalization*. EBV is intermittently shed asymptomatically in oropharyngeal secretions about 15% of the time in healthy individuals and accounts for the bulk of its transmission in the human population. The shedding rate increases significantly in patients with defects in cellular immunity although the virus is not highly contagious. EBV has been implicated in a variety of other disorders, including Burkitt lymphoma, nasopharyngeal carcinoma, oral hairy leukoplakia, and a variety of B cell and possibly T-cell lymphoproliferative disorders. Neurologic complications occur in lesser than 1% of patients with infectious mononucleosis and include encephalitis, aseptic meningitis, transverse myelitis, Guillain-Barré syndrome, optic neuritis, and peripheral neuropathies.

Epstein–Barr Virus Ocular Disease

EBV does not commonly cause eye disease but is the most common cause of acute dacryoadenitis, characterized by inflammatory enlargement of one or both lacrimal glands. Acute follicular conjunctivitis, Parinaud oculoglandular syndrome, and bulbar conjunctival nodules have been reported in patients with acute infectious mononucleosis and may be the result of EBV infection. EBV epithelial keratitis may occur with punctate epithelial keratitis, dendritic keratitis, or stromal keratitis. Forms of EBV unilateral or bilateral stromal keratitis include multifocal subepithelial infiltrates (that resemble adenoviral keratitis), multifocal, blotchy, pleomorphic infiltrates with active inflammation in the anterior to mid stroma, or multifocal deep or full-thickness peripheral infiltrates, with or without vascularization (that resemble interstitial keratitis due to syphilis) (23,24).

Diagnosis and Prevention of Epstein–Barr Virus Disease

Laboratory studies of EBV are hampered by the lack of a fully permissive cell system able to propagate the virus. Highly sensitive real-time PCR assays are now available for detection of primary EBV infection and infectious mononucleosis (25). Because of difficulty in viral isolation, the diagnosis of EBV infection depends on the detection of antibodies to various viral components. During acute infection, first IgM and then IgG antibodies to viral capsid antigens (VCAs) appear. Anti-VCA IgG may persist for the life of the patient. Antibodies to early antigens (EAs) also rise during the acute phases of the disease and subsequently decrease to low or undetectable levels in most individuals. Antibodies to EBV nuclear antigens (EBNAs) appear weeks to months later, providing serologic evidence of past infection (26).

No vaccine is currently available against EBV, but research is ongoing toward developing a cytotoxic T-cell–based vaccine.

CYTOMEGALOVIRUS

CMV was initially isolated from patients with congenital cytomegalic inclusion disease and replication is associated with a characteristic production of large intranuclear inclusions and smaller cytoplasmic inclusions. Like other herpesviruses, it shares a complex balance with the host and has the ability to establish a long-lived latent infection, and most of the clinical disease results from reactivation of latent virus in immune-impaired patients. In addition to inducing severe birth defects, CMV causes a wide spectrum of disorders in older children and adults. About 50% of adults in developed countries harbor antibodies, which are usually acquired during the first five years of life. Multiple mechanisms account for the person-to-person spread of this virus, including vertical transmission (in utero, during vaginal delivery, and by breast milk) and horizontal contact (saliva, genital, urine). The virus is also carried in circulating white blood cells. Oral and respiratory spread is probably the dominant routes of CMV transmission but clinically important mechanisms of transmission include blood transfusion and organ transplantation. Reinfection with a different strain of CMV may occur in a CMV-seropositive person.

Most CMV infections occurring in immunocompetent persons are asymptomatic. In some patients a clinical illness resembling infectious mononucleosis develops with fever, myalgia, asthenia, and lymphadenopathy. Although the syndrome occurs at all ages, it most often involves sexually active young adults. Complications are infrequent but can include a retinitis. In contrast to the generally benign course of CMV infection in healthy persons, CMV is a major cause of morbidity and mortality in immunocompromised patients with the clinical spectrum dependent upon the cause and degree of immunosuppression. Disseminated CMV infection was the most common opportunistic infection in AIDS before highly active antiretroviral therapy (HAART).

Cytomegalovirus in Congenital and Perinatal Infections

CMV is the most common cause of congenital infection in the United States and may result from either primary or recurrent CMV infection in a pregnant woman. The newborn may also acquire CMV at delivery by passage through an infected birth canal or by postnatal contact with infected breast milk or other maternal secretions. The majority of infants infected at or after delivery remains asymptomatic. Fetal infections range from inapparent to severe and disseminated,

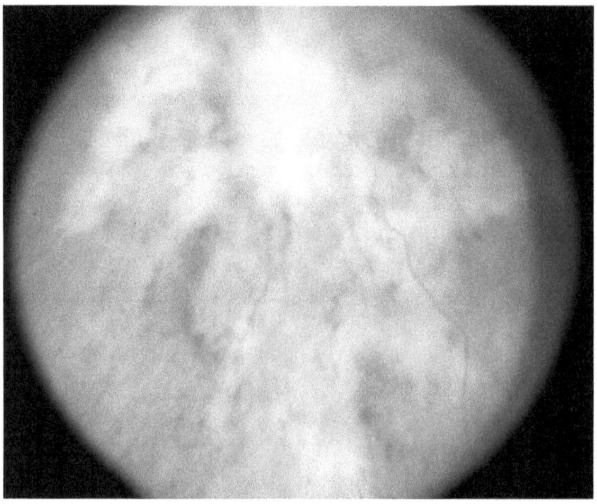

Figure 4 CMV retinitis in a patient with AIDS demonstrating opaque, white areas of granular retinal necrosis that spread in a centrifugal manner accompanied by hemorrhages, vessel sheathing, and retinal edema. The yellow-white retinal lesions often follow a vascular distribution. In the absence of therapy, involvement of the contralateral eye will often occur.

including growth retardation, hepatosplenomegaly and thrombocytopenic purpura, and less frequently jaundice, microcephaly, and chorioretinitis (27). The prognosis is poor for these infants. Some neonates who are asymptomatic at birth develop late sequelae, particularly mental retardation and sensorineural hearing loss.

Cytomegalovirus Ocular Disease in the Immunosuppressed

CMV retinitis is the most common ocular opportunistic infection in patients with AIDS and occurred in up to a third of AIDS patients before the era of HAART. CMV retinitis is occasionally the first AIDS-defining infection for an individual and most commonly occurs in people whose $CD4^+$ T-lymphocyte counts are below 50 cells/mm^3. CMV retinitis results from hematogenous spread of CMV and usually begins unilaterally with visual blurring, floaters, decreased acuity, and loss of visual fields and progresses to blindness if untreated, particularly in advanced AIDS. Early retinal lesions consist of small, opaque, white areas of granular retinal necrosis that spread in a centrifugal manner and are later accompanied by hemorrhages, vessel sheathing, and retinal edema (Fig. 4). The yellow-white retinal lesions often follow a vascular distribution because the virus initially infects the endothelium of the blood vessels. In the absence of therapy, involvement of the contralateral eye will often occur. Since the introduction of HAART, the incidence of serious CMV infections (e.g., retinitis) has decreased by 80% in the United States and Europe and CMV now only occurs in patients with advanced immunosuppression who are either not receiving or have failed to respond to antiretroviral therapy (28,29). The initiation of HAART may also lead to different clinical presentations of symptomatic CMV disease in the setting of immune reconstitution. During the first few weeks after institution of HAART, acute flare-ups of CMV retinitis may occur secondary to an immune reconstitution inflammatory syndrome and manifests with anterior uveitis, vitritis, and cystoid macular edema (30). The pathogenesis of this immune restitution is not well understood, but seems to relate to T-cell–mediated immune reconstitution to latent CMV intraocular antigens.

Laboratory Diagnosis and Prevention of Cytomegalovirus Disease

For the nonocular disease, the diagnosis of CMV infection usually cannot be made reliably on clinical grounds; however, the ocular findings are distinctive alone in the proper setting of immunosuppression. Culture is a "gold standard" for detecting CMV infection, but false-negative culture results and the slow culture isolation undermines its value. More rapid culture results within one to two days may be possible with the shell vial culture technique involving centrifugation and an immunocytochemical detection technique employing monoclonal antibodies to an immediate-early CMV antigen. Monoclonal antibodies can be used to detect CMV antigens directly in peripheral blood leukocytes (antigenemia) as well as offer the possibility of a quantitative assay. Assays of viral DNA by PCR can also detect viral load and is more sensitive than culture and at least comparable to antigenemia assays. Seroconversion is an excellent marker for primary CMV infection, but rises in IgG titers, even fourfold or greater, are not diagnostic of newly acquired infection because IgG titers may reappear during reactivation of latent CMV. The presence of IgG antibody is a sensitive marker of past infection and is used to screen blood or transplant patients. Detection of viral IgM antibodies suggests a current infection. Microscopically, the hallmark of CMV infection is a large (cytomegalic), 25- to 35-μm cell containing a large central, basophilic intranuclear inclusion, referred to as an "owl's eye."

Because CMV is transmitted by exchange of secretions or excretions, infection can be diminished by reducing exposure to body fluids and in selecting appropriate donors for blood or organs. Filters that remove leukocytes from blood products are also effective in reducing transmission of CMV. Although live attenuated CMV vaccines induce antibody formation and CMI, there are possible oncogenic properties so that nonviable vaccines are being evaluated in clinical trials.

PAPILLOMAVIRUS

Human papillomaviruses (HPVs) are widely distributed in nature and transmitted primarily by direct contact. They are strictly epitheliotropic and have a high degree of species and tissue specificity for either the skin or the mucous membranes. More than 100 individual HPV types have been described to date. Persistent HPV infection of susceptible epithelial cells induces cellular proliferation and can lead to malignant transformation. HPV subtypes 6 and 11 are maintained in a latent state within basal epithelial cells as circular episomes. Early viral gene products stimulate cell growth and lead to skin warts or conjunctival papillomas. Some become permissive for complete viral gene expression and produce infectious virus. Neoplastic transformation due to HPV 6 or 11 is very rare. In contrast, HPV 16 and 18 stereotypically integrate their viral genome into host chromosomal DNA, and may be associated with malignant transformation and squamous cell carcinoma. HPV is closely associated with condylomata (genital warts), cervical intraepithelial neoplasia and cervical cancer, conjunctival intraepithelial neoplasia, some cases of head and neck squamous cell carcinoma, and possibly lung adenocarcinoma. HPV is probably the most common sexually transmitted viral infection in humans. Infected persons are usually asymptomatic. The course of HPV infection is altered profoundly by HIV-induced immunosuppression

Human Papillomavirus Ocular Disease

HPV initiates a neoplastic growth of epithelial cells with vascular proliferation giving rise to a grayish verruca vulgaris (wart) of the eyelid skin or a reddish pedunculated papilloma of the conjunctiva (Fig. 5). The reported frequency of HPV DNA in conjunctival papillomas varies from 50% to 100% (31). The subtypes found in the conjunctiva are usually 6 or 11 and less frequently 18, 33, and 16. In older adults, the papillomas are usually subtype 16 or they may not have a viral origin. The transmission route to the conjunctiva is not known. Conjunctivial papillomas may be unilateral or bilateral with multiple lesions of variable size usually in the fornices (32). The lesions can be sessile or pedunculated with multiple finger-like projections that may regress spontaneously but some lesions in the elderly can undergo malignant transformation.

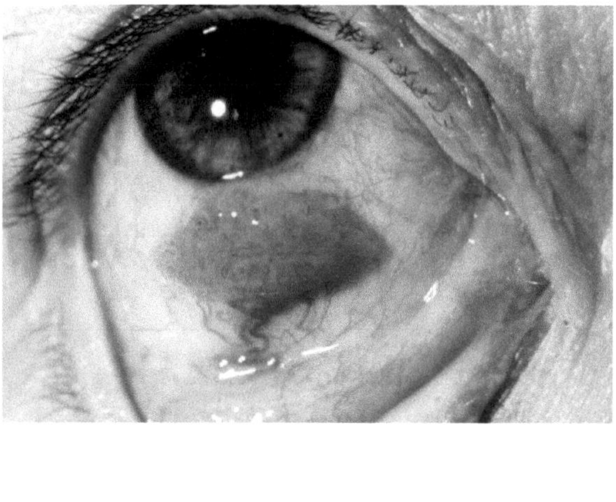

Figure 5 Conjunctival papilloma demonstrated as a reddish pedunculated lesion with multiple finger-like projections. Human papillomovirus can be detected in most of these lesions, especially subtypes 6 or 11 and less frequently 18, 33, and 16. The lesions may be unilateral or bilateral with multiple lesions of variable size usually in the fornices.

Papillomavirus-associated conjunctival intraepithelial neoplasia and squamous cell carcinoma share many histologic features with similar lesions in the uterine cervix.

Diagnosis and Prevention of Human Papilloma Virus Disease

Most cutaneous and anogenital warts are diagnosed on the basis of history and clinical inspection but histologic examination of a biopsy specimen is diagnostic. Since HPV complete their life cycle only in terminally differentiated epithelial cells they are difficult to grow in cell culture. The presence of HPV in a tissue is ascertained by nucleic acid hybridization assays or PCR amplification (33). Immunologic assays can be used for type-specific HPV diagnosis.

Prevention of HPV infection depends on avoidance of contact with infectious lesions and reduction of susceptibility through immunization. Women at high risk for HPV cervical disease should receive HPV testing at the time of the Papanicolaou test. A newly licensed quadrivalent vaccine (types 6, 11, 16, 18) given in three doses is highly effective in susceptible persons.

VARIOLA VIRUS

The Poxviridae encompasses a family of enveloped DNA viruses, with a distinctive brick or ovoid shape and a complex capsid structure. Poxviruses are large viruses (200 to 320 nm) and replicate in cell cytoplasm producing eosinophilic cytoplasmic inclusion bodies. The poxviruses of ocular significance are smallpox (variola) virus, vaccinia virus, and the molluscum contagiosum virus (MCV). Smallpox, caused by variola virus, was eradicated through widespread immunization with vaccinia virus and meticulous epidemiologic investigation to find all cases and immunize all contacts. However, the possibility of smallpox bioterrorism is a continuing issue. Smallpox is spread from person to person through aerosolized droplets from respiratory discharges, direct skin contact, and through fomites. Once inhaled, variola virus invades the oropharyngeal or the respiratory mucosa, migrates to regional lymph nodes, and begins to multiply. The initial or prodromal symptoms are similar to other viral diseases and as the digestive tract is commonly involved, nausea and vomiting and backache often occur. Later skin involvement occurs and has several variants of severity. The case-fatality rates reach 20–35% among unvaccinated individuals.

Smallpox Virus Ocular Disease

Both variola virus and vaccinia virus are associated with serious ocular complications with smallpox reportedly causing more than one-third of the blindness in Europe prior to immunization, and even as late as the 1960s, smallpox remained a significant cause of blindness in Africa. The eyelids are usually involved as an extension of the generalized pustular rash. Conjunctival smallpox pustules are accompanied by pain, photophobia, and lacrimation with occasional conjunctival phlyctenules reported. Corneal involvement usually occurs through contiguous spread of a pustule at the limbus. Corneal ulceration is the most common serious complication and may result in perforation, iris prolapse, or endophthalmitis, and a late corneal scar (leukoma). Disciform keratitis tends to appear several weeks after the rash. Less common complications include secondary glaucoma, retinitis, chorioretinitis, optic neuritis, paralysis of accommodation, paralysis of extraocular muscles, retrobulbar hemorrhage with proptosis, and dacryocystitis.

VACCINIA VIRUS

Multiple-puncture vaccinia virus infection through a bifurcated needle is the current smallpox vaccination regimen used for the US military, public health care personnel, and laboratory personnel working with orthopoxviruses. Most commonly, the infection progresses through a standard course of events from vesicle to pustule. Of all vaccines used today, the vaccinia virus vaccine, which is composed of live, replicative virus, has one of the highest rates of adverse events. Major complications include progressive vaccinia, eczema vaccinatum, generalized vaccinia, postvaccinial encephalitis, accidental infection, and carditis (34).

Vaccinia Virus Ocular Disease

About 10 to 20 patients develop ocular complications per 1 million smallpox immunizations, usually through autoinoculation although mishandling of the vaccine container or needle by

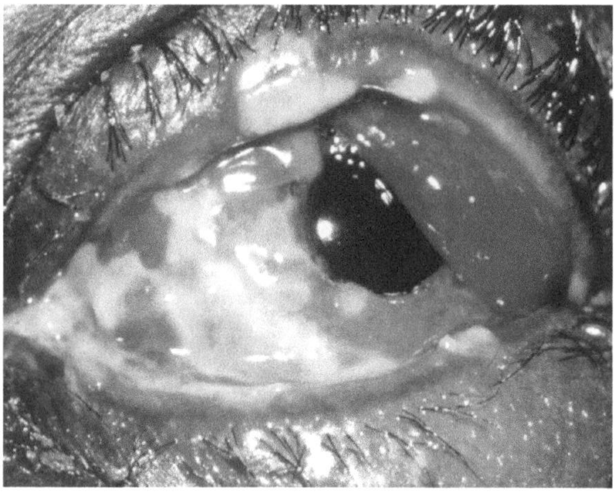

Figure 6 Vaccinia blepharoconjunctivitis from autoinoculation in a recently smallpox-vaccinated individual. There is intense painful preoribital cellulitis with lid and conjunctival ulcers and purulent discharge. There is high risk for corneal stromal involvement with abscesses.

health care personnel is also a potential source of infection. The severity of the infection varies with the level of immunity of the new host. The nonimmune patient develops an ocular reaction that resembles the primary vaccination "take" reaction of the skin, including fever, malaise, with orbital cellulitis, blepharitis, conjunctivitis, and keratitis. Vaccinial blepharoconjunctivitis is the most common reaction with coalescing vesicles accompanied by intense painful preoribital or orbital cellulites (Fig. 6). The vesicles evolve from pustules that umbilicate to scab that resolve with red pitted eyelid scars and loss of eyelashes. Vaccinia conjunctivitis is a follicular reaction with conjunctival ulcers, purulent discharge, and inflammatory membranes and later conjunctival scarring. Vaccinial keratitis results from live viral invasion of the cornea, causing a superficial punctate keratitis and later stromal involvement that may consist of either subepithelial opacities or deeper abscesses within the corneal stroma. A delayed disciform and necrotizing stromal keratitis may occur and an immune-mediated corneal perforation is possible. Semba reviewed much of the international literature on ocular complications of vaccinia and found cases of iritis, central retinal artery occlusion, pigmentary retinopathy, chorioretinitis, central serous retinopathy, exudative retinitis, optic neuritis, retrobulbar optic neuritis with encephalomyelitis, and transient strabismus (35,36).

Laboratory Diagnosis and Prevention of Vaccinia Virus Disease

Ocular vaccinia is a clinical diagnosis based on history, timing, and presentation. Diagnosis can be confirmed by obtaining scrapings and swabs of lesions and ocular discharge. Smears show numerous polymorphonuclear cells with epithelial cells containing Guarnieri bodies. Antigens can be detected by direct and indirect immunofluorescent methods, DNA can be detected PCR, and the variola virus can be isolated on chick chorioallantoic membrane cultures.

Ocular vaccinia infection can be reduced by avoiding contact with the immunization site and the eye and frequent hand washing.

MOLLUSCUM CONTAGIOSUM VIRUS

MCV appears worldwide but is often more generalized, severe, and persistent in AIDS patients than in other groups, frequently involving the face and upper body. Traditional modes of transmission are associated with mild skin trauma such as abrasions, direct contact with a lesion, fomites (e.g., shared towels), or sexually transmitted. Occasionally, MCV has been associated with outbreaks, but it usually occurs sporadically. MCV disease is characterized by multiple smooth painless, pearly white nodules 2 to 5 mm in diameter with a central umbilication. There are usually 1 to 20 lesions, but hundreds may occasionally be present. Incubation periods vary from several days to several weeks, and lesions may clear rapidly or persist for up to 18 months. The highest incidence is reported in children younger than five years, particularly in hot climates and crowded conditions. More severe and prolonged infection tends to occur in individuals with impaired CMI, including persons with HIV infection.

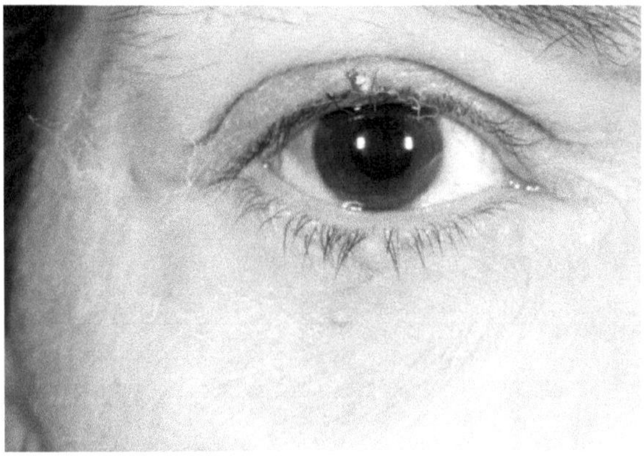

Figure 7 Molluscum contagiosum virus lesions of the skin characterized by multiple smooth painless, pearly white nodules 2 to 5 mm in diameter with a central umbilication. Eyelid umbilicated nodules release viral particles into the tear film causing a toxic follicular conjunctivitis and corneal punctate epithelial erosions. Extensive facial and eyelid molluscum lesions may occur in association with AIDS.

Molluscum Contagiosum Virus Ocular Disease

Eyelid umbilicated nodules release viral particles into the tear film causing a toxic follicular conjunctivitis (Fig. 7). Punctate epithelial erosions and, in rare cases, a corneal pannus (fibrosis), ulceration, or perforation may occur (37). Extensive facial and eyelid molluscum lesions occur in association with AIDS (38). The conjunctivitis may require weeks to resolve after elimination of the skin lesion.

Diagnosis and Prevention of Molluscum Contagiosum Virus Disease

Diagnosis is based on the characteristic skin or eyelid lesions. MCV can be diagnosed readily by examination of a skin scraping or biopsy specimen, revealing a circumscribed epidermal pseudotumor with epidermal hyperplasia and characteristic large eosinophilic intracytoplasmic inclusion bodies (Henderson–Patterson bodies, also called Molluscum bodies). Although other human poxviruses can be isolated in culture, MCV cannot be cultured using standard tissue culture techniques. Serology generally is not useful. DNA detection with direct hybridization or restriction endonuclease analysis or PCR permits rapid identification and differentiation.

Atypical numerous confluent MCV lesions may be an indicator of systemic immunocompromise and HIV status should be determined.

MUMPS VIRUS

Mumps virus is highly infectious and spreads rapidly among susceptible people living in close quarters. It is typically transmitted by respiratory droplets, direct contact, or fomites. Prior to the widespread use of an effective vaccine, mumps primarily occurred in young children frequently accompanied by a nonspecific prodrome consisting of low grade fever, malaise, headache, myalgias, and anorexia followed by an acute, self-limited, viral syndrome. Within 48 hours a parotitis develops and is the classic feature of mumps infection. Some patients develop epididymo-orchitis and CNS disease. Symptomatic infection in adults is usually more severe than in children.

Mumps Virus Ocular Disease

Ocular findings in mumps include photophobia, acute mucoid follicular conjunctivitis, and epithelial and stromal keratitis (39). Inflammation of the lacrimal gland (dacryoadenitis) sometimes occurs concurrent with parotid gland involvement. Less commonly iritis, trabeculitis, scleritis, ocular motor palsies, neuroretinitits, and optic neuritis occur within the first two weeks after onset of parotitis (40).

Diagnosis and Prevention of Mumps Virus Disease

In patients with classic symptoms of mumps, laboratory confirmation is not required. Supportive laboratory diagnosis includes a positive IgM mumps antibody, rise in titers between acute and convalescent specimens, or detection of virus nucleic acid by PCR from a clinical specimen. The isolation culture techniques are time consuming.

Despite the impressive decline in the incidence of mumps cases since the introduction of live attenuated MMR vaccine, sporadic mumps outbreaks still occur. Prevention of transmission is dependent on early diagnosis, isolation of the infected patient, and immunization of susceptible exposed individuals.

RUBEOLA (MEASLES) VIRUS

Rubeola (Measles) was a universal childhood disease before immunization, confirming that it is the most infectious of microbial agents. In undernourished children in developing countries, measles has high case-fatality rates. Before immunization, epidemics occurred every two to three years in developed countries. By 2004, measles vaccination coverage had dramatically reduced measles cases and measles deaths in most areas of the world, with exceptions being in sub-Saharan Africa and certain areas in Southern and East Asia. Molecular epidemiologic studies suggest that most cases in the United States now result from importation of virus. Measles is spread by respiratory droplet aerosols produced by sneezing and coughing. The portals of entry for measles virus include cells of the respiratory tract and possibly the conjunctiva. The disease may be contagious from several days before the onset of rash and up to five days after lesions appear. Measles is clinically manifested by symptoms of fever, malaise, myalgia, and headache. The classic triad of acquired measles includes cough, coryza, and follicular conjunctivitis.

Measles Disease Ocular Disease

Within hours of the onset of measles symptoms, photophobia and conjunctival injection occur. The palpebral and, to a lesser extent, the bulbar conjunctivae are involved with watery conjunctivitis (41). Tiny white (Koplik's) spots may involve the palpebral conjunctiva or the conjunctiva may have a papillary or follicular conjunctivitis. Mild epithelial keratitis may be present. Measles keratopathy, a major source of blindness in the developing world, typically presents as corneal ulceration or opacification in malnourished, vitamin A–deficient children (42). Less common are optic neuritis and retinal vascular occlusion. Measles retinopathy presents with profound visual loss one to two weeks after the appearance of the characteristic exanthema and is characterized by attenuated arterioles, diffuse retinal edema, macular star formation, scattered retinal hemorrhages, blurred disc margins, and a clear media. With resolution of systemic symptoms and of the acute retinopathy, arteriolar attenuation with or without perivascular sheathing, optic disc pallor, and a secondary pigmentary retinopathy with either a bone spicule or salt-and-pepper appearance may evolve. It may be accompanied by encephalitis.

Congenital Measles Ocular Disease

Ocular manifestations of congenital measles infection include cataract, optic nerve head drusen, and bilateral diffuse pigmentary retinopathy involving both the posterior pole and retinal periphery. The retinopathy may also be associated with either normal or attenuated retinal vessels, retinal edema, and macular star formation. The differential diagnosis of congenital measles retinopathy includes entities comprising the TORCHES syndrome (toxoplasmosis; rubella; cytomegalic inclusion disease; herpesviruses, including EBV; and syphilis).

Diagnosis and Prevention of Measles Virus Disease

The observation of the characteristic rash, fever, coryza, and conjunctivitis in an epidemic setting is sufficient to establish the diagnosis. Multinucleated giant cells can often be detected in stained smears of nasal secretions. The virus can be isolated from nasal secretions or conjunctiva by cultivation on primate cell monolayers but is technically demanding. Measles antigen or RNA can be demonstrated by immunofluorescence or reverse-transcriptase PCR methods, respectively. A rise in hemagglutination inhibition antibodies during a period of two to three weeks confirms the diagnosis. Confirmation by measles-specific IgM enzyme immunoassay is also available.

Despite the existence of a safe, effective, and inexpensive vaccine since 1963, measles remains a leading cause of mortality worldwide among young children; in the United States, however, measles is now quite rare. Measles vaccine is available in monovalent form, but also combined with mumps and rubella vaccine (MMR). Uncomplicated measles is managed symptomatically but in areas where vitamin A deficiency may be present, an oral dose of vitamin A can reduce the morbidity and mortality of measles.

HUMAN IMMUNODEFICIENCY VIRUS

Retroviruses encode a viral enzyme, reverse transcriptase, which assists in conversion of the single-stranded RNA genome into a circular double-stranded DNA molecule that then integrates into host cell chromosomal DNA. The retrovirus of greatest medical importance is human immunodeficiency virus (HIV) which causes a spectrum of disease, including an asymptomatic carrier state, the acquired immunodeficiency syndrome (AIDS), and AIDS-related complex (ARC). HIV preferentially infects T cells, especially T-helper (CD4+) lymphocytes, identified by monoclonal antibodies OKT4 and Leu 3. HIV enters the human host through sexual contact at mucosal surfaces, through breast feeding, or through blood-contaminated needles. Sexually transmitted infection is facilitated by uptake of HIV by dendritic cells at mucosal surfaces. Primary viremia results in an infectious mononucleosis-like HIV prodrome, followed by seeding of the peripheral lymphoid organs and development of a measurable immune response. Infected patients may remain otherwise asymptomatic for several years, but CD4+ T lymphocytes are progressively depleted. Clinical immunodeficiency eventually develops. Alternatively some develop acute retroviral syndrome, the typical symptoms being fever, fatigue, weight loss, myalgias, headache, pharyngitis, and nausea.

Human Immunodeficiency Virus Ocular Disease

HIV itself has been isolated from tears, conjunctiva, cornea, aqueous humor, iris, sclera, vitreous humor, and retina, and has been suspected of causing intraocular inflammation, in the absence of other pathogens, in occasional patients. Patients have been described with anterior or posterior uveitis with HIV cultured from either the aqueous humor or vitreous humor, although the definitive cause from HIV remains suspect. Another inflammatory condition, termed multifocal punctate retinal infiltrates (also called HIV microvasculopathy or HIV retinopathy), can have associated anterior segment inflammation and may be a direct effect of HIV infection or a result of ischemia.

Most of the ocular findings in HIV-infected patients are related to other infections associated with the profoundly depressed immune state, and the eye findings may be the first sign of disseminated systemic infection (43,44) These HIV AIDS-related ocular diseases include HZO, MCV disease, keratoconjunctivitis sicca, microsporidial keratoconjunctivitis, HIV neuropathy, cryptococcal optic neuritis, retinal microvasculopathy, choroiditis, and retinitis caused by a variety of infections including syphilis, mycobacteria, pneumocystis, toxoplasmosis, CMV, HSV, and VZV (45,46). Some of these entities were well described, but uncommon, before the HIV epidemic and some have presentations that differ between affected individuals with HIV disease and those from the general population who are immunocompetent. The treatment of many of these diseases is challenging because of host immunodeficiency. The longer survival of patients with HIV has also led to a change in some of the ocular manifestations or frequency of ocular manifestations of HIV infection. HAART causes the presentation of certain diseases to be different than earlier in the AIDS epidemic, as improved immune function may limit the severity of infections, but allows increased inflammatory reactions.

CMV retinitis remains the most prevalent of the blinding ocular disorders in HIV patients. CMV is transmitted by blood, saliva, breast milk, and mucous membrane contact but the infection is usually asymptomatic in immunocompetent individuals or produces a self-limited, mononucleosis-like syndrome. CMV dissemination and tissue-destructive infection of the retina occurs with severe immunosuppression as in the late manifestation of AIDS and, in the pre-HAART era, patients rarely survived longer than one to two years after the diagnosis of CMV retinitis. The most distinctive feature of CMV retinitis is a dry granular border with multiple dot-like satellite lesions caused by advancement of infection into normal retina in the absence of prominent inflammatory reactions. Epithelial keratitis or stromal keratitis has been reported probably correlating with virus in tears, conjunctiva, and cornea.

MCV disease is more common in people with AIDS than in the general population and lesions of the eyelid may be the initial clinical manifestation of HIV disease. In HIV+ patients, MCV disease produces a more aggressive course with extensive dissemination of lesions especially on the eyelids and the face although the conjunctival and cornea sequelae of MCV disease found in immunocompetent patients do not develop, most likely due to an inability to mount an inflammatory reaction. Reconstitution of immune function with HAART can result in resolution of MCV disease without therapy specifically directed toward the virus

HPV infection is very common in all stages of HIV infection and patients with significant immunosuppression may have particularly extensive warts that are recalcitrant to standard treatment.

HZ occurs more frequently in HIV-infected individuals and the possibility of HIV infection should be considered in young adults who present with HZO, and who have no other known risk factors for immune suppression. HZO can be the initial clinical manifestation of HIV infection and HZO associated with HIV infection can be especially severe. A chronic VZV infection of the corneal epithelium has been described in ophthalmic HZ in HIV patients. HZ may be more frequently multidermatomal, disseminated, or have a chronic verrucous cutaneous form in more severely immunosuppressed patients.

HSV infection in HIV+ patients seems to be the same as in non-HIV infected patients except for an increased risk of recurrence and a relative lack of corneal stromal inflammation.

Acute necrotizing herpetic retinitis has several variants (47). PORN syndrome is a unique variant of VZV retinitis seen in patients with AIDS and is distinct from acute retinal necrosis (10) syndrome, the form of VZV retinitis usually seen in immunocompetent patients. In its early stages, PORN may be difficult to differentiate from peripheral CMV retinitis. However, PORN's characteristic rapid progression and relative absence of vitreous inflammation usually allows this entity to be distinguished from CMV retinitis and from the ARN syndrome.

Laboratory Diagnosis and Prevention of HIV Infection

HIV infection is defined by viral culture or the presence of HIV-specific antibodies, HIV antigens, or HIV RNA in serum. HIV can be cultured from lymphocytes in peripheral blood and occasionally from specimens of other sites. The numbers of circulating infected cells vary with the stage of disease. Test kits are commercially available for measuring antibodies by enzyme-linked immunoassay (EIA). If properly performed, these tests have a very high sensitivity and specificity. Amplification assays such as the reverse transcription–polymerase chain reaction (RT-PCR), DNA PCR, and bDNA tests are commonly used to detect viral RNA in clinical specimens. The ELISA is used for screening, and Western blot, immunofluorescent antibody studies, p24 antigens determination, qualitative DNA PCR, viral culture, and other techniques are used to evaluate suspected false-positive or false-negative test results.

Despite HIV having been demonstrated in tears, conjunctival epithelial cells, corneal epithelial cells, aqueous, retinal vascular endothelium, and retina, the risk of HIV transmission during most ophthalmic examinations and procedures is extremely low or nonexistent. Because of the possibility of virus transmission during ophthalmic examinations or through corneal transplantation and other procedures, however, the CDC has issued a series of recommendations to prevent the spread of HIV, as well as other pathogens; they include careful handwashing, use of gloves, and sterilization of any equipment that comes into direct contact with patient tissues or fluids, including tears. Because HIV-infected patients may show no overt manifestations of disease, universal precautions should be employed in all patient encounters. Instruments that come into direct contact with external surfaces of the eyes should be wiped clean and disinfected by a 5- to 10-minute exposure to one of the following: (*i*) a fresh solution of 3% hydrogen peroxide; (*ii*) a fresh solution containing 5000 parts per million (ppm) free available chlorine—a one-tenth dilution of common household bleach (sodium hypochlorite); (*iii*) 70% ethanol; or (*iv*) 70% isopropanol. The device should be thoroughly rinsed in tap water and dried before use. There have been no reported cases of HIV transmission by corneal transplantation but potential corneal donors are screened for antibodies against HIV; those with a positive ELISA test and Western blot analysis or evidence of high-risk behaviors for HIV infection are excluded from use in corneal transplantation procedures. Transmission of HIV by contact lens fitting sets is also a concern, as HIV has been isolated from soft contact lenses worn by individuals with AIDS and rinsing alone is ineffective for disinfecting lenses. Contact lenses used in trial fitting should be disinfected between fittings with a commercially available hydrogen peroxide contact lens disinfecting system or with the standard heat disinfection regimen (78–80°C for 10 minutes).

ENTEROVIRUS

The enteroviruses are transmitted from person to person through fecal-oral contact. The enteroviruses are traditionally divided into five subgenera: Polioviruses, group A Coxsackieviruses, group B Coxsackieviruses, Echoviruses, and "newer" Enteroviruses, based on differences in host range and pathogenic profile. Members of ocular significance include the coxsackievirus and the enterovirus.

Enterovirus and Coxsackievirus Ocular Disease

AHC is an extremely contagious, self-limited ocular infection occurring predominantly as epidemics in crowded coastal areas of tropical countries during the hot, rainy season (48). Infection is transmitted from fingers or fomites directly to the eye, and contagion is encouraged by crowded unsanitary living conditions. Large-scale epidemics have been caused by enterovirus 70 and coxsackievirus A24 and less commonly, adenovirus type 11. Enterovirus 70, first recognized in 1969 and hence called "Apollo 11 disease" (49), caused a global pandemic during the early 1970s. These viruses have caused epidemics throughout Southeast Asia and the Indian subcontinent whereas disease in the West has been confined to seasonal outbreaks in the Caribbean, Central America, and south Florida. The reasons for the appearance and disappearance of particular serotypes are not known. AHC is characterized by an explosive onset and quick peak within 24 hours associated with ocular burning, foreign body sensation, photophobia, swelling of the eyelids, and watery discharge. The most distinctive sign is a subconjunctival hemorrhage, which is present in almost 90% of patients with enterovirus 70 but less frequently in cases caused by coxsackievirus A24 (50). Other ocular features include small follicles that appear on the tarsal conjunctiva within a few days of onset and a fine punctate epithelial keratitis. The preauricular lymph nodes are often tender and enlarged. Recovery is usually quick but the keratitis may persist for several weeks. Acute motor paralysis indistinguishable from poliomyelitis has been reported in association with enterovirus 70.

Diagnosis and Prevention of Enterovirus and Coxsachieviruses Disease

Both enterovirus 70 and coxsackievirus A24 are readily isolated from tears, but only infrequently from other sites. Cell culture from conjunctival swabs or scrapings is labor intensive and expensive but permits typing of the isolate for clinical and epidemiologic research. The specimens should be transported to the laboratory under cooled conditions and, if virus cultures cannot be performed immediately, the specimens should be stored at −4°F (−20°C). The enteroviruses produce a characteristic cytopathic effect in cultured cells. PCR is more sensitive than culture for identification of enteroviruses. A microneutralization test is usually used on both acute and convalescent sera for the determination of antibodies to enteroviruses.

Crowding and poor sanitation increases the risk of spread. Reuse of water for bathing and sharing of towels contribute to the spread of infection so simple hygienic measures should be reinforced. In contrast to infection caused by polioviruses, active immunization against the nonpolio enteroviruses is not practical because of the large number of serotypes.

DENGUE VIRUS

The dengue virus is an arthropod-borne virus, with Dengue fever transmitted through the bite of an infected female Aedes aegypti/albopictus mosquito. It has a large geographic distribution including the tropics and subtropics. With air travel and tourism, the incidence and geographic distribution of dengue is increasing. Dengue viral infection can be classified into five clinical presentations: nonspecific febrile illness, classic dengue, dengue hemorrhagic fever, dengue hemorrhagic fever with dengue shock syndrome, and other unusual syndromes such as encephalopathy and hepatitis. Dengue hemorrhagic fever and dengue shock syndrome are the most severe manifestation but are rare. The severity of dengue disease correlates with both the level and quality of the dengue virus-specific T lymphocyte responses. Viremia is detectable 6 to 18 hours before the onset of symptoms, and ends as the fever resolves. The immunopathologic mechanisms involved in dengue fever seem to comprise a complex series of immune responses initiated by direct infection with the virulent dengue virus, followed by antibody-dependent enhancement.

Dengue Virus Ocular Disease

Ocular manifestations in dengue fever are uncommon. Patients may present with blurred vision, central scotomas, floaters, photopsia, and haloes (51). The interval between the onset of the disease and the appearance of ocular symptoms varies from days to two weeks. Findings include retinal pigment epithelial disturbance, localized retinal and retinal pigment epithelial thickening, macular and retinal hemorrhages, peripapillary hemorrhage, Roth's spot, diffuse retinal edema, vitreous cells, blurring of the optic disc margin, serous retinal detachment, choroidal effusions, and nonspecific maculopathy. The prognosis for vision is variable related to the degree of macular involvement. The ocular manifestations associated with dengue fever, as with the general disease, seem to be an immune-mediated process rather than a direct viral infection, with the time interval corresponding to the time of onset of antibody production, immune complex deposition, or production of autoantibodies.

Diagnosis and Prevention of Dengue Virus Disease

Isolation of virus and detection of viral genomic sequences by RT-PCR are equally sensitive in identifying dengue virus in blood taken during the acute phase of illness. Virus-specific IgM can be detected within seven days after onset in most cases by enzyme immunoassay or with dot-blot kits.

Monitoring of vital signs and fluid balance of patients with dengue hemorrhagic shock syndrome is essential. Live attenuated dengue vaccines are under evaluation.

WEST NILE VIRUS

West Nile virus (WNV) is maintained in an enzootic cycle; birds are the natural host of the virus, which is transmitted from them to humans and other vertebrates through the bite of an infected mosquito of the Culex genus. WNV was isolated in the West Nile district of Uganda in 1937 and is related to the viruses that cause encephalitis and are endemic to Europe, Australia, Asia, and Africa. WNV first appeared in the United States during an outbreak in New York City in 1999 and has subsequently spread throughout the country. The clinical presentation of WNV infection is marked by the acute onset of a febrile illness, often accompanied by myalgias, arthralgias, headache, conjunctivitis, and lymphadenopathy. A maculopapular or roseolar rash appears in some, and WNV can occasionally result in meningitis or encephalitis.

West Nile Virus Ocular Disease

Ocular involvement secondary to WNV has recently been described with a host of ophthalmic signs and symptoms (52–54), although the precise prevalence of ocular involvement is not known. Presenting ocular symptoms include ocular pain, photophobia, conjunctival hyperemia, retrobulbar pain, and blurred vision. A characteristic multifocal chorioretinitis is seen in the majority of patients, together with nongranulomatous anterior uveitis and vitreous cellular infiltration. Chorioretinal lesions are distributed most often in the retinal periphery in a random pattern or in linear arrays, following the course of the choroidal blood vessels, or, less frequently, in the posterior pole. Other findings include intraretinal hemorrhages, optic disc swelling, and, less commonly, focal retinal vascular sheathing. The pathogenesis of chorioretinal lesions is unknown but may be analogous to the hematogenous dissemination to the choroidal circulation, and multifocal granulomatous chorioretinitis seen in presumed ocular histoplasmosis syndrome and idiopathic multifocal choroiditis. The majority of patients experience a self-limiting course without sequelae after several months.

Diagnosis and Prevention of West Nile Virus Disease

IgM antibody to the virus using the IgM antibody-capture ELISA is the most commonly used laboratory method for diagnosis and can be confirmed by plaque reduction neutralization testing.

HEPATATIS C VIRUS

Hepatitis C virus (HCV), a small enveloped RNA virus, is the most common cause of chronic viral hepatitis. Approximately 20–40% of acute viral hepatitis cases reported in the United States are of the non-A, non-B type; of these, the majority are caused by HCV. Of all the hepatitis viruses, HCV causes the most damage in immunocompetent hosts because of direct hepatocyte cytotoxicity, which may result in cirrhosis, fulminant hepatitis, and hepatocellular carcinoma. Although the liver is the primary target of the virus, HCV infections are associated with disorders of various organs through immunologic mechanisms.

Hepatitis C Virus Ocular Disease

Viral infections such as hepatititis C, along with the EBV, type I human T-cell lymphotropic virus (HTLV-I), and HIV, have been associated with the development of Sjögren-like syndromes in both animal models and humans. This viral infection may contribute to chronic autoimmune destruction of lacrimal and salivary glands; the presence of actual virus persistent in the glands is not clear (55). A possible pathogenetic association between Mooren ulcer, recurrent keratitis, and chronic hepatitis C infection has been proposed (56). Mooren ulcer is a chronic, progressive, painful, idiopathic ulceration of the peripheral corneal stroma and epithelium. The eye is inflamed and pain can be intense, with photophobia and tearing. Perforation may occur with minor trauma. Although the etiology of this peripheral ulcerative keratitis is unknown, evidence is mounting that autoimmunity plays a key role and the peripheral cornea has distinct morphologic and immunologic characteristics that predispose it to inflammatory reactions and immune complex formation. Some patients have an underlying chronic HCV infection although the majority of cases are idiopathic. Other forms of ocular involvement in patients with chronic HCV include episcleritis, retinopathy, and retinal vasculitis. HCV RNA has been detected in aqueous humor and tears.

Diagnosis and Prevention of Hepatitis C Virus Disease

A sensitive enzyme immunoassay can detect and quantify total HCV core antigen in anti-HCV-positive or anti-HCV-negative sera. PCR assays are available to detect HCV RNA and provide HCV genotyping.

RUBELLA VIRUS

Rubella virus is surrounded by a lipid envelope, or "toga," and hence its inclusion in the Togaviridae family. It was first isolated from army recruits in 1962. Rubella (German measles) is usually a benign febrile exanthem, but when it occurs in pregnant women it can produce major congenital malformations. Humans are the only known natural hosts. Before the introduction of a rubella vaccine in 1969, epidemics occurred in the United States at six- to nine-year intervals, predominantly in children. Rubella has now almost disappeared in the United States, although outbreaks have occurred, primarily in young adults. Acquired infection with rubella presents with a nonspecific prodrome of malaise and fever in adolescents and adults followed by the rubella exanthem. Rubella is moderately contagious and most likely transmitted by aerosolized particles from the respiratory secretions. The erythematous, maculopapular rash appears first on the face, spreads toward the hands and feet, involves the entire body within 24 hours, and disappears by the third day. Once the virus invades the bloodstream, it may spread to the skin and distal organs or, transplacentally, to the developing fetus.

Rubella Virus Ocular Disease

Ocular manifestations of acquired rubella infection include a mild follicular conjunctivitis, an epithelial keratitis, and a retinitis. Rubella retinitis presents with acute onset of decreased vision and multifocal chorioretinitis, large areas of bullous neurosensory detachment, underlying pigment epithelial detachment involving the entire posterior pole, anterior chamber and preretinal vitreous cells, and dark gray atrophic lesions of the retinal pigment epithelium. Most recently, chronic rubella virus infection has been implicated in the pathogenesis of Fuchs heterochromic iridocyclitis, as evidenced by the presence of rubella-specific intraocular antibody production and intraocular persistence of the virus (57).

Congenital Rubella Virus Disease

The association between maternal rubella and congenital defects was described in 1941 by an ophthalmologist. Up to 25% of women of childbearing age lack rubella antibodies and are susceptible to primary infection. Rubella is the prototypical teratogenic viral agent as the fetus is infected with the rubella virus transplacentally, secondary to maternal viremia during the course of primary infection. Although obvious maternal infection during the first trimester of pregnancy may end in spontaneous abortion, stillbirth, or severe fetal malformations, asymptomatic maternal rubella may also result in severe fetal disease. The frequency of fetal infection is highest during the first 10 weeks and during the final month of pregnancy, with the rate of congenital defects varying inversely with gestational age. The classic features of the congenital rubella syndrome (CRS) include cardiac malformations (patent ductus arteriosus, interventricular septal defects, and pulmonic stenosis), ocular findings (chorioretinitis, cataract, corneal clouding, microphthalmia, strabismus, and glaucoma), and deafness.

Congenital Rubella Virus Ocular Disease

The ocular findings in CRS can be early or late (58,59). Transient, early ocular clinical manifestations of CRS include generalized cloudy corneas that may resolve in weeks. Permanent or progressive ocular manifestations of CRS include "salt-and-pepper" retinopathy and cataracts. Retinopathy occurs in about half of infants with symptomatic infection and is usually unilateral. The "salt-and-pepper fundus" shows considerable variation, ranging from finely stippled, bone spicule–like, small black irregular masses to gross pigmentary irregularities with coarse, blotchy mottling. Cataracts occur in up to a third of infants with retinopathy; they are bilateral in about 50% of affected infants and are associated with microphthalmia (small eye) in about 60%. Late-onset ocular manifestations of congenital rubella may not be identified until two years or longer after birth (10). Glaucoma as well as abnormalities of the cornea and lens can also have late onset. Retinal neovascularization secondary to congenital retinal vascular atrophy can result in visual disturbances. Permanent ocular manifestations of CRS include progressive visual disturbances from severe myopia (near sightedness), especially in children with retinopathy in whom progressive macular scarring develops. The rubella virus can persist in the fetus and newborn and can be excreted for months to years after birth; the virus can be isolated from the lens contents at the time of cataract surgery.

Diagnosis and Prevention of Rubella Virus Disease

The presence of acquired rubella can be confirmed by virus isolation, by PCR detection, or by demonstration of seroconversion in response to rubella antigens (60). Virus isolation is often difficult because rubella virus does not cause cytopathic effects in the cell lines that are generally employed in diagnostic laboratories. Serologic criteria for rubella infection include a fourfold increase in rubella-specific IgG in paired sera one to two weeks apart or the new appearance of rubella-specific IgM. The diagnosis of CRS is confirmed by the pathognomonic retinal findings, associated systemic findings, and a history of maternal exposure to rubella. Because the fetus is capable of mounting an immune response to rubella virus, specific IgM or IgA antibodies to rubella in the cord blood confirms the diagnosis. Elevated IgM antibodies may return to nondiagnostic levels by three to six months, and persistence of IgG antibodies beyond this period may also help diagnose neonatal infection. Unlike in acquired infection, rubella virus can be isolated for up to a year or more from the nasopharynx, buffy coat of the blood, cerebrospinal fluid, and urine of infants with congenital infection. The differential diagnosis of congenital rubella retinitis consists of those entities comprising the TORCHES syndrome (toxoplasmosis; rubella; cytomegalic inclusion disease; herpesviruses, including EBV; and syphilis).

Rubella immunization programs in the United States include two strategies, universal immunization of all infants and targeted vaccination of susceptible prepubertal girls and women of childbearing age. CRS still occurs in areas of the world that lack routine rubella vaccination programs.

CHIKUNGUNYA VIRUS

Chikungunya virus is spread by the bite of the infected Aedes mosquitoes, primarily *A. aegypti*. Molecular characterization has demonstrated two distinct lineages of strains that caused epidemics in Africa and Asia. Chikungunya fever is a re-emerging viral disease with a recent epidemic in India after a period of quiescence (61). Chikungunya fever is characterized by abrupt onset of fever, headache, fatigue, nausea, vomiting, muscle pain, rash, and severe arthralgia lasting for one to seven days. The systemic manifestation may be related to viremia, and the joint involvement may be immune mediated from the viral antigen and antibody reactions. The disease is usually self-limiting and rarely fatal although can be debilitating.

Chikungunya Virus Ocular Disease

Iiridocyclitis and retinitis are the most common ocular manifestations associated with chikungunya fever, with a typically benign clinical course (62). Chikungunya retinitis occurs several weeks after the primary illness. Less frequent ocular lesions include episcleritis and optic neuritis (63). The optic neuritis may be retrobulbar, retrochiasm, or involve the optic nerve head (papillitis) or nerve fibers (neuroretinitis); partial or complete recovery usually occurs.

Diagnosis and Prevention of Chikungunya Virus Disease

Chikungunya is clinically difficult to differentiate from dengue fever; however, correlation with parameters such as serology (IgM antibody) and platelet count can differentiate.

RABIES VIRUS

Rabies is a neurotrophic virus transmitted from the saliva or neural tissue of infected animals. Rabies has a worldwide distribution but is rare in the United States. In North America, certain wild animals are major reservoirs of infection, since most domesticated animals have been vaccinated. The virus can enter through skin or mucous membranes and there is a 9 to 90 day incubation period before the virus enters the myoneuronal junction with access to the peripheral nervous system and then the central nervous system. Symptoms include pain and itching around the wound site and later headache, fever, vomiting, and loss of appetite. An anxious or agitated state becomes manifest with spasmodic contractions while attempting to swallow. Seizures and paralysis are common. Rabies is almost always fatal once symptoms begin.

Rabies Virus Ocular Disease

Ocular findings include photophobia, loss of corneal reflexes, pupil irregularities, disc edema, retinal hemorrhages, and paralysis of extraocular muscles. Rabies virus can be transmitted via corneal transplant (64).

Diagnosis and Prevention of Rabies

Fluids such as saliva or CNS, or tissue can be cultured. Neutralizing antibodies to rabies can be measured. Rabies can be detected in skin or other tissues by many techniques, including PCR. Corneal biopsy and impression cytology have been useful in assisting in the early diagnosis of rabies (65).

Suspected infected animals should be captured and quarantined. If suspected, the animal should be euthanized and the brain analyzed for rabies virus or Negri bodies (cytoplasmic inclusion bodies). Animal bites should be washed vigorously, followed by povidone iodine. Tetanus prophylaxis should be provided. Rabies can be prevented by immune globulin and rabies vaccine.

REFERENCES

1. El-Sayed Zaki M, Abd-El Fatah GA. Rapid detection of oculopathogenic adenovirus in conjunctivitis. Curr Microbiol 2008; 56(2):105–109.
2. Sambursky R, Tauber S, Shirra F, et al. The RPS adeno detector for diagnosing adenoviral conjunctivitis. Ophthalmology 2006; 113(10):1758–1764.
3. Wiley LA, Roba LA, Kowalski RP, et al. A 5-year evaluation of the adenoclone test for the rapid diagnosis of adenovirus from conjunctival swabs. Cornea 1996; 15(4):363–367.
4. Koidl C, Bozic M, Mossbock G, et al. Rapid diagnosis of adenoviral keratoconjunctivitis by a fully automated molecular assay. Ophthalmology 2005; 112(9):1521–1528.

5. Elnifro, E.M, Cooper RJ, Klapper, PE, et al. Multiplex polymerase chain reaction for diagnosis of viral and chlamydial keratoconjunctivitis. Invest Ophthalmol Vis Sci 2000; 41(7):1818–1822.
6. Gottsch JD, Froggatt JW, Smith DM, et al. Prevention and control of epidemic keratoconjunctivitis in a teaching eye institute. Ophthalmic Epidemiol 1999; 6(1):29–39.
7. Liesegang TJ. Herpes simplex virus epidemiology and ocular importance. Cornea 2001; 20(1):1–13.
8. Hodge WG, Margolis TP. Herpes simplex virus keratitis among patients who are positive or negative for human immunodeficiency virus: an epidemiologic study. Ophthalmology 1997; 104(1):120–124.
9. Streilein JW, Dana MR, Ksander BR. Immunity causing blindness: five different paths to herpes stromal keratitis. Immunol Today 1997; 18(9):443–449.
10. Arnold JJ, McIntosh ED, Martin EJ, et al. A fifty-year follow-up of ocular defects in congenital rubella: late ocular manifestations. Aust N Z J Ophthalmol 1994; 22(1):1–6.
11. Walters G, James TE. Viral causes of the acute retinal necrosis syndrome. Curr Opin Ophthalmol 2001; 12(3):191–195.
12. Farhatullah S, Kaza S, Athmanathan S, et al. Diagnosis of herpes simplex virus-1 keratitis using Giemsa stain, immunofluorescence assay, and polymerase chain reaction assay on corneal scrapings. Br J Ophthalmol 2004; 88(1):142–144.
13. El-Aal AM, El Sayed M, Mohammed E, et al. Evaluation of herpes simplex detection in corneal scrapings by three molecular methods. Curr Microbiol 2006; 52(5):379–382.
14. Herpetic Eye Disease Study Group. Acyclovir for the prevention of recurrent herpes simplex virus eye disease. N Engl J Med 1998; 339(5):300–306.
15. Liesegang TJ. The Varicella-Zoster Virus Diseases. Contemp Ophthalmol 2006; 5:1–7.
16. Liesegang TJ. Herpes zoster ophthalmicus natural history, risk factors, clinical presentation, and morbidity. Ophthalmology 2008; 115(suppl 2):S3–S12.
17. Liesegang TJ. Herpes zoster virus infection. Curr Opin Ophthalmol 2004; 15(6):531–536.
18. Ganatra JB, Chandler D, Santos C, et al. Viral causes of the acute retinal necrosis syndrome. Am J Ophthalmol 2000; 129(2):166–172.
19. Holland GN. Standard diagnostic criteria for the acute retinal necrosis syndrome. Executive Committee of the American Uveitis Society. Am J Ophthalmol 1994; 117(5):663–667.
20. Engstrom RE Jr., Holland GN, Margolis TP, et al. The progressive outer retinal necrosis syndrome. A variant of necrotizing herpetic retinopathy in patients with AIDS. Ophthalmology 1994; 101(9):1488–1502.
21. Weidmann M, Meyer-Konig U, Hufert FT. Rapid detection of herpes simplex virus and varicella-zoster virus infections by real-time PCR. J Clin Microbiol 2003; 41(4):1565–1568.
22. Davis JL, Feuer W, Culbertson WW, et al. Interpretation of intraocular and serum antibody levels in necrotizing retinitis. Retina 1995; 15(3):233–240.
23. Chodosh J. Epstein-Barr virus stromal keratitis. Ophthalmol Clin North Am 1994; 7:549–556.
24. Matoba AY. Ocular disease associated with Epstein-Barr virus infection. Surv Ophthalmol 1990; 35(2):145–150.
25. Pitetti RD, Laus S, Wadowsky RM. Clinical evaluation of a quantitative real time polymerase chain reaction assay for diagnosis of primary Epstein-Barr virus infection in children. Pediatr Infect Dis J 2003; 22(8):736–739.
26. Chodosh J, Gan YJ, Sixbey JW. Detection of Epstein-Barr virus genome in ocular tissues. Ophthalmology 1996; 103(4):687–690.
27. Coats DK, Demmler GJ, Paysse EA, et al. Ophthalmologic findings in children with congenital cytomegalovirus infection. J AAPOS 2000; 4(2):110–116.
28. Jabs DA, Van Nattan NL, Kempen JH, et al. Characteristics of patients with cytomegalovirus retinitis in the era of highly active antiretroviral therapy. Am J Ophthalmol 2002; 133(1):48–61.
29. Hodge WG, Boivin JF, Shapiro SH, et al. Clinical risk factors for cytomegalovirus retinitis in patients with AIDS. Ophthalmology 2004; 111(7):1326–1333.
30. Hoover DR, Peng Y, Saah, A, et al. Occurrence of cytomegalovirus retinitis after human immunodeficiency virus immunosuppression. Arch Ophthalmol 1996; 114(7):821–827.
31. Sjo NC, Heegaard S, Prause JU, et al. Human papillomavirus in conjunctival papilloma. Br J Ophthalmol 2001; 85(7):785–787.
32. Shields CL, Shields JA. Tumors of the conjunctiva and cornea. Surv Ophthalmol 2004; 49(1):3–24.
33. Nakamura Y, Mashima Y, Kameyama K, et al. Detection of human papillomavirus infection in squamous tumours of the conjunctiva and lacrimal sac by immunohistochemistry, in situ hybridisation, and polymerase chain reaction. Br J Ophthalmol 1997; 81(4):308–313.
34. Neff JM, Lane JM, Fulginiti VA, et al. Contact vaccinia–transmission of vaccinia from smallpox vaccination. JAMA 2002; 288(15):1901–195.
35. Semba RD. The ocular complications of smallpox and smallpox immunization. Arch Ophthalmol 2003; 121(5):715–719.

36. Fillmore GL, Ward TP, Bower KS, et al. Ocular complications in the Department of Defense Smallpox Vaccination Program. Ophthalmology 2004; 111(11):2086–2093.
37. Charteris DG, Bonshek RE, Tullo AB. Ophthalmic molluscum contagiosum: clinical and immunopathological features. Br J Ophthalmol 1995; 79(5):476–481.
38. Charles NC, Friedberg DN. Epibulbar molluscum contagiosum in acquired immune deficiency syndrome. Case report and review of the literature. Ophthalmology 1992; 99(7):1123–1126.
39. Riffenburgh RS. Ocular manifestations of mumps. Arch Ophthalmol 1961; 66:739–743.
40. Foster RE, Lowder CY, Meisler DM, et al. Mumps neuroretinitis in an adolescent. Am J Ophthalmol 1990; 110(1):91–93.
41. Kayikcioglu O, et al. Ocular findings in a measles epidemic among young adults. Ocul Immunol Inflamm 2000; 8(1):59–62.
42. Whitcher JP, Srinivasan M, Upadhyay MP. Corneal blindness: a global perspective. Bull World Health Organ 2001; 79(3):214–221.
43. Holland GN. AIDS and ophthalmology: the first quarter century. Am J Ophthalmol 2008; 145(3): 397–408.
44. Belfort R Jr. The ophthalmologist and the global impact of the AIDS epidemic LV Edward Jackson Memorial Lecture. Am J Ophthalmol 2000; 129(1):1–8.
45. Cunningham ET Jr., Margolis TP. Ocular manifestations of HIV infection. N Engl J Med 1998; 339(4):236–244.
46. Jeng BH, Holland GN, Lowder CY, et al. Anterior segment and external ocular disorders associated with human immunodeficiency virus disease. Surv Ophthalmol 2007; 52(4):329–368.
47. Vrabec TR. Posterior segment manifestations of HIV/AIDS. Surv Ophthalmol 2004; 49(2):131–157.
48. Wairagkar NS, Gogate SS, Labhsetwar AS. Investigation of an epidemic of acute haemorrhagic conjunctivitis in Pune, India. J Commun Dis 1999; 31(1):41–43.
49. Kono R. Apollo 11 disease or acute hemorrhagic conjunctivitis: a pandemic of a new enterovirus infection of the eyes. Am J Epidemiol 1975; 101(5):383–390.
50. Park SW, Lee CS, Jang HC, et al. Rapid identification of the coxsackievirus A24 variant by molecular serotyping in an outbreak of acute hemorrhagic conjunctivitis. J Clin Microbiol 2005; 43(3):1069–1071.
51. Lim WK, Mathur R, Koh A, et al. Ocular manifestations of dengue fever. Ophthalmology 2004; 111(11):2057–2064.
52. Garg S, Jampol LM. Systemic and intraocular manifestations of West Nile virus infection. Surv Ophthalmol 2005; 50(1):3–13.
53. Chan CK, Limstrom SA, Tarasewicz DG, et al. Ocular features of west nile virus infection in North America: a study of 14 eyes. Ophthalmology 2006; 113(9):1539–1546.
54. Priestley Y, Koevary S. Systemic and ophthalmic manifestations of West Nile Virus infection. Expert Rev Ophthalmol 2008; 3(3):279–292.
55. Jacobi C, Wenkel H, Jacobi A, et al. Hepatitis C and ocular surface disease. Am J Ophthalmol 2007; 144(5):705–711.
56. Wilson SE, Lee WM, Murakami C, et al. Mooren's corneal ulcers and hepatitis C virus infection. N Engl J Med 1993; 329(1):62.
57. Quentin CD, Reiber H. Fuchs heterochromic cyclitis: rubella virus antibodies and genome in aqueous humor. Am J Ophthalmol 2004; 138(1):46–54.
58. Givens KT, Lee DA, Jones T, et al. Congenital rubella syndrome: ophthalmic manifestations and associated systemic disorders. Br J Ophthalmol 1993; 77(6):358–363.
59. Wolff SM. The ocular manifestations of congenital rubella. Trans Am Ophthalmol Soc 1972; 70:577–614.
60. Rajasundari TA, Sundaresan P, Vijaylakshmi P, et al. Laboratory confirmation of congenital rubella syndrome in infants: an eye hospital based investigation. J Med Virol 2008; 80(3):536–546.
61. Chhabra M, Mittal V, Bhattacharya D, et al. Chikungunya fever: a re-emerging viral infection. Indian J Med Microbiol 2008; 26(1):5–12.
62. Mahendradas P, Ranganna SK, Shetty R, et al. Ocular manifestations associated with chikungunya. Ophthalmology 2008; 115(2):287–291.
63. Mittal A, Mittal S, Bharati MJ, et al. Optic neuritis associated with chikungunya virus infection in South India. Arch Ophthalmol 2007; 125(10):1381–1386.
64. Gode GR, Bhide NK. Two rabies deaths after corneal grafts from one donor. Lancet 1988; 2(8614):791.
65. Zaidman GW, Billingsley A. Corneal impression test for the diagnosis of acute rabies encephalitis. Ophthalmology 1998; 105(2):249–251.

26 Arthropod-Borne Viruses

Matthias Niedrig, Andreas Nitsche, and Oliver Donoso-Mantke
Robert Koch-Institut, Berlin, Germany

The arthropod-borne viruses (arboviruses) comprise a rather large and heterogeneous number of different virus families like Togaviridae, Flaviviridae, Bunyaviridae, Reoviridae, Rhabdoviridae, and Orthomyxoviridae. More than hundred viruses are currently classified as arboviruses from which the Togaviridae and Flaviviridae are the best known (1). As arthropod borne, all of these viruses share the common feature of the transmission by mosquitoes, ticks, or phlebotomes. Since the frequent accidental transmission of these viruses by an arthropod vector can cause severe clinical and subclinical infections in humans, a quick and reliable diagnosis of the viral infection is very important.

In order to provide clear and state-of-the-art information, this chapter concentrates on the most important pathogens with high impact on public health. These viruses include Dengue Virus (DENV), Yellow Fever Virus (YFV), Tick-borne Encephalitis Virus (TBEV), and West Nile (WN) for the Flaviviruses, Chikungunya Virus (CHIKV) as an Alphavirus and the Sandfly Fever Virus (SFV), Crimean Congo hemorrhagic fever virus (CCHFV), and Rift valley Fever virus (RVF) for the Bunyaviruses.

Diagnostic testing for arboviruses is mainly limited to patients who are suspected to be infected by these viruses. For example, DENV, displaying four serotypes, has a global distribution in nearly all subtropical areas and have become a major international public health concern in recent years, with about 2.5 billion people at risk for the disease (2). The clinical features of dengue fever vary according to the age of the patient and the DEN strain, ranging from a mild fever to the Dengue hemorrhagic fever (DHF) with potentially deadly complications, or to the Dengue shock syndrome (DSS) that is characterized by rapid signs of circulatory failure and where the patient may die within 12 to 24 hours without proper treatment. Depending on the cause of disease, different measures for virus diagnosis must be initiated to confirm the clinical diagnosis.

During acute infection, the direct detection of the viral pathogen itself is the only possibility for a successful diagnosis. This can be achieved only with a limited number of diagnostic assays focusing either on the detection of virus-specific proteins or the virus genome (Fig. 1). For particle detection by electron microscopy, virus loads have to exceed 10^6/mL, which is largely dependent on the virus, the course of infection, and the time point of investigation. Table 1 gives a brief overview on the sensitivity and specificity of the different diagnostic methods. The estimated time required for the different diagnostic methods also gives important information on the practicality of these diagnostic methods in the acutely infected patient.

Because of low commercial impact for most of the flaviviruses, specific assays for protein detection do not exist, and such assays are available as in-house assays in only a few laboratories (Table 2). Some assays may involve immunohistochemistry staining of tissue samples with monoclonal antibodies directed against the suspected virus (3). Other assays applied for the detection of virus antigens are specific capture enzyme immunoassays (EIA), but these are rather uncommon and not very sensitive. Owing to the unique genetic code of RNA and DNA, the detection of virus-specific sequences by PCR methods has been the major technology for diagnostic testing for many years. In general, PCR approaches are very rapid, enable extremely low detection limits, and provide genetic material for further characterization (for example, by sequencing). On the other hand, the high specificity due to the exact base pairing of primer and template may lead to false-negative results even after minor changes in the target sequence. Since flaviviruses are prone to frequent sequence aberrations the PCR assays used should be selected carefully. Finally, the exact knowledge of the target virus sequences has to be available for the proper design of reliable assays, a prerequisite that is hard to achieve for flaviviruses. However, nearly all recent cases of acute viral hemorrhagic fevers imported to the western world

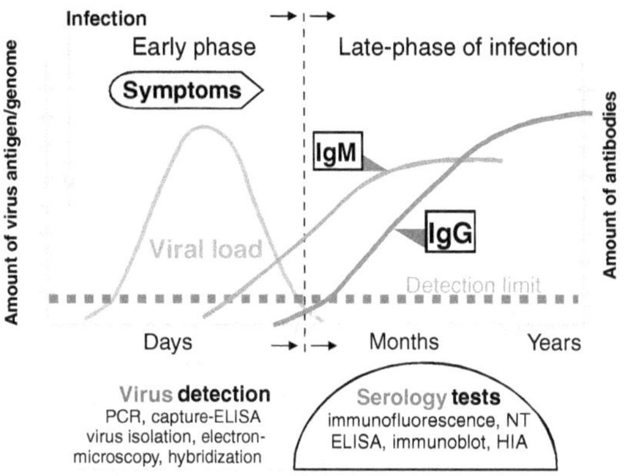

Figure 1 (*See color insert*). Schema for diagnosis of viral infections.

were diagnosed by PCR, before a confirmation by virus cultivation was performed successfully. This virus cultivation on suitable cells is an additional option for virus detection in an acute phase of disease. Cultivation is laborious and can take one to seven days depending on the susceptible cells used and amount of virus in the patient's serum. However, a specific method for the unequivocal identification of samples showing even typical cytopathic effects is essential. Since nearly all of these viruses (except the SFV) are highly pathogenic and have to be handled in a class 3 or 4 biosafety containment laboratory, virus cultivation is limited to a few highly experienced facilities.

For the diagnosis of Dengue (DEN) infection in the early phase, rapid assays are commercially available. These indirect virus detection assays are based on identification of specific IgM shown by diffusion of blood or serum on blot paper spotted with Dengue protein and an antibody control. The staining of the respective band including the control gives indication of existing IgM. However, these rapid tests, which are available in different formats from different companies, have some pros and cons like all other serological assays. Despite the very quick performance and easy handling, which does not require skilled personnel, the rate of false-positive results caused by unspecific reactions seems higher and the sensitivity of these assays seems to be lower than other commercially available tests (4,5).

Table 1 Basic Features of Different Diagnostic Methods

	Time for diagnosis	Sensitivity	Specificity
Virus detection			
virus isolation	1–7 day	High[a]	High[b]
hybridization	3–4 hr	High[c]	Good
PCR	2–4 hr	High[d]	High
Pyrosequencing	1 hr	High[e]	High
Electron microscopy	30 min	Low[f]	High[g]
capture ELISA	3–5 hr	Good[h]	High
Serology			
ELISA	3–4 hr	High	Low
Immunofluorescence	2–4 hr	Good	Good
Immunoblot	2–4 hr	Good	Good
Neutralization	4–7 day	Good	High
HIA	2–4 hr	Low	Good

[a] Depending on cultivation system.
[b] Depending on detection system.
[c] ca. 10^4 particle/mL.
[d] ca. 200 genome equivalent/mL.
[e] Requires PCR.
[f] $\geq 10^6$ particle/mL.
[g] Detection of the virus family.
[h] ca. 0.01 μg antigen/mL.

Table 2 Overview of the Most Used Diagnostic Methods for Selected Arbovirus Infections in Patients or Environmental Samples

	Diagnostic		
Viral pathogen	**Acute/early phase**	**Acute/late phase**	**Convalescent phase**
Dengue Virus 1–4	PCR[a,b], VI[b], IHA, EIA[a] (NS-1)	IgM EIA[a], IFA[a], HIA, rapid tests	IgM/IgG EIA[a], IgM/IgG IFA[a]
Yellow Fever Virus	PCR[b], VI[b], IHA	IgM EIA , IFA[a]	IgM/IgG EIA, IgM/IgG IFA[a]
West Nile Virus	PCR[a,b], VI[b], IHA	IgM EIA[a]; IgM IFA[a], HIA	IgM/IgG EIA[a], IgM/IgG IFA[a], HIA
Tick-borne Encephalitis Virus	PCR[b], VI[b], IHA	IgM EIA[a]; IFA[a], HIA	IgM/IgG EIA[a], IgM/IgG IFA[a], HIA
Chikungunya Virus	PCR[b], VI[b], IHA	IgM/IgG EIA, IgM IFA[a], HIA	IgM/IgG IFA[a]; HIA
Sandfly Fever Virus	PCR[a,b], VI[b]	IB[a]; IgM IFA[a]	IB[a], IgM IFA[a], IgM/IgG EIA[a]
Crimean Congo hemorrhagic fever virus	PCR[b], VI[b]	IgM/IgG EIA[a], IgM IFA	IgM/IgG EIA, IgG IFA
Rift Valley Fever virus	PCR[b], VI[b]	IgM/IgG EIA[a], IgM IFA, HIA, inhibition EIA	IgM/IgG EIA, IgG IFA, HIA

[a]Commercial assays available.
[b]Methods useful for analysis of enviromental or vector (mosquitoes, ticks, phlebotomes) samples.
Abbreviations: VI, virus isolation on cells; PCR, polymerase chain reaction; IHA, immunhistolological analysis of tissue samples; c-EIA, capture enzyme immune assay; EIA, enzyme immune assay for detection of IgM and/or IgG; IFA, immunofluorescence assay; HIA, hemagglutination inhibition assay; IB, immunoblot.

Many attempts were made to analyze the serotype specific immune response for DEN 1–4 by type-specific serological assays. One rational reason is the discussed risk for a more severe disease progression for DHF in secondary infections caused by a further serotype. The underlying mechanism is called antibody-dependent enhancement (2). Unfortunately, all attempts to select suitable serotype-specific epitopes and the generation of monoclonal antibodies specific to recombinant expressed proteins or peptides have failed up to now (6). The reason for these difficulties, beside the high homology in the sequence of the surface E-protein of the flaviviruses, might be the conformational presentation of these specific epitopes, which cannot be simulated by short recombinant proteins or peptides.

Only the plaque reduction neutralization test (PRNT) is, so far, capable of differentiating the four DEN serotypes, but does not offer a practical alternative considering the time and workload required for performance. It was also proven that IgG ELISA and hemagglutination inhibition assay (HIA) were less sensitive than PRNT 50 in detecting low levels of antibodies (7). While the diagnosis of dengue is routinely accomplished by serologic assays like IgM/IgG ELISAs, as well as the hemagglutination inhibition assay (HIA), these tests are rarely used for diagnosis during the early symptomatic phase. In addition, antibodies against DEN are broadly reactive with other flaviviruses. In recent investigation, it turned out that the DEN NS-1 antigen testing is a useful tool for an early diagnosis of dengue infections after onset of fever and a good alternative to the PCR in clinical laboratories (5,8). Real-time RT-PCR has also become a more and more important tool for an early and specific detection of DEN virus genomes in human serum samples (9–12). Generic DEN assays, detecting viruses of all four serotypes, as well as DEN serotype-specific assays, differentiating the serotypes by specific primer or probes, have been established. A recent promising approach was based on the generic amplification and detection of all four DEN serotypes, with subsequent typing by pyrosequencing. Pyrosequencing is a new technique in which the enzymatic incorporation of the four different nucleotides, which are added separately and are complementary to the sequenced DNA strand, is monitored in real time. For each incorporated nucleotide, a clear light signal is generated and presented as a peak histogram, called a pyrogram. Pyrosequencing has its strength in the sequencing of fragments of up to 80 bases and the identification of single nucleotide polymorphisms (SNPs). As shown previously, this maximal length is sufficient for the identification and typing of several pathogens, like the highly diverse group of Hantaviruses. Here, the generation of short sequences of up to 40 bases enables an attribution to already known serotypes and the identification of still unknown genotypes.

Even though a very efficient vaccine against Yellow Fever (YF) exists, the diagnosis of a YF infection is still an important and often urgent matter. Yellow Fever is transmitted by mosquitoes to humans in the tropical regions of Africa and South America and causes endemic/epidemic disease in approximately 200,000 cases per year (13). Significant progress was made by integrating the YF vaccine into the childhood immunization programs in Africa. However, the great movement of the population requires regular serological analysis of the vaccination coverage to avoid a decrease of the protective immunity in parts of the population living in endemic areas. These measures require analysis by YF-specific EIA for detecting IgG antibodies in the African population. In order to evaluate the specificity of the serological EIA results, a more specific test such as the neutralization assay is required (14). For the analysis of acute infection in patients either from endemic areas or in unvaccinated tourists returning from endemic regions, RT-PCR is the most sensitive and advanced method for a rapid diagnosis (15,16). The later isolation of YFV in cell culture is a useful confirmation, but it is too time consuming for a quick diagnosis. In addition, to ensure a timely decision about the requirement of YFV vaccinations as preventive countermeasure, the evaluation of suspected cases in cities in endemic areas relies on RT-PCR.

The relatively short viremia of only a few days in patients recovering from YF infection limits the chance for a positive PCR test, and requires further investigation by serological assays. Because of the immense cross-reactivity between the different flaviviruses, the analysis of consecutive sera with a clear fourfold increase in titer is recommended for a positive result by WHO. In patients with severe YF, very high virus titers can be found, and virus is even detectable in liver tissue by electron microscopy and immunohistochemical staining with specific antibodies (3,16). The high specificity and sensitivity of the recently developed IFA makes it a useful tool for rapid diagnosis of yellow fever during outbreaks, for epidemiological studies, and for serosurveillance after vaccination (17).

The introduction of West Nile Virus (WNV) into North America in 1999 as a new pathogen caused serious problems, including being misdiagnosed as Saint Louis Encephalitis (SLE) in the beginning. Up to then SLE was a common infection in the United States, which is caused by SLEV, a virus also belonging to the flavivirus family with high potential for cross-reactivity with other flavivirus assays. It turned out very soon that this new pathogen easily spread by infected migrating birds and can be also occasionally transmitted to humans by infected mosquito vectors. This new emerging infection made it necessary to develop diagnostic tools for detecting WNV in acute encephalitis patients and for providing safe blood and plasma for medical care. Accidental transmission by transplanting organs of an infected donor clearly shows that biological safety issues require intensive diagnostic measures (18). Tremendous efforts were made by introduction of routine screening of blood and blood products by EIA and RT-PCR to avoid any further transmission. Such a routine testing is always based on evaluated and standardized diagnostic assays.

However, the use of certified EIAs does not overcome the general problem of cross-reactivity within the flavivirus family. As found in evaluation studies for different commercially available EIAs for WNV-specific IgG, all of them showed positive reactivity with anti-YF and -TBE sera as well (19–21). Even when applying the PRNT, usually resulting in the highest specificity, certain high titer sera displaying anti-YF and/or anti-TBE reactivity showed cross-neutralization activity to WNV as well. In an external quality assurance for the serological detection of WN positive sera, it turned out, that only 8/27 (29.6%) of the participating laboratories reached the proficiency criteria for correct IgM and IgG detection. These data demonstrate the challenge of a reliable serological diagnosis of a WN infection (22). To overcome these cross-reactivity problems, the analysis of antibody avidity might be a rapid and simple option. As shown for differentiation of primary from previous WN infection, the IgG avidity assay provides an additional diagnostic certainty (23). This needs to be evaluated also for cross-reactivity with other flavivirus-reactive sera. A reverse ELISA based on the B domain of the E-protein seems to be more specific (24). With this assay, it was possible to avoid the frequent cross-reaction caused by sera directed against other flaviviruses.

For the diagnosis of WNV by PCR, several assays have been published (25,26). In studies, it could be shown that the two lineages found in Africa have a very distinct pattern of distribution (27). In the United States, due to the exclusive introduction of one virus strain, only variants of lineage 1 originally from an Israeli prototype are present. The introduction of WNV lineage 2 into Hungary was demonstrated recently (28). This knowledge has immediate consequences for the design of WNV-specific RT-PCR systems to be used in Europe, which should react with sequences of both lineages. As shown in an external quality assurance, only 11/30 laboratories were able to detect lineage 2 in high concentration, which demonstrates the necessity for improving their assays by adaptation of primers to lineage 2 WNV (29).

The tick-borne encephalitis virus (TBEV) is the most important flavivirus infection in Europe and Russia transmitted by infected *Ixodes* ticks. The highest incidence is found in the area of Kemerovo in Siberia but ticks are also present in central and Eastern Europe, with high incidences causing a great number of encephalitis cases (30). The biphasic course of the disease starts with uncharacteristic influenza-like symptoms followed by a symptom-free interval before a meningitis phase develops in about 20–30% of the patients, with an occasionally fatal outcome (ca. 2%) (31). Very often the patients do not recognize or remember a tick bite that results in a late diagnosis for the patient. Serology is the most used diagnostic approach for suspected cases, although some retrospective studies have shown that the TBEV could be detected by PCR in the early phase of the disease (32,33). In the majority of cases, the first-line diagnosis is performed by testing for IgM using commercial TBE EIAs available from different manufacturers with reasonable quality. Although these EIAs cannot overcome the problem of serological cross-reactivity between the different flaviviruses, their quality improved after an extensive assay evaluation some years ago (33,34,35). In an external quality assessment study for TBE serology, it turned out that correct results were obtained for at least 90% of the samples by 33/40 (83%) participating laboratories for IgM and 16/42 (38%) laboratories for IgG, with testing often based on commercial EIAs (36). In contrast it was shown that only 35–44% of the laboratories detected the far eastern and/or the Siberian TBE subtypes correctly despite the fact that the far eastern subtype is present in the Baltic countries already (37,38). It can be speculated that with the increasing requirement for the diagnosis of suspected acute TBE infections, rapid diagnosis by PCR will become more important for the analysis of serum and/or liquor specimens. Unfortunately, because of the short viremia, a negative PCR result is not predictive for a negative TBEV infection (39). One argument for physicians not requesting PCR testing is that no specific medical treatment is available for TBE patients. However, one can hope that this will change; in addition, other treatments such as antibiotics are often given, which might be harmful for the patient.

The chikungunya virus (CHIKV), geographically distributed in Africa, India, and South-East Africa, belongs to the alphavirus family and is transmitted by mosquitoes to humans. Since 2005 outbreaks of chikungunya fever have involved several countries in Africa, the southwestern Indian Ocean region, Asia, and recently Europe (40,41). The 266,000 CHIK cases of La Reunion and the 1.3 million cases in India highlight the importance for a rapid and reliable diagnosis (42,43). The majority of infections were diagnosed by in-house EIAs for detection of IgM and/or IgG (44). As demonstrated in a study on returning CHIKV-infected travelers by IFA, IgM and IgG antibodies appeared nearly at the same time on the second day after the onset of symptoms (42). Along with RT-PCR analysis, which gave positive results in all patients from the first until the fourth day post onset, other assays are available for the diagnosis of acute cases. As demonstrated in this investigation, the rate for successful virus isolation correlates quite well with the number of genome copies/mL found in the patient's sera.

Furthermore, as shown in an external quality assurance study for the serological CHIKV detection, the HIA seems to be the most sensitive assay used by the participating laboratories compared to other tests like EIA or IFA (45). Many laboratories had suboptimal performance for IgM detection, regardless of whether they used commercial or in-house assays. Potential problems with cross-reactive antibodies in patients caused by other alphavirus pathogens such as Eastern-, Western-, or Venezuelan equine encephalitis virus, O'nyong-nyong, Ross River, or Semliki Forest virus seem to be less important due to the different distribution patterns of these diseases. The recent development of commercial serological assays will help to improve the diagnostic quality for this newly emerging disease (46).

The detection of the broadly divergent CHIKV strains distributed between West Africa and Asia by PCR seems not to be a major problem, as genome sequences with high homology for PCR primer binding could be selected (42). Nevertheless, the sensitivity of some testing methods needs to be improved, as demonstrated in an EQA study in which 87% of the laboratories could not detect 1000 genome copies/mL (47).

The Sandfly fever virus (SFV) transmitted by Phlebotomus flies is widely distributed in all countries of the Mediterranean region. The Sandfly fever group comprises several different isolates named after the places where they were first isolated during local outbreaks: Toscana, Naples, Sicily, Corfu, and Cyprus (48). While the Sandfly fever disease is normally very mild and aseptic, meningitis, meningoencephalitis, and encephalitis are rare complications. The number of reported infections is rather low (49). However, in the recent years several studies demonstrated the presence of Sandfly fever in the South of France, Portugal, Spain, Greece, Italy, Cyprus, and Algeria, thus underlining the importance of the infection (50–52). Most of the diagnoses were performed by in-house assays, although commercial assays for serology such as EIA, IFA, and IB are available. The recently developed RT-PCRs for virus genome detection support the commercially available nested PCR assay, and can be used for screening for Phlebotomus as well as for analysis of patient specimens in acute infection (53,54). The introduction of new and more sensitive PCR-based tools will help to improve the diagnosis of viral meningitis patients in Sandfly endemic areas.

The name Crimean Congo hemorrhagic fever virus (CCHFV) refers to the two locations where the virus was isolated, demonstrating the wide distribution of this disease (55). CCHFV can be found in East-, West-, and South Africa, the Middle East, Asia, and South-East Europe. Recently an outbreak in Turkey and sporadic cases in northern Greece highlight that these areas belong to the endemic regions (56–58). The transmission is caused mainly by *Hyalomma* ticks in people having close contact to animals, like shepherds, farmers, or butchers. The diagnosis of acute cases is mainly performed by in-house PCR and or IgM-specific capture EIA. In severely ill patients with intense viremia, the detection of virus antigen by antigen capture is possible, but rarely used (59). The comparison of a capture EIA and RT-PCR displayed one to two orders of magnitude lower sensitivity for the antigen detection assay (60).

For serology, IgM and IgG EIA and IFA assays are available. The problem of handling this highly infectious human pathogen in serological platforms, usually requiring a biological class 4 containment laboratory for virus propagation, can be overcome by the use of nonhazardous recombinant nucleoprotein (rNP) antigen as shown recently (61). Recombinant antigens provide a novel, sensitive, and specific tool with equivalent results for CCHF diagnosis. With the introduction of new quantitative PCRs, measurement of CCHF virus genome load in clinical samples was possible to compare with the severity of the CCHF infection in the patient (62). In summary, the RNA genome copy number can be considered as a predictive value for a more or less severe outcome of an infected patient. In eight of nine patients with fatal outcome, viral loads of $\geq 1 \times 10^9$ copies/mL were detected, whereas in 25 of 26 patients with nonfatal outcomes, viral loads were $<1 \times 10^9$ copies/mL ($P < 0.001$).

Rift Valley fever (RVF) virus is a mosquito-borne RNA virus responsible for large outbreaks of acute febrile disease in livestock and humans throughout Africa and Saudi Arabia. Virus infections in animals often result in abortion, with significant mortality in newborn livestock and high economic impact. Human infections result in a flu-like illness, with 1% to 2% of patients developing severe complications, including encephalitis or hemorrhagic fever with high fatality rates. In a cohort study summarizing laboratory characteristics of the RVF epidemic that occurred in Saudi Arabia from August 2000 through September 2001, laboratory results of 834 reported cases were available. From these cases 81.9% were laboratory confirmed, of which 51.1% were positive for only RVF IgM, 35.7% were positive for only RVF antigen, and 13.2% were positive for both (63). Since then, and as demonstrated in the recent outbreaks in Kenya in 2006, several studies have been published demonstrating the improvement of the diagnosis of acute infections by RT-PCR (64–66). An inhibition enzyme-linked immunosorbent assay for detection of antibody to RVFV was developed for the serological evaluation of humans, domestic animals, and wild ruminants in disease-surveillance and control programs (67). This validated in-house EIA was more sensitive in detection of the earliest immunological responses compared to virus neutralization and hemagglutination–inhibition tests and can be used by experienced laboratories as a safe, robust, and highly accurate diagnostic tool. In a further development, it was demonstrated that inactivated whole virus antigen could be efficiently replaced by the recombinant N-protein that has the potential to complement the traditional assays for serodiagnosis of RVF (68). It is very likely that the improved diagnostic assays for serology and genome detection by PCR will have well an impact on disease surveillance and control programs, and on therapeutic intervention in diseased RVF patients, as previously shown for other viral hemorrhagic fever infections.

REFERENCES

1. Hayes E, Machenzie J, Shope R. American Public Health Association. Arthropod-borne viral diseases. In: Heyman DL, ed. Control of Communicable Diseases Manual. 19th ed. Washington, DC: United Book Press, Inc, 2008:34–39.
2. Halstead SB. Dengue. Lancet 2007; 370(9599):1644–1652. Review.
3. Bae HG, Drosten C, Emmerich P, et al. Analysis of two imported cases of Yellow Fever infection from Ivory Coast and The Gambia to Germany and Belgium. J Clin Virol 2005; 33:274–280.
4. Blacksell SD, Bell D, Kelley J, et al. Prospective study to determine accuracy of rapid serological assays for diagnosis of acute dengue virus infection in Laos. Clin Vaccine Immunol 2007; 14(11):1458–1464.
5. Dussart P, Petit L, Labeau B, et al. Evaluation of Two New Commercial Tests for the Diagnosis of Acute Dengue Virus Infection Using NS1 Antigen Detection in Human Serum. PLoS Negl Trop Dis 2008; 2(8):280.
6. Wichmann O, Stark K, Shu PY, et al. Clinical features and pitfalls in the laboratory diagnosis of dengue in travellers. BMC Infect Dis 2006; 6:120.
7. Pengsaa K, Yoksan S, Limkittikul K, et al. Maternally transferred neutralising dengue antibodies in Thai infants: A pilot study. Ann Trop Paediatr 2003; 23(3):159–165.
8. Chuansumrit A, Chaiyaratana W, Pongthanapisith V, et al. The use of dengue nonstructural protein 1 antigen for the early diagnosis during the febrile stage in patients with dengue infection. Pediatr Infect Dis J 2008; 27(1):43–48.
9. Das S, Pingle MR, Muñoz-Jordán J, et al. Detection and serotyping of dengue virus in serum samples by multiplex reverse transcriptase PCR-ligase detection reaction assay. J Clin Microbiol 2008; 46(10):3276–3284.
10. Domingo C, Palacios G, Niedrig M, et al. A new tool for the diagnostic and molecular surveillance of Dengue infections in clinical samples. WHO Dengue Bull 2004; 28:87–95.
11. Teichmann D, Göbels K, Niedrig M, et al. Virus isolation for diagnosing dengue virus infections in returning travellers. Eur J Clin Microbiol Infect Dis 2003; 22(11):697–700.
12. Teichmann D, Göbels K, Niedrig M, et al. Clinical and laboratory features of Dengue virus infected traveller previously vaccinated against Yellow Fever. Scan J Infect Dis 2003; 35(6–7):427–429.
13. Monath TP. Yellow fever as an endemic/epidemic disease and priorities for vaccination. Bull Soc Pathol Exot 2006; 99(5):341–347.
14. Koraka P, Zeller H, Niedrig M, et al. Reactivity of serum samples from patients with a Flavivirus infection measured by IFA and ELISA. Microbes Infect 2003; 4:1209–1215.
15. Bae HG, Nitsche A, Teichmann A, et al. Detection of Yellow Fever: A comparison of quantitative real time and plaque assay. J Virol Methods 2003; 110(2):185–191.
16. Teichmann D, Grobusch MP, Wesselmann H, et al. A haemorrhagic fever from the Côte d'Ivoire. Lancet 1999; 354 (9190):1608.
17. Niedrig M, Kürsteiner O, Herzog C, et al. Evaluation of an Indirect Immunofluorescence Assay for detection of IgM and IgG Antibodies against Yellow Fever Virus. Clin Vaccine Immunol 2008; 15(2):171–181.
18. Iwamoto M, Jernigan DB, Guasch A,et al. West Nile Virus in Transplant Recipients Investigation Team Transmission of West Nile virus from an organ donor to four transplant recipients. N Engl J Med 2003; 348(22):2196–2203.
19. Pfleiderer C, König C, Chudy M, et al. Molecular epidemiology of West Nile Virus in humans. Dev Biol (Basel) 2006; 126:197–201.
20. Linke S, Muehlen M, Niedrig M, et al. Assessing the exposure of German and Austrian bird ringers to west Nile virus (Flavivirus) and evaluating their potential risk to acquire an infection. J Ornithol 2008; 149:271–275.
21. Malan AK, Martins TB, Hill HR, et al. Evaluations of commercial West Nile virus immunoglobulin G (IgG) and IgM enzyme immunoassays show the value of continuous validation. J Clin Microbiol 2004; 42(2):727–733.
22. Niedrig M, Donoso-Mantke O, Altmann D, et al. External quality assurance study on serology diagnosis of West Nile. BMC Infect Dis.2007; 7:72.

23. Levett PN, Sonnenberg K, Sidaway F, et al. Use of IgG avidity assays for differentiation of primary from previous infections with West Nile virus. J Clin Microbiol 2005; 43(12) 5873–5875.
24. Ludolfs D, Linckh S, Niedrig M, et al. Reverse ELISA for the detection of anti West Nile IgG antibodies in humans. Clin Microbiol Infect 2007; 26(7):467–473.
25. Lanciotti RS, Kerst AJ. Nucleic acid sequence-based amplification assays for rapid detection of West Nile and St. Louis encephalitis viruses. J Clin Microbiol 2001; 39(12):4506–4513.
26. Linke S, Ellerbrok H, Niedrig M, et al. Detection of West Nile virus lineage 1 and 2 by real-time PCR. J Virol Methods 2007; 146:355–358.
27. Campbell GL, Marfin AA, Lanciotti RS, et al. West Nile virus. Lancet Infect Dis 2002; 2(9):519–529.
28. Bakonyi T, Hubálek Z, Rudolf I, et al. Novel flavivirus or new lineage of West Nile virus, central Europe. Emerg Infect Dis 2005; 11(2):225–231.
29. Niedrig M, S Linke, H Zeller,et al. First international proficiency study on West Nile virus molecular detection. J Clin Chem 2006; 10(12):2200–2203.
30. Korenberg E, Likhacheva T Analysis of the long-term dynamics of tick-borne encephalitis (TBE) and ixodid tick-borne borrelioses (ITBB) morbidity in Russia. Int J Med Microbiol 2006; 296(suppl 40):54–58.
31. Poponnikova TV. Specific clinical and epidemiological features of tick-borne encephalitis in Western Siberia. Int J Med Microbiol 2006; 296(40):59–62.
32. Schultze D, Dollenmaier G, Rohner A, et al. Benefit of detecting tick-borne encephalitis viremia in the first phase of illness. J Clin Virol 2007; 38(2):172–175.
33. Donoso Mantke O, Achazi K, Niedrig M. Serological versus PCR methods for the detection of tick-borne encephalitis virus infections in humans. Future Virol 2007; 2(6):565–572.
34. Niedrig M, Vaisviliene D, Klockmann U, et al. Comparison of six commercial IgG-ELISA Kits for the detection of TBEV-antibodies. J Clin Virol 2000; 20:179–182.
35. Jääskeläinen A, Xiuqi Han X, Niedrig M, et al. Diagnostics of tick-borne encephalitis with μ-capture IgM-EIA based on secreted recombinant antigen produced in insect cells. J Clin Microbiol 2003; 41(9):4336–4342.
36. Niedrig M, Avšič T, Aberle SW, et al. Quality control assessment for the serological diagnosis of tick borne encephalitis virus infections. J Clin Virol 2007; 38:260–264.
37. Donoso Mantke O, Schädler R, Niedrig M. Survey on cases of tick borne encephalitis in European countries. Euro Surveill 2008; 13(17):18848.
38. Golovljova I, Vene S, Sjölander KB, et al. Characterization of tick-borne encephalitis virus from Estonia. J Med Virol 2004; 74(4):580–588.
39. Donoso Mantke O, Aberle SW, Avšič T, et al. Quality control assessment for the PCR diagnosis of tick borne encephalitis virus infections. J Clin Virol 2007; 38:73–77.
40. Schuffenecker I, Iteman I, Michault A, et al. Genome microevolution of chikungunya viruses causing the Indian Ocean outbreak. PLoS Med 2006; 3(7):263.
41. Brisse S, Iteman I, Schuffenecker I. Chikungunya outbreaks. N Engl J Med 2007; 356(25):2650–2652.
42. Panning M, Grywna K, van Esbroeck M, et al. Chikungunya fever in travelers returning to Europe from the Indian Ocean region, 2006. Emerg Infect Dis 2008; 14(3):416–422.
43. Laurent P, Le Roux K, Grivard P, et al. Development of a sensitive real-time reverse transcriptase PCR assay with an internal control to detect and quantify chikungunya virus. Clin Chem 2007; 53(8):1408–1414.
44. Gérardin P, Guernier V, Perrau J, et al..Estimating Chikungunya prevalence in La Réunion Island outbreak by serosurveys: Two methods for two critical times of the epidemic. BMC Infect Dis 2008; 8:99.
45. Niedrig M, Zeller H, Schuffenecker I, et al. First international diagnostic accuracy study for the serological detection of chikungunya virus infection. Clin Microbiol Infect 2009; 15(9):880–884.
46. Litzba N, Schuffenecker I, Zeller H, et al. Evaluation of the first commercial chikungunya virus indirect immunofluorescence test. J Virol Methods 2008; 149:175–179.
47. Panning M, Charrel RN, Landt O, et al. Coordinated implementation and external quality assessment of diagnostic Chikungunya virus RT-PCR. Emerg Infect Dis 15(3):469–471.
48. Xu F, Chen H, Travassos da Rosa AP, et al. Phylogenetic relationships among sandfly fever group viruses (Phlebovirus: Bunyaviridae) based on the small genome segment. J Gen Virol 2007; 88(8):2312–2319.
49. De Lamballerie X, Tolou H, Durand JP, et al. Prevalence of Toscana virus antibodies in volunteer blood donors and patients with central nervous system infections in southeastern France. Vector Borne Zoonotic Dis 2007; 7(2):275–277.
50. Konstantinou GN, Papa A, Antoniadis A. Sandfly-fever outbreak in Cyprus: Are phleboviruses still a health problem? Travel Med Infect Dis 2007; 5(4):239–242.
51. Santos L, Simões J, Costa R, et al. Toscana virus meningitis in Portugal, 2002–2005. Euro Surveill 2007; 12(6):E3–E4.

52. Izri A, Temmam S, Moureau G, et al. Sandfly fever Sicilian virus, Algeria. Emerg Infect Dis 2008; 14(5):795–797.
53. Pérez-Ruiz M, Collao X, Navarro-Marí JM, et al. Reverse transcription, real-time PCR assay for detection of Toscana virus. J Clin Virol 2007; 39(4):276–281.
54. Weidmann M, Sanchez-Seco MP, Sall AA, et al. Rapid detection of important human pathogenic Phleboviruses. J Clin Virol 2008; 41(2):138–142.
55. Whitehouse CA. Crimean-Congo hemorrhagic fever. Antiviral Res 2004; 64(3):145–160.
56. Gozalan A, Esen B, Fitzner J, et al. Crimean-Congo haemorrhagic fever cases in Turkey. Scand J Infect Dis 2007; 39(4):332–336.
57. Papa A, Christova I, Papadimitriou E, et al.. Crimean-Congo hemorrhagic fever in Bulgaria. Emerg Infect Dis 2004; 10(8):1465–1467.
58. Papa A, Maltezou HC, Tsiodras S, et al. A case of Crimean-Congo haemorrhagic fever in Greece, June 2008. Euro surveill 2008; 13:33.
59. Smirnova SE, Karavanov AS. Detection of Crimean-Congo haemorrhagic fever virus antigen by solid phase enzyme immunosorbent assay. Acta Virol 1985; 29:87–90.
60. Garcia S, Chinikar S, Coudrier D, et al. Evaluation of a Crimean-Congo hemorrhagic fever virus recombinant antigen expressed by Semliki Forest suicide virus for IgM and IgG antibody detection in human and animal sera collected in Iran. J Clin Virol 2006; 35(2):154–159.
61. Larichev VF, Manzeniuk IN, Naĭdenova EV, et al. ELISA and PCR test systems used to detect Crimean-Congo hemorrhagic fever virus. Vopr Virusol 2007; 52(4):43–46.
62. Cevik MA, Erbay A, Bodur H, et al. Viral load as a predictor of outcome in Crimean-Congo hemorrhagic fever. Clin Infect Dis 2007; 45(7):e96–e100.
63. Madani TA, Al-Mazrou YY, Al-Jeffri MH, et al. Rift Valley fever epidemic in Saudi Arabia: Epidemiological, clinical, and laboratory characteristics. Clin Infect Dis 2003; 37(8):1084–1092.
64. Centers for Disease Control and Prevention (CDC). Rift Valley fever outbreak–Kenya, November 2006-January 2007. MMWR Morb Mortal Wkly Rep 2007; 56(4):73–76.
65. Bird BH, Bawiec DA, Ksiazek TG, et al. Highly sensitive and broadly reactive quantitative reverse transcription-PCR assay for high-throughput detection of Rift Valley fever virus. J Clin Microbiol 2007; 45(11):3506–3513.
66. Bird BH, Khristova ML, Rollin PE, et al. Complete genome analysis of 33 ecologically and biologically diverse Rift Valley fever virus strains reveals widespread virus movement and low genetic diversity due to recent common ancestry. J Virol 2007; 81(6):2805–2816.
67. Paweska JT, Mortimer E, Leman PA, et al. An inhibition enzyme-linked immunosorbent assay for the detection of antibody to Rift Valley fever virus in humans, domestic and wild ruminants. J Virol Methods 2005; 127(1):10–18.
68. Fafetine JM, Tijhaar E, Paweska JT, et al. Cloning and expression of Rift Valley fever virus nucleocapsid (N) protein and evaluation of a N-protein based indirect ELISA for the detection of specific IgG and IgM antibodies in domestic ruminants. Vet Microbiol 2007; 121(1–2):29–38.

27 Viral Infections of the Immunocompromised Host

Steven A. Pergam

Department of Medicine, University of Washington, and Vaccine and Infectious Disease Institute, Fred Hutchinson Cancer Research Center, Seattle, Washington, U.S.A.

Keith R. Jerome

Department of Laboratory Medicine, University of Washington, and Vaccine and Infectious Disease Institute, Fred Hutchinson Cancer Research Center, Seattle, Washington, U.S.A.

INTRODUCTION

Immunocompromised patients present unique diagnostic challenges for the clinical virology laboratory. The host immune response plays a critical role in controlling viral infections, and in the absence of such control, viruses can cause unique and serious manifestations. Thus, immunocompromised patients require diligent and frequent monitoring for viral infections, and the clinical virology laboratory is an important component to managing these patients.

Immunosuppression as a general term encompasses a broad spectrum of conditions that vary in severity, but are all at an increased risk for infection. Normal hosts are by definition immunocompetent, but at specific time points during life they are also more susceptible to viral infections. Young children, particularly newborns and premature infants, are at higher risk for developing life-threatening disease from viral pathogens (1). The development of immune senescence in the aged leads to a weakening of T-cell responsiveness to viral infections such as varicella zoster virus (VZV) for example and can lead to increased rates of clinical disease (2,3). Women also develop a physiologic decline in immune function during pregnancy (4), which can lead to more severe complications from particular viral infections (5,6).

Many underlying clinical diseases can also lead to immunosuppression. Human immunodeficiency virus (HIV), covered extensively in chapter 23, is one of the leading causes of immunosuppression worldwide. Moderate to profound immune compromise is also a common manifestation of advanced malignancy, particularly patients with leukemia, lymphoma, or multiple myeloma. Other examples can be found in patients with organ dysfunction, such as kidney and liver failure, whose altered immunity can make them more susceptible to infections and less responsive to vaccinations (7–9). There also exist a large number of hereditary immunodeficiencies, which depending on their specific abnormalities, may also be associated with an increased susceptibility to viral infections (10).

One of the most frequent causes of immunosuppression in the United States is the use of immunomodulatory drugs for the treatment of various medical conditions. Immunosuppressive regimens used in hematopoietic stem cell transplantation (HSCT) and solid organ transplantation (SOT) predispose patients to viral infections for which the clinical virology laboratory plays an important role in diagnosis and management. This chapter will focus on specific viral infections in immunocompromised patients, and will focus on transplantation. This chapter is meant to provide an overview, as additional information regarding the viruses discussed here can be found in other chapters throughout this book.

IMMUNOSUPPRESSION IN TRANSPLANT PATIENTS

Hematopoietic Stem Cell Transplantation

Patients undergoing allogeneic HSCT receive stem cells from one of three main graft sources: harvested donor bone marrow, peripheral blood, or umbilical cord blood (UCB) stem cells. In addition to allogeneic transplants, autotransplantation of a patient's own stem cells is a common therapeutic option used to deliver high-dose chemotherapy to a diverse group of conditions including non-Hodgkin's lymphoma and multiple myeloma (11,12). In preparation for the infusion of stem cells various conditioning regimens are used to ablate the recipient's marrow; a common combination regimen includes high-dose cyclophosphamide with fractionated total-body irradiation (TBI) or busulfan. Nonmyeloablative (reduced intensity) allogeneic transplant regimens have become more popular for treating older patients and those with other serious comorbidities (13,14). Regardless of conditioning regimen, the goal behind most regimens is to destroy the patient's immune system to prevent graft rejection, leaving patients severely immunocompromised in the immediate post-transplant period (15).

Following conditioning, immune recovery is markedly delayed. Initially patients have no peripheral neutrophils or lymphocytes, and are at a high risk for opportunistic infections. Monocytes are the first cells to engraft, followed by granulocytes and natural killer cells (NK) (15). Lymphocyte function recovers less quickly, and although total lymphocyte count usually returns to normal within the second month of transplant, patients remain at high risk for viral infections. It is thought that limited T-cell clonality from the donor limits the diversity of antigen specificity in the early post-transplant period (16). Furthermore, virus-specific T-cell responses are often delayed, particularly in patients who develop graft-versus-host disease (GVHD) or those that require steroid therapy (17). Patients who undergo T-cell depleted and umbilical cord transplants have a more pronounced delay in T-cell recovery (18). B-cell recovery is also delayed by chronic GVHD (19,20), and may be further delayed due to treatment with anti-B-cell monoclonal antibodies such as rituximab (21). Together this delay in cellular immunity leads to high rates of viral infections during the post-transplant period.

Following transplantation, HCST recipients are placed on immunosuppressive therapy as prophylaxis for GVHD. The amount of immunosuppression is a balance: broader protection portends an increase in the number and severity of infections, and insufficient levels of immunosuppression leave patients at risk for severe GVHD complications. Typical agents used for GVHD prevention include methotrexate, calcineurin inhibitors, and mycophenolate mofetil (MMF), and all are associated with higher rates of viral disease (22). Regardless of prophylaxis, patients who do develop GVHD are also given high-dose glucocorticoid therapy and additional immunosuppression that can further exacerbate the risk of viral complications (23). Depleting anti-T cell therapies such as antithymocyte globulin (ATG) and alemtuzumab (monoclonal antibody for CD52) are used to treat grafts prior to transplant in some centers to prevent GVHD, but can also be used as therapy in cases of severe GVHD; both are associated with significant post-transplant infections (24,25).

Solid Organ Transplantation

Solid organ transplant patients also undergo induction therapy prior to receipt of their donor graft. Although very center specific, induction regimens usually are made up of high-dose standard immunosuppressive medications (e.g., calcineurin inhibitor, glucocorticoids, and MMF) or a combination of low-dose immunosuppressive therapy and biologic agents such as antibodies directed toward T-cell antigens. Since data indicate an improved graft survival in patients receiving anti-T-cell antibody therapy (26), most centers use these agents for induction in combination with low-dose immunosuppression and steroids (27). These biologic agents can be classified into three groups, primarily polyclonal antibody therapy (ATG), focused IL-2 receptor (anti-CD25) specific antibodies, and alemtuzumab that affects both B and T-cell lineages. ATG and alemtuzumab are highly associated with infectious complications (28,29), however nondepleting antibodies, such as basiliximab and daclizumab, are associated with a lower risk of infection (30–32).

Following organ transplantation, patients remain on a cocktail of immunosuppressive agents to prevent the development of acute and chronic organ rejection. These antirejection agents are often used in combination and include: low-dose steroids, calcineurin inhibitors (sirolimus/cyclosporine), MMF, azathioprine, or mammalian target of rapamycin (mTOR) inhibitors (sirolimus/everolimus). Steroid-free regimens have also been used to limit systemic side effects of glucocorticoids and have been shown to be effective (33).

Immunosuppressive Drugs in Transplant

The various immunosuppressive compounds used in transplantation have different mechanisms by which they exert their effects upon the immune system. Corticosteroids inhibit transcription of genes encoding cytokines and other immune mediators, downregulate adhesion molecules, and generally dampen the inflammatory response (34). In contrast, tacrolimus and cyclosporine have more targeted effects via inhibition of calcineurin, which leads to inhibition of T-cell signal transduction and IL-2 transcription (35). MMF acts by inhibiting inosine monophosphate dehydrogenase, thus suppressing de novo purine synthesis required during T- and B-cell proliferation (34). Newer agents such as everolimus and sirolimus, which inhibit mTOR, have both antiproliferative and immunosuppressive effects (36). Monoclonal antibodies target specific components of the immune system, such as T-cells (through IL-2 receptor/anti-CD25 antibodies, e.g., basiliximab) and B-cells (anti-CD20, e.g., rituximab). ATG and alemtuzumab, which more broadly target lymphocytes, can also be added to standard immunosuppressive agents as treatment during acute rejection in SOT and with severe GVHD in HSCT. Newer agents such as tumor necrosis factor (TNF) inhibitors (e.g., infliximab) are less commonly used in transplant recipients but are associated with decreased immune function (37).

To optimize therapeutic benefit of standard agents while minimizing risk of infectious complications, providers routinely measure the physiologic concentration of immunosuppressive agents. Since multiple factors can alter drug metabolism, including individual age (38), genetics (39), and drug–drug interactions (40–42), monitoring drug levels can help prevent the development of over-immunosuppression and drug toxicity. Drug monitoring for cyclosporine has been available for some time, and most centers rely on the drug concentration two hours after dosing as a surrogate marker for total cyclosporine exposure (43). The utility of such monitoring has been demonstrated by multiple groups, and has also been shown to be superior to predose monitoring. Tacrolimus and sirolimus, on the other hand, are generally measured immediately before dosing, and there appears to be no advantage to monitoring at other times. The utility of drug monitoring for MMF remains controversial (43,44).

VIRAL INFECTIONS IN TRANSPLANTATION

Viral pathogens remain a persistent problem in transplant patients. Herpesviruses and respiratory viruses in particular are leading causes of morbidity and mortality in transplantation, and viruses are leading causes of meningitis, encephalitis, pneumonia, gastritis/colitis, among other infectious complications in transplantation (Fig. 1). The clinical virology laboratory provides diagnostic specificity, and allows for management, monitoring, and treatment of these viral infections.

Cytomegalovirus (CMV)

CMV is the most common viral infection in transplantation. After primary infection in a normal host, CMV establishes lifelong latency. Once the immune system is compromised by the transplant process, CMV can escape immune control and replicate leading to major complications. The risk of CMV after transplantation depends largely upon the serostatus of both the recipient and the donor. The lowest risk population in transplant settings are CMV seronegative patients who receive transplants from seronegative donors, a situation denoted D−/R− (52). In HSCT, the highest risk group are seropositive recipients (R+) (52,53), of whom 60% to 70% will develop CMV during the post-transplant period (54). Alternatively, seronegative recipients (R−) receiving a seropositive organ (D+) are the highest risk in SOT. Patients who are D+/R− HSCT or D−/R+ SOT recipients are at moderate risk. These differences are not trivial; as compared to seronegative recipients, seropositive recipients have been shown to be at increased risk of transplant-related mortality in HSCT (52). High-risk patients have also been reported to have an increased incidence of GVHD, organ rejection, opportunistic infections, and respiratory failure (55–57). However, since serostatus is only one determining factor in deciding the optimal donor for a given recipient, and other factors such as HLA mismatch and age appear to be stronger determinants of overall survival (58), CMV risk often cannot be avoided.

CMV reactivation during the post-transplant period can lead to life-threatening invasive complications such as pneumonitis, enteritis, hepatitis, retinitis, encephalitis, or as disseminated multiorgan disease (23,59,60). Invasive CMV disease is still a major cause of infectious mortality in transplantation. CMV disease occurs in approximately 6% to 9% of seropositive HSCT recipients (23). Temporally, CMV disease occurs in two forms after HSCT. Early CMV disease is typically seen within the first three months post-transplantation, during the period of greatest

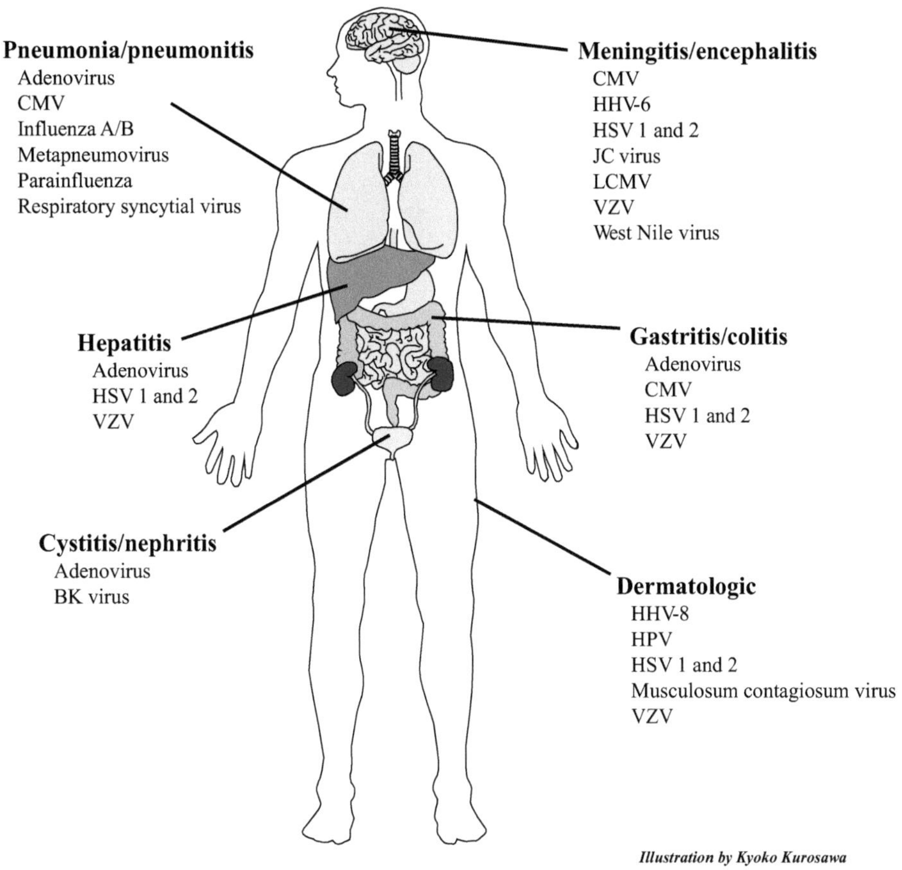

Figure 1 Schema of viral disease in post-transplant patients. *Abbreviations*: CMV, cytomegalovirus; HPV, human papilloma virus; HSV, herpes simplex virus; LCMV, lymphocytic choriomeningitis virus; VZV, varicella zoster virus.

immunosuppression. Late CMV disease occurs more than three months after HSCT, and results from defects in cell-mediated immunity that persist even after engraftment (23,61,62). Risk factors for late CMV disease include detectable CMV in plasma or whole blood in the early post-transplant period (≤100 days), persistent lymphopenia, and deficient CMV-specific T-cell immunity (60). In SOT, risk of post-transplant CMV disease is dependent on the level of immunosuppression and the organ transplanted (63). Rates of late disease are increasing due to increased use of primary CMV antiviral prophylaxis in the early post-transplant period (64,65).

Clinically, the most common manifestations of CMV disease in immunocompromised patients are enteritis and pneumonia (59). In addition, in the solid organ setting, CMV disease can manifest in the allograft and lead to graft failure (57). Prompt diagnosis of CMV in these settings is essential, as a delay in diagnosis can lead to serious consequences or even death. CMV pneumonia is diagnosed through the detection of CMV in bronchoalveolar lavage (BAL) or biopsy specimens (66). The diagnosis of CMV pneumonia can be made by shell-vial testing, direct fluorescent antigen (DFA), cytology, or by immunohistochemistry from biopsy samples (59). Similarly, CMV gastrointestinal disease is diagnosed by detection of CMV in biopsy specimens. A high index of suspicion is needed even in patients who have no evidence of viremia, as gastrointestinal CMV disease (and to a lesser extent other localized CMV disease) can occur in the absence of detectable CMV in the blood. CMV disease is normally treated with ganciclovir (plus intravenous immunoglobulin for pneumonia) as a first-line agent, and foscarnet as an alternative in patients with ongoing neutropenia (Table 1).

Data and clinical experience demonstrate that waiting for symptoms before initiating anti-CMV treatment is suboptimal, since by that time high-level viral replication and tissue invasive

Table 1 Selected Antiviral Agents with Activity Against Transplant-Associated Viral Infections

	Agent	Virus
Herpesviruses	Acyclovir/valacyclovir/famciclovir	HSV 1 and 2
		VZV
	Ganciclovir/valganciclovir	CMV
		HHV-6[a]
		KSHV/HHV-8[b]
		VZV
	Foscarnet	CMV
		HSV 1 and 2
		HHV-6[a]
		VZV
	Cidofovir	CMV
		HSV 1 and 2
		HHV-6
		KSHV/HHV-8[c]
Polyomaviruses	Cidofovir	BK virus
		JC virus
	Leflunomide	BK virus[d]
Respiratory viruses	Adamantines	Influenza A
	Cidofovir	Adenovirus[e]
	Neurominidase inhibitors	Influenza A and B
	Ribavirin	Adenovirus[e]
		Influenza A and B
		Metapneumovirus[f]
		RSV

Abbreviations: CMV, cytomegalovirus; HSV, herpes simplex virus; KSHV, Kaposi's sarcoma associated herpesvirus; RSV, respiratory syncytial virus; VZV, varicella zoster virus.
[a]See Ref. 45.
[b]See Ref. 46.
[c]See Refs. 47,48.
[d]See Ref. 49.
[e]See Ref. 50.
[f]See Ref. 51.

disease may have already been established. Studies have demonstrated survival of only 30% to 50% after the onset of symptoms of CMV pneumonia even with aggressive therapy (67,68). To prevent such dire consequences, two major prophylactic strategies are currently used in practice to prevent the development of CMV disease: primary antiviral prophylaxis and pre-emptive therapy. Recent meta-analyses have compared the efficacy of these two approaches, and found them both to be effective in preventing CMV disease, although the prophylactic approach appeared to be superior in the highest risk patients (69,70).

Several clinical trials of ganciclovir prophylaxis have been performed and shown a decrease in CMV disease; however, overall survival was not improved (59). Prophylactic approaches with ganciclovir can also lead to prolonged neutropenia, increasing susceptibility to bacterial or fungal infections (71). Additionally, extended exposure to antiviral agents, particularly in the setting of a weak or nonexistent immune response, can lead to the selection of drug-resistant viral mutants (72,73). Primary prophylaxis is used at numerous SOT centers in the early post-transplant period, and due to its improved bioavailability and lower risk of developing resistant CMV, valganciclovir is the agent of choice at most centers in the United States (74).

At many transplant centers, pre-emptive therapy is used for CMV prevention in HCST and SOT recipients. In HSCT transplant, recipients are monitored weekly for CMV viremia by quantitative real-time PCR for the first 100 days post-transplant; other centers use the CMV antigenemia assay. Not all patients with low-level CMV viremia detectable by PCR will progress to clinical disease; thus, individual centers and clinicians need to develop viral load thresholds that will minimize unnecessary treatment. After 100 days post-transplant, PCR surveillance may be discontinued in low-risk patients (59). However, high-risk patients continue weekly surveillance for late CMV disease until they reach minimal immunosuppressant levels and have at least three consecutive negative weekly tests (59). During surveillance for late CMV disease in HSCT a typical threshold for initiation of therapy is 1000 copies/mL or a greater than five-fold increase in viral load above the patient's established baseline (59).

Finally, it should be noted that CMV infection is itself immunosuppressive, and CMV is associated with an increased risk of other infections, including other viruses (especially EBV), fungal infections, and bacteremia (75). Thus, control of CMV viremia is critical in the management of transplant patients, not only to prevent the direct pathogenic effects of the virus, but also to minimize the synergistic effects on other pathogenic infections.

Epstein–Barr Virus (EBV)

EBV infection in the normal host is generally a self-limited process, but similar to CMV, develops lifetime latency. Patients who develop severe immunosuppression are at risk for the development of EBV-related complications post-transplant. EBV in post-transplant recipients has been associated with hepatitis, hemophagocytic syndrome, as well as EBV-associated lymphoproliferative disease (EBV-LPD), also referred to as post-transplant lymphoproliferative disorder (PTLD). In EBV-LPD, EBV infects resting B-cells, inducing their transformation into proliferating blasts. These blasts then differentiate into resting memory B-cells, which constitute the long-term reservoir for latent EBV (reviewed in Ref. 76). Infected B-cells that are unable to establish or maintain the latent state express viral antigens, and are therefore targeted and destroyed by EBV-specific T-cells. However, in the setting of immunosuppression, the host T-cell response may be unable to effectively control EBV replication in B-cells, leading to high-level EBV viremia and virus-driven B-cell proliferation.

The clinical presentation of EBV-LPD in transplantation is variable (77). Some patients may be asymptomatic, others may have nonspecific symptoms such as fever, lethargy, and weight loss; and some present with a mononucleosis-like syndrome (78). Lymphadenopathy is common, as are lymphoid proliferations in extranodal sites such as gums, subcutaneous tissue, liver, lungs, CNS, kidneys, intestines, and spleen (78).

Primary EBV infection post-transplant is a major risk factor for the development of EBV-LPD (79). EBV seronegative patients receiving organs from seropositive donors (D+/R−) are at especially high risk for EBV-LPD. The risk of EBV-LPD is also increased by higher intensity immunosuppression and may be associated with coinfection by CMV (80). These factors combine to cause a much higher incidence of EBV-LPD in children compared to adults (78). In addition, certain immunosuppressive agents may have direct proneoplastic effects that contribute to the development of EBV-LPD independent of their immune modulatory effects (81).

If detected early, EBV-LPD is responsive to a number of therapeutic options (reviewed in Refs. 81,82). Interestingly, antivirals have not proven effective against this disease, presumably because the tumor cells in EBV-LPD have acquired secondary mutations allowing virus-independent proliferation (76). Initial therapy is focused on reducing immunosuppression, in hopes of restoring effective T-cell control of EBV-positive B cells, although this must be balanced against the risk of graft rejection or GVHD. For early disease, reduction of immunosuppression combined with local control by excision or radiotherapy can be curative in many patients (83). Most commonly, however, reduction of immunosuppression is combined with anti-B-cell therapy using the chimeric anti-CD20 antibody, rituximab (81,82). Rituximab therapy is well tolerated and highly effective, especially when given early in the course of disease, and leads to a rapid and complete response in most patients. For patients failing rituximab therapy or those with diffuse involvement, combination chemotherapy such as CHOP (cyclophosphamide, doxorubicin, vincristine, and prednisone) can be effective, although such regimens are associated with significant toxicity.

Early detection of EBV-LPD is critical for the best therapeutic outcomes with the least toxicity. Because of the importance of prompt diagnosis of EBV-driven T-cell proliferation, some centers follow post-transplant patients by PCR. Various groups have advocated testing whole blood, PBMCs, or cell-free plasma. In general, all of these specimens provide good sensitivity, although comparisons suggest that cell-free plasma may provide superior specificity for EBV-LPD (84). Many authors favor pre-emptive therapy, in which detection of EBV DNA levels above a certain threshold triggers reduction of immunosuppression +/− rituximab therapy.

There are no current guidelines for the management of EBV viremia, however a threshold of 500 to 1000 viral copies has been suggested for initiating therapy, as this appears to predict the development of clinical complications related to EBV (85,86). Other groups have advocated the so-called "prompt" therapy, in which treatment is triggered only when detection of EBV DNA is accompanied by clinical signs of EBV-LPD. Prompt therapy has shown similar efficacy to pre-emptive therapy (87). However, based on the concern that the delay in treatment while waiting for symptoms may lead to more severe disease, many groups still favor pre-emptive approaches (82).

Herpes Simplex Virus 1 and 2 and Varicella Zoster Virus

Herpes simplex viruses (1 and 2) and VZV can reactivate during the post-transplant period. Laboratory testing for VZV and HSV is important in transplant recipients, and essential for patients with atypical presentations (88). Typical orolabial and genital HSV lesions are often recurrent and prolonged during periods of immunosuppression. Mucocutaneous lesions can involve the esophagus, and breakdown from oral lesions can increase the risk of bacterial superinfection. Rare complications, such as pneumonitis, encephalitis, and hepatitis can also occur. The diagnosis of HSV hepatitis must be considered in patients who present with abdominal pain, elevated liver function tests, and consumptive coagulopathy, even in the absence of skin or mucocutaneous lesions (89). High-dose IV acyclovir should be given presumptively until a diagnosis can be made (Table 1). Orolabial and genital reactivations can be treated with oral acyclovir, unless there is evidence of major spread or dissemination.

Approximately 90% of older adults are seropositive for VZV, and immunity has increased in younger patients due to the institution of routine childhood vaccination in the United States. VZV reactivations are common in transplant recipients occurring at a median of five months post-HSCT (90) and nine months after SOT (91). Patients who develop primary varicella zoster during transplant are at high risk for major complications (92), so serologic testing prior to transplant is necessary; particularly since 2% to 3% of adult transplant recipients may be seronegative (93). The most common form of VZV post-transplant is herpes zoster, which usually presents as a painful pruritic rash along a dermatomal distribution. Transplant recipients may also present with dissemination similar to that seen in primary VZV infection (94). Postherpetic neuralgia can occur in up to 25% to 35% of patients following an episode of herpes zoster, which can cause chronic pain and debilitation (95,96). In patients who develop herpes zoster, high-dose valacyclovir is often used for therapy due to its infrequent dosing (Table 1). Transplant patients are also at risk for developing severe invasive disease, including pneumonitis, visceral VZV, and encephalitis. VZV encephalitis and visceral VZV, a life-threatening condition that presents with severe abdominal pain, markedly elevated liver function tests and syndrome of inappropriate antidiuretic hormone (SIADH), may present either prior to or without the development of a rash (97,98). Such unusual presentations are rare, but when suspected, patients should undergo early VZV PCR testing; presumptive high-dose IV acyclovir should not be delayed while awaiting the results of diagnostic testing (Table 1).

While severe disease is rare, HSV and VZV lead to significant morbidity. In order to avoid complications, both prophylactic and symptomatic treatment approaches have been advocated. Studies using both moderate and low-dose acyclovir during the post-transplant period have demonstrated a marked decrease in both VZV and HSV complications (99,100). The use of acyclovir or valacyclovir is recommended for at least one year post-transplant, and through six months after cessation of all immunosuppressive therapy (101). Long-term prophylaxis also appears to prevent the emergence of acyclovir-resistant HSV (102). Studies evaluating VZV and HSV antiviral prophylaxis in SOT are lacking, and therefore are not currently recommended for long-term use (103).

HHV-6

HHV-6 is a human herpesvirus that was found to be the etiologic agent for roseola infantum (exanthema subitum), one of the common febrile rash causing illnesses in children (104). There are two subtypes of HHV-6: type A and type B. The two subtypes share certain biological properties and a high level of sequence homology, but are clearly two distinct viruses both virologically and epidemiologically (105). The majority of childhood illnesses and transplant complications are due to HHV-6B. Similar to other herpesviruses, after primary infection HHV-6 establishes latency within the host. A benign disease in the normal host, HHV-6 can lead to delayed engraftment after HSCT, skin rash, and devastating encephalitis in transplant recipients (106,107). Patients with encephalitis can present with confusion that can range from short-term memory loss to global dysfunction with frank seizures. Magnetic resonance imaging (MRI) can be normal, can demonstrate abnormalities in the mesotemporal lobe, or can be diffusely abnormal; patients with changes on imaging tend to have a poorer prognosis (108). Cerebrospinal fluid (CSF) for HHV-6 can be used to confirm the diagnosis. While not all patients with CNS disease have HHV-6 viremia, patients with high levels may be at higher risk for developing disease (45,106). High-dose ganciclovir or foscarnet is considered the treatment of choice (Table 1). Unfortunately, despite aggressive treatment, there is high associated morbidity and mortality (105).

Kaposi's Sarcoma Associated Herpesvirus (KSHV/HHV-8)

KSHV is the cause of Kaposi's sarcoma, and is associated with multicentric Castleman's disease and primary effusion lymphoma (109). KSHV is an uncommon cause of clinical illness in transplant, but Kaposi's sarcoma (KS) makes up approximately 5.7% of all post-transplant malignancies in SOT (110). Rates may be higher in areas of KSVH endemnicity. There appear to be regional differences in presentation, as US SOT recipients appear to present with more cutaneous KS, where internationally born patients have higher rates of visceral disease (111). HSCT patients are also at risk for the development of post-transplant KS (112), and KSHV may also lead to nonmalignant complications including bone marrow failure, hepatitis, and febrile illness (112–115). It has been suggested that KSHV transmission from donor grafts in HSCT is possible as CD34+ hematopoietic progenitor cells may be a reservoir for the virus (116). Cidofovir and valganciclovir have been shown to provide potential antiviral options for therapy, though neither is routinely recommended (Table 1) (48).

Respiratory Viruses

Respiratory infections are as common in immunocompromised patients, as they are in the community at large. The syndromes caused by these infections show a large degree of clinical overlap, and in general it is not possible to definitively identify the causative virus on clinical grounds alone. Unfortunately, in the immunocompromised population, these viruses can cause a spectrum of disease, including upper respiratory infection, pneumonia, and/or airflow obstruction (117), and may be causes of significant morbidity and mortality (118). Influenza A and B, parainfluenza virus, metapneumovirus, respiratory syncytial virus, and adenovirus can all be associated with severe and often fatal lower tract disease, and they are the leading respiratory viruses responsible for major complications in transplant recipients (Table 2) (118). Rarely other respiratory viruses have been associated with severe life-threatening pulmonary complications (119).

Table 2 Respiratory Viruses and Risk of Severe Lower Tract Disease

Highest risk
Adenovirus
Influenza A
Influenza B
Metapneumovirus
Parainfluenza virus
Respiratory syncytial virus
Lowest risk
Bocavirus
Coronaviruses
Rhinovirus[a]

[a]Has rarely been associated with lower tract disease (119).

While mild respiratory illness in immunocompetent persons rarely requires laboratory evaluation, definitive diagnosis is important in the management of immunosuppressed patients, and this need has led to many of the advances in respiratory virus detection. Effective antiviral therapy is now available for several respiratory viruses (Table 1), increasing the importance of specific diagnostics. For example, respiratory syncytial virus (RSV) is a common cause of fatal pneumonia in immunosuppressed patients (120). Fortunately, RSV infection is responsive to aerosolized ribavirin therapy, which reduces the mortality to approximately 20% (121). Survival may be improved when aerosolized ribavirin is used in combination therapy with intravenous immunoglobulin or the RSV-specific monoclonal antibody palivizumab (118). Similarly, a diagnosis of influenza virus infection allows prompt initiation of neuraminidase inhibitor therapy (e.g., oseltamivir), while adenovirus pneumonia may potentially respond to cidofovir (122). Although approved therapeutics are not available for most respiratory viruses, a specific diagnosis can spare patients exposure to antivirals that are unlikely to be effective, can guide the use of new and experimental therapies, and can provide critical information for infection control and hospital epidemiology.

Traditional virologic testing can be performed on immunocompromised patients, but molecular assays provide the greatest yield and have become the standard of care in this setting. The greater sensitivity of molecular techniques has demonstrated high rates of respiratory viral coinfection in this setting (123). As noted above, the clinical syndromes caused by the respiratory viruses overlap, and therefore comprehensive PCR panels or multiplex testing approaches have been established in several laboratories (123). Commercial panels and multiplex approaches are under development by multiple manufacturers. Currently, only a few such tests have received FDA approval, including the xTAG Respiratory Viral Panel, manufactured by Luminex Molecular Diagnostics, which detects 12 different respiratory viruses, and the ProFlu+ Assay, from Prodesse, which detects RSV and influenza A and B. New techniques are currently being used to detect resistant strains of influenza virus, and these may become more available as resistance continues to increase (124).

Adenovirus

The adenovirus family is associated with multiple transplant-related complications including hemorrhagic cystitis (HC), enterocolitis, pneumonia, hepatitis, encephalitis, and disseminated disease. Specific adenovirus subtypes and serotypes are associated with specific clinical illness (reviewed in Ref. 125). Definitive adenoviral disease is identified when a patient with appropriate clinical symptoms has documented adenovirus from the target organ, and disseminated disease is defined as clinical symptoms consistent with adenovirus and documentation of adenovirus in two or more organs (126). Risk factors for the development of invasive adenovirus include T-cell depletion, pediatric recipient, and graft mismatching (50,127). Disseminated disease, pneumonia, and hepatitis generally have a poor prognosis even with antiviral therapy (127–129). Intravenous cidofovir has been the most commonly used agent for therapy, but other less typical agents have been used to treat patients with severe adenovirus (Table 1) (125). Patients treated with cidofovir have been shown to have a decrease in mortality in retrospective studies (130–132), and prospective studies have suggested that virologic response to the drug is associated with clinical responses (133,134). There are also in vitro studies that suggest ribavirin may be somewhat effective for treatment (135); however, this remains controversial, as nonrandomized studies in humans have demonstrated varied results (reviewed in Ref. 136). Pre-emptive monitoring for adenovirus is advocated by some experts (126,137), but this remains controversial.

Polyoma Viruses

BK and JC viruses are common viruses, and usually lead to asymptomatic infection in childhood. Following transplant, both viruses are significant threats in the setting of immunosuppression. BK virus frequently reactivates in HSCT and SOT, and can lead to the development of HC (138,139). HC presents in patients as bladder irritation with frequent urination, pain, frank hematuria, and in severe cases the development of clotting in the bladder that can lead to urinary obstruction. Viral shedding can be detected by detecting cytologically abnormal cells in urine (also called "decoy cells"), culture, or through PCR tests on urine. Levels of BK viruria do not appear to predict the development of cystitis in HSCT, but BK viremia >10,000 copies does appear to have an increased risk of developing HC (140). Beyond HC, BK nephropathy is an important cause of graft dysfunction and failure in kidney transplant recipients. Kidney transplant recipients are screened for BK viruria and or viremia at most centers routinely to prevent the development of BK nephropathy (141). While BK nephropathy can occur in HSCT, it is much less frequently observed. Reduction in immunosuppression is effective in many patients with viruria, but in patients with severe disease cidofovir, leflunomide, IVIG, and flourquinolones have all been suggested as possible therapeutic options (Table 1) (49).

JC virus is a less common pathogen in transplantation, but can be associated with devastating consequences. JC virus has been associated with HC and nephropathy, and is the causative agent of progressive multifocal leukoencephalopathy (PML). Patients with PML typically present with new focal neurologic findings, and or altered mental status. Neuroimaging demonstrates scattered, multifocal areas of white matter demyelination that do not enhance with addition of contrast. However, transplant recipients may also develop atypical radiologic presentations (142). The gold standard for diagnosis is brain biopsy, but JC virus PCR from CSF is a useful diagnostic test in many patients. A negative JC virus PCR from CSF does not rule out PML, but a positive CSF sample confirms the diagnosis and may be useful for determining long-term prognosis (143,144). Therapeutic options are limited and must include reduction in immunosuppressive therapy if possible. Cidofovir has been used successfully in case reports (145), but others have shown no benefit (146,147). New in vitro data suggests that serotonin reuptake inhibitors that block $5HT_{2A}R$, the receptor the JC virus uses to infect glial cells, may also provide potential options for future therapy (148). Although a rare complication, many transplant recipients progress to death even after a decrease in immunosuppression and aggressive antiviral therapy (149,150).

Human Papillomavirus (HPV)

HPV is associated with nonmelanoma skin cancer, cervical cancer, and anal cancer, and immunosuppression as a result of transplantation can increase the risk of developing any of these HPV-related cancers. Studies of post-transplant women demonstrate an increased risk of cervical dysplasia and a higher risk for cervical cancer (151–153). Post-transplant nonmelanoma skin cancers, such as squamous cell carcinoma, are felt to be strongly associated with HPV, and HPV is more commonly detected in these cancers in transplant patients when compared to normal hosts (154–157). Risks of HPV-associated anal cancer are also increased in transplant patients (158,159). Therapy for HPV-related cancer is limited to regional resection, radiation, and systemic chemotherapy. While antiviral therapy is not effective in HPV-associated cancer, topical cidofovir for early treatment of premalignant lesions holds some promise as a therapeutic option (160). Since HPV disease can be prevented, post-transplant patients should have yearly skin exams and routine pap smears. Furthermore, the HPV vaccine may become a future option for primary prevention in transplant recipients (151,161).

Emerging Viral Infections

The majority of infections seen in transplant patients are well described, but these patients are also at risk for atypical and emerging infections. For example, with the emergence of West Nile virus (WNV) as a novel pathogen in the United States, transplant patients were found to be an at-risk group for the development of severe neuroinvasive disease (162,163). Donor screening also became important for prevention, as it was noted that transplant patients were also at increased risk due to potential exposure from blood products and donor grafts (163–165). The potential for viral transmission through donor grafts is an important issue, as it is possible that donors have a viral pathogen that either is asymptomatic in the normal host, or in the case of solid organ transplants, may be an unidentified cause of death of the donor. Recent examples of such situations have involved WNV, lymphocytic choriomeningitis (LCMV) (166), and rabies viruses (167). Additionally, transplant recipients are an important population that can serve to identify new pathogens—serving as a "canary in the coal mine" for emerging pathogens. An example is the newly identified Arenavirus, discovered in a virology laboratory following the death by acute febrile illness in all organ transplant recipients from a single donor. The donor died from unknown causes and was later determined to have had the same virus (168).

As new pathogens spread across the globe, these viruses can be expected to lead to disease in transplant recipients. It is important for clinical virology laboratories to remain up-to-date on new emerging viruses, and to be aware of new testing and technical considerations for these pathogens.

SUMMARY

Viral infections are a major cause of morbidity and mortality in immunocompromised patients, and the clinical virology lab is essential for the diagnosis and treatment of these pathogens. The major viral threats in a given patient are typically determined by the underlying disease process and the cause of immunosuppression. Since prompt therapy is especially important in immunosuppressed patients, close monitoring of patients for viral infection, typically by molecular testing, is an essential component of appropriate management.

REFERENCES

1. Neuzil KM, Zhu Y, Griffin MR, et al. Burden of interpandemic influenza in children younger than 5 years: a 25-year prospective study. J Infect Dis 2002; 185:147–152.
2. Gnann JW Jr., Whitley RJ. Clinical practice. Herpes zoster. N Engl J Med 2002; 347:340–346.
3. Jumaan AO, Yu O, Jackson LA, et al. Incidence of herpes zoster, before and after varicella-vaccination-associated decreases in the incidence of varicella, 1992-2002. J Infect Dis 2005; 191:2002–2007.
4. Aagaard-Tillery KM, Silver R, Dalton J. Immunology of normal pregnancy. Semin Fetal Neonatal Med 2006; 11:279–295.
5. Navaneethan U, Al Mohajer M, Shata MT. Hepatitis E and pregnancy: understanding the pathogenesis. Liver Int 2008; 28:1190–1199.
6. McNeil S, Halperin B, MacDonald N. Influenza in pregnancy: the case for prevention. Adv Exp Med Biol 2009; 634:161–183.
7. Yoon JW, Gollapudi S, Pahl MV, et al. Naive and central memory T-cell lymphopenia in end-stage renal disease. Kidney Int 2006; 70:371–376.
8. Pesanti EL. Immunologic defects and vaccination in patients with chronic renal failure. Infect Dis Clin North Am 2001; 15:813–832.
9. Cheong HJ, Song JY, Park JW, et al. Humoral and cellular immune responses to influenza vaccine in patients with advanced cirrhosis. Vaccine 2006; 24:2417–2422.
10. Casanova JL, Fieschi C, Zhang SY, et al. Revisiting human primary immunodeficiencies. J Intern Med 2008; 264:115–127.
11. Hunault-Berger M, Ifrah N, Solal-Celigny P. Intensive therapies in follicular non-Hodgkin lymphomas. Blood 2002; 100:1141–1152.
12. Harousseau JL, Moreau P. Autologous hematopoietic stem-cell transplantation for multiple myeloma. N Engl J Med 2009; 360:2645–2654.
13. Baron F, Storb R. Hematopoietic cell transplantation after reduced-intensity conditioning for older adults with acute myeloid leukemia in complete remission. Curr Opin Hematol 2007; 14:145–151.
14. Sorror ML, Storer BE, Sandmaier BM, et al. Five-year follow-up of patients with advanced chronic lymphocytic leukemia treated with allogeneic hematopoietic cell transplantation after nonmyeloablative conditioning. J Clin Oncol 2008; 26:4912–4920.
15. Storek J, Geddes M, Khan F, et al. Reconstitution of the immune system after hematopoietic stem cell transplantation in humans. Semin Immunopathol 2008; 30:425–437.
16. Roux E, Helg C, Chapuis B, et al. T-cell repertoire complexity after allogeneic bone marrow transplantation. Hum Immunol 1996; 48:135–138.
17. Hakki M, Riddell SR, Storek J, et al. Immune reconstitution to cytomegalovirus after allogeneic hematopoietic stem cell transplantation: impact of host factors, drug therapy, and subclinical reactivation. Blood 2003; 102:3060–3067.
18. Komanduri KV, St John LS, de Lima M, et al. Delayed immune reconstitution after cord blood transplantation is characterized by impaired thymopoiesis and late memory T-cell skewing. Blood 2007; 110:4543–4551.
19. Abrahamsen IW, Somme S, Heldal D, et al. Immune reconstitution after allogeneic stem cell transplantation: the impact of stem cell source and graft-versus-host disease. Haematologica 2005; 90:86–93.
20. Storek J, Ferrara S, Ku N, et al. B cell reconstitution after human bone marrow transplantation: recapitulation of ontogeny? Bone Marrow Transplant 1993; 12:387–398.
21. Buser A, Stern M, Arber C, et al. Impaired B-cell reconstitution in lymphoma patients undergoing allogeneic HSCT: an effect of pretreatment with rituximab? Bone Marrow Transplant 2008; 42: 483–487.

22. Hambach L, Stadler M, Dammann E, et al. Increased risk of complicated CMV infection with the use of mycophenolate mofetil in allogeneic stem cell transplantation. Bone Marrow Transplant 2002; 29:903–906.
23. Boeckh M, Nichols WG, Papanicolaou G, et al. Cytomegalovirus in hematopoietic stem cell transplant recipients: Current status, known challenges, and future strategies. Biol Blood Marrow Transplant 2003; 9:543–558.
24. Issa NC, Fishman JA. Infectious complications of antilymphocyte therapies in solid organ transplantation. Clin Infect Dis 2009; 48:772–786.
25. Mueller NJ. New immunosuppressive strategies and the risk of infection. Transpl Infect Dis 2008; 10:379–384.
26. Charpentier B, Rostaing L, Berthoux F, et al. A three-arm study comparing immediate tacrolimus therapy with antithymocyte globulin induction therapy followed by tacrolimus or cyclosporine A in adult renal transplant recipients. Transplantation 2003; 75:844–851.
27. Meier-Kriesche HU, Li S, Gruessner RW, et al. Immunosuppression: evolution in practice and trends, 1994-2004. Am J Transplant 2006; 6:1111–1131.
28. Schadde E, D'Alessandro AM, Knechtle SJ, et al. Alemtuzumab induction and triple maintenance immunotherapy in kidney transplantation from donors after cardiac death. Transpl Int 2008; 21:625–636.
29. Taherimahmoudi M, Ahmadi H, Baradaran N, et al. Cytomegalovirus infection and disease following renal transplantation: preliminary report of incidence and potential risk factors. Transplant Proc 2009; 41:2841–2844.
30. Brennan DC, Daller JA, Lake KD, et al. Rabbit antithymocyte globulin versus basiliximab in renal transplantation. N Engl J Med 2006; 355:1967–1977.
31. Husain S, Singh N. The impact of novel immunosuppressive agents on infections in organ transplant recipients and the interactions of these agents with antimicrobials. Clin Infect Dis 2002; 35:53–61.
32. Mattei MF, Redonnet M, Gandjbakhch I, et al. Lower risk of infectious deaths in cardiac transplant patients receiving basiliximab versus anti-thymocyte globulin as induction therapy. J Heart Lung Transplant 2007; 26:693–699.
33. Filipe R, Mota A, Alves R, et al. Kidney transplantation with corticosteroid-free maintenance immunosuppression: a single center analysis of graft and patient survivals. Transplant Proc 2009; 41:843–845.
34. Rosen HR. Transplantation immunology: what the clinician needs to know for immunotherapy. Gastroenterology 2008; 134:1789–1801.
35. Liu J, Farmer JD Jr., Lane WS,et al. Calcineurin is a common target of cyclophilin-cyclosporin A and FKBP-FK506 complexes. Cell 1991; 66:807–815.
36. Morath C, Arns W, Schwenger V, et al. Sirolimus in renal transplantation. Nephrol Dial Transplant 2007; 22(suppl 8):viii61–viii65.
37. Strangfeld A, Listing J, Herzer P, et al. Risk of herpes zoster in patients with rheumatoid arthritis treated with anti-TNF-alpha agents. JAMA 2009; 301:737–744.
38. Lill J, Bauer LA, Horn JR, et al. Cyclosporine-drug interactions and the influence of patient age. Am J Health Syst Pharm 2000; 57:1579–1584.
39. Provenzani A, Notarbartolo M, Labbozzetta M, et al. The effect of CYP3A5 and ABCB1 single nucleotide polymorphisms on tacrolimus dose requirements in Caucasian liver transplant patients. Ann Transplant 2009; 14:23–31.
40. Marty FM, Lowry CM, Cutler CS, et al. Voriconazole and sirolimus coadministration after allogeneic hematopoietic stem cell transplantation. Biol Blood Marrow Transplant 2006; 12:552–559.
41. Nowack R. Herb-drug interactions in nephrology: documented and theoretical. Clin Nephrol 2008; 69:319–325.
42. Kuypers DR. Influence of interactions between immunosuppressive drugs on therapeutic drug monitoring. Ann Transplant 2008; 13:11–18.
43. Wavamunno MD, Chapman JR. Individualization of immunosuppression: concepts and rationale. Curr Opin Organ Transplant 2008; 13:604–608.
44. Sanchez Fructuoso AI, de la Higuera MA, Garcia-Ledesma P, et al. Graft outcome and mycophenolic acid trough level monitoring in kidney transplantation. Transplant Proc 2009; 41:2102–2103.
45. Zerr DM, Gupta D, Huang ML, et al. Effect of antivirals on human herpesvirus 6 replication in hematopoietic stem cell transplant recipients. Clin Infect Dis 2002; 34:309–317.
46. Casper C, Krantz EM, Corey L, et al. Valganciclovir for suppression of human herpesvirus-8 replication: a randomized, double-blind, placebo-controlled, crossover trial. J Infect Dis 2008; 198:23–30.
47. De Clercq E, Naesens L, De Bolle L, et al. Antiviral agents active against human herpesviruses HHV-6, HHV-7 and HHV-8. Rev Med Virol 2001; 11:381–395.
48. Mazzi R, Parisi SG, Sarmati L, et al. Efficacy of cidofovir on human herpesvirus 8 viraemia and Kaposi's sarcoma progression in two patients with AIDS. AIDS 2001; 15:2061–2062.

49. Ramos E, Drachenberg CB, Wali R, et al. The decade of polyomavirus BK-associated nephropathy: state of affairs. Transplantation 2009; 87:621–630.
50. Echavarria M. Adenoviruses in immunocompromised hosts. Clin Microbiol Rev 2008; 21:704–715.
51. Deffrasnes C, Hamelin ME, Boivin G. Human metapneumovirus. Semin Respir Crit Care Med 2007; 28:213–221.
52. Boeckh M, Nichols WG. The impact of cytomegalovirus serostatus of donor and recipient before hematopoietic stem cell transplantation in the era of antiviral prophylaxis and preemptive therapy. Blood 2004; 103:2003–2008.
53. Bacigalupo A, Tedone E, Sanna MA, et al. CMV infections following allogeneic BMT: risk factors, early treatment and correlation with transplant related mortality. Haematologica 1992; 77:507–513.
54. Boeckh M. Current antiviral strategies for controlling cytomegalovirus in hematopoietic stem cell transplant recipients: prevention and therapy. Transpl Infect Dis 1999; 1:165–178.
55. Cornelissen JJ, Carston M, Kollman C, et al. Unrelated marrow transplantation for adult patients with poor-risk acute lymphoblastic leukemia: strong graft-versus-leukemia effect and risk factors determining outcome. Blood 2001; 97:1572–1577.
56. Broers AE, van der Holt R, van Esser JW, et al. Increased transplant-related morbidity and mortality in CMV-seropositive patients despite highly effective prevention of CMV disease after allogeneic T-cell-depleted stem cell transplantation. Blood 2000; 95:2240–2245.
57. Kanter J, Pallardo L, Gavela E, et al. Cytomegalovirus infection renal transplant recipients: risk factors and outcome. Transplant Proc 2009; 41:2156–2158.
58. Kollman C, Howe CW, Anasetti C, et al. Donor characteristics as risk factors in recipients after transplantation of bone marrow from unrelated donors: the effect of donor age. Blood 2001; 98:2043–2051.
59. Boeckh M, Ljungman P. How we treat cytomegalovirus in hematopoietic cell transplant recipients. Blood 2009; 113:5711–5719.
60. Boeckh M, Leisenring W, Riddell SR, et al. Late cytomegalovirus disease and mortality in recipients of allogeneic hematopoietic stem cell transplants: importance of viral load and T-cell immunity. Blood 2003; 101:407–414.
61. Zaia JA, Gallez-Hawkins GM, Tegtmeier BR, et al. Late cytomegalovirus disease in marrow transplantation is predicted by virus load in plasma. J Infect Dis 1997; 176:782–785.
62. Krause H, Hebart H, Jahn G, et al. Screening for CMV-specific T cell proliferation to identify patients at risk of developing late onset CMV disease. Bone Marrow Transplant 1997; 19:1111–1116.
63. Fisher RA. Cytomegalovirus infection and disease in the new era of immunosuppression following solid organ transplantation. Transpl Infect Dis 2009; 11:195–202.
64. Limaye AP, Bakthavatsalam R, Kim HW, et al. Late-onset cytomegalovirus disease in liver transplant recipients despite antiviral prophylaxis. Transplantation 2004; 78:1390–1396.
65. San Juan R, Aguado JM, Lumbreras C, et al. Impact of current transplantation management on the development of cytomegalovirus disease after renal transplantation. Clin Infect Dis 2008; 47:875–882.
66. Ljungman P, Griffiths P, Paya C. Definitions of cytomegalovirus infection and disease in transplant recipients. Clin Infect Dis 2002; 34:1094–1097.
67. Ljungman P, Engelhard D, Link H, et al. Treatment of interstitial pneumonitis due to cytomegalovirus with ganciclovir and intravenous immune globulin: experience of European Bone Marrow Transplant Group. Clin Infect Dis 1992; 14:831–835.
68. Enright H, Haake R, Weisdorf D, et al. Cytomegalovirus pneumonia after bone marrow transplantation. Risk factors and response to therapy. Transplantation 1993; 55:1339–1346.
69. Kalil AC, Levitsky J, Lyden E, et al. Meta-analysis: the efficacy of strategies to prevent organ disease by cytomegalovirus in solid organ transplant recipients. Ann Intern Med 2005; 143:870–880.
70. Small LN, Lau J, Snydman DR. Preventing post-organ transplantation cytomegalovirus disease with ganciclovir: a meta-analysis comparing prophylactic and preemptive therapies. Clin Infect Dis 2006; 43:869–880.
71. Salzberger B, Bowden RA, Hackman RC, et al. Neutropenia in allogeneic marrow transplant recipients receiving ganciclovir for prevention of cytomegalovirus disease: risk factors and outcome. Blood 1997; 90:2502–2508.
72. Eid AJ, Arthurs SK, Deziel PJ, et al. Emergence of drug-resistant cytomegalovirus in the era of valganciclovir prophylaxis: therapeutic implications and outcomes. Clin Transplant 2008; 22:162–170.
73. Limaye AP. Antiviral resistance in cytomegalovirus: an emerging problem in organ transplant recipients. Semin Respir Infect 2002; 17:265–273.
74. Perrottet N, Decosterd LA, Meylan P, et al. Valganciclovir in adult solid organ transplant recipients: pharmacokinetic and pharmacodynamic characteristics and clinical interpretation of plasma concentration measurements. Clin Pharmacokinet 2009; 48:399–418.

75. Boeckh M, Nichols WG. Immunosuppressive effects of beta-herpesviruses. Herpes 2003; 10:12–16.
76. Thorley-Lawson DA, Gross A. Persistence of the Epstein-Barr virus and the origins of associated lymphomas. N Engl J Med 2004; 350:1328–1337.
77. Taylor AL, Marcus R, Bradley JA. Post-transplant lymphoproliferative disorders (PTLD) after solid organ transplantation. Crit Rev Oncol Hematol 2005; 56:155–167.
78. Penn I. Post-transplant malignancy: the role of immunosuppression. Drug Saf 2000; 23:101–113.
79. Epstein-Barr virus and lymphoproliferative disorders after transplantation. Am J Transplant 2004; 4(suppl)10:59–65.
80. Manez R, Breinig MK, Linden P, et al. Factors associated with the development of post-transplant lymphoproliferative disease (PTLD) in Epstein-Barr virus (EBV)-seronegative adult liver transplant recipients. Transpl Int 1994; 7(suppl 1):S235–S237.
81. Everly MJ, Bloom RD, Tsai DE, et al. Posttransplant lymphoproliferative disorder. Ann Pharmacother 2007; 41:1850–1858.
82. Meijer E, Cornelissen JJ. Epstein-Barr virus-associated lymphoproliferative disease after allogeneic haematopoietic stem cell transplantation: molecular monitoring and early treatment of high-risk patients. Curr Opin Hematol 2008; 15:576–585.
83. Dror Y, Greenberg M, Taylor G, et al. Lymphoproliferative disorders after organ transplantation in children. Transplantation 1999; 67:990–998.
84. Wagner HJ, Wessel M, Jabs W, et al. Patients at risk for development of posttransplant lymphoproliferative disorder: plasma versus peripheral blood mononuclear cells as material for quantification of Epstein-Barr viral load by using real-time quantitative polymerase chain reaction. Transplantation 2001; 72:1012–1019.
85. Frias C, Lauzurica R, Bayes B, et al. Prospective follow-up of Epstein-Barr virus load in adult kidney transplant recipients by semiquantitative polymerase chain reaction in blood and saliva samples. Eur J Clin Microbiol Infect Dis 2001; 20:892–895.
86. Meerbach A, Wutzler P, Hafer R, et al. Monitoring of Epstein-Barr virus load after hematopoietic stem cell transplantation for early intervention in post-transplant lymphoproliferative disease. J Med Virol 2008; 80:441–454.
87. Wagner HJ, Cheng YC, Huls MH, et al. Prompt versus preemptive intervention for EBV lymphoproliferative disease. Blood 2004; 103:3979–3981.
88. Young JA. Infectious complications of acute and chronic GVHD. Best Pract Res Clin Haematol 2008; 21:343–356.
89. Basse G, Mengelle C, Kamar N, et al. Disseminated herpes simplex type-2 (HSV-2) infection after solid-organ transplantation. Infection 2008; 36:62–64.
90. Steer CB, Szer J, Sasadeusz J, et al. Varicella-zoster infection after allogeneic bone marrow transplantation: incidence, risk factors and prevention with low-dose aciclovir and ganciclovir. Bone Marrow Transplant 2000; 25:657–964.
91. Gourishankar S, McDermid JC, Jhangri GS, et al. Herpes zoster infection following solid organ transplantation: incidence, risk factors and outcomes in the current immunosuppressive era. Am J Transplant 2004; 4:108–115.
92. Pandya A, Wasfy S, Hebert D, et al. Varicella-zoster infection in pediatric solid-organ transplant recipients: a hospital-based study in the prevaricella vaccine era. Pediatr Transplant 2001; 5:153–159.
93. Geel AL, Landman TS, Kal JA, et al. Varicella zoster virus serostatus before and after kidney transplantation, and vaccination of adult kidney transplant candidates. Transplant Proc 2006; 38:3418–3419.
94. Fehr T, Bossart W, Wahl C, et al. Disseminated varicella infection in adult renal allograft recipients: four cases and a review of the literature. Transplantation 2002; 73:608–611.
95. Onozawa M, Hashino S, Haseyama Y, et al. Incidence and risk of postherpetic neuralgia after varicella zoster virus infection in hematopoietic cell transplantation recipients: Hokkaido Hematology Study Group. Biol Blood Marrow Transplant 2009; 15:724–729.
96. Koc Y, Miller KB, Schenkein DP, et al. Varicella zoster virus infections following allogeneic bone marrow transplantation: frequency, risk factors, and clinical outcome. Biol Blood Marrow Transplant 2000; 6:44–49.
97. Rau R, Fitzhugh CD, Baird K, et al. Triad of severe abdominal pain, inappropriate antidiuretic hormone secretion, and disseminated varicella-zoster virus infection preceding cutaneous manifestations after hematopoietic stem cell transplantation: utility of PCR for early recognition and therapy. Pediatr Infect Dis J 2008; 27:265–268.
98. Pergam S, Erard V, Delaney C, et al. Varicella zoster virus CNS disease in hematopoietic cell transplantation: a single center experience. Int J Infect Dis 2008; 12:S21–S21.
99. Erard V, Guthrie KA, Varley C, et al. One-year acyclovir prophylaxis for preventing varicella-zoster virus disease after hematopoietic cell transplantation: no evidence of rebound varicella-zoster virus disease after drug discontinuation. Blood 2007; 110:3071–3077.

100. Asano-Mori Y, Kanda Y, Oshima K, et al. Long-term ultra-low-dose acyclovir against varicella-zoster virus reactivation after allogeneic hematopoietic stem cell transplantation. Am J Hematol 2008; 83:472–476.
101. Tomblyn M, Chiller T, Einsele H, et al. Guidelines for preventing infectious complications among hematopoietic cell transplantation recipients: a global perspective. Biol Blood Marrow Transplant 2009; 15:1143–1238.
102. Erard V, Wald A, Corey L, et al. Use of long-term suppressive acyclovir after hematopoietic stem-cell transplantation: impact on herpes simplex virus (HSV) disease and drug-resistant HSV disease. J Infect Dis 2007; 196:266–270.
103. Miller GG, Dummer JS. Herpes simplex and varicella zoster viruses: forgotten but not gone. Am J Transplant 2007; 7:741–747.
104. Zerr DM, Meier AS, Selke SS, et al. A population-based study of primary human herpesvirus 6 infection. N Engl J Med 2005; 352:768–776.
105. Zerr DM. Human herpesvirus 6 and central nervous system disease in hematopoietic cell transplantation. J Clin Virol 2006; 37(suppl 1):S52–S56.
106. Zerr DM, Corey L, Kim HW, et al. Clinical outcomes of human herpesvirus 6 reactivation after hematopoietic stem cell transplantation. Clin Infect Dis 2005; 40:932–940.
107. Yoshikawa T, Asano Y, Ihira M, et al. Human herpesvirus 6 viremia in bone marrow transplant recipients: clinical features and risk factors. J Infect Dis 2002; 185:847–853.
108. Noguchi T, Mihara F, Yoshiura T, et al. MR imaging of human herpesvirus-6 encephalopathy after hematopoietic stem cell transplantation in adults. AJNR Am J Neuroradiol 2006; 27:2191–2195.
109. Casper C. New approaches to the treatment of human herpesvirus 8-associated disease. Rev Med Virol 2008; 18:321–329.
110. Penn I. Kaposi's sarcoma in transplant recipients. Transplantation 1997; 64:669–673.
111. Woodle ES, Hanaway M, Buell J, et al. Kaposi sarcoma: an analysis of the US and international experiences from the Israel Penn International Transplant Tumor Registry. Transplant Proc 2001; 33:3660–3661.
112. Luppi M, Barozzi P, Rasini V, et al. HHV-8 infection in the transplantation setting: a concern only for solid organ transplant patients? Leuk Lymphoma 2002; 43:517–522.
113. Luppi M, Barozzi P, Schulz TF, et al. Bone marrow failure associated with human herpesvirus 8 infection after transplantation. N Engl J Med 2000; 343:1378–1385.
114. Luppi M, Barozzi P, Schulz TF, et al. Nonmalignant disease associated with human herpesvirus 8 reactivation in patients who have undergone autologous peripheral blood stem cell transplantation. Blood 2000; 96:2355–2357.
115. Palencia SI, Rodriguez-Peralto JL, Castano E, et al. Kaposi's sarcoma after allogeneic peripheral blood stem cell transplantation. Int J Dermatol 2003; 42:647–649.
116. Wu W, Vieira J, Fiore N, et al. KSHV/HHV-8 infection of human hematopoietic progenitor (CD34+) cells: persistence of infection during hematopoiesis in vitro and in vivo. Blood 2006; 108:141–151.
117. Erard V, Chien JW, Kim HW, et al. Airflow decline after myeloablative allogeneic hematopoietic cell transplantation: the role of community respiratory viruses. J Infect Dis 2006; 193:1619–1625.
118. Boeckh M. The challenge of respiratory virus infections in hematopoietic cell transplant recipients. Br J Haematol 2008; 143:455–467.
119. Gutman JA, Peck AJ, Kuypers J, et al. Rhinovirus as a cause of fatal lower respiratory tract infection in adult stem cell transplantation patients: a report of two cases. Bone Marrow Transplant 2007; 40:809–811.
120. Ebbert JO, Limper AH. Respiratory syncytial virus pneumonitis in immunocompromised adults: clinical features and outcome. Respiration 2005; 72:263–269.
121. Boeckh M, Englund J, Li Y, et al. Randomized controlled multicenter trial of aerosolized ribavirin for respiratory syncytial virus upper respiratory tract infection in hematopoietic cell transplant recipients. Clin Infect Dis 2007; 44:245–249.
122. Yusuf U, Hale GA, Carr J, et al. Cidofovir for the treatment of adenoviral infection in pediatric hematopoietic stem cell transplant patients. Transplantation 2006; 81:1398–1404.
123. Kuypers J, Campbell AP, Cent A, et al. Comparison of conventional and molecular detection of respiratory viruses in hematopoietic cell transplant recipients. Transpl Infect Dis 2009; 11:298–303.
124. Oseltamivir-resistant novel influenza A (H1N1) virus infection in two immunosuppressed patients - Seattle, Washington, 2009. MMWR Morb Mortal Wkly Rep 2009; 58:893–896.
125. Ison MG. Adenovirus infections in transplant recipients. Clin Infect Dis 2006; 43:331–339.
126. Chakrabarti S, Milligan DW, Moss PA, et al. Adenovirus infections in stem cell transplant recipients: recent developments in understanding of pathogenesis, diagnosis and management. Leuk Lymphoma 2004; 45:873–885.

127. Symeonidis N, Jakubowski A, Pierre-Louis S, et al. Invasive adenoviral infections in T-cell-depleted allogeneic hematopoietic stem cell transplantation: high mortality in the era of cidofovir. Transpl Infect Dis 2007; 9:108–113.
128. Wang WH, Wang HL. Fulminant adenovirus hepatitis following bone marrow transplantation. A case report and brief review of the literature. Arch Pathol Lab Med 2003; 127:e246–e248.
129. Bertheau P, Parquet N, Ferchal F, et al. Fulminant adenovirus hepatitis after allogeneic bone marrow transplantation. Bone Marrow Transplant 1996; 17:295–298.
130. Bordigoni P, Carret AS, Venard V, et al. Treatment of adenovirus infections in patients undergoing allogeneic hematopoietic stem cell transplantation. Clin Infect Dis 2001; 32:1290–1297.
131. Nagafuji K, Aoki K, Henzan H, et al. Cidofovir for treating adenoviral hemorrhagic cystitis in hematopoietic stem cell transplant recipients. Bone Marrow Transplant 2004; 34:909–914.
132. Ljungman P, Ribaud P, Eyrich M, et al. Cidofovir for adenovirus infections after allogeneic hematopoietic stem cell transplantation: a survey by the Infectious Diseases Working Party of the European Group for Blood and Marrow Transplantation. Bone Marrow Transplant 2003; 31:481–486.
133. Leruez-Ville M, Minard V, Lacaille F, et al. Real-time blood plasma polymerase chain reaction for management of disseminated adenovirus infection. Clin Infect Dis 2004; 38:45–52.
134. Hoffman JA, Shah AJ, Ross LA, et al. Adenoviral infections and a prospective trial of cidofovir in pediatric hematopoietic stem cell transplantation. Biol Blood Marrow Transplant 2001; 7:388–394.
135. Morfin F, Dupuis-Girod S, Frobert E, et al. Differential susceptibility of adenovirus clinical isolates to cidofovir and ribavirin is not related to species alone. Antivir Ther 2009; 14:55–61.
136. Lenaerts L, Naesens L. Antiviral therapy for adenovirus infections. Antiviral Res 2006; 71:172–180.
137. Anderson EJ, Guzman-Cottrill JA, Kletzel M, et al. High-risk adenovirus-infected pediatric allogeneic hematopoietic progenitor cell transplant recipients and preemptive cidofovir therapy. Pediatr Transplant 2008; 12:219–227.
138. Erard V, Storer B, Corey L, et al. BK virus infection in hematopoietic stem cell transplant recipients: frequency, risk factors, and association with postengraftment hemorrhagic cystitis. Clin Infect Dis 2004; 39:1861–1865.
139. de Bruyn G, Limaye AP. BK virus-associated nephropathy in kidney transplant recipients. Rev Med Virol 2004; 14:193–205.
140. Erard V, Kim HW, Corey L, et al. BK DNA viral load in plasma: evidence for an association with hemorrhagic cystitis in allogeneic hematopoietic cell transplant recipients. Blood 2005; 106:1130–1132.
141. Hirsch HH, Brennan DC, Drachenberg CB, et al. Polyomavirus-associated nephropathy in renal transplantation: interdisciplinary analyses and recommendations. Transplantation 2005; 79:1277–1286.
142. Lima MA, Hanto DW, Curry MP, et al. Atypical radiological presentation of progressive multifocal leukoencephalopathy following liver transplantation. J Neurovirol 2005; 11:46–50.
143. Yiannoutsos CT, Major EO, Curfman B, et al. Relation of JC virus DNA in the cerebrospinal fluid to survival in acquired immunodeficiency syndrome patients with biopsy-proven progressive multifocal leukoencephalopathy. Ann Neurol 1999; 45:816–821.
144. Bossolasco S, Calori G, Moretti F, et al. Prognostic significance of JC virus DNA levels in cerebrospinal fluid of patients with HIV-associated progressive multifocal leukoencephalopathy. Clin Infect Dis 2005; 40:738–744.
145. Viallard JF, Lazaro E, Lafon ME, et al. Successful cidofovir therapy of progressive multifocal leukoencephalopathy preceding angioimmunoblastic T-cell lymphoma. Leuk Lymphoma 2005; 46:1659–1662.
146. Marra CM, Rajicic N, Barker DE, et al. A pilot study of cidofovir for progressive multifocal leukoencephalopathy in AIDS. 2002; 16:1791–1797.
147. Osorio S, de la Camara R, Golbano N, et al. Progressive multifocal leukoencephalopathy after stem cell transplantation, unsuccessfully treated with cidofovir. Bone Marrow Transplant 2002; 30:963–966.
148. Nukuzuma S, Nakamichi K, Nukuzuma C, et al. Inhibitory effect of serotonin antagonists on JC virus propagation in a carrier culture of human neuroblastoma cells. Microbiol Immunol 2009; 53:496–501.
149. Kharfan-Dabaja MA, Ayala E, Greene J, et al. Two cases of progressive multifocal leukoencephalopathy after allogeneic hematopoietic cell transplantation and a review of the literature. Bone Marrow Transplant 2007; 39:101–107.
150. Shitrit D, Lev N, Bar-Gil-Shitrit A, et al. Progressive multifocal leukoencephalopathy in transplant recipients. Transpl Int 2005; 17:658–665.
151. Savani BN, Stratton P, Shenoy A, et al. Increased risk of cervical dysplasia in long-term survivors of allogeneic stem cell transplantation–implications for screening and HPV vaccination. Biol Blood Marrow Transplant 2008; 14:1072–1075.
152. Paternoster DM, Cester M, Resente C, et al. Human papilloma virus infection and cervical intraepithelial neoplasia in transplanted patients. Transplant Proc 2008; 40:1877–1880.

153. Ozsaran AA, Ates T, Dikmen Y, et al. Evaluation of the risk of cervical intraepithelial neoplasia and human papilloma virus infection in renal transplant patients receiving immunosuppressive therapy. Eur J Gynaecol Oncol 1999; 20:127–130.
154. Stockfleth E, Nindl I, Sterry W, et al. Human papillomaviruses in transplant-associated skin cancers. Dermatol Surg 2004; 30:604–609.
155. Harwood CA, Surentheran T, McGregor JM, et al. Human papillomavirus infection and non-melanoma skin cancer in immunosuppressed and immunocompetent individuals. J Med Virol 2000; 61:289–297.
156. Purdie KJ, Surentheran T, Sterling JC, et al. Human papillomavirus gene expression in cutaneous squamous cell carcinomas from immunosuppressed and immunocompetent individuals. J Invest Dermatol 2005; 125:98–107.
157. Meyer T, Arndt R, Nindl I, et al. Association of human papillomavirus infections with cutaneous tumors in immunosuppressed patients. Transpl Int 2003; 16:146–153.
158. Albright JB, Bonatti H, Stauffer J, et al. Colorectal and anal neoplasms following liver transplantation [published online ahead of print March 26, 2009]. Colorectal Dis.
159. Patel HS, Silver AR, Northover JM. Anal cancer in renal transplant patients. Int J Colorectal Dis 2007; 22:1–5.
160. Coremans G, Snoeck R. Cidofovir: clinical experience and future perspectives on an acyclic nucleoside phosphonate analog of cytosine in the treatment of refractory and premalignant HPV-associated anal lesions. Expert Opin Pharmacother 2009; 10:1343–1352.
161. Savani BN, Goodman S, Barrett AJ. Can routine posttransplant HPV vaccination prevent commonly occurring epithelial cancers after allogeneic stem cell transplantation? Clin Cancer Res 2009; 15:2219–2221.
162. DeSalvo D, Roy-Chaudhury P, Peddi R, et al. West Nile virus encephalitis in organ transplant recipients: another high-risk group for meningoencephalitis and death. Transplantation 2004; 77: 466–469.
163. Hong DS, Jacobson KL, Raad II, et al. West Nile encephalitis in 2 hematopoietic stem cell transplant recipients: case series and literature review. Clin Infect Dis 2003; 37:1044–1049.
164. Centers for Disease Control and Prevention (CDC). West Nile virus infections in organ transplant recipients–New York and Pennsylvania, August-September, 2005. MMWR Morb Mortal Wkly Rep 2005; 54:1021–1023.
165. Stramer SL, Fang CT, Foster GA, et al. West Nile virus among blood donors in the United States, 2003 and 2004. N Engl J Med 2005; 353:451–459.
166. Brief report: lymphocytic choriomeningitis virus transmitted through solid organ transplantation–Massachusetts, 2008. MMWR Morb Mortal Wkly Rep 2008; 57:799–801.
167. Lapierre V, Tiberghien P. Transmission of rabies from an organ donor. N Engl J Med 2005; 352:2552; author reply 2552.
168. Palacios G, Druce J, Du L, et al. A new arenavirus in a cluster of fatal transplant-associated diseases. N Engl J Med 2008; 358:991–998.

Index